Collected Works of
Jaroslav Hájek

WILEY SERIES IN PROBABILITY AND STATISTICS

Established by WALTER A. SHEWHART and SAMUEL S. WILKS

A complete list of the titles in this series appears at the end of this volume

Collected Works of Jaroslav Hájek - With Commentary

Compiled by

M. Hušková
Charles University, Czech Republic

R. Beran
University of California, USA

and

V. Dupač
Charles University, Czech Republic

John Wiley & Sons
Chichester • New York • Weinheim • Brisbane • Singapore • Toronto

National 01243 779777
International (+44) 1243 779777

e-mail (for orders and customer service enquires): cs-books@wiley.co.uk

Visit our Home Page on http://www.wiley.co.uk or http://www.wiley.com

Other Wiley Editorial Offices

John Wiley & Sons, Inc., 605 Third Avenue,
New York, NY 10158-0012, USA

WILEY-VCH Verlag GmbH
Pappelallee 3, D-69469 Weinheim, Germany

Jacaranda Wiley Ltd, 33 Park Road, Milton,
Queensland 4064, Australia

John Wiley & Sons (Asia) Pte Ltd, 2 Clementi Loop #02-01,
Jin Xing Distripark, Singapore 129809

John Wiley & Sons (Canada) Ltd, 22 Worcester Road,
Rexdale, Ontario, M9W 1L1, Canada

British Library Cataloguing in Publication Data

A catalogue record for this book is available from the British Library

ISBN 0 471 97586 9

Produced from camera-ready copy supplied by the editors.
Printed and bound in Great Britain by Biddles Ltd, Guildford and King's Lynn.
This book is printed on acid-free paper responsibly manufactured from sustainable forestry in which at least two trees are planted for each one used for paper production.

Contents

Preface

The year 1998 marks the 25th anniversary of the Prague Symposium on Asymptotic Statistics, first convened by Jaroslav Hájek, and the 650th anniversary of Charles University, his home as a scholar. However, these historical resonances, gracenotes of time, are not the reason for this volume. Though Hájek died in 1974 at the age of 48, the vitality of his work endures. That is why we have collected his papers.

Not too long ago, one of us noticed that Hájek's (1971) convolution theorem – a staple in courses on the asymptotic theory of estimation – also characterizes when standard bootstrap distributions are consistent.[1] Ordinary bootstrap distributions fail to converge correctly at any superefficiency point in the parameter space. This technical consequence of the convolution theorem has awkward implications. Modern biased estimation techniques, including model-selection or smoothing procedures or shrinkage methods, succeed by creating 'good' superefficiency points. By so doing, such modern estimators improve, for instance, on the maximum likelihood estimator of a trend in Gaussian noise, which is the raw data itself. How unpleasant that ordinary bootstrap distributions or risk estimators cannot be trusted for sophisticated estimators. How pleasant that Hájek's thought penetrated this question before it arose.

Sharp insight infuses Hájek's contributions to Statistics and Probability. He came to prominence in the mid-1960s through a brilliant series of papers on the asymptotic distributions of rank statistics. The conditions for his theorems were elegantly minimal, improving substantially on their predecessors. His treatment introduced a generation of statisticians to Le Cam's results on contiguous probability measures and to Prohorov's theory of weak convergence of empirical processes. In the preceding decade, Hájek had already made a profound contribution to the theory of stochastic processes. The famous Dichotomy Theorem, stating that two Gaussian processes are either perpendicular or equivalent, was published independently by Hájek and by Feldman in the year 1958. Visiting Professor appointments at Berkeley in 1965–66 and at Florida State in 1969–70 were a direct outcome of his growing reputation. Up until then, Hájek had worked in relative isolation, more than once discovering results that he later learned were known in western journals unavailable to him.

Hájek's work on asymptotically efficient estimation marked the climax of one historical development and the early stages of another that is still under

[1] Beran, R. J. (1997) *Ann. Inst. Statist. Math. 49*, 1–24.

way. In an influential 1925 paper,[2] R. A. Fisher conjectured that the mean squared error of maximum likelihood estimators converges asymptotically to the information bound, pointwise in the parameter space, and that this is the best possible performance in large samples. Discoveries in the mid-1950s showed that maximum likelihood and related estimators may overfit the model when the parameter space is of high or infinite dimension, paying too much attention to bias and not enough to variance. C. Stein's result on the inadmissibility of the sample mean in three or more dimensions was one such finding. Kernel smoothing techniques for consistent estimation of spectral densities and of probability densities were two more. The Hodges estimator, which is a simple model-selection procedure, revealed another difficulty: the possible existence of 'bad' superefficiency points in the parameter space. These unexpected discoveries, which are not unconnected, forced rethinking of Fisher's conjecture.

Hajek's (1972) paper formulated elegantly what one can expect from the local asymptotic minimax risk of any estimator sequence, whether or not it possesses superefficiency points. While giving generous credit to earlier work in this direction, the paper made crucial innovations, such as bringing to bear on the problem Le Cam's theory of local asymptotic normality. Hájek's local asymptotic minimax bound gave precise meaning to the claim that maximum likelihood estimators, or corrections thereof, are asymptotically efficient. At the same time, it did not deny the existence of estimators that may perform better in high dimensional parameter spaces. For instance, asymptotic admissibility was claimed only for one-dimensional estimators of the maximum likelihood type. The intense development of asymptotic minimax theory spurred by Hájek's work soon yielded the 1980 asymptotic minimax bound of M.S. Pinsker[3] for estimation of trend in a Gaussian process. Pinsker's result explained how biased shrinkage estimators can improve on the risk of maximum likelihood estimators. So began the mainstream of modern statistical trend recovery, whose implications for efficient estimation in high-dimensional parameter spaces remain to be plumbed.

Hájek's earliest papers and last, posthumous book were on the subject of survey sampling. He was among the pioneers of unequal probability sampling. The name 'Hájek predictor' now labels his contributions to the use of auxiliary data in estimating population means. In a different direction, he proved a very general central limit theorem for the possible limiting distributions of the sample mean in finite population sampling. This theorem also connects finite population asymptotics with those of linear rank statistics.

Many now would regard Hájek as a theoretical statistician with diverse interests. However, his undergraduate training at the Technical University in Prague was applied in nature. To earn the degree of Statistical Engineer in 1949, he studied not only mathematics, probability, and statistics, but also

[2] Fisher, R. A. (1925) *Proc. Cambridge Math. Soc. 22*, 700–725.

[3] Pinsker, M. S. (1980) *Problems of Information Transmission 16*, 120–133.

computational methods, financial mathematics, some economics and law, and specific statistical methods useful in transport, insurance, agriculture, and industry. This broad training seems not to have harmed Hájek. His subsequent career illustrates how talented individuals define their own categories of achievement.

Hájek published papers in Czech and in Russian as well as in English. Some have already been translated. We include in this volume the first translations from Czech of several early papers on sample surveys. In particular, after almost 40 years of citation, Hájek's paper on sampling with probability proportional to size (representative cluster sampling in his terminology) is now accessible to more readers. Whenever possible, we have made corrections that Hájek marked on his own copies of papers. Several distinguished colleagues have contributed essays placing the papers in perspective. Inevitably, these essays throw light on Hájek's personality – how gracefully he lived in the shadow of the kidney disease that ended his life so early.

Many individuals have assisted greatly in the preparation of this volume. Our editor at Wiley, Helen Ramsey, and her collaborators gave us much useful advice. Nick Fisher helped clarify some vital communications. Jana Jurečková, Josef Machek, Lucien Le Cam, Emanuel Parzen, Zuzana Prášková, and Pranab Kumar Sen wrote review essays. Jitka Dupačová, Josef Machek, and Ivo Müller translated papers originally published in Czech or in Russian. Many staff members and PhD students in the Department of Probability and Mathematical Statistics, Charles University, read the proofs carefully. Iva Marešová did excellent typing in LaTeX and patiently made our repeated changes and corrections. Pavel Boček and Mirek Hušek prepared figures. To all of these people we extend our sincere thanks.

We are also indebted to the following individuals and organizations for permission to reproduce Jaroslav Hájek's papers: Hájek family (Alžběta Hájková, Helena Kazárová, Alžběta Slámová), V.P. Godambe, Lucien Le Cam, D.A. Sprott, Akadémiai Kiadó, American Mathematical Society, Cambridge University Press, International Statistical Society, Institute of Mathematics of the Polish Academy of Sciences, Springer-Verlag GmbH& Co. KG, Institute of Mathematical Statistics, and University of California Press.

January 1998 Marie Hušková, Rudolf Beran, Václav Dupač

Jaroslav Hájek presenting his famous Dichotomy Theorum
(Reproduced by permission of the Hájek family: Al. Hájková, H. Kazárová and Al. Slámová)

Part I: Historical Overview

Part I starts with résumé of Jaroslav Hájek's life and professional career and shows also his personality as viewed by his colleagues and former students. The essays that follow appraise his role in the development of modern statistical science, particularly his fundamental results and methods in statistical inference in stochastic processes, in finite population sampling, in the theory of rank tests and in the asymptotics of parameter estimators. Recollected as well are his less known early papers. This part concludes with a complete list of Hájek's publications and a list of his PhD students and their theses.

Biography of Jaroslav Hájek

Jaroslav Hájek was born on February 4th, 1926 in Poděbrady, a quiet town in Central Bohemia well known as a spa and rehabilitation center for heart diseases. There he spent the first years of his life and received his primary school education. Later he studied in Prague at a Gymnasium (which is roughly an equivalent of high school and is a condition for admission to university). Like nearly all young men of his generation, he did not escape assignment under the Nazi regime to 'Totaleinsatz,' a species of forced work in the armaments industry. Having finished the Gymnasium in 1945, he enrolled in the Czech Technical University where he studied so-called Statistical Engineering, which at that time also included insurance mathematics. During the last two years of his studies, Hájek was already working as Assistant Lecturer in the Institute of Mathematics at the Czech Technical University. In June 1949 he received his diploma as a Statistical Engineer; in the same year his first paper, on a problem in the theory of sampling surveys, appeared. The two years that followed he spent in obligatory military service.

During the period 1951 to 1954, Hájek undertook graduate studies in the Mathematical Institute of the Czechoslovak Academy of Sciences. Having finished these graduate studies and obtained the degree of PhD, he continued his career in the Mathematical Institute as Senior Research Worker for a full 12 years — from 1954 to 1966. It can be said that this was a happy, successful and fruitful period of his life, both private and scientific. In 1954 he married Alžběta (Betty) Galambos, who proved to be his reliable, patient and stimulating companion for all the remaining days of his life. Within the first five years of the marriage, she gave him two daughters and shared with him all the difficulties and pleasures of everyday life (which was not always easy: he travelled a lot, accepted many scientific tasks, and maintained cooperations with many institutions). Hájek was, however, an attentive husband to her and a loving father to their daughters.

During these years, Hájek continued his scientific work indefatigably. In addition to more than 20 papers, he published his first book, *The Theory of Probability Sampling with Applications to Sample Surveys*. In this period also, a small book appeared, *Probability in Science and Engineering* (written jointly with V. Dupač). During these years, Hájek established himself as a foremost international authority in several fields of statistics, particularly in nonparametric methods and their asymptotic theory. His reputation led to an increasing number of invitations to international statistical events, to service on editorial boards of international journals, and to longer stays in foreign universities as a Visiting Professor. The most important of these longer stays were, perhaps, his visits at the University of California in Berkeley, the Michigan State University in East Lansing, and the Florida State University in Tallahassee. During these stays abroad Hájek established many contacts, both professional and friendly, with many leading statisticians.

In 1963 he was declared Doctor of Physico–Mathematical Sciences after having defended a thesis on statistical problems in stochastic processes. In the same year, he started his cooperation with the Faculty of Mathematics and Physics at the Charles University in Prague. From 1964 until his death in 1974, he held the Chair of the Department of Probability and Statistics in this Faculty.

From the moment Hájek assumed the Chair, he directed great efforts towards reforming both the content and the method of teaching, which were then, admittedly, somewhat old-fashioned and obsolete; and he succeeded. In the course of the years, he brought together a group of graduate students whom he tutored and oriented in such a superb manner that the majority of them have become renowned researchers in various fields of statistical theory. Many of these now form the kernel of the Department. At this point, it may be worthwhile to mention Hájek's 'technique' of tutoring his graduate students. It consisted of two parts: individual consultations in which special topics, befitting the student's specializations, were discussed; and seminars held for the entire group in which advanced problems of common interest were solved collectively.

One of the ways through which Hájek tried to increase the level of statistical education was the following. Every year or two, Hájek would select young members of the Department as his 'aides-de-camp' for some of the important courses in the curriculum. He then taught each such course during one or two terms; the task of the 'aide' was to participate, take detailed notes, and then coauthor the corresponding mimeographed text. In the next term, such a collaborator took over the class, naturally with liberty to introduce modifications or innovations. In this way, several texts arose.

As a teacher, Hájek was rather exigent but at the same time comprehensive, fair and friendly. Years later, one of his former students (undergraduate) told me: 'I didn't mind being suspended by Hájek, because, during the exam, he taught me something, explained to me my deficiencies and did not make me feel humiliated'.

J. Hájek also encouraged social life in the Department. He introduced regular 'informal meetings' — collective trips to the surroundings of Prague with students participating — whose organization has continued to this date.

In addition to his time- and energy-absorbing teaching activities, he continued his own research work. Besides 16 more original papers published in international journals, he wrote two books (*Theory of Rank Tests*, jointly with Z. Šidák, and *A Course in Nonparametric Statistics*).

For many years, Hájek had suffered from a serious kidney disease. In the early 1970s, his illness worsened month by month. With admirable strength of will, Hájek tried to continue his work. In about 1971, he proposed the organization of an international statistical conference in Prague. He himself suggested as its key topic the asymptotic methods of statistics and addressed himself to several prominent statisticians, asking them to act as invited speakers.

In 1973 the event took place under the heading 'Prague Symposium on Asymptotic Statistics.' The majority of statisticians abroad who were addressed by Hájek responded favourably; and thus, thanks to Hájek's popularity, the Symposium became a highlighted scientific event. But this was, for Hájek, the last opportunity to meet his friends from abroad. On June 10th, 1974 he succumbed to his fatal disease and died at the age of 48 years. His untimely death was a loss for statistics in general and even more so for our country, because he founded the modern Czech statistical school of the post-war period.

In August 1998, the 6th Prague Symposium on Asymptotic Statistics will take place. It will form a part of the celebrations of the 650th Anniversary of the foundation of the Charles University. It will also provide an opportunity to pay due tribute to Professor Hájek as one of those who most contributed to the University's international prestige.

ESSAY 1

Hájek and the Superefficiency Breakthrough

Rudolf Beran
Department of Statistics
University of California
Berkeley, CA 94720
U.S.A.

INTRODUCTION

In the first part of the twentieth century, statistical theory studied the asymptotic efficiency of maximum likelihood estimators when the dimension of the parameter space is finite and is held fixed as the number of observations increases. Influential papers on this topic include Fisher (1925), Le Cam (1953), Chernoff (1956), Bahadur (1964), and Wolfowitz (1965). Earlier contributions to the topic were reviewed by Le Cam (1953), 277–280, Pratt (1976), Stigler (1986), and Pfanzagl (1994), 207–208. This historical development climaxed in the local asymptotic minimax bound of Hájek (1972) and Le Cam (1972), in the convolution theorems of Hájek (1970) and Inagaki (1970), and in Le Cam's (1970) paper on the asymptotic normality of maximum likelihood estimators.

Mid-century brought growing interest in statistical models whose parameter spaces have high or infinite dimension. Advances in computation and in mathematical technique undoubtedly played a role. It began to be seen that maximum likelihood or related estimators may 'overfit' the model, paying too much attention to bias and not enough to variance, when the dimension of the parameter space is substantial. Blackman and Tukey (1959) reviewed smoothing methods for consistent estimation of spectral densities. Rosenblatt (1956) treated consistent nonparametric estimation of probability densities. After Fourier transformation, a kernel probability density estimator shrinks the empirical characteristic function while a kernel spectral density estimator

Collected Works of Jaroslav Hájek – With Commentary
Edited by M. Hušková, R. Beran and V. Dupač

shrinks the lagged sample covariances. The shrinkage factor varies with the frequency or lag, respectively.

Stein (1956) demonstrated, in the Gaussian case, the inadmissibility of maximum likelihood estimators when the dimension of the parameter space is not too small. Better shrinkage estimators were proposed by James and Stein (1961) and by Stein (1966). Mallows (1973) discussed model-selection in least squares regression, a restricted type of shrinkage estimation. Pinsker (1980) proved asymptotic minimaxity, over ellipsoids in the parameter space, of certain shrinkage estimators for the mean of a Gaussian time-series. Shrinking Fourier or wavelet or other orthonormal coefficients in Pinsker's problem amounts to a form of smoothing. Subsequent work by many authors has transformed these pioneering results into effective techniques for nonparametric regression and curve estimation. Yet much remains unknown about estimation in general parametric models with high-dimensional parameter spaces.

Did classical asymptotics for maximum likelihood estimators detect the possibility of improved estimation by shrinkage or smoothing? The answer seems to be enigmatically (through a glass darkly). Le Cam (1953) treated superefficient estimators whose asymptotic risk at certain parameter values is less than that determined by the information bound. However, the initial examples of superefficiency, such as the Hodges estimator, offered poor global risk properties. Only later and by other paths were the possible benefits of superefficiency realized. Through his convolution theorem, Hájek (1970) characterized estimators that are *not* superefficient at a given parameter value. How the convolution theorem distinguishes between maximum-likelihood-type estimators, on the one hand, and shrinkage or smoothing estimators, on the other hand, is the theme of this essay.

CONVOLUTION STRUCTURE PERVADES THE PARAMETER SPACE

Suppose that the sample X_n consists of n independent, identically distributed (iid) random variables whose joint distribution is $P_{\theta_0,n}$. The value of θ_0 is unknown but is restricted to a parameter space Θ that is an open subset of R^k. The problem is to estimate $\tau(\theta_0)$, where $\tau : \Theta \to R^m$ is a given function differentiable at every $\theta \in \Theta$, the derivative matrix $\nabla\tau(\theta)$ has full rank, and $m \leq k$.

Let $P^c_{\theta,n}$ denote the part of $P_{\theta,n}$ that is absolutely continuous with respect to $P_{\theta_0,n}$. For $h \in R^k$, let $L_n(h, \theta_0)$ designate the log-likelihood ratio of $P^c_{\theta_0+n^{-1/2}h,n}$ with respect to $P_{\theta_0,n}$. In the terminology of Le Cam (1960), the model $P_{\theta,n}$ is *locally asymptotically normal* (LAN) at θ_0 if there exist a column vector $Y_n(\theta_0)$ and a nonsingular matrix $I(\theta_0)$ such that, under $P_{\theta_0,n}$,

$$L_n(h_n, \theta_0) = h'Y_n(\theta_0) - 2^{-1}h'I(\theta_0)\,h + o_p(1)$$

for every $h \in R^k$ and for every sequence $\{h_n \in R^k\}$ converging to h, and if the distribution of $Y_n(\theta_0)$ under $P_{\theta_0,n}$ converges weakly to $N(0, I(\theta_0))$ as $n \to \infty$. The matrix $I(\theta_0)$ updates the classical definition of the information matrix through second derivatives.

In LAN models, the log-likelihood ratio behaves asymptotically like the log-likelihood ratio of $N(h, I^{-1}(\theta_0))$ with respect to $N(0, I^{-1}(\theta_0))$. This suggests that good statistical procedures for the normal location model should have counterparts for any LAN model. Smoothly parametrized exponential families are the most familiar LAN models. But certain models where maximum likelihood estimators fail, such as the three-parameter lognormal (see Hill (1963)), are also LAN at every θ_0 and thereby motivate theory for LAN models.

Let $\{T_n : n \geq 1\}$ be any sequence of estimators of $\tau(\theta)$ and let $H_n(\theta)$ denote the distribution of $n^{1/2}(T_n - \tau(\theta))$ under $P_{\theta,n}$. Suppose that $H_n(\theta_0)$ converges weakly to a limit distribution $H(\theta_0)$ as $n \to \infty$. The estimators $\{T_n\}$ are called *locally asymptotically equivariant* (LAE) at θ_0 if, for every $h \in R^k$ and for every sequence $\{h_n \in R^k\}$ that converges to h, the distributions $\{H_n(\theta_0 + n^{1/2}h_n)\}$ converge weakly to $H(\theta_0)$. The LAE property at θ_0 asserts that the distribution $H(\theta_0)$ asymptotically approximates the true distribution $H_n(\theta)$ whenever θ is at or sufficiently near θ_0. Consistent estimators for weakly continuous functionals of $H_n(\theta)$, such as risks, are a by-product of LAE.

The LAE property is very common. Suppose that $\{H_n(\theta)\}$ converges weakly to a limit distribution $H(\theta)$ for every $\theta \in \Theta$. Then, under conditions on the model weaker than LAN at every θ, there exists a Lebesgue null set E such that the $\{T_n\}$ are LAE at every $\theta \in \Theta - E$ (see Le Cam (1973), Le Cam and Yang (1990), 76–77). In practice, standard maximum likelihood or minimum distance or method-of- moments estimators are usually LAE at every θ. On the other hand, shrinkage estimators may lack the LAE property at key points in the parameter space, as illustrated in the following examples:

Hodges estimator. Suppose that the observations are iid random variables, each with $N(\theta, 1)$ distribution, where $\theta \in R$. Let $\bar{X}_n$ denote the sample mean. The Hodges estimator $T_{n,H}$ of θ equals 0 whenever $|\bar{X}_n| \leq n^{-1/4}$ and equals $\bar{X}_n$ otherwise (Le Cam (1953)). This estimator is a model-selection method that chooses between fitting the $N(0, 1)$ model and the $N(\theta, 1)$ model to the data. The Hodges estimator is LAE at every $\theta \neq 0$ but is not LAE at $\theta = 0$.

James–Stein estimator. Let I_k denote the k-dimensional identity matrix. Suppose that the observations are iid random k-vectors, each distributed according to $N(\theta, I_k)$, where $\theta \in R^k$. Let $[\cdot]^+$ represent the positive-part function on the real line. For $k \geq 3$, the James–Stein (1961) estimator of θ is $T_{n,JS} = [1 - (k-2)/(n|\bar{X}_n|^2)]^+ \bar{X}_n$. While this estimator operates more smoothly on the data than does the Hodges estimator, it too is LAE at every $\theta \neq 0$ and is not LAE at $\theta = 0$.

In LAN models, estimation methods such as maximum likelihood, or one-step maximum likelihood when maximum likelihood fails, generate estimators $T_{n,E}$ of $\tau(\theta)$ whose asymptotic structure under $P_{\theta_0,n}$ is

$$T_{n,E} = \tau(\theta_0) + n^{-1/2}\nabla\tau(\theta_0)I^{-1}(\theta_0)Y_n(\theta_0) + o_p(n^{-1/2})$$

for every $\theta_0 \in \Theta$. Estimators with this property are LAE at θ_0 and are usually called *asymptotically efficient* at θ_0. Information bounds discussed in the next section justify this terminology within the class of all estimators that are LAE at θ_0.

Let $K_n(\theta)$ denote the joint distribution of $n^{1/2}(T_n - T_{n,E})$ and $Y_n(\theta)$ under $P_{\theta,n}$. Let $\Sigma_\tau(\theta_0) = \nabla\tau(\theta_0)I^{-1}(\theta_0)\nabla'\tau(\theta_0)$. Hájek's (1970) convolution theorem may be expressed as follows (see Beran (1997)):

Suppose that the model is LAN at θ_0 and that the distribution $H_n(\theta_0)$ of $n^{1/2}(T_n - \tau(\theta_0))$ under $P_{\theta_0,n}$ converges weakly to $H(\theta_0)$ as $n \to \infty$. Then the following two statements are equivalent:

(a) *The estimators $\{T_n\}$ are LAE at θ_0 with limit distribution $H(\theta_0)$.*

(b) *For every $h \in R^k$ and every sequence $\{h_n \in R^k\}$ that converges to h,*

$$K_n(\theta_0 + n^{-1/2}h_n) \Rightarrow D(\theta_0) \times N(0, I(\theta_0))$$

for some distribution $D(\theta_0)$ such that $H(\theta_0) = D(\theta_0) * N(0, \Sigma_\tau(\theta_0))$.

The distribution $D(\theta_0)$ is a point mass at the origin if and only if $\{T_n\}$ has the asymptotic structure of $\{T_{n,E}\}$.

Statement (b) above implies that $n^{1/2}(T_n - T_{n,E})$ and $Y_n(\theta_0)$ are asymptotically independent under $P_{\theta_0,n}$, with marginal limit distributions $D(\theta_0)$ and $N(0, I(\theta_0))$. Therefore, the limit distribution $H(\theta_0)$ of $n^{1/2}(T_n - T_{n,E}) + n^{1/2}(T_{n,E} - \tau(\theta_0))$ is the convolution of $D(\theta_0)$ with $N(0, \Sigma_\tau(\theta_0))$. This structure in $H(\theta_0)$ gives the result its name and reveals the surprising strength of the LAE property in LAN models. If the hypotheses of the theorem are satisfied for every θ_0, then the LAE property, and so convolution structure, must hold almost everywhere in the parameter space.

SUPEREFFICIENCY IS IN TWO KINDS

Let l be any real-valued, symmetric, subconvex, non-negative function defined on R^k. Suppose that the model is LAN at θ_0 and that the estimators $\{T_n\}$ are

LAE at θ_0. Then, the second part of the convolution theorem, Fatou's lemma, and Anderson's lemma (Ibragimov and Has'minskii (1981), p. 155) imply

$$\liminf_{n\to\infty} \mathrm{E}_{\theta_0} l[n^{1/2}(T_n - \tau(\theta_0))] \geq \mathrm{E}l[\Sigma_\tau^{1/2}(\theta_0)Z_k],$$

where Z_k has a standard normal distribution on R^k. Moreover, if l is continuous a.e. and bounded, then

$$\lim_{n\to\infty} \mathrm{E}_{\theta_0} l[n^{1/2}(T_{n,E} - \tau(\theta_0))] = \mathrm{E}l[\Sigma_\tau^{1/2}(\theta_0)Z_k].$$

This last equality justifies the claim that the estimators $\{T_{n,E}\}$ of $\tau(\theta)$ are asymptotically efficient among all estimators that are LAE at θ_0.

If an estimator sequence $\{T_n\}$ is such that

$$\limsup_{n\to\infty} \mathrm{E}_{\theta_0} l[n^{1/2}(T_n - \tau(\theta_0))] < \mathrm{E}l[\Sigma_\tau^{1/2}(\theta_0)Z_k],$$

then θ_0 is called a *superefficiency* point of $\{T_n\}$. The preceding discussion shows that points of superefficiency must be non-LAE points, thereby recovering the result of Le Cam (1953) and Bahadur (1964) that superefficiency points form a Lebesgue null set. In general, the set of non-LAE points is larger than the set of superefficiency points.

If θ_0 is not an LAE point, the limiting risk computed at θ_0 need not approximate the actual risk at values of θ near θ_0. This means that superefficiency at θ_0 does not determine whether an estimator has good risk or poor risk near θ_0, except in special circumstances. Consider our two examples:

Hodges estimator. Here $\lim_{n\to\infty} n\mathrm{E}_\theta(T_{n,H} - \theta)^2$ is 0 when $\theta = 0$ and is 1 otherwise. The estimator is superefficient at the origin under quadratic loss and is best LAE at all other values of θ. However, for each n, the normalized risk $n\mathrm{E}_\theta(T_{n,H} - \theta)^2$ is less than the information bound 1 in a neighborhood of $\theta = 0$, then climbs quickly above 1, and finally falls slowly towards 1 as $|\theta|$ increases (see Lehmann (1983), Chapter 6). As n increases, the neighborhood of the origin where risk is less than 1 narrows while the maximum risk tends to infinity.

James–Stein estimator. If $k \geq 3$, $\lim_{n\to\infty} n\mathrm{E}_\theta|T_{n,JS} - \theta|^2$ is 2 when $\theta = 0$ and is k otherwise. The estimator is superefficient at the origin and is best LAE at every other value of θ. Unlike the risk of the Hodges estimator, the normalized risk $n\mathrm{E}_\theta|T_{n,JS} - \theta|^2$ is strictly less than the information bound for every value of n and θ and decreases monotonically with the signal-to-noise ratio $n|\theta|^2/k$. Historically, this small k phenomenon induced exaggerated mistrust of superefficiency. When k is not too small, the James–Stein estimator and the more useful multiple shrinkage estimators in Stein (1966) demonstrated tacitly the global reductions in risk possible through superefficiency. Further developments had to await wider understanding of how parameter dimension affects estimation. That superefficiency points form a set of Lebesgue measure zero does not preclude beneficial influence on the entire risk function when n is finite and k is not too small.

A VIEW OF THE BORDER

The ambiguity of the superefficiency phenomenon in classical asymptotics, seen above, can be resolved through alternative analyses. Stein (1956) touched on 'time-series' asymptotics, in which the dimension of the multivariate normal model increases. Taken further, this approach links the James–Stein estimator to Pinsker's (1980) asymptotic minimax bounds for estimating the mean of a Gaussian process (see Beran (1996) and Nussbaum (1996)) and to risk calculations for model selection estimators (see the discussion by Shao (1997)). The discovery of superefficiency, both breakthrough and enigma, foreshadowed estimators that dominate maximum likelihood in high or infinite dimensional models. Hájek's convolution theorem elegantly drew the border between the old world of maximum likelihood estimators and the new world of smoothing or shrinkage estimators.

REFERENCES

Bahadur, R. (1964). A note on Fisher's bound on asymptotic variance. *Ann. Math. Statist.* *35*, 1545–1552.

Beran, R. (1996). Stein estimation in high dimensions: a retrospective. *Festschrift in Honor of Madan L. Puri* (eds. E. Brunner and M. Denker). VSP, Zeist., 91–110,

Beran, R. (1997). Diagnosing bootstrap success. *Ann. Inst. Statist. Math.* *47*, 1–24.

Blackman, R. B. and Tukey, J. W. (1959). *The Measurement of Power Spectra*. Dover, New York.

Chernoff, H. (1956). Large sample theory: parametric case. *Ann. Math. Statist.* *27*, 1–22.

Fisher, R. A. (1925). Theory of statistical estimation. *Proc. Cambridge Math. Soc.* *22*, 700–725.

Hájek, J. (1970). A characterization of limiting distributions of regular estimators. *Z. Wahrsch. Verw. Gebiete* *14*, 323–330.

Hájek, J. (1972). Local asymptotic minimax and admissibility in estimation. *Proc. Fifth Berkeley Symp. Math. Statist. Prob.* (eds. L. M. Le Cam, J. Neyman and E. M. Scott), Vol. 1. University of California Press, Berkeley, 175–194.

Hill, B. M. (1963). The three-parameter lognormal distribution and Bayesian analysis of a point source epidemic. *J. Amer. Statist. Assoc.* *58*, 72–84.

Ibragimov, I. A. and Has'minskii, R. Z. (1981). *Statistical Estimation: Asymptotic Theory*. Springer, New York.

Inagaki, N. (1970). On the limiting distribution of a sequence of estimators with uniformity property. *Ann. Inst. Statist. Math.* *22*, 1–13.

James, W. and Stein, C. (1961). Estimation with quadratic loss. *Proc. Fourth Berkeley Symp. Math. Statist. Prob.* (ed. J. Neyman), Vol. 1. University of California Press, Berkeley, 361–380.

Le Cam, L. (1953). On some asymptotic properties of maximum likelihood estimates and related Bayes estimates. *Univ. of California Publ. in Statistics* *1*, 277–330.

Le Cam, L. (1960). Locally asymptotically normal families of distributions. *Univ. of California Publ. in Statistics* *3*, 37–98.

Le Cam, L. (1970). On the assumptions used to prove asymptotic normality of maximum likelihood estimates. *Ann. Math. Statist.* *41*, 802–828.

Le Cam, L. (1972). Limits of experiments. *Proc. Fifth Berkeley Symp. Math. Statist. Prob.* (eds. L. M. Le Cam, J. Neyman and E. M. Scott), Vol. 1. University of California Press, Berkeley, 245–261.

Le Cam, L. (1973). Sur les contraintes imposées par les passages à limite usuels en statistique. *Bull. Int. Statist. Inst.* *45*, 169–180.

Le Cam, L. (1974). *Notes on Asymptotic Methods in Statistical Decision Theory*. Centre de Recherches Mathématiques, Université de Montréal.

Le Cam, L. and Yang, G. L. (1990). *Asymptotics in Statistics*. Springer-Verlag, New York.

Lehmann, E. L. (1983). *Theory of Point Estimation*. Wiley, New York.

Mallows, C. (1973). Some comments on C_p. *Technometrics* *15*, 661–675.

Nussbaum, M. (1996). The Pinsker bound: a review. Unpublished manuscript.

Pfanzagl, J. (1994). *Parametric Statistical Theory*. Walter de Gruyter, Berlin.

Pinsker, M. S. (1980). Optimal filtration of square–integrable signals in Gaussian white noise. *Problems Inform. Transmission* *16*, 120–133.

Pratt, J. (1976). F. Y. Edgeworth and R. A. Fisher on the efficiency of maximum likelihood estimation. *Ann. Statist.* *4*, 501–514.

Rosenblatt, M. (1956). Remarks on some nonparametric estimates of a density function. *Ann. Math. Statist.* *27*, 832–837.

Shao, J. (1997). An asymptotic theory for model selection. *Statistica Sinica* *7*, 221–264.

Stein, C. (1956). Inadmissibility of the usual estimator for the mean of a multivariate normal distribution. *Proc. Third Berkeley Symp. Math. Statist. Prob.* (ed. J. Neyman), Vol. 1. University of California Press, Berkeley, 197–206.

Stein, C. (1966). An approach to the recovery of inter-block information in balanced incomplete block designs. *Research Papers in Statistics: Festschrift for Jerzy Neyman* (ed. F. N. David). Wiley, New York, 351–366.

Stigler, S. M. (1986). *The History of Statistics: The Measurement of Uncertainty before 1900*. Belknap Harvard, Cambridge, Massachusetts.

Wolfowitz, J. (1965). Asymptotic efficiency of the maximum likelihood estimator. *Theory Probab. Appl. 10*, 247–260.

ESSAY 2

Jaroslav Hájek and his Impact on the Theory of Rank Tests

Marie Hušková and Jana Jurečková
Department of Probability and Statistics
Charles University
Sokolovská 83, 186 00 Prague
Czech Republic

Rank tests were one of Hájek's beloved subjects. He joined the Department of Mathematical Statistics of Charles University in 1964, soon after the publication of his two basic asymptotic papers (Hájek (1961, 1962)) on the permutational CLT and on the asymptotically optimal rank test. Just at that time, while he intensively worked on the monograph (Hájek and Šidák (1967)), he accepted his first PhD students. Naturally, his students of that period wrote theses on rank-based procedures.

Hájek enriched the asymptotic theory of rank tests not only for several basic results but also for several original methods and tools which were later used by many authors. These include the *permutational central limit theorem* in Hájek (1961), the application of Le Cam's concept of *contiguity* and Le Cam's lemmas (Hájek (1962)), the use of the *weak convergence of empirical processes* (Hájek (1965)), the *projection method* and the *variance inequality* for the linear rank statistics (Hájek (1968 c)), fully and partially *adaptive rank tests* (Hájek (1962, 1970)), among others.

Hájek systematically studied asymptotic properties of linear and of some nonlinear rank tests. The basic components of linear rank tests are *the simple linear rank statistics*, while nonlinear rank tests Hájek expressed as functionals of the rank scores processes. In the series of papers: Hájek (1961, 1962, 1968 a, b, c, 1970) and Hájek and Dupač (1969 a, b), Hájek systematically investigated the asymptotic properties of linear rank statistics under null hypotheses, under local (contiguous) alternatives and some non-local alternatives. Besides that, in Hájek (1965) he proposed a test which represents a straightforward extension of the Kolmogorov–Smirnov test to the regression alternatives and in Hájek (1968 a) he derived the rank test of

Collected Works of Jaroslav Hájek – With Commentary
Edited by M. Hušková, R. Beran and V. Dupač

independence in a bivariate distribution, locally most powerful against specific dependence alternatives.

The results published before 1967 were then included, unified and complemented, in the famous monograph, Hájek and Šidák (1967), written jointly with Z. Šidák. Hájek's textbook (Hájek (1969)) of rank tests also deserves attention. This collection of papers, though not large, represents a substantial contribution to the asymptotic theory of rank tests; it was a starting point of the research of many authors and it is a rich source of ideas even today.

The typical form of the simple linear rank statistic corresponding to the random vector $X_{N1}, \ldots, X_{NN}$ is

$$S_N = \sum_{i=1}^{N} c_{Ni} a_N(R_{Ni}),$$

where $R_{N1}, \ldots, R_{NN}$ are the ranks of $X_{N1}, \ldots, X_{NN}$, while X_{Ni} has the distribution function F_{Ni}, $i = 1, \ldots, n$; the c_{Ni} are known regression constants, and the $a_N(i)$ are the scores. Hájek (1961) gave necessary and sufficient conditions of the Lindeberg type for the asymptotic normality of S_N as $N \to \infty$ for independent $X_{N1}, \ldots, X_{NN}$, under the null hypothesis $F_1 = \ldots = F_N = F$, where F is a continuous *df*, otherwise unspecified; this extends the results of Wald and Wolfowitz (1944), Noether (1949), Hoeffding (1951), Dwass (1953, 1955), and Motoo (1957). If both the a_{Ni}'s and c_{Ni}'s are infinitesimal when normalized to zero sums and unit sums of squares, then the condition reads

$$\lim_{N\to\infty} \sum \sum_{|c_{Ni}a_{Nj}|>\epsilon} c_{Ni}^2 a_{Nj}^2 = 0$$

for every $\epsilon > 0$. This *permutational CLT* is applicable not only to linear rank statistics, but also, e. g., in sampling from a finite population. Under very mild conditions, Hájek (1961) also derived the first order asymptotic representation for S_N by a sum of independent summands; this result was later extended by many authors in various contexts.

Making use of the Le Cam *concept of contiguity*, Hájek (1962) proved the asymptotic normality of S_N under contiguous alternatives, restricting consideration to absolutely continuous underlying distributions with finite Fisher information. For example, under location alternatives with likelihood function $\prod_{i=1}^{N} f(x_i - d_{Ni})$ compared with $\prod_{i=1}^{N} f(x_i)$, the conditions $\max_{1\leq i\leq N}(d_{Ni} - \bar{d}_N)^2 \to 0$ and $I(f)\sum_{i=1}^{N}(d_{Ni} - \bar{d}_N)^2 \to b^2$, $0 < b < \infty$, where $I(f)$ is the Fisher information, establish the asymptotic normality of S_N with simple expressions for asymptotic mean and variance. This result enables finding asymptotically most powerful rank tests against specific contiguous alternatives. Hájek derived similar results for other situations, e. g., for scale alternatives, for paired comparisons, and for the one-way layout with χ^2-type statistics.

In the paper, Hájek (1962), Hájek also constructed a histogram type estimator $\widehat{\varphi}$ of the optimal score-generating function φ_F and showed that the rank test based on $\widehat{\varphi}$ is asymptotically efficient. The proposed test is probably one of the first fully adaptive tests. Hájek (1970) proposed also some partially *adaptive rank-based procedures*; the procedure selects one of a finite set of scores functions and hence the pertaining rank test. These partially adaptive procedures were later considered by many authors.

In the paper, Hájek (1965), Hájek extended the Kolmogorov–Smirnov test to verify the hypothesis of randomness against regression alternatives and derived the limit distribution of the test criterion. Expressing the criterion as a functional of a special empirical process, Hájek attacked this problem using the *weak convergence of empirical processes*, a pioneering approach in 1965. To prove the tightness, Hájek extended the Kolmogorov inequality to dependent summands; this inequality is interesting in itself.

Chernoff and Savage (1958) and Govindarajulu, Le Cam and Raghavachari (1966) proved the asymptotic normality of two-sample rank statistics under some nonlocal alternatives for some classes of score-generating functions. Later Hájek (1968 c) and Hájek and Dupač (1969 a) gave a far-reaching generalizaton of the Chernoff–Savage theorem. Hájek proved the results with the aid of the L_2 projection of S_N on the space of sums of N independent summands. The *projection method*, together with an ingenious *variance inequality*,

$$\operatorname{var}\left\{\sum_{i=1}^{N} c_{Ni} a_N(R_{Ni})\right\} \leq 21 \max_{1\leq i\leq N} (c_{Ni} - \bar{c}_N)^2 \sum_{i=1}^{N} (a_N(R_{Ni}) - \bar{a}_N)^2,$$

has found wide applications in various contexts.

Hájek and Dupač (1969 a), using the projection method and a more elaborate treatment of the residual variance, extended the above results to possibly discontinuous score functions, under slightly more restrictive conditions on the distributions. The same authors then, in Hájek and Dupač (1969 b), specialized the results to the two-sample Wilcoxon statistic under divergent alternatives.

In his last paper on rank tests (Hájek (1974)), Hájek demonstrated that not only the best Pitman efficiency but also the best exact Bahadur slope is attainable by rank statistics; in other words, he proved the asymptotic sufficiency of the vector of ranks in the Bahadur sense for testing randomness against a general class of two-sample alternatives.

Hájek's methods were later used in various contexts and his results were extended to general (univariate and multivariate) linear models, to time series and to other models. Practically every test now has a nonparametric rank-based counterpart. The class of R-estimators, based on ranks, is considered as one of the basic classes of robust estimators; their asymptotic properties are also derived from those of the pertaining rank tests. Hájek's students were some of the first to extend his results in their PhD theses (see p. 51 for the list of students).

It is impossible to review all further developments and directions of rank test theory after Hájek. Instead, we shall briefly describe a recent application of Hájek's 1962 paper in the change point problem and the concept of *regression rank scores* which surprisingly extends Hájek's results at a time when we were sure that little of importance could be added to his ideas.

The results (both the CLT and Lemma 2.1 for simple linear rank statistics) in Hájek (1961)) together with the martingale property of a class of two-sample rank statistics are crucial tools in investigation of limit behavior of test statistics and estimators for the change point problem based either on ranks or on M-statistics.

The regression rank scores of observations $Y_1, \ldots, Y_n$ in the linear model $\boldsymbol{Y} = \boldsymbol{X}\boldsymbol{\beta} + \boldsymbol{e}$, $\boldsymbol{\beta} \in \boldsymbol{r}^p$, with an intercept $[x_{i1} = 1,\ i = 1, \ldots, n]$ are defined as the optimal solution $\widehat{\boldsymbol{a}}_n(\alpha) \in \boldsymbol{r}^n$ of the parametric linear programming problem

$$\boldsymbol{Y}'\widehat{\boldsymbol{a}}(\alpha) := \max$$
$$\boldsymbol{X}'\widehat{\boldsymbol{a}}(\alpha) = (1-\alpha)\,\boldsymbol{X}'\mathbf{1}_n$$
$$\widehat{\boldsymbol{a}}(\alpha) \in [0,1]^n, \quad 0 < \alpha < 1.$$

The regression rank scores first appeared in Gutenbrunner (1986) as dual in the linear programming sense to the *regression quantiles* of Koenker and Bassett (1978) and were later studied in Gutenbrunner and Jurečková (1992). When specialized to the location model $[\boldsymbol{X} = \mathbf{1}_n]$, the process $\{\widehat{\boldsymbol{a}}_n(\alpha) : 0 \leq \alpha \leq 1\}$ coincides with the Hájek rank scores process studied in Hájek (1965). The linear programming duality also extends the duality of ranks and order statistics to the linear regression model. Moreover, $\widehat{\boldsymbol{a}}_n(\alpha)$ are invariant to $\boldsymbol{\beta}$ in the sense that $\widehat{\boldsymbol{a}}_n(\alpha, \boldsymbol{Y} + \boldsymbol{X}\boldsymbol{b}y = \widehat{\boldsymbol{a}}_n(\alpha, \boldsymbol{Y}) + \boldsymbol{b}\ \forall \boldsymbol{b} \in \boldsymbol{r}^p$. Hence, the inference based on $\widehat{\boldsymbol{a}}_n(\alpha)$ is also invariant and could be used in the presence of a nuisance regression. The tests based on regression rank scores successfully replace the aligned rank tests and their asymptotic behavior is the same as in the case of known $\boldsymbol{\beta}$, hence it follows directly from Hájek's asymptotic theory. For the theory and practical applications of regression rank scores we refer to Gutenbrunner et al. (1993), Koenker and d'Orey (1987, 1994), Hušková (1994), Koul and Saleh (1995), Jurečková (1997), Hallin and Jurečková (1997) and Hallin et al. (1997, a, b) among others.

REFERENCES

Chernoff, H. and Savage, I. R. (1958). Asymptotic normality and efficiency of certain nonparametric test statistics. *Ann. Math. Statist. 29*, 972–994.

Dwass, M. (1953). On the asymptotic normality of certain rank order statistics. *Ann. Math. Statist. 24*, 303–306.

Dwass, M. (1955). On the asymptotic normality of some statistics used in non-parametric setup. *Ann. Math. Statist. 26*, 334–339.

Gombay, E. and Hušková, M. (1997). Rank based estimators of the change-problem. *J. Statist. Plann. and Infer.*, to appear.

Govindarajulu, L., Le Cam, L. and Raghavachari, M. (1966). Generalizations of theorems of Chernoff and Savage on the asymptotic normality of test statistics. In *Proc. 5th Berkeley Symp. Math. Statist. Probab. 1*, 609–638.

Gutenbrunner, C. (1986). Zur Asymptotik von Regressionquantileprozessen und daraus abgeleiteten Statistiken. *PhD Thesis*, Universität Freiburg.

Gutenbrunner, C. and Jurečková, J. (1992). Regression rank scores and regression quantiles. *Ann. Statist. 20*, 305–330.

Gutenbrunner, C., Jurečková, J., Koenker, R. and Portnoy, S. (1993). Tests of linear hypotheses based on regression rank scores. *J. Nonpar. Statist. 2*, 307–331.

Hájek, J. (1961). Some extensions of the Wald – Wolfowitz – Noether theorem. *Ann. Math. Statist. 32*, 506–523.

Hájek J. (1962). Asymptotically most powerful rank order tests. *Ann. Math. Statist. 33*, 1124–1147.

Hájek, J. (1965). Extension of the Kolmogorov – Smirnov test to the regression alternative. In *Bernoulli – Bayes – Laplace*, Proc. Internat. Research Seminar (eds. J. Neyman and L. Le Cam). Springer Verlag, 45–60.

Hájek, J. (1968 a). Locally most powerful tests of independence. In *Studies in Math. Statist.* (eds. K. Sarkadi and I. Vincze). Akadémiai Kiadó, Budapest, 45-51.

Hájek, J. (1968 b). Some new results in the theory of rank tests. In *Studies in Math. Statist.* (eds. K. Sarkadi and I. Vincze). Akadémiai Kiadó, Budapest, 53–55.

Hájek, J. (1968 c). Asymptotic normality of simple linear rank statistics under alternatives. *Ann. Math. Statist. 39*, 325–346.

Hájek, J. (1969). *A Course in Nonparametric Statistics*. Holden–Day, San Francisco.

Hájek, J. (1970). Miscellaneous problems of rank test theory. *Nonparametric Techniques in Statist. Inference.* (ed. M. L. Puri). Cambridge Univ. Press, 3–19.

Hájek, J. (1974). Asymptotic sufficiency of the vector of ranks in the Bahadur sense. *Ann. Statist. 2*, 75–84.

Hájek, J. and Dupač, V. (1969 a). Asymptotic normality of simple linear rank statistics under alternatives II. *Ann. Math. Statist. 40*, 1992–2017.

Hájek, J. and Dupač, V. (1969 b). Asymptotic normality of the Wilcoxon statistic under divergent alternatives. *Zastos. Mat. 10*, 171–178.

Hájek, J. and Šidák, Z. (1967). *Theory of Rank Tests.* Academia, Prague and Academic Press, New York.

Hallin, M. and Jurečková, J. (1997). Optimal tests for autoregressive models based on regression rank scores. Submitted.

Hallin, M., Jurečková, J., Picek, J. and Zahaf, T. (1997). Nonparametric tests of independence of two autoregressive time series based on autoregression rank scores. *J. Statist. Plann. Infer.*, in print.

Hallin, M., Zahaf, T., Jurečková, J., Kalvová, J. and Picek, J. (1997). Nonparametric tests in AR models with applications in climatic data. *Environmetrics 8*, to appear.

Hoeffding, W. (1951). A combinatorial central limit theorem. *Ann. Math. Statist. 22*, 558–566.

Hušková M. (1994). Some sequential procedures based on regression rank scores. *J. Nonpar. Statist. 3*, 277–298.

Hušková M. (1997a). Limit theorems for rank statistics. *Statist. & Probab. Letters 32*, 45–55.

Hušková M.(1997b). Limit theorems for M–processes via rank statistics processes. In *Advances in Combinatorial Meth. and Appl. to Probability and Statistics* (ed. N. Balakrishnan). Birkhäuser Boston, 521–534.

Jurečková, J. (1997). Regression rank scores tests against heavy-tailed alternatives. *Bernoulli* (to appear).

Koenker, R. and Bassett, R. (1978). Regression quantiles. *Econometrica 46*, 33–50.

Koenker, R. and d'Orey, V. (1987). Computing regression quantiles. *Appl. Statist. 36*, 383–393.

Koenker, R. and d'Orey, V. (1994). A remark on algorithm AS 229: Computing dual regression quantiles and regression rank scores. *Appl. Statist. 42*, 410–414.

Koul, H. L. and Saleh, A. K. Md. E. (1995). Autoregression quantiles and related rank scores processes. *Ann. Statist. 23*, 670–689.

Motoo, M. (1957). On the Hoeffding's combinatorial central limit theorem. *Ann. Inst. Statist. Math. 8*, 145–154.

Noether, G. E. (1949). On a theorem of Wald and Wolfowitz. *Ann. Math. Statist. 20*, 455–458.

Raghavachari, M. (1970). On a theorem of Bahadur on the rate of convergence of test statistics. *Ann. Math. Statist. 41*, 1695–1699.

Wald, A. and Wolfowitz, J. (1944). Statistical tests based on permutations of the observations. *Ann. Math. Statist. 15*, 358–372.

Woodworth, G. (1970). Large deviatons and Bahadur efficiency in linear rank statistic. *Ann. Math. Statist. 41*, 251–283.

ESSAY 3

Recollections on my Contacts With Jaroslav Hájek

Lucien Le Cam
Department of Statistics
University of California
Berkeley, CA 94720
U.S.A.

INTRODUCTION

Jaroslav Hájek was one of the most inspired and inspiring statisticians of our times. His scientific path crossed mine on a number of subjects. One can mention the introduction of contiguity, the use of derivatives in quadratic mean for square roots of densities (instead of the classical Cramér type conditions), and some improvements on the handling of Chernoff–Savage statistics. There are even some results to which both our names have become attached: the Hájek–Le Cam asymptotic minimax theorem and the Hájek–Le Cam convolution theorem. This happened in spite of the fact that we had little opportunity for collaboration and came to the problems from very different directions.

I shall recount below, to the best of my recollection, how these matters developed.

PERSONAL CONTACTS

Our personal contacts were very limited. The first occurred during the Academic Year 1961–1962, part of which Hájek spent at Berkeley. This was not a particularly propitious time. It was my first year as Chairman of the Berkeley Department. In spite of the fact that chairmanship at that time was much less demanding than now, it was still a drain on my time. Our system was not as structured as it is now. I inherited a situation where I was

Collected Works of Jaroslav Hájek – With Commentary
Edited by M. Hušková, R. Beran and V. Dupač

chairman, graduate adviser, admission committee, personnel committee, and whatever else. I had a full teaching load and the ordinary crises did not fail to occur. In addition to that, Hájek, being extremely courteous, did not want to 'bother' me. So we had precious few contacts. Still they brought forth some results that will be described below.

Our second chance came in Spring and Summer 1966. I had left the chairmanship, but my time was taken by the editing of the Fifth Berkeley symposium, a rather monumental task. That year I had the occasion to witness how our University administration took advantage of visitors from Iron Curtain countries by paying them miserly salaries computed on some version of the official exchange rates. Having asked Hájek to stay in the Summer, I had to battle fiercely to secure him a decent compensation on my NSF Grant!

We met again in 1970 at the occasion of the Sixth Symposium, but only for a short period. Since we hardly corresponded by mail, it is astonishing that our efforts were at all linked. I shall now describe some specifics of our interactions.

CONTIGUITY AND RELATED SUBJECTS

In his ' Asymptotically most powerful rank order tests' of 1962, Hájek says: 'The notion of contiguity had been developed independently by Le Cam and the present author.' He then explains that a previous version of his paper, based on a criterion of de la Vallée Poussin, was modified to take into account some of my ideas.

I can certainly claim priority for introducing the *name* 'contiguity'. This was done in my office around 1956, with the help of J. D. Esary. For the concept itself, the matter is more complicated. Contiguity is a property of sequences (or nets) of pairs (P_ν, Q_ν) of probability measures. The pair (P_ν, Q_ν) is carried by a space $(\mathcal{X}_\nu, \mathcal{A}_\nu)$ that may vary as ν changes. In the terminology used by de la Vallée Poussin, it could have been called some sort of uniform asymptotic absolute continuity. However, in de la Vallée Poussin's case, the spaces $(\mathcal{X}_\nu, \mathcal{A}_\nu)$ were fixed, equal to a given $(\mathcal{X}, \mathcal{A})$. Also one of the measures, say, P_ν, was fixed, equal to P.

The jump from a fixed $(\mathcal{X}, \mathcal{A}, \mathcal{P})$ to totally variable entities was an easy one for me, as it must have been for Hájek. Yet it bothered some of my more mathematically inclined friends, in particular Michel Loève, who would not hear of such heresy. In his 1962 paper Hájek indicates that he had made the jump and adapted to that situation an integral criterion of de la Vallée Poussin that, however, gave him weaker results than my own analysis of the situation. I had not thought of looking at de la Vallée Poussin.

In late 1961 or early 1962, I gave Hájek a set of handwritten notes. They contained what Hájek and Šidák were kind enough to baptize 'Le Cam's second lemma' in their 1967 book. It is a lemma that allows, in the independent case and under some restrictions, passage from sums of square roots of densities to the sums of their logarithms. It differed from previously available results of Loève (1947) by the fact that expectations of truncated variables do not appear in its statement or in its applications. (It should be noted that in the so-called i.i.d. case the proof of the lemma can be greatly simplified. Also a lemma of a similar nature remains valid for martingale situations. See Le Cam (1986), Section 5, Chapter 10).

What Hájek and Šidák called 'Le Cam's first lemma' and 'Le Cam's third lemma' had already been published in my 1960 LAN paper. What my 'second lemma' does in the i.i.d. case is to make it almost obvious that conditions of the Cramér type can be replaced by a condition of differentiability in quadratic mean on the square roots of the densities. Hájek uses that forcefully in his 1962 paper, albeit only for densities parametrized by a shift parameter.

Which one of the two of us first thought of the possibility of replacing Cramér's conditions by a condition of differentiability in quadratic mean is unclear. Both of us certainly knew of the possibility in early Spring 1962. I had come to it through considerations about the asymptotic normality of experiments but it is clear that I did not know exactly what the situation was when, in December 1957, I wrote my LAN paper, eventually published in 1960 in a series intended for fast publication! Hájek had come to it through his study of rank order tests. For these he could restrict himself to shift families, and did so. In that case, differentiability in quadratic mean of square roots of densities boils down to the finiteness of 'Fisher's information', a particularly pleasant happenstance. I have put 'Fisher's information' in quotation marks because one has to be careful about exactly what that means. See the remarks in my 1986 book, page 585.

Later, in a paper on 'The assumptions used to prove asymptotic normality of maximum likelihood estimates', I was to emphasize the difference between differentiability in quadratic mean when that mean is taken under a dominating measure contrasted with taking means under the probability at a particular parameter value. I stated there a Proposition 1, page 808. My 'proof' is all messed up. Hájek pointed out a very simple proof. It is recorded in my 1986 book, page 570.

I should add that at the time Hájek wrote his 1962 paper I was preoccupied by my 'distance between experiments' and not yet able to see what it could say about rank tests. My paper 'Sufficiency and approximate sufficiency' had been written in 1960. It was obvious to me at the time that it could clarify and simplify many of the considerations of my LAN paper. Indeed, these considerations were what spurred my introduction of distances between experiments. However, I had not had the time to write up the connection.

THE CHERNOFF–SAVAGE STATISTICS

Hájek and I had a second chance to interact in 1966. By that time I had been involved in some non-parametric test problems because of a paper with Raju and Chari (= Govindarajulu and Raghavachari). Hájek had also been involved with the problem treated by Chernoff and Savage, as can be seen in the Hájek–Šidák book, page 233. (Note that the book appeared in 1967, but that it was written in 1963!) Hájek had previously used his projection theorem, for which see his 1968 paper, Lemma 4.1 and Theorem 4.1. In late January 1966, Hájek presented me with a magnificent result: a bound on the variance of statistics that are linear combinations of functions of individual ranks. See his Theorem 3.1 in the 1968 paper mentioned above. The distributions of the observable variables could be anything they pleased, subject to independence. He knew, and I could see, that this could drastically modify and simplify proofs of theorems on the Chernoff–Savage statistics.

Being preoccupied by other matters, I did not react promptly and properly. On the other hand, Neyman did react, in a way that made Hájek's life and mine a bit difficult: Neyman insisted that Hájek should not interfere with the work I had done with Raju and Chari! I believe that Neyman visualized us as competing about something or other. There was nothing of the sort. We were just not built that way. Further, to be 'competing' about the Chernoff–Savage statistics would have been ridiculous: my involvement with them was distant, accidental and incidental.

ASYMPTOTIC MINIMAXITY AND ADMISSIBILITY

There is a theorem, often called the Hájek–Le Cam asymptotic minimax theorem, that is extremely useful in asymptotic considerations. It was stated very clearly in Hájek's paper given in the 1970 Berkeley Symposium for an asymptotically normal situation, a weakened form of the LAN conditions. Hájek's theorem has two parts. The first is an asymptotic minimax assertion. The second is a statement of asymptotic uniqueness and admissibility. Except for the consideration of nuisance parameters, the argument is restricted to one-dimensional parameters.

As it turns out I had obtained a related result in my 1952 thesis, published in

1953. Theorem 13, page 326, of that paper is a theorem about admissibility. However, my paper was only about the i.i.d. case under conditions of the Cramér type. (This was not so bad for those times!) The admissibility result was for one-dimensional parameters with a remark that:'In the case of an r-dimensional parameter, the problem becomes much more complicated. The difficulties are conceptual as well as mathematical.' (That does not mean that I had anticipated Charles Stein's result on the inadmissibility of means in the Gaussian case when $r \geq 3$. I was simply a bit cautious.) I did not say much about asymptotic uniqueness.

Rereading my 1953 paper, I was puzzled and chagrined by the fact that it does not contain an asymptotic minimax theorem for the estimation of multi-dimensional parameters. Such a theorem follows in a few lines from my Theorem 11, page 322. When, in 1954, Herman Chernoff presented a minimax theorem adapted from unpublished statements of H. Rubin and C. Stein (see his 1956 paper), I could argue that my theorem was 'better' than his because it was multi-dimensional and covered a large variety of loss functions. Chernoff's was one dimensional and for possibly truncated quadratic loss. From that point on, I behaved under the illusion that the multi-dimensional theorem was actually stated in my paper. I should have reread it!

The essential parts of that paper had precious little impact anyway. It is usually remembered only for the part about 'super-efficiency'. It is probably because of my illusions that, when I first heard about Hájek's theorem, I automatically interpreted its 'minimax' part as applying to the multi-dimensional case. We know now that the minimax part is valid for any finite number of dimensions but that the asymptotic uniqueness holds only in one and two dimensions. Hájek must have attached to it more importance than I did.

After mulling over the situation and Hájek's 1970 paper on the convolution theorem, I had some idea of what lies behind the result. I expressed it in 1972 in the Proceedings of the Sixth Berkeley Symposium. That paper of mine is marred by several untenable 'proofs', but the results are correct. Later, in the Hájek Memorial Volume, I gave what I think are correct proofs. Thus the eponym 'Hájek–Le Cam' for the asymptotic minimax theorem is probably justified. The term 'minimax' could have been advantageously changed to one recognizing that it is an asymptotic lower semi-continuity result.

Perhaps I should explain why I kept working on the asymptotic minimax theorem: my mind works in a strange manner. It does not really accept things unless they are codified in a general preconstructed structure. Here the preconstructed structure happened to be that of convergence of experiments and, more particularly, weak convergence of experiments. It turned out to be helpful for the understanding of Hájek's asymptotic minimax result as well as for his convolution theorem (for which see the next section). Note that Hájek's 1972 theorem has two parts. Part 4.1 is about minimax. Part 4.2 is about admissibility. The first part was easy to imbed in my preconceived notions. The second part took more time and work. It was finally written in 1974 for the Hájek Memorial Volume published in 1979.

After some excursions in blind alleys, I reached the following formulation. Take a statistical experiment $\mathcal{F} = \{Q_\theta; \theta \in \Theta\}$ and a loss function W, bounded from below for each value of θ. Let $\mathcal{R}(\mathcal{F}, W)$ be a space defined as follows. One takes those functions of θ that are possible risk functions for the experiment $\mathcal{F}$ and W or larger than such functions. One closes it for pointwise convergence on Θ. For an experiment $\mathcal{E}$ indexed by Θ, let $\mathcal{E}[S]$ be the experiment obtained by restricting θ to a subset $S \subset \Theta$. For $b < \infty$, let W_b denote the infimum $W_b = b \wedge W$. The theorem then says the following.

Theorem 1. *Take a function $f \notin \mathcal{R}(\mathcal{F}, W)$. Then there is a finite set $S \subset \Theta$, an $\epsilon > 0$, an $\alpha > 0$ and a $b < \infty$ such that if the distance between $\mathcal{E}[S]$ and $\mathcal{F}[S]$ does not exceed ϵ then $f + \alpha$ restricted to S does not belong to $\mathcal{R}(\mathcal{E}[S], W_b)$.*

Given this, it takes just a bit of fiddling to obtain the first assertion in Theorem 4.1 of Hájek, 1972. The admissibility part of that theorem was not so obvious, but I finally obtained a statement that implies Hájek's. It singles out those elements $g \in \mathcal{R}(\mathcal{F}, W)$ that are uniquely admissible and produced by a non-randomized procedure. See my 1979 paper or my 1986 book, Chapter 7, Section 5, pages 112–117.

HÁJEK'S CONVOLUTION THEOREM

The theorem appeared in *Z. Wahrschein. verw. Gebiete* (1970). Peter Bickel brought it to my attention and we discussed it briefly some time before the start of the Sixth Berkeley Symposium. Peter thought that Hájek's proof was too complicated. He devised his own. That proof still used analyticity properties of Gaussian characteristic functions. See Bickel et al. (1993). I decided on the spot that I could construct a proof that did not depend on special properties of Gaussian shift experiments except that they are obtained by shifting a measure that is absolutely continuous with respect to the Lebesgue measure. That became my contribution to the Sixth Symposium. It runs along the following lines:

(a) Take a sequence $\mathcal{E}_n$ of experiments $\mathcal{E}_n = \{P_{\theta,n}; \theta \in \Theta\}$ and statistics S_n defined on $\mathcal{E}_n$. Suppose that the distributions of S_n for $P_{\theta,n}$ converge to limits, say, G_θ. Suppose also that the experiments $\mathcal{E}_n$ converge weakly in the sense of experiments to a limit $\mathcal{F}$. Then the experiment $\mathcal{G} = \{G_\theta; \theta \in \Theta\}$ is weaker than $\mathcal{F}$. I called 'distinguished' those sequences S_n such that $\mathcal{F}$ and $\mathcal{G}$ are equivalent.
(b) The next step was to note that if the pair $(\mathcal{F}, \mathcal{G})$ is invariant under the action of an amenable group, there is a transition A from $\mathcal{F}$ to $\mathcal{G}$ that has the resulting invariance properties.

(c) Finally, if $\mathcal{F}$ and $\mathcal{G}$ are generated by the group shifts acting on measures, say F and G, carried by the same group, if the group is locally compact and if F is dominated by the Haar measure, then the invariant transition A is a convolution by a certain probability measure.

Parts (b) and (c), or something very close, were known from C. Boll's thesis of 1955. I gave a different argument for (b) in my 1964 paper on 'Sufficiency and approximate sufficiency.' That part (c) was a theorem of J. Wendel published in 1952 was unknown to all of us! The fact that the 'centerings' that occur in the locally asymptotically normal expansions of log likelihoods ratios give 'distinguished' sequences of statistics was rather obvious. Thus I had a proof of the convolution theorem built on considerations rather different from those of Hájek. It had the merit that it could be conveyed to an interested listener without writing anything down.

It was only a few years later that I noted that the invariance properties involved in (b) are automatic almost everywhere for the usual passages to the limit in parametric problems. My theorem had also the added feature that it could be applied to non-Gaussian limits. They do occur, but not frequently enough that this would be a main virtue. However, the fact that the argument in part (c) involved (i) locally compact groups, and (ii) absolute continuity with respect to Haar measure was a puzzle. This later part has something to do with the possibility of representation of the dual of the L-space of $\mathcal{F}$ by a space of measurable functions, but the situation is not clear, at least to me. The involvement of 'local compactness' had to be remedied for applications to non-parametric problems. This was done eventually by P. W. Millar (1985) and by A. van der Vaart (1988, 1991). I think that further generalizations are possible, but my 1994 paper is not entirely convincing.

It is clear that in the case of the convolution theorem, Hájek did the pioneering work. I added a little bit and feel honored that my name was coupled to Hájek's.

CONCLUSION

Hájek and I had precious little chance for collaboration. In spite of that our work, originating from different considerations, overlapped to a noticeable extent. I clearly benefited from his ideas and he used some of mine.

He was a very good friend. I deeply regret his untimely demise.

REFERENCES

Bickel, P. , Klaassen, C., Ritov, Y. and Wellner, J. (1993). *Efficient and Adaptive Estimation for Semiparametric Models.* Johns Hopkins.

Boll, C. (1955). *Comparison of experiments in the infinite case and the use of invariance in establishing sufficiency.* Unpublished Ph.D. thesis, Stanford.

Chernoff, H. (1956). Large sample theory: Parametric case. *Ann. Math. Statist. 27*, 1–22.

Govindarajulu, Z., Le Cam, L. and Raghavachari, M. (1967). Generalizations of theorems of Chernoff and Savage on asymptotic normality of nonparametric test statistics. *Proc. Fifth Berkeley Symp. Math. Statist. Probab. 1*, 609–638.

Hájek, J. (1962). Asymptotically most powerful rank order tests. *Ann. Math. Statist. 33*, 1124–1147.

Hájek, J. (1968). Asymptotic normality of simple linear rank statistics under alternatives. *Ann. Math. Statist. 39*, 325–346.

Hájek, J. (1970). A characterization of limiting distributions of regular estimates. *Z. Wahrsch. verw. Gebiete. 14*, 323–330.

Hájek, J. (1972). Local asymptotic minimax and admissibility in estimation. *Proc. Sixth Berkeley Symp. Math. Statist. Probab. 1*, 175–194.

Hájek, J. and Šidák, Z. (1967). *Theory of rank tests.* Academia Praha and Academic Press.

Le Cam, L. (1953). On some asymptotic properties of maximum likelihood estimates and related Bayes' estimates. *Univ. California Publ. Statist. 1*, 277–330.

Le Cam, L. (1960). Locally asymptotically normal families of distributions. *Univ. California Publ. Statist. 3*, 37–98.

Le Cam, L. (1972). Limits of experiments. *Proc. Sixth Berkeley Symp. Math. Statist. Probab. 1*, 245–261.

Le Cam, L. (1979). On a theorem of J. Hájek. *The Jaroslav Hájek Memorial Volume.* (ed. J. Jurečková). Academia Prague, 119–135.

Le Cam, L. (1986). *Asymptotic Methods In Statistical Decision Theory.* Springer–Verlag.

Le Cam, L. (1994). An infinite dimensional convolution theorem. *Statistical Decision Theory and Related Topics 5.* (eds. S. Gupta and J. Berger). Springer–Verlag, New York, 401-411.

Loève, M. (1947). A l'interieur du problème limite central. *Publ. Inst. Statist. Univ. Paris 6*, 313-325.

Millar, P. W. (1985). Nonparametric applications of an infinite dimensional convolution theorem. *Z. Wahrsch. verw. Gebiete 68*, 545–556.

van der Vaart, A. (1988). *Statistical Estimation in Large Parameter Spaces.* CWI Tracts 44, Center for Mathematics and Computer Sciences, Amsterdam.

van der Vaart, A. (1991). An asymptotic representation theorem. *Internat. Statist. Review 58*, 97–121.

Wendel, J. G. (1952). Left centralizers and isomorphisms of group algebras. *Pacific J. Math. 2*, 251–261.

ESSAY 4

On Some Early Papers of Jaroslav Hájek

Josef Machek
Department of Probability and Statistics
Charles University
Sokolovská 83, 186 00 Prague
Czech Republic

The papers will be surveyed here in chronological order.

Before proceeding to review Hájek's first papers, a few words about the conditions under which they had been written, about the environment in which the corresponding research had taken place, seem worthwhile.

J. Hájek studied 'statistical and actuarial engineering' at the Czech Technical University in Prague in the years 1945 to 1949. It should be borne in mind that this was exactly the period in which the consequences of the long gap in the development of Czech universities and science, caused by World War II, were still strongly felt: for the six years of the war there had been no contact with the advances made abroad, important bibliography was not accessible and these shortcomings were only gradually being overcome. No wonder then that in his first research papers Hájek was somewhat 'too eloquent', according to the criteria of the time, in his explications and that from time to time he coined his own terms for some concepts since the Czech terminology was not yet fully established.

The sources of inspiration for his research were, accordingly, two: (i) older statistical literature, and (ii) cooperation with institutions that applied statistical methods.

In spite of these unfavourable circumstances, already in Hájek' last year of university studies there appeared two papers written by him. Of those especially one – Hájek (1949 a) – deserves attention. In it the following problem is considered: to estimate the mean per element of a finite population of N elements subdivided into clusters of different sizes, on the basis of a probabilistic sample of n clusters. J. Neyman (1934) found an unbiased estimator for the population mean assuming a simple random sample of

Collected Works of Jaroslav Hájek – With Commentary
Edited by M. Hušková, R. Beran and V. Dupač

clusters. His estimator has a rather large variance when the cluster sizes vary considerably. Hájek in his paper proposed a modification of the sampling design consisting in first selecting one cluster with probability proportional to size and then a simple random sample of $n-1$ from the remaining $N-1$ clusters. This design provides an unbiased estimator of the population mean that is generally considerably less variable than Neyman's (1943) estimator. Hájek in his (1949 a) paper also gives a good approximation to the estimate of the variance of this estimator.

The procedure described above is, in fact, a special case of a procedure proposed two years later by Midzuno (1951) in relation to ratio estimates, frequently referred to as 'Midzuno's sampling scheme'.

In the other paper from the same year (Hájek (1949 b)), the author develops further the ideas of Hotelling's now classical (1944) paper on 'weighing designs'. Hájek's contribution here consists in finding optimal 'weighing designs' applying only fundamental theorems of probability theory without making recourse to the theory of the Least Squares Method. The results are illustrated in an experiment performed at the National Gauging Office.

After 1949 there is a surprising gap in Hájek's publishing activities and it is only in 1955 when new papers appear. One of these is not included in this volume. It is an invited lecture (co-authored by F. Fabian) read at a Meeting of Czechoslovak Statisticians. In this paper the authors deal with what may be called 'the foundations of (mathematical) statistics: its development, its relation to probability theory and, above all, the role of statistical methods both in science and engineering'. Three most common approaches to statistical inference are discussed, namely through 'inverse probability' (Bayes), decision-theoretic (Neyman–Pearson) and likelihood (R. A. Fisher). Special attention is paid to R. A. Fisher's ideas concerning the design of experiments.

In 1956 Hájek published two papers not included in this collection. The first of them (Hájek (1956 a)) is concerned with stratified sampling. In it the reader finds some refinements of the construction of confidence intervals for the mean or for the total of a stratified population and some remarks on the efficiency of what is called 'stratification after selection'. An interesting detail in this paper is the derivation of the well known 'principle of optimum allocation' through the Cauchy–Schwartz inequality.

From the same year we have a short note (Hájek (1956 b)) on a paper by Kosmák (1955). Kosmák in his paper finds the limit of a sequence of numbers defined by a rather complex recursive relation. In his note Hájek simplifies the proof and also suggests an alternative approach: the recursive relation in question defines a stationary Markov chain and the limit follows directly from the ergodic theorem and the special structure of the corresponding transition matrix. This brief note is mentioned here since it demonstrates Hájek's interest in – and aptitude for – 'matching probability problems to purely mathematical ones', e. g. to prove algebraic identities through statistical arguments.

Hájek was also involved in applied statistical work; this fact does not make itself palpable in his publications since his role generally was that of a consultant. However, there is one exception, namely his 1959 paper. This contains a report on a study of the distribution of the age at which a certain phenomenon characterizing the growth and development of man occurs. Among the biostatisticians it is known as 'the problem of milestones'. Hájek describes in this paper the sampling design applied and uses the techniques of analyzing biological assays with quantal response. We believe that the reading and discussion of this paper could, with advantage, be included in seminars for the students of applied statistics.

It is already Hájek's early work that manifests the main features of his personality as a statistician. First, multisidedness – from the very beginning of his activity he was interested in nearly all fields of statistics, such as sampling surveys, experimental design, parameter estimation, nonparametric methods, statistics in random processes, etc. Second, the originality of ideas – Hájek generally did not 'copy' some authority with an endeavour to obtain some slight generalization of results already known, but rather used to grasp the problem as a new one and to seek new approaches to its solution.

REFERENCES

Fabian, F. and Hájek, J. (1955). On some fundamental questions of mathematical statistics. (In Czech) *Časopis Pěst. Mat.* *80*, 387–399.

Hájek, J. (1949 a). Representative cluster sampling by a method of two phases. (In Czech) *Statistický Obzor 29*, 384–394.

Hájek, J. (1949 b). The use of factorial design and of confidence intervals in weighing. (In Czech) *Statistický Obzor 29*, 258–273.

Hájek, J. (1956 a). Stratified sampling. (In Czech) *Apl. Matem. 1*, 149–161.

Hájek, J. (1956 b). Remark on the article 'On certain sequences of points on a circle'. (In Czech) *Časopis Pěst. Mat. 81*, 77–78.

Hájek, J. and Poncová, V. (1959). The age of eruption of permanent teeth in children in Czechoslovakia. (In Czech) *Československá Stomatologie 2*, 104–113.

Hotelling, H. (1944). Some improvements in weighing and other experimental techniques. *Ann. Math. Stat. 15*, 297–306.

Kosmák, L. (1955). On certain sequences of groups of points on a circle. (In Czech) *Časopis Pěst. Mat. 80*, 299–309

Midzuno, H. (1951). On the sampling system with probability proportional to sum of sizes. *Ann. Inst. Statist. Math. 2*, 99–108.

Neyman, J. (1934). On the two different aspects of the representative method. *J. Roy. Statist. Soc. 97*, 558–623.

ESSAY 5

Contributions of Jaroslav Hájek to Statistical Inference on Stochastic Processes

Emanuel Parzen
Department of Statistics
Texas A & M University
College Station, Texas 77843–3143
U.S.A.

A few mathematical statisticians (including Whittle, Grenander, Rosenblatt, Parzen, Yaglom, and Hájek) in the 1950s pioneered the theory of statistical inference for stochastic processes and time series, which provides the foundations for statistical communication theory, which is applied by engineers in billion dollar information age industries, to design modern sophisticated communication devices. In his few papers in this research area Hájek made profound and long lasting contributions.

A famous, surprising, and extremely important theorem was published independently in 1958 by both Hájek and Feldman; called the *dichotomy* theorem, it states that Gaussian processes P_1 and P_2 are either perpendicular ($P_1 \perp P_2$) or equivalent ($P_1 \equiv P_2$). This theorem is important in the formulation of models for hypothesis testing problems to discriminate which Gaussian measure, P_1 or P_2, best fits a data sample (often P_1 represents noise and P_2 represents signal plus noise).

Statistical inference for stationary time series with *convex* correlation functions is a problem to which Hájek made profound contributions. He found useful lower bounds for the efficiency of predictors and estimators.

Among Hájek's papers on 'linear statistical problems in stochastic processes' his 1962 paper is acclaimed by statistical communication engineers in Weinert (1982) as a remarkable and profound contribution. The concepts and notation of RKHS (Reproducing Kernel Hilbert Space) were introduced by Parzen in 1958 'to problems in statistical signal processing–in particular to estimation (including regression) and detection problems. Parzen clearly demonstrated

Collected Works of Jaroslav Hájek – With Commentary
Edited by M. Hušková, R. Beran and V. Dupač

that an RKHS provides an elegant, general, unified framework' (quoting Weinert, p. 102). Hájek (1962) independently introduced applications of RKHS to statistical signal processing. It is reprinted in a collection of papers edited by Weinert (1982), who writes (p. 103):

> In a remarkable article presented here as Paper 4, Hájek establishes the basic congruence relation and shows that the estimation and detection problems mentioned previously can be solved by inverting the basic congruence map. Detailed procedures are given in carrying out this inversion for stationary processes with rational spectral densities. The reader will recognize in Hájek's Section 2 the reproducing property and other aspects of RKHS theory, even though Aronszajn's terminology is not used.

The theory of RKHS was developed by Aronszajn (1943, 1950) as a unification, using Hilbert space, of various concepts of kernel functions developed in the 1920s and 1930s to study various partial differential equations. Loève (1948) pioneered the link between RKHS and stochastic processes, and stated the basic congruence theorem between the Hilbert space $L_2(X)$ of random variables spanned by a stochastic process X and the RKHS $H(R)$ whose reproducing kernel R is the covariance kernel R of the process X. As a PhD student of Loève at Berkeley, Parzen was fortunate to study in 1951 the mathematics of RKHS (and harmonic analysis) which he began applying in 1953 to develop methods of time series estimation, detection, and statistical spectral analysis.

Parzen's motivation for introducing RKHS to statistical signal processing was to obtain a unified theory which avoids choosing a countable coordinate system to do the theory of statistical inference on continuous time processes. In the 1950s the usual coordinates were sampled data, Fourier series, or eigenfunction expansions. Parzen proposed that by using appropriate RKHS inner products one could do coordinate free statistical inference and solve statistical inference problems abstractly. Then one would choose coordinate systems for concrete computations with data (see Parzen (1961, 1967, 1970, 1979)).

Hájek accomplished all this, with many detailed solutions, without having available to him the formal theory and notation of RKHS. We can only marvel at the originality and hard work (genius) that Hájek demonstrates in his 1962 paper. Let me outline in RKHS notation its contributions.

Let $(X_t, t \epsilon T)$ be a time series with covariance kernel $R_{ts} = R(x_t, x_s) = \mathsf{E}[x_t \bar{x}_s]$; $L_2(X)$ denotes the Hilbert space spanned by X. Let $H(R)$ denote the Hilbert space of functions $(\phi_t, t \epsilon T)$ where $\phi_t = R(X_t, V_\phi)$ for some unique V_ϕ in $L_2(X)$ denoted $V_\phi = U^{-1}\phi = (\phi, X)_{\sim H(R)}$, called the congruence inner product; the inner product on $H(R)$ is defined by

$$(\phi, \psi)_{H(R)} = \mathsf{Q}(\phi, \psi) = R(U^{-1}\phi, U^{-1}\psi) = R(V_\phi, V_\psi).$$

We call U^{-1} a congruence mapping from $H(R)$ to $L_2(X)$ because it preserves inner products. $H(R)$ is an RKHS with reproducing kernel R, since for every

$t \epsilon T$ the function $R_{t.}$ belongs to $H(R)$ and (Hájek's equation (3.4))

$$\phi_t = R(X_t, V_\phi) = \mathsf{Q}(R_{t.}, \phi) = (R_{t.}, \phi)_{H(R)}.$$

The value of a function ϕ at a point t is a continuous linear functional.

We express Hájek's Application 2.1 in our notation. The conditional expectation $Y^* = \mathsf{E}[Y|X_t, t\epsilon T]$ satisfies $\mathsf{E}[Y^* X_t] = \mathsf{E}[YX_t] = \rho_t$. By the definition of $H(R)$, $Y^* = U^{-1}\rho = (\rho, X)\tilde{}_{H(R)}$. This observation generates computational algorithms because we interpret it: Y^* is the same linear combination of X_t, $t\epsilon T$, that ρ is of $R_{t.}$, $t\epsilon T$.

We express Hájek's Application 2.2 in our notation. A continuous parameter time series regression model

$$Y_t = \sum_{j=1}^{r} \beta_j \phi_j(t) + X_t,$$

where X_t has zero mean and covariance kernel R has minimum variance linear unbiased estimators of regression coefficients

$$\hat{\beta}_j = \left\{(\phi_j, \phi_k)_{H(R)}\right\}^{-1} \left\{(\phi_j, Y)\tilde{}_{H(R)}\right\}$$

assuming regression functions ϕ_j belong to $H(R)$.

There is an immense literature on RKHS methods for numerical analysis, approximation theory, time series estimation and detection and statistical signal processing. The diversity of applications of RKHS is represented by the work of Kailath (in Weinert, 1982), Sacks and Ylvisaker (1970), Wahba (1974) and Eubank (1982).

REFERENCES

Aronszajn, N. (1943). La théorie générale des noyaux réproduisants et ses applications. *Proc. Cambridge Philos. Soc.* *39*, 133–153.

Aronszajn, N. (1950). Theory of reproducing kernels. *Trans. Amer. Math. Soc.* *68*, 337–404.

Eubank, R. L. (1982). On the computation of optimal designs for certain time series models with applications to optimal quantile selection for location and scale parameter estimation. *SIAM J. Sci. Statist. Comput.* *3*, 238–249. With P. L. and P. W. Smith.

Feldman, J. (1958). Equivalence and perpendicularity of Gaussian processes. *Pacific J. Math.* *8*, 699–708; correction, Vol. 9 (1959), 1295–1296.

Grenander, U. (1950). Stochastic processes and statistical inference. *Ark. Mat.* *1*, 195–277.

Hájek, J. (1958). On a property of normal distribution of any stochastic process. (In Russian) *Czechoslovak Math. J. 8*, 610–618.

Hájek, J. (1962). On linear statistical problems in stochastic processes. *Czechoslovak Math. J. 12*, 404-444.

Loève, M. (1948). Fonctions alléatoires du second ordre, supplement to P. Lévy. *Processus Stochastiques et Mouvement Brownien*, Paris, Gauthier–Villars.

Parzen, E. (1958). On asymptotically efficient consistent estimate of the spectral density function of a stationary time series. *J. Roy. Statist. Soc. Ser. B 20*, 303 –322.

Parzen, E. (1961). An approach to time series analysis. *Ann. Math. Stat. 32*, 951–989.

Parzen, E. (1967). *Time Series Analysis Papers*. Holden–Day, San Francisco.

Parzen, E. (1970). Statistical inference on time series by RKHS methods. *12th Biennial Seminar Canadian Mathematical Congress Proc.* (ed. R. Pyke). Canadian Mathematical Congress, Montreal, pp. 1–37.

Parzen, E. (1979). Nonparameteric statistical data modeling. *J. Amer. Statist. Assoc. 74*, 105–131.

Sacks, J. and Ylvisaker, D. (1970). Designs for regression problems with correlated errors III. *Ann. Math. Statist. 41*, 2057–2074.

Wahba, C. (1974). Regression design for some equivalence classes of kernels. *Ann. Statist. 2*, 925–934.

Weinert, H.L. (1982). *Reproducing Kernel Hilbert Spaces*. Hutchinson Ross, Stroudsburg, PA.

ESSAY 6

The Hájek Perspectives in Finite Population Sampling

Zuzana Prášková
Department of Probability and Statistics
Charles University
Sokolovská 83, 186 00 Prague
Czech Republic

Pranab Kumar Sen
Department of Biostatistics
University of North Carolina
Chapel Hill, NC 27599–7400
U.S.A.

THE PREAMBLE

The late Professor Jaroslav Hájek laid down the theoretical foundations of objective or probabilistic sampling from a finite population covering both the design (planning) and statistical analysis (inference) aspects. He developed and unified the essential probability structures underlying finite population sampling plans, incorporated the much needed asymptotics pertaining to a variety of contemporary sampling designs, covering equal as well as unequal probability sampling, with and without replacement schemes, random systematic sampling and successive sampling, among others, and also introduced the notion of rejective sampling and Poisson sampling schemes that play a vital role in the subsequent developments in this field. Hájek's outstanding contributions to finite population sampling (along with the theory of rank tests and regular estimates) not only induced statisticians from all over the world to follow his steps but also endowed him with the right leadership to form an active group of researchers in his own school with some of his former advisees and colleagues who carried out the task even after his

Collected Works of Jaroslav Hájek – With Commentary
Edited by M. Hušková, R. Beran and V. Dupač

premature death in 1974. Based on these observations, it seems more logical to look into the state of the art of the developments with finite population sampling preceding Professor Hájek's entrance to this field, continuing on to the spontaneous developments during the 1960s when he emerged as one of the most outstanding researchers of our time, and finally the aftermath of developments following his demise. We shall depict the scenario in this order but putting more emphasis on the contributions made by Jaroslav Hájek.

THE SRSWOR AND HÁJEK ASYMPTOTICS

Simple random sampling with replacement (SRSWR) is the precursor of other relatively more complex (though useful) sampling designs that are in current use in finite population sampling (FPS) for drawing valid and efficient statistical conclusions based on a representative sample. Simple random sampling without replacement (SRSWOR) retains the equal probability sampling (EPS) structure but violates the stochastic independence of the sample observations; nevertheless, there is an inherent interchangeability (or exchangeability) structure which enables one to draw conclusions in a probabilistic approach with greater precision than in SRSWR. There are various unequal probability sampling (UPS) schemes that may have distinct advantages over SRSWR/SRSWOR, but their structural complexities may often make them unadoptable when the sample size is not small. For these reasons, 40 years ago, there was a pressing need for the development of probability bases in FPS that would justify the statistical conclusions in finite sample sizes and would facilitate the incorporation of large sample theory to minimize the complexities arising therein. For the finite as well as asymptotic treatise in FPS, considerable credit goes to Professor Hájek.

A systematic study of FPS was made in the seminal article of Hansen and Hurwitz (1943), though not much was known at that time about the asymptotics in SRS, not to mention more complex sampling designs. While the case of SRSWR did not pose any serious problem, the lack of independence in SRSWOR stood in the way of the development of general methodology including asymptotics. Madow (1948), inspired by the structural equivalence of SRSWOR and the permutational exchangeability of ranks in conventional randomization tests, incorporated the classical Wald–Wolfowitz (1944) permutational central limit theorem (PCLT) for the study of the large sample distribution (i. e. asymptotic normality) of the sample mean under SRSWOR; similar asymptotic results hold for a larger class of linear statistics in SRSWOR. Yet the PCLT went through a succession of generalizations wherein the stringent Wald–Wolfowitz regularity conditions were relaxed to some extent. The ingenuity of Hájek lay in the effective imputation of rank test asymptotics in FPS in a rigorous and yet lucid manner, and in this respect he made most spectacular contributions to both the fields. Hájek (1960 a) started connecting the general FPS asymptotics with nonparametrics in a more effective manner. The developments in PCLT culminated with Hájek (1961) that pertains to the asymptotic permutational laws for linear statistics in their most general forms, and his treatise covered the FPS in a more general form. Even after almost 36 years, Hájek (1961) is regarded as a milestone in the area of PCLTs for linear rank statistics in nonparametrics, as well as for linear statistics in SRSWOR. There is one other point that merits an appraisal. Even in SRSWR, the number of distinct units in a sample conveys some statistical information that can be incorporated to enhance the efficiency of an estimator of the population total or mean. This simple observation is due to D. Basu (1958), and a lot of related work appeared in the literature during the past 30 years. For some reason this was not pursued very systematically during the prime period enriched by the Hájek contributions.

Besides the asymptotic theory of FPS, Hájek (1959) dealt with the general problem of estimation of population parameters and with the problem of the choice of a sampling design that would be optimal in balancing accuracy and the cost of sampling. In this respect the seminal paper of Mahalanobis (1944) deserves mention too. Hájek developed a general formula for the variance and estimated variance of a linear estimator of the population total under a representative condition explained below. A linear estimator of the total (say, Y_N) of a finite population of size N is representative if it is identical to Y_N when the sample includes all the N observations in the population. Thus, in a sense, it is the FPS analogue of the classical Fisher consistency criterion in the theory of estimation. He also explained the importance of his notion of representativeness in the optimality of sampling strategies, and thereby introduced the notion of representative sampling in FPS. In Hájek (1960 b), a monograph on sample surveys written in Czech, using the notion of representativeness, a very simple non-negative unbiased variance estimate of the ratio estimator in PPS (probability proportional to size) sampling was introduced along the lines of Hájek (1949), Lahiri (1951) and Midzuno (1952), where the case of $n = 2$ was considered. Rao and Vijayan (1977) considered the same problem much later on based on a more complex approach. Unfortunately, Hájek's notion of representativeness has not received due appreciation in the literature; he intended to unify and revitalize this concept in the early 1970s, but his unexpected death left the task at least partially incomplete; nevertheless the posthumous monograph [Hájek (1981)] contains an updated and slightly reconstructed (by his colleagues) version of this notion. This monograph of Hájek was in preparation in the early 1970s, but left partially incomplete by his illness and death, was in circulation for quite some time in several places in the USA and Canada before appearing in print with some updating. In our opinion this contains a lucid, original and rigorous treatment of the probability bases underlying FPS, and indeed it deserves considerable appreciation from theoreticians as well as practitioners in this field. Unfortunately, it would not be possible to have a reproduction of this monograph in this collected work of Hájek. Coming back to SRSWOR schemes, we note that Nandi and Sen (1963) extended some of the results (on unbiasedness, variance estimation and asymptotic normality) to general symmetric (i. e. U-)statistics (of arbitrary degree) under SRSWOR, while a few years later, Sen (1970) established the reverse martingale property of sample means and U-statistics, in general, under SRSW(O)R, which was incorporated in the derivation of deeper asymptotic as well as finite sample results for SRSWOR.

BEYOND THE SRS

Hájek's contributions to UPS schemes are most noteworthy; he introduced the notion of rejective sampling and Poisson sampling schemes, and contributed very generously towards the development of the much needed asymptotics. In a general without replacement UPS scheme, the main complication arises due to the precise formulation of second and higher order inclusion probabilities. While Hájek (1981) contains a nice formulation of the underlying probability space and these inclusion probabilities, the ingenuity of Hájek lay in the incorporation of novel asymptotics that are very much needed in the simplification of the variance functions and their estimators, and in distributions of the estimators when the sample size is not small. Hájek (1964) is the most significant publication in this respect. Prior to that Hartley and Rao (1962) formulated some asymptotic results that deserve mention; their formulation is based on a basic postulation that the π_{ij}, the second order inclusion probabilities, and the π_i, the first order ones, satisfy the condition that as $N, n \to \infty$,

$$n\{\pi_i\pi_j - \pi_{ij}\}/\{\pi_i\pi_j\} \to 1, \quad \forall\, i, j.$$

Hájek (1964) demonstrated that the above condition may not generally hold under UPS when both n and $N - n$ tend to ∞, or more precisely, when $\sum_{i=1}^{N} \pi_i(1 - \pi_i) \to \infty$. Therefore to handle the general asymptotic case, without the above postulation, Hájek (1964) introduced two basic tools – Poisson sampling and rejective sampling – presented the underlying probability spaces clearly, and rigorously formulated the general asymptotics. The notion of successive sampling with varying probabilities also cropped up in this context, and Rosén (1972 a, b), led primarily by the motivation and ingenuity of Hájek (1964), has made a fundamental contribution to the general asymptotics in this context. Some martingale characterizations [viz. Sen (1979, 1980)] have added more flexibility to the asymptotics in this field. The *Coupon collector's* problem that cropped up in this context indeed opened up a connection with some general *occupancy* problems, and Hájek's contributions are fundamental in this respect too. Successive subsampling schemes are also allied to these general asymptotics, and Hájek's influence is clearly visible in later works [see Sen (1980) for some details].

Hájek's work in UPS, especially new definitions of probability spaces and ideas of conditioning by the sample size, influenced and motivated also his colleagues and students, who have extended Hájek's results in various directions, whether giving solutions of problems formulated and/or conjectured by Hájek [Dupačová (1979), Prášková (1995)], setting up a unified approach to the asymptotics in rejective, Sampford and successive sampling schemes [Víšek (1979)] or contributing to the higher order asymptotics in UPS [Prášková (1982, 1984)] as well as in SRSWOR from a finite set of random variables [Prášková (1989)].

CONCLUDING REMARKS

Asymptotics in FPS, in EPS as well as in UPS, have been largely nurtured by the poineering work of Jaroslav Hájek. Besides, he also worked on sampling designs, optimal allocation and other practical aspects. Most of these publications in original form with some editorial comments appear in this collection of his published work. For various reasons, not all of his publications could be included in this collection, and hence we take special care in citing some of his works in finite population sampling that are not reproduced in this volume; they appear in the following reference list with asterisks to make this point clear.

Besides papers on applications of sample survey to stomatology and other research papers not mentioned here, Hájek (1956) considered various problems arising with stratified sampling like confidence intervals for ratio estimates or optimal and random allocation. In Hájek (1962) a general optimization problem in multiparameter estimation is solved which can be applied to find an optimum design in stratified, two-stage and other sampling schemes. The mathematical rigour of the paper is balanced well by practical computing from a real sampling situation; both great theoretical as well as practical interest are remarkable features of Professor Hájek's scientific work in general.

REFERENCES

Basu, D. (1958). On sampling with and without replacement. *Sankhyā, Ser. A 20*, 287–294.

Dupačová, J. (1979). A note on rejective sampling. *Contribution to Statistics* (J. Hájek Memorial Volume, ed. J. Jurečková). Academia, Prague and Reidel, Dordrecht, 71–78.

Hájek, J. (1949). Representative cluster sampling by a method of two phases. (In Czech) *Statistický Obzor 29*, 384–394.

* Hájek, J. (1956). Stratified sampling. (In Czech) *Appl. Math. 1*, 149–161.

Hájek, J. (1959). Optimum strategy and other problems in probability sampling. *Časopis Pěst. Mat. 84*, 387–423.

Hájek, J. (1960 a). Limiting distributions in simple random sampling from a finite population. *Publ. Math. Inst. Hungar. Acad. Sci. 5*, 361–374.

Hájek, J. (1960 b). *Theory of Probability Sampling with Applications to Sample Survey.* (In Czech) Nakladatelství ČSAV, Prague.

Hájek, J. (1961). Some extensions of the Wald-Wolfowitz-Noether theorem. *Ann. Math. Statist. 32*, 506–532.

* Hájek, J. (1962). Cost minimization in multiparameter estimation. (In Czech) *Aplikace Matem. 7*, 405–425.

Hájek, J. (1964). Asymptotic theory of rejective sampling with varying probabilities from a finite population. *Ann. Math. Statist. 35*, 1491–1523.

Hájek, J. (1981). *Sampling from a Finite Population.* Marcel Dekker, New York.

Hansen, M. H. and Hurwitz, W. N. (1943). On the theory of sampling from finite populations. *Ann. Math. Statist. 14*, 333–362.

Hartley, E. O. and Rao, J. N. K. (1962). Sampling with unequal probabilities and without replacement. *Ann. Math. Statist. 33*, 350–374.

Lahiri, D. B. (1951). A method of sample selection providing unbiased ratio estimates. *Bull. Internat. Statist. Inst. 33*, 133–140.

Madow, W. G. (1948). On the limiting distributions of estimates based on samples from finite universes. *Ann. Math. Statist. 19*, 535–545.

Mahalanobis, P. C. (1944). On large scale sample surveys. *Phil. Trans. Roy. Soc. London, Ser. B 231*, 329–451.

Midzuno, H. (1952). On the sampling system with probability proportionate to sum of sizes. *Ann. Inst. Statist. Math. 3*, 99–107.

Nandi, H. K. and Sen, P. K. (1963). On the properties of U-statistics when the observations are not independent. Part II. Unbiased estimation of the parameters of a finite population. *Calcutta Statist. Assoc. Bull. 12*, 125–143.

Prášková, Z. (1982). Rate of convergence for simple estimate in the rejective sampling. *Probability and Statistical Inference* (eds. W. Grossmann et al.). Reidel, Dordrecht, 307–317.

Prášková, Z. (1984). On the rate of convergence in Sampford–Durbin sampling from a finite population. *Statist. Decisions 2*, 339–350.

Prášková, Z. (1989). Sampling from a finite set of random variables: The Berry–Essén bound for the studentized mean. *Proc. 4th Prague Symp. on Asympt. Statistics* (eds. P. Mandl and M. Hušková). Charles University, Prague, 67–81.

Prášková, Z. (1995). On Hájek's conjecture in stratified sampling. *Kybernetika 31,* 303–314.

Rao, J.N.K. and Vijayan, K. (1977). On estimating the variance in sampling with probability proportional to aggregate size. *J. Amer. Statist. Assoc. 72,* 579–584.

Rosén, B. (1972 a,b). Asymptotic theory for successive sampling with varying probabilities without replacement I, II. *Ann. Math. Statist. 43,* 373–397, 748–776.

Sen, P.K. (1970). The Hájek-Rényi inequality for sampling from a finite population. *Sankhyā A 32,* 181–188.

Sen, P.K. (1972). Finite population sampling and weak convergence to a Brownian bridge. *Sankhyā A 34,* 85–90.

Sen, P.K. (1979). Invariance principles for the coupon collector's problem: A martingale approach. *Ann. Statist. 7,* 372–380.

Sen, P.K. (1980). Limit theorems for an extended coupon collector's problem and for successive sampling with varying probabilities. *Calcutta Statist. Assoc. Bull. 29,* 113–132.

Sen, P.K. (1995). The Hájek asymptotics for finite population sampling and their ramifications. *Kybernetika 31,* 251–268.

Víšek, J.Á. (1979). Asymptotic distribution of sample estimate for rejective, Sampford and succesive sampling. *Contribution to Statistics.* (J. Hájek Memorial Volume, ed. J. Jurečková). Academia, Prague and Reidel, Dordrecht, 263–275.

Wald A. and Wolfowitz J. (1944). Statistical tests based on permutations of the observations. *Ann. Math. Statist. 15,* 358–372.

Publications of Jaroslav Hájek

Books

1960

Theory of Probability Sampling with Applications to Sample Surveys. (In Czech) Nakladatelství ČSAV, Prague.

1962

Probability in Science and Ingineering (with V. Dupač). (In Czech) Nakladatelství ČSAV, Prague.

1967

Probability in Science and Ingineering (with V. Dupač). (In Czech) Academia, Prague and Academic Press, New York.

Theory of Rank Tests (with Z. Šidák). Academia, Prague and Academic Press, New York. (Russian translation: Izd. Nauka, Moscow 1971).

1969

A Course in Nonparametric Statistics. Holden–Day, San Francisco.

1981

Sampling from a Finite Population. Marcel Dekker, New York.

Articles

1949

The use of factorial design and of confidence intervals in weighing. (In Czech) *Statistický Obzor 29*, 258–273.

Representative cluster sampling by a method of two phases. (In Czech) *Statistický Obzor 29*, 384–394.

1955

Some rank distributions and their applications. (In Czech) *Časopis Pěst. Mat. 80*, 17–31. (English translation in *Selected Transl. Math. Statist. Prob. 2*, (1962), 27–40.)

Generalization of an inequality of Kolmogorov (with A. Rényi). *Acta Math. Acad. Sci. Hungar. 6*, 281–283.

On some fundamental questions of mathematical statistics (with F. Fabian). (In Czech) *Časopis Pěst. Mat. 80*, 387–399.

1956

Asymptotic efficiency of a certain sequence of tests. (In Russian) *Czechoslovak Math. J. 6*, 26–30.

Linear estimation of the mean value of a stationary random process with convex correlation function. (In Russian) *Czechoslovak Math. J. 6*, 94–117. (English translation in *Selected Transl. Math. Statist. Prob.* 2 (1962), 41–61).

Stratified sampling. (In Czech) *Apl. Mat. 1*, 149–161.

Remark on the article 'On certain sequences of sets of points on a circle'. (In Czech) *Časopis Pěst. Mat. 81*, 77–78.

1957

Inequalities for the Student's distribution and their applications. (In Czech) *Časopis Pěst. Mat. 82*, 182–194. (English translation in *Selected Transl. Math. Statist. Prob. 2*, (1962), 63–74.)

1958

Predicting a stationary process when the correlation function is convex. *Czechoslovak Math. J. 8*, 150–154.

A property of J-divergence of marginal probability distributions. *Czechoslovak Math. J. 8*, 460–463.

On a property of normal distribution of any stochastic process. (In Russian) *Czechoslovak Math. J. 8*, 610–618. (English translation in *Selected Transl. Math. Statist. Prob. 1*, (1961), 245–252.)

On the distribution of some statistics in the presence of intraclass correlation. (In Czech) *Časopis Pěst. Mat. 83*, 327–329. (English translation in *Selected Transl. Math. Statist. Prob. 2*, (1962), 75–77.)

On the theory of ratio estimates. *Apl. Mat. 3*, 384–398.

Some contributions to the theory of probability sampling. *Bull. Inst. Internat. Statist. 36*, 127–134.

1959

Optimum strategy and other problems in probability sampling. *Časopis Pěst. Mat. 84*, 387–423.

The age of eruption of permanent teeth in children in Czechoslovakia (with V. Poncová). (In Czech) *Československá Stomatologie 2*, 104–113.

1960

On a simple linear model in Gaussian processes. *Transactions of the 2nd Prague Conference on Information Theory, Statistical Decision Functions, Random Processes.* (ed. J. Kožešník), Liblice 1959, 185–197.

Limiting distributions in simple random sampling from a finite population. *Publ. Math. Inst. Hungar. Acad. Sci. 5*, 361–374.

1961

On plane sampling and related geometrical problems (with T. Dalenius and S. Zubrzycki). *Proc. Fourth Berkeley Symp. Math. Statist. Prob. 1* (ed. L. Le Cam), Univ. of California Press, 125–150.

Some extensions of the Wald–Wolfowitz–Noether theorem. *Ann. Math. Statist. 32*, 506–523.

On linear estimation theory for an infinite number of observations. *Teor. Verojatnost. i Primenen. 6*, 182–193.

Concerning relative accuracy of stratified and systematic sampling in a plane. *Colloq. Math. 8* , 133–134.

1962

On linear statistical problems in stochastic processes. *Czechoslovak Math. J. 12*, 404–444.

An inequality concerning random linear functionals on a linear space with a random norm and its statistical applications. *Czechoslovak Math. J. 12*, 486–491.

Asymptotically most powerful rank-order tests. *Ann. Math. Statist. 33*, 1124–1147.

Cost minimization in multiparameter estimation. (In Czech) *Apl. Mat. 7*, 405–425.

1964

Asymptotic theory of rejective sampling with varying probabilities from a finite population. *Ann. Math. Statist. 35*, 1491–1523.

Strictly equivalent Gaussian measures. (In Russian) *Trudy 4-go Vses. Matem. C'ezda 1961, 2*, Nauka, Leningrad, 341–345.

1965

Extension of the Kolmogorov–Smirnov test to regression alternatives. *Bernoulli–Bayes–Laplace; Proceedings of an International Research Seminar*. (eds. J. Neyman and L. Le Cam). Springer–Verlag, Berlin 1963, 45–60.

1966

On the foundations of statistics. *Conf. on Math. Methods in Economic Research*, Smolenice. (ed. Economic–Mathematical Laboratory in Economic Institute of ČSAV), 1–17.

1967

On basic concepts of statistics. *Proc. Fifth Berkeley Symp. Math. Statist. Prob. 1*, Univ. of California Press 1967, 139–162.

1968

Locally most powerful rank tests of independence. *Studies in Math. Statist.* (eds. K. Sarkadi and I. Vincze). Akadémiai Kiadó, Budapest, 45–51.

Asymptotic normality of simple linear rank statistics under alternatives. *Ann. Math. Statist. 39*, 325–346.

1969

Asymptotic normality of simple linear rank statistics under alternatives II (with V. Dupač). *Ann. Math. Statist. 40*, 1992–2017.

Asymptotic normality of the Wilcoxon statistic under divergent alternative (with V. Dupač). *Zastos. Mat. 10*, 171–178.

On the calculus of probability. *Matematika ve Škole 19*, 449–462.

1970

Miscellaneous problems of rank test theory. *Nonparametric Techniques in Statistical Inference.* (ed. M. L. Puri). Cambridge Univ. Press, 3–19.

A characterization of limiting distributions of regular estimates. *Z. Wahrscheinlichkeitstheorie und Verw. Gebiete 14*, 323–330.

1971

Limiting properties of likelihood and inference. *Foundations of Statistical Inference.* (eds. V. P. Godambe and D. A. Sprott). Holt, Rinehart and Winston, Toronto, 142–162.

1972

Local asymptotic minimax and admissibility in estimation. *Proc. Sixth Berkeley Symp. Math. Statist. Prob. 1*, Univ. of California Press 1972, 175–194.

1974

Asymptotic sufficiency of the vector of ranks in the Bahadur sense. *Ann. Statist. 2*, 75–83.

Regression design in autoregressive stochastic processes (with G. Kimeldorf). *Ann. Statist. 2*, 520–527.

Asymptotic theories of sampling with varying probabilities without replacement. *Proc. Prague Symp. on Asymptotic Statistics.* (ed. J. Hájek). 1, 127–138.

Texts

1955

Theory of Sample Surveys. (In Czech) SPN, Praha. (Text for School of Economics).

1961

Algebra for 3rd Grade of Grammar Schools. (In Czech) SPN, Praha. (Author of chapters on probability and statistics, further joint authors of the book A. Hyška, A. Robek, A. Hustá.)

1963

Survey of Applicable Mathematics. (In Czech) SNTL, Praha. 2nd edition 1968, 3rd edition 1973. (Author of chapters on probability and statistics, joint authorship of the book in a group of 18 members headed by K. Rektorys.) (English translation Iliffe Bok, Ltd., London 1969.)

1966

Mathematics for 3rd Grade of Grammar Schools, Science Branch. (In Czech) SPN, Praha. (Author of chapters on probability and statistics, further joint authors of the book E. Kraemer, F. Veselý, J. Voříšek, M. Zöldy.)

Nonparametric Methods. (In Czech) SPN, Praha. (Text for Faculty of Math. and Phys., Charles Univ., jointly with D. Vorlíčková.)

1969

Stationary Processes. (In Czech) SPN, Praha. (Text for Faculty of Math. and Phys., Charles Univ., jointly with J. Anděl.)

1977

Mathematical Statistics. (In Czech) SPN, Praha. (Text for Faculty of Math. and Phys., Charles Univ., jointly with D. Vorlíčková.)

Hájek's PhD Students

— J. Anděl (1965) defended a thesis on problems related to the asymptotic power and efficiency of tests of Kolmogorov – Smirnov type. He evaluated, in particular, the local power of one-sided tests of Kolmogorov – Smirnov and of Rényi types and found an approximation for the asymptotic power. The main results of the thesis were published in the paper Anděl (1967). Later on Anděl oriented himself to the area of time series, where he is now a well-known specialist.

— Nguyen Van Ho in his thesis (1971) studied the problem of large deviations and of Bahadur efficiency for signed rank statistics, in parallel with Hájek's (1974) prepared paper on Bahadur efficiency of two-sample rank statistics. The main result of the thesis was published in Nguyen Van Ho (1974).

— J. Hurt derived results on estimation of reliability in some particular models. He started under the guidance of J. Hájek and finished with V. Dupač. He defended his thesis in 1978, the main results of which are published in two papers (1976, 1978). He continued to work in reliability theory. His current interests are in financial mathematics.

— M. Hušková in her thesis (1968) extended some of Hájek's results on simple linear rank statistics to signed rank statistics and their multivariate analogs. She studied their limit behavior under the hypothesis of symmetry and under local as well as non-local alternatives. The results were published in the papers Hušková (1970, 1971). Her research then continued in this direction. Later she studied the rate of convergence of the distributions of various rank statistics to the normal distribution and dealt with adaptive rank processes. Recently, in Hušková (1997), she successfully applied rank-based procedures to the change point problem.

— The thesis of J. Jurečková (1967) proposed a class of R-estimators for the regression parameter vector in the linear model and studied its asymptotic properties. As the main tool, she proved *uniform asymptotic linearity* of the linear rank statistic in the regression parameter; this property was later found useful not only for R-estimators but also for *aligned rank tests* and in various other contexts. These results were published in Jurečková (1969) and Jurečková (1971). Later Jurečková (1977) dealt with robust procedures and proved an asymptotic equivalence between R-estimators and M-estimators in the linear regression model. Recently she returned to the area of ranks, dealing with the regression rank score process, a concept which surprisingly extends Hájek's (1965) rank score process to the linear regression model. More details of this concept are given in the essay by Hušková and Jurečková (1998).

— Z. Prášková originally received a subject for her thesis Prášková (1974) from Hájek but because of his poor health she finished the thesis with Václav Dupač. In her thesis she developed local limit theorems and established asymptotic expansions for integer-valued rank statistics; some results were published in Prášková (1976 a, b). Later on she continued her research in the theory of sampling methods.

— J. Štěpán studied in his thesis (1968 a) geometrical and topological structures of sets of probability distributions with given properties. He published related results in three papers, Štěpán (1968 b, 1969, 1970). He continued to work in the area; his results on marginal probability problems are well known.

— J.Á. Víšek started to work on his thesis with Hájek and finished with V. Dupač in 1975. Víšek derived some asymptotic results useful in sampling theory; some of them were published in Víšek (1979). He continued his research mostly in robust statistics.

— I. Volný in his thesis obtained a general set of conditions implying LAN in a series of independent experiments. He defended his thesis in 1978 and published part of the results in Volný (1978 b).

— The thesis of D. Vorlíčková (1968) deals with linear rank statistics in a situation when the observations have a discrete distribution or are rounded off. She studied the limit conditional and unconditional distributions of simple linear rank statistics in the presence of ties, under the null hypothesis as well as under contiguous alternatives. The main results were published in the papers Vorlíčková (1970, 1972); a similar situation is considered in several other papers of hers.

REFERENCES

Anděl, J. (1965). Local asymptotic power of the tests of Kolmogorov - Smirnov type. (In Czech) *PhD Thesis, Charles University,* Prague.

Anděl, J. (1967). Local asymptotic power and efficiency of tests of Kolmogorov - Smirnov type. *Ann. Math. Statist.* *38*, 1705–1725.

Hájek, J. (1974). Asymptotic sufficiency of the vector of ranks in the Bahadur sense. *Ann. Statist.* *2*, 75–84.

Ho, Nguyen Van (1971).Probability of large deviations and asymptotic efficiency in the Bahadur sense for the signed rank tests. *PhD Thesis, Charles University,* Prague.

Ho, Nguyen Van (1974). Asymptotic efficiency in the Bahadur sense for the signed rank tests. In *Proc. Prague Symp. on Asymptotic Statistics 2,* (ed. J. Hájek), 127–156.

Hurt, J. (1976). On estimation of reliability in the exponential case. *Aplikace Matematiky 21*, 263–272.

Hurt, J. (1978). Estimation of reliability in the exponential and normal distributions. (In Czech) *PhD Thesis, Charles University*, Prague.

Hurt, J. (1980). Estimates of reliability for normal distributions. *Aplikace Matematiky 25*, 432–444.

Hušková, M. (1968). Asymptotic Distributions of Simple Linear Rank Statistics. (In Czech) *PhD Thesis, Charles University*, Prague.

Hušková, M. (1970). Asymptotic distribution of simple linear rank statistics for testing symmetry. *Z.Wahrsheinlichktheorie und Verw. Gebiete 14*, 308–322.

Hušková, M. (1971). Asymptotic distribution of rank statistics used for testing multivariate symmetry. *J. Multivar. Anal. 1*, 461–484.

Hušková, M. (1997). Limit theorems for rank statistics. *Stat. & Probab. Letters 30*, 45–55.

Hušková, M. and Jurečková, J. (1998). Jaroslav Hájek and his impact on the theory of rank tests. In *Collected Works of Jaroslav Hájek with Commentary*, Wiley, 51–54.

Jurečková, J. (1967). Nonparametric estimate of regression coefficients. (In Czech) *PhD Thesis*, Mathematical Institute of the Czechoslovak Academy of Sciences.

Jurečková, J. (1969). Asymptotic linearity of a rank statistic in regression parameter. *Ann. Math. Statist. 40*, 1889–1900.

Jurečková, J. (1971). Nonparametric estimate of reression coefficients. *Ann. Math. Statist. 42*, 1328–1338.

Jurečková, J., (1977). Asymptotic relation of M-estimates and R-estimates in linear regression model. *Ann. Statist. 5*, 464–472.

Prášková-Vízková, Z. (1974). Local limit theorems for some rank tests statistics. (In Czech) *PhD Thesis, Charles University,* Prague

Prášková-Vízková, Z. (1976 a). Asymptotic expansion and a local limit theorem for a function of the Kendall rank correlation coefficient. *Ann. Statist. 4*, 597–606.

Prášková, Z. (1976 b). Asymptotic expansions and a local limit theorem for the signed Wilcoxon statistic. *CMUC 17*, 335–347.

Štěpán, J. (1968 a). On convex hulls of sets of probability measures. (In Czech) *PhD Thesis, Charles University*, Prague.

Štěpán, J. (1968 b). On convex sets of probability measures. (In Czech) *Časopis Pěst.Mat. 93*, 73–79. (English translation in *Selected Transl. Math.Statist Prob. 13*, 1–7).

Štěpán, J. (1969). Some notes on the convolution semigroup of probabilities on a metric group. *CMUC 10*, 612–613.

Štěpán, J. (1970). On the family of translations of a tight probability measure on topological group. *Z.Wahrsheinlichktheorie und Verw.Gebiete 15*, 131 – 138.

Víšek, J.Á. (1975). Two contributions to the asymptotic theory of some sampling designs. (In Czech) *PhD Thesis, Charles University*, Prague.

Víšek, J.Á. (1979). Asymptotic distribution of simple estimate for rejective, Sampford and successive sampling. In *Contributions to Statistics* (ed. J. Jurečková), Academia, Prague.

Volný, I. (1978 a). Differential conditions of local asymptotic normality for sequences of experiments. (In Czech) *PhD Thesis, Charles University*, Prague.

Volný, I. (1978 b). Differential conditions of local asymptotic normality for sequences of experiments. *CMUC 19*, 281 –290.

Vorlíčková, D. (1968). Rank Tests for Discrete Distributions. (In Czech) *PhD Thesis, Charles University*, Prague.

Vorlíčková, D. (1970). Asymptotic properties of rank tests under discrete distributions. *Wahrsheinlichktheorie und Verw. Gebiete 14*, 275–289.

Vorlíčková, D. (1972). Asymptotic properties of rank tests of symmetry under discrete distributions. *Ann. Math. Statist. 43*, 2013–2018.

Part II: Collected Works of Jaroslav Hájek

Part II consists of 36 papers that, excluding his monographs, make up the overwhelming part of Hájek's scientific heritage. The papers are arranged in chronological order. Thematically, they belong to the fields of inference in stochastic processes $(5, 7, 8, 9, 14, 18, 20, 21, 35)$, rank tests $(2, 15, 17, 22, 25, 27, 28, 29, 30, 34)$, finite population sampling $(1, 11, 12, 13, 15, 19, 24, 36)$, asymptotics of parameter estimators $(31, 32, 33)$ and other topics $(3, 4, 6, 10, 16, 23, 26)$.

CHAPTER 1

Representative Cluster Sampling by a Method of Two Phases

Statistický Obzor 29 (1949), 384–394. (In Czech.)
Translated by J. Machek.
Reproduced by permission of Hájek family: Al. Hájková, H. Kazárová and Al. Slámová.

1 INTRODUCTION

In the applications of the representative method it is frequently impossible to select the individual single sampling units (elements of the population under study) but only some groups of them, called clusters (e. g., not individual households, but apartment buildings, not individual workers, but rather factories, not individual farms, but whole districts, etc.). These clusters generally do not comprise the same number of elements, which makes the problem of unbiased estimation of the population mean per element rather difficult.

First attempts to solve this problem were based on the so called 'purposive sampling': it was assumed that the sample will be representative with respect to the variable under study if it is selected in such a way that it be representative with respect to some other 'control variables' whose distribution in the population is known and which are correlated to the main variable. More information can be found in Bowley (1926). However, the analysis of data collected by means of such a purposive sample had to be based – to a considerable extent – on intuition. Finally, J. Neyman (1934) provided definitive arguments against this method when he showed that full representativity could be attained through a suitable stratification

Collected Works of Jaroslav Hájek – With Commentary
Edited by M. Hušková, R. Beran and V. Dupač
Published in 1998 by John Wiley & Sons Ltd.

and that the samples within the strata could be obtained by means of random mechanisms providing unbiased estimates of population mean whose precision could be assessed.

However, the procedure proposed by Neyman has one drawback: the variance of the estimators depends not only on the variability of cluster means (group means) but also on the variation of group sizes (cluster sizes). In this chapter we will present a modification of the sampling design which eliminates the effects of the variability of cluster sizes and produces unbiased estimators that have, in most usual situations, smaller variance than those of J. Neyman.

2 NOTATION

Let us have a finite population of N elements subdivided in K groups, called clusters in what follows. Denote the averages (per element) of the variable under study in these clusters by u_1, $u_2, \ldots, u_K$ and the sizes of the clusters (i. e., the numbers of elements in them) by h_1, $h_2, \ldots, h_K$. Thus the total population size N is

$$N = h_1 + h_2 + \cdots + h_K$$

and the population mean $\overline{z}$ per element is

$$\overline{z} = \frac{h_1 u_1 + h_2 u_2 + \cdots + h_k u_k}{h_1 + h_2 + \cdots + h_K}.$$

Out of the K clusters we now select by means of some random procedure k clusters. Let us denote them $S_1, S_2, \ldots, S_k$, their means $y_1, y_2, \ldots, y_k$, their sizes $r_1, r_2, \ldots, r_k$ and their weighted mean, to be called sample mean in the sequel,

$$\overline{y} = \frac{r_1 y_1 + r_2 y_2 + \cdots + r_k y_k}{r_1 + r_2 + \cdots + r_k}. \tag{2.1}$$

The difference between the values u_i and h_i in the population and their counterparts y_i, r_i in the sample consists, naturally, in the following fact: while u_i, h_i are some fixed numbers, y_i and r_i depend on the outcome of the sampling. If the sampling design is once given, y_i and r_i are, before the sample is taken, random variables which take on some of the values u_i, h_i with probabilities given by the design of the sample.

In what follows, our goal will be to find a sampling design under which the sample mean $\overline{y}$ is an unbiased estimate of the population mean $\overline{z}$, i. e.

$$\mathsf{E}(\overline{y}) = \overline{z}.$$

J. Neyman used simple random sampling of clusters, i. e., the design that gives to any possible combination of k clusters the same probability to be selected. Then, of course, the population mean cannot be estimated by the sample mean (2.1) since the latter is biased. The unbiased estimator of $\overline{z}$ for simple random sampling of clusters is

$$\overline{y}_0 = (r_1 y_1 + \cdots + r_k y_k)\,\frac{K}{kN}; \tag{2.2}$$

but this, as already mentioned above, usually has a larger variance.

3 THE SAMPLING DESIGN

To define a sampling design (procedure) means to assign to any possible combination of k clusters $S_1, \ldots, S_k$ the probability that precisely this combination of clusters will be selected. If this probability is denoted by $\mathsf{P}[S_1, \ldots, S_k]$, then the expectation (mean value) of the sample mean (2.1) can be written as

$$\mathsf{E}(\overline{y}) = \sum_C \frac{r_1 y_1 + \cdots + r_k y_k}{r_1 + \cdots + r_k}\,\mathsf{P}[S_1, \ldots, S_k], \tag{3.1}$$

where the symbol C under the summation sign indicates that the sum extends over all $\binom{K}{k}$ subsets of k clusters that can be formed. The main difficulty in the study of (3.1) consists in the fact that the denominators of the summands differ when the sizes of clusters are not all equal. However, this difficulty can be overcome if we make the probability $\mathsf{P}[S_1, \ldots, S_k]$ proportional to the sum $r_1 + \cdots + r_k$, i. e.,

$$\mathsf{P}[S_1, \ldots, S_k] = \text{const}(r_1 + \cdots + r_k). \tag{3.2}$$

Then the sum (3.1) reduces to

$$\mathsf{E}(\overline{y}) = \text{const} \sum_C (r_1 y_1 + \cdots + r_k y_k) \tag{3.3}$$

and we will show later that this is equal to a multiple of the population mean $\overline{z}$. Thus if we succeed in finding a sampling design such that (3.2) holds, a multiple of $\overline{y}$ will be an unbiased estimator of $\overline{z}$.

The constant in (3.2) is easily determined from the fact that the probabilities $\mathsf{P}[S_1, \ldots, S_k]$ must sum up to 1,

$$\sum_C \mathsf{P}[S_1, \ldots, S_k] = \text{const} \sum_C (r_1 + \cdots + r_k) = 1. \tag{3.4}$$

Since in the sum $\sum_C (r_1 + \cdots + r_k)$ every cluster size r_i appears exactly as many times as there are combinations that include S_i, i. e. $\binom{K-1}{k-1}$ times, we have

$$\sum_C (r_1 + \cdots + r_k) = \binom{K-1}{k-1}(h_1 + \cdots + h_k) = \binom{K-1}{k-1} N, \tag{3.5}$$

whence

$$\text{const} = \left(\binom{K-1}{k-1} N\right)^{-1}$$

and

$$\mathsf{P}[S_1, \ldots, S_k] = \frac{r_1 + \cdots + r_k}{N\binom{K-1}{k-1}}. \tag{3.6}$$

By inspection of this last expression it may be easily seen what sampling procedure should be used to satisfy (3.6). It will consist in two phases. In the first phase one cluster is selected by a mechanism which gives to every cluster the probability of selection proportional to its size. In the second phase, $(k-1)$ groups are selected from the remaining $K-1$ by simple random sampling without replacement, i. e. assigning to every of the possible combinations of $k-1$ out of $K-1$ the same probability. Thus the probability that the two phases result in the selection of k given clusters $S_1, \ldots, S_k$ is equal to the product of two probabilities; first the probability that in the first phase one of the given clusters is selected, and second the probability that at the second phase precisely $k-1$ remaining clusters are drawn. The former probability is equal to $(r_1 + r_2 + \cdots + r_k)/N$ and the latter to $1/\binom{K-1}{k-1}$, so that their product is (3.6).

Let us now determine the expectation of the sample mean $\overline{y}$. Inserting (3.6) into (3.1) and applying the same reasoning as that used to evaluate the sum (3.5) we obtain

$$\begin{aligned} \mathsf{E}(\overline{y}) &= \frac{1}{N\binom{K-1}{k-1}} \sum_C (r_1 y_1 + \cdots + r_k y_k) \qquad (3.7) \\ &= \frac{1}{N\binom{K-1}{k-1}} \binom{K-1}{k-1} (h_1 u_1 + \cdots + h_K u_K) = \overline{z}. \end{aligned}$$

Hence, the sample mean $\overline{y}$ constructed by the proposed method is an unbiased estimator of the population mean.

4 EXECUTION OF THE FIRST PHASE

Of the two phases mentioned above only the first one requires some attention as to the technique of selection. The problem here is to ensure that every cluster is selected with probability proportional to its size.

A very primitive method could be the following: to prepare a card for every *element* in the population, to put the cards in an urn and then draw one card completely at random, and then to select that cluster to which the selected cards belonged. Obviously, this procedure would make the probability of selection of any cluster proportional to its size. If the sizes of all the clusters were large, it could be sufficient for practical purposes to assign one card to every subgroup of, say, ten elements.

However, the simplest and most expedient way to perform the first phase is by means of a table of random numbers. Let us illustrate it by a simple example. Suppose that the population is composed of three clusters which contain 37, 45 and 20 elements, respectively. The first phase is carried out so that we assign to every cluster as many numbers as it has elements. Thus the first cluster will be assigned numbers from 0 to 36, the second the numbers 37–81, and the third the numbers 82–101. Then we select a triplet of digits from a table of random digits, omitting numbers higher than 101. The first cluster is selected if the number obtained is from the interval [000, 036], the second, if it is from the interval [037, 081] and the third if it is within [082, 101].

5 SOME COMPARISONS

The method of J. Neyman mentioned above could be called 'the method of totals', since the estimator (2.2) applied in it is, in fact, equal to the total of all the values in the sample divided by the fixed number Nk/K. To illustrate the weak point of this method we can use the following example. Consider a population of $N = 6$ elements on all of which the variable under study takes the value 1, so that the population mean per element also is $\overline{z} = 1$. Let us subdivide this population into three clusters ($K = 3$) in the following way: the first cluster contains one value equal to 1, the second two such values and the third three of them. Thus we have $u_1 = u_2 = u_3 = 1$, $h_1 = 1$, $h_2 = 2$, $h_3 = 3$. Let us select at random (by simple random sampling without replacement) two clusters, i. e. $k = 2$. There are three possible outcomes here and any of them has, under Neyman's sampling scheme, the same probability equal to 1/3. Applying (2.2) we obtain, according to which of the three possible pairs of clusters is selected, the following values of the estimator:

$$\begin{aligned}
\overline{y}_0 &= (h_1u_1 + h_2u_2)\,K/kN = (1+2)\cdot 3/(2\cdot 6) = 3/4 \\
\overline{y}_0 &= (h_1u_1 + h_3u_3)\,K/kN = (1+3)\cdot 3/(2\cdot 6) = 1 \\
\overline{y}_0 &= (h_2u_2 + h_3u_3)\,K/kN = (2+3)\cdot 3/(2\cdot 6) = 5/4.
\end{aligned}$$

It is seen that the method gives different values of the estimator although the values of the variable are the same for all the elements in the population. The reason for this is that the cluster totals h_iu_i differ from one another due to differences in cluster sizes rather than as a consequence of the variability of the cluster means u_i. In exceptional cases these two factors can cancel out mutually and then, of course, the method of totals would be the most indicated. However, in most cases the cluster means and cluster sizes are mutually independent and, consequently, large variation in cluster sizes can produce considerable losses in precision.

On the other hand, we can see that the sample mean $\overline{y}$ computed according to (2.1) does not suffer from this difficulty and its value is – in the present example – equal to 1 in all the cases.

Among the advantages of Neyman's method we should mention the following two: first, it is not necessary to know in advance the sizes of the individual clusters and second, that the average number of sampled elements is smaller than with the method of two phases. This last fact is intuitively clear: the larger sampling units (clusters) are assigned a larger probability to be included in the sample. A formal mathematical proof follows.

Let us denote by n_1 the total number of elements in the sample when Neyman's method is used. Its expectation is

$$\mathsf{E}(n_1) = \sum_C (r_1 + \cdots + r_k) \Big/ \binom{K}{k} = N\binom{K-1}{k-1} \Big/ \binom{K}{k} = Nk/K. \quad (5.1)$$

(We can see that the difference between the estimators (2.1) and (2.2) consists in the fact that in the former case we are dividing the sample total by the *real size* of the sample but in the latter by its *expected size.*)

Denoting by n_2 the total number of elements sampled when the method of two phases is used we obtain

$$\begin{aligned}\mathsf{E}(n_2) &= \sum_C (r_1 + \cdots + r_k)^2 \Big/ \left(N\binom{K-1}{k-1}\right) \\ &= \left\{\sum_C \sum_1^k r_i^2 + \sum_c \sum_{i \neq j}^k r_i r_j\right\} \Big/ \left(N\binom{K-1}{k-1}\right)\end{aligned}$$

The first of the two summations is evaluated by means of the technique already applied above (in the derivation of (3.5))

$$\sum_C \sum^k r_i^2 = \binom{K-1}{k-1} \sum^K h_i^2.$$

The second term is evaluated through a similar process, only it must be kept in mind that any of the products $h_i h_j$ appears as many times as there are combinations with the corresponding two clusters, i. e. $\binom{K-2}{k-2}$times:

$$\begin{aligned}\sum_C \sum_{i \neq j}^k r_i r_j &= \binom{K-2}{k-2} \sum_{i \neq j}^K h_i h_j = \binom{K-2}{k-2} \left\{\left(\sum^K h_i\right)^2 - \sum^K h_i^2\right\} \\ &= \binom{K-2}{k-2} \left(N^2 - \sum^K h_i^2\right).\end{aligned}$$

Combining the results we have

$$\mathsf{E}\, n_2 = \frac{k-1}{K-1} N + \frac{K-k}{(k-1)\, N} \sum^{K} h_i^2.$$

To obtain a form convenient for the comparison with $\mathsf{E}(n_2)$, we apply the identity

$$\sum^{K} h_i^2 = \sum^{K} \left(h_i - \frac{N}{K} \right)^2 + N^2/K,$$

so that finally

$$\mathsf{E}(n_2) = Nk/K + \frac{K-k}{(K-1)\, N} \sum^{K} \left(h_i - \frac{N}{K} \right)^2 .$$

According to (5.1), the first term on the right-hand side is exactly equal to $\mathsf{E}(n_1)$, so that $\mathsf{E}(n_2)$ always exceeds a multiple of the variance of cluster sizes. If all the clusters were the same size, then both methods would coincide and so would $\mathsf{E}(n_1)$ and $\mathsf{E}(n_2)$.

6 VARIANCE OF THE SAMPLE MEAN

In order to be able to evaluate the merits of the sample mean we need to know its variance (which will be denoted by $\mathsf{var}(\overline{y})$ in the sequel) – eventually a suitable estimator of the same (which we denote by $s^2_{\overline{y}}$) based on the sample.

Since a straightforward calculation of $\mathsf{var}(\overline{y})$ from its definition

$$\mathsf{var}(\overline{y}) = \mathsf{E}(\overline{y} - \overline{z})^2 \tag{6.1}$$

presents rather serious difficulties, some round about way must be chosen.

To begin with, we introduce the *unweighted* sample mean $\overline{y}'$:

$$\overline{y}' = (y_1 + \cdots + y_k)/k.$$

We start from the identity

$$\mathsf{E}(\overline{y} - \overline{z})^2 = \mathsf{E}(\overline{y} - \overline{z})\,(\overline{y}' - \overline{z}) + \mathsf{E}(\overline{y} - \overline{y}')^2 - \mathsf{E}(\overline{y}' - \overline{y})\,(\overline{y}' - \overline{z}). \tag{6.2}$$

Next we show that the first term on the right-hand side of (6.2) can be expressed in a simple form. The second term does not present any difficulty: it is based only on the sample values and the very square $(\overline{y} - \overline{y}')^2$ is its unbiased estimator. And as far as the third term is concerned, we will show that its value is very small in comparison with the other two, moreover generally negative, so that it may be neglected.

Now let us analyze the first term:

$$
\begin{aligned}
& \mathsf{E}(\overline{y} - \overline{z})\,(\overline{y}' - \overline{z}) \\
= \; & \sum_C (\overline{y} - \overline{z})\,(\overline{y}' - \overline{z})\,\frac{r_1 + \cdots + r_k}{N\binom{K-1}{k-1}} \\
= \; & \frac{1}{N\binom{K-1}{k-1}\,k} \sum_C \left\{ \sum^k (y_i - \overline{z})\, r_i \sum^k (y_i - \overline{z}) \right\} \qquad (6.3) \\
= \; & \frac{1}{N\binom{K-1}{k-1}\,k} \left\{ \sum_C \sum^k (y_i - \overline{z})^2\, r_i + \sum_C \sum_{i \neq j}^k (y_i - \overline{z})\,(y_j - \overline{z})\, r_i \right\}
\end{aligned}
$$

Proceeding in a similar manner as in the evaluation of $\mathsf{E}(n_2)$ we obtain

$$
\sum_C \sum^k (y_i - \overline{z})^2\, r_i - \binom{K-1}{k-1} \sum^K (u_i - \overline{z})^2\, h_i = N\binom{K-1}{k-1}\,\sigma^2,
$$

where σ^2 is the population variance,

$$
\sigma^2 = \frac{1}{N} \sum^K (u_i - \overline{z})^2\, h_i .
$$

Further we have

$$
\begin{aligned}
& \sum_C \sum_{i \neq j}^k (y_i - \overline{z})\,(y_j - \overline{z})\, r_i = \binom{K-2}{k-2} \sum_{i \neq j}^K (u_i - \overline{z})\,(y_i - \overline{z})\, h_i \\
= \; & -\binom{K-2}{k-2} \sum (u_i - \overline{z})^2\, h_i = -N\binom{K-2}{k-2}\,\sigma^2 .
\end{aligned}
$$

Introducing these results into (6.3) we obtain

$$
\mathsf{E}(\overline{y} - \overline{z})\,(\overline{y}' - \overline{z}) = \frac{\sigma^2}{k}\,\frac{K-k}{K-1}. \qquad (6.4)
$$

Thus, according to (6.1) and (6.2), the variance of the sample mean is

$$
\mathsf{var}(\overline{y}) = \frac{\sigma^2}{k}\,\frac{K-k}{K-1} + \mathsf{E}(\overline{y}' - \overline{y})^2 - \mathsf{E}(\overline{y}' - \overline{y})\,(\overline{y}' - \overline{z}). \qquad (6.5)
$$

7 ESTIMATION OF THE VARIANCE OF THE SAMPLE MEAN

Let us start from the sample variance s^2:

$$
s^2 = \left\{ \sum^k (y_i - \overline{y})^2\, r_i \right\} \Big/ \left(\sum^k r_i \right) = \left(\sum^k y_i^2\, r_i \right) \Big/ \left(\sum^k r_i \right) - \overline{y}^2. \qquad (7.1)
$$

The expectation of s^2 is equal to

$$\mathsf{E}(s^2) = \mathsf{E}\left\{\frac{\sum^k y_i^2\, r_i}{\sum^k r_i}\right\} - \mathsf{E}(\overline{y}^2). \tag{7.2}$$

Since s^2 is invariant with respect to the translation, we can assume, without loss of generality, that $\overline{z} = 0$. Then we obtain, following the way already known, that the first term on the right-hand side of (7.2) is equal directly to $\mathsf{var}(\overline{y})$ as expressed in (6.5). Introducing these results into (7.2) we obtain

$$\mathsf{E}(s^2) = \sigma^2 \frac{K(k-1)}{k(K-1)} - \mathsf{E}(\overline{y} - \overline{y}')^2 + \mathsf{E}(\overline{y}' - \overline{y})\,(\overline{y}' - \overline{z}). \tag{7.3}$$

Now we will find a pair of constants p and q such that $p\,s^2 + q(\overline{y}' - \overline{y})^2$ be an unbiased estimate of $\mathsf{var}(\overline{y})$, up to a multiple of $\mathsf{E}(\overline{y}' - \overline{y})\,(\overline{y}' - \overline{z})$, i. e.

$$\mathsf{E}\left\{p\,s^2 + q(\overline{y}' - \overline{y})^2\right\} = \mathsf{var}(\overline{y}) + \text{const}\,\mathsf{E}\left\{(\overline{y}' - \overline{y})\,(\overline{y}' - \overline{z})\right\}.$$

Substituting from (7.3) and (6.5) we should thus have

$$\begin{aligned} p\,\sigma^2 \frac{K(k-1)}{k(K-1)} &- p\,\mathsf{E}(\overline{y}' - \overline{y})^2 + p\,\mathsf{E}\left\{(\overline{y}' - \overline{y})\,(\overline{y}' - \overline{z}) + q\,\mathsf{E}(\overline{y}' - \overline{y})^2\right\} \\ = \;& \frac{\sigma^2}{k}\frac{K-k}{K-1} + \mathsf{E}(\overline{y}' - \overline{y})^2 - \mathsf{E}\left\{(\overline{y}' - \overline{y})\,(\overline{y}' - \overline{z})\right\} \\ & + \text{const}\,\mathsf{E}\left\{(\overline{y}' - \overline{y})\,(\overline{y}' - \overline{z})\right\}. \end{aligned}$$

Comparing the coefficients of σ^2 we obtain

$$p = \frac{K-k}{(k-1)\,K}.$$

Substituting for p and comparing the coefficients of the terms containing $\mathsf{E}(\overline{y}' - \overline{y})^2$ and $\mathsf{E}(\overline{y}' - \overline{y})\,(\overline{y}' - \overline{z})$ we obtain for q and for the constant

$$q = \text{const} = \frac{k(K-1)}{K(k-1)}.$$

Accordingly, as an estimator for $\mathsf{var}(\overline{y})$, we choose

$$s_{\overline{y}}^2 = s^2\,\frac{K-k}{K(k-1)} + (\overline{y}' - \overline{y})^2\,\frac{k}{K}\,\frac{K-1}{k-1}.$$

This estimator is 'nearly unbiased', since its expectation is

$$\mathsf{E}(s_{\overline{y}}^2) = \mathsf{var}(\overline{y}) + \frac{k(K-1)}{K(k-1)}\,\mathsf{E}\left\{(\overline{y}' - \overline{y})\,(\overline{y}' - \overline{z})\right\}. \tag{7.4}$$

Substituting in (7.1) the mean $\overline{y}$ by the mean $\overline{y}'$, we obtain the variance $s'^2 = s^2 + (\overline{y}' - \overline{y})^2$, so that the estimator $s_{\overline{y}}^2$ can also be written in the form

$$s_{\overline{y}}^2 = s'^2\,\frac{K-1}{(k-1)\,K} + (\overline{y}' - \overline{y})^2.$$

8 STRATIFIED SAMPLING

Now let the population of N elements be divided in L subpopulations called strata. Denote the number of elements in the ith stratum by N_i, the number of clusters K_i and the stratum mean z_i per element $\overline{z}_i$ $(i = 1, 2, \ldots, L)$. The population mean per element is thus

$$\overline{z} = (N_1\overline{z}_1 + \cdots + N_L\overline{z}_L)/N.$$

Now let us take, from every stratum, a sample of K_i clusters by the method of two phases described above and denote the sample means $\overline{y}_i$. The samples from different strata are supposed to be independent of each other. The overall sample mean calculated as

$$\overline{y} = (N_1\overline{y}_1 + \cdots + N_L\overline{y}_L)/N$$

is then an unbiased estimate of the population mean. This is easily proved by straightforward calculation:

$$\mathsf{E}(\overline{y}) = (N_1\,\mathsf{E}(\overline{y}_1) + \cdots + N_L\,\mathsf{E}(\overline{y}_L))/N = (N_1\overline{z}_1 + \cdots + N_L\overline{z}_L)/N = \overline{z}.$$

The variance of $\overline{y}$ can be expressed through the variances of individual sample means

$$\mathsf{var}(\overline{y}) = \left(N_1^2\mathsf{var}(\overline{y}_1) + \cdots + N_L^2\mathsf{var}(\overline{y}_L)\right)/N^2$$

(due to the assumed independence of sampling from different strata). Thus the analysis of $\mathsf{var}(\overline{y})$ and its estimation reduces to the analysis and estimation of the individual $\mathsf{var}(\overline{y}_i)$s carried out in Sections 6 and 7.

It may be noted that the relations just shown do not depend on the numbers k_i, i.e. on the numbers of clusters selected in different strata. However, as in the case of sampling individual elements, it is generally recommendable[1] to select the same proportion of clusters in all strata.

9 APPENDIX

Before concluding we will indicate briefly the derivation of the term $\mathsf{E}(\overline{y}' - \overline{y})(\overline{y}' - \overline{z})$ neglected in the evaluation of $\mathsf{var}(\overline{y})$.

Introduce first the following auxiliary quantities:

$$\begin{aligned}\overline{r} &= (r_1 + \cdots + r_k)/k \\ \overline{z}' &= (u_1 + \cdots + u_K)/K\end{aligned}$$

[1] This recommendation may be doubtful when the strata differ considerably in average cluster size. (Translator's note.)

$$\sigma_1^2 = \sum^{K} (u_i - \overline{z}')^2 / K \tag{9.1}$$
$$\sigma_2^2 = \sum^{K} (u_i - \overline{z}')^2 \, h_i / N.$$

We start with the decomposition

$$\mathsf{E}(\overline{y}' - \overline{y})\,(\overline{y}' - \overline{z}) = \mathsf{E}(\overline{y}' - \overline{y})\,(\overline{y}' - \overline{z}') + \mathsf{E}(\overline{y}' - \overline{y})\,(\overline{z}' - \overline{z}) \tag{9.2}$$

and then calculate separately the two terms on the right-hand side.

First we write

$$\mathsf{E}(\overline{y}' - \overline{y})\,(\overline{y}' - \overline{z}') = \frac{1}{N\binom{K-1}{k-1}\,k}\left\{\sum_{C}\sum^{k}(y_i - \overline{z}')^2\,(\overline{r} - r_i) \right. \tag{9.3}$$
$$\left. + \sum_{C}\sum_{i \neq j}^{k}(y_i - \overline{z}')\,(y_j - \overline{z}')\,(\overline{r} - r_i)\right\}.$$

The first summation in the brackets is

$$\sum_{C}\sum^{k}(y_i - \overline{z}')^2\,(\overline{r} - r_i) = \sum^{K}(u_i - \overline{z}')^2 \sum_{C_i}(\overline{r} - h_i), \tag{9.4}$$

where the symbol C_i under the summation sign indicates that only combinations including h_i are summed, so that

$$\sum_{C_i}(\overline{r} - h_i) = \frac{1}{k}\left\{\binom{K-2}{h-2}(N - h_i) + \binom{K-1}{k-1} h_i\right\} - \binom{K-1}{k-1} h_i$$
$$= \frac{1}{k}\binom{K-2}{k-2}(N - K\,h_i).$$

Introducing this expression in (9.4) we obtain for (9.4) the value

$$(9.4) = \frac{1}{k}\binom{K-2}{k-2}\left\{N\sum^{K}(u_u - \overline{z}')^2 - K\sum^{K}(u_i - \overline{z}')^2\,h_i\right\}$$
$$= \frac{NK}{k}\binom{K-2}{k-2}(\sigma_1^2 - \sigma_2^2).$$

The second sum in the braces is

$$\sum_{C}\sum_{i \neq j}^{k}(y_i - \overline{z}')\,(y_j - \overline{z}')\,(\overline{r} - r_i) = \sum_{i \neq j}^{K}(u_i - \overline{z}')\,(u_j - \overline{z}')\sum_{C_{ij}}(\overline{r} - h_i), \tag{9.5}$$

where C_{ij} under the summation sign indicates that the summation extends over the combinations that include both h_i and h_j so that

$$\sum_{C_{ij}}(\overline{r}-r_i) = \frac{1}{k}\left\{\binom{K-3}{k-3}(N-h_i-h_j)+\binom{K-2}{k-2}(h_i+h_j)\right\}$$
$$-\binom{K-2}{k-2}h_i = \frac{1}{k}\binom{K-3}{k-3}\left(N+\frac{K-k}{k-2}h_j-\frac{Kk-k-K}{k-2}h_i\right).$$

Introducting this in (9.5) leads to

$$(9.5) = \frac{1}{k}\binom{K-3}{k-3}\left\{N\sum_{i\neq j}^{K}(u_i-\overline{z}')(u_j-\overline{z}')+\frac{K-k}{k-2}\sum_{i\neq j}^{K}(u_i-\overline{z}')(u_j-\overline{z}')h_j\right.$$
$$\left.-\frac{Kk-k-K}{k-2}\sum_{i\neq j}^{K}(u_i-\overline{z}')(u_j-\overline{z}')h_i\right\}$$
$$= -\frac{NK}{k}\binom{K-3}{k-3}(\sigma_1^2-\sigma_2^2).$$

Adding the two summations in brackets we obtain

$$(9.4)+(9.5) = \frac{NK}{k}\binom{K-2}{k-2}\frac{K-k}{K-2}(\sigma_1^2-\sigma_2^2)$$

and thus finally according to (22)

$$\mathsf{E}(\overline{y}'-\overline{y})(\overline{y}-\overline{z}) = \frac{1}{N\binom{K-1}{k-1}k}\{(23)+(24)\} = \frac{K(k-1)}{k(K-1)}\frac{\sigma_1^2-\sigma_2^2}{k}\frac{K-k}{K-2}. \quad (9.6)$$

Now we will evaluate the second term on the right-hand side of (9.2):

$$\mathsf{E}(\overline{y}'-\overline{y})(\overline{z}'-\overline{z}) = (\overline{z}'-\overline{z})(\mathsf{E}(\overline{y}')-\overline{z}). \quad (9.7)$$

The expectation of $\overline{y}'$ is

$$\mathsf{E}(\overline{y}') = \frac{1}{N\binom{K-1}{k-1}k}\sum_{C}\sum^{k}y_i(r_1+\cdots+r_k), \quad (9.8)$$

where

$$\sum_{C}\sum^{k}y_i(r_1+\cdots+r_k) = \sum^{K}\sum_{C_i}(r_1+\cdots+r_k)$$
$$= \sum^{K}u_i\left\{\binom{K-2}{k-2}(N-h_i)+\binom{K-1}{k-1}h_i\right\}$$

$$= \binom{K-2}{k-2} \sum^{K} u_i \left(N + \frac{K-k}{k-1} \, h_i \right)$$
$$= \binom{K-2}{k-2} \left(N \, K \, \overline{z}' + N \, \frac{K-k}{k-1} \, \overline{z} \right),$$

so that (9.8) reduces to

$$\mathsf{E}(\overline{y}') = \frac{K(k-1)}{k(K-1)} \, \overline{z}' + \frac{K-k}{k(K-1)} \, \overline{z}.$$

Hence

$$\mathsf{E}(\overline{y}') - \overline{z} = \frac{K(k-1)}{k(K-1)} \, (\overline{z}' - \overline{z})$$

and according to (9.7)

$$\mathsf{E}(\overline{y}' - \overline{y}) \, (\overline{z}' - \overline{z}) = \frac{K(k-1)}{k(K-1)} \, (\overline{z}' - \overline{z})^2. \tag{9.9}$$

Introducing (9.6) and (9.7) into (9.2) we finally obtain

$$\mathsf{E}(\overline{y}' - \overline{y}) \, (\overline{y}' - \overline{z}) = \frac{K(k-1)}{k(K-1)} \left\{ \frac{\sigma_1^2 - \sigma_2^2}{k} \, \frac{K-k}{K-2} + (\overline{z}' - \overline{z})^2 \right\}. \tag{9.10}$$

This expression could only be negative if σ_2^2 exceeded σ_1^2 considerably. Since the definition (9.1) implies that

$$\sigma_2^2 - \sigma_1^2 = \frac{1}{N} \sum^{K} (u_i - \overline{z}')^2 \, (h_i - N/K) = \frac{K}{N} \, \mathsf{cov} \left\{ (u_i - \overline{z}')^2, h_i \right\}$$

this would be possible only if there were a strong positive correlation between the squares of $u_i - \overline{z}'$ and the cluster sizes h_i.

From Equations (7.4) and (9.10) it may be seen that the usually positive bias of $s_{\overline{y}}^2$ is equal to

$$\frac{\sigma_1^2 - \sigma_2^2}{k} \, \frac{K-k}{k-z} + (\overline{z}' - \overline{z})^2.$$

REFERENCES

Bowley, A. L. (1926). Measurement of the precison attained in sampling. *Bulletin de l'Institut International de Statistique 22*.

Neyman, J. (1934). On the two different aspects of the representative method. *J. Roy. Statist. Soc. 97*, 558–625.

CHAPTER 2

Some Rank Distributions and their Applications

Časopis Pěst. Mat. 80 (1955), 17–31. (In Czech).
English translation in Selected Transl. Math. Statist. Prob. 2 (1962), 27–40.
Translated by L. Rosenblatt.
Reproduced by permission of American Mathematical Society.

Two rank tests, capable of replacing the t-test, with asymptotic efficiency 0.955, are derived, by application of the results of combinatorial considerations about the set of numbers $1, 2, \ldots, N$.

1 INTRODUCTION AND SUMMARY

This work is, with a few changes, one of the three parts of the dissertation which I presented in 1949. Three laws of distribution are derived in it, following from simple considerations about the set of numbers $1, 2, \ldots, N$, and further, the application of these distributions to testing the null hypothesis is indicated. The numbers $1, 2, \ldots, N$ play the role of ranking in these tests, and thus they are commonly called rank tests. The distribution and the test denoted by the letter α occur here probably for the first time. However the original idea on which this test is based should be credited to R. A. Fisher (1935), who introduces, in Chapter III of his *The Design of Experiments*, a test based on the same principle, only with the difference that it is not based on the rank of the values but on the values themselves. It is obvious that from the arithmetical point of view it is very laborious to construct the test, since a special law of distribution must be calculated for every concrete event. With the α-test, on the other hand, as soon as the distribution α is calculated and tabulated, the performance of the test is only a few minutes' work. An

Collected Works of Jaroslav Hájek – With Commentary
Edited by M. Hušková, R. Beran and V. Dupač
Published in 1998 by John Wiley & Sons Ltd.

extensive tabulation of critical values of α for five significant levels and for 10 to 50 pairs of observations is given in this chapter.

The distribution and test denoted by the letter β was developed by Wilcoxon (1945) in an article which was inaccessible to me. Therefore I shall concern myself only with the areas where my method appears to be original. The third distribution and test, denoted by the letter γ, formed from the previous system, is good in statistics by comparison with the rank correlation coefficient r.

For future use of the tests introduced in this work, we determine their advantages and disadvantages with respect to the present Student's t-test, which is ranked by determination of significance for

(i) the difference between the average of two joint samples
(ii) the difference between the average of two disjoint samples
(iii) the regression coefficient.

In case (i) I was able to replace the t-test by the α-test, in case (ii) by the β-test (i. e., the Wilcoxon test or the Mann–Whitney U-statistic), and in case (iii) by the γ-test.[1] The chief advantages introduced by the rank tests are simplicity and speed, and the fact that, being more general, the assumptions are more realistic. Also their efficiency is not bad, since, as shown by van der Waerden (1952), the asymptotic efficiency of the β-test is $3/\pi = 0.955$. A similar result is also valid for the α-test, which we would like to show in a separate section. In using rank tests, however, we must not forget the importance of the fact that the cases where it is necessary to carry out only the test of significance are relatively rare. Always as soon as significance is shown it is necessary to determine the confidence interval simultaneously, and while the t-test presents it directly the rank tests are able to present it only with great arithmetical difficulty.

2 α-DISTRIBUTION

Our point of departure is the set of numbers $1, 2, \ldots, N$ whose subsets we examine, assuming each of them to be an elementary event with equal probability 2^{-N}. The sum of the numbers in the separate subsets, denoted by α, will then be a random variable acquiring an integral value in the interval

$$0 \leq \alpha \leq \frac{1}{2}N(N+1).$$

We easily find the generating polynomial for the probability distribution of the random variable α if we take into account that α is the sum of n independent random variables:

$$\alpha = \alpha_1 + \alpha_2 + \cdots + \alpha_N,$$

[1] For brevity we denote by one letter both the form of the law of distribution and the random variable governed by that law of distribution, and likewise the respective test.

where

$$\begin{aligned} \alpha_k &= k \text{ (when the number } k \text{ is in the given subset)} \\ &= 0 \text{ (in the opposite case).} \end{aligned}$$

Here both possibilities are equally probable, since the number of subsets in which the number k is contained is equal to that in which it is not contained. Decomposition of α into the sum $\alpha_1 + \alpha_2 + \cdots + \alpha_N$ has a real counterpart in that we are able to divide the generation of the typical subset into N independent steps. In the first we include those not included as a subset of the number 1, in the second for the number 2, and similarly for all N numbers. The generating polynomial of the random variables α_k is equal to

$$\frac{1}{2}\left(1 + t^k\right)$$

and the generating polynomial of α is given by the relation

$$P_N(t)\,2^N = (1+t)\,\left(1+t^2\right)\cdots\left(1+t^N\right).$$

The coefficients of the polynomials $P_N(t)\,2^N$, given 2^N, give the probabilities of the respective values of α, and can be calculated with the help of the obvious recurrence relation

$$P_N(t)\,2^N = P_{N-1}(t)\,2^{N-1}\left(1+t^N\right).$$

That is, we substitute for the sequence of coefficients of $P_{N-1}(t)\,2^{N-1}$ the same sequence shifted back about N places, and sum to obtain the coefficient of $2^N\,P_N(t)$. Since the distributions of the random variables α_k are symmetric, this is also valid for the distribution of α. With the help of the corresponding characteristic variables α_k we easily calculate the mean value, the variance and the kurtosis of the distribution of α:

$$\begin{aligned} \mathsf{E}(\alpha_k) &= \frac{1}{2}k \\ \mathsf{var}(\alpha_k) &= \frac{1}{4}k^2 \\ \mu_4(\alpha_k) &= \frac{1}{16}k^4 \\ \kappa_4(\alpha_k) &= \mu_4(\alpha_k) - 3\mathsf{var}^2(\alpha_k) \;=\; -\frac{1}{8}k^4 \end{aligned}$$

and hence

$$\begin{aligned} \mathsf{E}(\alpha) &= \frac{1}{4}N(N+1) \\ \mathsf{var}(\alpha) &= \frac{1}{24}N(N+1)\,(2N+1) \\ \kappa_4(\alpha) &= -\frac{1}{240}N(N+1)\,(2N+1)\,(3N^2+3N-1) \\ \gamma_2(\alpha) &= \frac{\kappa_4(\alpha)}{\mathsf{var}^2(\alpha)} \;\sim\; -\frac{18}{5N}. \end{aligned}$$

Convergence of the α-distribution to the normal distribution follows directly from Liapunov's theorem: the third absolute moments of the variables α_k are equal to $k^3\, 2^{-3}$ and their sum is

$$\sum_{k=1}^{N} k^3\, 2^{-3} = \frac{1}{32} N^2 (N+1)^2.$$

The ratio of the third root of this sum to the standard deviation $\sqrt{\mathsf{var}(\alpha)}$ converges to zero as $N \to \infty$.

3 γ-DISTRIBUTION

We again start with the collection of numbers $1, 2, \ldots, N$ and consider all permutations, assuming each of them to be elementary events with equal probability $1/N!$. Then the number of inversions – we denote it by γ – will be a random variable. The random variable γ assumes integral values of the interval

$$0 \le \gamma \le \frac{1}{2} N(N-1),$$

and again can be decomposed into the sum of $N-1$ independent terms

$$\gamma = \gamma_1 + \gamma_2 + \cdots + \gamma_{N-1},$$

where γ_k is the number of inversions arising from the fact that the number $(k+1)$ precedes some smaller number, i. e., some number from $1, 2, \ldots, k$. This decomposition of the random variable γ has a real counterpart in that the generation of the typical permutation can be decomposed into $N-1$ independent steps. In the first we put the number 2 either after or before 1; in the second we put 3 either after both preceding numbers or between them or before them both, etc. till on the $(N-1)$th step we put the number N either after $(N-1)$ preceding numbers or in one of the $(N-2)$ positions between them or before them all.

Since the random variables γ_k assume the values $0, 1, 2, \ldots, k$ with equal probabilities, and since each eventuality corresponds to an equal number of the permutations (elementary events), their generating polynomials are equal to

$$\frac{1}{k+1}\left(1 + t + t^2 + \cdots + t^k\right) = \frac{1}{k+1}\left(1 - t^{k+1}\right)(1-t)^{-1}$$

and the generating polynomial of γ is given by the relation

$$R_N(t)\, N!(1-t)^N = (1-t)\,(1-t^2)\cdots(1-t^N). \tag{3.1}$$

We could deduce the convergence of the distribution of γ to the normal distribution, its symmetry, as well as the main moments, in exactly the same way as for the distribution of α. We give here only

$$\begin{aligned} \mathsf{E}(\gamma) &= \frac{1}{4}N(N-1), \\ \mathsf{var}(\gamma) &= \frac{1}{72}N(N-1)(2N+5). \end{aligned}$$

4 β-DISTRIBUTION

Again we start with the numbers $1, 2, \ldots, N$ and consider all possible divisions into two groups, one of n elements and the other of $(N-n)$ complementary elements considering each group of n elements as elementary events with equal probabilities $1 \Big/ \binom{N}{n}$. Then the number of inversions between the groups and their complements – denoted by β – will be a random variable. We shall consider the inversions between the groups as the occurrence in the group of n elements of a number greater than some number from the complementary group, and we determine the total number of inversions, if we investigate all $n(N-n)$ pairs of numbers, of which the first is from the group of n elements and the others from the complementary group. The random variable β then assumes an integral value in the interval

$$0 \leq \beta \leq n(N-n).$$

We return now to the random variable γ considered in the preceding paragraph. If we divide each permutation into two parts – in the first n and in the other $(N-n)$ elements – then we see that the random variable γ can be decomposed into the sum of three independent variables

$$\gamma = \gamma_1 + \gamma_2 + \beta,$$

where γ_1 is the number of inversions within the first part, γ_2 is the number of inversions within the second part and β is the number of inversions between both parts. This decomposition of the random variable γ has a real counterpart in that the resulting common permutation can be divided into three independent steps: in the first we determine the numbers which will be in the first n spaces, and hence also the numbers of the following $(N-n)$ spaces; in the second step we permute between themselves the numbers determined for the first n spaces; and in the third step we carry this out for the remaining $(N-n)$ numbers.

The generating polynomials of γ, γ_1 and γ_2 are given by (3.1). If we denote the generating polynomial of β by the symbol $Q_{N,n}(t)$, then from the relation

$$R_N(t) = R_n(t)\, R_{N-n}(t)\, Q_{N,n}(t)$$

immediately follows:

$$Q_{N,n}(t)\binom{N}{n} = \frac{(1-t^N)(1-t^{N-1})\cdots(1-t^{N-n+1})}{(1-t)(1-t^2)\cdots(1-t^n)}. \tag{4.1}$$

If we write formal analogies of the factorials

$$(1-t)(1-t^2)\cdots(1-t^k) = (1-t^k)!,$$

we have

$$Q_{N,n}(t)\binom{N}{n} = \frac{(1-t^N)!}{(1-t^n)!\,(1-t^{N-n})!},$$

so that the generating polynomial of β is analogous to the combination numbers between the polynomials. We see that it does not change if we put $(N-n)$ in place of n:

$$Q_{N,n}(t) = Q_{N,\,N-n}(t).$$

However $Q_{N,\,N-n}(t)$ is the generating polynomial for the distribution of the random variable $n(N-n)-\beta$, from which it is easy to see that the distribution of β is symmetric.

We find the coefficients of $Q_{N,n}(t)\binom{N}{n}$, and with their help also the probabilities of the corresponding values of β, easily from the recursion relation

$$R_{N,n}(t)\binom{N}{n} = R_{N-1,\,n}(t)\binom{N-1}{n} + t^{N-n}\,R_{N-1,\,n-1}(t)\binom{N-1}{n-1},$$

which the reader, using (4.1), can easily confirm. The real meaning of this relation is that the division of the numbers $1, 2, \ldots, N$ into two groups can be divided into two cases, according to whether the number N is in the complementary group or in the group of n elements. In the first case there will be just as many inversions as when correspondingly only the numbers $1, 2, \ldots, N-1$ were divided, while in the second case there will be $(N-n)$ more of them.

If we call the sum of numbers in the group of n elements S, it is easy to confirm that

$$\beta = S - \frac{1}{2}n(n+1).$$

S is the sum of n numbers chosen without replacement and with equal probabilities from the basic set $1, 2, \ldots, N$ in which the basic variance is $\sigma^2 = (1/12)(N^2-1)$. Therefore

$$\mathsf{var}(S) = n\,\sigma^2\frac{N-n}{N-1} = \frac{1}{12}n(N-n)(N+1)$$

and then also

$$\mathsf{var}(\beta) = \frac{1}{12}n(N-n)(N+1).$$

The mean value of β we find by symmetry as the average of the extreme values 0 and $n(N-n)$:

$$\mathsf{E}(\beta) = \frac{1}{2}n(N-n).$$

Finally let us investigate to what distribution the distribution of β converges as $N \to \infty$. Here we must consider two cases:

(a) n remains constant. In this case we will derive the distribution of the variable β/N. The characteristic function $\phi(t)$ of the distribution of the variable β/N can be found if we replace t by the expression $e^{it/N}$ in the suitable polynomial (4.1):

$$\phi(t) = \frac{\left(1-e^{it\frac{N}{N}}\right)\left(1-e^{it\frac{N-1}{N}}\right)\cdots\left(1-e^{it\frac{N-n+1}{N}}\right)1\cdot 2\cdots n}{\left(1-e^{it\frac{1}{N}}\right)\left(1-e^{it\frac{2}{N}}\right)\cdots\left(1-e^{it\frac{n}{N}}\right)N(N-1)\cdots(N-n+1)}.$$

Since

$$\lim_{N\to\infty}\left(1-e^{it\frac{N-k+1}{N}}\right) = 1-e^{it}, \quad k = 1, 2, \ldots, n$$

$$\lim_{N\to\infty}\left(1-e^{it\frac{k}{N}}\right)\frac{N-k+1}{k} = -it,$$

we obtain

$$\lim_{N\to\infty}\phi(t) = \left\{\frac{e^{it}-1}{it}\right\}^n.$$

This is the characteristic function of the sum of n independent random variables having uniform distribution on the interval $(0,1)$.

(b) $n \to \infty$ and $N-n \to \infty$. Since the distribution of the sum of n independent values chosen from the same distribution converges as $n \to \infty$ to the normal distribution we can easily see from this that for some manner of joint convergence of n and $N-n$ to infinity the distribution of β will converge to the normal distribution. However it has been shown (see Mann and Whitney (1947)) that convergence to the normal distribution occurs for any convergence $n \to \infty$ and $N-n \to \infty$.

5 α-TEST

Now we show how we can use the results derived in the above paragraphs in testing null hypotheses. Let us have N pairs of observations (x_1, y_1), (x_2, y_2), $\ldots, (x_n, y_n)$ mutually independent and such that the observations x_i are obtained under condition [treatment] A, and observations y_i under condition [treatment] B. Still other conditions may be effective that are the same within each pair but different between the pairs (see the example below). Now we test the null hypothesis: that the change from conditions A to conditions

B did not have influence on the size of the observed values, and that all differences within the pairs can be considered random. In other words, we test the null hypothesis that each pair of observations represented two values independently chosen from the same continuous distribution which, however, can change from pair to pair.

We form the differences

$$d_i = x_i - y_i, \quad i = 1, 2, \ldots, N$$

and arrange them according to absolute value. The continuity of the distribution excludes the case $|d_i| = |d_j|$ for $i \neq j$ and also the case $d_i = 0$. The differences arranged according to absolute value we call $d^{(1)}, d^{(2)}, \ldots, d^{(N)}$, so that

$$\left|d^{(1)}\right| < \left|d^{(2)}\right| < \cdots < \left|d^{(N)}\right|.$$

Then we can associate each batch of pairs of observations (x_i, y_i) with a subset of the numbers $1, 2, \ldots, N$ such that the number k will be present in it if and only if the difference $d^{(k)}$ is positive. From the null hypothesis it immediately follows that each such subset will have equal probability and the sum of the ranks of positive differences in their arrangement according to absolute value will have the α-distribution.

We will carry out this test as follows: we calculate the difference between pairs of observations, arrange them according to absolute value, sum the rank indices of the positive differences and then compare the obtained α with the tabulated critical value for the respective level of significance. It stands to reason that if we expect violation of the null hypothesis only in one direction, then we reduce the level of significance to one half. In order not to have to tabulate the upper critical values for α, we can take advantage of the fact that it is up to us whether we add rank indices of the positive or negative differences. If we denote α for positive differences by α_+ and α for negative differences by α_-, then it is valid that

$$\alpha_+ + \alpha_- = \frac{1}{2}N(N+1),$$

so that α_+ exceeds the upper critical value just when α_- exceeds the lower critical value. This enables us to tabulate only the lower critical value with which we compare α based on the differences of that sign which occurs less frequently.

Table 1

i, k	x_i	y_i	d_i	$d^{(k)}$
1	188	139	49	6
2	96	163	–67	8
3	168	160	8	14
4	176	160	16	16
5	153	147	6	23
6	172	149	23	24
7	177	149	28	28
8	163	122	41	29
9	146	132	14	41
10	173	144	29	–48
11	186	130	56	49
12	168	144	24	56
13	177	102	75	60
14	184	124	60	–67
15	96	144	–48	75

Example. We use the α-test for the analysis of Darwin's classical experiment, given in *The Design of Experiments* by R. A. Fisher. Here we find an investigation of whether the descendants of plants engendered by cross-breeding differ significantly as to height from descendants of self-fertilized parentage. Fifteen pairs of different varieties of plants were individually exposed as far as possible to the same attention and the same original conditions, so that each respective pair, for all but random influences, differed only in that one originated from cross-breeding and the second from self-fertilization. In Table 1 in the second column the results of height attainments are given for the individuals of cross-bred parentage and in the third column the parallel results for individuals of self-fertilized parentage, where the unit is 1/8 inch. The differences are in the fourth column, and in the fifth these differences are arranged according to absolute value. The numbers in the first column determine the rank d_i as well as $d^{(k)}$. Since there are only two negative differences, we base the α-test on them. Since -48 has rank 10 and -67 has rank 14 we get

$$\alpha = 10 + 14 = 24.$$

In Table 2 we find that for $N = 15$ the 5 % critical value is 25.3, hence if we are satisfied with a 5 % level of significance, the obtained value of α is significantly small and we judge that the manner of origin of the plant has an influence on its height. If we use the t-test, we get $t = 2.148$, which is also a value barely larger than the 5 % value, so both tests give essentially the same result.

Table 2 Lower critical values for α-test

No. of pairs N	Levels of significance				
	10 %	5 %	2 %	1 %	0.1 %
10	10.8	8.0	4.9	3.0	
11	13.9	10.7	7.1	4.8	
12	17.5	13.8	9.6	7.0	0.8
13	21.4	17.2	12.6	9.6	2.4
14	25.7	21.0	15.8	12.4	4.2
15	30.5	25.3	19.5	15.7	6.3
16	35.6	29.9	23.5	19.3	8.8
17	41.2	34.9	27.9	23.3	11.6
18	47.1	40.3	32.6	27.6	14.7
19	53.6	46.1	37.8	32.3	18.1
20	60.4	52.3	43.3	37.3	21.9
21	67.6	58.9	49.1	42.7	25.9
22	75.2	66.0	55.4	48.6	30.3
23	83.3	73.4	62.2	54.7	35.1
24	91.8	81.2	69.3	61.4	40.3
25	100.8	89.5	76.7	68.3	45.8
26	110.1	98.2	84.6	75.7	51.6
27	119.9	107.3	92.9	83.4	57.7
28	130.2	116.8	101.6	91.6	64.2
29	140.8	126.8	110.7	100.0	71.2
30	151.9	137.1	120.2	108.9	78.6
31	163.4	147.9	130.2	118.2	86.3
32	175.5	159.1	140.5	128.0	94.3
33	187.9	170.7	151.2	138.2	102.7
34	200.7	182.8	162.3	148.8	111.5
35	214.0	195.3	173.9	159.8	120.7
36	227.7	208.2	185.9	171.0	130.2
37	241.9	221.5	198.3	182.7	140.1
38	256.5	235.3	211.1	195.0	150.4
39	271.5	249.6	224.4	207.6	161.1
40	287.0	264.2	238.1	220.6	172.1
41	303.3	279.2	252.1	233.8	183.6
42	319.3	294.7	266.5	247.5	195.4
43	336.2	310.6	281.5	261.8	207.7
44	353.4	327.0	296.9	276.5	220.4
45	371.2	343.8	312.7	291.6	233.5
46	389.3	361.1	328.9	307.1	247.0
47	408.0	378.7	345.5	322.8	260.8
48	427.0	396.8	362.5	339.0	275.0
49	446.5	415.4	380.0	355.8	289.7
50	466.5	434.4	397.9	373.0	304.7

In Table 2 the lower critical values for 10 %, 5 %, 2 %, 1 % and 0.1 % levels of significance are calculated. For small N actual α-distributions were used. For larger N it was possible to use an approximation with the help of the normal distribution, corrected by a term of the Edgeworth series, taking into consideration the kurtosis.

Remark 1. If the number $\alpha_{0.05}$ stands in the column of the 5 % critical values in Table 2, it means that

$$\mathsf{P}(\alpha_+ \leq \alpha_{0.05}) + \mathsf{P}(\alpha_- \leq \alpha_{0.05}) = 0.05.$$

If the number $\alpha_{0.05}$ is a non-integer, for example $\alpha_{0.05} = 25.3$, it means that with probability 0.3 we can also consider as significant the value 26, without the probability of rejecting the null hypothesis, if it is correct, rising above 0.05.

Remark 2. In actual experiments it will happen that some differences will equal zero and others again will be equal as to their absolute values. This case can be carried over to the previous one if each zero difference is set with probability 0.5 to be positive and with probability 0.5 to be negative, and if, at the same time, the groups of differences of the same absolute value are arranged randomly according to absolute value so that each arrangement has equal probability. After this artificial 'completion' of the experiment, the α-test can be used precisely as in the case of sampling from continuous distributions. Its sensitivity, of course, changes thus. For application, however, it will be sufficient to use in place of the 'completed' experiment the following rules of thumb:

(i) if there occur k zero differences, then we add to the value of α the number $(1/2)(1 + 2 + \cdots + k)$;
(ii) if there occur in the positions $r + 1, \ldots, r - k + 1$ k differences of the same absolute value, among them k_+ positive and k_- negative, $k_+ + k_- = k$, then we add to the value of α $\left(r + \frac{k-1}{2}\right) k_+$ and $\left(r + \frac{k-1}{2}\right) k_-$, respectively, according to whether we use α_+ or α_-.

In those groups, where all differences are of equal sign, it is obviously not necessary to take any measures. The statistic α, obtained from the rule of thumb, will indeed not have exactly the distribution which we tabulated, but we can expect that in the case where there are not many zero and absolutely equal differences, the tables will remain applicable.

Remark 3. If the null hypothesis is rejected, we have to find the confidence interval for unknown mean value μ. Here we can proceed in the following way. We take μ arbitrarily and assume that the mean value is actually equal to that μ. Then, obviously, with the corresponding probability (namely 0.95),

α calculated from the values $(d_1 - \mu)$, $(d_2 - \mu), \ldots, (d_N - \mu)$ must not fall into the critical region. Those μ for which this is true will form the confidence interval. Calculation of the end points of this interval would, however, be very difficult.

Remark 4. Application of the α-test is not restricted only to paired samples. We can also test with it a somewhat more general hypothesis, that the mean value of n observations having symmetric distributions is equal to a fixed constant. (In paired samples we test the hypothesis that the mean value of the differences is equal to zero.) Another generalization would be the non-central α-test, where it would, however, be necessary to specify the form of the distribution function.

Table 3

m \ n	3	4	5	6	7	8	9	10
4	—	0.7		Lower 5 % critical values for β-test				
5	0.4	1.6	2.8					
6	1.0	2.4	3.9	5.4				
7	1.5	3.3	5.1	6.9	8.8			
8	2.0	4.1	6.2	8.4	10.7	13.0		
9	2.5	5.0	7.4	10.0	12.6	15.3	18.0	
10	3.0	5.7	8.5	11.5	14.5	17.6	20.6	23.7

Table 4

m \ n	2	3	4	5	6	7	8	9	10
3	—	0.0			Lower 10 % critical values for β-test				
4	—	0.7	1.8						
5	0.1	1.4	2.8	4.1					
6	0.4	2.1	3.7	5.4	7.2				
7	0.8	2.7	4.7	6.8	8.9	11.1			
8	1.2	3.3	5.7	8.2	10.7	13.2	15.8		
9	1.6	4.0	6.8	9.6	12.5	15.4	18.3	21.3	
10	1.9	4.6	7.8	11.0	14.2	17.5	21.0	24.3	27.6

6 β-TEST

If we have two unpaired independent samples $x_1, x_2, \ldots, x_n$ and $y_1, y_2, \ldots, y_m$, we can test the null hypothesis that all $n + m$ values were samples from the same distribution by means of the β-test. It is sufficient to assertain the number of inversions between both samples, i. e., the number of cases where $y_i < x_j$ for $i = 1, 2, \ldots, m$ and $j = 1, 2, \ldots, n$. Under the assumption that the null hypothesis is valid, the number of inversions will have a β-distribution, where $N-n=m$. The lower 5 % and 10 % critical values for n, $m \leq 10$ are given for the β-test in Tables 3 and 4. Again, β calculated on the basis of the smaller number of inversions will be compared with the tabulated critical values.

7 γ-TEST

We have N independent pairs of observations (x_1, y_1), $(x_2, y_2), \ldots, (x_N, y_N)$ from a two-dimensional distribution. The null hypothesis, that x and y are independent, can now be tested with the help of the γ-test. It is sufficient to ascertain the number of inversions in the sequence $y^{(1)}, y^{(2)}, \ldots, y^{(N)}$ formed such that the corresponding values x_i form an increasing sequence $x^{(1)} < x^{(2)} < \cdots < x^{(N)}$. Under the assumption that the null hypothesis is valid, the number of inversions will have a γ-distribution.

REFERENCES

Fisher R. A. (1935). *The Design of Experiments.* Edinburgh.

Mann M. B. and Whitney D. R. (1947). On a test of whether one of two random variables is stochastically larger than the other. *Ann. Math. Statist. 18*, 50–60.

van der Waerden B. L. (1952). Order tests of the two-sample problem and their power. *Nederl. Akad. Wetensch. Proc. Ser. A 55*, 453–458.

van der Waerden B. L. (1953a). Order tests for the two-sample problem, II. *Nederl. Akad. Wetensch. Proc. Ser. A 56*, 303–310.

van der Waerden B. L. (1953b). Order tests for the two-sample problem, III. *Nederl. Akad. Wetensch. Proc. Ser. A 56*, 311–316.

Wilcoxon F. (1945). Individual comparisons by ranking methods. *Biometrics Bull. 1*, 80–83.

CHAPTER 3

Generalization of an Inequality of Kolmogorov

Co-author A. Rényi
Acta Math. Acad. Sci. Hungar. *6* (1955), 281–283.
Reproduced by permission of Akadémiai Kiadó, Budapest.

In what follows $\mathsf{P}(A)$ denotes the probability of the event A, $\mathsf{E}(\xi)$ the mean value, and $\mathsf{E}(\xi|A)$ the conditional mean value, under the condition A, of the random variable ξ.

The inequality of A. N. Kolmogorov (1928) in question states that if $\xi_1, \xi_2, \ldots, \xi_k, \ldots$ is a sequence of mutually independent random variables with mean values $\mathsf{E}(\xi_k) = 0$ and finite variances $\mathsf{E}(\xi_k^2) = D_k^2$ $(k = 1, 2, \ldots)$, we have, for any $\varepsilon > 0$,

$$\mathsf{P}\left(\max_{1 \le k \le m} |\xi_1 + \xi_2 + \cdots + \xi_k| \ge \varepsilon\right) \le \frac{1}{\varepsilon^2} \sum_{k=1}^{m} D_k^2. \tag{1.1}$$

This inequality is extremely useful in proving the strong law of large numbers and related theorems. In what follows the inequality (1.2) will be proved which contains (1.2) as a special case. The use of (1.2) instead of (1.1) makes it possible to simplify the proofs mentioned.

The inequality (1.2) has been found in Hájek (1953) by the first named author; the proof of the inequality given below is due to Rényi.

The following theorem will be proved.

Theorem. *If $\xi_1, \xi_2, \ldots, \xi_k, \ldots$ is a sequence of mutually independent random variables with mean values $\mathsf{E}(\xi_k) = 0$ and finite variances $\mathsf{E}(\xi_k^2) = D_k^2$ $(k = 1, 2, \ldots)$, and c_k $(k = 1, 2, \ldots)$ is a non-increasing sequence of positive numbers, we have for any $\varepsilon > 0$ and any positive integers n and*

Collected Works of Jaroslav Hájek – With Commentary
Edited by M. Hušková, R. Beran and V. Dupač
Published in 1998 by John Wiley & Sons Ltd.

m $(n < m)$

$$\mathsf{P}\left(\max_{n\le k\le m} c_k|\xi_1+\xi_2+\cdots+\xi_k|\ge\varepsilon\right)\le\frac{1}{\varepsilon^2}\left(c_n^2\sum_{k=1}^{n}D_k^2+\sum_{k=n+1}^{m}c_k^2\,D_k^2\right). \tag{1.2}$$

Proof. Let us put

$$\zeta=\sum_{k=n}^{m-1}(\xi_1+\xi_2+\cdots+\xi_k)^2\,(c_k^2-c_{k+1}^2)+c_m^2(\xi_1+\cdots+\xi_m)^2. \tag{1.3}$$

It follows that

$$\mathsf{E}(\zeta)=c_n^2\sum_{k=1}^{n}D_k^2+\sum_{k=n+1}^{m}c_k^2\,D_k^2. \tag{1.4}$$

Denoting by A_r $(r=n, n+1,\ldots,m)$ the event consisting in the simultaneous validity of the inequalities[1]

$$c_s|\xi_1+\cdots+\xi_s|<\varepsilon\quad(n\le s<r)\quad\text{and}\quad c_r|\xi_1+\cdots+\xi_r|\ge\varepsilon,$$

the inequality (1.2) can be written in the form

$$\sum_{r=n}^{m}\mathsf{P}(A_r)\le\frac{1}{\varepsilon^2}\,\mathsf{E}(\zeta). \tag{1.5}$$

The inequality (??) is the consequence of (1.6a), (1.6b) and (1.6c) of which the first two are evident:

$$\mathsf{E}(\zeta)\ge\sum_{r=n}^{m}\mathsf{E}(\zeta|A_r)\,\mathsf{P}(A_r), \tag{1.6a}$$

$$\mathsf{E}(\zeta|A_r)\ge\sum_{k=r}^{m-1}\mathsf{E}\left((\xi_1+\cdots+\xi_k)^2|A_r\right)\cdot(c_k^2-c_{k+1}^2)$$
$$+c_m^2\,\mathsf{E}\left((\xi_1+\cdots+\xi_m)^2|A_r\right), \tag{1.6b}$$

$$\mathsf{E}\left((\xi_1+\cdots+\xi_k)^2|A_r\right)\ge\mathsf{E}\left((\xi_1+\cdots+\xi_r)^2|A_r\right)\ge\frac{\varepsilon^2}{c_r^2},\quad r\le k\le m. \tag{1.6c}$$

In verifying the inequality (1.6c), one has to use the fact that according to the definition of the event A_r the random variables ξ_k $(k>r)$ are mutually independent of each other also under the condition A_r, and further that they are independent also under the condition A_r of the variables $\xi_1,\xi_2,\ldots,\xi_r$. It should be mentioned that the same fact is utilized in the proof of (1.1) due to Kolmogorov (1928). □

[1] If $r=n$, only the second inequality is supposed.

Remark 1. If we choose $n = 1$ and $c_1 = c_2 = \cdots = c_m = 1$, we obtain from (1.2) as a special case the inequality (1.1). If we choose $c_k = 1/k$ ($k = n, n+1, \ldots, m$), we obtain the inequality

$$\mathsf{P}\left(\max_{n\le k\le m}\frac{|\xi_1+\xi_2+\cdots+\xi_k|}{k}\ge\varepsilon\right)\le\frac{1}{\varepsilon^2}\left(\frac{\sum_{k=1}^n D_k^2}{n^2}+\sum_{k=n+1}^m\frac{D_k^2}{k^2}\right). \quad (1.7)$$

Remark 2. By means of passing to the limit $m \to \infty$, it is easy to deduce from (1.2) the following inequality:

$$\mathsf{P}\left(\sup_{n\le k} c_k|\xi_1+\cdots+\xi_k|\ge\varepsilon\right)\le\frac{1}{\varepsilon^2}\left(c_n^2\sum_{k=1}^n D_k^2+\sum_{k=n+1}^\infty c_k^2 D_k^2\right). \quad (1.8)$$

In particular, if $c_k = 1/k$ ($k = 1, 2, \ldots$), we obtain the inequality

$$\mathsf{P}\left(\sup_{n\le k}\frac{|\xi_1+\xi_2+\cdots+\xi_k|}{k}\ge\varepsilon\right)\le\frac{1}{\varepsilon^2}\left(\frac{\sum_{k=1}^n D_k^2}{n^2}+\sum_{k=n+1}^\infty\frac{D_k^2}{k^2}\right). \quad (1.9)$$

Remark 3. It follows from (1.9) immediately that the strong law of large numbers holds for the sequence of mutually independent random variables $\xi_1, \xi_2, \ldots, \xi_k, \ldots$ if the ξ_ks have mean values 0, finite variances $D_k^2 = \mathsf{E}(\xi_k^2)$ and

$$\sum_{k=1}^\infty\frac{D_k^2}{k^2} \quad (1.10)$$

converges (see Kolmogorov (1930)).

As a matter of fact, it follows from (1.9) and (1.10) that for any $\varepsilon > 0$

$$\lim_{n\to\infty}\mathsf{P}\left(\sup_{n\le k}\frac{|\xi_1+\cdots+\xi_k|}{k}\ge\varepsilon\right)=0 \quad (1.11)$$

and therefore we have

$$\mathsf{P}\left(\lim_{n\to\infty}\frac{\xi_1+\xi_2+\cdots+\xi_n}{n}=0\right)=1. \quad (1.12)$$

REFERENCES

Kolmogorov, A. (1928). Über die Summen durch den Zufall bestimmten unabhängigen Größen. *Math. Ann. 99*, 309–319. (See also the corrections, ibid. *102* (1929), 484–488).

Kolmogorov, A. (1930). Sur la loi forte des grands nombres. *Comptes Rendus Acad. Sci. Paris 191*, 910–912. (See also A. Kolmogorov (1933), Grundbegriffe der Wahrschenlichkeitsrechnung, *Ergebnisse der Math. 2*, no. 3.)

CHAPTER 4

Asymptotic Efficiency of a Certain Sequence of Tests

Czechoslovak Math. J. 6 (1956), 26–30. (In Russian).
Translated by I. Müller.
Reproduced by permission of Hájek family: Al. Hájková, H. Kazárová and Al. Slámová.

Let H_μ be a hypothesis that random variables $D_1, D_2, \ldots, D_n$ are independent and normal $(\mu\sigma, \sigma^2)$. It is shown in this paper that the test of the null hypothesis H_0 against the alternative $H_\mu (\mu \neq 0)$ based on the statistic (1.1) has in the sense of Definition 5.1 asymptotic efficiency (5.2).

1 STATISTIC α_k

Let k be an arbitrary natural number less than n. We define the statistic

$$\alpha_k = \sum_{1 \leq i_1 < \ldots < i_k \leq n} \phi(D_{i_1}, \ldots, D_{i_k}), \tag{1.1}$$

where the function $\phi(d_1, \ldots, d_k)$ takes values

$$\begin{aligned} \phi(d_1, \ldots, d_k) &= 1 \quad \text{if } d_1 + \ldots + d_k > 0, \\ &= 0 \quad \text{otherwise.} \end{aligned} \tag{1.2}$$

2 MEAN VALUE OF α_k

Under H_μ, $D_1 + \ldots + D_k$ has normal distribution $(k\mu\sigma, k\sigma^2)$. Hence, the mean value can be written in the form

$$\mathsf{E}(\alpha_k | H_\mu) = \binom{n}{k} \mathsf{E}[\phi(D_1, \ldots, D_k) | H_\mu] = \binom{n}{k} \mathsf{P}(D_1 + \ldots + D_k > 0 | H_\mu)$$

Collected Works of Jaroslav Hájek – With Commentary
Edited by M. Hušková, R. Beran and V. Dupač
Published in 1998 by John Wiley & Sons Ltd.

$$= \binom{n}{k} \int_{-\infty}^{\mu\sqrt{k}} e^{-\frac{1}{2}x^2} \mathrm{d}x \Big/ \sqrt{2\pi} \,. \tag{2.1}$$

3 VARIANCE OF α_k

From (1.1) and (1.2) it follows by Hoefding (1948) that:

$$\mathrm{var}(\alpha_k | H_\mu) \tag{3.1}$$

$$= \binom{n}{k} \sum_{r=1}^{k} \binom{k}{r} \binom{n-k}{k-r} \mathrm{cov}[\phi(D_1, \ldots, D_k), \phi(D_{k-r+1}, \ldots, D_{2k-r}) | H_\mu].$$

Covariances on the right hand side are continuous functions of μ. Hence, for an arbitrary sequence $\{\mu_n\}, \mu_n \to 0$,

$$\lim_{n\to\infty} \mathrm{D}^2(\alpha_k | H_{\mu_n}) \frac{(k-1)!^2}{n^{2k-1}} = \mathrm{cov}[\phi(D_1, \ldots, D_k), \phi(D_k, \ldots, D_{2k-1}) | H_0]. \tag{3.2}$$

Under H_0, $D_1 + \ldots + D_k, D_k + \ldots + D_{2k-1}$ have normal distribution $(0, k\sigma^2)$, with covariance σ^2, so that

$$\mathrm{P}(D_1 + \ldots + D_k > 0, D_k + \ldots + D_{2k-1} > 0 | H_0)$$

$$= \iint_{\substack{x>0 \\ y>0}} e^{-\frac{x^2 - 2xy/k + y^2}{2k(1-1/k^2)\sigma^2}} \frac{\mathrm{d}x\mathrm{d}y}{2k\pi\sigma^2\sqrt{1-1/k^2}}.$$

Using the substitution

$$\begin{aligned} x &= u\sigma\sqrt{k} \\ y - \frac{1}{k}x &= v\sigma\sqrt{1-1/k^2}, \end{aligned}$$

the integral above can be expressed in the form of

$$\iint_{\substack{u>0 \\ u+v\sqrt{k^2-1}>0}} e^{-\frac{1}{2}(u^2+v^2)} \frac{\mathrm{d}u\mathrm{d}v}{2\pi},$$

and, finally, by means of the substitution

$$u = \rho\cos\omega, \quad v = \rho\sin\omega,$$

in the form of

$$\frac{1}{2\pi} \int_{-\arcsin 1/k}^{\frac{1}{2}\pi} \mathrm{d}\omega \int_0^\infty \rho e^{-\frac{1}{2}\rho^2} \mathrm{d}\rho = \frac{1}{4} + \frac{\arcsin 1/k}{2\pi}.$$

Bringing these equations together, we get

$$\begin{aligned}
&\operatorname{cov}[\phi(D_1,\ldots,D_k),\phi(D_k,\ldots,D_{2k-1})|H_0] \\
= \; & \mathsf{P}(D_1+\ldots+D_k>0, D_k+\ldots+D_{2k-1}>0|H_0)-\frac{1}{4} \\
= \; & \frac{\arcsin 1/k}{2\pi}.
\end{aligned}$$

Plugging this into (3.2), we get

$$\lim_{n\to\infty} \mathsf{D}^2(\alpha_k|H_{\mu_n})\,\frac{(k-1)!^2}{n^{2k-1}} = \frac{\arcsin 1/k}{2\pi}, \quad (\mu_n \to 0). \tag{3.3}$$

4 LIMITING DISTRIBUTION OF α_k

Through a minor modification[1] of a general procedure used in Hoefding (1948), it can be shown that the distribution of $(\alpha_k - \mathsf{E}\alpha_k)/\mathsf{D}\alpha_k$, conditioned on the hypothesis H_{μ_n}, is asymptotically normal (0,1) for $n \to \infty$ and $\mu_n \to 0$.

5 ASYMPTOTIC EFFICIENCY OF AN α_k-TEST

As the T-test of the hypothesis, H_0 will be called a test based on a statistic $T = T(D_1,\ldots,D_n)$ in such a way that H_0 is rejected if $T \le T_1$, $T \ge T_2$ resp., where T_1 and T_2 are critical values on a specific significance level. The asymptotic efficiency of the T-test in our case[2] is defined in the following way.

Definition 5.1. *Let the distribution of $T = T(D_1,\ldots,D_n)$ under the hypothesis H_{μ_n} be asymptotically normal for $n \to \infty$, $\mu_n \to 0$. Then the number*

$$e = \lim_{n\to\infty} \frac{\left[\frac{\partial}{\partial\mu}\mathsf{E}(T|H_\mu)|_{\mu=0}\right]^2}{n\mathsf{D}^2(T|H_{\mu_n})} \; (\mu_n \to 0), \tag{5.1}$$

if it exists, is called the asymptotic efficiency of the T-test of the hypothesis H_0 against the alternative H_μ ($\mu \neq 0$).

---

[1] The main point at issue here is that instead of the limiting distribution of partial sums of a sequence of i.i.d. random variables, one has to examine the limiting distribution of sums $\sum_{i=1}^{n} \xi_{in}$, where $\xi_{1n},\ldots,\xi_{nn}$ have the same distribution that is, however, in general different for different n. This difficulty can be overcome, for example, by using Theorem 11 from Chapter 9 of the book by B. V. Gnedenko: 'A Course in Probability Theory', Moscow, 1950.

[2] A more general definition is not given, since it doesn't seem to have been adopted in the literature yet.

As was shown in Noether (1955), the essence of asymptotic efficiency e is that given data, the T-test on n observations has approximately the same power as the Student t-test on ne observations if n is sufficiently large and $|\mu|$ sufficiently small.

Let us now calculate the asymptotic efficiency e_k of the α_k-test. According to (2.1)

$$\frac{\partial}{\partial\mu}\mathsf{E}(\alpha_k|H_\mu)\bigg|_{\mu=0} = \binom{n}{k}\frac{\partial}{\partial\mu}\int_{-\infty}^{\mu\sqrt{k}} e^{-\frac{1}{2}x^2}\,\mathrm{d}x/\sqrt{2\pi}\bigg|_{\mu=0} = \binom{n}{k}\sqrt{\frac{k}{2\pi}}.$$

Applying this result to (5.1), and with a view of (3.3), we get

$$e_k = \frac{1}{k \arcsin 1/k}, \qquad k = 1, 2, \ldots. \tag{5.2}$$

6 APPLICATIONS

The statistic α_1 equals the number of events when $D_i > 0, i = 1, \ldots, n$, so that the α_1-test is identical to the well-known 'sign' test. According to (5.2), the asymptotic efficiency of this test will be

$$e_1 = \frac{2}{\pi} \doteq 0.637. \tag{6.1}$$

The α_2-test is markedly more efficient:

$$e_2 = \frac{3}{\pi} \doteq 0.955. \tag{6.2}$$

The most important test for applications is the α-test based on the statistic $\alpha = \alpha_2 + \alpha_1$, which is introduced in Hájek (1955). It is calculated in such a way that the observations $d_1, d_2, \ldots, d_n$ of random variables $D_1, D_2, \ldots, D_n$ are ordered in absolute value

$$|d^{(1)}| < |d^{(2)}| < \ldots < |d^{(n)}|$$

and the upper rank indices of positive (negative) observations are summed. Since $\lim \alpha_1/\mathsf{D}\alpha_2 = 0$, the asymptotic efficiency of the α-test equals the asymptotic efficiency of the α_2-test (6.2).

Comparing (6.1) and (6.2), it can be seen that the increase in computational effort when using the α-test instead of the 'sign' test is fully outweighed by the increase in asymptotic efficiency. Moreover, in testing a more general null hypothesis, one can make use not only of the α_1-test but also of the α-test. The α-test can be used, for example, in the case of the null hypothesis stating that the random variables $D_1, D_2, \ldots, D_n$ be independent and each have zero-symmetric distribution. This can occur, for example, if $D_i = X_i - Y_i$, with the vectors (X_i, Y_i) having the same distribution as $(Y_i, X_i), i = 1, \ldots, n$.

It follows from (6.2) that the asymptotic efficiency of the α_k-test increases with k, and

$$\lim_{k \to \infty} e_k = 1. \tag{6.3}$$

In practice, however, α_k-tests ($k > 2$) are less convenient, because they are less efficient and at the same time more computationally demanding than the Student t-test. Besides, analogously to the t-test, they depend on the specific type of distribution of $D_1, D_2, \ldots, D_n$.

REFERENCES

Hájek, J. (1955). Some rank distributions and their applications (in Czech). English translation: in *Selected Transl. Math. Statist. Prob.* *2* (1962), 27–40. *Časopis Pěst. Mat. 80*, 17–31

Hoefding, W. (1948). A class of statistics with asymptotically normal distributions. *Ann. Math. Stat. 19*, 293–325.

Noether, G. E. (1955). On a theorem of Pitman. *Ann. Math. Stat. 26*, 64–68

CHAPTER 5

Linear Estimation of the Mean Value of a Stationary Random Process with Convex Correlation Function

Czechoslovak Math. J. 6 (1956), 94–117. (In Czech.)
English translation in *Selected Transl. Math. Statist. Prob.* *2* (1962), 41–6.
Translated by P. L. Zador.
Reproduced by permission of American Mathematical Society.

With the aid of the special spectral resolution (3.2) of convex correlation functions we obtain certain results concerning the linear estimation of the mean value of a stationary process; the essence of these results is contained in Theorems 5.1, 5.2 and 5.5.

1 INTRODUCTION AND SUMMARY

By the random process $x(t)$ we mean a collection of random variables labeled by the real time parameter t. The probability distribution of this process is given if for every n and for all sets of parameter values $t_1, \ldots, t_n$ the probability distribution of the random vector $\{x(t_1), \ldots, x(t_n)\}$ is known. The random process is stationary if the probability distribution of the vectors $\{x(t_1 + u), \ldots, x(t_n + u)\}$ is independent of translation u.

The most important characteristic of a stationary random process is its correlation function

$$R(\tau) = \frac{\mathsf{E}x(t)\,x(t+\tau) - [\mathsf{E}x(t)]^2}{\mathsf{E}x^2(t) - [\mathsf{E}x(t)]^2}.$$

Collected Works of Jaroslav Hájek – With Commentary
Edited by M. Hušková, R. Beran and V. Dupač
Published in 1998 by John Wiley & Sons Ltd.

The correlation function of continuous stationary processes – and only such processes will be considered in this chapter – is a continuous function in the variable τ.

Theorem 5.1 gives a lower bound for the variance of the linear estimate of the mean value of a stationary process with convex correlation function; Theorem 5.2 establishes the fact that the ordinary integral estimate (5.15) is asymptotically effective in the sense of Definition 5.1 whenever the correlation function is convex; the content of Theorem 5.5 is the assertion that for every correlation function there exists a uniquely determined distribution function $\Phi(t)$ which will furnish the best linear estimation of the form of (4.3).

2 CONVEX CORRELATION FUNCTIONS

We shall say that the correlation function $R(\tau)$ is convex if it satisfies the inequality[1]

$$R(\tau_1) + R(\tau_2) \geq 2R\left(\frac{\tau_1 + \tau_2}{2}\right); \quad \tau_1 \geq 0, \ \tau_2 \geq 0. \tag{2.1}$$

This implies immediately that every convex correlation function is non-increasing. Therefore the following limit exists:

$$\lim_{\tau \to \infty} R(\tau) = R(\infty). \tag{2.2}$$

This limit cannot be negative, for that would contradict the following lemma.

Lemma 2.1. *For each correlation function $R(\tau)$ and for every T it is true that*

$$\sup_{T \leq \tau} \{R(\tau)\} \geq 0. \tag{2.3}$$

Proof. $R(\tau)$ is positive definite:

$$\sum_{j=1}^{n} \sum_{i=1}^{n} R(t_i - t_j) \geq 0.$$

It must be all the more true that

$$n + n(n-1) \max \{R(t_i - t_j);\, i \neq j\} \geq 0,$$

[1] The statistical meaning of inequality (2.1) is the following: if we want to place between two observations on the t axis a third observation so that the variance of the sum of all three observations should become minimal it turns out that the best we can do is to put it in the middle.

or, equivalently,

$$\max\{R(t_i - t_j);\ i \neq j\} \geq -\frac{1}{n-1},$$

from which (2.3) follows readily. □

Thus, in fact, $R(\infty) \geq 0$. However it is sufficient to investigate the case when $R(\infty) = 0$. Namely if $R(\infty) > 0$, then it is possible to decompose the corresponding stationary process as the sum of dynamic and non-dynamic components in such a way that, for the dynamic component, $R(\infty) = 0$ will hold. This circumstance is stated more precisely in the following theorem.

Theorem 2.1. *Every stationary random process $x(t)$ with convex correlation function $R(\tau)$ and with the property that $R(\infty) < 1$ can be represented in the form*

$$x(t) = y(t) + z, \tag{2.4}$$

where $y(t)$ is a stationary random process with convex correlation function, $Q(\tau)$, and $Q(\infty) = 0$; z is a random variable independent of t and uncorrelated with $y(t)$ for all t.

Proof.For the sake of simplicity we shall assume that the mean value of the process $x(t)$ vanishes and that its variance is equal to unity. Set

$$z = \lim_{T\to\infty} \frac{1}{T}\int_0^T x(t)\,\mathrm{d}t. \tag{2.5}$$

The existence of this limit in the sense of mean square convergence is guaranteed by the celebrated ergodic theorem; see Khintchine (1934). From the convexity of $R(\tau)$ it follows that

$$\lim_{t\to\infty} R(t-u) = R(\infty),$$

and then

$$\lim_{T\to\infty} \frac{1}{T}\int_0^T R(t-u)\,\mathrm{d}t = R(\infty). \tag{2.6}$$

Therefore, also

$$\lim_{\substack{T\to\infty\\ U\to\infty}} \frac{1}{TU}\int_0^U\int_0^T R(t-u)\,\mathrm{d}t\mathrm{d}u = R(\infty).$$

But the last expression is nothing but the variance of the random variable z defined by equation (2.5), so that

$$\mathsf{D}^2 z = R(\infty). \tag{2.7}$$

Set

$$y(t) = x(t) - z. \tag{2.8}$$

We shall prove first of all that this process is not correlated with the variable z. Since, according to (2.8),

$$zy(t) = zx(t) - z^2,$$

it is sufficient to prove that

$$\mathsf{E}zx(t) = R(\infty). \tag{2.9}$$

However, owing to (2.5),

$$\mathsf{E}zx(t) = \lim_{T\to\infty} \frac{1}{T} \int_0^T R(t-u)\,\mathrm{d}u,$$

from which, noting (2.6), we get (2.9).

Thus it remains to find the correlation function $Q(\tau)$ of the process $y(t)$ and to prove that it is convex and that it satisfies the relation $Q(\infty) = 0$. Since we have just proved that the correlation between $y(t)$ and z is zero,

$$\mathsf{D}^2\, y(t) = \mathsf{D}^2\, x(t) - \mathsf{D}^2\, z = 1 - R(\infty),$$

and in addition, in view of (2.8), (2.7) and (2.9),

$$\mathsf{E}\, y(t)\, y(t+\tau) = \mathsf{E}x(t)\, x(t+\tau) - \mathsf{E}zx(t) - \mathsf{E}zx(t+\tau) + \mathsf{E}z^2 = R(\tau) - R(\infty).$$

But this means that

$$Q(\tau) = \frac{R(\tau) - R(\infty)}{1 - R(\infty)},$$

which implies first that $Q(\tau)$ is a convex function and second that $Q(\infty) = 0$. Since by hypothesis $R(\infty) < 1$, $Q(\tau)$ always exists. This completes the proof. □

A further consequence of the convexity and boundedness of $R(\tau)$ is that the right and left derivatives exist at each point; see Littlewood, Hardy and Pólya (1934). (For our purposes it will be satisfactory to employ the same symbol $R'(\tau)$ for both these derivatives.) This being so, for $\tau > 0$, the right derivative is not smaller than the left derivative, both these derivatives are non-decreasing functions, and $-R'(\tau)$ does not approach zero faster than τ^{-1}. The last assertion follows from the following lemma.

Lemma 2.2. *If $R(\tau)$ is a convex correlation function, then*

$$\lim_{\tau\to\infty} \tau R'(\tau) = \lim_{\tau\to 0} \tau R'(\tau) = 0. \tag{2.10}$$

Proof. $R(\tau)$ is absolutely continuous for $\tau > \epsilon > 0$, because $\tau > \epsilon > 0 \Rightarrow -R'(\tau) < -R'(\epsilon) < \infty$. Furthermore, $-R(\tau)$ does not increase since

$$R\left(\frac{1}{2}\tau\right) - R(\tau) = -\int_{\frac{1}{2}\tau}^{\tau} R'(u)\,\mathrm{d}u \geq -\frac{1}{2}\tau R'(\tau) \geq 0.$$

Since $R(\tau)$ is a convex function $R(\infty)$ exists and, consequently, the left, and a fortiori also the right side of this relation converges to zero for $\tau \to \infty$. The existence of the second limit follows in a similar manner from the continuity of $R(\tau)$. □

3 THE SPECTRAL RESOLUTION OF A CONVEX CORRELATION FUNCTION

The simplest convex correlation function is the following:

$$\begin{aligned} r(\tau) &= 1 - |\tau| \quad \text{for} \quad |\tau| \leq 1, \\ &= 0 \quad\quad\quad \text{for} \quad |\tau| > 1. \end{aligned} \tag{3.1}$$

It is easy to show that this correlation function corresponds, besides others, to the process $x_0(t)$ which can be constructed in the following way: take (i) mutually independent random variables y_k ($k = 0, \pm 1, \pm 2, \ldots$) with arbitrary,[2] but identical, distributions; (ii) a random variable ϕ which is independent of the variables y_k and has a uniform (rectangular) distribution over the interval $(0, 1)$, and then for t in the (random varying) intervals

$$\phi + k - 1 \leq t < \phi + k,$$

set

$$x_0(t) = y_k.$$

The process constructed this way represents a step function which is constant on intervals of length 1. For fixed ϕ two observation $x(t_1)$ and $x(t_2)$ will be identical or independent according to whether t_1 and t_2 lie in the same interval $[\phi + k - 1, \phi + k)$ or not.

However, the most important property of the correlation function $r(\tau)$ is that all convex correlation functions with $R(\infty) = 0$ possess spectral resolutions in terms of this function. This is the content of the following fundamental theorem.

---

[2] We assume that y_k have finite variances.

Theorem 3.1. *Let $R(\tau)$ be a convex correlation function with $R(\infty) = 0$. Then*

$$R(\tau) = \int_0^\infty r\left(\frac{\tau}{\lambda}\right) \mathrm{d}F(\lambda), \tag{3.2}$$

where $F(\lambda)$ is a positive[3] *distribution function; more precisely,*

$$F(\lambda) = \int_0^\lambda \tau \,\mathrm{d}R'(\tau), \quad 0 \le \lambda < \infty, \tag{3.3}$$

and $r\left(\frac{\tau}{\lambda}\right)$ is defined by equations (3.1).

Proof.We have to prove first of all that the function $F(\lambda)$, as defined by (3.3), indeed possesses all the properties of positive distribution functions, i. e., that it does not decrease, and that $F(0) = 0,\ F(\infty) = 1$. That $F(\lambda)$ is a non-decreasing function follows immediately from the fact that $R'(\tau)$ is a non-decreasing function, and the equations $F(0) = 0,\ F(\infty) = 1$ are consequences of the identity

$$F(\lambda) = \lambda R'(\lambda) - R(\lambda) + 1, \tag{3.4}$$

Lemma 2.2 and hypothesis $R(\infty) = 0$.

The validity of equation (3.2) for $\tau = 0$ follows immediately from the fact that $F(\lambda)$ is a positive distribution function. In the case $\tau > 0$ we have only to recall (3.3) and consider the following chain of identities:

$$\begin{aligned} \int_0^\infty r\left(\frac{\tau}{\lambda}\right) \mathrm{d}F(\lambda) &= \int_{|\tau|}^\infty \frac{\lambda - |\tau|}{\lambda} \mathrm{d}F(\lambda) = \int_{|\tau|}^\infty (\lambda - |\tau|) \,\mathrm{d}R'(\lambda) \\ &= -\int_{|\tau|}^\infty R'(\lambda) \,\mathrm{d}\lambda = R(|\tau|) = R(\tau). \end{aligned}$$

□

Remark 3.1. If $R(\tau)$ has second derivative $R''(\tau)$ for $\tau > 0$ then the spectral density $f(\lambda) = F'(\lambda)$ exists and

$$f(\lambda) = \lambda R''(\lambda).$$

For example, we get for the correlation function $e^{-a\tau}$

$$f(\lambda) = a^2 \,\lambda \,e^{-a\lambda}.$$

---

[3] We call a distribution function positive if $F(0) = 0$.

Remark 3.2. It follows from Theorem 3.1 that any convex correlation function $R(\tau)$ will, besides others, correspond to the process $x_0(t)$ which is constructed by taking (i) mutually independent random variables y_k ($k = 0, \pm 1, \pm 2, \ldots$) with arbitrary but identical distributions, (ii) a random variable ϕ which is independent of y_k and has uniform (rectangular) distribution over the interval $(0, 1)$, (iii) another random variable λ which is independent of ϕ and of y_k and has the distribution function $F(\lambda)$ which appears in equation (3.3), and then finally, for each t in the (random varying) intervals,

$$\lambda(\phi + k - 1) \leq t < \lambda(\phi + k), \quad k = 0, \pm 1, \pm 2, \ldots$$

setting

$$x_0(t) = y_k. \tag{3.5}$$

The usefulness of this representation lies in the fact that for any fixed ϕ and λ it replaces the stationary random process by the sequence of independent random variables y_k, each occupying an interval of length λ on the t axis. Important applications of this process are found in the proofs of Theorems 5.1 and 5.4.

Remark 3.3. It follows immediately from spectral representation (3.2) that every convex function $R(\tau)$ which is continuous at point 0 and satisfies $R(0) = 1$, $R(\infty) = 0$, is positive definite, and hence is the correlation function of some continuous stationary process. Indeed, the only properties of function $R(\tau)$ that were used in the proof of representation (3.2) were its convexity and the assumptions concerning its values at points 0 and ∞. In this, the condition $R(\infty) = 0$ has a secondary importance.

4 LINEAR ESTIMATION OF THE MEAN VALUE OF A STATIONARY PROCESS

By the linear estimate of the mean value

$$m = \mathsf{E}x(t)$$

of a stationary random process $x(t)$ over the interval $(0, T)$ we shall mean the Stieltjes[4] integral

$$\overline{x} = \int_0^T x(t)\, \mathrm{d}\Phi(t), \tag{4.1}$$

[4] Since in this paper we shall not leave the framework of correlation theory it is sufficient if we take here the integral $\int_0^T x(t)\, \mathrm{d}\Phi(t)$ in the sense of mean square convergence.

where the weighting function $\Phi(t)$ may be any function of bounded variation which satisfies the conditions

$$\Phi(0) = 0, \quad \Phi(T) = 1. \tag{4.2}$$

Besides this we shall assume for the sake of definiteness that $\Phi(t)$ is continuous to the left in $0 < t < T$. We shall call, as it is commonly done, the linear estimate with the smallest variance, only, of course, if it exists, the *best linear estimate*.

The weighting function yielding the best linear estimate possesses the important property established in the following theorem.

Theorem 4.1. *The weighting function $\Phi(t)$ gives the best linear estimate of the form (4.3) if and only if*

$$f(u) = \int_0^T R(t-u)\,\mathrm{d}\Phi(t) = \text{const}, \quad 0 \le u \le T. \tag{4.3}$$

If none of the admissible weighting functions satisfies this relation then the best linear estimate does not exist.

Proof.Regard the stationary random process $x(t)$ as a continuous curve in the Hilbert space the origin of which has been shifted to the point $x(0)$, so that the norm of $x(t)$ will be $\mathsf{D}^2[x(t)-x(0)]$. Then take the smallest linear subspace containing the segment of curve $x(t)$ which corresponds to $0 \le t \le T$. This subspace will, of course, contain all points of the form (4.3). It is known that this subspace contains one and only one point $\tilde{x}$ at which the distance from the initial origin becomes minimal. This means that for this unique point $\tilde{x}$

$$\mathsf{D}^2\,\tilde{x} = \inf_{\Phi}\left\{\mathsf{D}^2\left[\int_0^T x(t)\,\mathrm{d}\Phi(t)\right]\right\}, \tag{4.4}$$

where the infimum is taken for all admissible weighting functions. Besides, point $\tilde{x}$, and only this point, will satisfy for all u in the interval $0 \le u \le T$ the relation

$$\operatorname{cov}[\tilde{x},\, x(u) - \tilde{x}] = 0, \quad 0 \le u \le T,$$

or, equally,

$$\operatorname{cov}[\tilde{x},\, x(u)] = \text{const}, \quad 0 \le u \le T. \tag{4.5}$$

If, however, $\tilde{x}$ can be represented in the form of (4.3) then (4.7) is equivalent to (4.5). But this simply means that (4.5) is a necessary and sufficient condition for the statistics (4.3) to be the best linear estimate. If, however, condition (4.5) is not satisfied by any of the weighting functions $\Phi(t)$, then none of the statistics (4.3) will be identical with $\tilde{x}$, and, consequently, the best linear estimate does not exist. This completes the proof. □

Using condition (4.5) it is possible to determine 'by selection' the best weighting function for some important concrete correlation functions.

Example 4.1. Let $R(\tau) = e^{-|\tau|}$. Then the best weighting function, if it exists at all, must satisfy condition (4.5), that is

$$\int_0^T e^{-|t-u|}\,\mathrm{d}\Phi(t) = \text{const}, \quad 0 \le u \le T. \tag{4.6}$$

We rewrite equation (4.8) in the form

$$e^{-2u}\int_0^u e^t\,\mathrm{d}\Phi(t) + \int_0^T e^{-t}\,\mathrm{d}\Phi(t) = e^{-u}\,\text{const}, \quad 0 \le u \le T,$$

and investigate the conditions $\Phi(t)$ must satisfy if it has a derivative at points u, for $0 < u < T$. By differentiating both sides of the equation we get, after simplifying the result,

$$\int_0^u e^t\,\mathrm{d}\Phi(t) = \frac{1}{2}\,e^u\,\text{const}, \quad 0 < u < T.$$

But this condition is satisfied by the 'weighting' function which has a jump of the size $(1/2)$const at $t = 0$, and is linear for $0 < t \le u$. If we want (4.6) to be valid even at $u = T$, i.e.,

$$\int_0^T e^t\,\mathrm{d}\Phi(t) = e^T\,\text{const},$$

then $\Phi(t)$ must have another jump of the same size at $t = T$. By this means we obtain the 'weighting' function

$$\begin{aligned} \Phi(0) &= 0 \\ \Phi(t) &= \frac{1}{2}(1+t)\,\text{const}, \quad 0 < t < T, \\ \Phi(T) &= \frac{1}{2}(2+T)\,\text{const}, \end{aligned} \tag{4.7}$$

which can easily be shown to satisfy condition (4.5). We deduce from the conditions $\Phi(T) = 1$ that

$$\text{const} = \frac{2}{2+T}; \tag{4.8}$$

hence the best linear estimate is of the form

$$\tilde{x} = \frac{x(0) + x(T) + \int_0^T x(t)\,\mathrm{d}t}{2+T}. \tag{4.9}$$

This expression represents the weighted average of the ordinary integral estimate and of the arithmetical mean of the end points.

From the identity

$$\mathsf{D}^2\,\tilde{x} = \sigma^2 \int_0^T \int_0^T R(t-u)\,\mathrm{d}\Phi(t)\,\mathrm{d}\Phi(u)$$

and from equations (4.5) and (4.10) we get that the variance of the best linear estimate for the case when $e^{-|\tau|}$ is the correlation function is equal to

$$\mathsf{D}^2\,\tilde{x} = \frac{2}{2+T}\,\sigma^2. \tag{4.10}$$

By analogy, for the more general function $e^{-a|\tau|}$, $a > 0$, we get

$$\tilde{x} = \frac{x(0) + x(T) + a\int_0^T x(t)\,\mathrm{d}t}{2+aT} \tag{4.11}$$

and

$$\mathsf{D}^2\,\tilde{x} = \frac{2}{2+aT}\,\sigma^2. \tag{4.12}$$

Remark 4.1. The best weighting function, if it exists at all, will be found among the functions which are symmetric with respect to the point $(1/2)T$ since for any weighting function $\Phi(t)$ we can define another weighting function $\Psi(t)$,

$$\Psi(t) = \frac{1}{2}[\Phi(t) + 1 - \Phi(T-t)],$$

which is obviously symmetrical with respect to the point $(1/2)T$ and which gives, at the worst, a linear estimate identical to the one yielded by $\Phi(t)$. To be more precise we have the equation

$$\mathsf{D}^2\,\overline{x}_\Phi = \mathsf{D}^2\,\overline{x}_\Psi + \mathsf{E}(\overline{x}_\Phi - \overline{x}_\Psi)^2$$

where $\overline{x}_\Phi$ and respectively $\overline{x}_\Psi$ is the linear estimate obtained by means of weighting function $\Phi(t)$, or respectively, $\Psi(t)$.

This circumstance, namely that we can restrict ourselves to symmetric weighting functions, very often makes it easier to find the best solution. The following example will demonstrate this clearly.

Example 4.2. We shall look for the best weighting function for the correlation function $r(\tau)$ defined by equations (3.1). It is not difficult to see that condition (4.5) is satisfied by the weighting function $\Phi(t)$ which corresponds to finding the estimate $\tilde{x}$ from the equation

$$\tilde{x} = \frac{(n+1)\,[x(0) + x(T)] + n[x(1) + x(T-1)] + \cdots + [x(n) + x(T-n)]}{(n+1)\,(n+2)}, \tag{4.13}$$

where n is the largest integer not exceeding T, $T = n+\tau$, $0 \le \tau < 1$. However, it will be better to indicate how function $\Phi(t)$ can be found by 'selection', for example, for $T = 2+\tau$, $0 \le \tau < 1$. We shall assume that the function $\Phi(t)$ has jumps $p_1, p_2, p_3, p_4, p_5, p_6$ in the points $0, \tau, 1, 1+\tau, 2, 2+\tau$ and that it is continuous at all other points. By Remark 4.1 we can restrict ourselves to the case when the weighting function is symmetrical, that is when

$$p_1 = p_6, \quad p_2 = p_5, \quad p_3 = p_4, \quad p_1 + p_2 + p_3 = p_4 + p_5 + p_6 = \frac{1}{2}. \tag{4.14}$$

For $u < \tau$ condition (4.5) can be transcribed in the form

$$\begin{aligned} \int_0^{u+1} (1 - |t-u|)\, \mathrm{d}\Phi(t) &= p_1(1-u) + p_2(1-\tau+u) + p_3 u \\ &= (p_3 + p_2 - p_1)\, u + p_1 + p_2(1-\tau) \;=\; \text{const}, \end{aligned}$$

or, equally

$$p_3 + p_2 - p_1 = 0. \tag{4.15}$$

In a similar fashion for $r < u < 1$ we get

$$\begin{aligned} \int_0^{u+1} (1 - |t-u|)\, \mathrm{d}\Phi(t) &= p_1(1-u) + p_2(1-u+\tau) + p_3 u + p_4(u-\tau) \\ &= (p_4 + p_3 - p_2 - p_1)\, u + p_1 + p_2(1+\tau) - p_4\tau \\ &= \text{const}, \end{aligned}$$

or, in view of (4.16),

$$2p_3 - p_2 - p_1 = 0. \tag{4.16}$$

Equations (4.16), (4.17) and (4.18) imply that

$$p_1 = p_6 = \frac{1}{4}, \quad p_2 = p_5 = \frac{1}{12}, \quad p_4 = p_2 = \frac{1}{6},$$

which does not contradict the general formula (4.15).

The variance of the best linear estimate (4.3) is equal to the common value of integrals (4.5). For example, we get easily for $u = 0$ that the value of integral (4.5) and, consequently, also of $\mathsf{D}^2\,\tilde{x}$, is equal to

$$\mathsf{D}^2\,\tilde{x} = \sigma^2 \int_0^1 (1-t)\, \mathrm{d}\Phi(t) = \left[\frac{1}{n+2} + \frac{1-\tau}{(n+1)\,(n+2)}\right] \sigma^2. \tag{4.17}$$

Introducing the function $\ell(T)$ which for T integer assumes the values

$$\ell(T) = \frac{1}{1+T}, \quad T = 0, 1, 2, \ldots, \tag{4.18}$$

and in the remaining points is defined to be linear, we may write (4.19) in the form

$$\mathsf{D}^2\,\tilde{x} = \sigma^2\,\ell(T). \tag{4.19}$$

For the more general correlation function $r\left(\frac{\tau}{\lambda}\right)$ we shall have

$$\mathsf{D}^2\,\tilde{x} = \sigma^2\,\ell\left(\frac{T}{\lambda}\right). \tag{4.20}$$

Remark 4.2. If the process $x_0(t)$, introduced in the beginning of Section 3, is considered as the representation of correlation function $r(\tau)$, then it is clear that for T integer and for any ϕ there will be $T+1$ independent random variables y_k with identical distributions in the interval $(0, T)$. The best linear estimate will then be, of course, their arithmetical mean, which is obtained for every ϕ if and only if we put

$$\tilde{x} = \frac{1}{T+1}\,[x(0) + x(1) + \cdots + x(T)]. \tag{4.21}$$

In accordance with this the general formula (4.15) becomes for T integer (4.21), which can easily be checked.

5 LINEAR ESTIMATE OF THE MEAN VALUE OF A STATIONARY RANDOM PROCESS WITH CONVEX CORRELATION FUNCTION

The variance of the linear estimate (4.3) depends, obviously, only on the weighting function $\Phi(t)$ and on the correlation function and variance of the corresponding stationary process $x(t)$. Still in the case of a convex correlation function $\Phi(t)$ it turns out to be useful to consider the specific stationary process $x_0(t)$ defined in Remark 3.2 and then deduce the general results with the aid of this process. As it was already stated in Remark 3.2 the process $x_0(t)$ in the (random varying) intervals,

$$\lambda(\phi + k - 1) \leq t < \lambda(\phi + k), \quad k = 0, \pm 1, \pm 2, \ldots \tag{5.1}$$

is equal to

$$x_0(t) = y_k \tag{5.2}$$

where λ, ϕ, y_k $(k = 0, \pm 1, \pm 2, \ldots)$ are mutually independent random variables such that the distribution function of λ is $F(\lambda)$, which depends on $R(\tau)$ in the manner given by (3.3), ϕ is uniformly distributed over the interval $(0, 1)$, and y_k are arbitrarily but identically distributed.

Let us first of all investigate the distribution of the random variable ν which is by definition equal to the number of indices k such that $\lambda(\phi+k-1) \in (0,T)$. The real meaning of this quantity is that for some fixed λ and ϕ (of which ν is a function) we can reconstruct the process $x_0(t)$ on the whole interval $(0,T)$ by means of $\nu+1$ independent random variables $y_0, y_1, \dots, y_\nu$. For a fixed λ and varying ϕ the random variable ν assumes obviously only the two values n and $n+1$, where n is the largest integer not exceeding T/λ:

$$T = \lambda(n+\tau), \quad 0 \le \tau < 1. \tag{5.3}$$

We have also

$$P(\nu = n \mid \lambda) = 1-\tau,$$

$$P(\tau = n+1 \mid \lambda) = \tau,$$

where T is now determined from equation (5.3). This means that

$$\mathsf{E}(\nu+1 \mid \lambda) = (n+1)(1-\tau) + (n+2)\tau = n+\tau+1,$$

or, in view of (5.3),

$$\mathsf{E}(\nu+1 \mid \lambda) = \frac{T}{\lambda} + 1. \tag{5.4}$$

Also

$$\mathsf{E}\left(\frac{1}{\nu+1}\middle|\lambda\right) = \frac{1-\tau}{n+1} + \frac{\tau}{n+2} = \ell\left(\frac{T}{\lambda}\right). \tag{5.5}$$

where the function $\ell(T)$ was determined by equation (4.20).

The distribution function of random variable λ is $F(\lambda)$, so that we get in accordance with (5.4)

$$\mathsf{E}(\nu+1) = 1 + T\int_0^\infty \frac{1}{\lambda}\,\mathrm{d}F(\lambda). \tag{5.6}$$

Since by definition (3.3)

$$F(\lambda) = \int_0^\lambda \tau\,\mathrm{d}R'(\tau),$$

we may write using the well-known properties of a Stieltjes integral

$$\int_0^\infty \frac{1}{\lambda}\,\mathrm{d}F(\lambda) = \int_0^\infty \mathrm{d}R'(\lambda) = |R'(0)|;$$

and substituting this into (5.6) we get

$$\mathsf{E}(\nu+1) = 1 + |R'(0)|\,T. \tag{5.7}$$

The linear estimate for the process $x_0(t)$ may be written in the form

$$\tilde{x} = \int_0^T x_0(t)\,\mathrm{d}\Phi(t) = \sum_{k=0}^{\nu} y_k\,P_k,$$

where the weights P_k are equal to

$$\begin{aligned} P_0 &= \Phi(\lambda\phi), \\ P_\nu &= 1 - \Phi[\lambda(\phi + \nu - 1)], \\ P_k &= \Phi[\lambda(\phi + k)] - \Phi[\lambda(\phi + k - 1)], \quad k = 1, 2, \ldots, \nu - 1. \end{aligned}$$

For some fixed λ and ϕ, therefore, $\overline{x}$ will be the weighted average of the independently and identically distributed random variables $y_0, y_1, \ldots, y_\nu$ and as such it cannot have a smaller variance than the arithmetical mean $\frac{1}{\nu+1}(y_0 + y_1 + \cdots + y_\nu)$. This means that

$$\mathsf{D}^2(\overline{x} \mid \lambda, \phi) \geq \frac{\sigma^2}{\nu + 1}$$

and consequently,

$$\mathsf{D}^2\,\overline{x} \geq \sigma^2\,\mathsf{E}\left(\frac{1}{\nu+1}\right). \tag{5.8}$$

On the other hand, according to (5.7)

$$\mathsf{E}\left(\frac{1}{\nu+1}\right) \geq \frac{1}{\mathsf{E}(\nu+1)} = \frac{1}{1 + |R'(0)|\,T};$$

in this way, for process $x_0(t)$ and, therefore, also for every other process with variance σ^2 and with convex correlation function $R(\tau)$, the following inequality holds, irrespective of the choice of the weighting function $\Phi(t)$:

$$\mathsf{D}^2\left[\int_0^T x(t)\,\mathrm{d}\Phi(t)\right] \geq \frac{\sigma^2}{1 + |R'(0)|\,T}. \tag{5.9}$$

Thus, for example, we have for correlation function $e^{-\tau}$ that

$$\mathsf{D}^2\,\overline{x} \geq \frac{\sigma^2}{1+T},$$

which is in complete agreement with result (4.12). Inequality (5.9) is interesting because of its simplicity but we need, however, the more exact result, which follows from equations (5.5) and (5.8):

$$\mathsf{D}^2\left[\int_0^T x(t)\,\mathrm{d}\Phi(t)\right] \geq \sigma^2 \int_0^\infty \ell\left(\frac{T}{\lambda}\right)\mathrm{d}F(\lambda). \tag{5.10}$$

Remark 5.1. For the correlation function $r(\tau)$ defined by equations (3.1) the inequality (5.10) becomes

$$\mathsf{D}^2\,\overline{x} \geq \sigma^2\,\ell(T).$$

As we have already seen in Example 4.2 (cf. equation (4.21)) equality will occur for the best linear estimate in this inequality. This means that for the stationary process $x_0(t)$, $\tilde{x}$ will always be the arithmetical mean of observations y_k irrespective of whether $n+1$ or $n+2$ among them fall into interval $(0, T)$ in dependence on the value of ϕ. This can be immediately checked from the defining equations (4.15).

Let us return to inequality (5.10). Since

$$\lim_{T\to\infty} T\,\ell\left(\frac{T}{\lambda}\right) = \lambda,$$

we may write

$$\liminf_{T\to\infty} T\,\mathsf{D}^2\left[\int_0^T x(t)\,\mathrm{d}\Phi_T(t)\right] \geq \sigma^2\int_0^\infty \lambda\,\mathrm{d}F(\lambda), \tag{5.11}$$

where for every T function $\Phi_T(t)$ may be any arbitrary weighting function. In order to express inequality (5.11) 'in the language of correlation functions' we have to prove the following lemma.

Lemma 5.1. *If the convex correlation function $R(\tau)$ is such that $R(\infty) = 0$, then*

$$\int_0^\infty \lambda\,\mathrm{d}F(\lambda) = 2\int_0^\infty R(\tau)\,\mathrm{d}\tau, \tag{5.12}$$

where the function $F(\lambda)$ is defined by equation (3.3).

Proof.We may write on the basis of Fubini's theorem

$$\begin{aligned}\int_0^\infty R(\tau)\,\mathrm{d}\tau &= \int_0^\infty\int_0^\infty r\left(\frac{\tau}{\lambda}\right)\mathrm{d}F(\lambda)\,\mathrm{d}\tau \\ &= \int_0^\infty\int_0^\infty r\left(\frac{\tau}{\lambda}\right)\mathrm{d}\tau\mathrm{d}F(\lambda) = \frac{1}{2}\int_0^\infty \lambda\,\mathrm{d}F(\lambda),\end{aligned}$$

which indeed completes the proof. □

In the following theorem we formulate our results.

Theorem 5.1. *If the correlation function $R(\tau)$ of a stationary random process $x(t)$ is convex then*

$$\liminf_{T\to\infty} T\,\mathsf{D}^2\left[\int_0^T x(t)\,\mathrm{d}\Phi_T(t)\right] \geq 2\sigma^2\int_0^\infty R(\tau)\,\mathrm{d}\tau \tag{5.13}$$

with $\Phi_T(t)$ being an arbitrary weighting function for every T.

Definition 5.1. If under the conditions of Theorem 5.1

$$\lim_{T\to\infty} T\,\mathrm{D}^2\left[\int_0^T x(t)\,\mathrm{d}\Phi_T(t)\right] = 2\sigma^2\int_0^\infty R(\tau)\,\mathrm{d}\tau < +\infty, \tag{5.14}$$

then we agree to call the corresponding linear estimate *asymptotically effective.*

The following theorem asserts that the ordinary 'integral' estimate

$$\frac{1}{T}\int_0^T x(t)\,\mathrm{d}t \tag{5.15}$$

is asymptotically effective.

Theorem 5.2. *If a convex correlation function satisfies*

$$\int_0^\infty R(\tau)\,\mathrm{d}\tau < +\infty, \tag{5.16}$$

then the linear estimate (5.15) *is asymptotically effective in the sense of Definition* 5.1.

Proof.Multiply by $(\sigma^{-2}T)$ the variance of estimate (5.15); then we get

$$\begin{aligned}\frac{1}{T}\int_0^T\int_0^T R(t-u)\,\mathrm{d}t\mathrm{d}u &= 2\int_0^T\left(1-\frac{t}{T}\right)R(t)\,\mathrm{d}t\\ &= 2\int_0^T R(t)\,\mathrm{d}t - \frac{2}{T}\int_0^T tR(t)\,\mathrm{d}t.\end{aligned}$$

It has to be proved, therefore, that the last integral converges to zero for $T\to\infty$. By assumption one can find for every $\varepsilon > 0$ some T_0 such that for all $T > T_0$

$$\int_{T_0}^T R(t)\,\mathrm{d}t < \frac{1}{2}\varepsilon.$$

Choose $T_1 > T_0$ so that

$$\frac{T_0}{T_1} < \frac{\varepsilon}{2\int_0^\infty R(t)\,\mathrm{d}t}.$$

Then for $T > T_1$ we shall have

$$\begin{aligned}\frac{1}{T}\int_0^T t\,R(t)\,\mathrm{d}t &= \frac{1}{T}\int_0^{T_0} t\,R(t)\,\mathrm{d}t + \frac{1}{T}\int_{T_0}^T t\,R(t)\,\mathrm{d}t\\ &\le \frac{T_0}{T}\int_0^{T_0} R(t)\,\mathrm{d}t + \int_{T_0}^T R(t)\,\mathrm{d}t < \frac{\varepsilon}{2}+\frac{\varepsilon}{2} = \varepsilon,\end{aligned}$$

which completes the proof. □

It was seen in Examples 4.1 and 4.2 that the best weighting functions corresponding to correlation functions $e^{-|\tau|}$ and $r(\tau)$ are non-decreasing; that is, essentially they are distribution functions. The following theorem asserts that the same is true for all convex correlation functions.

Theorem 5.3.

(i) *Let a convex correlation function $R(\tau)$ be given which corresponds to the stationary random process $x(t)$ and which satisfies the condition $R(\infty) < 1$.*

(ii) *Take an arbitrary positive integer n and some set of values $t_1, t_2, \dots, t_n$ and consider the set of linear estimates of the form*

$$\sum_{i=1}^{n} p_i\, x(t_i), \quad \sum_{i=1}^{n} p_i = 1. \tag{5.17}$$

Then the weights $p_1, p_2, \dots, p_n$ minimizing the variance of estimate (5.17) are all non-negative, i. e.,

$$p_i \geq 0. \tag{5.18}$$

Proof.Since

$$\mathsf{D}^2\left[\sum_{i=1}^{n} p_i\, x(t_i)\right] = \sigma^2 \sum_{i=1}^{n}\sum_{j=1}^{n} R(t_i - t_j)\, p_i p_j, \tag{5.19}$$

then the best weighting factors are solutions to the set of linear equations[5]

$$p_1\, R(t_i - t_1) + p_2\, R(t_i - t_2) + \cdots + p_n\, R(t_i - t_n) = \text{const}, \quad i = 1, 2, \dots, n. \tag{5.20}$$

Let us assume that the numbers $t_1, t_2, \dots, t_n$ are arranged in an increasing order, $t_1 < t_2 < \cdots < t_n$; take the weights $p_1, p_2, \dots, p_n$ which solve the above equations and divide the indices $1, 2, \dots, n$ into three mutually disjoint sets A, B, C as follows:

$$\left.\begin{aligned} j \in A &\Longleftrightarrow p_j > 0 \\ j \in B &\Longleftrightarrow p_j < 0 \\ j \in C &\Longleftrightarrow p_j = 0 \end{aligned}\right\} j = 1, 2, \dots, n. \tag{5.21}$$

Now introduce the functions

$$f(t) = \sum_{j \in A} p_j\, R(t - t_j), \tag{5.22}$$

[5] Equations (5.21) are the non-continuous analogues of condition (4.5).

and

$$g(t) = \sum_{j \in B} (-p_j) \, R(t - t_i). \tag{5.23}$$

With these functions we may write the system of equations (5.21) in the form

$$f(t_i) - g(t_i) = \text{const}, \quad i = 1, 2, \ldots, n. \tag{5.24}$$

We shall now show that the assumption that $p_1, p_2, \ldots, p_n$ satisfy equations (5.25) but at the same time do not satisfy inequalities (5.19), i. e., that set B derived from them is not empty, leads to contradiction.

Set A can never be empty because $\sum p_i = 1$. If B is also a non-empty set, then an index $r > 1$ can be found such that either

$$\{1, 2, \ldots, r-1\} \subset B \cup C \quad \text{and} \quad r \in A, \tag{5.25}$$

or

$$\{1, 2, \ldots, r-1\} \subset A \cup C \quad \text{and} \quad r \in B. \tag{5.26}$$

If (5.26) corresponds to the situation then we may write the following chain of implications:

$$\{j \in A\} \Rightarrow \{j \geq r\} \Rightarrow \{|t_r - t_j| < |t_{r-1} - t_j|\}.$$

Hence, in view of the fact that $R(\tau)$ is a non-increasing function, follows the inequalities

$$R(t_r - t_j) \geq R(t_{r-1} - t_j), \quad j \in A. \tag{5.27}$$

By hypothesis, $R(\infty) < 1$, implying also $R(\tau) < 1$ whenever $\tau > 0$. This means that in the special case when $j = r$ we get the sharp inequality

$$1 = R(t_r - t_r) > R(t_{r-1} - t_r). \tag{5.28}$$

An immediate consequence of inequalities (5.28) and (5.29) is the inequality

$$\sum_{j \in A} p_j \, R(t_r - t_j) > \sum_{j \in A} p_j \, R(t_{r-1} - t_j),$$

or, equivalently,

$$f(t_r) > f(t_{r-1}). \tag{5.29}$$

By this assumption we have equation (5.25) and therefore inequality (5.30) is equivalent to the inequality

$$g(t_r) > g(t_{r-1}). \tag{5.30}$$

If the second alternative (5.27) occurs we can also obtain inequalities (5.30) and (5.31). The only difference is that in that case we first prove inequality (5.31) and then, as its consequence, (5.30).

For the same reasons there exists an $s < n$, such that either

$$\{s+1, \ldots, n\} \subset B \cup C \quad \text{and} \quad s \in A, \tag{5.31}$$

or

$$\{s+1, \ldots, n\} \subset A \cup C \quad \text{and} \quad s \in B, \tag{5.32}$$

and in the same way as for index r it can be proved that we always have

$$f(t_s) > f(t_{s+1}), \tag{5.33}$$

together with

$$g(t_s) > g(t_{s+1}). \tag{5.34}$$

It can be seen from the definition of r and s that $r \leq s$.[6] Therefore, there exists an index m such that $r \leq m \leq s$ and

$$f(t_m) > f(t_{m-1}), \quad f(t_m) \geq f(t_{m+1}), \tag{5.35}$$

and also

$$g(t_m) > g(t_{m-1}), \quad g(t_m) \geq g(t_{m+1}). \tag{5.36}$$

If no index of this kind existed, then at least one of the inequalities (5.30) and (5.34), and correspondingly at least one of (5.31) and (5.35), could not be satisfied.

$R(\tau)$ is assumed to be a convex function over every interval not containing the origin. This means that function $f(t)$, or $g(t)$, defined by equation (5.23), respectively, (5.24), is a convex function in every interval which does not contain any of the points t_j such that $j \in A$, respectively, $j \in B$. However this property of functions $f(t)$ and $g(t)$ is compatible with inequalities (5.36) and (5.37) if and only if $m \in A$ and $m \in B$. In other words, we must have simultaneously $m \in A$ and $m \in B$, i. e., $p_m > 0$ and $p_m < 0$. In this way we found a contradiction, which completes our proof. □

Remark 5.2. From a geometrical point of view the best linear estimate $\tilde{x}$ is the centre of the sphere passing through the points $x(t_1), \ldots, x(t_n)$. In this way Theorem 5.3 is equivalent to the following assertion. A simplex with vertices at points $x_1, x_2, \ldots, x_n$ contains the centre $\tilde{x}$ of the sphere passing through these points if there exist numbers $t_1, t_2, \ldots, t_n$ and a function $R(\tau)$ which is convex over the interval $(0, \infty)$ and satisfies

$$(x_i - \tilde{x}, x_j - \tilde{x}) = \sigma^2 R(|t_i - t_j|), \quad i, j = 1, \ldots, n.$$

In the following theorem we shall prove that the solutions obtained in Examples 4.1 and 4.2 are the only possible solutions – more precisely, that for a convex correlation function there is at most one best weighting function.

[6] Translator's note: This last assertion needs some elucidation. If r is defined by (5.26) and s by (5.33) then we have $r > s$. For example, for $p_1 < 0$, $p_2 > 0$, $p_3 > 0$, $r = 2$ and $s = 1$. However it can easily be seen that in the typical case also $r = s + 1$ and then (5.30) and (5.34) contradict.

Theorem 5.4. *Let the convex correlation function $R(\tau)$ of a stationary random process $x(t)$ satisfy the condition $R(\infty) < 1$. Then the weighting functions $\Phi(t)$ and $\Psi(t)$ may satisfy the equation*

$$\int_0^T x(t)\,\mathrm{d}\Phi(t) = \int_0^T x(t)\,\mathrm{d}\Psi(t) \tag{5.37}$$

if and only if

$$\Phi(t) \equiv \Psi(t). \tag{5.38}$$

Proof.If $R(\infty) > 0$, then by Theorem 2.1 we may write $x(t) = y(t) + z$ where the correlation function of $y(t)$ satisfies $Q(\infty) = 0$. But equation (5.38), since z is independent of t, is equivalent to the equation

$$\int_0^T y(t)\,\mathrm{d}\Phi(t) = \int_0^T y(t)\,\mathrm{d}\Psi(t).$$

Therefore we may restrict ourselves to the case when $R(\infty) = 0$.

Let us assume that equation (5.38) holds but equation (5.39) is not true. In this case there exists a point t such that

$$\Delta(t) = \Phi(t) - \Psi(t) \neq 0, \quad t \in (0, T). \tag{5.39}$$

In accordance with the convention accepted in Section 4 we shall consider only those weighting functions which are continuous to the left throughout interval $(0, T)$, and thus relation (5.40) guarantees us the existence of some ε, t_0 such that $\varepsilon > 0$, $0 < t_0 < t$ and

$$\Delta^2(u) > \varepsilon \quad \text{whenever } t_0 < u < t. \tag{5.40}$$

To prove the incompatibility of this assumption with equation (5.38) we have only to show that under these conditions

$$\mathsf{D}^2\left[\int_0^T x(t)\,\mathrm{d}\Delta(t)\right] > 0. \tag{5.41}$$

Let us again construct for our correlation function $R(\tau)$ the process $x_0(t)$ according to Remark 3.2. Then choose some $\lambda_0 > 0$ in such a way that $F(\lambda_0) < 1$, and denote by S the event which consists of choosing λ and ϕ in accordance with the following constraints:

(a) $\lambda \geq \lambda_0$,

(b) t_0 and t are not contained in any of the intervals (5.1) at the same time.

The probability of event S is equal to

$$P(S) = \int_{\lambda_0}^{\infty} h(\lambda)\,\mathrm{d}F(\lambda),$$

where

$$h(\lambda) = \min\left[1, \frac{t - t_0}{\lambda}\right].$$

Since λ_0 was fixed in such a way that $F(\lambda_0) < 1$ we shall have $P(S) > 0$, and therefore inequality (5.42) for $x(t) = x_0(t)$ will follow from the inequality

$$\mathsf{D}^2 \left[\int_0^T x_0(t)\, \mathrm{d}\Delta(t) \mid S\right] > 0. \tag{5.42}$$

Take now any arbitrary λ and ϕ which are compatible with the occurrence of event S. Since whenever event S occurs the numbers t and t_0 do not fall simultaneously into any of the intervals (5.1), it is possible to find some number u such that $t_0 < u < t$, and the random variables $\int_0^u x_0(t)\, \mathrm{d}\Delta(t)$ and $\int_0^T x_0(t)\, \mathrm{d}\Delta(t)$ have independent conditional distributions. However this means that

$$\mathsf{D}^2 \left[\int_0^T x_0(t)\, \mathrm{d}\Delta(t) \mid \lambda, \phi\right] \geq \mathsf{D}^2 \left[\int_0^u x_0(t)\, \mathrm{d}\Delta(t) \mid \lambda, \phi\right]. \tag{5.43}$$

By hypothesis (5.41) we have $\Delta(u) \neq 0$ and therefore the random variable $[\Delta(u)]^{-1} \int_0^u x_0(t)\, \mathrm{d}\Delta(t)$ exists. For fixed λ and ϕ this quantity is the weighted average of the random variables y_k occuring according to (5.2) in interval $(0, u)$. However, the number of random variables y_k in interval $(0, u)$ can be no more than $u/\lambda + 2$ and certainly no more than $t/\lambda_0 + 2$, since $u < t$ and $\lambda \geq \lambda_0$. From this we can conclude that

$$[\Delta(u)]^2\, \mathsf{D}^2 \left[\int_0^u x_0(t)\, \mathrm{d}\Delta(t) \mid \lambda, \phi\right] \geq \frac{\sigma^2}{t/\lambda_0 + 2}. \tag{5.44}$$

Since $t_0 < u < t$, (5.41) is true and consequently

$$\mathsf{D}^2 \left[\int_0^u x_0(t)\, \mathrm{d}\Delta(t) \mid \lambda, \phi\right] \geq \frac{\varepsilon\sigma^2}{t/\lambda_0 + 2}. \tag{5.45}$$

Further, we must have according to (5.44)

$$\mathsf{D}^2 \left[\int_0^T x_0(t)\, \mathrm{d}\Delta(t) \mid \lambda, \phi\right] \geq \frac{\varepsilon\sigma^2}{t/\lambda_0 + 2}. \tag{5.46}$$

The last inequality holds for any λ and ϕ in S, and therefore (5.43) must also be true. This proves the theorem because as we have already said inequality (5.43) implies inequality (5.42), which in turn leads to contradiction. It is true that we proved the theorem only for process $x_0(t)$ but owing to the linearity of the problem this proves it for an arbitrary process with the same correlation function. □

Remark 5.3. Theorem 5.4 implies, among other things, that under the conditions of Theorem 5.3 the random variables $x(t_1), \dots, x(t_n)$ will be independent; from a geometrical point of view they can be considered as vertices of a simplex.

Theorem 5.5. *If the convex correlation function $R(\tau)$ of the stationary random process $x(t)$ satisfies the condition $R(\infty) < 1$, then there exists one and only one weighting function $\Phi(t)$ realizing the best linear estimate of the mean value of process $x(t)$ over the interval $(0, T)$. The function $\Phi(t)$ is non-decreasing.*

Proof.Put

$$t_i^{(n)} = T\frac{i}{n}, \qquad \begin{array}{l} i = 0, 1, \dots, n \\ n = 1, 2, \dots, \end{array} \tag{5.47}$$

and find the weights $p_i^{(n)}$ for the best linear estimate $\sum_{i=1}^n p_i^{(n)} x(t_i^{(n)})$. This means that for any set of numbers $q_i^{(n)}$, $\sum_{i=1}^n q_i^{(n)} = 1$, we shall have the inequality

$$\mathsf{D}^2 \left[\sum_{i=1}^n p_i^{(n)} x(t_i^{(n)}) \right] \leq \mathsf{D}^2 \left[\sum_{i=1}^n q_i^{(n)} x(t_i^{(n)}) \right]. \tag{5.48}$$

According to Theorem 5.3, for every n, $p_i^{(n)} \geq 0$. By Helly's theorem there exists therefore a subsequence of indices $n_1 < n_2 < \dots$ such that

$$\lim_{j \to \infty} \sum_{t_i(n_j) < t} p_i^{(n_j)} = \Phi(t), \quad 0 \leq t < T, \tag{5.49}$$

where $\Phi(t)$ is a non-decreasing function with $\Phi(0) = 0$. Put $\Phi(T) = 1$; if in $0 < t < T$, $\Phi(t)$ is not continuous to the left then we change it in such a way as to make it left continuous and take it for the weighting function. From the other Helly theorem it follows that

$$\mathsf{D}^2 \left[\int_0^T x(t)\, \mathrm{d}\Phi(t) \right] = \lim_{j \to \infty} \mathsf{D}^2 \left[\sum_{i=0}^{n_j} p_i^{(n_j)} x(t_i^{(n_j)}) \right]. \tag{5.50}$$

Take some other arbitrary weighting function $\Psi(t)$ and put

$$q_i^{(n)} = \Psi\left(t_{i+1}^{(n)}\right) - \Psi\left(t_i^{(n)}\right), \qquad \begin{array}{l} i = 0, 1, \dots, n, \\ n = 1, 2, \dots, \end{array}$$

where the points $t_i^{(n)}$ are determined from equations (5.47). Then we have

$$\mathsf{D}^2 \left[\int_0^T x(t)\, \mathrm{d}\Psi(t) \right] = \lim_{j \to \infty} \mathsf{D}^2 \left[\sum_{i=0}^{n_j} q_i^{(n_j)} x(t_i^{(n_j)}) \right]. \tag{5.51}$$

It follows from equations (5.51) and (5.52) and from inequality (5.48) that

$$\mathsf{D}^2\left[\int_0^T x(t)\,\mathrm{d}\Phi(t)\right] \le \mathsf{D}^2\left[\int_0^T x(t)\,\mathrm{d}\Psi(t)\right], \tag{5.52}$$

which is equivalent to the assertion that $\Phi(t)$ is the best weighting function; as it has already been stated, $\Phi(t)$ does not decrease, and from Theorem 5.4 it follows that there are no other best weighting functions. Indeed, the best linear estimate represents a definite point in Hilbert space and as such it cannot be represented, according to Theorem 5.4, by means of two different weighting functions. □

6 LINEAR ESTIMATION OF THE VARIANCE

The variance of a stationary random process $x(t)$ is the mean value of the process

$$[x(t) - \mathsf{E}x(t)]^2. \tag{6.1}$$

If $x(t)$ is a Gaussian process[7] then we have the following relation between its correlation function and the correlation function of process (6.1):

Theorem 6.1. *If the correlation function $R(\tau)$ of a Gaussian stationary process $x(t)$ is convex then the correlation function $Q(\tau)$ of process (6.1) will also be convex; more precisely,*

$$Q(\tau) = R^2(\tau). \tag{6.2}$$

Proof. The proof follows from the following well-known property of the two-dimensional normal distribution:

$$\frac{1}{2\pi(1-R^2(\tau))^{1/2}} \int_{-\infty}^{\infty}\int_{-\infty}^{\infty} u^2\,v^2\,e^{-\frac{1}{2}\,\frac{u^2-2R(\tau)\,uv+v^2}{1-R^2(\tau)}}\,\mathrm{d}u\mathrm{d}v = 1 + 2R^2(\tau).$$

□

Example 6.1. In the important case when $R(\tau) = e^{-a\tau}$ we get therefore $Q(\tau) = e^{-2a\tau}$ and thus the best linear estimate of the variance will, by analogy with equation (4.13), be

$$s^2 = \frac{1}{1+aT}\left\{\frac{1}{2}[x(0)-m]^2 + \frac{1}{2}[x(T)-m]^2 + a\int_0^T [x(t)-m]^2\,\mathrm{d}t\right\}, \tag{6.3}$$

[7] A stationary process is said to be Gaussian if the vectors $\{x(t_1), \ldots, x(t_n)\}$ are normally distributed.

where $m = \mathsf{E}x(t)$. Its variance, according to (4.14), is

$$\mathsf{D}^2\, s^2 = \frac{2\sigma^2}{1+aT}. \tag{6.4}$$

7 LINEAR ESTIMATION OF THE DISTRIBUTION FUNCTION

The value of the distribution function $G(h)$ of a stationary process $x(t)$ at point h is equal to the mean value of the process $\chi(t)$ defined by the following equations:

$$\begin{aligned} \chi(t) &= 1, \quad \text{if } x(t) < h, \\ &= 0, \quad \text{otherwise.} \end{aligned} \tag{7.1}$$

If $x(t)$ is a Gaussian process then we have the following relation between its correlation function and the correlation function of process (7.1).

Theorem 7.1. *If the correlation function $R(\tau)$ of a stationary Gaussian random process $x(t)$ satisfies the relation $R(\tau) \geq 0$ then the correlation function $Q(\tau)$ of the stationary random process $\chi(t)$, defined by (7.1), satisfies the inequalities*

$$0 \leq Q(\tau) \leq R(\tau). \tag{7.2}$$

Proof.Without restricting generality we may assume that $\mathsf{E}x(t) = 0$ and $\mathsf{E}\,x^2(t) = 1$, so that

$$G(h) = \frac{1}{\sqrt{2\pi}} \int_{-\infty}^{h} e^{-\frac{v^2}{2}}\, \mathrm{d}v. \tag{7.3}$$

Put

$$\begin{aligned} B(\tau) &= \mathsf{E}\,\chi(t)\,\chi(t-\tau) \\ &= \frac{1}{2\pi(1-R^2(\tau))^{1/2}} \int_{-\infty}^{h}\int_{-\infty}^{h} e^{-\frac{1}{2}\frac{u^2-2R(\tau)\,uv+v^2}{1-R^2(\tau)}}\, \mathrm{d}u\mathrm{d}v \\ &= 2\int_{-\infty}^{h} \left[\frac{1}{\sqrt{2\pi(1-R^2(\tau))}} \int_{-\infty}^{u} e^{-\frac{1}{2}\frac{(v-R(\tau)\,u)^2}{1-R^2(\tau)}}\, \mathrm{d}v\right] \frac{e^{-\frac{u^2}{2}}}{\sqrt{2\pi}}\, \mathrm{d}u, \end{aligned} \tag{7.4}$$

or using (7.3)

$$B(\tau) = 2\int_{-\infty}^{h} G\left(u\sqrt{\frac{1-R(\tau)}{1+R(\tau)}}\right) \mathrm{d}G(u). \tag{7.5}$$

Since

$$1 - R(\tau) < \sqrt{\frac{1-R(\tau)}{1+R(\tau)}},$$

and since function $G(z)$ is convex for $z \leq 0$, we have for $u \leq 0$ the following inequalities:

$$G\left(u\sqrt{\frac{1-R(\tau)}{1+R(\tau)}}\right) \leq G\left(u[1-R(\tau)]\right) \leq [1-R(\tau)]\,G(u) + R(\tau)\,G(0).$$

Hence, by substituting into (7.5) we get for $h \leq 0$

$$\begin{aligned} B(\tau) &\leq 2\int_{-\infty}^{h} \left([1-R(\tau)]\,G(u) + R(\tau)\,G(0)\right) \mathrm{d}G(u) \\ &= [1-R(\tau)]\,G^2(h) + R(\tau)\,G(h), \end{aligned} \tag{7.6}$$

or, equivalently,

$$Q(\tau) = \frac{B(\tau) - G^2(h)}{G(h)\,(1-G(h))} \leq \frac{[1-R(\tau)]\,G^2(h) + R(\tau)\,G(h) - G^2(h)}{G(h)\,(1-G(h))} = R(\tau). \tag{7.7}$$

On the other hand for $h \leq 0$ we also have

$$B(\tau) \geq 2\int_{-\infty}^{h} G(u)\,\mathrm{d}G(u) = G^2(h),$$

whence it follows that

$$Q(\tau) \geq 0. \tag{7.8}$$

Inequalities (7.7) and (7.8) give (7.2) for $h \leq 0$. For $h > 0$ it is sufficient to recall that the process $\chi(t \mid h)$ and the process $1 - \chi(t \mid -h)$ have the same distribution and the same correlation function. □

REFERENCES

Hardy G. H., Littlewood E. J. and Pólya G. (1934). *Inequalities.* Cambridge Univ. Press.

Khintchine A. (1934). Korrelationstheorie der stationären stochastischen Prozesse. *Math. Ann.* 109, 604–615.

CHAPTER 6

Inequalities for the Generalized Student's Distribution and their Applications

Časopis Pěst. Mat. 82 (1957), 182–194. (In Czech).
English translation *Selected Transl. Math. Statist. Prob. 2* (1962), 63–74.
Translated by L. Rosenblatt.
Reproduced by permission of American Mathematical Society.

The generalization addresses the fact that the estimate of the variance does not have the usual structure (1.1) but has a more complicated structure (1.2). This is the case for k independent samples whose variances differ in an unknown manner. In this chapter inequalities are found introducing in this way the generalized Student's distribution in relation to ordinary Student's distributions. The applications are concerned with the usual testing of the null hypothesis, and the construction of a confidence interval.

1 INTRODUCTION AND SUMMARY

If we know that a certain statistic (random variable)x is normally distributed, then to calculate the deviation $x-\mu$ (of x) from its mean value μ it is sufficient to know either its variance σ^2 or at least some estimate of it, s^2. 'Student', in his classic work, found the distribution of the ratio $(x-\mu)/s$ for the case where s^2 has the structure

$$s^2 = \sigma^2 \frac{1}{m} \chi^2(m), \tag{1.1}$$

where $\chi^2(m)$ denotes a random variable independent of x and having a chi-square distribution with m degrees of freedom. This is called Student's

Collected Works of Jaroslav Hájek – With Commentary
Edited by M. Hušková, R. Beran and V. Dupač
Published in 1998 by John Wiley & Sons Ltd.

distribution with m degrees of freedom. Since that time many statisticians have been occupied with the problem of how $(x - \mu)/s$ behaves under assumptions which – as was often the case in practice – differ somewhat from those investigated by Student. It was ascertained in this regard that Student's distribution of the ratio $(x-\mu)/s$ is not too sensitive to the eventual deviation of the initial distribution law from the normal one. In this chapter we will be concerned with another deviation from Student's assumptions – that s^2 has a somewhat more complicated structure, and that

$$s^2 = \sigma^2 \sum_{j=1}^{k} \frac{\lambda_j}{m_j} \chi_j^2(m_j), \quad \lambda_j \geq 0, \quad \sum_{j=1}^{k} \lambda_j = 1, \tag{1.2}$$

where λ_j are unknown constants and $\chi_j^2(m_j)$ are independent of each other as well as of x. As a result of the additivity of the independent χ^2-variables, (1.2) is identical to (1.1) as soon as $m_1 + \cdots + m_k = m$ and $\lambda_j/m_j = 1/m$.

We graduate to the estimate (1.2) by means of k independent samples, whose variance differs in an unknown manner. In some cases, for instance, we proceed in such a way that first of all for each sample we calculate a certain statistic x_j and the estimate s_j^2 for its variance σ_j^2; then we put $x = x_1 + \cdots + x_k$ and along with that

$$s^2 = s_1^2 + \cdots + s_k^2. \tag{1.3}$$

If we have each s_j^2 of the structure (1.1) with m_j degrees of freedom, i. e.,

$$s_j^2 = \sigma_j^2 \frac{1}{m_j} \chi_j^2(m_j),$$

then s^2 has structure (1.2), where $\lambda_j = \sigma_j^2/(\sigma_1^2 + \cdots + \sigma_k^2)$.[1]

A classical example of this type is the difference $\overline{x} - \overline{y}$ of two sample averages, when the x observations have a variance different from that of the ys. This problem was investigated by R. A. Fisher (1935), where he proposed a fiducial solution which gave rise, till the present time, to unfinished polemics. The Fisher and the later Welch (1947) solutions are based on ascertainment, that the probability of the inequality $x - \mu < ts$, where s has structure (1.2), can be made independent of the numbers λ_j in that manner, and that the number

[1] The estimates $s_1^2 + \cdots + s_k^2$ however we use only when the ratios λ_j are really unknown. If we know them then the estimate

$$s^2 = \sum_{j=1}^{k} \frac{s_j^2 m_j}{\lambda_j m} = (\sigma_1^2 + \cdots + \sigma_k^2) \sum_{j=1}^{k} \frac{m_j}{m} \frac{s_j^2}{\sigma_j^2}, \tag{1.4}$$

which has structure (1.1) with $m = m_1 + \cdots + m_k$ degrees of freedom, is better. On the other hand, as soon as the ratios $\sigma_j^2/(\sigma_1^2 + \cdots + \sigma_k^2)$ are not properly known, estimate (1.4) can be false, because then it is not even obliged to be unbiased.

t is also a random variable – a fixed function of the ratios $s_j^2/(s_1^2 + \cdots + s_k^2)$, where s_j^2 are known components of the estimate (1.3).

Fisher is of course concerned with fiducial probability, while B. L. Welch is concerned with probability in the usual sense of the word. Numerically both these solutions are almost undistinguishable; the corresponding tables presented in Sukhatme (1938) and Aspin (1949) for $k = 2$ have three entries – for m_1, m_2 and $s_1^2/(s_1^2 + s_2^2)$. Already for $k = 3$ there would have to be five entries, which is practically unrealizable. Therefore Welch proposed approximate solutions, for which t is looked up from the Student's distribution with a random number of degrees of freedom,[2]

$$m' = \frac{(s_1^2 + \cdots + s_k^2)^2}{\sum_{j=1}^{k} s_j^4/(m_j + 2)} - 2, \tag{1.5}$$

which is called the *effective* number of degrees of freedom. Welch showed numerically in Welch (1949) that the use of the effective number of degrees of freedom gives good results even for small numbers of degrees of freedom.

Structure (1.2) is assumed in this chapter, and it is shown that the probability that $(x - \mu)/s$ falls into a fixed interval containing zero is, for any numbers λ_j, smaller than the analogous probability for the Student's distribution with $m = m_1 + \cdots + m_k$ degrees of freedom,[3] but is always greater than for the Student's distribution with number of degrees of freedom not exceeding any of the numbers m_j/λ_j – for example, being equal to the smallest of the numbers m_j. This result is used for testing null hypotheses, since in accordance with it these hypotheses must be rejected if rejected for numbers of degrees of freedom equal to the smallest of the numbers m_j, and it should not be rejected if not rejected for $m = m_1 + \cdots + m_k$ degrees of freedom; similarly it is used for constructing confidence intervals. This is also a further argument for the use of effective numbers of degrees of freedom.

[2] Welch hesitates between formula (1.5) and the simpler formula $(s_1^2 + \cdots + s_k^2)^2/\left(\sum_{j=1}^{k} s_j^4/m_j\right)$. For not very small m_j, as for the biological experiments, the two formulas are equivalent. We have, however, in mind also applications for stratified samples where in the case that in each stratum we choose only two units, we obtain $m_j = 1$, and then probably (1.5) is better. Reasoning of formula (1.5) will be given in Section 4.

[3] Hence it follows that, in the case where the λ_j are known, estimate (1.4) is actually always better than estimate (1.3).

2 CONSTRUCTION OF AUXILIARY RANDOM VARIABLES α_j

In the following paragraph we will require random variables α_j which have prescribed mean value

$$\mathsf{M}\,\alpha_j = \lambda_j, \quad j = 1, \ldots, m, \quad \lambda_j > 0, \quad \sum_{j=1}^{m} \lambda_j = 1 \tag{2.1}$$

and at the same time satisfy the condition that for each concrete realization

$$\begin{gathered} \nu \text{ of the random variables } \alpha_1, \ldots, \alpha_m \text{ are equal to } \frac{1}{\nu} \\ \text{and the remaining } m - \nu \text{ equal } 0. \end{gathered} \tag{2.2}$$

In other words, the random variables α_j must be a random permutation of m numbers, ν of them equalling $1/\nu$ and the remaining $m - \nu$ equaling 0, where the probabilities of the separate permutations must be such that (2.1) is valid. It is obviously not possible to do this for all numbers $\lambda_1, \lambda_2, \ldots, \lambda_m$ and ν. For example, from condition (2.2), it follows directly that the maximum value of each variable α_j equals $1/\nu$, and this cannot be less than its mean value, i. e.,

$$\lambda_j \leq \frac{1}{\nu}, \quad j = 1, \ldots, m \tag{2.3}$$

must be valid. However, as we will show, condition (2.3) is not only necessary but also sufficient, in order that the desired random variables α_j exist.

We divide the right open interval $\langle 0, 1)$ into m right open intervals with lengths $\lambda_1, \lambda_2, \ldots, \lambda_m$, for example into the intervals

$$\Delta_j = \langle \lambda_1 + \cdots + \lambda_{j-1}, \lambda_1 + \cdots + \lambda_j), \quad j = 1, \ldots, m. \tag{2.4}$$

Then with each point $\omega \in \langle 0, \frac{1}{\nu})$ we associate a value of the variables $\alpha_j = \alpha_j(\omega)$ such that $\alpha_j = \frac{1}{\nu}$ when some of the points $\omega, \omega + \frac{1}{\nu}, \ldots, \omega + \frac{\nu-1}{\nu}$ fall in the interval Δ_j and $\alpha_j = 0$ in the opposite case. Each of the ν points

$$\omega, \omega + \frac{1}{\nu}, \ldots, \omega + \frac{\nu - 1}{\nu}$$

falls in *another* interval Δ_j, since in accordance with assumption (2.3) the distances of these points are greater than or at most equal to the lengths λ_j of the intervals Δ_j. That, however, means that for each point $\omega \in \langle 0, \frac{1}{\nu})$ condition (2.2) is satisfied.

It remains to show that the points $\omega \in \langle 0, \frac{1}{\nu})$ can be given such a probability distribution as to make (2.1) valid. A rectangular distribution obviously satisfies this claim, where the probability of the event $\omega \in \Delta$ for

$\Delta \subset \langle 0, \frac{1}{\nu})$ equals $\nu\, d(\Delta)$ where $d(\Delta)$ denotes the length of the interval Δ. Actually, then the probability that one of the defined points $\omega + \frac{h-1}{\nu}$ falls in the given interval Δ_j will be equal to a ν-multiple of the length of the common part of this interval with the range of the random variable $\omega + \frac{h-1}{\nu}$, i. e. with the interval $\langle \frac{h-1}{\nu}, \frac{h}{\nu})$; this may be expressed as

$$\mathsf{P}\left(\omega + \frac{h-1}{\nu} \in \Delta_j\right) = \nu\, d\left\{\Delta_j \cap \left\langle \frac{h-1}{\nu}, \frac{h}{\nu}\right)\right\}.$$

The events $\omega + \frac{h-1}{\nu} \in \Delta_j$, $h = 1, 2, \ldots, \nu$, as was ascertained above, are mutually exclusive, and therefore the probability of occurrence of at least one of them, i. e., the probability of the event $\alpha_j = \frac{1}{\nu}$, equals

$$\begin{aligned}\mathsf{P}\left(\alpha_j = \frac{1}{\nu}\right) &= \nu \sum_{h=1}^{\nu} d\left\{\Delta_j \cap \left\langle \frac{h-1}{\nu}, \frac{h}{\nu}\right)\right\} \\ &= \nu\, d\left\{\Delta_j \cap \langle 0, 1)\right\} = \nu\, d\{\Delta_j\} = \nu\, \lambda_j, \quad j = 1, \ldots, m.\end{aligned}$$

Hence it follows that $\mathsf{M}\, \alpha_j = \frac{1}{\nu}\, \mathsf{P}\left(\alpha_j = \frac{1}{\nu}\right) = \lambda_j$, $j = 1, 2, \ldots, m$, which is what we set out to prove.

3 MAIN RESULT

Proof of the main proposition will be based on the Jensen inequality, which can be expressed as follows. If f is a concave real-valued function in the range of values of a random variable y and if M denotes the mean value, then

$$\mathsf{M}\, f(y) \leq f(\mathsf{M}\, y). \tag{3.1}$$

In our case we will consider the function

$$\Phi(t\sqrt{y}) = \frac{1}{\sqrt{2\pi}} \int_0^{t\sqrt{y}} e^{-\frac{x^2}{2}}\, dx. \tag{3.2}$$

Since its derivative

$$\frac{\partial}{\partial y}\, \Phi\left(t\sqrt{y}\right) = \frac{t\, e^{-\frac{1}{2}t^2 y}}{2\sqrt{2\, \pi y}}$$

is for arbitrary $t > 0$ a decreasing function of y in the region $0 < y < \infty$, this function is concave in the region $0 < y < \infty$. To this region we can obviously add also the point $y = 0$, since $\Phi(t\sqrt{y})$ is continuous at the point $y = 0$.

From the Jensen inequality we can write

$$\mathsf{M}\, \Phi\left(t\sqrt{y}\right) \leq \Phi\left(t\sqrt{\mathsf{M}\, y}\right) \tag{3.3}$$

for each non-negative variable y. Now we show the main result.

Let x be a normally distributed random variable with mean value μ and variance σ^2. For variance σ^2 we have the estimate s^2 which is independent of x and has structure (1.2); we choose arbitrary limits $t' \leq 0 \leq t''$. Then the probability of the event

$$t' < \frac{x-\mu}{s} < t'', \quad t' \leq 0 \leq t'' \tag{3.4}$$

lies in the interval

$$P_\nu \leq P \leq P_m, \tag{3.5}$$

where P_m and P_ν are the probability of the event (3.4) given that $(x-\mu)/s$ has Student's distribution with m and ν degrees of freedom, respectively, where

$$m = m_1 + m_2 + \cdots + m_k, \tag{3.6}$$

and for ν we can take any arbitrary integer number satisfying

$$\nu \leq \min_{1 \leq j \leq k} \frac{m_j}{\lambda_j}, \tag{3.7}$$

for example

$$\nu = \min_{1 \leq j \leq k} m_j. \tag{3.8}$$

Before we get to the p r o o f, we take note of the following two simple facts. First, we can decompose each random variable $\chi_j^2(m_j)$ into the sum of m_j chi-square random variables with one degree of freedom, and in accordance with that rewrite structure (1.2) in the form

$$s^2 = \sigma^2 \sum_{j=1}^{m} \lambda_j \, \chi_j^2(1), \quad m = m_1 + \cdots + m_k, \; \lambda_j \geq 0, \; \sum_{j=1}^{m} \lambda_j = 1. \tag{3.9}$$

This form allows us to omit the numbers m_j and keep in mind only that the constants λ_j in (3.9) correspond to the constants λ_j/m_j in (1.2). In the second place, since the distribution of the ratio $(x-\mu)/s$ may already be assumed to be symmetrical around zero,[4] both for the usual or generalized s^2, it is sufficient to prove inequalities (3.6) only for the probability of the event

$$0 < \frac{x-\mu}{s} < t \tag{3.10}$$

for arbitrary $t > 0$.

The conditional distribution of $(x-\mu)/s$ for given s^2 is normal $(0, \sigma^2/s^2)$, so that, using definition (3.3), we can rewrite the conditional probability of

[4] For symmetry around zero it suffices only that s^2 be independent of x.

the event (3.10) in the form $\Phi\left(t\sqrt{s^2/\sigma^2}\right)$. However it is known that the unconditional probability of the event (3.10) equals

$$\mathsf{P}\left(0 < \frac{x-\mu}{s} < t\right) = \mathsf{E}\,\Phi\left(t\sqrt{s^2/\sigma^2}\right), \tag{3.11}$$

where E denotes the mean value with respect to s^2. The value of probability (3.11) depends upon which is the form of the estimate s^2. If the estimate s^2 has structure (1.1), then we obtain the probability

$$P_m = \mathsf{E}\,\Phi\left(t\sqrt{\frac{1}{m}\chi^2(m)}\right) \tag{3.12}$$

corresponding to the Student's distribution with m degrees of freedom, while to the general structure (3.9) corresponds the probability

$$P = \mathsf{E}\,\Phi\left(t\sqrt{\sum_{j=1}^{m}\lambda_j\,\chi_j^2(1)}\right). \tag{3.13}$$

Since the random variables $\chi_j^2(1)$ are independent and are identically distributed, the value of P does not change if we replace the constants $\lambda_1, \lambda_2, \ldots, \lambda_m$ by their arbitrary permutation $\beta_1, \beta_2, \ldots, \beta_m$: $P = \mathsf{E}\Phi\left(t\sqrt{\sum_{j=1}^{m}\beta_j\,\chi_j^2(1)}\right)$ for arbitrary permutation $\beta_1, \beta_2, \ldots, \beta_m$ of the numbers $\lambda_1, \lambda_2, \ldots, \lambda_m$. If this identity is valid for arbitrary permutation, it is valid (in the sense of mean value) also for the randomly chosen permutation $\beta_1, \beta_2, \ldots, \beta_m$. If we let the probability of the different permutations $\beta_1, \beta_2, \ldots, \beta_m$ be unspecified at the moment and if we use the symbol E for the mean value with respect to $\chi_j^2(1)$ and the symbol M for the mean value with respect to β_j, we can rewrite P in the form of the following iterated mean values:

$$P = \mathsf{M}\,\mathsf{E}\,\Phi\left(t\sqrt{\sum_{j=1}^{m}\beta_j\,\chi_j^2(1)}\right) = \mathsf{E}\,\mathsf{M}\,\Phi\left(t\sqrt{\sum_{j=1}^{m}\beta_j\,\chi_j^2(1)}\right).$$

If we use the Jensen inequality (3.4) on the interior mean value in the last expression, we get

$$P \leq \mathsf{E}\,\Phi\left(t\sqrt{\sum_{j=1}^{m}(\mathsf{M}\,\beta_j)\,\chi_j^2(1)}\right). \tag{3.14}$$

Now it is sufficient to give each permutation $\beta_1, \beta_2, \ldots, \beta_m$ equal probability $1/m!$. Then obviously it is $\mathsf{M}\,\beta_j = \frac{1}{m}\sum_{j=1}^{m}\lambda_j = \frac{1}{m}$, which substituted in

(3.14) gives

$$P \leq \mathsf{E}\,\Phi\left(t\sqrt{\sum_{j=1}^{m}\chi_j^2(1)}\right) = \mathsf{E}\,\Phi\left(t\sqrt{\frac{1}{m}\chi^2(m)}\right) = P_m.$$

Thus the inequality $P \leq P_m$ is proved. The inequality $P_\nu \leq P$ we will prove analogously with the aid of the random variables $\alpha_1, \alpha_2, \ldots, \alpha_m$ constructed in the last paragraph. In accordance with (2.2) ν of these variables equal $1/\nu$ and $m - \nu$ equal 0, so that $\sum_{j=1}^{m} \alpha_j\, \chi_j^2(1) = \frac{1}{\nu}\,\chi^2(\nu)$ always, and as a result of this for each realization of the random variables $\alpha_1, \alpha_2, \ldots, \alpha_m$ we have

$$P_\nu = \mathsf{E}\,\Phi\left(t\sqrt{\sum_{j=1}^{m}\alpha_j\,\chi_j^2(1)}\right),$$

where P_ν denotes the probability of the event (3.11) given that $(x - \mu)/s$ has Student's distribution with ν degrees of freedom. If we carry through again the same chain of reasoning as before, and if we take into account that in accordance with (2.1) $\mathsf{M}\,\alpha_j = \lambda_j$ is valid, we get

$$\begin{aligned} P_\nu &= \mathsf{M}\,\mathsf{E}\,\Phi\left(t\sqrt{\sum_{j=1}^{m}\alpha_j\,\chi_j^2(1)}\right) = \mathsf{E}\,\mathsf{M}\,\Phi\left(t\sqrt{\sum_{j=1}^{m}\alpha_j\,\chi_j^2(1)}\right) \\ &\leq \mathsf{E}\,\Phi\left(t\sqrt{\sum_{j=1}^{m}(\mathsf{M}\,\alpha_j)\,\chi_j^2(1)}\right) = \mathsf{E}\,\Phi\left(t\sqrt{\sum_{j=1}^{m}\lambda_j\,\chi_j^2(1)}\right) = P. \end{aligned}$$

The only condition for the existence of the random variables $\alpha_1, \ldots, \alpha_m$ with the properties (2.1) and (2.2) just used is the satisfaction of inequalities (2.3). These inequalities are however identical with inequality (3.8), since, as has been mentioned, for λ_j in formula (3.10) it is necessary to take λ_j/m_j from formula (1.2). Thus the proof is completed. □

The proved theorem could also be rephrased, saying that the generalized Student's distribution is less concentrated around zero than the usual Student's distribution with $m = m_1 + \cdots + m_k$ degrees of freedom, but, on the other hand, that it is more concentrated than the usual Student's distribution with $\min_{1 \leq j \leq k}\{m_j\}$ degrees of freedom.

Generally it can be said that the distribution corresponding to the constants $\lambda_1, \lambda_2, \ldots, \lambda_m$ in structure (3.9) is more concentrated around zero than the distribution corresponding to the other constants $\kappa_1, \kappa_2, \ldots, \kappa_m$ as soon as the numbers $\lambda_1, \lambda_2, \ldots, \lambda_m$ can be expressed as the mean value of the permutation of the numbers $\kappa_1, \kappa_2, \ldots, \kappa_m$, randomized with the help of suitable probabilities.

4 APPLICATIONS

We will show first of all in some examples how we can use the derived results for testing null hypotheses. Specifically, we will be concerned with the following, that with the help of the t-test we test the hypothesis that a fixed statistic whose variance has an estimate of structure (1.2) has a given (without loss of generality, zero) mean value.

Example 1. With the help of the usual t-test two independent sample averages were compared, each formed from 12 observations, and there resulted

$$t = \frac{\overline{x} - \overline{y}}{\sqrt{s_1^2 + s_2^2}} = 2.52.$$

(The null hypothesis: $\overline{x} - \overline{y}$ has mean value 0; s_1^2 and s_2^2 are estimates of variances of $\overline{x}$ and $\overline{y}$.) Since it was carried at the 5 % level of significance and with $12 + 12 - 2 = 22$ degrees of freedom, this value was recognized as significant, since it is greater than the corresponding critical value $t = 2.09$. Later, however, there was found to be an objection, that the variances in both samples arose might not be the same. In other words, a suspicion arose that t behaves not according to the usual, but to the generalized, Student's distribution. Since, however, the smallest of the numbers $m_1 = m_2 = 11$ and the critical value for 11 degrees of freedom is equal to 2.20, which is still smaller than 2.52, $t = 2.52$ must be recognized as significant irrespective of the equality of the inequality of the variances.

Example 2. We have the same case as in Example 1, only with the difference that $t = 2.16$. This value is no doubt significant for 22 degrees of freedom, but it is not significant for 11 degrees of freedom. Let us admit, however, that the assumption is given that none of the variances σ_1^2 and σ_2^2 is more than three times as large as the other, i. e., that the maximum of the numbers $\lambda_j = \sigma_j^2/(\sigma_1^2 + \sigma_2^2)$, $j = 1, 2$, is not greater than 3/4. Then,

$$\min_{j=1,2} \frac{m_j}{\lambda_j} \geq 11 \cdot \frac{4}{3} > 14,$$

and therefore in accordance with condition (3.8) we are able to use the critical value for 14 degrees of freedom $t = 2.14$, with respect to which the observed value $t = 2.16$ remains significant.

In the first example we were concerned with the simplest case, where $k = 2$, $m_1 = m_2$ and $\sigma_1^2 = \sigma_2^2$ and also $\lambda_1 = \lambda_2$. This case is remarkable in that the estimates (1.3) and (1.4) are identical, so that a suitable error in the assumption $\sigma_1^2 = \sigma_2^2\,(\lambda_1 = \lambda_2)$ does not have an influence on the unbiasedness of s^2, but only on the fact that structure (1.1) changed to structure (1.2). If we use, however, estimate (1.4) in the situation where σ_j^2 and m_j are different from each other, the error in the ratios $\lambda_j = \sigma_j^2/(\sigma_1^2 + \cdots + \sigma_j^2)$ has more complicated consequences. Estimate (1.4) can be rewritten in the form

$$s^2 = (\sigma_1^2 + \cdots + \sigma_k^2) \sum_{j=1}^{k} \frac{\lambda_j^0}{m_j} \chi_j^2(m_j),$$

where

$$\lambda_j^0 = \frac{\sigma_j^2 m_j}{\lambda_j(\sigma_1^2 + \cdots + \sigma_k^2) m}, \quad j = 1, 2, \ldots, m,$$

where λ_j are chosen arbitrarily. If $\lambda_j = \sigma_j^2/(\sigma_1^2 + \cdots + \sigma_k^2)$ is not valid, then neither can $\sum_{j=1}^{k} \lambda_j^0 = 1$ be valid, i. e., the estimate (1.4) not only does not have structure (1.1), but is not even unbiased. If $\sum_{j=1}^{k} \lambda_j^0 < 1$ then the mean value of s^2 is shifted towards zero, and therefore not only the number of degrees of freedom but also the value of t itself is overestimated. Here, then, the danger of hasty rejection of null hypotheses is a maximum. If

$$\sum_{j=1}^{k} \lambda_j^0 > 1$$

then the mean value of s^2 is shifted upwards, and then while the number of degrees of freedom is overestimated, the values of t are underestimated. For this reason, as soon as doubts have arisen about the (at least approximate) accuracy of the choice of the numbers λ_j, it is better to drop estimate (1.4) and use estimate (1.3).

Example 3. From two independent samples of 10 and 15 observations we calculated averages $\overline{x} = 213$ and $\overline{y} = 155$ and the estimates of the variances $s_1^2 = 526$ and $s_2^2 = 94$. It was assumed that the separate observations in both samples have the same variance, so that the variances of the sample averages vary inversely with the number of observations from which they were calculated, i. e., $\sigma_1^2 : \sigma_2^2 = 15 : 10$ and then $\lambda_1 = 3/5$ and $\lambda_2 = 2/5$. The estimate of the variance

$$s^2 = \frac{s_1^2 m_1}{\lambda_1 m} + \frac{s_2^2 m_2}{\lambda_2 m} = \frac{526 \cdot 9}{\frac{3}{5} \cdot 23} + \frac{95 \cdot 14}{\frac{2}{5} \cdot 23} = 486$$

was calculated in accordance with formula (1.4) and then

$$t = \frac{x - y}{s} = \frac{213 - 155}{\sqrt{486}} = 2.63.$$

This value is clearly significant at the 5 % level with 23 degrees of freedom. However the excessive difference between s_1^2 and s_2^2 gives rise to doubts about the accuracy of the supposition that $\sigma_1^2 : \sigma_2^2 = 15 : 10$, and therefore it was decided to estimate the variance in accordance with (1.3), i. e., to take $s^2 = s_1^2 + s_2^2 = 526 + 94 = 620$ and $t = \frac{213-155}{\sqrt{620}} \doteq 2.33$. This value is significant at the 5 % level even if we use $9 = \min[9, 14]$ degrees of freedom since the corresponding critical point is $t = 2.26$. Then we can consider the difference $\overline{x} - \overline{y}$ as significantly large without a specific inequality for the variances.

Example 4. We have the same case as in Example 3, only with the difference that $\overline{y} = 165$, so that we first calculate $t = \frac{48}{\sqrt{486}} \doteq 2.18$, which is still significant for 23 degrees of freedom. The revised value, however, will be $t = \frac{48}{\sqrt{620}} \doteq 1.93$, which is not significant for 23 degrees of freedom, and therefore we cannot reject the null hypothesis at the 5 % level.

The obtained inequalities suggest the idea of approximating the generalized Student's distribution by the usual Student's distribution with suitable choice of the number of degrees of freedom. If we choose this number of degrees of freedom m' such that s^2 with structure (1.1) and m' degreees of freedom has the same variance as if s^2 had structure (1.2), then we arrive by easy calculation to

$$m' = \frac{1}{\sum_{j=1}^{k} \lambda_j^2 / m_j}. \tag{4.1}$$

However the numbers λ_j, as mentioned, are unknown, because otherwise we could use formula (1.4) and the whole problem would vanish. It is inevitable then that we must estimate λ_j from the samples. If s^2 is given as in (1.3), then the estimate for (24) is just (1.5). This estimate is not unbiased, but is a simple function of two unbiased estimates, which has practically the same effect as unbiasedness itself:

$$\begin{aligned} \frac{\mathsf{E}(s_1^2 + \cdots + s_k^2)^2}{\mathsf{E} \sum_{j=1}^{k} s_j^4/(m_j+2)} - 2 &= \frac{(\sigma_1^2 + \cdots + \sigma_k^2)^2 + 2\sum \sigma_j^4/m_j}{\sum_{j=1}^{k} \sigma_j^4/m_j} - 2 \\ &= \frac{(\sigma_1^2 + \cdots + \sigma_k^2)^2}{\sum_{j=1}^{k} \sigma_j^4/m_j} \end{aligned}$$

which just equals (24) since $\lambda_j = \sigma_j^2/(\sigma_1^2 + \cdots + \sigma_k^2)$.

Example 5. Three independent samples give us the following statistics, their variance estimates and corresponding degrees of freedom:

$$\begin{aligned} x_1 &= 11.8, & s_1^2 &= 1.02, & m_1 &= 10, \\ x_2 &= 16.4, & s_2^2 &= 0.94, & m_2 &= 10, \\ x_3 &= 14.1, & s_3^2 &= 0.52, & m_3 &= 4. \end{aligned}$$

Our task is to find the 95 % confidence interval around the statistic

$$x = 0.5\,x_1 + 0.2\,x_2 + 0.3\,x_3 = 13.41.$$

If we put the variance of the separate sums and the corresponding degree of freedom into (1.5), we get

$$\begin{aligned} m' &= \frac{[(0.5)^2 \times 1.02 + (0.2)^2 \times 0.94 + (0.3)^2 \times 0.52]^2}{\frac{[(0.5)^2\times 1.02]^2}{10+2} + \frac{[(0.2)^2\times 0.94]^2}{10+2} + \frac{[(0.3)^2\times 0.52]^2}{4+2}} - 2 \\ &= \frac{(0.3394)^2}{0.0059} - 2 \doteq 17.5. \end{aligned}$$

If we round off downwards, the effective number of degrees of freedom is $m' = 17$ and the corresponding $t = 2.11$. The estimate of the variance of x is found by computing m' to $s^2 = 0.3394$, so that the confidence interval $x \pm ts$ equals 13.41 ± 1.23.

Remark. If the minimum m_j is not too small, we can take it to be the number of degrees of freedom and we would be aware that we obtained a confidence interval which covers the unknown parameter with somewhat greater probability than that which we claimed.

REFERENCES

Aspin A. A. (1949). Tables for use in comparisons whose accuracy involves two variances, separately estimated. *Biometrika 36*, 290–293.

Fisher R. A. (1935). The fiducial argument in statistical inference. *Ann. Eugenics 6*, 391–398.

Sukhatme P. V. (1938). On Fisher and Behrens' test of significance for the difference in means of two normal samples. *Sankhyā 4*, 39–48.

Welch B. L. (1947). The generalization of 'Student's' problem when several different population variances are involved. *Biometrika 34*, 28–35.

Welch B. L. (1949). Further note on Mrs. Aspin's table and on certain approximations to the tabulated function. *Biometrika 36*, 293–296.

CHAPTER 7

Predicting a Stationary Process when the Correlation Function is Convex

Czechoslovak Math. J. 8 (1958), 150–154.
Reproduced by permission of Hájek family: Al. Hájková, H. Kazárová and Al. Slámová.

By the method worked out in the paper by Hájek (1956) it is proved that, in the case of a convex correlation function, it suffices to base the linear prediction on the last observation only, because the relative reduction of the residual variance, attainable by making use of any number of preceding observations, cannot exceed 50 %.

1 INTRODUCTION AND SUMMARY

Let us consider a wide-sense stationary process $\{x_t, -\infty < t < \infty\}$ and suppose that its mean value μ, variance σ^2 and correlation function R_τ are known. The best linear prediction of the state of the process at the moment $t + \Delta$, say Pred $x_{t+\Delta}$, based on the single observation x_t at the moment t, viz.

$$\text{Pred } x_{t+\Delta} = \mu + R_\Delta(x_t - \mu), \quad \Delta > 0, \tag{1.1}$$

is known to possess the residual variance

$$\mathsf{D}\{x_{t+\Delta} - \text{Pred } x_{t+\Delta}\} = \sigma^2\left(1 - R_\Delta^2\right). \tag{1.2}$$

This variance may be reduced by making use of a certain number of preceding observations, i. e. by putting

$$\text{Pred } x_{t+\Delta} = \mu + \sum_{i=1}^{n} c_i\left(x_{t_i} - \mu\right), \quad t_1 < \cdots < t_n = t < t + \Delta, \ n \geq 1. \tag{1.3}$$

Collected Works of Jaroslav Hájek – With Commentary
Edited by M. Hušková, R. Beran and V. Dupač
Published in 1998 by John Wiley & Sons Ltd.

We know, however, that if the correlation function is exponential, $R_\tau = e^{-a\tau}$ $(a > 0)$, then the last observation contains all 'linear' information, so that no reduction is possible in this way. The theorem below shows that a somewhat weaker result is valid for all other convex correlation functions, namely to the effect that the residual variance of (1.3) cannot be less than half of the right side of (1.2). In fact, the right side of (1.2) is only $1 + R_\Delta$ times greater than the lower bound indicated by (2.1).

2 THEOREM

Theorem. *Let $\{x_t, -\infty < t < \infty\}$ be a wide-sense stationary process whose correlation function R_τ is convex. Let $\mathrm{Pred}\, x_{t+\Delta}$ be any linear prediction of $x_{t+\Delta}$ defined by* (1.3)*. Then*

$$\mathsf{D}\{x_{t+\Delta} - \mathrm{Pred}\, x_{t+\Delta}\} \geq \sigma^2(1 - R_\Delta), \tag{2.1}$$

where $\sigma^2 = \mathsf{D}x_t$ denotes the variance of x_t.

Proof. Let us first suppose that R_τ is continuous and $R_\infty = 0$. In Hájek (1956) it is proved that such a convex correlation function is possessed by the process $x_t^0 = x_t^0(\lambda, \varphi, \ldots, y_{-1}, y_0, y_1, \ldots)$ defined by

$$x_t^0 = y_{\left[\frac{t}{\lambda}+\varphi\right]}, \quad -\infty < t < \infty, \tag{2.2}$$

where $\left[\frac{t}{\lambda} + \varphi\right]$ denotes the integral part of $\frac{t}{\lambda} + \varphi$, and $\lambda, \varphi, \ldots, y_{-1}, y_0, y_1, \ldots$ are mutually independent random variables; the distribution function of λ is given by

$$F(\lambda) = \begin{cases} \int_0^\lambda \tau\, \mathrm{d}R'(\lambda) = \tau\, R'(\lambda) - R(\lambda) + 1, & \lambda > 0, \\ 0, & \lambda \leq 0, \end{cases} \tag{2.3}$$

φ is distributed rectangularly over $(0, 1)$, and $\ldots, y_{-1}, y_0, y_1, \ldots$ have each the same but otherwise arbitrary distributions with mean value μ and variance σ^2. As the validity of the inequality (2.1) for given instants $t_1 < \cdots < t_n = t < t + \Delta$ and constants $c_1, \ldots, c_n$ depends only on the correlation function R_τ, it suffices to prove it for any particular process possessing this correlation function, e. g. for the process x_t^0 as defined above. This is the main idea of our proof.

Let us choose an arbitrary $n \geq 1$ and instants $t_1 < t_2 < \cdots < t_n = t < t_{n+1} = t + \Delta$, and keep them fixed in the course of all further considerations. Under the condition that

$$\left[\frac{t_i}{\lambda} + \varphi\right] = k_i, \quad i = 1, \ldots, n+1, \tag{2.4}$$

we may write

$$x^0_{t+\Delta} - \operatorname{Pred} x^0_{t+\Delta} = y_{k_{n+1}} - \mu - \sum_{i=1}^{n} c_i \left(y_{k_i} - \mu \right). \tag{2.5}$$

Owing to the independence of the random variables $\lambda, \varphi, \ldots, y_{-1}, y_0, y_1, \ldots$, the conditional distribution of $\{y_{k_1}, \ldots, y_{k_{n+1}}\}$ under the condition (2.4) will be identical to the non-conditional distribution, namely, the variance of $y_{k_{n+1}}$ will equal σ^2 and $y_{k_{n+1}}$ will be independent of $\{y_{k_1}, \ldots, y_{k_n}\}$ when $k_1 \le \cdots \le k_n < k_{n+1}$. Consequently, denoting the conditional variance under the condition A by $\mathsf{D}\{\cdot \mid A\}$, we have

$$\begin{aligned}
&\mathsf{D}\left\{ x^0_{t+\Delta} - \operatorname{Pred} x^0_{t+\Delta} \,\middle|\, \left[\frac{t_i}{\lambda} + \varphi\right] = k_i,\, i = 1, \ldots, n+1 \right\} \\
&= \mathsf{D}\left\{ y_{k_{n+1}} - \mu - \sum_{i=1}^{n} c_i (y_{k_i} - \mu) \right\} = \mathsf{D}\left\{ y_{k_{n+1}} \right\} + \mathsf{D}\left\{ \sum_{i=1}^{n} c_i\, y_{k_i} \right\} \\
&\ge \mathsf{D}\left\{ y_{k_{n+1}} \right\} \;=\; \sigma^2.
\end{aligned} \tag{2.6}$$

From the inequality (2.6) it follows that

$$\begin{aligned}
&\mathsf{D}\left\{ x^0_{t+\Delta} - \operatorname{Pred} x^0_{t+\Delta} \right\} \\
&\ge \sum_{k_1 \le \cdots \le k_n < k_{n+1}} \mathsf{P}\left\{ \left[\frac{t_i}{\lambda} + \varphi\right] = k_i,\, i = 1, \ldots, n+1 \right\} \\
&\qquad \times \mathsf{D}\left\{ x^0_{t+\Delta} - \operatorname{Pred} x^0_{t+\Delta} \,\middle\|\, \left[\frac{t_i}{\lambda} + \varphi\right] = k_i,\; i = 1, \ldots, n+1 \right\} \\
&\ge \sigma^2 \sum_{k_1 \le \cdots \le k_n < k_{n+1}} \mathsf{P}\left\{ \left[\frac{t_i}{\lambda} + \varphi\right] = k_i,\, i = 1, \ldots, n+1 \right\} \\
&= \sigma^2 \mathsf{P}\left\{ \left[\frac{t}{\lambda} + \varphi\right] < \left[\frac{t+\Delta}{\lambda} + \varphi\right] \right\}.
\end{aligned} \tag{2.7}$$

The probability $\mathsf{P}\left\{ \left[\frac{t}{\lambda} + \varphi\right] < \left[\frac{t+\Delta}{\lambda} + \varphi\right] \right\}$ may be bound by means of the conditional probability $\mathsf{P}\left\{ \left[\frac{t}{\lambda} + \varphi\right] < \left[\frac{t+\Delta}{\lambda} + \varphi\right] \middle| \lambda \right\}$ with respect to λ. As λ and φ are independent, the probability $\mathsf{P}\left\{ \left[\frac{t}{\lambda} + \varphi\right] < \left[\frac{t+\Delta}{\lambda} + \varphi\right] \middle| \lambda \right\}$ can for all $\lambda > 0$ be chosen equal to the ordinary probability of $\left[\frac{t}{\lambda} + \varphi\right] < \left[\frac{t+\Delta}{\lambda} + \varphi\right]$ when λ is fixed and φ is distributed rectangularly over $(0, 1)$. The latter probability, however, as may easily be seen, equals 1 or $\frac{\Delta}{\lambda}$ according to whether $\Delta \ge \lambda$ or $\Delta < \lambda$, respectively. Hence, in accordance with (2.3),

$$\begin{aligned}
&\mathsf{P}\left\{ \left[\frac{t}{\lambda} + \varphi\right] < \left[\frac{t+\Delta}{\lambda} + \varphi\right] \right\} \\
&= \int_0^\infty \mathsf{P}\left\{ \left[\frac{t}{\lambda} + \varphi\right] < \left[\frac{t+\Delta}{\lambda} + \varphi\right] \middle| \lambda \right\} \mathrm{d}F(\lambda)
\end{aligned}$$

$$
\begin{aligned}
&= \int_0^{\Delta} \mathrm{d}F(\lambda) + \int_{\Delta}^{\infty} \frac{\Delta}{\lambda} \mathrm{d}F(\lambda) = F(\Delta) + \Delta \int_{\Delta}^{\infty} \mathrm{d}R'(\tau) \\
&= \Delta R'(\Delta) - R(\Delta) + 1 - \Delta R'(\Delta) = 1 - R(\Delta).
\end{aligned}
$$

Inserting this result in (2.7) we get (2.1).

Let us complete our proof by examining an arbitrary convex correlation function. Every convex correlation function is continuous for $0 < \tau < \infty$,[1] non-negative and non-increasing (see Hájek (1956)), so that there exist limits $0 \leq R_\infty \leq R_{0+} \leq 1$.

Let us introduce mutually independent stationary processes y_t^0, v_t, $z_t = z$ such that

$$
\begin{aligned}
&\mathsf{D}y_t = \sigma^2 \left(R_{0+} - R_\infty\right), \quad -\infty < t < \infty, \\
&\mathsf{D}v_t = \sigma^2 \left(1 - R_{0+}\right), \\
&\mathsf{D}z = \sigma^2 R_\infty, \\
&\mathsf{cov}\{y_t, y_{t+\tau}\} = \sigma^2 \left(R_\tau - R_\infty\right), \\
&\mathsf{cov}\{v_t, v_{t+\tau}\} = 0, \quad \tau \neq 0,
\end{aligned}
\tag{2.8}
$$

and put $x_t^0 = y_t + v_t + z$, $-\infty < t < \infty$. Process x_t^0 is obviously stationary and has the variance σ^2 and correlation function R_τ. Furthermore, owing to the independence of y_t, v_t and $z_t = z$, the relation

$$
\begin{aligned}
&\mathsf{D}\left\{x_{t+\Delta}^0 - \mathrm{Pred}\, x_{t+\Delta}^0\right\} \\
= \;&\mathsf{D}\left\{y_{t+\Delta} - \mathrm{Pred}\, y_{t+\Delta}\right\} + \mathsf{D}\left\{v_{t+\Delta} - \mathrm{Pred}\, v_{t+\Delta}\right\} + \mathsf{D}\left\{z - \mathrm{Pred}\, z\right\}
\end{aligned}
\tag{2.9}
$$

holds. The component y_t possesses the correlation function $\frac{R_\tau - R_\infty}{R_{0+} - R_\infty}$ which is clearly continuous and tends to 0 when $\tau \to 0$. Applying the first part of our proof, we may therefore write

$$
\mathsf{D}\left\{y_{t+\Delta} - \mathrm{Pred}\, y_{t+\Delta}\right\} \geq \sigma^2 \left(R_{0+} - R_\infty\right) \left(1 - \frac{R_\Delta - R_\infty}{R_{0+} - R_\infty}\right) = \sigma^2 \left(R_{0+} - R_\infty\right). \tag{2.10}
$$

Further, as the component v_t is a stationary process with uncorrelated component random variables, $v_{t+\Delta}$ is uncorrelated with every prediction of the form (1.3), and accordingly

$$
\mathsf{D}\left\{v_{t+\Delta} - \mathrm{Pred}\, v_{t+\Delta}\right\} \geq \mathsf{D}\left\{v_{t+\Delta}\right\} = \sigma^2 \left(1 - R_{0+}\right). \tag{2.11}
$$

Now, inserting (2.10), (2.11) and $\mathsf{D}\{z - \mathrm{Pred}\, z\} \geq 0$ in (2.9), we can see that the inequality (2.1) is proved for all convex correlation functions. □

[1] The possible discontinuity may occur in the point $\tau = 0$ only.

Remark 1. The result applies obviously also to processes with discontinuous parameter $t = \ldots, -1, 0, 1 \ldots$.

Remark 2. The lower bound established in (2.1) is exact. Indeed, if we take the convex correlation function

$$r(\tau) = \begin{cases} 1 - \tau, & \tau < 1, \\ 0, & \tau \geq 1, \end{cases}$$

and $\Delta < 1$, then we obtain for

$$\operatorname{Pred} x_{t+\Delta} = \mu + \sum_{i=0}^{n} x_{t-i} \left[\frac{n+2}{n+1} \left(1 - \frac{\Delta}{n+2} \right) - \frac{i+1}{n+1} \right] - \sum_{i=0}^{n} x_{t+\Delta-i} \frac{n-i}{n+1}$$

the residual variance $\mathsf{D}\{x_{t+\Delta} - \operatorname{Pred} x_{t+\Delta}\} = \sigma^2 \frac{n+2}{n+1} \Delta \left[1 - \frac{\Delta}{n+2}\right]$, which tends to $\sigma^2 \Delta = \sigma^2(1 - r_\Delta)$ as $n \to \infty$; for $\Delta > 1$ the right side of (2.1) is exactly attained when we put $\operatorname{Pred} x_{t+\Delta} = \mu$.

REFERENCES

Hájek J. (1956). Linear estimation of the mean value of a stationary random processes with convex correlation function. *Czechoslovak Math. J. 6*, 94–117.

CHAPTER 8

A Property of J–Divergence of Marginal Probability Distributions

Czechoslovak Math. J. 8 (1958), 460–463.
Reproduced by permission of Hájek family: Al. Hájková, H. Kazárová and Al. Slámová.

It is proved that the J-divergence of any two probability distributions of any stochastic process equals the supremum of J-divergences of finite-dimensional marginal distributions. If this supremum is finite then the distributions are absolutely continuous with respect to each other.

Let us have two arbitrary probability distributions, P and Q, on a Borel field $\mathcal{F}$ of subsets Λ of a space $\Omega = \{\omega\}$. Let P_a and Q_a, $a \in A$, be corresponding 'marginal' distributions on Borel subfields $\mathcal{F}_a \subset \mathcal{F}$, defined by $\mathsf{P}_a(\Lambda) = \mathsf{P}(\Lambda)$ and $\mathsf{Q}_a(\Lambda) = \mathsf{Q}(\Lambda)$ for $\Lambda \in \mathcal{F}_a$, $a \in A$.

Definition. (See Jeffreys (1948), page 158, and Kullback (1951).) J-divergence J_a between distributions P and Q on the Borel field $\mathcal{F}_a \subset \mathcal{F}$ is the number

$$J_a = \int \left(\frac{p_a}{q_a} - 1\right) \log \frac{p_a}{q_a} \, \mathrm{d}\mathsf{Q} \quad \text{if} \quad \mathsf{P}_a \equiv \mathsf{Q}_a, \tag{1.1}$$

and

$$J_a = \infty, \quad \text{if } \mathsf{P}_a \not\equiv \mathsf{Q}_a, \tag{1.2}$$

where $\mathsf{P}_a \equiv \mathsf{Q}_a$ denotes that $[\mathsf{Q}(\Lambda) = 0] \Leftrightarrow [\mathsf{P}(\Lambda) = 0]$ for $\Lambda \in \mathcal{F}_a$, and $\frac{p_a}{q_a} = \frac{\mathrm{d}\mathsf{P}_a}{\mathrm{d}\mathsf{Q}_a}$ is the likelihood ratio (Radon–Nikodym's derivative) of P_a w.r.t. Q_a, i.e. such a function of ω that

$$\mathsf{P}(\Lambda) = \int_\Lambda \frac{p_a(\omega)}{q_a(\omega)} \, \mathrm{d}\mathsf{Q}, \quad \Lambda \in F_a. \tag{1.3}$$

Our definition is that of Kullback (1951) extended to the case when $\mathsf{P}_a \not\equiv \mathsf{Q}_a$.

Collected Works of Jaroslav Hájek – With Commentary
Edited by M. Hušková, R. Beran and V. Dupač
Published in 1998 by John Wiley & Sons Ltd.

Divergence J_a is symmetrical in P and Q and possesses certain valuable properties. It may be easily shown that $J_a \geq 0$, where the sign of equality holds if and only if $\mathsf{P}_a = \mathsf{Q}_a$. Furthermore, $\mathcal{F}_b \subset \mathcal{F}_a$ implies $J_b \leq J_a$, where the sign of equality holds if and only if either $J_b = \infty$ or $\frac{p_a}{q_a} = \frac{p_b}{q_b}[\mathsf{P}]$; in the latter case F_b is a sufficient Borel field for distinguishing between P_a and Q_a.[1]

Theorem 1. *Let $J_1 \leq J_2 \leq \cdots \leq J_\infty$ be a sequence of J-divergences (see definition) between distributions* P *and* Q *on the Borel fields $\mathcal{F}_1 \subset \mathcal{F}_2 \subset \cdots \subset \mathcal{F}_\infty$, where $\mathcal{F}_\infty$ is the smallest Borel field containing $\bigcup_1^\infty \mathcal{F}_n$. Then*

$$J_\infty = \lim_{n\to\infty} J_n. \tag{1.4}$$

If $\lim_{n\to\infty} J_n < \infty$, then $\mathsf{P}_\infty \equiv \mathsf{Q}_\infty$.

Proof. If $\lim_{n\to\infty} J_n = \infty$, then (1.4) follows from $J_n \leq J_\infty$, $n \geq 1$. Hence we may restrict ourselves to the case when

$$\lim_{n\to\infty} J_n < \infty. \tag{1.5}$$

First let us prove that (1.5) implies $\mathsf{P}_\infty \equiv \mathsf{Q}_\infty$. We shall suppose that $\mathsf{P}_\infty \not\equiv \mathsf{Q}_\infty$ and deduce a contradiction. If, for example, $\mathsf{P}_\infty \ll \mathsf{Q}_\infty$ does not hold, then there exists an event $\Lambda \in \mathcal{F}_\infty$ such that $\mathsf{P}_\infty(\Lambda) = \varepsilon > 0$ and $\mathsf{Q}_\infty(\Lambda) = 0$. Consequently (Halmos (1950), Exercise 8, § 13), to each $k \geq 1$ we may choose $n_k \geq 1$ such that there exists an event Λ_k in $\mathcal{F}_{n_k}$ satisfying the inequalities

$$\frac{3}{4}\varepsilon < \mathsf{P}(\Lambda_k), \quad \mathsf{Q}(\Lambda_k) < \frac{\varepsilon}{4(k-1)}. \tag{1.6}$$

Bearing (1.6) in mind and denoting

$$\Lambda_k^* = \Lambda_k \cap \left\{\omega : \frac{p_{n_k}}{q_{n_k}} \geq k\right\} \tag{1.7}$$

we may write

$$\begin{aligned} \frac{\varepsilon}{2} &< \mathsf{P}(\Lambda_k) - \mathsf{Q}(\Lambda_k) = \int_{\Lambda_k} \left(\frac{p_{n_k}}{q_{n_k}} - 1\right) \mathrm{d}\mathsf{Q} \\ &= \int_{\Lambda_k^*} \left(\frac{p_{n_k}}{q_{n_k}} - 1\right) \mathrm{d}\mathsf{Q} + \int_{\Lambda_k - \Lambda_k^*} \left(\frac{p_{n_k}}{q_{n_k}} - 1\right) \mathrm{d}\mathsf{Q} \end{aligned}$$

[1] It may be shown, however, that J-divergence does not possess the triangle property of a metric. Let us consider three normal distributions on the real line having variances $\sigma_1^2 = 0.1$, $\sigma_2^2 = 1$, $\sigma_3^2 = 2$ and mean values $\mu_1 = \mu_2 = \mu_3 = 0$. The J-divergence of any two of them turns out to be $J_{ik} = \frac{\sigma_i^2}{\sigma_k^2}\left(\frac{\sigma_k^2}{\sigma_i^2} - 1\right)^2$, $1 \leq i \neq k \leq 3$, from which we get $J_{12} = 8.1$, $J_{13} = 18.05$, $J_{23} = 0.5$, i. e. $J_{12} + J_{23} < J_{13}$.

$$\leq \int_{\Lambda_k^*} \left(\frac{p_{n_k}}{q_{n_k}} - 1\right) \mathrm{dQ} + (k-1)\,\mathrm{Q}(\Lambda_k - \Lambda_k^*)$$

$$\leq \int_{\Lambda_k^*} \left(\frac{p_{n_k}}{q_{n_k}} - 1\right) \mathrm{dQ} + \frac{\varepsilon}{4},$$

i.e.

$$\int_{\Lambda_k^*} \left(\frac{p_{n_k}}{q_{n_k}} - 1\right) \mathrm{dQ} \geq \frac{\varepsilon}{4}. \tag{1.8}$$

From (1.7) and (1.8) it follows that

$$\begin{aligned} J_{n_k} &= \int \left(\frac{p_{n_k}}{q_{n_k}} - 1\right) \log \frac{p_{n_k}}{q_{n_k}} \mathrm{dQ} \geq \int_{\Lambda_k^*} \left(\frac{p_{n_k}}{q_{n_k}} - 1\right) \log \frac{p_{n_k}}{q_{n_k}} \mathrm{dQ} \\ &\geq \log k \int_{\Lambda_{n_k}^*} \left(\frac{p_{n_k}}{q_{n_k}} - 1\right) \mathrm{dQ} \geq \frac{\varepsilon}{4} \log k, \quad k = 1, 2, \ldots . \end{aligned}$$

This last inequality contradicts the supposition (1.5) and thereby proves $\mathrm{P}_\infty \equiv \mathrm{Q}_\infty$.

Now, from $\mathrm{P}_\infty \equiv \mathrm{Q}_\infty$ it follows that there exists $\frac{p_\infty}{q_\infty}$ and

$$J_\infty = \int \left(\frac{p_\infty}{q_\infty} - 1\right) \log \frac{p_\infty}{q_\infty} \mathrm{dQ}.$$

Moreover, $\frac{p_n}{q_n} = M\left\{\frac{p_\infty}{q_\infty}\middle| \mathcal{F}_n\right\}$, which implies (J. L. Doob, Theorem 4.3, Ch. VII) that

$$\frac{p_\infty}{q_\infty} = \lim_{n\to\infty} \frac{p_n}{q_n} \quad [\mathrm{Q}]. \tag{1.9}$$

By means of (1.9) and using Fatou's lemma we get from (1.1) that

$$\varliminf_{n\to\infty} J_n \geq \int \left(\frac{p_\infty}{q_\infty} - 1\right) \log \frac{p_\infty}{q_\infty} \mathrm{dQ} = J_\infty. \tag{1.10}$$

Inequality (1.10) combined with the obvious opposite inequality, $\lim J_n \leq J_\infty$, gives (1.4). The theorem is thus proved.[2] □

The following version of Theorem 1 is useful for stochastic processes.

---

[2] We may write $J_n = -H_n(\mathrm{P},\mathrm{Q}) - H_n(\mathrm{P},\mathrm{Q})$, where $H_n(\mathrm{P},\mathrm{Q}) = -\int \frac{p_n}{q_n} \log \frac{p_n}{q_n} \mathrm{dQ}$ is the entropy of P w.r.t. Q on $\mathcal{F}_n$, $\mathrm{P}_n \ll \mathrm{Q}_n$, see Perez (1957). The corresponding theorem for entropies, namely that (i) $H_n(\mathrm{P},\mathrm{Q}) \to H_\infty(\mathrm{P},\mathrm{Q})$ and (ii) $\lim H_n(\mathrm{P},\mathrm{Q}) > -\infty \Rightarrow \mathrm{P}_\infty \ll \mathrm{Q}_\infty$, could be proved without any essential change in our method.
A related but considerably weaker result is contained in Perez (1957), Theorem 7, part (ii), where $H_n(\mathrm{P},\mathrm{Q}) \to H_\infty(\mathrm{P},\mathrm{Q})$ is proved under supposition that $\lim_{n\to\infty} H_n(\mathrm{P},\mathrm{Q}) > -\infty$ and $\mathrm{P}_\infty \ll \mathrm{Q}_\infty$; the supposition $\mathrm{P}_\infty \ll \mathrm{Q}_\infty$, being implied by $\lim H_n(\mathrm{P},\mathrm{Q}) > -\infty$, is superfluous.

Theorem 2. *Let $\{x_t,\, t \in T\}$ be an arbitrary system of random variables. Let J_K be the J-divergence between distributions* P *and* Q *on the Borel field $\mathcal{F}$ generated by a subsystem $\{x_t,\, K \subset T\}$. Then*

$$J_T = \sup_{K \in \mathcal{K}} J_K, \tag{1.11}$$

where $\mathcal{K}$ is the class of all finite subsets of T. If $\sup_{K\in\mathcal{K}} J_K < \infty$, *then* $\mathsf{P}_T \equiv \mathsf{Q}_T$.

Proof. If T is countable, then it is possible to choose finite subsets $K_1 \subset K_2 \subset \ldots$ such that $\mathcal{F}_T$ is the smallest Borel field containing $\bigcup_{n=1}^{\infty} \mathcal{F}_{K_n}$, and the theorem is reduced to Theorem 1.

If T fails to be countable, it may be easily shown that $J_T = J_S$ for some countable subset $S \subset T$. When $\mathsf{P}_T \not\equiv \mathsf{Q}_T$ then $\mathsf{P}_S \not\equiv \mathsf{Q}_S$ for at least one countable $S \subset T$, so that $J_T = J_S = \infty$. When $\mathsf{P}_T \equiv \mathsf{Q}_T$, then there exists $\frac{p_T}{q_T}$ which is measurable with respect to $\mathcal{F}_T$. However, we know that every function measurable w.r.t. $\mathcal{F}_T$ is measurable w.r.t. $\mathcal{F}_S$ for at least one countable $S \subset T$, so that $\frac{p_T}{q_T} = \frac{p_S}{q_S}$, which implies $J_T = J_S$. □

REFERENCES

Doob, J. L. (1953). *Stochastic Processes*. London, New York.

Halmos, P. R. (1950). *Measure Theory*. New York.

Jeffreys H. (1948). *Theory of Probability*. Oxford.

Kullback, S. and Leibler, R. A. (1951). On information and sufficiency. *Ann. Math. Stat.*, *22*, 79–86.

Perez, A. (1957). Notions generalisées d'incertitude, d'entropie et d'information du point de vue de la théorie de martingales. In *Transactions of the First Prague Conference on Information Theory, Statistical Decision Functions, Random Processes*, Prague.

CHAPTER 9

On a Property of Normal Distributions of any Stochastic Process

Czechoslovak Math. J. 8 (1958), 610–618 (in Russian).
English translation in *Selected Transl. Math. Statist. Prob. 1* (1961), 245–252.
Translated by B. Kuvshinoff.
Reproduced by permission of American Mathematical Society.

Proof is given that two arbitrary normal distributions of a stochastic process are either equivalent or singular.

1 INTRODUCTION AND SUMMARY

Two probability distributions $\mathsf{P}\{\cdot\}$ and $\mathsf{Q}\{\cdot\}$ given on some Borel field $\mathcal{F}$ are equivalent (symbolically $\mathsf{P} \equiv \mathsf{Q}$) if the equations $\mathsf{P}\{\Lambda\} = 0$ and $\mathsf{Q}\{\Lambda\} = 0$ follow from each other for all $\Lambda \in \mathcal{F}$. On the other hand, if an event Λ can be found such that

$$\mathsf{P}\{\Lambda\} = 0, \quad \mathsf{Q}\{\Lambda\} = 1, \tag{1.1}$$

then we say that P and Q are singular (symbolically $\mathsf{P}\perp\mathsf{Q}$). Arbitrarily chosen distributions P and Q need be neither equivalent nor singular, as, for example, the uniform distributions on the intervals $(0,1)$ and $(0,2)$. Nevertheless, as proved in Section 3 of this paper, any two normal distributions on a Borel field which is generated by an arbitrary stochastic process cannot be other than equivalent or singular. This means that the normal distributions of an arbitrary process can be divided into classes in such a manner that every two distributions are either equivalent or singular[1], depending on whether they belong to the same class or to different classes. In order to verify

Collected Works of Jaroslav Hájek – With Commentary
Edited by M. Hušková, R. Beran and V. Dupač
Published in 1998 by John Wiley & Sons Ltd.

these hypotheses, we find the following important result. If a given system of allowable normal distributions of the process at hand is such that it does not contain singular distributions, then all of these distributions are equivalent, and consequently can be given by probability densities, and any one of them can be dominant.

Theorem 3 also indicates a criterion which enables us to establish whether two normal distributions are singular or equivalent. This criterion is utilized in Section 4 in the solution of two examples.

2 THE RATIO AND THE J-DIVERGENCE OF TWO n-DIMENSIONAL NORMAL DENSITIES

Let us examine the following two special normal densities:

$$p(x_1, \ldots, x_n) = (2\pi)^{-\frac{n}{2}} e^{-\frac{1}{2}\sum_{i=1}^n x_i^2} \tag{2.1}$$

$$q(x_1, \ldots, x_n) = (2\pi)^{-\frac{n}{2}} e^{-\frac{1}{2}\sum_{i=1}^n \left[\left(\frac{x_i-\mu_i}{\sigma_i}\right)^2 - \log\sigma_i^2\right]}. \tag{2.2}$$

For

$$\log\frac{p}{q} = \frac{1}{2}\sum_{i=1}^n \left[\left(\frac{x_i-\mu_i}{\sigma_i}\right)^2 - x_i^2 - \log\sigma_i^2\right]$$

we will find the mean and the dispersion with respect to both densities p and q. The above mentioned means and dispersions will be denoted by E_p, E_q, var_p, var_q. By carrying out a simple calculation we will get:

$$\mathsf{E}_p\left\{\log\frac{p}{q}\right\} = \frac{1}{2}\sum_{i=1}^n \left[\frac{1+\mu_i^2}{\sigma_i^2} - 1 - \log\sigma_i^2\right], \tag{2.3}$$

$$\mathsf{E}_q\left\{\log\frac{p}{q}\right\} = \frac{1}{2}\sum_{i=1}^n \left[1 - \mu_i^2 - \sigma_i^2 - \log\sigma_i^2\right], \tag{2.4}$$

$$\mathsf{var}_p\left\{\log\frac{p}{q}\right\} = \frac{1}{2}\sum_{i=1}^n \left[\left(\frac{1}{\sigma_i^2} - 1\right)^2 + 2\frac{\mu_i^2}{\sigma_i^4}\right], \tag{2.5}$$

$$\mathsf{var}_q\left\{\log\frac{p}{q}\right\} = \frac{1}{2}\sum_{i=1}^n \left[(\sigma_i^2 - 1)^2 + 2\mu_i^2\sigma_i^2\right]. \tag{2.6}$$

[1] Singular distributions can be distinguished without error by only realizing and using one process with a probability of 1; thus, in non-degenerate statistical problems they are not encountered together.

Now we will determine the J-divergence J of the densities p and q by the relation

$$J = \mathsf{E}_p\left\{\log\frac{p}{q}\right\} - \mathsf{E}_q\left\{\log\frac{p}{q}\right\} \tag{2.7}$$

(see Hájek (1958), Jeffreys (1948) or Kullback and Leibler (1951)).

By using (2.3) and (2.4), we will get, for the special densities (2.1) and (2.2),

$$J = \frac{1}{2}\sum_{i=1}^{n}\left[\frac{(\sigma_i^2-1)^2}{\sigma_i^2} + \mu_i^2\left(1+\frac{1}{\sigma_i^2}\right)\right]. \tag{2.8}$$

Lemma 2.1. *For any given $\epsilon > 0$ we can find a number K_ϵ possessing the following property: if n is an arbitrary natural number and $p(x_1,\ldots,x_n)$ and $q(x_1,\ldots,x_n)$ are arbitrary (regular) n-dimensional normal densities, whose J-divergence given by (2.7) satisfies the condition $J > K_\epsilon$, then we can find an event Λ such that*

$$\int_\Lambda p(x_1,\ldots,x_n)\,\mathrm{d}x_1\cdots\mathrm{d}x_n < \epsilon, \qquad \int_\Lambda q(x_1,\ldots,x_n)\,\mathrm{d}x_1\cdots\mathrm{d}x_n > 1-\epsilon. \tag{2.9}$$

First, we will carry out the proof for densities (2.1) and (2.2). We will introduce the following events:

$$A: \quad \log\frac{p}{q} < \mathsf{E}_p\left\{\log\frac{p}{q}\right\} - \frac{1}{2}J = \mathsf{E}_q\left\{\log\frac{p}{q}\right\} + \frac{1}{2}J, \tag{2.10}$$

$$B_i: \quad x_i^2 > \sigma_i, \quad i=1,\ldots,n, \tag{2.11}$$

$$C_i: \quad (x_i-\mu_i)^2 < \sigma_i, \quad i=1,\ldots,n. \tag{2.12}$$

(Two expressions for the right hand side of inequality (2.10) are obtained from (2.7)). According to the following considerations, we can write

$$\mathsf{P}\{A\} \le \mathsf{var}_p\frac{\left\{\log\frac{p}{q}\right\}}{\left(\frac{1}{2}J\right)^2}, \quad \mathsf{Q}\{A\} \ge 1-\mathsf{var}_q\frac{\left\{\log\frac{p}{q}\right\}}{\left(\frac{1}{2}J\right)^2}, \tag{2.13}$$

$$\mathsf{P}\{B_i\} = 2\left[1-\Phi\left(\sqrt{\sigma_i}\right)\right], \quad \mathsf{Q}\{B_i\} \ge 2\left[1-\Phi\left(\frac{1}{\sqrt{\sigma_i}}\right)\right], \quad i=1,\ldots,n, \tag{2.14}$$

$$\mathsf{P}\{C_i\} \le 1-2[1-\Phi(\sqrt{\sigma_i})], \quad \mathsf{Q}\{C_i\} = 1-2\left[1-\Phi\left(\frac{1}{\sqrt{\sigma_i}}\right)\right], \quad i=1,\ldots,n, \tag{2.15}$$

where $\mathsf{P}\{\cdot\}$ and $\mathsf{Q}\{\cdot\}$ denote probabilities given by the densities p and q, and $\Phi(\lambda) = (2\pi)^{-\frac{1}{2}}\int_{-\infty}^{\lambda} e^{-\frac{x^2}{2}}\,\mathrm{d}x$. Actually, inequalities (2.13) follow from (2.10) and from the Chebyshev inequality. Equation (15a) follows immediately from (2.11) and (2.1). In deriving inequalities (15b) it is necessary to use the obvious inequality

$$\int_{x^2>a} e^{-\frac{(x-\mu)^2}{2\sigma^2}}\,\mathrm{d}x \ge \int_{x^2>a} e^{-\frac{x^2}{2\sigma^2}}\,\mathrm{d}x;$$

it also follows directly from (2.11) and (2.2). Relations (2.15) are obtained analogously to (2.14).

Now let us fix an arbitrary $\epsilon > 0$, and let us find a positive λ_ϵ such that

$$2[1 - \Phi(\lambda_\epsilon)] < \epsilon, \quad 2\left[1 - \Phi\left(\frac{1}{\lambda_\epsilon}\right)\right] > 1 - \epsilon. \tag{2.16}$$

If for some $i = 1, \ldots, n$, inequality $\sigma_i^{1/2} > \lambda_\epsilon$ or $\sigma_i^{-1/2} > \lambda_\epsilon$ holds, then from (2.14) and (2.15) it follows that relation (2.9) is satisfied for $\Lambda = B_i$, or that $\Lambda = C_i$. On the other hand, if

$$\frac{1}{\lambda_\epsilon^4} \leq \sigma_i^2 \leq \lambda_\epsilon^4, \quad i = 1, \ldots, n, \tag{2.17}$$

then from (2.5), (2.6) and (2.8) we can easily obtain inequality

$$J \geq \frac{1}{2\lambda_\epsilon^4}\left[\mathsf{var}_p\left\{\log\frac{p}{q}\right\} + \mathsf{var}_q\left\{\log\frac{p}{q}\right\}\right]. \tag{2.18}$$

Applying this result in (2.13), we obtain

$$\mathsf{P}\{A\} \leq \frac{8\lambda_\epsilon^4}{J}, \quad \mathsf{Q}\{A\} \geq 1 - \frac{8\lambda_\epsilon^4}{J}. \tag{2.19}$$

This means that for $\Lambda = A$, relation (2.9) is satisfied as soon as

$$J \geq \frac{8\lambda_\epsilon^4}{\epsilon}. \tag{2.20}$$

Thus, the lemma is proven for densities of the form (2.1) and (2.2). But as we will now show, any pair of densities can be brought to this form by appropriate transformations. If the densities p and q are determined by non-singular covariance matrices C and D, and by vectors of the means c and d, then, evidently,

$$\begin{aligned} p(x) &= (2\pi)^{-\frac{n}{2}} e^{-\frac{1}{2}\left[(x-c)' C^{-1}(x-c) + \log|C|\right]}, \\ q(x) &= (2\pi)^{-\frac{n}{2}} e^{-\frac{1}{2}\left[(x-d)' D^{-1}(x-d) + \log|D|\right]}, \end{aligned}$$

where $x = (x_1, \ldots, x_n)'$, C^{-1} and D^{-1} are inverse matrices, and $|C|$ and $|D|$ are determinants. Now let us take a matrix V such that $V'V = C$ (Jacobi's transformation) and an orthogonal matrix U such that $U'V'D^{-1}VU$ is a diagonal matrix. We will determine vectors y and u by the equations

$$x - c = VUy, \quad d - c = VU\mu. \tag{2.21}$$

Then, on the basis of the indicated properties of the matrices U and V, it is evident that

$$(x-c)'\,C^{-1}(x-c) = y'U'V'C^{-1}VUy = y'y = \sum_{i=1}^{n} y_i^2, \tag{2.22}$$

$$(x-d)'\,D^{-1}(x-d) = (y-\mu)'U'V'C^{-1}VU(y-\mu) = \sum_{i=1}^{n} \frac{(y_i-\mu_i)^2}{\sigma_i^2}, \tag{2.23}$$

where σ_i^2 are the diagonal elements of the matrix $U'V'D^{-1}VU$. It follows from (2.22) and (2.23) that the densities p and q have the form of (2.1) and (2.2) as functions of the one-to-one mapping of the vector $y = (y_1, \ldots, y_n)'$. Thus, the lemma is proved completely.

We present the following assertion without proof.

Lemma 2.2. *If the distribution of the vector $(x_1, \ldots, x_n)$ is normal (regular or singular), then the equation*

$$x_n = \sum_{i=1}^{n-1} c_i\, x_i + c_0 \tag{2.24}$$

holds with probability 0 or 1. Here c_i are arbitrary constants.

3 THEOREM

Theorem. *Every two normal distributions $\mathsf{P}\{\cdot\}$ and $\mathsf{Q}\{\cdot\}$ of a stochastic process $\{x_t,\, t \in T\}$ are either equivalent or singular*[1]*, depending on whether their J-divergence, as determined in Hájek (1958), is finite or infinite.*

Proof. From Theorem 2 in Hájek (1958) it follows that the J-divergence of P and Q is equal to the supremum of the J-divergences of finite-dimensional distributions $\mathsf{P}_{t_1,\ldots,t_n}$ and $\mathsf{Q}_{t_1,\ldots,t_n}$ of vectors $\{x_{t_1}, \ldots, x_{t_n}\}$. Symbolically:

$$J_T = \sup_{t_1,\ldots,t_n \in T} J_{t_1,\ldots,t_n}. \tag{3.1}$$

If $J_T < \infty$, then $\mathsf{P} \equiv \mathsf{Q}$, because according to the definition in Hájek (1958), only equivalent distributions have a finite J-divergence. If $J_T = \infty$, then as evident from (3.1), we can select $t_1, \ldots, t_n$ such that $J_{t_1,\ldots,t_n}$ is greater than any previously given number. Hence, and from Lemma 2.1, it follows that, if all finite-dimensional distributions $\mathsf{P}_{t_1,\ldots,t_n}$ and $\mathsf{Q}_{t_1,\ldots,t_n}$ are regular, then for any m we can find an event A_m such that

$$\mathsf{P}\{A_m\} < 2^{-m}, \quad \mathsf{Q}\{A_m\} > 1 - 2^{-m}, \quad m = 1, 2, \ldots. \tag{3.2}$$

But this means that $\mathsf{P}\perp\mathsf{Q}$, because equations (1.1) are satisfied for $A = \bigcap_{k=1}^{\infty}\bigcup_{m=k}^{\infty} A_m$.

It remains to examine the case when the finite-dimensional distributions are singular – more precisely, when they are singular relative to the n-dimensional Lebesgue measure. We will assume that $P_{t_1,\dots,t_n}$ is singular. Then some x_{t_1}, assuming x_{t_n}, is a linear function of the remaining random variables; more exactly, with the $P_{t_1,\dots,t_n}$-probability equal to one,

$$x_{t_n} = \sum_{i=1}^{n-1} c_i\, x_{t_i} + c_0, \quad [\mathsf{P}_{t_1,\dots,t_n}]. \tag{3.3}$$

But on the other hand, because of Lemma 2.2, the $\mathsf{Q}_{t_1,\dots,t_n}$-probability of (3.3) is either 0 or 1. If it is equal to 0, then obviously $\mathsf{P}_{t_1,\dots,t_n}\perp\mathsf{Q}_{t_1,\dots,t_n}$, and, moreover, $\mathsf{P}\perp\mathsf{Q}$. If it is equal to 1, then $\{x_{t_1},\dots,x_{t_{n-1}}\}$ is the sufficient statistic for the pair $P_{t_1,\dots,t_n}$, $\mathsf{Q}_{t_1,\dots,t_{n-1}}$, so that according to Theorem 4.1 and Section 5 in Černý (1957), $J_{t_1,\dots,t_n} = J_{t_1,\dots,t_{n-1}}$. This means that only the following two cases can occur:

(a) for certain $t_1,\dots,t_n$, $\mathsf{P}_{t_1,\dots,t_n}\perp\mathsf{Q}_{t_1,\dots,t_n}$,

(b) for each $t_1,\dots,t_n \in T$ either $J_{t_1,\dots,t_n} = J_{t'_1,\dots,t'_p}$,

where $P_{t'_1,\dots,t'_p}$ are regular, or both distributions $\mathsf{P}_{t_1,\dots,t_n}$ and $\mathsf{Q}_{t_1,\dots,t_n}$ are concentrated at one point, i. e.,

$$\mathsf{P}_{t_1,\dots,t_n}\{x_{t_i} = \mu_{t_i},\ i = 1,\dots,n\} = \mathsf{Q}_{t_i,\dots,t_n}\{x_{t_i} = \mu_{t_1},\ i = 1,\dots,n\} = 1.$$

(In the latter case, of course, $J_{t_1,\dots,t_n} = 0$ so that such $t_1,\dots,t_n$ can be ignored, but nevertheless, $\sup J_{t_1,\dots,t_n} = \infty$ will remain.)

If case (a) occurs, $\mathsf{P}\perp\mathsf{Q}$ is obvious. If case (b) occurs, we can find an arbitrarily large $J_{t_1,\dots,t_n}$, even if we restrict ourselves to such $t_1,\dots,t_n$ for which $\mathsf{P}_{t_1,\dots,t_n}$ and $Q_{t_1,\dots,t_n}$ are regular, so that we can again use Lemma 2.1 as in the first part of the proof. Thus the theorem is proved. □

4 APPLICATIONS

The criterion of equivalence of two normal distributions, presented in the preceding theorem, can be applied directly to the solution of several problems.

Example 1. C h a n g e o f s c a l e. Let two normal distributions $\mathsf{P}\{\cdot\}$ and $\mathsf{Q}\{\cdot\}$ of the process $\{x_t,\, t \in T\}$ be given with the following covariances and means:

$$\mathsf{E}_p\{x_t\} = \mu_t, \quad \mathrm{Cov}_p\{x_t, x_s\} = \sigma_{ts},$$
$$\mathsf{E}_q\{x_t\} = \mu_t, \quad \mathrm{Cov}_q\{x_t, x_s\} = \lambda\,\sigma_{ts},\ \lambda > 0.$$

Let us take $t_1, \dots, t_n \in T$. If the rank of the matrix $\|\sigma_{t_i t_j}\|$ is m, we can find sufficient statistics in the form of m variables $y_1, \dots, y_m$ which are independent and normally distributed with parameters $(0,1)$ or $(0,\lambda)$, depending on whether P or Q holds. From here and from (2.8) it follows that

$$J_{t_1,\dots,t_n} = \frac{1}{2} m \frac{(\lambda-1)^2}{\lambda}.$$

Thus, for $\lambda \neq 1$, the following statement holds. Distributions P and Q are singular or regular, depending on whether or not it is possible to find $t_1, \dots, t_n$ for any n such that the distribution $\mathsf{P}_{t_1,\dots,t_n}$ (or matrix $\|\sigma_{t_i t_j}\|$) is regular. In particular, it is easy to see from this that we will obtain a singular distribution for the Wiener process when $\lambda \neq 1$, as it is shown by Cameron and Martin (1947).

Example 2. (Petr Mandl) The drift of the Gauss–Markov process. Let us examine the normal distributions $\mathsf{P}\{\cdot\}$ and $\mathsf{Q}\{\cdot\}$ of the process $\{x_t,\ 0 \le t \le T\}$, which are determined by the following covariances and means:

$$\begin{aligned} \mathsf{E}_p\{x_t\} &= 0, \quad \mathrm{cov}_p\{x_t, x_s\} = \sigma_t \sigma_s \frac{h_t}{h_s}, \quad t \le s, \\ \mathsf{E}_q\{x_t\} &= \mu_t, \quad \mathrm{cov}_q\{x_t, x_s\} = \sigma_t \sigma_s \frac{h_t}{h_s}, \quad t \le s \end{aligned}$$

where $\sigma_t > 0$ for $t > 0$, $|h_t|$ increases, and $\mu_0 = 0$ if $\sigma_0 = 0$. The J-divergence of the distributions $\mathsf{P}_{t_1,\dots,t_n}$ and $\mathsf{Q}_{t_1,\dots,t_n}$ can be found by means of the substitution:

$$y_i = \left(\frac{x_{t_i}}{\sigma_{t_i}} - \frac{h_{t_i-1}}{h_{t_i}} \cdot \frac{x_{t_i-1}}{\sigma_{t_i-1}}\right) \left[1 - \left(\frac{h_{t_i-1}}{h_{t_i}}\right)^2\right]^{-\frac{1}{2}},$$

where for $\sigma_0 = 0$ we set $x_0/\sigma_0 = 0$. Using formulas (2.8) and (3.1) we will get

$$J = \sup_{0 \le t_1,\dots,t_n \le T} \sum_{k=1}^{n} \frac{\left(h_{t_i} \frac{\mu_{t_i}}{\sigma_{t_i}} - h_{t_i-1} \frac{\mu_{t_i-1}}{\sigma_{t_i-1}}\right)^2}{h_{t_i}^2 - h_{t_i-1}^2},$$

where $t_0 = 0$. If $J < \infty$, then J is equal to the Hellinger integral

$$\int_0^T \frac{\left(\mathrm{d}h_t \frac{\mu_t}{\sigma_t}\right)^2}{\mathrm{d}h_t^2},$$

which is the necessary and sufficient condition for distributions $\mathsf{P}\{\cdot\}$ and $\mathsf{Q}\{\cdot\}$ to be equivalent. In the case of the Wiener process, $h_t = \sigma_t = \sqrt{t}$ and, consequently, the integral has the form

$$\int_0^T \frac{(\mathrm{d}\mu_t)^2}{\mathrm{d}t}.$$

Hence we can also conclude that the drift of the Wiener process is represented precisely by those functions μ_t for which

$$\int_0^1 \frac{(\mathrm{d}\mu_t)^2}{\mathrm{d}t}$$

exists. For another condition see Maruyama (1950).

In the case of the stationary Gauss–Markov process, $\sigma_t = \sigma$, $h_t = a^t$ and, consequently, the integral has the form

$$\sigma^{-2} \int_0^T \frac{(\mathrm{d}a^t \mu_t)^2}{\mathrm{d}a^{2t}}.$$

Note. If $0 < |\rho(x_t, x_s)| < 1$ for $t \neq s$, the covariance function of the Gauss–Markov process can always be written as shown, because the random variables x_{t_1} and x_{t_3} for $t_1 < t_2 < t_3$ and under the condition that $x_{t_2} = x$ are linearly independent. Consequently, for a particular correlation coefficient we must have

$$0 = \rho_{13,2} = (\rho_{13} - \rho_{12}\rho_{23})(1 - \rho_{12}^2)^{-\frac{1}{2}}(1 - \rho_{23}^2)^{-\frac{1}{2}}.$$

Hence, $\rho(x_{t_1}, x_{t_3}) = \rho(x_{t_1}, x_{t_2})\,\rho(x_{t_2}, x_{t_3})$. Consequently we can set $h_t = \rho(x_0, x_t)^{-1}$, where $|h_t|$ increases.

References

Cameron R. H. and Martin W. T. (1947). The behavior of measure and measurability under change of scale in Weiner space. *Bull. Amer. Math. Soc. 53*, 130–137.

Černý I. (1957). Das Hellinger Integral, *Časopis Pěst. Mat. 82*, 24–43.

Hájek J. (1958). A property of J–divergence of marginal probability distributions. *Czechoslovak Math. J. 8 (83)* 460–463.

Jeffreys H. (1948). *Theory of Probability*. Oxford.

Kullback S. and Leibler R. A. (1951). On information and sufficiency. *Ann. Math. Statist. 22*, 79–86.

Maruyama G. (1950). Notes on Wiener integrals. *Kōdai Math. Sem. Rep. 1950*, 41–44.

CHAPTER 10

On the Distribution of Some Statistics in the Presence of Intraclass Correlation

Časopis Pěst. Mat. 83 (1958), 327–329 (in Czech).
English translation *Selected Transl. Math. Statist. Prob. 2* (1962), 75–77.
Translated by L. Rosenblatt.

Summary. In this article it is shown that the probability density (1.6) corresponds to an n-dimensional normal distribution with constant variance σ^2 and correlation coefficient ρ. From its form it follows that the maximum-likelihood estimates of the regression coefficients β_j in the relation $\mathsf{E}y_i = a + \sum \beta_j(x_{ji} - \overline{x}_j)$ do not depend on ρ. Some results contained in Halperin (1951) and Walsh (1947) are derived simultaneously.

In a very simple way it is shown that the presence of intraclass correlation does not have influence on the distribution of certain important statistics. The results obtained include results arrived at by Halperin (1951).

Theorem. *Given normal random variables $y_1, y_2, \ldots, y_n$ with arbitrary mean values. We assert that the following two hypotheses H_1 and H_2 about variance and correlation coefficients*

$$\begin{aligned} H_1: \quad \operatorname{var} y_i &= \sigma^2 & & i, j = 1, \ldots, n; \\ \rho_{ij} &= \rho & & i \neq j, -\frac{1}{n-1} < \rho < 1; \end{aligned} \tag{1.1}$$

$$\begin{aligned} H_2: \quad \operatorname{var} y_i &= (1-\rho)\,\sigma^2 & & i, j = 1, \ldots, n; \\ \rho_{ij} &= 0 & & i \neq j, \end{aligned} \tag{1.2}$$

Collected Works of Jaroslav Hájek - With Commentary
Edited by M. Hušková, R. Beran and V. Dupač
Published in 1998 by John Wiley & Sons Ltd.

lead to the same distribution of the vector $(y_1 - \overline{y}, \dots, y_n - \overline{y})$, *where* $\overline{y} = \frac{1}{n}\sum_{i=1}^{n} y_i$, *and therefore of all statistics formed from* $s = s(y_1 - \overline{y}, \dots, y_n - \overline{y})$; *besides, for both hypotheses the average* $\overline{y}$ *is independent of the vector* $(y_1 - \overline{y}, \dots, y_n - \overline{y})$.

Proof. Let the normal random variables $z_1, \dots, z_n$ behave according to the hypothesis H_2, i. e., they are independent with variance $(1-\rho)\,\sigma^2$. By direct calculation it is easy to show that then the random variables

$$y_i = z_i - \frac{1}{n}\left(1 - \sqrt{\frac{1+(n-1)\,\rho}{1-\rho}}\right)\sum_{i=1}^{n}(z_i - \mathsf{E}z_i) \tag{1.3}$$

behave according to hypothesis H_1, having variance σ^2 and correlation coefficients $\rho_{ij} = \rho$; besides, the mean values are equal, $\mathsf{E}y_i = \mathsf{E}z_i$. From (1.4) it follows that $y_i - \overline{y} = z_i - \overline{z}$, so that $(y_1 - \overline{y}, \dots, y_n - \overline{y}) = (z_1 - \overline{z}, \dots, z_n - \overline{z})$ whence the first assertion follows. Further, from (1.4) and from $\mathsf{E}y_i = \mathsf{E}z_i$ it follows that

$$\overline{y} - \mathsf{E}\overline{y} = \sqrt{\frac{1+(n-1)\,\rho}{1-\rho}}\,(\overline{z} - \mathsf{E}\overline{z}), \tag{1.4}$$

so that the independence of $\overline{z}$ and $(z_i - \overline{z}, \dots, z_n - \overline{z})$, which is well known, implies the independence of $\overline{y}$ and $(y_1 - \overline{y}, \dots, y_n - \overline{y})$. Thus the proof is completed. □

The theorem enables us to carry over the results worked out for the usual hypothesis H_2 signifying independence to the somewhat more general case described by the hypothesis H_1 signifying the presence of intraclass correlation, i. e., correlation for all pairs of random variables.

Example 1. Special cases of statistics of the form $s = s(y_1 - \overline{y}, \dots, y_n - \overline{y})$ are the best linear estimates b_j of the regression coefficients β_j in the expression

$$\mathsf{E}y_i = \mu + \sum_{j=1}^{k}\beta_j(x_{ji} - \overline{x}_j), \tag{1.5}$$

further the residual variance $\sum_{i=1}^{n}\left(y_i - \overline{y} - \sum b_j(x_{ji} - \overline{x}_j)\right)^2$, as well as the vector whose components are just named statistics, etc. The presence of intraclass correlation has on these and similar statistics equal effect as changing the variance from σ^2 to $(1-\rho)\,\sigma^2$ and keeping the assumption of independence.

Example 2. For a random choice of n from N elements without replacement we have $\rho = -\frac{1}{N-1}$. With the assumption that the distribution of the observations is roughly normal it follows from our theorem that Student's ratio

$$t = \frac{\overline{y} - \overline{\overline{y}}}{\sqrt{\frac{\left(1-\frac{n}{N}\right)}{n(n-1)} \sum_{i=1}^{n} (y_i - \overline{y})^2}}$$

where $\overline{y}$ and $\overline{\overline{y}}$ are the average value of the sampled and of all elements, respectively, has roughly Student's distribution with $n-1$ degrees of freedom rather than with $\frac{n-1}{1-n/N}$, as is recommended in Cramér's (1946) book, p. 523. We arrive at this conclusion if we assume that the values $y_1, \ldots, y_N$ form a random sample from the normal distribution and the values $y_1, \ldots, y_n$ a part of this sample.

Remark. Representation (1.4), which is somewhat simpler than the one used by Walsh (1947), permits us immediately to find the density of the observations $(y_1, \ldots, y_n)$ when H_1 is valid. Indeed the density of the vector $(z_1, \ldots, z_n)$ is equal to

$$K_1\, e^{-\frac{1}{2}\sum_{i=1}^{n} \frac{(z_i - \mathsf{E}z_i)^2}{(1-\rho)\,\sigma^2}} = K_1\, e^{-\frac{1}{2}\left(\sum_{i=1}^{n} \frac{(z_i - \mathsf{E}z_i - \overline{z} + \mathsf{E}\overline{z})^2}{(1-\rho)\,\sigma^2} + \frac{n(\overline{z} - \mathsf{E}\overline{z})^2}{(1-\rho)\,\sigma^2}\right)}.$$

From the linear transformation (1.4), from which follows $z_i - \mathsf{E}z_i - \overline{z} + \mathsf{E}\overline{z} = y_i - \mathsf{E}y_i - \overline{y} + \mathsf{E}\overline{y}$, and (1.5), we arrive at the density of $(y_1, \ldots, y_n)$ in the form

$$K_2\, e^{-\frac{1}{2}\left[\sum_{i=1}^{n} \frac{(y_i - \overline{y} - \mathsf{E}y_i + \mathsf{E}\overline{y})^2}{(1-\rho)\,\sigma^2} + \frac{n}{\sigma^2[1+(n-1)\,\rho]}(\overline{y} - \mathsf{E}\overline{y})^2\right]}. \tag{1.6}$$

The determinant of the covariance matrix under hypothesis H_1 equals

$$\sigma^{2n}\,[1 + (n-1)\,\rho]\,(1-\rho)^{n-1},$$

so that

$$K_2 = (2\pi)^{-\frac{n}{2}}\,\sigma^{-n}(1-\rho)^{-\frac{n-1}{2}}\,[1 + (n-1)\,\rho]^{-\frac{1}{2}}.$$

From the density (1.6) it follows that the maximum likelihood estimates of the regression coefficients in (1.5) are the same for both hypotheses H_1 and H_2. Thus Halperin is wrong when he claims in Halperin (1951) that calculating maximum likelihood estimates under hypothesis H_1 is disproportionately complicated. In fact, the estimates (3.3) introduced in Halperin (1951) are actually maximum likelihood estimates.

REFERENCES

Cramér H. (1946). *Mathematical Methods of Statistics*. Princeton Univ. Press, Princeton, N.J.

Halperin M. (1951). Normal regression theory in the presence of intra-class correlation. *Ann. Math. Statist.* *22*, 573–580.

Walsh J. E. (1947). Concerning the effect of intraclass correlation on certain significance tests. *Ann. Math. Statist.* *18*, 88–96.

CHAPTER 11

On the Theory of Ratio Estimates

Apl. Mat. 3 (1958), 384–398.
Reproduced by permission of Hájek family: Al. Hájková, H. Kazárová and Al. Slámová.

Estimated variances, yielded by the large sample approach, are adjusted by a proportional regression approach; subsequently, under the assumption of normality, exact statements on confidence intervals are arrived at. The chapter deals, too, with complex types of ratio estimates, as well as with modifications needed, when stratification, multiple stages, or some special methods of first-stage sampling are present.

1 INTRODUCTION

Ratio estimates belong to the most efficient techniques of modern sample survey practice. Some of them are very ingenious (see, for example Hansen et al. (1953), vol. I, p. 413), and make use of very elaborate supplementary information. Nevertheless, the present theory of ratio estimates has not gone beyond approximating their variances. At the same time, the validity of estimated variances is inferred from the discrepancy between the confidence interval, yielded by them, and Fieller's confidence interval. We shall show, however, that, under conditions typical for sample surveys, the Fieller confidence interval is less advantageous than the usual one, because it is longer for any sample outcome, and despite this, covers the true value with a smaller probability. In addition, Fieller's device cannot be used for ratio estimates of more complex type. Thus the Fieller method, though very useful outside the sample surveys domain (because of the generality of conditions under which it works), cannot be considered as preferable in sample survey conditions.

Let us select n elements from a population of N elements by simple random sampling without replacement, observe values y_i, x_i, z_i of some variables, and

Collected Works of Jaroslav Hájek – With Commentary
Edited by M. Hušková, R. Beran and V. Dupač
Published in 1998 by John Wiley & Sons Ltd.

estimate the total

$$Y = \sum_{i=1}^{N} y_i, \tag{1.1}$$

under the supposition that we know the totals[1] and some subtotals of related values x_i and z_i, $i = 1, \ldots, N$. Now, let each value y_i, x_i, z_i split into a sum of values y_{aci}, x_{aci}, z_{aci}, respectively, so that

$$y_i = \sum_a \sum_c y_{aci}, \quad x_i = \sum_a \sum_c x_{aci}, \quad z_i = \sum_a \sum_c z_{aci}, \quad a \in A, \ c \in C. \tag{1.2}$$

As a rule, y_{aci}, x_{aci}, z_{aci} will refer to a two-way classification of conditions within the ith element; for example, y_{aci} may be the number of workers in the ath age–sex group, within the cth rural–urban part of the ith country; x_{aci} and z_{aci} may refer to the present and last census number of all persons, respectively, again in the ath age–sex group within the cth rural–urban part of the ith county.

Symbols Y, Y_a, Y_{ac}, X_a, X_{ac}, Z_c and y, y_a, y_c, y_{ac}, x_a, x_{ac}, z_c, will denote population and sample totals, respectively, extended over all subscript letters that are not indicated. For example,

$$X_a = \sum_c \sum_{i=1}^{N} x_{aci}, \quad y_{ac} = \sum_i y_{aci}, \text{ etc.,}$$

where $\sum_i$ denotes the sum extended over subscripts of sampled elements, $i = i_1, \ldots, i_n$.

We shall consider the following three types of ratio estimates:

$$X \frac{y}{x}, \tag{1.3–1}$$

$$\sum_a X_a \frac{y_a}{x_a}, \tag{1.3–2}$$

$$\sum_a X_a \frac{\sum_c Z_c y_{ac}}{\sum_c Z_c \frac{x_{ac}}{z_c}}. \tag{1.3–3}$$

The estimates (1.3–1) and (1.3–2) are obviously special cases of the estimate (1.3–3).

2 LARGE-SAMPLE APPROACH

The large-sample approach rests upon approximating estimates (1.3) by a linear function of the sample totals y, x, y_a, x_a, x_c, z_c, y_{ac} involved; the latter

[1] If we only estimate the ratio of totals, $\frac{Y}{X}$, we need not know the total X.

function can be obtained from the usual Taylor expansion about corresponding expectations. Carrying out this operation, we get the following approximate expressions for estimates (1.3):

$$Y + \frac{N}{n}\left(y - \frac{Y}{X}x\right), \tag{2.1–1}$$

$$Y + \frac{N}{n}\sum_a \left(y_a - \frac{Y_a}{X_a}x_a\right), \tag{2.1–2}$$

$$Y + \frac{N}{n}\sum_a\sum_c \left[y_{ac} - \frac{Y_{ac}}{Z_c}z_c - \frac{Y_a}{X_a}\left(x_{ac} - \frac{X_{ac}}{Z_c}z_c\right)\right]. \tag{2.1–3}$$

Estimated variances of 'concomitant' linear estimates (2.1) can be calculated in the usual way, with the only modification that

$$y_i - \frac{Y}{X}x_i, \tag{2.2–1}$$

$$\sum_a \left(y_{ai} - \frac{Y_a}{X_a}x_{ai}\right), \tag{2.2–2}$$

$$\sum_a\sum_c \left[y_{aci} - \frac{Y_{ac}}{Z_c}z_{ci} - \frac{Y_a}{X_a}\left(x_{aci} - \frac{X_{ac}}{Z_c}z_{ci}\right)\right], \tag{2.2–3}$$

are treated as single observations, and that sample ratios $\frac{y_a}{x_a}$, $\frac{y_{ac}}{z_c}$, $\frac{x_{ac}}{z_c}$ are substituted for population ratios $\frac{Y_a}{X_a}$, $\frac{Y_{ac}}{Z_c}$, $\frac{X_{ac}}{Z_c}$, respectively. In this manner we get estimated variances of estimates (2.1) in the general form

$$\frac{(N-n)\,n}{N(n-1)}\sum_i \Delta_i^2 \tag{2.3}$$

where, for individual estimates (1.3),

$$\Delta_i = \frac{N}{n}\left(y_i - \frac{y}{x}x_i\right) \tag{2.4–1}$$

$$\Delta_i = \sum_a \frac{N}{n}\left(y_{ai} - \frac{y_a}{x_a}x_{ai}\right) \tag{2.4–2}$$

$$\Delta_i = \sum_a\sum_c \frac{N}{n}\left[y_{aci} - \frac{y_{ac}}{z_c}z_{ci} - \frac{y_a}{x_a}\left(x_{aci} - \frac{x_{ac}}{z_c}z_{ci}\right)\right] \tag{2.4–3}$$

respectively. See, too, equations (3.2).

We shall not enlarge upon well-known asymptotical properties of estimates (1.3) and their estimated variances (2.3); reference is made to the paper Madow (1948). The main point of this section was to show how to get estimated variances in a simple form even for complex ratio estimates.

Remark 2.1. Inserting (2.4–3) into (2.3), we obtain the estimated variance which is much simpler but otherwise equivalent to that recommended in Hansen et al. (1953), vol. II, p. 226.

3 PROPORTIONAL REGRESSION APPROACH

The expression (2.2) suggests that the corresponding ratio estimates (1.3) are fitted to the following regression models:

$$\mathsf{E}(y_i \mid x_i) = \frac{Y}{X}\, x_i, \tag{3.1–1}$$

$$\mathsf{E}\left(y_{ai} \mid x_{a'i},\ a' \in A\right) = \frac{Y_a}{X_a}\, x_{ai}, \quad a \in A, \tag{3.1–2}$$

$$\begin{aligned} &\mathsf{E}(y_{aci} \mid x_{a'c'i},\ z_{c'i},\ a' \in A,\ c' \in C) \\ &= \frac{Y_{ac}}{Z_c}\, z_{ci} + \frac{Y_a}{X_a}\left(x_{aci} - \frac{X_{ac}}{Z_c}\, z_{ci}\right), \quad a \in A,\ c \in C. \end{aligned} \tag{3.1–3}$$

Here, and in what follows, $\mathsf{E}(y \mid x_\alpha,\ \alpha \in H)$ and $\mathsf{var}(y \mid x_\alpha,\ \alpha \in H)$ denote the conditional expectation and variance of y with respect to a family of variables $(x_\alpha,\ \alpha \in H)$. Linear regression, which passes through the origin, we shall call, for brevity, proportional regression.[2] All regressions given by (3.1) are proportional regressions. Relation (1.3–2) implies that y_{ai} given x_{ai} is linearly independent of $x_{a'i}$, $a' \neq a$. Relation (1.3–3) means that y_{ai} and x_{ai} are both proportional to z_{ci}, and moreover that the residual $y_{aci} - \frac{Y_{ac}}{Z_c}(z_{ci} - \overline{z}_c)$ is proportional to the residual $x_{aci} - \frac{X_{ac}}{Z_c}(z_{ci} - \overline{z}_c)$.

When sampling is from finite populations, (3.1) must, in general, be reinterpreted (see Remark 3.1). Let us consider, therefore, the parallel problem when the sample consists of n independent observations from a multidimensional distribution governed by one of the relations (3.1). Under this condition, the conditional expectations of ratio estimates (1.3), for any fixed x_{aci} and z_{aci}, will equal Y. It suffices to show it for the most general estimate (1.3–3):

$$\begin{aligned} &\mathsf{E}\left(\sum_a X_a \frac{\sum_c Z_c \frac{y_{ac}}{z_c}}{\sum_c Z_c \frac{x_{ac}}{z_c}} \,\middle|\, x_{a'c'1}, \ldots, x_{a'c'n}, z_{c'1}, \ldots, z_{c'n}, a' \in A, c' \in C\right) \\ &= \sum_a \sum_c \frac{X_a \frac{Z_c}{z_c}}{\sum_c Z_c \frac{x_{ac}}{z_c}} \sum_{i=1}^{n} \mathsf{E}\left(y_{aci} \mid x_{a'c'i},\ z_{c'i},\ a' \in A,\ c' \in C\right) \end{aligned}$$

---

[2] The proportionality relation is irreversible: if $\mathsf{E}(y \mid x) = \varphi x$ and $\mathsf{E}(x) \neq 0$, then $\mathsf{E}(x \mid y) \neq \psi y$, unless $y = \varphi x$ with probability 1.

$$
\begin{aligned}
&= \sum_a \sum_c \frac{X_a \frac{Z_c}{z_c}}{\sum_c Z_c \frac{x_{ac}}{z_c}} \sum_{i=1}^{n} \left[\frac{Y_{ac}}{Z_c} z_{ci} + \frac{Y_a}{X_a} \left(x_{aci} - \frac{X_{ac}}{Z_c} z_{ci} \right) \right] \\
&= \sum_a \sum_c \frac{Y_a Z_c \frac{x_{ac}}{z_c}}{\sum_c Z_c \frac{x_{ac}}{z_c}} + \sum_a \sum_c X_a \frac{Z_c}{z_c} \left(\frac{Y_{ac} \frac{z_c}{Z_c} - \frac{Y_a}{X_a} X_{ac} \frac{z_c}{Z_c}}{\sum_c Z_c \frac{x_{ac}}{z_c}} \right) \\
&= \sum_a Y_a + \sum_a \frac{X_a Y_a - Y_a X_a}{\sum_c Z_c \frac{x_{ac}}{z_c}} = Y.
\end{aligned}
$$

For this reason, the adequate confidence interval should be based on estimated conditional variance $\frac{n}{n-1} \sum_i \Delta_i^2$, where, for individual estimates (1.3), Δ_i equals

$$\Delta_i = \frac{X}{x} \left(y_i - \frac{y}{x} x_i \right) \tag{3.2–1}$$

$$\Delta_i = \sum_a \frac{X_a}{x_a} \left(y_{ai} - \frac{y_a}{x_a} x_{ai} \right) \tag{3.2–2}$$

$$\Delta_i = \sum_a \sum_c \frac{X_a}{\sum_c Z_c \frac{x_{ac}}{z_c}} \frac{Z_c}{z_c} \left[y_{aci} - \frac{y_{ac}}{z_c} z_{ci} - \frac{y_a}{x_a} \left(x_{aci} - \frac{x_{ac}}{z_c} z_{ci} \right) \right] \tag{3.2–3}$$

Returning, now, to sampling without replacement from finite populations, we may, by analogy, hope that the Δ_is given by (3.2), when inserted into (2.3), will give better estimated variances than the Δ_is given by (2.4). Asymptotically both considered methods of estimating the variance are equivalent, because for large n and $(N - n)$

$$\frac{X}{x} \approx \frac{N}{n}, \quad \frac{X_a}{x_a} \approx \frac{N}{n}, \quad \frac{X_a}{\sum_c Z_c \frac{x_{ac}}{z_c}} \approx 1, \quad \frac{Z_c}{z_c} \approx \frac{N}{n}.$$

For small samples, however, the improvement may be essential, generally in the sense that (3.2) will yield considerably greater estimated variance, on the average, than (2.4). Furthermore, without using (3.2), we cannot detect the possible negative effect of splitting x_i into too many components x_{ai} (see Hansen et al. (1953), vol. II, p. 139).

Remark 3.1. In finite populations, relations (3.1) are reflected in such a manner that the regression of y_{aci} upon x_{aci} and z_{ci} passes through the origin. Exceptionally, they may be fulfilled precisely, for example if x_i equals 0 or 1, and if $x_i = 0$ implies $y_i = 0$.

Remark 3.2. Inserting (3.2–2) into (2.3), we obtain the estimated variance, which for the special case $n = 2$, and stratified sampling, has been derived in Keyfitz (1957).

4 NORMAL THEORY APPROACH

Assuming normality and (3.1), we shall prove that the confidence intervals

$$X\,\frac{y}{x} \pm \frac{X}{x}\,t_\alpha \sqrt{\frac{n}{n-1}\sum_i \left(y_i - \frac{y}{x}\,x_i\right)^2}, \tag{4.1-1}$$

$$\sum X_a\,\frac{y_a}{x_a} \pm t_\alpha \sqrt{\frac{n}{n-1}\sum_i \left[\sum_a \frac{X_a}{x_a}\left(y_{ai} - \frac{y_a}{x_{ai}}\,x_i\right)\right]^2}, \tag{4.1-2}$$

$$\sum_a X_a\,\frac{\sum_c Z_c \frac{y_{ac}}{z_c}}{\sum_c Z_c \frac{x_{ac}}{z_c}}$$

$$\pm t_\alpha \sqrt{\frac{n}{n-1}\sum_i \left\{\sum_a \sum_c \frac{X_a Z_c}{z_c \sum_{c'} Z_{c'} \frac{x_{ac'}}{z_{c'}}}\left[y_{aci} - \frac{y_{ac}}{z_c} z_{ci} - \frac{y_a}{x_a}\left(x_{aci} - \frac{x_{ac}}{z_c} z_{ci}\right)\right]\right\}^2}, \tag{4.1-3}$$

where t_α denotes the critical value of Student's distribution with $n-1$ degrees of freedom for significance level α, will cover Y with probability greater than $1-\alpha$. In other words, confidence intervals (4.1) are 'conservative'.

The statistic $\sum_{i=1}^n \left(y_i - \frac{y}{x}\,x_i\right)^2$ does not have a chi-square distribution and is not independent of (y, x). Nevertheless, the following general inequality holds.

Theorem. *If $(y_{aci}, x_{aci}, z_{ci}, a \in A, c \in C)$, $i = 1, \ldots, n$, is a sample from a multidimensional normal distribution, then*

$$\sum_{i=1}^n \left\{\sum_a \sum_c K_{ac}\left[y_{aci} - \frac{y_{ac}}{z_c}\,z_{ci} - \frac{y_a}{x_a}\left(x_{aci} - \frac{x_{ac}}{z_c}\,z_{ci}\right)\right]\right\}^2 \geq \chi^2_{n-1}\,\sigma^2, \tag{4.2}$$

where

$$y_{ac} = \sum_{i=1}^n y_{aci}, \quad x_{ac} = \sum_{i=1}^n x_{aci}, \quad z_c = \sum_{i=1}^n z_{ci}, \quad a \in A,\ c \in C,$$

K_{ac} are arbitrary constants, which may depend on y_{ac}, x_{ac}, z_c, $a \in A$, $c \in C$, χ^2_{n-1} is a random variable which is independent of $\{y_{ac}, x_{ac}, z_c, a \in A, c \in C\}$ and has a chi-square distribution with $n-1$ degrees of freedom, and σ^2 is the conditional variance of $\sum_a \sum_c K_{ac}\, y_{aci}$ w.r.t. $\{x_{aci'}, z_{ci'}, a \in A, c \in A, i' = 1, \ldots, n\}$.

Proof. Let us choose an orthogonal transformation, given by the matrix $\{b_{ij}\}$, such that $b_{ij} = \frac{1}{\sqrt{n}}$, $j = 1, \ldots, n$, and denote

$$u_{aci} = \sum_{j=1}^{n} b_{ij}\, y_{acj}, \quad v_{aci} = \sum_{j=1}^{n} b_{ij}\, x_{acj}, \quad w_{ci} = \sum_{j=1}^{n} z_{cj}, \quad a \in A,\ c \in C.$$

Obviously $z_{ac1} = \frac{y_{ac}}{\sqrt{n}}$, $v_{ac1} = \frac{x_{ac}}{\sqrt{n}}$, $w_{c1} = \frac{z_c}{\sqrt{n}}$. Then, first owing to

$$\begin{aligned} &\sum_a \sum_c K_{ac} \left[u_{aci} - \frac{y_{ac}}{z_c} w_{ci} - \frac{y_a}{x_a} \left(v_{aci} - \frac{x_{ac}}{z_c} w_{ci} \right) \right] \\ &= \sum_{j=1}^{n} b_{ij} \left\{ \sum_a \sum_c K_{ac} \left[y_{aci} - \frac{y_{ac}}{z_c} z_{ci} - \frac{y_a}{x_a} \left(x_{aci} - \frac{x_{ac}}{z_c} z_{ci} \right) \right] \right\}, \\ &\qquad i = 1, \ldots, n, \end{aligned}$$

and to

$$\sum_{i=1}^{n} \sum_a \sum_c K_{ac} \left[y_{aci} - \frac{y_{ac}}{z_c} z_{cj} - \frac{y_a}{x_a} \left(x_{acj} - \frac{x_{ac}}{z_c} z_{cj} \right) \right] = 0,$$

the following well-known algebraic identity (see (Cramér (1946), p. 116)):

$$\begin{aligned} &\sum_{i=1}^{n} \left\{ \sum_a \sum_c K_{ac} \left[y_{aci} - \frac{y_{ac}}{z_c} z_{ci} - \frac{y_a}{x_a} \left(x_{aci} - \frac{x_{ac}}{z_c} z_{ci} \right) \right] \right\}^2 \\ &= \sum_{i=2}^{n} \left\{ \sum_a \sum_c K_{ac} \left[u_{aci} - \frac{y_{ac}}{z_c} w_{ci} - \frac{y_a}{x_a} \left(v_{aci} - \frac{x_{ac}}{z_c} w_{ci} \right) \right] \right\}^2 \end{aligned} \tag{4.3}$$

will hold, and, secondly, random vectors $\{u_{aci}, v_{aci}, w_{ci}, a \in A, c \in C\}$, $i = 1, \ldots, n$:

(i) will be mutually independent,
(ii) will be independent of $\{y_{ac}, x_{ac}, z_c, a \in A, c \in C\}$,
(iii) will possess mean values

$$\mathsf{E}(u_{aci}) = \mathsf{E}(v_{aci}) = \mathsf{E}(w_{ci}) = 0, \ a \in A, \ c \in C, \ i = 2, \ldots, n,$$

(iv) will be governed by the same variance covariance matrix as random vectors $\{y_{aci}, x_{aci}, z_{ci}, a \in A, c \in C\}$, $i = 1, \ldots, n$.

From (i) (ii) and (iii) it follows that

$$\sum_{i=2}^{n} \left\{ \sum_a \sum_c K_{ac} \left[u_{aci} - \frac{y_{ac}}{z_c} w_{ci} - \frac{y_a}{x_a} \left(v_{aci} - \frac{x_{ac}}{z_c} w_{ci} \right) \right] \right\}^2 \tag{4.4}$$

$$= \chi^2_{n-1} \,\mathsf{E}\left\{\left(\sum_a \sum_c K_{ac}\left[u_{ac2} - \frac{y_{ac}}{z_c} w_{c2} - \frac{y_a}{x_a}\left(v_{ac2} - \frac{x_{ac}}{z_c} w_{c2}\right)\right]\right)^2 \right.$$
$$\left. \middle| y_{ac}, x_{ac}, z_c, a \in A, c \in C\right\}.$$

It remains to be proved that the latter conditional mean value is always greater than or equal to σ^2. As, according to (ii), random vectors $\{u_{ac2}, v_{ac2}, w_{c2}, a \in A, c \in C\}$ and $\{y_{ac}, x_{ac}, z_c, a \in A, c \in C\}$ are independent, we may write

$$\mathsf{E}\left\{\left(\sum_a \sum_c K_{ac}\left[u_{ac2} - \frac{y_{ac}}{z_c} w_{c2} - \frac{y_a}{x_a}\left(v_{ac2} - \frac{x_{ac}}{z_c}\right)\right]\right)^2 \right.$$
$$\left. \middle| y_{ac}, x_{ac}, z_c, a \in A, c \in C\right\}$$
$$= \mathsf{E}^*\left\{\left(\sum_a \sum_c K_{ac}\left[u_{ac2} - \frac{y_{ac}}{z_c} w_{c2} - \frac{y_a}{x_a}\left(v_{ac2} - \frac{x_{ac}}{z_c} w_{c2}\right)\right]\right)^2\right\} \tag{4.5}$$

where the asterisk denotes that $y_{ac}, x_{ac}, z_c, a \in A, c \in C$ are treated as constants. The mean square of any normal random variable cannot be smaller than its conditional variance; i. e.

$$\mathsf{E}^*\left\{\left(\sum_a \sum_c K_{ac}\left[u_{ac2} - \frac{y_{ac}}{z_c} w_{c2} - \frac{y_a}{x_a}\left(v_{ac2} - \frac{x_{ac}}{z_c} w_{c2}\right)\right]\right)^2\right\}$$
$$\geq \mathsf{var}^*\left\{\sum_a \sum_c K_{ac}\left[u_{ac2} - \frac{y_{ac}}{z_c} w_{c2} - \frac{y_a}{x_a}\left(v_{ac2} - \frac{x_{ac}}{z_c} w_{c2}\right)\right]\right\}$$
$$\geq \mathsf{var}^*\left\{\sum_a \sum_c K_{ac}\, u_{ac2} \,|\, v_{ac2}, w_{c2}, a \in A, c \in C\right\}. \tag{4.6}$$

Now, according to (iv),

$$\mathsf{var}^*\left\{\sum_a \sum_c K_{ac}\, u_{ac2} \,|\, v_{ac2}, w_{c2}, a \in A, c \in C\right\} \tag{4.7}$$
$$= \mathsf{var}^*\left\{\sum_a \sum_c K_{ac}\, y_{aci} \,|\, x_{aci}, z_{ci}, a \in A, c \in C\right\}.$$

Finally, as vectors $(y_{aci}, x_{aci}, z_{ci}, a \in A, c \in C)$, $i = 1, \ldots, n$, are independent, we may write

$$\text{var}^*\left\{\sum_a\sum_c K_{ac}\,y_{aci}\,|\,x_{aci},\, z_{ci}\; a\in A,\; c\in C\right\}$$

$$= \text{var}\left\{\sum_a\sum_c K_{ac}\,y_{aci}\,|\,x_{aci'},\, z_{ci'},\, a\in A,\, c\in C,\, i'=1,\ldots,n\right\}=\sigma^2. \quad (4.8)$$

By the chain of relations (4.4) – (4.8) the proof is completed. □

In the special case, when conditions (1.3–3) are fulfilled and

$$K_{ac} = \frac{X_a\, Z_c}{z_c \sum_{c'} Z_{c'} \frac{x_{ac'}}{z_{c'}}},$$

σ^2 becomes the conditional variance of

$$\sum_a X_a \frac{Z_c \frac{y_{aci}}{z_c}}{\sum_c Z_c \frac{x_{aci}}{z_c}} \quad (4.9)$$

for fixed $\{x_{aci},\, z_{ci},\, a\in A,\, c\in C,\, i=1,\ldots,n\}$. The estimate given in (1.3–3) and (4.1–3) is the sum of the observations (4.9). Observations (4.9) are independent (for fixed $\{x_{aci},\, z_{ci},\, a\in A,\, c\in C,\, i=1,\ldots,n\}$) and, as we have seen in Section 3, their sum is an unbiased estimator of Y. Inequality (4.2) shows that the estimated variance used in (4.1–3) is greater than or equal to an estimator of variance that is (a) unbiased, (b) chi-square distributed, and (c) independent of the sum of observations (4.9). Consequently, the above statement concerning the confidence interval (4.1–3) is really true. Intervals (4.1–1) and (4.1–2) need no specific consideration, as they are special cases of the interval (4.1–3).

Remark 4.1. Our results could be generalized for the case where the same intraclass correlation is present, as in sampling without replacement. The only modification needed is to replace the coefficients t_α in (4.1) by $t_\alpha\sqrt{1-\frac{n}{N}}$. Thus we obtain confidence intervals corresponding to simple random sampling without replacement.

5 COMPARISON WITH FIELLER'S METHOD

The confidence interval (4.1–1), corresponding to the simple ratio estimate, can be compared with that provided by Fieller's method. We shall show that Fieller's interval is always longer.

Let us have a random sample $(x_1, y_1), \ldots, (x_n, y_n)$ from a two-dimensional normal distribution, and put $\xi = \mathsf{E}x_i$, $\eta = \mathsf{E}y_i$. Fieller's solution rests upon the obvious fact that the random variables $y_i - \frac{\eta}{\xi}\, x_i$ have mean values 0, so that

$$t = \frac{-\frac{\eta}{\xi}\, x}{\sqrt{\frac{n}{n-1} \sum_{i=1}^n \left[y_i - \overline{y} - \frac{\eta}{\xi}\,(x_i - \overline{x})\right]^2}} \tag{5.1}$$

is governed by Student's distribution with $n-1$ degrees of freedom. Thus for any $\frac{\eta}{\xi}$, with probability $1-\alpha$, it holds that $|t| \leq t_\alpha$, where t_α is the critical value of Student's distribution with $n-1$ degrees of freedom for significance level α. By the inequality $|t| \leq t_\alpha$, where t is given by (5.1), a certain confidence region for $\frac{\eta}{\xi}$ is defined, which covers the true value of $\frac{\eta}{\xi}$ with probability $1-\alpha$. Fieller has shown (1944) that, under the condition

$$1 - t_\alpha^2\, c_1^2 > 0 \tag{5.2}$$

the confidence region is equal to the interval with end-points

$$\frac{y}{x}\, \frac{1 - t_\alpha^2\, c_{12} \pm \sqrt{(1 - t_\alpha^2\, c_{12})^2 - (1 - t_\alpha^2\, c_1^2)\,(1 - t_\alpha^2\, c_2^2)}}{1 - t_\alpha^2\, c_1^2} \tag{5.3}$$

where

$$\begin{aligned} c_1^2 &= \left(\frac{1}{x}\right)^2 \frac{n}{n-1} \sum_{i=1}^n (x_i - \overline{x})^2, \\ c_2^2 &= \left(\frac{1}{y}\right) \frac{n}{n-1} \sum_{i=1}^n (y_i - \overline{y})^2, \\ c_{12} &= \frac{1}{xy}\, \frac{n}{n-1} \sum_{i=1}^n (x_i - \overline{x})\,(y_i - \overline{y}). \end{aligned}$$

If (5.2) does not hold, it may be proved that Fieller's confidence region equals the whole line, or half-line, or complement of a finite interval. The square of the length of the interval with end-points (5.3), say d_F, equals

$$\begin{aligned} d_F &= 4\left(\frac{y}{x}\right)^2 \frac{(1 - t_\alpha^2\, c_{12})^2 - (1 - t_\alpha^2\, c_1^2)\,(1 - t_\alpha^2\, c_2^2)}{(1 - t_\alpha^2\, c_1^2)^2} \\ &= 4\left(\frac{y}{x}\right)^2 t_\alpha^2\, \frac{(1 - t_\alpha^2\, c_1^2)\left(c_2^2 - \frac{c_{12}^2}{c_1^2}\right) + \left(c_1 - \frac{c_{12}}{c_1}\right)^2}{(1 - t_\alpha^2\, c_1^2)^2}. \end{aligned} \tag{5.4}$$

Let us compare this length with the length of the interval obtained from (4.1–1) after dividing by X:

$$\frac{y}{x} \pm \frac{1}{x}\, t_\alpha \sqrt{\frac{n}{n-1} \sum_{i=1}^n \left(y_i - \frac{y}{x}\, x_i\right)^2}. \tag{5.5}$$

The square of the length of the interval (5.5), say d_A, is

$$d_A = 4\left(\frac{t_\alpha}{x}\right)^2 \frac{n}{n-1}\sum_{i=1}^{n}\left(y_i - \frac{y}{x}x_i\right)^2 = 4\left(\frac{y}{x}\right)^2 t_\alpha^2(c_2^2 - 2c_{12} + c_2^2)$$

$$= 4\left(\frac{y}{x}\right)^2 t_\alpha^2\left[c_2^2 - \frac{c_{12}^2}{c_1^2} + \left(c_1 - \frac{c_{12}}{c_1}\right)^2\right]$$

$$\leq 4\left(\frac{y}{x}\right)^2 t_\alpha^2 \frac{(1 - t_\alpha^2 c_1^2)\left(c_2^2 - \frac{c_{12}^2}{c_1^2}\right) + \left(c_1 - \frac{c_{12}}{c_1}\right)^2}{(1 - t_\alpha^2 c_1^2)^2} = d_F, \tag{5.6}$$

i.e. the Fieller interval (5.3) is always longer than the usual interval (5.5).

6 SOME OTHER METHODS OF SAMPLING–ESTIMATING

Ratio estimates are generally biased. In the paper (Hájek (1949)), however, it is shown that the bias of the simple ratio estimate $X\frac{y}{x}$ can be completely removed by the following method of sampling. In the first phase, we select one element with probabilities $\frac{x_i}{X}$, $i = 1, \ldots, N$, and in the second phase we select $n-1$ elements from the remaining $N-1$ elements by simple random sampling. The estimated variance, of course, at least for small n, must be changed accordingly. Following Yeates and Grundy (see (1953)) we get the estimated variance in the form

$$\left(\frac{X}{x}\frac{n}{N}\right)^2 \sum_i\sum_j\left(y_i - \frac{y}{x}x_i - y_j + \frac{y}{x}x_j\right)^2\left(\frac{\pi_i\pi_j}{\pi_{ij}} - 1\right), \tag{6.1}$$

where π_i denotes the probability that the ith element will be included in the sample, and π_{ij} denotes the probability that the ith and jth elements will both be included in the sample simultaneously. For the 'two-phase' sampling described above, it may easily be shown that

$$\pi_i = \frac{x_i}{X}\frac{N-n}{N-1} + \frac{n-1}{N-1}, \quad i = 1, \ldots, N, \tag{6.2}$$

$$\pi_{ij} = \frac{x_i + x_j}{X}\frac{n-1}{N-2}\frac{N-n}{N-1} + \frac{n-1}{N-1}\frac{n-2}{N-2}, \quad i, j = 1, \ldots, N. \tag{6.3}$$

Let us mention briefly two other methods of sampling–estimating. First method: We choose, n-times independently, one element with probabilities $\frac{x_i}{X}$, $i = 1, \ldots, N$, and use the following point and interval estimate for Y:

$$\frac{X}{n}\left[\sum_i^{n_0}\right]\frac{y_i}{x_i} + (n - n_0)\frac{\sum_i^{n_0} y_i}{\sum_i^{n_0} x_i} \pm t_\alpha\sqrt{\frac{X^2}{n(n-1)}\sum_i^{n}\left[\frac{y_i}{x_i} - \frac{1}{n}\sum_i^{n}\frac{y_i}{x_i}\right]^2}, \tag{6.4}$$

where $\sum_i^n$ and $\sum_i^{n_0}$ denote the sums extended over all sampled elements and over distinct sampled elements, respectively, and n_0 denotes the number of distinct elements in the sample; (some of the elements may appear twice or more times in the sample).

Second method: We carry out N independent experiments, the ith of which decides with probability $c\,\frac{x_i}{X}$ and $1 - c\,\frac{x_i}{X}$ whether or not the ith element will be included in the sample, so that the number n of elements included in the sample will be a random variable with $\mathsf{E}(n) = c$. Then we use the following point and interval estimates of Y:

$$\frac{X}{n}\sum_i \frac{y_i}{x_i} \pm t_\alpha \sqrt{\frac{X^2}{n(n-1)}\sum_i \left(\frac{y_i}{x_i} - \frac{1}{n}\sum_i \frac{y_i}{x_i}\right)^2 \left(1 - \frac{cx_i}{X}\right)}. \tag{6.5}$$

7 STRATIFICATION AND SUBSAMPLING

We shall only touch this topic very briefly. Let us label the strata, and statistics referred to the strata, by the subscript $h = 1, \ldots, L$. If the stratified sampling is proportionate, i.e. $\frac{N_h}{n_h} = \frac{N}{n}$, $h = 1, \ldots, L$, then formulas (1.3) need no change, and the estimated variance (2.3) should be replaced by

$$\sum_{h=1}^{L} \frac{(N_h - n_h)\, n_h}{N_h(n_h - 1)} \left[\sum_i \Delta_{hi}^2 - \frac{1}{n_h}\left(\sum_i \Delta_{hi}\right)^2\right], \tag{7.1}$$

where

$$\Delta_{hi} = \frac{N_h}{n_h}\left(y_{hi} - \frac{y}{x}\, x_{hi}\right), \quad \text{etc.},$$

in accordance with (2.4).

If the stratified sampling fails to be proportionate, then the sample totals y, x, y_a etc. in formulas (1.3) should be replaced by

$$\sum_{h=1}^{L} \frac{N_h}{n_h}\sum_i y_{hi}, \quad \sum_{h=1}^{L} \frac{N_h}{n_h}\sum_i x_{hi}, \quad \sum_{h=1}^{L} \frac{N_h}{n_h}\sum_i y_{hai}, \quad \text{etc.}, \tag{7.2}$$

respectively. Formula (7.1) needs no change. The theorem of Section 4 could be generalized for stratified sampling in an obvious manner. Corresponding statements concerning confidence, of course, will denote the critical value based on the generalized Student distribution (see Hájek (1957)).

On some occasions, other ratio estimates may be useful in the case of stratified sampling:

$$\sum_{h=1}^{L} X_h\, \frac{y_h}{x_h}, \tag{7.3–1}$$

$$\sum_{h=1}^{L}\sum_{a} X_{ha}\frac{y_{ha}}{x_h}, \tag{7.3–2}$$

$$\sum_{a} X_a \frac{\sum_{h=1}^{L}\sum_c Z_{hc}\frac{y_{hac}}{z_{hc}}}{\sum_{h=1}^{L}\sum_c Z_{hc}\frac{x_{hac}}{z_{hc}}}. \tag{7.3–3}$$

The corresponding estimated variances equal

$$\sum_{h=1}^{L}\frac{(N_h - n_h)n_h}{N_h(n_h-1)}\sum_i \Delta_{hi}^2, \tag{7.4}$$

where

$$\Delta_{hi} = \frac{X_h}{x_h}\left(y_{hi} - \frac{y_h}{x_h}x_{hi}\right), \quad \text{etc.},$$

in accordance with (3.2). The estimates (7.3) are advantageous in such cases, where the ratios $\frac{Y_h}{X_h}$, etc., vary substantially from stratum to stratum. On the other hand, they may be heavily biased, unless 'two-phase' sampling, described in Section 6, is used in each stratum. Moreover, if $\frac{Y_h}{X_h}$ does not vary very much, and if the number of strata is large and n_hs are small, then the variance of estimates (7.3) may even be greater than the variance of estimates (1.3).

Finally, when more stages are involved in our sampling plan, then the only modification needed is to replace the values y_i, y_{ai}, y_{aci}, and, possibly, values x_i, x_{ai}, x_{aci}, z_{ci} by their estimates obtained from subsampling. After this adaptation formulas (4.1) are also valid, in a conservative manner, for multistage sampling. Subsampling rates are often chosen so that the estimate turns out to be an unweighted sum of the ultimate observations. For example, let the hith element contain M_{hi} subsampling elements, m_{hi}, which we select by subsampling. If we put

$$m_{hi} = \frac{1}{k}\frac{M_{hi}X_h}{x_h}, \quad h = 1,\ldots,L;\ i = 1,\ldots,N_h, \tag{7.5}$$

then the estimate (7.3–1) turns out to equal ky', where y' denotes the sum of ultimate observations.

8 SOME CONCLUDING REMARKS

The above discussion, based on a combination of the large-sample, regression and normal theory approaches, will also apply to other possible types of ratio estimates. Some important problems, however, remain. Two of these are as follows: (a) For what set of normal distributions is it true that the confidence intervals (4.1) cover Y with a probability greater than or equal to $1 - \alpha$? (b) How can we simplify the computations of estimated variances considerably? (The estimated variances, derived in this paper, despite their relative simplicity, are too complicated for daily practice, particularly when hundreds of items are tabulated simultaneously.)

REFERENCES

Cramér, H. (1946). *Mathematical Methods of Statistics.* Princeton.

Fieller, E. C. (1944). A fundamental formula in the statistics of biological assay, and some applications. *Quart. Journ. Pharm. 17*, 117–123.

Hájek, J. (1949). Representative cluster sampling by a method of two phases. *Statistický obzor 29*, 384–394.

Hájek, J. (1957). Inequalities for the generalized Student's distribution and their applications. *Časopis Pěst. Mat. 83*, 182–194.

Hansen, M. H. and Hurwitz, W. N. (1943). On the theory of sampling from finite populations. *Ann. Math. Stat. 14*, 393.

Hansen, M. H., Hurwitz, W. N. and Madow, W. G. (1953). *Sample Survey Methods and Theory.* Vol. I, II, New York, London.

Keyfitz, W. (1957). Estimates of sampling variance where two units are selected from each stratum. *Journ. Am. Stat. Assoc. 52*, 503–510.

Madow, W. G. (1948). On the limiting distributions of estimates based on samples from finite universes. *Ann. Math. Stat. 19*, 535–545.

Yeates, F. and Grundy, P. M. (1953). Selection without replacement from within strata with probability proportional to size. *Journ. Royal. Stat. Soc. B, 15*, 253–261.

CHAPTER 12

Some Contributions to the Theory of Probability Sampling

Bull. Inst. Internat. Statist. *36* (1958), 127–134.
Reproduced by permission of International Statistical Institute.

1 CONFIDENCE INTERVAL FOR THE RATIO OF TWO MEANS

There are two methods in use, one based on Fieller's device (see Fieller (1944)) and one based on the usual theory of large-sample approximations. The large-sample method, however, can be justified on the grounds of a 'small-sample' theorem, and, moreover, it may be shown that under the very conditions of probability sampling which are often met with in practice, the large-sample 'approximation' is superior to Fieller's 'exact' method.

In the most simple case when we deal with n independent and identically distributed pairs of observations (x_i, y_i), Fieller's and the large-sample confidence intervals for $\varphi = \mathsf{E}y/\mathsf{E}x$ are solutions of the following inequalities in φ, respectively:

$$\left(\sum y_i - \varphi \sum x_i\right)^2 \leq t^2 \frac{n}{n-1} \sum (y_i - \varphi x_i - \overline{y} + \varphi \overline{x})^2 \tag{1.1}$$

$$\left(\sum y_i - \varphi \sum x_i\right)^2 \leq t^2 \frac{n}{n-1} \sum (y_i - f x_i)^2, \quad f = \frac{\sum y_i}{\sum x_i}. \tag{1.2}$$

Now, the following assertions are valid:

Collected Works of Jaroslav Hájek – With Commentary
Edited by M. Hušková, R. Beran and V. Dupač
Published in 1998 by John Wiley & Sons Ltd.

(i) From algebraic considerations, Fieller's confidence interval given by (1.1) is always longer than the large-sample interval given by (1.2).

(ii) Despite this fact, the large-sample confidence interval may contain the true value of φ with a greater probability than Fieller's, namely in the case when the pairs (x_i, y_i) are normally distributed and the regression line of y on x passes through the origin.

(iii) In the latter case the probability that the interval given by (1.2), i.e. interval

$$f \pm t \frac{\sqrt{\frac{n}{n-1} \sum (y_i - f x_i)^2}}{\sum x_i}, \tag{1.3}$$

contains the true value of φ is greater than the probability which corresponds to the used t (with $n-1$ d.f.).

Remark 1.1. If the regression line of y on x passes through the origin, then φ equals the regression coefficient $b = \sigma_{xy}/\sigma_x^2$, and, consequently, it would be possible to construct a better confidence interval than (1.3). But this interval would not be sufficiently 'robust' to the possible violation of the supposition that $\varphi = b$. On the other hand, the assertions (i) – (iii) remain true, even in cases when $\varphi \simeq b$ only approximately.

2 UPPER ESTIMATE OF SAMPLING ERROR IN MULTISTAGE SAMPLING

Let us denote by y_i and Q_i the y-value and the probability of being included in the sample, respectively, corresponding to the ith element (elementary unit). The unbiased estimate of the total $Y = \sum y_i$ is known to be

$$\hat{Y} = S\left(\frac{y_i}{Q_i}\right), \tag{2.1}$$

where S stands for summation over the sample. The variance of $\hat{Y}$ equals

$$\mathsf{var}\,\hat{Y} = \sum y_i^2 \left(\frac{1}{Q_i} - 1\right) + \sum_{i \neq j}\sum y_i y_j \left(\frac{Q_{ij}}{Q_i Q_j} - 1\right), \tag{2.2}$$

where Q_{ij} denotes the probability of the simultaneous inclusion of both elements i and j in the sample. In many cases, especially in multistage sampling, the unbiased estimation of $\mathsf{var}\,\hat{Y}$ is unduly complicated. We may, however, find a much simpler upper estimate $v(\hat{Y})$ satisfying the condition $\mathsf{E}\{v(\hat{Y})\} \geq \mathsf{var}\,\hat{Y}$.

Now, let us denote by Y_j and P_j the total of the y-values and the probability of being included in the sample, respectively, for the jth (primary) sampling unit. In addition, let $\hat{Y}_j$ be the unbiased estimate of Y_j obtained from the subsampling (further sampling stages). Then $\hat{Y}$ may be written in an equivalent form,

$$\hat{Y} = S\left(\frac{\hat{Y}_j}{P_j}\right),$$

where S stands for summation over the sample of primary sampling units. If the subsampling in each of the primary sampling units is conducted independently, then, irrespective of the further properties of the subsampling, there is the following upper estimate of variance of $\hat{Y}$,

$$v(\hat{Y}) = S\left(\frac{\hat{Y}_j}{P_j}\right)^2 + \underset{j\neq k}{SS}\ \hat{Y}_j\,\hat{Y}_k\left(\frac{1}{P_j\,P_k} - \frac{1+C_{jk}}{P_{jk}}\right), \qquad (2.3)$$

where the C_{jk}s may be any constants such that for any λ_js the relation

$$\sum_{j=1}^{M}\lambda_j^2 - \sum\sum_{j\neq k}\lambda_j\,\lambda_k\,C_{jk} \geq 0$$

holds; for example, $C_{jk} = 0$ or $C_{jk} = 1/(M-1)$, where M is the number of primary sampling units.

Example 2.1. If the first stage of sampling is random with sampling rate m/M, then one of the upper estimates of variance of $\hat{Y} - \varphi\hat{X}$ equals

$$v(\hat{Y} - \varphi\hat{X}) = \left(\frac{M}{m}\right)^2 \frac{m}{m-1} S(\hat{Y}_j - f\,\hat{X}_j)^2, \qquad (2.4)$$

where

$$f = \frac{S(\hat{Y}_j)}{S(\hat{X}_j)}.$$

3 POISSON'S SAMPLING

If the ratio estimate is used, there is no necessity to keep fixed the number of elements or even of sampling units in the sample. On the other hand, if the size of the sample may vary, it is easy to find such a sampling procedure which (i) is without replacement, (ii) guarantees to elements (or primary sampling units) varying probabilities of being included in the sample, and (iii) has easily calculable and estimable variance. One of the simplest procedures of this kind I call Poisson's sampling owing to the apparent connection with Poisson's process. Poisson's sampling simply consists of N independent experiments, the ith of which decides with probabilities Q_i and $1 - Q_i$ whether or not the ith element will be included in the sample. Then, obviously, $Q_{ij} = Q_i, Q_j$, and the variances (2.2) of $\hat{Y}$ and $\hat{Y} - \varphi\hat{X}$ have these simple forms:

$$\mathsf{var}\,\hat{Y} = \sum_{i=1}^{N} y_i^2 \left(\frac{1}{Q_i} - 1\right), \tag{3.1}$$

$$\mathsf{var}\{\hat{Y} - \varphi\hat{X}\} = \sum_{i=1}^{N} (y_i - \varphi x_i)^2 \left(\frac{1}{Q_i} - 1\right). \tag{3.2}$$

Example 3.1. Let the first stage of sampling be Poisson's sampling with probabilities P_j and the second stage random with sampling rates $n_j/N_j = Q/P_j$, so that every element has the same probability Q of being included in the sample. Under these conditions we have for the ratio estimate of $\varphi = Y/X$, the variance of $\hat{Y} - \varphi\hat{X}$ and the confidence interval for φ (based on the upper estimate of $\mathsf{var}\{\hat{Y} - \varphi\hat{X}\}$) the following formulae:

$$f = \frac{S(y_i)}{S(x_i)}$$

$$\mathsf{var}\{\hat{Y} - \varphi\hat{X}\} = \sum_{j=1}^{M} \left[(Y_j - \varphi X_j)^2 \left(\frac{1}{P_j} - 1\right) + N_j\,\sigma_j^2 \left(\frac{1}{Q} - \frac{1}{P_j}\right)\right] \tag{3.3}$$

$$f \pm t \frac{\sqrt{S[S_j(y) - f\,S_j(x)]^2}}{S(x)},$$

where M is the number of sampling units, Y_j and X_j are totals over the jth sampling unit and S_j stands for summation over the subsample within the jth primary sampling unit.

Remark 3.1. There are many useful modifications of Poisson's sampling. For example, we may, in the first phase, select randomly r elements and then, from this 'wide' sample, conduct Poisson's sampling with probabilities NQ_i/r (supposition that $NQ_i \leq r$). This device makes the sampling procedure easy. The variance remains almost the same:

$$\mathsf{var}\{\hat{Y} - \varphi\hat{X}\} = \sum_{i=1}^{N} (y_i - \varphi x_i)^2 \left(\frac{1}{Q_i} - \frac{N(r-1)}{r(N-1)}\right). \tag{3.4}$$

Remark 3.2. The sampling of only one or a few elements with probabilities proportional to the numbers $a_1, \ldots, a_N$ may be conveniently carried out in the following manner: we choose an arbitrary number $a \geq a_i$ $(i = 1, \ldots, N)$ and then select one element randomly and accept it definitely (with probability a_i/a) if $a_i \geq \xi$, where ξ is a random number chosen between 1 and a; if $a_i < \xi$, the selected element is not accepted and the procedure is repeated until one of the randomly selected elements is accepted. The average number of steps needed equals $a/\overline{a}$, where $\overline{a} = \sum_{i=1}^{N} a_i/N$. One of the possible applications of this method is sampling which makes the ratio estimate $S(y)/S(x)$ unbiased; it consists of the sampling of one element with probabilities proportional to $x_1, \ldots, x_N$ to which are added $n-1$ further elements selected from the remaining $N-1$ elements at random.

4 OPTIMUM SAMPLING STRATEGY

Sometimes, the method of sampling and estimating is predetermined with the exception of parameters $\nu_1, \ldots, \nu_h$, whose values should be chosen so that either

(a) the variance V is minimum for given costs C, or
(b) the costs C are minimum for given variance V.

If V and C can be – as is usually the case – expressed in the following 'canonical' forms:

$$V = \frac{V_1}{\nu_1} + \cdots + \frac{V_h}{\nu_h} + V_0, \qquad V_i \geq 0, \tag{4.1}$$

$$i = 1, \ldots, h$$

$$C = C_1 \nu_1 + \cdots + C_h \nu_h + C_0, \quad C_i \geq 0, \tag{4.2}$$

then both problems (a) and (b) can be solved simultaneously by minimizing the product $(V - V_0)(C - C_0)$. According to Schwartz's inequality

$$(V - V_0)(C - C_0) \geq \left(\sum \sqrt{C_i V_i}\right)^2$$

and the minimum value is reached when

$$\nu_i = \lambda \sqrt{\frac{V_i}{C_i}}, \quad i = 1, \ldots, h \tag{4.3}$$

(see Stuart (1954)).

Example 4.1. Using the above-mentioned method, one can easily solve the optimum allocation in stratified sampling, the optimum arrangement of multistage sampling and so on. As a less trivial example, let us consider the 'without replacement' analogue of the problem solved by Hansen and Hurwitz (1949). Suppose we have a sampling design described in Example 3.1 and let us rewrite the variance (3.3) in the form

$$V = \sum_{j=1}^{M} \frac{(Y_j - \varphi X_j)^2 - N_j \sigma_j^2}{P_j} + \frac{1}{Q} \sum_{j=1}^{M} N_j \sigma_j^2 - \sum_{j=1}^{M} (Y_j - \varphi X_j)^2. \tag{4.4}$$

If, in addition, the cost function C has the Hansen–Hurwitz form

$$C = \sum_{j=1}^{M} (C_1 + C_2 N_j) P_j + C_3 Q N \tag{4.5}$$

we can see that both V and C are 'canonical' in P_js and Q. Consequently, according to (4.3), the optimum probabilities are

$$P_j = \lambda \sqrt{\frac{(Y_j - \varphi X_j)^2 - N_j \sigma_j^2}{C_1 + C_2 N_j}}, \quad Q = \lambda \sqrt{\frac{\sum_{j=1}^{M} N_j \sigma_j}{C_3 N}}, \tag{4.6}$$

where λ is determined from either of the equations (4.4) or (4.5).

Remark 4.1. If $V_i < 0$, then the optimum ν_i assumes its minimum possible value. For example, if $(Y_j - \varphi X_j)^2 - N_j \sigma_j^2 < 0$ in (4.4), then optimum $P_j = Q$ (i. e. $n_j = N_j$).

In some papers an attempt has been made to give an optimum solution of sampling strategy (i. e. sampling design + estimating method) with respect to some *a priori* distribution of parameters $y_1, \ldots, y_N$. This 'Bayes solution' is quite justified, since we usually have some *a priori* knowledge about the sampled population and, moreover, the accepted assumptions about *a priori* distribution of $y_1, \ldots, y_N$ affect only the choice of sampling strategy and have no influence on the validity of probability statements about constructed confidence limits. (We note that the solution (4.6) has no practical values unless some 'expected value' is substituted for $(Y_j - \varphi X_j)^2$.) We shall consider the two simplest cases when $y_1, \ldots, y_N$ (i) are independent or (ii) form a stationary sequence with convex (= concave upward) correlation function (see Cochran (1946) and Godambe (1955)).

Now, suppose the y_is are independent with expected means $E_i = \mathsf{E} y_i$ and variances $D_i = \mathsf{var} y_i$. Let us consider all possible linear unbiased estimates $\hat{Y}$ of the total $Y = \sum y$,

$$\hat{Y} = S(y_i\, w_{is}),$$

where S stands for summation over the sample and weights w_{is} may depend on the sample s. In addition, let the cost function be

$$C = C_1 Q_1 + \cdots + C_N Q_N,$$

where Q_i is the probability that the ith element is included in the sample. Under these conditions, the expected variance of $\hat{Y}$ is minimum at given costs C (or conversely) if

$$w_{is} \quad = \quad 1/Q_1 \quad \text{for all samples,} \tag{4.7}$$

$$Q_i = \lambda\sqrt{\frac{D_i}{C_i}}, \tag{4.8}$$

$$S\left(\frac{E_i}{Q_i}\right) = \sum_{i=1}^{N} E_i \quad \text{with probability 1.} \tag{4.9}$$

For example, if the expected means E_i and costs C_i are constant over some strata, then the above conditions are fulfilled by optimum allocation of a stratified sample, or, if the expected means E_i are proportional to some numbers x_i, then the condition (4.7) is fulfilled approximately and condition (4.9) exactly by the ratio estimate

$$\hat{Y} = \frac{S\,(y_i/Q_i)}{S\,(x_i/Q_i)} \sum_{i=1}^{N} x_i$$

where the Q_is are calculated from (4.8).

If $y_i, \ldots, y_N$ form a stationary sequence with convex correlation function, and if the probabilities Q_i of including the ith element in the sample are given, then among all possible sampling designs the systematic design yields the minimum variance of $S(y)$. This is a generalization of Cochran's result contained in Cochran (1946).

5 THE NUMBER OF DEGREES OF FREEDOM OF STUDENT'S t FOR FINITE POPULATION

Sometimes, it is misleadingly suggested that, in the case of sampling without replacement from a finite population, Student's t is distributed not with $n-1$ but with $(n-1)/(1-n/N)$ degrees of freedom, approximately. In reality, however, in spite of the fact that the estimate of variance has approximately chi-square distribution with $(n-1)/(1-n/N)$ d.f., the number of d.f. of Student's t should still remain $n-1$. This apparent paradox is due to the interdependence of the sample mean and sample variance. Indeed, the distribution of Student's t remains unchanged when the condition of independence of observations is replaced by a weaker condition that all pairs of (normal and equidistributed) observations have the same covariance $\sigma_{ij} = \rho\sigma^2$ (in the case of random sampling from a finite population we have $\sigma_{ij} = -\sigma^2/(N-1)$) (see Walch (1947)).

6 DEFINITIONS OF FUNDAMENTAL CONCEPTS

The fundamental concepts of the theory of probability sampling may be introduced in the following formal way:

A population (universe) U is a set of couples (i, y_i), $U = \{(i, y_i)\}$.

A probability sample is a subset $s \subset U$ selected by a random technique defining a probability distribution $P(s)$ over all possible s, $s \subset U$.

An estimate (statistic) t is any function of the probability sample s, $t = t(s)$.

Thus the characteristic feature of the 'sampling from a finite population' is not the finiteness of the population but the *real existence* and *identificability* of its elements. The estimates, consequently, may depend not only on the y-values but also on the elements on which they were observed.

It would be possible (and more usual) to define the probability sample as a *sequence* $\{(i_1, y_{i1}), \ldots, (i_n, y_{in})\}$. Such a definition is, however, (a) inconvenient, since the size of the sample n may be itself a random variable, and (b) unnecessary, since each 'good' estimate does not depend on the order or on the number of times an element is included in the sample (i. e. 'good' estimates are functions of the sample s in the sense of the above definition). Each estimate depending on the order or multiplicity of sampled elements may be adjusted by taking its average value over all samples which differ only in order and multiplicity of sampled elements. This adjusted estimate has the same mean value but a smaller variance.

Example 6.1. Let us draw independently n times one of N elements with probabilities $q_1, \ldots, q_N$. Then the unbiased estimate of the total $Y = \sum y$ is known to be

$$\hat{Y}_1 = S_1 \left(\frac{y_i}{Q_i} \right),$$

where $Q_i = n\, q_i$ and S_1 stands for summation over the selected *sequence* (some elements may appear in the sum more than once). If there are $n - 1$ different elements in this sequence (precisely one element was sampled twice), then the average value $\hat{Y}$ of $\hat{Y}_1$ over all sequences formed from those $n - 1$ elements equals

$$\hat{Y}_1 = S \left(\frac{y_i}{Q_i} \right) + \frac{S(y_i)}{S(Q_i)}$$

where S stand for summation over the sample as defined above.

REFERENCES

Cochran, W. G. (1946). Relative accuracy of systematic and stratified random sample for certain class of populations. *Ann. Math. Stat. 17*, 164–177.

Godambe, V. P. (1955). A unified theory of sampling from finite population. *J. Roy. Stat. Soc., Ser. B, 17*, 269–278.

Hansen, M. H. and Hurwitz, W. N. (1949). On the determination of optimum probabilities in sampling. *Ann. Math. Stat., 20*, 426–432.

Fieller, E. C. (1944). A fundamental formula in the statistics of biological assay, and some applications. *Quart. J. Pharm. 17*, 117–123.

Stuart, A. (1954). A simple presentation of optimum sampling results. *J. Roy. Stat. Soc., Ser. B 16*, 239–241.

Walch, J. E. (1947). Concerning the effect of interclass correlation on certain significance tests. *Ann. Math. Stat. 18*, 88–96.

CHAPTER 13

Optimum Strategy and Other Problems in Probability Sampling

Časopis Pěst. Mat. 84 (1959), 387–423.
Reproduced by permission of Hájek family: Al. Hájková, H. Kazárová and Al. Slámová.

The statistician's strategy in probability sampling consists in the choice of the sampling design (plan) and of the estimation method (procedure). A strategy may be called optimum if it solves the conflict between cost and accuracy in the best way. In this chapter the Bayes approach is accepted, i. e. the accuracy is measured by the expected variance with respect to a certain a priori distribution of ascertained values. A general solution of the problem is derived for a rather wide class of admissible sampling designs, estimators, cost functions, and for the following two most important assumptions concerning the a priori distribution: (a) the ascertained values are realizations of noncorrelated random variables; (b) the ascertained values are realizations of a random sequence with stationary convex correlation function and stationary coefficients of variations.

In the introductory sections the conceptions of 'sample' and 'estimate' are defined, and a general formula for the variance and estimated variance of linear estimates is derived; furthermore, a method of improving estimates based on sufficient statistics is presented, and two sampling designs with varying probabilities are discussed.

Collected Works of Jaroslav Hájek – With Commentary
Edited by M. Hušková, R. Beran and V. Dupač
Published in 1998 by John Wiley & Sons Ltd.

A. INTRODUCTORY SECTIONS (1 – 5)

1 DEFINITIONS

Let us have a population S consisting of N elements of arbitrary nature, so that they may be represented by integers $1, \dots, N$, $S = \{1, \dots, N\}$. From S we select a subset s in such a way that any subset $s \subset S$ possesses a probability $\mathsf{P}(s)$ of being selected. The selected subset s will be called the *sample*.

Let us denote by $y_1, \dots, y_N$ values of a certain variable associated with elements $1, \dots, N$, respectively. We try to estimate the total

$$Y = \sum_{i=1}^{N} y_i \tag{1.1}$$

by an estimate $\hat{Y}$ having the form

$$\hat{Y} = \sum_{i \in S} y_i \, w_i(s), \tag{1.2}$$

where $w_i(s)$ are arbitrary weights, $i \in s$, $s \subset S$, and $\sum_{i \in S}$ extends over the elements i included in the sample s. The estimate (1.2) will be called *a linear estimate*.

The necessary and sufficient condition for the estimate (1.2) to be *unbiased* is, obviously, that

$$\sum_{s \ni i} w_i(s) \, \mathsf{P}(s) = 1, \quad i = 1, \dots, N, \tag{1.3}$$

where the sum $\sum_{s \ni i}$ extends over all samples containing the element i.

The probability of selecting a sample s which contains the element i, say π_i, equals

$$\pi_i = \sum_{s \ni i} \mathsf{P}(s). \tag{1.4}$$

Similarly, the probability of selecting a sample which contains both elements i and j, say π_{ij}, equals

$$\pi_{ij} = \sum_{\substack{s \ni i \\ s \ni j}} \mathsf{P}(s), \quad i, j = 1, \dots, N. \tag{1.5}$$

If we put $w_i(s) = 1/\pi_i$, $i \in s$, $s \subset S$, we get the *simple linear estimate*:

$$\hat{Y} = \sum_{i \in s} \frac{y_i}{\pi_i}. \tag{1.6}$$

It is easily seen that the simple linear estimate is unbiased, i.e. that (1.3) holds.

We have defined the 'sample' as a *subset* of the population. However, one could think of a more detailed specification of the 'sample'. For example, it is possible to define it as *an ordered subset* of the population, or, still more distinctively, as *a sequence*, all members of which belong to the population. For brevity, let us use the following symbols:

$$\text{subset } s, \quad \text{ordered subset } s', \quad \text{sequence } s''. \tag{1.7}$$

The sample s only tells us what elements have been selected, while s' and s'' comprise further information. The sample s' fully describes the element-by-element sampling without replacement, and s'' fully describes the element-by-element sampling with replacement.

If we delete in the sequence s'' all members which appear in some of the preceding places we get an ordered set s', $s' = s'(s'')$. If we dispense with the ordering in s', we get a set s, $s = s(s')$. Symbolically,

$$s = s(s') = s(s'(s'')), \tag{1.8}$$

i.e. s' is an abstract function of s'' and s is an abstract function of s', and, naturally, of s'' too.

The collection (1.7) of possible definitions of the sample is naturally not exhaustive. For example, we shall use, in Section 3, the sample s^* which tells how many times each element has been included in the sample. Clearly, if we dispense with the ordering in s'', we get s^*. Thus it holds that

$$s^* = s^*(s''), \quad s = s(s^*). \tag{1.9}$$

If we deal with double sampling, we may define the sample as a couple of subsets (s_1, s), where s_1 is the 'larger' sample and s is the ultimately selected sample. As we can see, the possible definitions of the sample might be continued as long as we wished.

Now, let us define the *observation* and the *estimate* in probability sampling. The observation, say (s, y), (s', y), (s'', y), etc., involves a knowledge of the sample s, s', s'', etc. and of the values y_i associated with elements in s, s', s'', etc., respectively. The estimate t is any function of (s, y), (s', y), (s'', y), etc.:

$$t = t(s, y), \quad t = t(s', y), \quad t = t(s'', y), \text{ etc.} \tag{1.10}$$

2 ESTIMATING SAMPLING ERROR OF LINEAR ESTIMATES

We begin with a definition.

Definition 2.1. Any estimate $t_i(s)$, which equals 0 if s does not contain the element i, will be called an (i)-estimate; any estimate $t_{ij}(s)$, which equals 0 if s does not contain the element i or j (or both), will be called an (i, j)-estimate.

If we complete the definition of $w_i(s)$, as a function of s, putting for s not containing the element i

$$w_i(s) = 0, \quad s \text{ non } \ni i, \tag{2.1}$$

then $w_i(s)$ becomes an (i)-estimate of 1 and (1.2) may be rewritten in the following form:

$$\hat{Y} = \sum_{i=1}^{N} y_i \, w_i(s). \tag{2.2}$$

Theorem 2.1. *If for the values $z_1, \ldots, z_N$ the equation*

$$\sum_{i=1}^{N} z_i \, w_i(s) = \sum_{i=1}^{N} z_i \quad (\text{i. e. } \hat{Z} = Z) \tag{2.3}$$

holds with probability 1, then the mean-square error of the estimate (2.2) equals

$$\mathsf{M}(\hat{Y} - Y)^2 = \frac{1}{2} \sum_{i=1}^{N} \sum_{j=1}^{N} \left(\frac{y_i}{z_i} - \frac{y_j}{z_j} \right)^2 z_i z_j (\mathsf{M}w_i + \mathsf{M}w_j - 1 - \mathsf{M}w_i w_j) \tag{2.4}$$

where $\mathsf{M}t = \sum t(s) \, P(s)$ denotes the mean value over all possible s. In the unbiased case ($\mathsf{M}w_i = 1$, $i = 1, \ldots, N$) we have

$$\mathsf{M}(\hat{Y} - Y)^2 = \frac{1}{2} \sum_{i=1}^{N} \sum_{j=1}^{N} \left(\frac{y_i}{z_i} - \frac{y_j}{z_j} \right)^2 z_i z_j (1 - \mathsf{M}w_i w_j). \tag{2.5}$$

The sum of the weights standing in (2.4) at the terms $\frac{1}{2} \left(\frac{y_i}{z_i} - \frac{y_j}{z_j} \right)^2$, $i \neq j$, equals

$$\sum\sum_{i \neq j} z_i z_j (\mathsf{M}w_i + \mathsf{M}w_j - 1 - \mathsf{M}w_i w_j) = \sum_{i=1}^{N} z_i^2 \, \mathsf{M}(w_i - 1)^2. \tag{2.6}$$

Proof is based on the following easy identities:

$$\mathsf{M}(\hat{Y} - Y)^2 = \mathsf{M} \left[\sum y_i (w_i - 1) \right]^2 = \sum_{i=1}^{N} \sum_{j=1}^{N} y_i y_j \, \mathsf{M}(w_i - 1)(w_j - 1)$$

$$
\begin{aligned}
&= \sum_{i=1}^{N}\sum_{j=1}^{N} \frac{y_i}{z_i}\frac{y_j}{z_j}\, z_i z_j\, \mathsf{M}(w_i-1)(w_j-1) \\
&= \sum_{i=1}^{N}\sum_{j=1}^{N} \left[-\frac{1}{2}\left(\frac{y_i}{z_i}-\frac{y_j}{z_j}\right)^2+\frac{1}{2}\left(\frac{y_i}{z_i}\right)^2+\frac{1}{2}\left(\frac{y_j}{z_j}\right)^2\right] z_i z_j\, \mathsf{M}(w_i-1)(w_j-1) \\
&= \frac{1}{2}\sum_{i=1}^{N}\sum_{j=1}^{N}\left(\frac{y_i}{z_i}-\frac{y_j}{z_j}\right)^2 [\mathsf{M}w_i+\mathsf{M}w_j-1-\mathsf{M}w_iw_j] \\
&\quad +\sum_{i=1}^{N}\left(\frac{y_i}{z_i}\right)^2 \mathsf{M}\left[z_i(w_i-1)\sum_{j=1}^{N} z_j(w_j-1)\right].
\end{aligned}
$$

The last term, however, vanishes, since, according to (2.3), $\sum_{i=1}^{N} z_i(w_i-1)=0$ with probability 1. The identity (2.6) is implied by the same fact: we get

$$
\begin{aligned}
0 = -\mathsf{M}\left(\sum z_i(w_i-1)\right)^2 &= \sum\sum_{i\neq j} z_i z_j\, [\mathsf{M}w_i+\mathsf{M}w_j-1-\mathsf{M}w_iw_j] \\
&\quad -\sum_{i=1}^{N} z_i^2\, \mathsf{M}(w_i-1)^2
\end{aligned}
$$

which is equivalent to (2.6). Formula (2.5) for a simple estimate was given in Yates and Grundy (1953). □

As regards the estimated mean-square error, we shall use the following.

Theorem 2.2. *On replacing* 1, $\mathsf{M}w_i$, $\mathsf{M}w_iw_j$ *in* (2.4) *or* (2.5) *by any of their* (i,j)*-estimates, we get an estimated mean square error of the estimate* (2.2). *Ones of the possible unbiased* (i,j)*-estimates of* 1, $\mathsf{M}w_i$ *and* $\mathsf{M}w_iw_j$ *are the following:*

Estimate of $1 =$

$$
\begin{aligned}
&1+\frac{1}{n-1}\sum_{k\,\mathrm{non}\in s}\frac{\mathsf{P}(s-\{j\}+\{k\})}{\mathsf{P}(s)}+\frac{1}{n-1}\sum_{k\,\mathrm{non}\in s}\frac{\mathsf{P}(s-\{i\}+\{k\})}{\mathsf{P}(s)} \\
&+\frac{1}{n(n-1)}\sum_{k\,\mathrm{non}\in s}\sum_{\substack{h\,\mathrm{non}\in s\\ k\neq h}}\frac{\mathsf{P}(s-\{i,j\}+\{k,h\})}{\mathsf{P}(s)}, \quad i,j\in s. \qquad (2.7)
\end{aligned}
$$

Estimate of $\mathsf{M}w_i =$

$$
w_i(s)+\frac{1}{n-1}\sum_{k\,\mathrm{non}\in s} w_i(s-\{j\}+\{k\})\frac{\mathsf{P}(s-\{j\}+\{k\})}{\mathsf{P}(s)}, \quad i,j\in s. \qquad (2.8)
$$

Estimate of $\mathsf{M}w_iw_j =$

$$w_i(s)\, w_j(s), \quad i, j \in s, \tag{2.9}$$

where $s-\{j\}+\{k\}$ *and* $s-\{i,j\}+\{k,h\}$ *denotes the subset obtained from the subset* s *by replacing the element* j *by the element* k *or the elements* $\{i,j\}$ *by the elements* $\{k,h\}$*, respectively, and* $\sum_{k\, \text{non}\in s}$ *denotes the summation over all elements* k *not contained in* s*.*

Proof. The first assertion of the Theorem only tells that on replacing 1, $\mathsf{M}w_i$ and $\mathsf{M}w_iw_j$ in (2.4) or (2.5) by any of their (i,j)-estimates, we get a function of (s,y), i.e. an estimate (see Section 1).

Now, in view of (2.1), we have

$$\sum_{s\ni i,j} w_i(s)\, w_j(s)\, \mathsf{P}(s) = \sum_{s\subset S} w_i(s)\, w_j(s)\, \mathsf{P}(s) = \mathsf{M}\, w_iw_j,$$

where $\sum_{s\ni i,j}$ denotes the sum extended over the subsets s containing both elements i and j, by which it is shown that w_iw_j is an unbiased (i,j)-estimate of $\mathsf{M}\, w_iw_j$.

In order to prove the assertion concerning (2.8), let us note that any subset z containing the element i but not containing the element j may be converted in a subset s containing both elements i and j by omitting an element, say k, different from i, and by replacing it by j, and that this may be done in $n-1$ ways. Consequently,

$$\begin{aligned}
&\sum_{s\ni i,j} \left[w_i(s) + \frac{1}{n-1} \sum_{k\,\text{non}\in s} w_i(s-\{j\}+\{k\})\, \frac{\mathsf{P}(s-\{j\}+\{k\})}{\mathsf{P}(s)} \right] \mathsf{P}(s) \\
= \;& \sum_{s\ni i,j} w_i(s)\, \mathsf{P}(s) + \sum_{s\ni i,j} \frac{1}{n-1} \sum_{k\,\text{non}\in s} w_i(s-\{j\}+\{k\})\, \mathsf{P}(s-\{j\}+\{k\}) \\
= \;& \sum_{s\ni i,j} w_i(s)\, \mathsf{P}(s) + \sum_{\substack{s\ni i \\ s\,\text{non}\ni j}} \frac{1}{n-1}(n-1)\, w_i(z)\, \mathsf{P}(z) \\
= \;& \sum_{s\ni i} w_i(s)\, \mathsf{P}(s) \;=\; \mathsf{M}\, w_i.
\end{aligned}$$

Before proceeding to (2.7), let us note that any subset z not containing either i or j may be converted in a subset s, which does contain both elements i and j, by omitting any two of its elements, say k and h, and by replacing them by i and j, and that this may be done in $n(n-1)$ ways. Consequently,

$$\sum_{s\ni i,j} \left[1 + \frac{1}{n-1} \sum_{k\,\text{non}\in s} \frac{\mathsf{P}(s-\{j\}+\{k\})}{\mathsf{P}(s)} + \frac{1}{n-1} \sum_{k\,\text{non}\in s} \frac{\mathsf{P}(s-\{i\}+\{k\})}{\mathsf{P}(s)} \right.$$

$$
\left. + \frac{1}{n(n-1)} \sum_{\substack{k \text{ non}\in s, h \text{ non}\in s \\ k \neq h}} \sum \frac{\mathsf{P}(s - \{i,j\} + \{k,h\})}{\mathsf{P}(s)} \right] \mathsf{P}(s)
$$

$$
= \sum_{s \ni i,j} \mathsf{P}(s) + \sum_{\substack{s \ni i \\ s \text{ non}\ni j}} \frac{1}{n-1}(n-1)\,\mathsf{P}(s) + \sum_{\substack{s \ni j \\ s \text{ non}\ni j}} \frac{1}{n-1}(n-1)\,\mathsf{P}(s)
$$

$$
+ \sum_{\substack{s \text{ non}\ni i \\ s \text{ non}\ni j}} \frac{1}{n(n-1)} n(n-1)\,\mathsf{P}(s) \;=\; \sum_{s \subset S} \mathsf{P}(s) = 1,
$$

which concludes the proof. □

Remark 2.1. The (i,j)-estimates shown in Theorem 2.2 are of use in situations where $w_i(s)$ does not depend greatly on s ($s \ni i$), and they will be used in Section 4. Their scope might be widened by the following device. We may, for each (i,j) separately, select several samples containing both elements i, j, say $s_1, \ldots, s_k$, and then replace the (i,j)-estimate, say $u_{ij}(s)$, by the arithmetic mean

$$
u_{ij}^*(s_1, \ldots, s_k) = \frac{1}{k} \sum_{\nu=1}^{k} u_{ij}(s_\nu). \tag{2.10}
$$

If the probabilities π_{ij} of including both elements i and j and mean values $\mathsf{M}w_iw_j$ and $\mathsf{M}w_i$ are simple, we may use the following unbiased (i,j)-estimates:

$$
\begin{aligned}
\textit{Estimate of}\,[\mathsf{M}w_i + \mathsf{M}w_j - 1 - \mathsf{M}w_iw_j] &= \frac{\mathsf{M}w_i + \mathsf{M}w_j - 1 - \mathsf{M}w_iw_j}{\pi_{ij}} \text{ if } \{i,j\} \subset s, \\
&= 0 \text{ otherwise,}
\end{aligned} \tag{2.11}
$$

or, in the unbiased case,

$$
\begin{aligned}
\textit{Estimate of}\,[1 - \mathsf{M}w_iw_j] &= \frac{1 - \mathsf{M}w_iw_j}{\pi_{ij}} \text{ if } \{i,j\} \subset s, \\
&= 0 \text{ otherwise.}
\end{aligned} \tag{2.12}
$$

Remark 2.2. The device of ratio estimation may be useful also in estimating the mean square error. For example, in view of (2.6), we may hope that the estimated mean square error

$$
m(\hat{Y} - Y)^2 = \tag{2.13}
$$

$$
\frac{\frac{1}{2} \sum_{i \in s} \sum_{j \in s} \left(\frac{y_i}{z_i} - \frac{y_j}{z_j} \right)^2 z_i z_j [\mathsf{M}w_i + \mathsf{M}w_j - 1 - \mathsf{M}w_iw_j]_{i,j}}{\sum_{i \in s} \sum_{j \in s} z_i z_j [\mathsf{M}w_i + \mathsf{M}w_j - 1 - \mathsf{M}w_iw_j]_{i,j}} \sum_{i=1}^{N} z_i^2\, \mathsf{M}(w_i - 1)^2,
$$

where $[\cdot]_{i,j}$ denotes a property (i,j)-estimate of $[\cdot]$, will be better than the estimated mean square error

$$\mathsf{m}(\hat{Y}-Y)^2 = \frac{1}{2}\sum_{i\in S}\sum_{j\in S}\left(\frac{y_i}{z_i}-\frac{y_j}{z_j}\right)^2 z_i z_j[\mathsf{M}w_i+\mathsf{M}w_j-1-\mathsf{M}w_iw_j]_{i,j}. \quad (2.14)$$

Remark 2.3. If the estimated mean square errors (2.13) and (2.14) are too laborious, we may select randomly with equal probabilities a subset M of k couples $\{i,j\}\subset s$ without common elements, when possible, and then replace (2.13) and (2.14) by

$$\mathsf{m}(\hat{Y}-Y)^2 = \quad (2.15)$$

$$\frac{\frac{1}{2}\sum\sum_{\{i,j\}\in M}\left(\frac{y_i}{z_i}-\frac{y_j}{z_j}\right)^2 z_iz_j[\mathsf{M}w_i+\mathsf{M}w_j-1-\mathsf{M}w_iw_j]_{i,j}}{\sum\sum_{\{i,j\}\in M} z_iz_j[\mathsf{M}w_i+\mathsf{M}w_j-1-\mathsf{M}w_iw_j]_{i,j}}\sum_{i=1}^N z_i^2\,\mathsf{M}(w_i-1)^2$$

and by

$$\mathsf{m}(\hat{Y}-Y)^2 = \frac{n(n-1)}{2k}\sum\sum_{\{i,j\}\in M}\left(\frac{y_i}{z_i}-\frac{y_j}{z_j}\right)^2 z_iz_j[\mathsf{M}w_i+\mathsf{M}w_j-1-\mathsf{M}w_iw_j]_{i,j}. \quad (2.16)$$

Remark 2.4. If not the values y_i directly but unbiased estimates of them are at hand, say $\hat{y}_i$, e. g. in the case of subsampling, and if estimates $\hat{y}_i$ are mutually independent, then the formula (1.2) is changed into

$$\hat{\hat{Y}} = \sum_{i\in s}\hat{y}_i w_i(s),$$

and the formula (2.3) into

$$\mathsf{M}(\hat{\hat{Y}}-Y)^2 = \quad (2.17)$$

$$\frac{1}{2}\sum_{i=1}^N\sum_{j=1}^N \underset{2}{\mathsf{M}}\left[\left(\frac{\hat{y}_i}{z_i}-\frac{\hat{y}_j}{z_j}\right)^2\right] z_{ij}[\mathsf{M}w_i+\mathsf{M}w_j-1-\mathsf{M}w_iw_j]+\sum_{i=1}^N \underset{2}{\mathsf{D}}(\hat{y}_i),$$

where $\underset{2}{\mathsf{M}}(\cdot)$ and $\underset{2}{\mathsf{D}}(\cdot)$ denote the mean value and the variance over the estimates $\hat{y}_i$ (for example, over the subsampling).

It means that the estimated mean square error has, generally, the form

$$\mathsf{m}(\hat{\hat{Y}} - Y)^2 = \frac{1}{2}\sum_{i\in s}\sum_{j\in s}\left(\frac{\hat{y}_i}{z_i} - \frac{\hat{y}_j}{z_j}\right)^2 [\mathsf{M}w_i + \mathsf{M}w_j - 1 - \mathsf{M}w_i w_j]_{i,j} + \hat{D}_2, \quad (2.18)$$

where $\hat{D}_2$ is an estimate of $\sum_{i=1}^{N} \underset{2}{\mathsf{D}}(\hat{y}_i)$.

3 IMPROVING ESTIMATES CONNECTED WITH ELEMENT–BY–ELEMENT SAMPLING WITH AND WITHOUT REPLACEMENT

We have denoted by (s'', y) the observation consisting in a knowledge of s'' and of values y_i associated with elements appearing in s'', and a similar meaning has been ascribed to (s, y) and (s', y). It is easily seen that the knowledge of (s'', y) enables us to establish (z'', y) for all z'' such that

$$s(z'') = s(s''), \quad (3.1)$$

or, similarly, knowledge of (s', y) enables us to establish (z', y) for all z' such that $s(z') = s(s')$. For example, if we know that on the sequence of elements $s''=(2, 4, 3, 2, 4)$ there were values $(13, 18, 15, 13, 16)$, respectively, we may infer that on the sequence $z'' = (3, 3, 2, 4)$ would be ascertained values (15, 15, 13, 18), respectively, because we simply know that $y_3 = 15$, $y_2 = 13$ and $y_4 = 18$. Conversely, knowing (z'', y) for any z'' such that (3.1) holds, we can establish (s'', y).

This trivial fact has an interesting application. Having an arbitrary estimate

$$t'' = t''(s'', y), \quad (3.2)$$

we may replace it by the estimate

$$t = t(s, y) = \frac{\sum_{[s]} t''(z'', y)\,\mathsf{P}(z'')}{\sum_{[s]} \mathsf{P}(z'')}, \quad (3.3)$$

where the sum $\sum_{[s]}$ extends over all z'' such that $s(z'') = s = s(s'')$. Indeed, if we know (s'', y), we can evaluate $t''(z'', y)$ for any z'' such that (3.1) holds, and hence we can evaluate (3.3).

A brief inspection of equation (3.3) shows that t is a conditional mean value of t'' with respect to (s, y). Consequently

$$\mathsf{M}t = \mathsf{M}t'' \quad (3.4)$$

$$\mathsf{D}t = \mathsf{D}t'' - \mathsf{M}(t - t'')^2. \quad (3.5)$$

In other words, t is at worst as equally good an estimate as t''. This means that the subset definition of a sample is fully satisfactory, since any good estimate is a function of (s, y) only. (Of course, it may happen that t'' and the estimated variance of t'' are easier to compute than t and the estimated variance of t.)

A similar conclusion may be drawn concerning the estimates $t' = t'(s', y)$, etc.

Now, let us show how the method works.

Example 3.1. Let us perform independent samples of one element with probabilities $\alpha_1, \ldots, \alpha_N$, $\sum_{i=1}^N \alpha_i = 1$, until n distinct elements have been selected. Let us choose any unbiased estimator of the total $Y = \sum_{i=1}^N y_i$, for example,

$$\hat{Y} = \frac{y_{i_1}}{\alpha_{i_1}} \tag{3.6}$$

where i_1 denotes the element selected as the first. Let us take a conditional mean value of $\hat{Y}$ with respect to (s, y), where s is the set of n distinct elements included in the sample. Using the formula (3.3) we see that

$$\tilde{Y} = \mathsf{M}(\hat{Y} \mid (s, y)) = \sum_{i \in s} \frac{y_i}{\alpha_i} \mathsf{P}_i(s), \tag{3.7}$$

where $\mathsf{P}_i(s)$ is the conditional probability that the element i will be included as the first in the sample under the condition that the distinct elements selected consist of the set s. Probabilities $\mathsf{P}_i(s)$ are not easy to compute, except when the sampling is uniform, i. e. $\alpha_1 = \cdots = \alpha_N = \frac{1}{N}$, or when $n = 2$.

In the case when the sampling is uniform, we clearly get $\mathsf{P}_i(S) = \frac{1}{n}$, so that

$$\tilde{Y} = \frac{N}{n} \sum_{i \in s} y_i.$$

Now, let us consider the case $n = 2$, denoting the two distinct elements included in the sample by i and j. The probability that the element i was selected first and the element j second is

$$P_{ij} = \mathsf{P}\{s' = [i, j]\} = \frac{\alpha_i \, \alpha_j}{1 - \alpha_i}. \tag{3.8}$$

Similarly, the probability that the element j was selected first and the element i second is

$$P_{ji} = \mathsf{P}\{s' = [j, j]\} = \frac{\alpha_i \, \alpha_j}{1 - \alpha_j}. \tag{3.9}$$

This means that the conditional probability that the element i has been selected first is

$$\mathsf{P}_i(s) = \frac{1 - \alpha_j}{2 - \alpha_i - \alpha_j}. \tag{3.10}$$

When substituting into (3.7) we get

$$\tilde{Y} = \frac{(1-\alpha_j)\,\frac{y_i}{\alpha_i} + (1-\alpha_i)\,\frac{y_j}{\alpha_j}}{2-\alpha_i-\alpha_j}. \tag{3.11}$$

As regards the variance of $\tilde{Y}$, we may use the formula (2.5) since (2.3) is clearly satisfied for $z_i = \alpha_i$. It means that

$$\mathsf{M}(\tilde{Y}-Y)^2 = \frac{1}{2}\sum_{i=1}^{N}\sum_{j=1}^{N}\left(\frac{y_i}{\alpha_i}-\frac{y_j}{\alpha_j}\right)\alpha_i\alpha_j[1-\mathsf{M}w_iw_j]. \tag{3.12}$$

In order to find $\mathsf{M}w_iw_j$ let us first compute π_{ij}:

$$\begin{aligned}\pi_{ij} &= \mathsf{P}\{i,j\in s\} = \mathsf{P}\{s=\{i,j\}\} = P_{ij}+P_{ji} = \frac{\alpha_i\,\alpha_j}{1-\alpha_i}+\frac{\alpha_j\,\alpha_i}{1-\alpha_j}\\ &= \frac{\alpha_i\alpha_j(2-\alpha_i-\alpha_j)}{(1-\alpha_i)\,(1-\alpha_j)}.\end{aligned} \tag{3.13}$$

Now, in view of (3.11),

$$w_i(s) = \frac{1-\alpha_i}{2-\alpha_i-\alpha_j}\,\frac{1}{\alpha_i} \quad \text{for } s=\{i,j\}$$

and, consequently,

$$\begin{aligned}\mathsf{M}w_iw_j &= \sum_{s\ni i,j} w_i(s)\,w_j(s)\,\mathsf{P}(s) = w_iw_j\pi_{ij}\\ &= \frac{\alpha_i\alpha_j(2-\alpha_i-\alpha_j)}{(1-\alpha_i)\,(1-\alpha_j)}\,\frac{(1-\alpha_j)\,(1-\alpha_i)}{(2-\alpha_i-\alpha_j)^2}\,\frac{1}{\alpha_i\alpha_j} = \frac{1}{2-\alpha_i-\alpha_j}.\end{aligned}$$

On substituting (3.13) into (3.12) we get that

$$\mathsf{M}(\tilde{Y}-Y)^2 = \frac{1}{2}\sum_{i=1}^{N}\sum_{j=1}^{N}\left(\frac{y_i}{\alpha_i}-\frac{y_j}{\alpha_j}\right)^2\alpha_i\alpha_j\frac{1-\alpha_i-\alpha_j}{2-\alpha_i-\alpha_j}. \tag{3.14}$$

Finally, we may use the (i,j)-estimate (2.12), where π_{ij} are given by (3.13), and get the following unbiased estimated variance:

$$d\tilde{Y} = \mathsf{m}(\tilde{Y}-Y)^2 = \left(\frac{y_i}{\alpha_i}-\frac{y_j}{\alpha_j}\right)^2\frac{(1-\alpha_i)\,(1-\alpha_j)\,(1-\alpha_i-\alpha_j)}{(2-\alpha_i-\alpha_j)^2}. \tag{3.15}$$

As $(1-\alpha_i)\,(1-\alpha_j) \le \left(1-\frac{\alpha_i+\alpha_j}{2}\right)^2$, we have that

$$d\tilde{Y} \le \frac{1}{4}\left(\frac{y_i}{\alpha_i}-\frac{y_j}{\alpha_j}\right)^2(1-\alpha_i-\alpha_j). \tag{3.16}$$

If the ordered-set sample s' consists of elements $s' = [i_1, \ldots, i_n]$, then there are the following unbiased estimates of Y:

$$\begin{aligned}\hat{Y}_1 &= \frac{y_i}{\alpha_{i_1}}, \\ \hat{Y}_2 &= y_{i_1} + \frac{y_{i_2}}{\alpha_{i_2}}(1 - \alpha_{i_1}), \\ &\vdots \\ \hat{Y}_n &= y_{i_1} + \cdots + y_{i_{n-1}} + \frac{y_{i_n}}{\alpha_{i_n}}(1 - \alpha_{i_1} - \cdots - \alpha_{i_{n-1}}),\end{aligned} \tag{3.17}$$

where i_k is the element included in the sample as the kth. These estimates are noncorrelated, and, moreover, the conditional mean value of $\hat{Y}_k$, given $\hat{Y}_1, \ldots, \hat{Y}_{k-1}$, equals Y. Consequently, any constants $c_1, \ldots, c_n$, $\sum_{i=1}^n c_i = 1$, generate an unbiased estimate $\hat{Y} = \sum_{i=1}^n c_i \hat{Y}_i$ such that

$$\mathsf{D}\left(\sum_{i=1}^n c_i \hat{Y}_i\right) = \sum_{i=1}^n c_i^2 \, \mathsf{D}\, \hat{Y}_i. \tag{3.18}$$

A further unbiased estimate might be

$$\frac{1}{v} \sum_{s''} \frac{y_i}{\alpha_i}, \tag{3.19}$$

where the sum $\sum_{s''}$ extends over all elements of the sequence s'' and v denotes the number of members of s'' (i. e. the number of independent selections of one element until n distinct elements have been selected).

The estimate (3.19) may be identified as a conditional mean value of (3.1) with respect to (s^*, y), where s^* is the orderless-sequence sample defined in Section 1:

$$\frac{1}{v} \sum_{s''} \frac{y_i}{\alpha_i} = \mathsf{M}\left(\frac{y_{i_1}}{\alpha_{i_1}} \middle| (s^*, y)\right). \tag{3.20}$$

This means, in view of (1.9), that the conditional mean of (3.19) with respect to (s, y) also equals (3.7). In addition, it is thereby proved that (3.19) is an unbiased estimate.

Example 3.2. Let us perform a fixed number, say m, of independent selections of one element always with probabilities $\alpha_1, \ldots, \alpha_N$ and consider the well-known estimate

$$\hat{Y} = \frac{1}{m} \sum_{s''} \frac{y_i}{\alpha_i} \tag{3.21}$$

where the sum $\sum_{s''}$ extends over the selected sequence $s'' = (i_1, \ldots, i_m)$. It may be easily shown that the estimate (3.21) is a conditional mean value of (3.1) with respect to (s^*, y), where s^* is again the orderless-sequence sample.

If the sampling is uniform, i.e. $\alpha_1 = \cdots = \alpha_N = \frac{1}{N}$, then the conditional mean value of (3.21) with respect to (s, y) is easily seen to be

$$\tilde{Y} = \mathsf{M}(\hat{Y} \mid (s, y)) = \frac{N}{d} \sum_{i \in s} y_i, \tag{3.22}$$

where d is the number of distinct elements in the sample and s is the set of the distinct elements in the sample.

If the probabilities α_i are varying we may get $\tilde{Y}$ as follows. First let us rewrite $\hat{Y}$ in the form

$$\hat{Y} = \frac{1}{m} \sum_{i \in s} \frac{y_i}{\alpha_i} + \frac{1}{m} \sum_{i \in s} (k_i - 1) \frac{y_i}{\alpha_i} \tag{3.23}$$

where k_i is the number of times the element i has been selected. The conditional probabilities of events $\{k_i = k_i^0,\ i \in s\}$ obviously equal

$$\text{const} \prod_{i \in s} \frac{\alpha_i^{k_i^0 - 1}}{k_i^0!}, \quad k_i^0 \geq 1, \quad \sum_{i \in s} k_i^0 = m. \tag{3.24}$$

If all elements in s'' are distinct then $\tilde{Y} = \hat{Y}$. If there are $m - 1$ or $m - 2$ distinct elements in the sample s'', then (3.24) generates the following conditional distribution of $\sum_{i \in s} (k_i - 1)\, y_i / \alpha_i$:

$$\mathsf{P}\left\{ \sum_{i \in s} (k_i - 1) \frac{y_i}{\alpha_i} = \frac{y_a}{\alpha_a} \,\middle|\, (s, y) \right\} = c\alpha_a, \ a \in s, \ \begin{bmatrix} m-1 \text{ distinct} \\ \text{elements} \end{bmatrix} \tag{3.25}$$

$$\left.\begin{aligned} &\mathsf{P}\left\{ \sum_{i \in s} (k_i - 1) \tfrac{y_i}{\alpha_i} = \tfrac{y_a}{\alpha_a} + \tfrac{y_b}{\alpha_b} \,\middle|\, (s, y) \right\} = c \tfrac{\alpha_a \alpha_b}{2!2!} \\ &\mathsf{P}\left\{ \sum_{i \in s} (k_i - 1) \tfrac{y_i}{\alpha_i} = 2 \tfrac{y_a}{\alpha_a} \,\middle|\, (s, y) \right\} = c \tfrac{\alpha_a^2}{3!} \end{aligned} \right\} a, b \in s, \ \begin{bmatrix} m-2 \text{ distinct} \\ \text{elements} \end{bmatrix}, \tag{3.26}$$

where c is a constant. From (3.25) and (3.26), after some computations, we get that

$$\tilde{Y} = \frac{1}{m} \left[\sum_{i \in s} \frac{y_i}{\alpha_i} + \frac{\sum_{i \in s} y_i}{\sum_{i \in s} \alpha_i} \right], \quad \begin{bmatrix} m-1 \text{ distinct} \\ \text{elements} \end{bmatrix}, \tag{3.27}$$

$$\tilde{Y} = \frac{1}{m} \left[\sum_{i \in s} \frac{y_i}{\alpha_i} + 2 \frac{\sum_{i \in s} y_i \left(1 + \frac{1}{3} \frac{\alpha_i}{\sum_{j \in s} \alpha_j} \right)}{\sum_{i \in s} \alpha_i \left(1 + \frac{1}{3} \frac{\alpha_i}{\sum_{j \in s} \alpha_j} \right)} \right], \quad \begin{bmatrix} m-2 \text{ distinct} \\ \text{elements} \end{bmatrix}. \tag{3.28}$$

If there are less then $m - 2$ distinct elements, then precise evaluation of $\tilde{Y}$ becomes too complicated. We may use, however, the approximation

$$\tilde{Y} = \frac{1}{m}\left[\sum_{i \in s} \frac{y_i}{\alpha_i} + r\,\frac{\sum_{i \in S} y_i}{\sum_{i \in s} \alpha_i}\right], \quad \left[\begin{matrix} m - r \text{ distinct} \\ \text{elements} \end{matrix}\right] \tag{3.29}$$

which seems to be a good one.

Example 3.3. Let us consider a double sampling design and denote by s_1 and s ($s \subset s_1$) the subsets of elements selected in the first and second phase, respectively. Let

$$\hat{Y} = \sum_{i \in s_1} y_i\, w_{i1}(s_1) \tag{3.30}$$

be an estimate based on the sample s_1. Now, we may judge the sample s_1 as a population with ascertained values $y_i\, w_i(s_1)$, $i \in s_1$, and construct a linear estimate

$$\hat{\hat{Y}} = \sum_{i \in s} y_i\, w_{i1}(s_1)\, w_{i2}(s_1, s) \tag{3.31}$$

whose weights depend on s_1. As the final observation (s, y) is, obviously, an abstract function of (s_1, s, y), and good estimate $\hat{\hat{Y}}$ must not depend on s_1, we must have

$$w_{i1}(s_1)\, w_{i2}(s_1, s) = w_i(s), \quad s \subset s_1 \subset S. \tag{3.32}$$

If (3.32) does not hold, the estimate may be improved by the method we have used in Examples 3.1 and 3.2.

If $\hat{Y} = \underset{2}{\mathsf{M}}\, \hat{\hat{Y}}$, where $\underset{2}{\mathsf{M}}\,(\cdot)$ denotes the mean value with respect to the second phase (or stage) of sampling, then

$$\mathsf{M}(\hat{\hat{Y}} - Y)^2 = \mathsf{M}(\hat{Y} - Y)^2 + \mathsf{M}(\hat{Y} - \hat{\hat{Y}})^2. \tag{3.33}$$

Example 3.4. The same considerations may be applied to sampling whose result is given by k interpenetrating samples $(s_1, \ldots, s_k)$. We come to the conclusion that the 'good' estimates must not depend on how many times an element has been selected, i. e. it must be a function of the set s of elements contained in at least one set $s_1, \ldots, s_k$ and of the observations ascertained thereon.

Now we shall leave this topic, since, as will be shown in the Section 4, there exists an exact theory of fixed-size sampling with varying probabilities without replacement.

Remark 3.1. The method could be formulated as an application of the well-known Rao–Blackwell theorem on improving estimates by taking their conditional mean value with respect to a sufficient statistic. In fact, in the space of all possible observations (s'', η) we may consider the system $\{\mathsf{P}_{y_1 \cdots y_N}(\cdot)\}$ of admissible distributions generated by all possible sequences of values $y_1, \ldots, y_N$ in such a way that

$$\begin{aligned} \mathsf{P}_{y_1 \cdots y_N}(s'', \eta) &= \mathsf{P}(s''), \quad \text{if } (s'', \eta) = (s'', y), \\ &= 0, \qquad\quad \text{otherwise.} \end{aligned}$$

Let us note that $(y_1, \ldots, y_N)$ plays the role of a parameter.

4 REJECTIVE SAMPLING

Let us perform n independent draws of one element always with probabilities $\alpha_1, \ldots, \alpha_N$ and accept or reject all the selected elements if, or if not, at no two draws the same elements have been selected, respectively. If the sample is rejected, let us repeat this procedure until we get an acceptable sample. In this well-known sampling scheme, we have, obviously

$$\mathsf{P}(s) = \lambda \prod_{i \in s} \alpha_i, \text{ for any } s \text{ consisting of } n \text{ elements,} \tag{4.1}$$

where

$$\lambda = \left[\sum_{s \in V_n} \prod_{i \in s} \alpha_i \right]^{-1},$$

where V_n denotes the class of all subsets of the population which consist of n elements. Our point is to show that the sampling design just described is capable of exact and easy treatment.

First, let us observe that any sample z not containing the element i may be converted in a sample which does contain the element i by omitting any one of its elements and replacing it by the element i. This may be done in n ways as z contains n elements. If the obtained sample is s, then, by (4.1),

$$\mathsf{P}(z) = \frac{\alpha_k}{\alpha_i} \mathsf{P}(s),$$

where k is the element which has been omitted. This means that

$$\begin{aligned} 1 &= \sum_{s \ni i} \mathsf{P}(s) + \sum_{z \text{ non} \ni i} \mathsf{P}(z) = \sum_{s \ni i} \mathsf{P}(s) + \frac{1}{n} \sum_{s \ni i} \mathsf{P}(s) \sum_{k \text{ non} \in s} \frac{\alpha_k}{\alpha_i} \\ &= \sum_{s \ni i} \mathsf{P}(s) \left[1 + \frac{1}{n\alpha_i} \sum_{k \text{ non} \in s} \alpha_k \right] = \sum_{s \ni i} \mathsf{P}(s) \frac{1 - \sum_{k \in s} \alpha_k + n\alpha_i}{n\alpha_i} \end{aligned} \tag{4.2}$$

where $\sum_{z\,\mathrm{non}\ni i}$ extends over all samples not containing the element i. Denoting

$$\alpha = \sum_{k\in s} \alpha_k, \quad w_i(s) = \frac{1-\alpha+n\alpha_i}{n\alpha_i}, \quad i \in s, \tag{4.3}$$

we may rewrite (4.2) in the form

$$\sum_{s\ni i} w_i(s)\,\mathsf{P}(s) = 1. \tag{4.4}$$

Equation (4.4), however, means that $w_i(s)$ are weights of the unbiased estimate

$$\hat{Y} = \sum_{i\in s} y_i \frac{1-\alpha+n\alpha_i}{n\alpha_i} = \frac{1-\alpha}{n}\sum_{i\in s}\frac{y_i}{\alpha_i} + \sum_{i\in s} y_i. \tag{4.5}$$

The estimate (4.5) equals Y identically when y_i are exactly proportional to numbers α_i, $i = 1,\ldots,N$, i.e. (2.3) is satisfied for $z_i = \alpha_i$. Consequently, according to (2.11), we have

$$\mathsf{D}\hat{Y} = \frac{1}{2}\sum_{i\in s}\sum_{j\in s}\left(\frac{y_i}{\alpha_i} - \frac{y_j}{\alpha_j}\right)^2 \alpha_i\alpha_j(1-\mathsf{M}w_iw_j). \tag{4.6}$$

We shall seek the (i,j)-estimates of 1 and of $\mathsf{M}w_iw_j$ in the form (2.7) and (2.9). Let us observe that, in view of (4.1),

$$\frac{\mathsf{P}(s-\{i\}+\{k\})}{\mathsf{P}(s)} = \frac{\alpha_k}{\alpha_i},$$
$$\frac{\mathsf{P}(s-\{i,j\}+\{k,h\})}{\mathsf{P}(s)} = \frac{\alpha_k\alpha_h}{\alpha_i\alpha_j},$$

and substitute these results into (2.7) and (2.9). We get:

Estimate of 1 $=$

$$1 + \frac{1}{n-1}\sum_{k\,\mathrm{non}\in s}\frac{\alpha_k}{\alpha_j} + \frac{1}{n-1}\sum_{k\,\mathrm{non}\in s}\frac{\alpha_k}{\alpha_i}$$

$$+\frac{1}{n(n-1)} + \sum_{\substack{k\,\mathrm{non}\in s,\, h\,\mathrm{non}\in s\\ k\neq h}}\sum \frac{\alpha_k\alpha_h}{\alpha_i\alpha_j}$$

$$= \frac{(1-\alpha+n\alpha_i)\,(1-\alpha+n\alpha_j) - n\alpha_i\alpha_j - \sum_{k\,\mathrm{non}\in s}\alpha_k^2}{n(n-1)\,\alpha_i\alpha_j}, \quad i,j\in s,$$

Estimate of $\mathsf{M}w_iw_j$ $=$

$$w_iw_j = \frac{(1-\alpha+n\alpha_i)\,(1-\alpha+n\alpha_j)}{n^2\alpha_i\alpha_j}.$$

This gives:
Estimate of $[1 - \mathsf{M}w_i w_j] =$

$$\frac{(1-\alpha+n\alpha_i)(1-\alpha+n\alpha_j)}{n^2(n-1)} - \frac{\alpha_i\alpha_j}{n-1} - \frac{\sum_{k\,\text{non}\in s}\alpha_k^2}{n(n-1)}. \tag{4.7}$$

On substituting (4.7) into (4.6) we obtain the following unbiased estimated variance:

$$\begin{aligned} d\hat{Y} &= \frac{1}{2}\frac{1}{n-1}\sum_{i\in s}\sum_{j\in s}\left(\frac{y_i}{\alpha_i}-\frac{y_j}{\alpha_j}\right)^2\left[\frac{(1-\alpha+n\alpha_i)}{n}\,\frac{(1-\alpha+n\alpha_j)}{n}\right. \\ &\quad \left. -\alpha_i\alpha_j - \frac{1}{n}\sum_{k\,\text{non}\in s}\alpha_k^2\right]. \end{aligned} \tag{4.8}$$

Now, remembering that, for any numbers p_i, $\sum_{i\in s} p_i = 1$, the identity

$$\frac{1}{2}\sum_{i\in s}\sum_{j\in s}\left(\frac{y_i}{\alpha_i}-\frac{y_j}{\alpha_j}\right)^2 p_i p_j = \sum_{i\in s}\left(\frac{y_i}{\alpha_i}-\sum_{i\in s}\frac{y_i}{\alpha_i}p_i\right)^2 p_i$$

holds, we can rewrite (4.8) in the following form:

$$\begin{aligned} d\hat{Y} &= \frac{1}{n-1}\left[\sum_{i\in s}\left(\frac{y_i}{\alpha_i}-\hat{Y}\right)^2\frac{1-\alpha+n\alpha_i}{n}\right. \\ &\quad -n\sum_{k\,\text{non}\in s}\alpha_k^2\sum_{i\in s}\left(\frac{y_i}{\alpha_i}-\frac{1}{n}\sum_{i\in s}\frac{y_i}{\alpha_i}\right)^2\frac{1}{n} \\ &\quad \left. -\alpha^2\sum_{i\in s}\left(\frac{y_i}{\alpha_i}-\frac{\sum_{k\in s}y_k}{\sum_{k\in s}\alpha_k}\right)^2\frac{\alpha_i}{\alpha}\right]. \end{aligned} \tag{4.9}$$

The relative magnitudes of the terms on the right-hand side are 1, $\frac{n}{N}\left(1-\frac{n}{N}\right)$ and $\left(\frac{n}{N}\right)^2$, respectively. If $\alpha_1 = \cdots = \alpha_N = \frac{1}{N}$, we get the well-known formula for simple random sampling.

If we wish to exploit the proportionality of values y_i to certain values x_i, we may use the ratio estimate

$$\breve{Y} = \frac{\frac{1-\alpha}{n}\sum_{i\in s} y_i/\alpha_i + \sum_{i\in s} y_i}{\frac{1-\alpha}{n}\sum_{i\in s} x_i/\alpha_i + \sum_{i\in s} x_i}\sum_{i=1}^{N} x_i = f\,X. \tag{4.10}$$

Alternatively, we may also use the unbiased ratio-type estimate

$$\breve{Y} = \sum_{i\in S} y_i + \frac{1-\alpha}{n-1}\sum_{i\in s}\frac{y_i}{\alpha_i} + \frac{1}{n}\sum_{i\in s}\frac{y_i}{x_i}\left[X - \sum_{i\in s} x_i - \frac{1-\alpha}{n-1}\sum_{i\in S}\frac{x_i}{\alpha_i}\right] \tag{4.11}$$

which is a variant of the estimate introduced by Goodman and Hartley in the case of simple random sampling (see Goodman and Hartley (1958)). To show the unbiasedness, let us rewrite (4.11) in the form

$$\begin{aligned}\breve{Y} &= \sum_{i\in s} y_i + \frac{1-\alpha}{n}\sum_{i\in s}\frac{y_i}{\alpha_i} \\ &\quad + \frac{1}{n}\sum_{i\in s}\frac{y_i}{x_i}\left[X - x_i - \sum_{j\in s-\{i\}} x_j - \frac{1-\alpha}{n-1}\sum_{j\in s-\{i\}}\frac{x_j}{\alpha_j}\right]. \quad (4.12)\end{aligned}$$

The first two terms on the right-hand side are nothing else but the unbiased estimate (4.5) of Y. Consequently, if we show that the conditional mean value of

$$\sum_{j\in s-\{i\}} x_j + \frac{1-\alpha}{n-1}\sum_{j\in s-\{i\}}\frac{x_j}{\alpha_j} \tag{4.13}$$

under the condition $s \ni i$ equals $X - x_i$, our proof will be completed. However, from (4.1) it is easily seen that conditional probabilities of s under the condition $s \ni i$, say $P(s \mid i)$, equal

$$\mathsf{P}(s \mid i) = \lambda_i \prod_{j\in s-\{i\}} \alpha_j, \quad s \ni i,$$

i. e. a conditional distribution has the same structure as the nonconditional one. Now the estimate (4.13) has the same form as the unbiased estimate (4.5) except that n is replaced by $n-1$, probabilities $\alpha_1, \ldots, \alpha_N$ are replaced by probabilities $\alpha_1/(1-\alpha_i), \ldots, \alpha_{i-1}/(1-\alpha_i)$, $\alpha_{i+1}/(1-\alpha_i), \ldots, \alpha_N/(1-\alpha_i)$, and the element i is omitted from the population. Consequently, the estimate (4.13) actually is an unbiased estimate of $X - x_i$ under the condition $s \ni i$.

The variances of (4.10) and (4.11) are naturally complicated but they may be estimated in lines with Theorem 2.2, since (2.3) is satisfied for $z_i = x_i$.

5 PERMUTATION SAMPLING

The rule of including and not including the element i in the sample is the following. We take a random permutation $R_1, \ldots, R_N$ of numbers $1, \ldots, N$, all permutations having the same probability, and then include or not include the element i in the sample if or if not

$$R_i \leq \pi_i N, \quad i = 1, \ldots, N. \tag{5.1}$$

If the numbers $\pi_i N$ are not integers, we may replace R_i by $R_i - \xi_i$, where ξ_i are independent random variables distributed uniformly over the interval $(0, 1)$.

It is easily seen that the numbers π_i used in (5.1) are directly probabilities of including the element in the sample. If $\pi_i \leq \pi_j$, and the element i has been included in the sample, than, because $R_j \neq R_i$ and $R_i \leq \pi_i N \leq \pi_j N$, R_j may take on $\pi_j N - 1$ integers not greater than $\pi_j N$ and $N - \pi_j N$ integers greater than $\pi_j N$, each of them with the same probability $\frac{1}{N-1}$. Consequently, the probability of including both elements i and j in the sample equals

$$\pi_{ij} = \pi_i \frac{\pi_j N - 1}{N - 1}, \quad [\pi_i \leq \pi_j]. \tag{5.2}$$

When we are using the above mentioned random variables ξ_i, the formula for π_{ij} must be slightly modified.

The permutation sampling is useful, for example, in connection with the ratio estimate

$$\breve{Y} = \frac{\sum_{i \in s} y_i/\pi_i}{\sum_{i \in s} x_i/\pi_i} \sum_{i=1}^{N} x_i = f\,X. \tag{5.3}$$

Practical performation of permutation sampling goes as follows. We decide that the π_is should be proportional to the numbers a_i and that the mean sample size should be $\overline{n}$. Then we compute the number

$$\psi = \frac{\sum_{i=1}^{N} a_i}{\overline{n} N} = \frac{\overline{a}}{\overline{n}}.$$

Now we select elements by simple random sampling element-by-element without replacement and the element selected as the kth, say i_k, we accept or reject if or if not $a_{i_k} \geq k\psi$. It is easily seen that, if $a_* = \min_{1 \leq i \leq N} a_i$ and $a^* = \max_{1 \leq i \leq N} a_i$, the first $k \leq k_* = a_*/\psi$ elements are accepted certainly, and, on the other hand, the sampling is certainly finished when $k \geq k^* = a^*/\psi$.

We may observe that in permutation sampling we need to know the a_is only for the first $k \leq a^*/\psi$ selected elements and that no sums of a_is are needed.

B. OPTIMUM STRATEGY. CASE I: THE ASCERTAINED VALUES ARE UNCORRELATED (6 – 8)

The assumption that the ascertained values are (a priori) noncorrelated is of fundamental importance and, as the reader will see in Section 8, of rather wide scope. For previous results see Godambe (1955).

6 BAYES APPROACH TO THE OPTIMUM STRATEGY

Sampling–estimating strategy is defined by probabilities $\mathsf{P}(s)$, $s \subset S$, and by weights $w_i(s)$, $i \in s$, $s \subset S$. The quality of this strategy will be judged, on the one side, by the mean square error

$$\mathsf{M}(\hat{Y} - Y)^2 = \sum_{s \subset S} \left(\sum_{i \in s} y_i \, w_i(s) - \sum_{i=1}^{N} y_i \right)^2 \mathsf{P}(s), \tag{6.1}$$

and, on the other side, by costs which generally have the form

$$C = \sum_{s \subset S} c(s, w) \, \mathsf{P}(s) \tag{6.2}$$

where $c(s, w)$, $s \subset S$, are the costs of ascertaining the values y_i on s, and of computing $\hat{Y}$ with the weights $w_i(s)$, $i \in s$; perhaps in $c(s, w)$ may be included the costs of computing the estimated sampling error, i. e. the estimated square root of (6.1). The right-hand side of (6.2) is called a cost function.

Sampling–estimating may be optimum in two senses: it either minimizes the mean square error (6.1) for a given expected total cost (6.2), or conversely it minimizes the expected total cost (6.2) for a given mean square error (6.1).

The notion of optimum strategy is useless when only one particular sequence $y_1, \ldots, y_N$ is considered, because the problem disappears if we know it, and has no solution if we do not know it. This difficulty may be overcome in various ways. In this chapter we prefer the Bayes approach. It means we shall suppose that there is a certain probability distribution in the space of sequences $(y_1, \ldots, y_N)$, and our criterion of accuracy will be

$$\mathsf{EM}(\hat{Y} - Y)^2 \tag{6.3}$$

where E denotes the mean value over the random sequence $(y_1, \ldots, y_N)$ and M denotes the mean value over the samples s.

As can be seen from (6.1), $\mathsf{M}(\hat{Y} - Y)^2$ is a quadratic form in y_js, so that $\mathsf{EM}(\hat{Y} - Y)^2$ depends only on the mean values, covariances and variances of $y_1, \ldots, y_N$. Consequently, specification of the distribution of $y_1, \ldots, y_N$ may only consist in the determination of the first- and second-order moments:

$$\mathsf{E} y_i = \mu_i, \quad \mathsf{V} y_i = d_{ii}, \quad \operatorname{Cov}(y_i, y_j) = d_{ij}, \quad i, j = 1, \ldots, N \tag{6.4}$$

where V denotes the variance in the $(y_1, \ldots, y_N)$-space.

The Bayes approach seems to be reasonable on these grounds: (a) in most cases, we really have some knowledge of conditions producing values $y_1, \ldots, y_N$, and we can express them in the form (6.4); (b) assumptions (6.4), if accepted, only influence our choice of sampling–estimating strategy and do not influence the validity of our estimated sampling errors, confidence intervals etc. Sampling error will be valid for any particular sequence $y_1, \ldots, y_N$ and, consequently, any mistake in assumptions (6.4) will only cause the sampling errors to be on the average greater than they would be if our assumptions were right.

There are, of course, cases when the sample is selected in such a way that the sampling error cannot be estimated on the basis of ascertained values $y_1, \ldots, y_N$ only (e. g. in systematic sampling). However, even in these cases, the Bayes approach is the only way of getting any estimated sampling error at all. We shall conclude with a few remarks.

Remark 6.1. The first to use the Bayes approach for the solution of sampling strategy was, to the author's knowledge, Cochran, in the pioneering paper Cochran (1946).

Remark 6.2. As emphasized by R. A. Fisher, any statistical work is a whole consisting of two aspects: *experimental design* and *statistical procedure*. This applies to probability sampling, too, the two aspects being the *sampling design* and the *estimation procedure*.

Remarks 6.3. Not only in probability sampling but in any statistical work there is a fundamental distinction between the case where the a priori distribution is an *organic part* of the statistical procedure (i. e. influences the validity of the probability statements) and the case where it only influences the *choice* of the experimental design and of the statistical procedure.

7 SUFFICIENT CONDITIONS FOR THE OPTIMUM STRATEGY

We shall restrict ourselves to unbiased linear estimators, i. e. $\sum_{s \subset S} w_i(s)\,\mathsf{P}(s) = 1$, $i = 1, \ldots, N$, and to the simple cost function

$$C = \sum_{i=1} c_i \pi_i. \tag{7.1}$$

(7.1) is based on the assumption that the cost associated with the element i equals c_i, i. e. does not depend on the estimation method and on which other elements were selected.

Supposing that the y_ss are noncorrelated and that the estimator $\hat{Y}$ is unbiased, let us evaluate (6.3). First, we change the order of the mean-value operators E and M, and then make use of the assumption that the y_ss are uncorrelated:

$$\mathsf{EM}(\hat{Y}-Y)^2 = \mathsf{ME}(\hat{Y}-Y)^2 = \mathsf{ME}\left(\sum_{i\in s} y_i w_i(s) - \sum_{i=1}^{N} y_i\right)^2$$

$$= \mathsf{M}\left\{\left(\sum_{i\in s}\mu_i w_i(s) - \sum_{i=1}^{N}\mu_i\right)^2 + \sum_{i\in s} d_{ii}(w_i(s)-1)^2 + \sum_{i\,\mathrm{non}\in s} d_{ii}\right\}$$

$$= \mathsf{M}\left(\sum_{i\in s}\mu_i w_i(s) - \sum_{i=1}^{N}\mu_i\right)^2$$

$$+\sum_{i=1}^{N} d_{ii}\left[\sum_{s\ni i}(w_i(s)-1)^2\,\mathsf{P}(s) + \sum_{s\,\mathrm{non}\ni i}\mathsf{P}(s)\right] \tag{7.2}$$

where $\mu_i = \mathsf{E}y_i$, $d_{ii} = \mathsf{V}y_i$. Bearing in mind (1.3) and (1.4) we get

$$\sum_{s\ni i}(w_i(s)-1)^2\,\mathsf{P}(s) + \sum_{s\,\mathrm{non}\ni i}\mathsf{P}(s)$$

$$\geq \left[\sum_{s\ni i}(w_i(s)-1)\,\mathsf{P}(s)\right]^2 \frac{1}{\sum_{s\ni i}\mathsf{P}(s)} + \sum_{s\,\mathrm{non}\ni i}\mathsf{P}(s)$$

$$= \frac{1}{\pi_i}(1-\pi_i)^2 + (1-\pi_i) = \frac{1}{\pi_i} - 1, \tag{7.3}$$

where the sign of equality holds if, and only if, $w_i(s)$ are independent of s, $s \ni i$. Independency of $w_i(s)$ of s, in connection with (1.3) and (1.4), implies that

$$w_i(s) = \frac{1}{\pi_i}, \quad i \in s,\ s \subset S. \tag{7.4}$$

The result just obtained together with (7.2) gives

$$\mathsf{EM}(\hat{Y}-Y)^2 \geq \sum_{i=1}^{N} d_{ii}\left(\frac{1}{\pi_i} - 1\right) \tag{7.5}$$

where the sign of equality holds if and only if, first, $\sum_{i\in s}\mu_i w_i(s) = \sum_{i=1}^{N}\mu_i$ for any s such that $\mathsf{P}(s) > 0$, and, second, (7.4) holds, i. e. $\hat{Y}$ is a simple linear estimate:

$$\hat{Y} = \sum_{i\in s}\frac{y_i}{\pi_i}. \tag{7.6}$$

Finally let us choose optimum $\pi_1, \ldots, \pi_N$ under the supposition that the expected total cost is given by (7.1). For brevity, let us denote the right-hand side of (7.5) by

$$D = \sum_{i=1}^{N} d_{ii} \left(\frac{1}{\pi_i} - 1 \right). \tag{7.7}$$

The use of Cauchy's inequality gives

$$C \left[D + \sum_{i=1}^{N} d_{ii} \right] = \left(\sum_{i=1}^{N} c_i \pi_i \right) \left(\sum_{i=1}^{N} \frac{d_{ii}}{\pi_i} \right) \geq \left(\sum_{i=1}^{N} \sqrt{c_i d_{ii}} \right)^2 \tag{7.8}$$

where the sign of equality holds if

$$c_i \pi_i = \lambda^2 \frac{d_{ii}}{\pi_i}, \quad i = 1, \ldots, N,$$

i. e.

$$\pi_i = \lambda \sqrt{\frac{d_{ii}}{c_i}}, \quad i = 1, \ldots, N, \tag{7.9}$$

where λ is a constant by which we may regulate the expected variance or cost when substituting into (7.7) or (7.1), respectively; of course, λ must be chosen so that $\pi_i \leq 1$.

Since the π_is satisfying (7.9) minimize the product $C \left[D + \sum_{i=1}^{N} d_{ii} \right]$ for any λ, (7.9) solves both problems simultaneously: minimization of C for given D, and of D for given C. The minimum value of C or D for given D or C, respectively, may be obtained from the equation

$$C \left[D + \sum_{i=1}^{N} d_{ii} \right] = \left(\sum_{i=1}^{N} \sqrt{c_i d_i} \right)^2 . \tag{7.10}$$

Let us summarize our results in a theorem.

Theorem 7.1. *Let us suppose that* $d_{ij} = 0,\ i \neq j, \sum_{s \subset S} w_i(s) \mathsf{P}(s) = 1,\ i = 1, \ldots, N$ *and* $C = \sum_{i=1}^{N} c_i \pi_i$. *Then any strategy for which the conditions*

$$\pi_i = \lambda \sqrt{\frac{d_{ii}}{c_i}}, \quad i = 1, \ldots, N, \tag{7.11}$$

$$w_i(s) = \frac{1}{\pi_i}, \quad i \in s,\ s \subset S, \tag{7.12}$$

$$\sum_{i \in s} \mu_i w_i(s) = \sum_{i=1}^{N} \mu_i, \quad P(s) > 0, \tag{7.13}$$

are fulfilled is the optimum one.

The condition (7.11) determines the expected frequency π_i of the element i in the sample and will be called the condition of *optimum allocation*. The condition (7.12) will be called the condition of *constancy of weights*. The condition (7.13) means that for any sample s which can really be selected the sampling error vanishes as soon as the y_i exactly equal their expected values μ_i, $i = 1, \ldots, N$. We can express this by words that the strategy is *representative* with respect to the values μ_i, $i = 1, \ldots, N$. In this way the vague notion of 'representativity' becomes well-defined.

We can see that the condition (2.3) assumed in Section 2 was nothing other than the condition of representativity with respect to the values z_i, $i = 1, \ldots, N$.

From the proof it is obvious that small violations of the above conditions are immaterial. Most frequently it is the condition (7.12) which is not fulfilled exactly; the weights are not constant, and, moreover, they are very often such that the estimate is not unbiased (e. g. in ratio estimation). Once we know that the above conditions are sufficient, it is irrelevant whether they are *approximately* fulfilled by an unbiased or biased estimate. Another question would be to choose an optimum solution within the whole class of linear estimates. However, this problem is too subtle and hardly fruitful.

Sometimes we may decide which of two possible systems of weights $w_i(s)$ and $w_i^*(s)$ is better in the following way. Suppose that the sampling design considered is such that (7.11) and (7.13) hold; then, according to (7.2),

$$\mathsf{EM}(\hat{Y} - Y)^2 = \mathsf{M}\left\{\sum_{i \in s} d_{ii}(w_i(s) - 1)^2 + \sum_{i\,\text{non}\in s} d_{ii}\right\}. \tag{7.14}$$

If it happens that for any s with $\mathsf{P}(s) > 0$

$$\sum_{i \in s} d_{ii}(w_i(s) - 1)^2 \leq \sum_{i \in s} d_{ii}(w_i^*(s) - 1)^2, \tag{7.15}$$

then, clearly, the weights $w(s)$ are the better of the two.

Example 7.1. Let us compare the estimates

$$\hat{Y}_1 = \sum_{i \in s} y_i \frac{1 - \alpha + n\alpha_i}{n\alpha_i}, \tag{7.16}$$

$$\hat{Y}_2 = \frac{1}{n} \sum_{i \in s} \frac{y_i}{\alpha_i}, \tag{7.17}$$

under the hypotheses that

$$d_{ii} = \lambda \alpha_i^\delta, \quad 0 \leq \delta \leq 2, \tag{7.18}$$

and that the sampling is rejective as described in Section 4. The estimate (7.16), as we have seen, is unbiased, and the estimate (7.17) is biased. We cannot, however, a priori say which of them is better. Putting

$$w_i(s) = \frac{1-\alpha+n\alpha_i}{n\alpha_i}, \quad w_i^*(s) = \frac{1}{n\alpha_i}, \quad i \in s,$$

we have

$$\begin{aligned}
&\sum_{i\in s} d_{ii}[(w_i(s)-1)^2 - (w_i^*(s)-1)^2] \\
&= \sum_{i\in s} d_{ii}(w_j(s)+w_i^*(s)-2)\,(w_i(s)-w_i^*(s)) \\
&= \lambda \sum_{i\in s} \alpha_i^{\delta-2}(2-\alpha-n\alpha_i)\,(n\alpha_i-\alpha).
\end{aligned}$$

The last expression, however, is nonpositive since it equals the covariance of values $\lambda'\alpha_i^{\delta-2}(2-\alpha-n\alpha_i)$ and values $n\alpha_i$ and since the relation

$$[n\alpha_i \le n\alpha_j] \implies \left[\alpha_i^{\delta-2}(2-\alpha-n\alpha_j) \ge \alpha_j^{\delta-2}(2-\alpha-n\alpha_j)\right], \quad 0 \le \delta \le 2,\ i, j \in s \tag{7.19}$$

holds. Consequently, under the hypothesis (7.18), the estimate (7.16) is better than the estimate (7.17).

The unbiased estimated mean square error of the estimate (7.17) can be derived along the lines of Sections 2 and 4:

$$m(\hat{Y}_2 - Y)^2 = \frac{1}{n-1}\sum_{i\in s}\left(\frac{y_i}{\alpha_i} - \frac{1}{n}\sum_{i\in s}\frac{y_i}{\alpha_i}\right)^2 \times \frac{(1-n\alpha_i)\,(1-\alpha)+n\sum_{k\,\text{non}\in s}\alpha_k^2}{n} + \alpha^2\left(\frac{1}{n}\sum_{i\in s}\frac{y_i}{\alpha_i} - \frac{\sum_{i\in s} y_i}{\sum_{i\in s}\alpha_i}\right)^2. \tag{7.20}$$

Now, we shall apply Theorem 7.1 to several typical examples.

Example 7.2. If $\mu_i = \text{const}$, $d_{ii} = \text{const}$, $c_i = \text{const}$, $i = 1, \ldots, N$, then all the requirements of Theorem 7.1 are fulfilled by simple random sampling together with the simple linear estimate $\frac{N}{n}\sum_{i\in s} y_i$. If μ_is, d_{ii}s and c_is are constant within some parts (strata) of the population, and when these strata are large enough, then the requirements of the optimum strategy are satisfied by the optimum allocated sampling together with the simple linear estimate. If the stratum S_h contains N_h elements from which we select n_h elements, then for an element i, belonging to the stratum S_h, we have

$$\pi_i = \frac{n_h}{N_h}, \quad i \in S_h. \tag{7.21}$$

According to (7.11) the n_hs must be chosen so that

$$\frac{n_h}{N_h} = \lambda\sqrt{\frac{d_{hh}}{c_h}}, \quad h = 1, \dots, H, \tag{7.22}$$

where d_{hh} and c_h are common values of d_{ii}s and c_is within stratum S_h. However, from (7.22) it is seen that whenever the numbers $\lambda N_h(d_{hh}/c_h)^{1/2}$ are too small, then (7.22) may not be fulfilled by any *integers* n_h with a reasonable degree of precision. Stratified sampling is ineffective in such situations.

Example 7.3. Very often we may suppose that μ_is are proportional to known numbers x_i:

$$\mu_i = \lambda x_i, \quad i = 1, \dots, N. \tag{7.23}$$

In such situations the ratio estimates (5.3), (4.10) or (4.11) meet the condition of representativity with respect to (7.23). The condition of constancy of weights, however, may not be fulfilled. When using the estimate (5.3), (4.10) or (4.11) we have weights

$$w_j^*(s) = \frac{1}{\pi_j}\frac{\sum_{i=1}^N x_i}{\sum_{i\in s} x_i/\pi_i}, \tag{7.24}$$

$$w_j^{**}(s) = \frac{1-\alpha+n\alpha_j}{n\alpha_j}\frac{\sum_{i=1}^N x_i}{\sum_{i\in s} x_i \frac{1-\alpha+n\alpha_i}{n\alpha_i}}, \tag{7.25}$$

$$w_j^{***}(s) = x_j\left[1+\frac{1-\alpha}{n\alpha_j}\right]+\frac{1}{nx_j}\left[X-\sum_{i\in s} x_i+\frac{1-\alpha}{n-1}\sum_{i\in s}\frac{x_i}{\alpha_i}\right], \tag{7.26}$$

respectively. The variability of the weights depends on the sampling design and on the relation between the values x_i and π_i or α_i. Generally, the weights (7.25) connected with the rejective sampling design described in Section 4 may be expected to vary least.

The weights (7.24) can be used in connection with the permutation sampling described in Section 5 with π_i satisfying the condition of optimum allocation (7.11). If the weights (7.25) are used, in connection with the rejective sampling described in Section 4, the α_is should be chosen so that the π_is again satisfy (7.11). Now from (1.4) and (4.3) it follows that

$$\frac{1}{\pi_i} = \frac{\sum_{i\in s}\frac{1-\alpha+n\alpha_i}{n\alpha_i}P(s)}{\sum_{i\in s} P(s)} \tag{7.27}$$

whence it is easily seen that

$$\frac{n\alpha_i}{1+(n-1)\max_j(\alpha_j-\alpha_i)} \le \pi_i \le b\frac{n\alpha_i}{1-(n-1)\max_j(\alpha_j-\alpha_i)}. \tag{7.28}$$

Consequently, (7.11) will be nearly fulfilled if simply

$$\alpha_i = \lambda'' \sqrt{\frac{d_{ii}}{c_i}}, \quad i = 1, \ldots, N. \tag{7.29}$$

If the π_is are not smaller than, say, 0.1, we may use the approximate relation

$$\pi_i \simeq \frac{n\alpha_i}{1 + \frac{n-1}{N-1} N \left(\alpha_i - \frac{1}{N}\right)}, \tag{7.30}$$

i. e.

$$\alpha_i \simeq \frac{\pi_i}{n} \frac{1 - \frac{n-1}{N-1}}{1 - \frac{(n-1)\,N}{(N-1)\,n} \pi_i}. \tag{7.31}$$

It may be of interest to modify the sampling design in order that the weights (7.24) vary as little as possible. Let us describe such a design for the uniform (self-weighting) case ($\pi_1 = \cdots = \pi_N = \pi$). In this case we try to select elements in such a way that $\pi_i = \pi$ and that the sum $\sum_{i \in s} x_i$ is nearly constant. Let us suppose that the x_is are integers (which causes no loss of generality) and consider the cyclical sequence

$$1,\, 2, \ldots, x_1 + \cdots + x_N, \quad 1,\, 2, \ldots, \omega.$$

Now we associate with the element i the segment of the sequence containing numbers $\{x_1 + \cdots + x_{i-1} + 1, \ldots, x_1 + \cdots + x_i\}$, $i = 1, \ldots, N$. Finally, we select an integer r with equal probabilities in the range $1 \leq r \leq x_1 + \cdots + x_N$, and include the element i in the sample with a probability $\frac{m_i(r)}{\omega}$, where $m_i(r)$ is the number of integers belonging simultaneously to both segments $\{r, \ldots, r+\omega-1\}$ and $\{x_1 + \cdots + x_{i-1} + 1, \ldots, x_1 + \cdots + x_i\}$. The probability of including an element in the sample is the same for all elements, namely

$$\pi_i = \frac{\omega}{x_1 + \cdots + x_N}. \tag{7.32}$$

This sampling may be preceded by any, random or nonrandom, ordering of elements.

Example 7.4. If the a priori hypothesis is expressed by a regression relation

$$\mu_i = \sum_{j=1}^{k} \beta_j\, x_i^{(j)} + \beta_0, \quad i = 1, \ldots, N, \tag{7.33}$$

where $\beta_0, \beta_1, \ldots, \beta_k$ are unknown constants, then the fulfillment of the condition of representativity is not easy. We may use some type of regression estimates, which are, however, difficult to compute, or, when the regression is passing through the origin (i. e. $\beta_0 = 0$), some type of ratio estimates.

Finally, we may use the so called acceptance–inspection method. This method consists in rejecting samples which give a bad result in estimating totals X_j of controlled (concomitant) variables used in the relation (7.31). For example, we repeat selections of a sample s until we get a sample for which

$$|\hat{X}_j - X_j| \leq \varepsilon_j \sqrt{\mathsf{D}\hat{X}_j}, \quad j = 1, \ldots, k, \tag{7.34}$$

where the ε_j are properly chosen constants, say $\varepsilon_j = 0.5$. This method of attaining to the representativity emphasizes the sampling-design aspect, and this is right in situations where we are dealing with extensive and fresh a priori data and ascertain a number of variables y_i', y_i'', ..., each of them related to variable $x_i^{(1)}, \ldots, x_i^{(k)}$. The theory of the acceptance–inspection method is based on the supposition that the random vector $(\hat{Y}, \hat{X}_1, \ldots, \hat{X}_k)$ has a $(k+1)$-dimensional normal distribution. However, no theoretical argument for this assumption is at our disposal in this time. Acceptance control is discussed in the paper by Koller (1957).

8 SOME REMARKS ON THE APPLICATIONS

8.1 Uniform Sampling

The cost function (7.1) does not reflect the jump in cost which arises when the sampling is uniform (self-weighting) so that the simple linear estimates are reduced to arithmetic sums, i. e. $\hat{Y} = \frac{1}{\pi} \sum_{i \in s} y_i$. Therefore, an optimum nonuniform sampling–estimating strategy is really acceptable only if it is better than the uniform sampling connected with the simple linear estimate or with the simple ratio estimate, etc. This device, consisting in dividing the possible strategies into classes within which the problem has a simple mathematical formulation, is of wide use.

8.2 Subsampling

The flexibility of the previous theory is well-illustrated by the subsampling problem. Let us suppose that the elements are primary sampling units consisting of N_i secondary units. We wish to select in any primary unit n_i secondary units by simple random sampling, where n_i is ascertained from the relation

$$\pi \frac{n_i}{N_i} = \tau, \quad i = 1, \ldots, M, \tag{8.1}$$

where π_i is the probability of including the primary unit in the sample and τ is the uniform overall probability of including a secondary unit in the sample.

A comparison of $\mathsf{EM}(\hat{\hat{Y}} - Y)^2$, where $\hat{\hat{Y}}$ is given by (2.30), with (7.2) shows that the effect of subsampling is the same as the effect of increasing the variances d_{ii} by the second-stage variance of $\hat{y}_i$, i. e. by

$$\frac{N_i^2}{n_i}\sigma_i^2\left(1 - \frac{n_i}{N_i}\right), \tag{8.2}$$

where σ_i^2 is the expected variance within the element (primary sampling unit) i. On the other hand, the cost associated with the element i is decreased by subsampling by

$$(N_i - n_i)\, e_i, \tag{8.3}$$

where e_i is the cost associated with a secondary unit within the element i.

If we insert the changed a priori variances and costs into (7.1) and (7.7), we get ($N \equiv M$)

$$C^* = \sum_{i=1}^{M} \pi_i\left[c_i - (N_i - n_i)\, e_i\right], \tag{8.4}$$

$$D^* = \sum_{i=1}^{M}\left[d_{ii} + \frac{N_i^2}{n_i}\sigma_i^2\left(1 - \left(\frac{n_i}{N_i}\right)\right)\right]\left(\frac{1}{\pi_i} - 1\right) \tag{8.5}$$

which by means of (8.1) gives

$$C^* = \sum_{i=1}^{M}\pi_i\,(c_i - N_i\, e_i) + \tau\sum_{i=1}^{M} e_i\, N_i, \tag{8.6}$$

$$D^* = \sum_{i=1}^{M}\frac{d_{ii} - N_i\,\sigma_i^2}{\pi_i} + \frac{1}{\tau}\sum_{i=1}^{M} N_i\,\sigma_i^2 - \sum_{i=1}^{M}(d_{ii} - N_i\,\sigma_i^2). \tag{8.7}$$

It means that, if $d_{ii} - N_i\,\sigma_i^2 \geq 0,\ c_i - N_i\, e_i \geq 0$, we have

$$C^*\left[D^* + \sum_{i=1}^{M}(d_{ii} - N_i\,\sigma_i^2)\right]$$

$$\geq \left(\sum_{i=1}^{M}\left((d_{ii} - N_i\,\sigma_i^2)\,(c_i - N_i\, e_i)\right)^{1/2} + \left(\sum_{i=1}^{M} N_i\,\sigma_i^2\sum_{j=1}^{M} N_j\, e_j\right)^{1/2}\right)^2,$$

where the sign of equality holds if

$$\frac{d_{ii} - N_i\,\sigma_i^2}{\pi_i} = \lambda^2\,\pi_i(c_i - N_i\, e_i) \quad \text{and} \quad \frac{\sum_{i=1}^{M} N_i\,\sigma_i^2}{\tau} = \lambda^2\,\tau\sum_{i=1}^{M} N_i\, e_i,$$

i. e.

$$\pi_i = \lambda \left(\frac{d_{ii} - N_i \sigma_i^2}{c_i - N_i e_i} \right)^{1/2}, \quad \tau = \lambda \left(\frac{\sum_{i=1}^M N_i \sigma_i^2}{\sum_{i=1}^M N_i e_i} \right)^{1/2}. \tag{8.8}$$

See also Hájek (1957 b, § 4).

8.3 Relative Emphasis on the Sampling Aspect

The conditions stated in Theorem 7.1 can be approximately fulfilled in many ways. When choosing a particular way, we may utilize some information which has not been included either in the assumptions about the a priori mean values and variances or in the cost function. A problem of this kind is how to distribute the effort between the sampling design and the estimation method. On some occasions it is better to pay attention to the sampling design (e. g. see Example 7.3) and on other occasions it is better to choose the estimation method carefully.

8.4 Connection Between the a priori Mean Values and the Assumption of Uncorrelation

The assumption of uncorrelation of ascertained values is the more realistic, the more specific are the a priori mean values. For example, in a population of areas on which total yields of cereal crops are ascertained, we may put

$$\mu_i = \mu \tag{8.9}$$

or

$$\mu_i = \mu \, x_i \tag{8.10}$$

or, finally,

$$\mu_i = b \, t_i \, x_i + a \, x_i \tag{8.11}$$

where x_i is the size of the area and t_i is an eye-estimate of the yield per unit area. The assumption (8.9) means that all influences are understood as random factors. The correlation of all these factors will cause the correlation of the values y_i. For example, in some regions there is a greater average size of areas than in others, and therefore a greater average total yield on these areas. If we know the sizes x_i and take the model (8.10), then this source of correlation will disappear. In model (8.11), there is excluded a further source of correlation, namely the correlation of fertility on neighbourhooding areas. When using this last model we may hope that the yields are a priori approximately uncorrelated.

8.5 Sampling Designs with Maximum Entropy

A simple way of avoiding unfavourable consequences of the possible violation of the hypothesis of the uncorrelation is to choose a sampling design which distributes the probabilities $P(s)$ as uniformly as possible. This requirement may be formulated in the form that we wish to maximize the entropy

$$E = -\sum_{s \subset S} \mathsf{P}(s) \log \mathsf{P}(s). \tag{8.12}$$

If the probabilities π_i, see (1.4), are fixed, then the sampling which maximizes (8.12) is the Poisson sampling described in Hájek (1957b). In fact, any sampling may be understood as N experiments having two possible outcomes: including or not including the element i in the sample. When the probability of including the element i in the sample, namely π_i, is given for each of these experiments, then the entropy will be maximum, as is well known, if all these experiments are independent. By the last requirement just the Poisson sampling is defined.

Now we shall show that the rejective sampling (Section 4) with fixed $\alpha_1, \ldots, \alpha_N$ has the greatest entropy in the class of sampling designs with the same $\pi_1, \ldots, \pi_N$ and with fixed sample size, n. In fact, bearing in mind (1.3) and using the usual method of Lagrange multiplicators λ_i, we get for $\mathsf{P}(s)$ and λ_i the following equations:

$$\begin{aligned} & \frac{\partial}{\partial P(s)} \left[\sum_{s \in V_n} -\mathsf{P}(s) \log \mathsf{P}(s) + \sum_{i=1}^{N} \lambda_i \sum_{s \ni i} \mathsf{P}(s) \right] \\ = \quad & -\log \mathsf{P}(s) - 1 + \sum_{i \in s} \lambda_i = 0, \quad s \in V_n, \end{aligned} \tag{8.13}$$

$$\sum_{s \ni i} \mathsf{P}(s) = \pi_i, \quad i = 1, \ldots, N, \tag{8.14}$$

where V_n is the set of all samples with sample size n. Since the function $-x \log x$ is strictly concave, there is a single maximum. If it happens that the solution is such that $0 \leq \mathsf{P}(s) \leq 1$, then it clearly coincides with the solution of the problem restricted to the domain $0 \leq \mathsf{P}(s) \leq 1$, $s \in V_n$. Now, it is easily seen that (8.13) is satisfied by the probabilities (4.1) and by

$$\lambda_i = \log \alpha_i + \frac{1}{n} \left(1 - \log \sum_{s \in V_n} \prod_{i \in s} \alpha_i \right). \tag{8.15}$$

As regards (8.14), it is fulfilled automatically, since we consider just those $\pi_1, \ldots, \pi_N$ which are yielded by the given $\alpha_1, \ldots, \alpha_N$.

C. OPTIMUM STRATEGY. CASE II: THE ASCERTAINED VALUES FORM A RANDOM SERIES WITH A STATIONARY CONVEX CORRELATION FUNCTION AND STATIONARY COEFFICIENTS OF VARIATIONS (9 – 10)

If the population is meaningfully ordered (in time or space), the ascertained values are very often governed by a convex correlation function. An empirical correlation function of this kind is shown in Figure 1, where also the estimated correlation function is drawn. The data refer to a population of plots where the area of forest land has been measured.[1]

Convex correlation functions are encountered so often that they deserve a detailed study. This has been undertaken in Cochran (1946), and in subsequent papers by Madow (1949) and Gautschi (1957), where, under the supposition of convexity, it is proved that systematic sampling is superior to simple random sampling, stratified sampling, and several independent systematic samplings. In this chapter, using quite a different method, we shall derive a general result that systematic sampling is better than any other sampling design under consideration. Moreover, we shall replace the supposition that $\mathsf{E}y_i$ and $\mathsf{V}y_i$ are stationary by a more general supposition that only the coefficients of variation, $\sqrt{\mathsf{V}y_i}/\mathsf{E}y_i$, are so. As a solution we shall obtain systematic sampling with generally unequal probabilities.

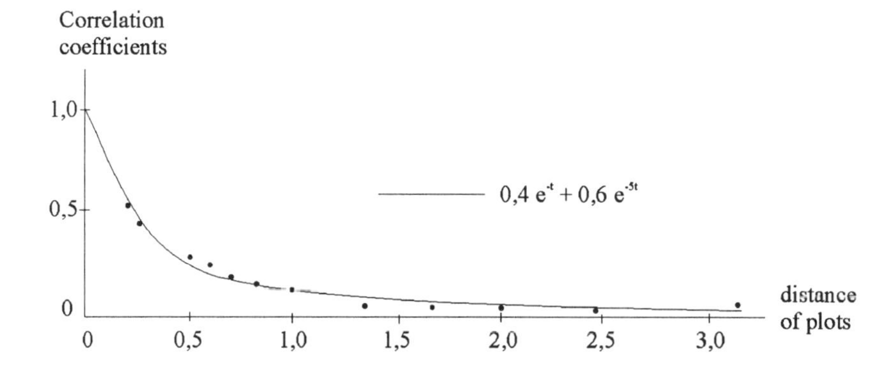

Figure 1

[1] Figure 1 taken from Matérn (1947) with the author's kind permission.

9 PRELIMINARIES

Lemma 9.1. *Among all integral-valued random variables x possessing a given mean value ξ the random variable which takes on only the values $[\xi]$ and $[\xi]+1$ is of least variance. Here $[\xi]$ denotes the greatest integral number not exceeding ξ.*

Proof. Putting $a = [\xi] + \frac{1}{2}$, we see that

$$\mathsf{D}x = \sum_{k=-\infty}^{\infty} (k-a)^2 \, \mathsf{P}\{x = k\} - (\xi - a)^2,$$

where

$$\begin{aligned}(k-a)^2 &= \frac{1}{4}, \quad \text{if } k = [\xi] \text{ or } [\xi]+1,\\ &> \frac{1}{4}, \quad \text{otherwise.}\end{aligned}$$

This concludes our lemma. □

Definition. A correlation function $R(i-j)$, $i, j = 1, \ldots, N$, that satisfies the condition

$$R(u) - 2R\left(\frac{u+v}{2}\right) + R(v) \geq 0, \quad 0 \leq u, v \leq N, \tag{9.1}$$

is called convex.

Lemma 9.2. *Any convex correlation function $R(i-j)$, $i, j = 1, \ldots, N$, of a discrete stationary random process may be expressed in the form*

$$R(i-j) = \sum_{\lambda=1}^{\infty} \sum_{\omega=1}^{N+\lambda-1} [R(\lambda+1) - 2R(\lambda) + R(\lambda-1)] \, q_{i\lambda\omega} q_{j\lambda\omega}, \quad 1 \leq i, j \leq N \tag{9.2}$$

where

$$\begin{aligned} q_{i\lambda\omega} &= 1, \quad \text{if } \omega - \lambda < i \leq \omega,\\ &= 0, \quad \text{otherwise}\end{aligned} \tag{9.3}$$

and

$$R(\lambda+1) - 2R(\lambda) + R(\lambda-1) \geq 0. \tag{9.4}$$

Proof. The inequality (9.4) follows from the assumption (9.1) when putting $u = \lambda+1$, $v = \lambda-1$. Equation (9.2) is a consequence of the following identities:

$$\begin{aligned}\sum_{\omega=1}^{N+\lambda-1} q_{i\lambda\omega} q_{j\lambda\omega} &= \lambda - |i-j|, \quad \text{if } \lambda > |i-j|\\ &= 0, \qquad\qquad \text{otherwise}\end{aligned}$$

and

$$
\begin{aligned}
&\sum_{\lambda=|i-j|+1}^{\infty} (\lambda - |i-j|)\,[R(\lambda+1) - 2R(\lambda) + R(\lambda-1)] \\
&= \sum_{\lambda=|i-j|+1}^{\infty} \sum_{\mu=|i-j|+1}^{\lambda} [R(\lambda+1) - 2R(\lambda) + R(\lambda-1)] \\
&= \sum_{\mu=|i-j|+1}^{\infty} \sum_{\lambda=\mu}^{\infty} [R(\lambda+1) - 2R(\lambda) + R(\lambda-1)] \\
&= \sum_{\mu=|i-j|+1}^{\infty} [R(\mu-1) - R(\mu)] \;=\; R(|i-j|) \;=\; R(i-j).
\end{aligned}
$$

□

10 AN OPTIMUM PROPERTY OF SYSTEMATIC SAMPLING

We shall suppose that the random sequence $y_1, \dots, y_N$ possesses stationary coefficients of variation and a stationary convex correlation function, i. e.

$$
\mathsf{E}y_i = \mu\, x_i, \quad \mathsf{V}y_i = \sigma^2\, x_i^2, \quad \operatorname{Cov}(y_i, y_j) = \sigma^2\, x_i\, x_j\, R(i-j), \quad i, j = 1, \dots, N, \tag{10.1}
$$

and shall choose a sampling design which

(a) yields samples of fixed size, n,

(b) gives the following probabilities π_i of including the element i in the sample:

$$
\pi_i = \frac{n x_i}{x_1 + \cdots + x_N}, \quad i = 1, \dots, N, \tag{10.2}
$$

(c) minimizes the expected variance of the estimator

$$
\hat{Y} = \frac{X}{n} \sum_{i \in s} \frac{y_i}{x_i}, \quad X = \sum_{i=1}^{N} x_i. \tag{10.3}
$$

Remark 10.1. The correlation function

$$
\begin{aligned}
R(i-j) &= 1, \quad \text{if } i = j, \\
&= 0, \quad \text{otherwise,}
\end{aligned}
$$

is convex, and, consequently, the scheme (10.1) incorporates the case of noncorrelated random variables with stationary coefficients of variation as a particular case. When supposing constant costs, i. e. $c_i = c,\ i = 1, \ldots, N$, then the condition (10.2) and the estimator (10.3) guarantee that the sampling–estimating strategy that we are choosing will be optimum for this particular case. In fact, then (10.2) is equivalent to (7.11) and the estimator (10.3) implies the fulfilment of (7.12) and (7.13).

Theorem 10.1. *The above problem is solved by systematic sampling defined as follows: The sample consists of those elements i for which the sum $x_1 + \cdots + x_i$ at first reaches or exceeds some of the numbers*

$$r,\ r + \frac{1}{n}X, \ldots, r + \frac{n-1}{n}X, \quad \left(X = \sum_{i=1}^{N} x_i\right),$$

where r is a random variable uniformly distributed over the interval $0 < r \leq \frac{1}{n}X$.

Proof. Unbiasedness of the estimator (10.3) for any particular sequence $y_1, \ldots, y_N$ justifies the first of the following identities (the remaining ones are obvious):

$$\begin{aligned}
\mathsf{EM}(\hat{Y} - Y)^2 &= \mathsf{EM}\left(\hat{Y} - \sum_{i=1}^{N} \mu x_i\right)^2 - \mathsf{E}\left(Y - \sum_{i=1}^{N} \mu x_i\right)^2 \\
&= \mathsf{ME}\left(\frac{X}{n} \sum_{i \in s} \frac{y_i}{x_i} - \mu X\right)^2 - \mathsf{E}(Y - \mu X)^2 \\
&= \left(\frac{X}{n}\right)^2 \sigma^2\, \mathsf{M}\left(\sum_{i \in s} \sum_{j \in s} R(i-j)\right) - \mathsf{E}(Y - \mu X)^2.
\end{aligned}$$

Consequently, it suffices to minimize $\mathsf{M}\left(\sum_{i \in s} \sum_{j \in s} R(i-j)\right)$. Using (9.2) we obtain

$$\begin{aligned}
&\mathsf{M}\left(\sum_{i \in s} \sum_{j \in s} R(i-j)\right) \\
&= \sum_{\lambda=1}^{\infty} \sum_{\omega=1}^{N-\lambda-1} [R(\lambda+1) - 2R(\lambda) + R(\lambda-1)]\, \mathsf{M}\left(\sum_{i \in s} q_{i\lambda\omega}\right)^2.
\end{aligned}$$

According to (9.4) it suffices to prove that $\mathsf{M}\left(\sum_{i \in s} q_{i\lambda\omega}\right)^2$ is minimized for each λ and ω. The definition (9.3) of $q_{i\lambda\omega}$ implies that

$$\sum_{i \in s} q_{i\lambda\omega} = \sum_{i=\omega-\lambda+1}^{\omega} p_i(s) \tag{10.4}$$

where p_i are random variables such that

$$\begin{aligned} p_i(s) &= 1, \quad \text{if } s \ni i, \\ &= 0, \quad \text{otherwise.} \end{aligned} \tag{10.5}$$

This means that

$$\mathsf{M}\left(\sum_{i\in s} q_{i\lambda\omega}\right) = \sum_{i=\omega-\lambda+1}^{\omega} \mathsf{M}[p_i(s)] = \sum_{i=\omega-\lambda+1}^{\omega} \pi_i = \frac{n}{X} \sum_{i=\omega-\lambda+1}^{\omega} x_i,$$

i. e., in accordance with the point (b), the mean value of $\sum_{i\in s} q_{i\lambda\omega}$ is constant for any considered sampling design. Thus minimizing $\mathsf{M}\left(\sum_{i\in s} q_{i\lambda\omega}\right)^2$ is equivalent to minimizing $\mathsf{D}\left(\sum_{i\in s} q_{i\lambda\omega}\right)$ for a given mean value $\frac{n}{X}\sum_{i=\omega-\lambda+1}^{\infty} x_i$. As $\sum_{i\in s} q_{i\lambda\omega}$ is an integral-valued random variable, it suffices, according to Lemma 9.1, to show that $\sum_{i\in s} q_{i\lambda\omega}$ only takes on values $\left[\frac{n}{X}\sum_{i=\omega-\lambda+1}^{\omega} x_i\right]$ and $\left[\frac{n}{X}\sum_{i=\omega-\lambda+1}^{\omega} x_i\right]+1$, where $[\xi]$ again denotes the greatest integral number not exceeding ξ. This may be done as follows: $\sum_{i\in s} q_{i\lambda\omega}$ is the number of members of the set $\{\omega-\lambda+1,\dots,\omega\}$ which have been included in the sample, and this number, according to the definition of the systematic sampling in Theorem 10.1, equals the number of integrals $k = 1,\dots,n$ for which

$$x_1 + \cdots + x_{\omega-\lambda} < r + \frac{k-1}{n} X \le x_1 + \cdots + x_\omega$$

i. e.

$$\frac{n}{X}(x_1 + \cdots + x_{\omega-\lambda} - r) + 1 < k \le \frac{n}{X}(x_1 + \cdots + x_\omega - r) + 1. \tag{10.6}$$

The number of integrals lying in an interval of fixed length ℓ, however, can only equal $[\ell]$ or $[\ell]+1$. The proof may thus be completed by observing that the length of the interval (10.6) equals $\frac{n}{X}\sum_{i=\omega-\lambda+1}^{\omega} x_i$. □

Remark 10.2. We have not proved that the sampling design described in Theorem 6.1 yields the probabilities of including the elements i in the sample given by (10.2). However, this is well known (see, for example, Hájek (1957 b, § 2).

Remark 10.3. In the most important particular case when both the expected values $\mathsf{E}y_i$ and variances $\mathsf{V}y_i$ are stationary, we get the uniform (or self-weighting) systematic sampling. From the point of view of practice, the case of stationary coefficients of variations which we have considered is only slightly more general, since there are only a few examples where the coefficients of variations are stationary even though the expected values and variances are not so. (Consider, as an example, the case where y_i and x_i denote the present and past city populations, respectively.)

In cases where the coefficients of variations are not stationary, the problem becomes more complicated, and a question arises whether a solution, common for all convex correlation functions, exists at all. We should minimize the expression

$$\sum_{\lambda=1}^{\infty} \sum_{\omega=1}^{N+\lambda-1} [R(\lambda+1) - 2R(\lambda) + R(\lambda-1)] \, \mathsf{M} \left(\sum_{i=\omega-\lambda+1}^{\omega} p_i(s) \, C_i \right)^2$$

where $p_i(s)$ are given by (10.5) and C_i are the coefficients of variations, $i = 1, \ldots, N$. Moreover, in such cases, the supposition (10.2) and the estimator (10.3) cannot any longer be considered as 'natural' ones and the whole problem needs to be reformulated.

REFERENCES

Cochran, W. G. (1946). Relative accuracy of systematic and stratified random sample for certain class of populations. *Ann. Math. Stat. 17*, 164–177.

Gautschi, W. (1957). Some remarks on systematic sampling. *Ann. Math. Stat. 28*, 385–394.

Godambe, V. P. (1955). A unified theory of sampling from finite population. *J. Roy. Stat. Soc. B 17*, 269–278.

Goodman, L. A. and Hartley, H. O. (1958). The precision of unbiased ratio-type estimators. *J. Amer. Statist. Assoc. 53*, 491–508.

Hájek, J. (1955). Some contributions to the theory of estimation. (In Czech) PhD Thesis, Czechoslovak Acad. of Sci. Prague.

Hájek, J. (1957 a). Inequalities for generalized Student distribution. (In Czech) *Časopis Pěst. Mat. 82*, 269–278.

Hájek, J. (1957 b). Some contributions to the theory of probability sampling. *Bull. Inst. Internat. Statist. 36*, 127–133.

Hansen, M. H. and Hurwitz, W. N. (1949). On the determination of optimum probabilities in sampling. *Ann. Math. Statist. 20*, 426–432.

Koller, S. (1957). On the problem of replicated sampling in German governmental statistics. *Bull. Inst. Internat. Statist. 36*, 135–143.

Madow, W. G. (1949). On the theory of systematic sampling, II. *Ann. Math. Statist. 20*, 333–354.

Matérn, B. (1947). Methods of estimating the accuracy of line and sample plot surveys. *Madd. Statens Skogsforskninginst. 36 (1)*, 138 pp.

Des Raj, Khamis, S. H. (1958). Some remarks on sampling with replacement. *Ann. Math. Statist. 29*, 550–557.

Yeates, F. and Grundy, P. M. (1953). Selection without replacement from within strata with probability proportional to size. *J. Roy. Statist. Soc. Ser. B* *15*, 253–261.

CHAPTER 14

On a Simple Linear Model in Gaussian Processes

In: Transactions of the 2nd Prague Conference on Information Theory, Statistical Decision Functions, Random Processes, 1960 (J. Kožešník, ed.), 185–197.
Reproduced by permission of Hájek family: Al. Hájková, H. Kazárová and Al. Slámová.

Summary. This chapter contains (a) proof of existence of a random variable which is a sufficient statistic for the linear model, (b) a criterion for the regular case, and (c) a method of finding in the regular case the sufficient statistic mentioned in (a). These results are used for processes with independent increments, Markov processes and stationary processes. As in Feldman (1958), Hilbert space considerations are widely used.

1 THE LINEAR MODEL

We shall be concerned with a space Ω and a family of random variables $\{x_t(\omega), t \in T\}$ defined thereon; T is an arbitrary set. We shall consider a system of probability measures $\{\mathsf{P}_\alpha(\cdot), -\infty < \alpha < \infty\}$, defined on the Borel field $\mathcal{F}$, generated by the random variables x_t, $t \in T$, such that the random process $\{x_t, t \in T\}$ is Gaussian via each probability measure P_α, $-\infty < \alpha < \infty$, namely with the following respective parameters:

$$\int x_t \,\mathrm{d}\mathsf{P}_\alpha = \alpha\varphi_t, \quad t \in T, \tag{1.1}$$

$$\int x_t x_s \,\mathrm{d}\mathsf{P}_\alpha = \int x_t x_s \,\mathrm{d}\mathsf{P}_0 + \alpha^2\varphi_t\varphi_s, \quad t,\ s \in T, \tag{1.2}$$

Collected Works of Jaroslav Hájek – With Commentary
Edited by M. Hušková, R. Beran and V. Dupač
Published in 1998 by John Wiley & Sons Ltd.

where φ_t is an a priori known non-vanishing real-valued function of $t \in T$. Thus we shall be dealing with a linear model with just one unknown parameter α.

2 CLOSED LINEAR MANIFOLD OF RANDOM VARIABLES

Taking all finite linear combinations $\sum c_t x_{t_\nu}$, $t_\nu \in T$, we obtain a linear manifold, say M. Any random variable $x \in M$ is normally distributed and well-defined in all points $\omega \in \Omega$. In defining closure of M, we have to overcome the difficulty arising from the fact that we are dealing not with a single probability measure but with a whole system of probability measures, and, consequently, any element of the closure of M must be well-defined with respect to any probability measure considered.

We shall proceed as follows. Consider the set, say $\mathcal{M}$, of random variables x such that, for some sequence $\{z_n\}$, $z_n \in M$,

$$\lim_{n\to\infty} \int (z_n - x)^2 \, \mathrm{dP}_\alpha = 0, \quad -\infty < \alpha < \infty \tag{2.1}$$

holds, where the z_ns do not depend on α.

The most important properties of the set $\mathcal{M}$ just defined are formulated in the following lemmas.

Lemma 2.1. *The P_α-distribution of random variables from $\mathcal{M}$ is Gaussian with the following parameters:*

$$\int x \, \mathrm{dP}_\alpha = \alpha \int x \, \mathrm{dP}_1, \quad y, x \in \mathcal{M}, \tag{2.2}$$

$$\int xy \, \mathrm{dP}_\alpha = \int xy \, \mathrm{dP}_0 + \alpha^2 \int x \, \mathrm{dP}_1 \int y \, d\mathrm{P}_1, \quad -\infty < \alpha < \infty. \tag{2.3}$$

Proof. For x, $y \in M$, (2.2) and (2.3) are direct consequences of (1.1) and (1.2). Generally, we choose sequences $\{z_n\}$ and $\{z'_n\}$ such that $z_n \to x$ and $z'_n \to y$ in the P_α-mean $(-\infty < \alpha < \infty)$, and then make a use of the well-known fact that an in-mean-limit of normally distributed random vectors is itself normally distributed with limiting parameters. □

Lemma 2.2. *$\mathcal{M}$ is a linear manifold.*

Proof. Obvious. □

Lemma 2.3. *The equation* $\mathsf{P}_1(x = y) = 1$ *implies the equations* $\mathsf{P}_\alpha(x = y) = 1$ $(-\infty < \alpha < \infty)$ *for any two elements* x, y *from* $\mathcal{M}$.

Proof. Since, by Lemma 2.2, $x - y \in \mathcal{M}$, it suffices to prove that

$$\int x^2 \, \mathrm{d}\mathsf{P}_\alpha \leq (1+\alpha^2) \int x^2 \, \mathrm{d}\mathsf{P}_1, \quad x \in \mathcal{M}. \tag{2.4}$$

But (2.4) follows easily from (2.3). □

Lemma 2.4. *$\mathcal{M}$, where random variables with $\mathsf{P}_1(x = y) = 1$ are considered as identical, and where the inner product is introduced by*

$$(x, y) = \int xy \, \mathrm{d}\mathsf{P}_1, \tag{2.5}$$

is a Hilbert space.

Proof. All properties needed are obvious with the only exception of completeness. Now, let us have a P_1-fundamental sequence $\{z_n\}$ and choose, for all $k \geq 1$, an n_k such that

$$\int (z_n - z_m)^2 \, \mathrm{d}\mathsf{P}_1 \leq \frac{1}{4^k}, \quad k \geq 1; \ m, n \geq n_k. \tag{2.6}$$

Then, clearly,

$$\int |z_n - z_m| \, \mathrm{d}\mathsf{P}_1 \leq \frac{1}{2^k}, \quad k \geq 1; \ m, n \geq n_k. \tag{2.7}$$

The inequalities (2.4) and (2.7) imply that

$$\sum_{k=1}^{\infty} \int |z_{n_{k+1}} - z_{n_k}| \, \mathrm{d}\mathsf{P}_\alpha \leq 1 + \alpha^2, \quad -\infty < \alpha < \infty. \tag{2.8}$$

Hence, according to a well-known theorem due to B. Lévy, the limit

$$x = \lim_{k \to \infty} z_{n_k} \tag{2.9}$$

exists with P_α-probability 1 for all $\alpha \in (-\infty, \infty)$. Further, in view of (2.6), we have

$$\int (z_n - x)^2 \, \mathrm{d}\mathsf{P}_\alpha \leq \lim_{m \to \infty} \int (z_n - z_m)^2 \, \mathrm{d}\mathsf{P}_\alpha$$
$$\leq (1+\alpha^2) \lim_{m \to \infty} \int (z_n - z_m)^2 \, \mathrm{d}\mathsf{P}_1 \leq (1+\alpha^2) \frac{1}{4^k}, \quad n \geq n_k \tag{2.10}$$

so that x satisfies the condition (2.1) and therefore belongs to $\mathcal{M}$. The completeness of $\mathcal{M}$ is proved. □

3 A SUFFICIENT STATISTIC

The basic result of this paper is contained in the following theorem.

Theorem 3.1. *There exists a random variable u in $\mathcal{M}$ which is a sufficient statistic for the system $\{\mathsf{P}_\alpha, -\infty < \alpha < \infty\}$ and for which*

$$\int u \, \mathrm{d}\mathsf{P}_\alpha = \alpha \quad (-\infty < \alpha < \infty); \tag{3.1}$$

u is uniquely characterized by the following relations:

$$\int u x_t \, \mathrm{d}\mathsf{P}_1 = \varphi_t \int u^2 \, \mathrm{d}\mathsf{P}_1, \quad t \in T. \tag{3.2}$$

Proof. Consider the space $\mathcal{M}$ as a subspace of the space $L_2(\mathsf{P}_1)$ of all P_1-quadratically integrable random variables measurable with respect to $\mathcal{F}'$, where $\mathcal{F}'$ consists of $\mathcal{F}$ and of subsets of random events $\Lambda \in \mathcal{F}$ with P_α-probability 0 for all the αs. The space $L_2(\mathsf{P}_1)$ certainly contains the random variable $y(\omega) \equiv 1$. Denote the projection of y on $\mathcal{M}$ by y^*. This projection is uniquely characterized by the following relations:

$$\int x \, \mathrm{d}\mathsf{P}_1 = \int y^* x \, \mathrm{d}\mathsf{P}_1, \quad x \in \mathcal{M}. \tag{3.3}$$

In particular,

$$\int y^* \, \mathrm{d}\mathsf{P}_1 = \int {y^*}^2 \, \mathrm{d}\mathsf{P}_1. \tag{3.4}$$

From (3.3) it follows that

$$\int {y^*}^2 \, \mathrm{d}\mathsf{P}_1 > 0, \tag{3.5}$$

since, if $\int {y^*}^2 \, \mathrm{d}\mathsf{P}_1 = 0$, we should have $\varphi_t = \int x_t \, \mathrm{d}\mathsf{P}_1 = \int y^* x_t \, \mathrm{d}\mathsf{P}_1 = 0$, $t \in T$, which contradicts the assumption that φ_t is non-vanishing. So, we may define the random variable

$$u = \frac{y^*}{\int {y^*}^2 \, \mathrm{d}\mathsf{P}_1}. \tag{3.6}$$

Let us show that u satisfies (3.1) and is characterized by (3.2). In accordance with (2.2) and (3.4), we have

$$\int u \, \mathrm{d}\mathsf{P}_\alpha = \alpha \int u \, \mathrm{d}\mathsf{P}_1 = \alpha \frac{\int y^* \, \mathrm{d}\mathsf{P}_1}{\int {y^*}^2 \, \mathrm{d}\mathsf{P}_1} = \alpha.$$

Further, u, being a multiple of y^*, is uniquely characterized by the same relations. On substituting for y^* into (3.3) we get that

$$\int x \, \mathrm{d}\mathsf{P}_1 = \int {y^*}^2 \, \mathrm{d}\mathsf{P}_1 \int u x \, \mathrm{d}\mathsf{P}_1. \tag{3.7}$$

On the other hand, from (3.6) it follows that

$$\int y^{*^2}\,\mathrm{dP}_1 = \frac{1}{\int u^2\,\mathrm{dP}_1}, \tag{3.8}$$

which together with (3.7) gives the result that u is uniquely characterized by

$$\int ux\,\mathrm{dP}_1 = \int u^2\,\mathrm{dP}_1 \int x\,\mathrm{dP}_1, \quad x \in \mathcal{M}. \tag{3.9}$$

It is well known, however, that (3.9) follows from (3.2), since any random variable $x \in \mathcal{M}$ may be approximated by finite linear combinations of x_t, $t \in T$.

It remains to prove that u is a sufficient statistic, which is the same as to prove that y^* is a sufficient statistic.

Any random variable $x \in \mathcal{M}$ may be represented in the form

$$x = cy^* + z \tag{3.10}$$

where $c = (x, y^*)$ and z satisfies the condition

$$\int y^* z\,\mathrm{dP}_1 = 0 \quad (\text{briefly } z \perp y^*). \tag{3.11}$$

The equation (3.10) holds with P_1-probability 1, and, hence, by Lemma 2.3, with P_α-probability 1 for all the αs. Applying (3.10) to x_t, $t \in T$, we can see that the Borel field, generated by the system of random variables $\{y^*, z', z'', \ldots\}$, where z', z'', ... are all random variables P_1-orthogonal to y^* and completed by subsets of random events with P_α-probability 0 for all the αs, contains the original Borel field $\mathcal{F}$ generated by the random variables x_t, $t \in T$. Hence, the system $\{y^*, z', z'', \ldots\}$ is a sufficient statistic.

(We have used the following obvious lemma. *Let $\mathcal{G}' \supset \mathcal{F}$, where $\mathcal{F}$ is the original Borel field, $\mathcal{G}$ is the Borel field generated by a statistic χ, and $\mathcal{G}'$ is the Borel field consisting of $\mathcal{G}$ and of subsets of all random events with zero-probability for all admissible probability measures. Then $\mathcal{G}$ is a sufficient Borel field or, equivalently, χ is a sufficient statistic.* Proof: the characteristic function $c_A(\omega)$ of any random event $A \in \mathcal{F}$ is measurable with respect to $\mathcal{G}'$ and may, therefore, serve as the conditional probability of A for any admissible probability measure.)

Equation (3.11), together with (3.3), implies that

$$\int z\,\mathrm{dP}_1 = 0 \quad (z \perp y^*). \tag{3.12}$$

From (3.12) it follows, in view of (2.2), that

$$\int z\,\mathrm{dP}_\alpha = 0 \quad (z \perp y^*, \ -\infty < \alpha < \infty), \tag{3.13}$$

and, in view of (2.3), that

$$\int zz'\,\mathrm{dP}_\alpha = \int zz'\,\mathrm{dP}_0 + \alpha^2 \int z\,\mathrm{dP}_1 \int z'\,\mathrm{dP}_1 = \int zz'\,\mathrm{dP}_0, \qquad (3.14)$$
$$(z \perp y^*,\ z' \perp y^*),$$

$$\int y^* z\,\mathrm{dP}_\alpha = \int y^* z\,\mathrm{dP}_0 + \alpha^2 \int y^*\,\mathrm{dP}_1 \int z\,\mathrm{dP}_1 = \int y^* z\,\mathrm{dP}_0, \qquad (3.15)$$
$$(z \perp y^*).$$

Since (3.15) must hold also for $\alpha = 1$, we have, in view of (3.11), that

$$\int y^* z\,\mathrm{dP}_\alpha = 0 \quad (z \perp y^*). \qquad (3.16)$$

Equations (3.13), (3.15) and (3.16) mean that the distribution of the system $\{z', z'', \ldots\}$ is (i) independent of y^* for all the αs, and (ii) the same for all the αs.

Now, the remainder of the proof is a purely formal matter. Denote the coordinate space corresponding to the system $\{y^*, z', z'', \ldots\}$ by $Y^* \times Z$, where Y^* and Z are the coordinate spaces corresponding to the systems $\{y^*\}$ and $\{z', z'', \ldots,\}$, respectively. The P_α-distribution in the space $Y^* \times Z$, as we have seen, is a direct product of a distribution Q_α in the Y^*-space and of a distribution R, independent of α, in the Z-space. Now, let us have a measurable subset B of the space $Y^* \times Z$ and prove that

$$\mathsf{P}_\alpha(A \cap B) = \int_A \mathsf{P}(B \mid y^*)\,\mathsf{Q}_\alpha(\mathrm{d}y^*) \qquad (3.17)$$

for any measurable subset A of Y^*, where $\mathsf{P}(B \mid y^*)$ is independent of α, and A on the left side denotes the respective cylinder-set. Denoting by χ_A and χ_B the respective characteristic functions, we have, by the theorem due to Fubini, that

$$\begin{aligned} \mathsf{P}_\alpha(A \cap B) &= \int_{Y^* \times Z} \chi_A(y^*)\,\chi_B(y^*, z', z'', \ldots)\,\mathsf{P}_\alpha(\mathrm{d}y^* \times \mathrm{d}(z', z'', \ldots)) \\ &= \int_{Y^*} \chi_A(y^*) \left[\int_Z \chi_B(y^*, z', z'', \ldots)\, R(\mathrm{d}(z', z'', \ldots)) \right] \mathsf{Q}_\alpha(\mathrm{d}y^*) \end{aligned}$$

which is the same as (3.12).

The theorem is thereby proved. □

The statistical problems concerning the system $\{\mathsf{P}_\alpha\}$ are thus reduced to the problem of one random variable normally distributed with known variance and unknown mean. This is true irrespective of the nature of the set T, of the function φ_t and of the covariance function $\int x_t x_s\,\mathrm{dP}_0$.

4 SINGULAR AND REGULAR CASES

If the equations $\mathsf{P}(\Lambda) = 0$ and $\mathsf{Q}(\Lambda) = 0$ each imply the other for any subset $\Lambda \in \mathcal{F}$, the probability measures are called equivalent, symbolically $\mathsf{P} \sim \mathsf{Q}$. If there exists a subset $\Lambda \in \mathcal{F}$ such that $\mathsf{P}(\Lambda) = 0$ and $\mathsf{Q}(\Lambda) = 1$, they are called perpendicular (or mutually singular), symbolically $\mathsf{P} \perp \mathsf{Q}$. In statistical problems concerning a system $\{\mathsf{P}_\vartheta,\ \vartheta \in \Theta\}$ of admissible probability measures we speak about a regular case, if $\mathsf{P}_{\vartheta_1} \sim \mathsf{P}_{\vartheta_2}$, and about a singular case, if $\mathsf{P}_{\vartheta_1} \perp \mathsf{P}_{\vartheta_2}$, both for all pairs ϑ_1, ϑ_2 from the set Θ.

Theorem 4.1. *In the linear problem considered we can have only the singular case or the regular case. The singular case takes place if and only if the random variable u mentioned in Theorem 3.1 has variance 0, which occurs if and only if y^* is P_1-equivalent to the random variable $y \equiv 1$.*

Proof. Since the random variable u is a sufficient statistic, and has normal distributions with mean values α and with a constant variance, the first assertion is clear. Now if $y^* \equiv 1(\mathsf{P}_1)$, then, in view of (3.4),

$$\int y^* \, \mathrm{d}\mathsf{P}_1 = \int {y^*}^2 \, \mathrm{d}\mathsf{P}_1 = 1 \tag{4.1}$$

so that

$$\operatorname{var} y^* = \int {y^*}^2 \, \mathrm{d}\mathsf{P}_1 - \left(\int y^* \, \mathrm{d}\mathsf{P}_1 \right)^2 = 0 \tag{4.2}$$

and, consequently, also $\operatorname{var} u = 0$. On the other hand, if $y^* \not\equiv 1(\mathsf{P}_1)$, then

$$\int y^* \, \mathrm{d}\mathsf{P}_1 = \int {y^*}^2 \, \mathrm{d}\mathsf{P}_1 < 1 \tag{4.3}$$

so that, in view of (4.3) and (3.5)

$$\operatorname{var} y^* = \left(\int y^* \, \mathrm{d}\mathsf{P}_1 \right) \left(1 - \int y^* \, \mathrm{d}\mathsf{P}_1 \right) > 0. \tag{4.4}$$

□

5 A USE OF LINEAR FUNCTIONALS

Theorem 3.1 gives a solution from a theoretical point of view only. An important practical problem is, however, how to find the aforementioned random variable u. The following theorem gives a criterion for the regular case, and also a method of finding the solution.

Theorem 5.1. *The regular case takes place if and only if*

$$\int x_t v \, \mathrm{dP}_0 = \varphi_t, \quad t \in T \tag{5.1}$$

for an element $v \in \mathcal{M}$.

If there exists a correspondence $x_t \leftrightarrow f_t$, where $f_t = f_t(\lambda)$ are quadratically integrable functions on a space $(\Lambda, \mathcal{S}, \mu)$, such that

$$\int x_t x_s \, \mathrm{dP}_0 = \int f_t f_s \, \mathrm{d}\mu, \quad t, s \in T, \tag{5.2}$$

then the regular case takes place if and only if in the closed linear space $\mathcal{N}(f_t, t \in T)$, generated by functions f_t, there exists an element g such that

$$\int f_t g \, \mathrm{d}\mu = \varphi_t \quad (t \in T). \tag{5.3}$$

For the sufficient statistic u, mentioned in Theorem 3.1, we have

$$u = \frac{v}{\int v \, \mathrm{dP}_1} \leftrightarrow \frac{g}{\int g^2 \, \mathrm{d}\mu} \quad \text{and} \quad \operatorname{var} u = \frac{1}{\int g^2 \, \mathrm{d}\mu}, \tag{5.4}$$

and for the statistic v from (5.1) we have

$$v \leftrightarrow g \quad \text{and} \quad \int v \, \mathrm{dP}_1 = \int v^2 \, \mathrm{dP}_0 = \int g^2 \, \mathrm{d}\mu. \tag{5.5}$$

Proof. According to Theorem 4.1 the regular case takes place if and only if

$$\int y^* \, \mathrm{dP}_1 < 1. \tag{5.6}$$

Now, in view of (3.3) and (2.3), we have

$$\varphi_t = \int x_t \, \mathrm{dP}_1 = \int y^* x_t \, \mathrm{dP}_1 = \int y^* x_t \, \mathrm{dP}_0 + \varphi_t \int y^* \, \mathrm{dP}_1, \quad t \in T,$$

which, in view of (5.6), results in

$$\varphi_t = \frac{1}{1 - \int y^* \, \mathrm{dP}_1} \int y^* x_t \, \mathrm{dP}_0, \quad t \in T.$$

Thus the first assertion is proved, where

$$v = \frac{y^*}{1 - \int y^* \mathrm{dP}_1}.$$

As can be easily seen, (5.1) extends to the whole space $\mathcal{M}$, i. e.

$$\int x \, \mathrm{dP}_1 = \int vx \, \mathrm{dP}_0, \quad x \in \mathcal{M}. \tag{5.7}$$

In other words, $\int x \, \mathrm{dP}_1$ is a linear functional on the space $\mathcal{M}$ with the inner product $[x, y] = \int xy \, \mathrm{dP}_0$, and $\varphi_t = \int x_t \, \mathrm{dP}_1$ is induced by this functional. This fact must hold also with respect to the space $\mathcal{N}$, because the correspondence (5.2) may be extended in a one-to-one way to the whole spaces $\mathcal{M}$ and $\mathcal{N}$. Then, clearly $v \leftrightarrow g$, where g is given by (5.3). As u is a multiple of y^*, we obtain $u \leftrightarrow cg$ for some constant c. This constant equals $1/\int g^2 \, \mathrm{d}\mu$, since, in view of (5.1),

$$1 = \int u \, \mathrm{dP}_1 = \int uv \, \mathrm{dP}_0 = c \int g^2 \, \mathrm{d}\mu.$$

Finally

$$\int u^2 \, \mathrm{dP}_0 = \int \left(\frac{g}{g^2 \, \mathrm{d}\mu} \right)^2 \mathrm{d}\mu = \frac{1}{\int g^2 \, \mathrm{d}\mu},$$

which completes the proof. □

6 APPLICATION TO PROCESSES WITH INDEPENDENT INCREMENTS

Consider a process with independent increments $\{x_t,\, 0 \le t \le T\}$, and denote

$$F(\lambda) = \begin{cases} \int x_\lambda^2 \, \mathrm{dP}_0, & 0 \le \lambda \le T \\ 0, & 0 > \lambda. \end{cases} \tag{6.1}$$

Make use of the well-known correspondence

$$x_t \leftrightarrow \chi_t \quad \text{where} \quad \begin{aligned} \chi_t(\lambda) &= 1 \quad \text{for } \lambda \le t, \\ &= 0 \quad \text{for } \lambda > t. \end{aligned} \tag{6.2}$$

So we get, in view of Theorem 5.1, the following condition for the regular case:

$$\begin{aligned} \varphi_t &= \int_{-0}^{T} \chi_t(\lambda)\, g(\lambda) \, \mathrm{d}F(\lambda) = \int_{-0}^{t} g(\lambda) \, \mathrm{d}F(\lambda) \\ &= g(0)\, F(0) + \int_0^t g(\lambda) \, \mathrm{d}F(\lambda) \end{aligned} \tag{6.3}$$

where

$$g_0^2 \, F_0 + \int_0^T g^2(\lambda) \, \mathrm{d}F(\lambda) < \infty. \tag{6.4}$$

In other words, φ_t must be absolutely continuous with respect to $F(t)$, and

$$\frac{\varphi_0^2}{F_0} + \int_0^T \left(\frac{\mathrm{d}\varphi_t}{\mathrm{d}F_t}\right)^2 \mathrm{d}F_t < \infty. \tag{6.5}$$

It is well known that to any function $g \in L_2(F)$ there corresponds the random variable (see Doob (1953), Chap. IX)

$$v = x_0 g_0 + \int_0^T g \, \mathrm{d}x_t \leftrightarrow g. \tag{6.6}$$

Inserting $\mathrm{d}\varphi_t / \mathrm{d}F_t$ for g in (6.6) and on using the formula (5.4) we obtain that the sufficient unbiased estimator u of α equals

$$u = \frac{x_0 \dfrac{\varphi_0}{F_0} + \displaystyle\int_0^T \frac{\mathrm{d}\varphi_t}{\mathrm{d}F_t} \, \mathrm{d}x_t}{\frac{\varphi_0^2}{F_0} + \int_0^T \left(\frac{\mathrm{d}\varphi_t}{\mathrm{d}F_t}\right)^2 \mathrm{d}F(t)} \tag{6.7}$$

and the variance of u is

$$\operatorname{var} u = \frac{1}{\frac{\varphi_0^2}{F_0} + \int_0^T \left(\frac{\mathrm{d}\varphi_t}{\mathrm{d}F_t}\right)^2 \mathrm{d}F_t}. \tag{6.8}$$

For $\varphi_t \equiv 1$, $\varphi_t \equiv t$ and $\varphi_t \equiv t^2$, and for $F(t) = F_0 + t$, we get:

$$\varphi_t \equiv 1: \quad u = x_0,$$

$$\varphi_t \equiv t: \quad u = \frac{x_T - x_0}{T},$$

$$\varphi_t \equiv t^2: \quad u = \frac{2\int_0^T t \, \mathrm{d}x_t}{\int_0^T (2t)^2 \, \mathrm{d}t} = \frac{T x_T - \int_0^T x_t \, \mathrm{d}t}{\frac{2}{3}T^3}$$

where we have used integration by parts, justified in Doob (1953), Chap. IX, formula (2.11).

Remark. It would be quite straightforward to generalize these results to arbitrary processes with independent increments defined as random set functions on an abstract measurable space.

7 APPLICATION TO MARKOV PROCESSES

This case can be reduced to the previous one by a simple transformation. In fact, if

$$\left| \int x_t x_s \, \mathrm{dP}_0 \right| > 0, \quad 0 \le t, \, s \le T \tag{7.1}$$

we may write that (see Hájek (1958), Example 2)

$$\int x_t x_s \, \mathrm{dP}_0 = \sigma_t \sigma_s \frac{h_t}{h_s}, \quad t \le s \tag{7.2}$$

where

$$\sigma_t^2 = \int x_t^2 \, \mathrm{dP}_0, \qquad 0 \le t \le T, \tag{7.3}$$

$$h_t = \frac{\sigma_0 \sigma_t}{\int x_0 x_t \, \mathrm{dP}_0}, \qquad 0 \le t \le T \tag{7.4}$$

and h_t^2 is non-decreasing. We shall for simplicity assume that $h_t > 0$. Putting

$$y_t = x_t \frac{h_t}{\sigma_t}, \quad 0 \le t \le T \tag{7.5}$$

we get that

$$\int y_s y_t \, \mathrm{dP}_0 = h_t^2, \quad 0 \le t \le s \le T, \tag{7.6}$$

i. e., because h_t^2 is non-decreasing, y_t is a process with independent increments. The corresponding parameter functions of y_t are (7.6) and

$$\int y_t \, \mathrm{dP}_\alpha = \frac{h_t}{\sigma_t} \int x_t \, \mathrm{dP}_\alpha = \alpha \frac{h_t}{\sigma_t} \varphi_t, \quad t \in T. \tag{7.7}$$

On substituting $\varphi_t h_t / \sigma_t$, $x_t h_t / \sigma_t$ and h_t^2 for φ_t, x_t and F_t, respectively, into (6.7) and (6.8), we get

$$u = \frac{\dfrac{\varphi_0 x_0}{\sigma_0^2} + \displaystyle\int_0^T \frac{\mathrm{d}\dfrac{\varphi_t h_t}{\sigma_t}}{\mathrm{d}h_t^2} \mathrm{d}\frac{x_t h_t}{\sigma_t}}{\dfrac{\varphi_0^2}{\sigma_0^2} + \displaystyle\int_0^T \left(\frac{\mathrm{d}\dfrac{\varphi_t h_t}{\sigma_t}}{\mathrm{d}h_t^2} \right)^2 \mathrm{d}h_t^2} \tag{7.8}$$

and

$$\mathsf{var}\, u = \frac{1}{\dfrac{\varphi_0^2}{\sigma_0^2} + \displaystyle\int_0^T \left(\frac{\mathrm{d}\dfrac{\varphi_t h_t}{\sigma_t}}{\mathrm{d}h_t^2} \right)^2 \mathrm{d}h_t^2} \tag{7.9}$$

For a stationary Markov process we have

$$\sigma_t^2 = \sigma^2, \quad h_t = e^{at} \quad (a > 0) \tag{7.10}$$

so that

$$u = \frac{\varphi_0 x_0 + \int_0^T \frac{\mathrm{d}\varphi_t e^{at}}{\mathrm{d}e^{2at}} \mathrm{d}x_t e^{at}}{\varphi_0^2 + \int_0^T \left(\frac{\mathrm{d}\varphi_t e^{at}}{\mathrm{d}e^{2at}}\right)^2 \mathrm{d}e^{2at}}. \tag{7.11}$$

If the function φ_t has an absolutely continuous first derivative φ_t', then

$$\frac{\mathrm{d}\varphi_t e^{at}}{\mathrm{d}e^{2at}} = \frac{\varphi_t' e^{at} + a\varphi_t e^{at}}{2ae^{2at}} = \frac{1}{2a}\,\varphi_t'\, e^{-at} + \frac{1}{2}\,\varphi_t\, e^{-at} \tag{7.12}$$

and

$$\left(\frac{\mathrm{d}\varphi_t e^{at}}{\mathrm{d}e^{2at}}\right)' = \frac{1}{2a}\,\varphi_t''\, e^{-at} - \frac{a}{2}\,\varphi_t\, e^{-at}. \tag{7.13}$$

On substituting (7.12) into (7.11) and on using again integration by parts (see Doob (1953), Chap. IX, formula (2.11)), we obtain, in view of (7.13),

$$u = \frac{\varphi_0 x_0 + \varphi_T x_T + \frac{1}{a}[\varphi_T' x_T - \varphi_0' x_0] + a\int_0^T x_t\left(\varphi_t - \frac{1}{a^2}\,\varphi''t\right)\mathrm{d}t}{2\varphi_0^2 + a\int_0^T \left(\varphi_t + \frac{1}{a}\,\varphi_t'\right)^2 \mathrm{d}t} \tag{7.14}$$

and

$$\operatorname{var} u = \frac{\sigma^2}{2\varphi_0^2 + a\int_0^T \left(\varphi_t + \frac{1}{a}\,\varphi_t'\right)^2 \mathrm{d}t}. \tag{7.15}$$

For $\varphi_t \equiv 1$ and $\varphi_t \equiv t$, (7.14) gives

$$\varphi_t \equiv 1: \qquad u = \frac{x_0 + x_T + a\int_0^T x_t\,\mathrm{d}t}{2 + aT} \tag{7.16}$$

$$\varphi_t \equiv t: \qquad u = \frac{Tx_T + \frac{1}{a}(x_T - x_0) + a\int_0^T t x_t\,\mathrm{d}t}{\frac{1}{a}T + T^2 + \frac{1}{3}aT^3}. \tag{7.17}$$

The formula (7.16) was first derived in Grenander (1949).

8 APPLICATION TO STATIONARY PROCESSES

On using the well-known correspondence

$$x_t \leftrightarrow e^{it\lambda}, \quad t \in T \tag{8.1}$$

we get from Theorem 5.1 that the regular case takes place if and only if

$$\varphi_t = \int e^{it\lambda} g(\lambda)\,\mathrm{d}F(\lambda), \quad t \in T \tag{8.2}$$

for some function $g(\lambda)$ quadratically integrable with respect to the spectral function $F(\lambda)$. This holds for any subset T of the real line. The necessity and sufficiency of the condition (8.2) for a subclass of functions φ_t is also proved in Grenander and Rosenblatt (1956), § 2.5. The statistical meaning of the regular case is that no consistent estimator for α exists. Let T_n be a sequence of subsets of parameter-points converging monotonically to T, and a_n a consistent estimator of α based on observations in points $t \in T_n$. Then for any pair $\alpha_1 \neq \alpha_2$ we would have $\lim_{n\to\infty} \mathsf{P}_{\alpha_1}(|a_n - \alpha_1| < |\alpha_1 - \alpha_2|/2) = 1$ and $\lim_{n\to\infty} \mathsf{P}_{\alpha_2}(|a_n - \alpha_1| < |\alpha_1 - \alpha_2|/2) = 0$, which implies the singular case.

Since $g \equiv 1$ always belongs to $L_2(F)$, $\varphi_t = r_t = \int e^{it\lambda}\, \mathrm{d}F(\lambda)$ always gives the regular case.

Lastly, let us show a connection between prediction and our linear model. Namely, if

$$\begin{aligned} \varphi_t &= 1, \qquad t = t_0, \\ &= 0, \qquad t \in T, \end{aligned} \tag{8.3}$$

where T is a subset of the real line not containing t_0, then we have that

$$u = x_{t_0} - x^*_{t_0} \tag{8.4}$$

where u is the sufficient statistic mentioned in Theorem 3.1 and $x^*_{t_0}$ is the best linear predictor of x_{t_0}.

In fact, $x^*_{t_0}$ is characterized by the relations

$$\int \left(x_{t_0} - x^*_{t_0}\right) x_t \,\mathrm{d}\mathsf{P}_0 = 0, \quad t \in T \tag{8.5}$$

from which, in view of (2.3) and (8.3), it follows that

$$\begin{aligned} &\int \left(x_{t_0} - x^*_{t_0}\right) x_t \,\mathrm{d}\mathsf{P}_1 \\ = &\int \left(x_{t_0} - x^*_{t_0}\right) x_t \,\mathrm{d}\mathsf{P}_0 + \int x_t \,\mathrm{d}\mathsf{P}_1 \int \left(x_{t_0} - x^*_{t_0}\right) \mathrm{d}\mathsf{P}_1 = 0, \quad t \in T. \end{aligned} \tag{8.6}$$

Further

$$\int \left(x_{t_0} - x^*_{t_0}\right) x_{t_0} \,\mathrm{d}\mathsf{P}_1 = \int \left(x_{t_0} - x^*_{t_0}\right)^2 \mathrm{d}\mathsf{P}_1. \tag{8.7}$$

Equations (8.6) and (8.7) may be written in the following unified form:

$$\int \left(x_{t_0} - x^*_{t_0}\right) x_t \,\mathrm{d}\mathsf{P}_1 = \varphi_t \int \left(x_{t_0} - x^*_{t_0}\right)^2 \mathrm{d}\mathsf{P}_1, \quad t \in T \cup \{t_0\}$$

which, in view of (3.2), proves (8.4).

REFERENCES

Doob J. L. (1953). *Stochastic Processes.* London, New York.

Feldman J. (1958). Equivalence and perpendicularity of Gaussian processes. *Pacific Journ. Math. 8*, 699–708.

Grenander U. (1949). Stochastic processes and statistical inference. *Arkiv för Matematik 1*, 195–277.

Grenander U. and Rosenblatt M. (1956). *Statistical Analysis of Stationary Time Series.* Stockholm.

Hájek J. (1958). On a property of normal distributions of any stochastic process. *Czech. Math. Jour. 8*, 610–618.

CHAPTER 15

Limiting Distributions in Simple Random Sampling from a Finite Population

Publ. Math. Inst. Hungar. Acad. Sci. 5 (1960), 361–374.
Reproduced by permission of Akadémiai Kiadó, Budapest.

1 INTRODUCTION AND SUMMARY

Sampling from a finite population may be considered as a random experiment whose outcomes are subsets s of the set $S = \{1, 2, \ldots, N\}$; s is called a sample and S is called a population. Denote an s consisting of k elements by s_k and the probability of s_k by $\mathsf{P}(s_k)$.

Simple random sampling of sample size n is defined by the following probabilities $\mathsf{P}(s_k)$:

$$\mathsf{P}(s_k) = \begin{cases} \binom{N}{n}^{-1} & \text{if } k = n \text{ [simple random sampling of size } n] \\ 0 & \text{otherwise.} \end{cases} \tag{1.1}$$

In this chapter we shall make use of so called Poisson sampling defined as follows:

$$\mathsf{P}(s_k) = \left(\frac{n}{N}\right)^k \left(1 - \frac{n}{N}\right)^{N-k} \quad \text{[Poisson sampling of mean size } n]. \tag{1.2}$$

Let us have a sequence $y_1, \ldots, y_N$ of real numbers, and put

$$\xi = \sum_{i \in s_n} y_i \tag{1.3}$$

where s_n is a simple random sample and $\sum_{i \in s_n}$ extends over all i contained in the sample s_n. Obviously $\xi = \xi(s_n)$ is a random variable with finite mean

---

Collected Works of Jaroslav Hájek – With Commentary
Edited by M. Hušková, R. Beran and V. Dupač
Published in 1998 by John Wiley & Sons Ltd.

value

$$\mathsf{E}\xi = \frac{n}{N}\sum_{i=1}^{N} y_i \tag{1.4}$$

and variance

$$\mathsf{D}\xi = \frac{n}{N}\,\frac{N-n}{N-1}\sum_{i=1}^{N}(y_i - \overline{Y})^2 \quad \left[\overline{Y} = \frac{1}{N}\sum_{i=1}^{N} y_i\right]. \tag{1.5}$$

Let us consider an infinite sequence of simple random sampling experiments, the νth of which is of size n_ν and refers to a population of size N_ν and with values $y_{\nu 1}, \ldots, y_{\nu N_\nu}$. Let ξ_ν be the random variable defined by (1.3) corresponding to the νth sampling experiment.

Now we ask about conditions concerning $\{y_{\nu i}, n_\nu, N_\nu\}$ under which

$$\lim_{\nu\to\infty} \mathsf{P}\left\{\xi_\nu - \mathsf{E}\xi_\nu < x\sqrt{\mathsf{D}\xi_\nu}\right\} = \frac{1}{\sqrt{2\pi}}\int_{-\infty}^{x} e^{-\frac{1}{2}t^2}\,\mathrm{d}t \tag{1.6}$$

or

$$\lim_{\nu\to\infty} \mathsf{P}\{\xi_\nu = k\} = e^{-\lambda}\,\frac{\lambda^k}{k!} \tag{1.7}$$

or generally, the distribution function of ξ_ν converges to a limiting distribution law. We naturally suppose that $n_\nu \to \infty$ and $N_\nu - n_\nu \to \infty$.

In the following sections we shall give a complete solution of this problem, i.e. we shall indicate necessary and sufficient conditions. As for convergence to the normal distribution law the necessary and sufficient condition coincides with that derived by Erdős and Rényi (1959), as could be expected.

2 THE FUNDAMENTAL LEMMA

We shall show that the above problem can be completely reduced to the same problem concerning sums of independent random variables.

First observe that Poisson sampling may be interpreted as simple random sampling of size k, where k is a binomial random variable attaining the value k with probability $\binom{N}{k}\left(\frac{n}{N}\right)^k\left(1-\frac{n}{N}\right)^{N-k}$. Actually, it suffices to consult (1.1) and (1.2) and notice that

$$\left(\frac{n}{N}\right)^k\left(1-\frac{n}{N}\right)^{N-k} = \left[\binom{N}{k}\left(\frac{n}{N}\right)^k\left(1-\frac{n}{N}\right)^{N-k}\right]\binom{N}{k}^{-1}.$$

Clearly $\mathsf{E}k = n$ and

$$\mathsf{E}(k-n)^2 = n\left(1-\frac{n}{N}\right). \tag{2.1}$$

Now, it is easy to define an experiment producing simultaneously a simple random sample s_n and a Poisson sample s_k such that $s_n \subset s_k$ if $n \leq k$, and $s_n \supset s_k$ if $n \geq k$. This may be done as follows:

(i) First we realize the binomial random variable k attaining the value k with probability

$$\binom{N}{k}\left(\frac{n}{N}\right)^k\left(1-\frac{n}{N}\right)^{N-k}, \quad 0 \leq k \leq N.$$

(ii) (a) When $k = n$, we select a simple random sample $s_n = s_k$ which is a simultaneous realization of both simple random sampling and Poisson sampling. (b) When $k > n$, then we select a simple random sample s_k, which is a realization of Poisson sampling; thereafter we select a simple random sample s_n of size n from s_k (s_k represents here a population), s_n being a realization of simple random sampling. (c) When $k < n$, then we select a simple random sample s_n, which is a realization of simple random sampling; thereafter we select a simple random sample s_k from s_n (s_n represents here a population), s_k being a realization of Poisson sampling.

Put

$$\eta = \sum_{i \in s_n}(y_i - \overline{Y}) = \xi - n\overline{Y} \tag{2.2}$$

and

$$\eta^* = \sum_{i \in s_k}(y_i - \overline{Y}) \tag{2.3}$$

where (s_n, s_k) are joined samples from the above experiments, s_n representing a simple random sample and s_k a Poisson sample. The number of summands is a constant n in (2.2) and is a binomial random variable in (2.3). Clearly,

$$\eta - \eta^* = \begin{cases} 0 & \text{if } k = n \\ \sum\limits_{i \in s_n - s_k} (y_i - \overline{Y}) & \text{if } k < n \\ \sum\limits_{i \in s_k - s_n} (y_i - \overline{Y}) & \text{if } k > n \end{cases} \tag{2.4}$$

since either s_k is a subset of s_n or, conversely, s_n is a subset of s_k.

Lemma 2.1. *The following inequality holds true:*

$$\frac{\mathsf{E}(\eta - \eta^*)^2}{\mathsf{D}\eta^*} \leq \sqrt{\frac{1}{n} + \frac{1}{N-n}}. \tag{2.5}$$

Proof. If k is fixed, then $s_n - s_k$ or $s_k - s_n$ represents a simple random sample of size $|k - n|$. Consequently, in view of (1.5), we have

$$\mathsf{E}\{(\eta - \eta^*)^2 \mid k\} = \mathsf{D}\{\eta - \eta^* \mid k\} \tag{2.6}$$

$$= \frac{|k-n|}{N}\,\frac{N-|k-n|}{N-1}\sum_{i=1}^{N}(y_i-\overline{Y})^2 \le |k-n|\,\frac{1}{N}\sum_{i=1}^{N}(y_i-\overline{Y})^2.$$

The inequality (2.6) together with (2.1) results in

$$\mathsf{E}\{(\eta-\eta^*)^2\} \le \mathsf{E}|k-n|\,\frac{1}{N}\sum_{i=1}^{N}(y_i-\overline{Y})^2 \tag{2.7}$$

$$\le \sqrt{\mathsf{E}(k-n)^2}\,\frac{1}{N}\sum_{i=1}^{N}(y_i-\overline{Y})^2 = \sqrt{n\left(1-\frac{n}{N}\right)}\,\frac{1}{N}\sum_{i=1}^{N}(y_i-\overline{Y})^2.$$

In a similar way we could derive

$$\mathsf{D}\eta^* = n\left(1-\frac{n}{N}\right)\frac{1}{N}\sum_{i=1}^{N}(y_i-\overline{Y})^2. \tag{2.8}$$

We prefer, however, to prove (2.8) by the following consideration. Any sampling experiment consists of N dichotomous experiments, the ith of which has the following two possible outcomes: including the element i in the sample s and not including the element i in the sample s. If all these experiments are mutually independent and the probability of including the element i is a constant n/N, $1 \le i \le N$, we can easily see that one gets Poisson sampling, i. e. that each sample s_k has the probability (1.2). This fact implies that η^* may be considered to be a sum of N independent random variables,

$$\eta^* = \sum_{i=1}^{N}\zeta_i \tag{2.9}$$

where

$$\zeta_i = \begin{cases} y_i - \overline{Y} & \text{with probability } n/N \text{ if } i \in s_k \\ 0 & \text{with probability } 1-n/N \text{ if } i \in S - s_k. \end{cases} \tag{2.10}$$

Clearly

$$\mathsf{D}\zeta_i = (y_i-\overline{Y})^2\,\frac{n}{N}\left(1-\frac{n}{N}\right) \quad (1 \le i \le N) \tag{2.11}$$

which proves (2.8). □

Combining (2.7) and (2.8) we obviously obtain the inequality (2.5), which was to be proved.

Let us consider a sequence of experiments of the above kind (i. e. producing joined simple random and Poisson samples) and denote by η_ν and η_ν^* the random variables (2.2) and (2.3) referring to the νth experiment. From Lemma 2.1 it follows that

$$\lim_{\nu\to\infty}\frac{\mathsf{E}(\eta_\nu-\eta_\nu^*)^2}{\mathsf{D}\eta_\nu^*} = 0 \quad \text{if} \quad \begin{cases} n_\nu \to \infty \\ N_\nu - n_\nu \to \infty. \end{cases} \tag{2.12}$$

Remark 2.1. The relation (2.12) implies that, provided $n_\nu \to \infty$ and $N_\nu - n_\nu \to \infty$, the limiting variances and distributions of random variables $A_\nu + B_\nu\,\eta_\nu$ and $A_\nu + B_\nu\,\eta_\nu^*$ exist under the same conditions, and if they exist, are the same. The random variable η_ν^*, however, is a sum of independent addends (2.10), so that when studying the limiting distributions of $A_\nu + B_\nu\,\eta_\nu^*$ we may simply apply the well-known theory of summation of independent random variables. See Gnedenko and Kolmogorov (1949).

3 CONVERGENCE TO THE NORMAL DISTRIBUTION

We shall prove that the condition derived by Erdős and Rényi (1959) is not only sufficient but also necessary provided that $n_\nu \to \infty$ and $N_\nu - n_\nu \to \infty$.

Theorem 3.1. *Let $S_{\nu\tau}$ be the subset of elements of $S_\nu = \{1, \ldots, N_\nu\}$ on which the inequality*

$$|y_{\nu i} - \overline{Y}_\nu| > \tau\sqrt{\mathsf{D}\xi_\nu} \tag{3.1}$$

holds; let $\mathsf{D}\xi_\nu$ denote the variance (1.5) referring to the νth experiment. Suppose that $n_\nu \to \infty$ and $N_\nu - n_\nu \to \infty$.

Then the random variable ξ_ν defined by (1.3) has an asymptotically normal distribution with parameters $(\mathsf{E}\xi_\nu, \mathsf{D}\xi_\nu)$ if and only if

$$\lim_{\nu\to\infty} \frac{\sum_{i\in S_{\nu\tau}} (y_{\nu i} - \overline{Y}_\nu)^2}{\sum_{i\in S_\nu} (y_{\nu i} - \overline{Y}_\nu)^2} = 0 \quad \text{for any } \tau > 0 \tag{3.2}$$

where $\sum_{i\in S_\nu}$ denotes the same summation as $\sum_{i=1}^{N_\nu}$.

Proof. In view of Remark 2.1, it suffices to establish sufficient and necessary conditions for asymptotic normality of the random variable η_ν^* defined by (2.9), namely with parameters $(0, \mathsf{D}\eta_\nu^*)$. Notice that $\mathsf{E}\eta_\nu^* = 0$, since

$$\mathsf{E}\eta^* = \sum_{i=1}^{N} \mathsf{E}\zeta_i = \sum_{i=1}^{N} \frac{n}{N}(y_i - \overline{Y}) = 0. \tag{3.3}$$

First suppose that the random variables $\zeta_{\nu i} = \zeta_i$ defined by (2.10) are infinitesimal, i. e. that

$$\lim_{\nu\to\infty} \frac{\max_{1\le i\le N} \mathsf{D}\zeta_{\nu i}}{\sum_{i=1}^{N_\nu} \mathsf{D}\zeta_{\nu i}} = 0. \tag{3.4}$$

In view of (2.11), (3.4) is equivalent to

$$\lim_{\nu\to\infty} \frac{\max_{1\le i\le N_\nu} (y_{\nu i} - \overline{Y}_\nu)^2}{\sum_{i=1}^{N_\nu} (y_{\nu i} - \overline{Y}_\nu)^2} = 0. \tag{3.5}$$

Condition (3.5) is clearly much weaker than condition (3.2); it is usually called the Noether condition.

Provided that (3.4) holds, the necessary and sufficient condition for asymptotic normality of η_ν^* with parameters $(0,\ \mathsf{D}\eta_\nu^*)$ is given by the Lindeberg condition. Since the random variables $\zeta_{\nu i} - \mathsf{E}\zeta_{\nu i}$ take on values $(y_{\nu i} - \overline{Y}_\nu)(1 - n_\nu/N_\nu)$ and $-(y_{\nu i} - \overline{Y}_\nu)\, n_\nu/N_\nu$ with respective probabilities n_ν/N_ν and $1 - n_\nu/N_\nu$ we have

$$\frac{1}{\mathsf{D}\eta_\nu^*}\sum_{i=1}^{N_\nu}\int_{|x|>\tau\sqrt{\mathsf{D}\eta_\nu^*}} x^2\,\mathrm{d}\,\mathsf{P}\{\zeta_{\nu i} - \mathsf{E}\zeta_{\nu i} < x\} \tag{3.6}$$

$$= \frac{\frac{n_\nu}{N_\nu}\left(1-\frac{n_\nu}{N_\nu}\right)^2 \sum_{i\in C_{\nu\tau}}(y_{\nu i}-\overline{Y}_\nu)^2 + \left(1-\frac{n_\nu}{N_\nu}\right)\left(\frac{n_\nu}{N_\nu}\right)^2 \sum_{i\in B_{\nu\tau}}(y_{\nu i}-\overline{Y}_\nu)^2}{\frac{n_\nu}{N_\nu}\left(1-\frac{n_\nu}{N_\nu}\right)\sum_{i\in S_\nu}(y_{\nu i}-\overline{Y}_\nu)^2}$$

$$= \frac{\left(1-\frac{n_\nu}{N_\nu}\right)\sum_{i\in C_{\nu\tau}}(y_{\nu i}-\overline{Y}_\nu)^2 + \frac{n_\nu}{N_\nu}\sum_{i\in B_{\nu\tau}}(y_{\nu i}-\overline{Y}_\nu)^2}{\sum_{i\in S_\nu}(y_{\nu i}-\overline{Y}_\nu)^2}$$

where $C_{\nu\tau}$ and $B_{\nu\tau}$ are subsets of elements of S_ν on which

$$C_{\nu\tau} : |y_{\nu i} - \overline{Y}_\nu|\left(1-\frac{n_\nu}{N_\nu}\right) > \tau\left(1-\frac{n_\nu}{N_\nu}\left(1-\frac{n_\nu}{N_\nu}\right)\sum_{i\in S_\nu}(y_{\nu i}-\overline{Y}_\nu)^2\right)^{1/2} \tag{3.7}$$

and

$$B_{\nu\tau} : |y_{\nu i} - \overline{Y}_\nu|\,\frac{n_\nu}{N_\nu} > \tau\left(\frac{n_\nu}{N_\nu}\left(1-\frac{n_\nu}{N_\nu}\right)\sum_{i\in S_\nu}(y_{\nu i}-\overline{Y}_\nu)^2\right)^{1/2} \tag{3.8}$$

respectively. In view of (1.5), we see that

$$C_{\nu\tau} = S_{\nu,\tau\sqrt{\frac{N_\nu(N_\nu-1)}{(N_\nu-n_\nu)^2}}} \quad\text{and}\quad B_{\nu\tau} = S_{\nu,\tau\sqrt{\frac{N_\nu(N_\nu-1)}{n_\nu^2}}}. \tag{3.9}$$

From (3.9) it follows that (3.2) is equivalent to the condition that the first member of (3.6) converges to 0, i. e. to the fulfilment of the Lindeberg condition for η_ν^*.

Thus it remains to prove that η_ν^* cannot have a limiting normal distribution with parameters $(\mathsf{E}\eta_\nu^*,\ \mathsf{D}\eta_\nu^*)$ if (3.5) does not hold.

We may suppose without any loss of generality that $n_\nu \le \frac{1}{2}N_\nu$ and

$$|y_{\nu 1} - \overline{Y}_\nu| \ge |y_{\nu 2} - \overline{Y}_\nu| \ge \cdots \ge |y_{\nu N_\nu} - \overline{Y}_\nu|. \tag{3.10}$$

If (3.5) is not satisfied, then there exists an $\varepsilon \neq 0$ such that

$$\lim_{\nu\to\infty}\frac{y_{\nu 1} - \overline{Y}_\nu}{\sqrt{\sum_{i\in S_\nu}(y_{\nu i}-\overline{Y}_\nu)^2}} = \varepsilon \neq 0 \tag{3.11}$$

for some subsequence of indices ν. Taking a new subsequence from this subsequence, we may assume that

$$\lim_{\nu} \frac{n_\nu}{N_\nu} = c \leq \frac{1}{2}. \tag{3.12}$$

For simplicity let us introduce no new symbols for denoting the subsequences.

Now the relations (3.11) and (3.12) mean that the distribution function of the random variable

$$\frac{\zeta_{\nu 1}}{\sqrt{\mathsf{D}\eta_\nu^*}} = \begin{cases} \frac{y_{\nu 1} - \overline{Y}_\nu}{\left[\left(\frac{n_\nu}{N_\nu}\right)\left(1-\frac{n_\nu}{N_\nu}\right)\sum_{i \in S_\nu}(y_{\nu i}-\overline{Y}_\nu)^2\right]^{1/2}} & \text{with probability } \frac{n_\nu}{N_\nu} \\ 0 & \text{with probability } 1 - \frac{n_\nu}{N_\nu} \end{cases} \tag{3.13}$$

converges to a distribution function which has a jump $1 - c$ at the point 0 and, if $c > 0$, a jump c at the point $\varepsilon(c(1-c))^{-1/2}$. Let us discuss each of the cases $c = 0$ and $c > 0$ separately.

If $c = 0$ the variance of (3.13) does not converge to the variance of the limiting distribution. Actually, (3.13) has a limiting variance ε^2 while the limiting distribution is degenerated to the single point 0 so that it has a variance 0. Hence if there existed a limiting distribution of the statistic

$$\frac{\eta_\nu^*}{\sqrt{\mathsf{D}\eta_\nu^*}} = \frac{\zeta_{\nu 1}}{\sqrt{\mathsf{D}\eta_\nu^*}} + \frac{\sum_{i=2}^{N_\nu} \zeta_{\nu i}}{\sqrt{\mathsf{D}\eta_\nu^*}} \tag{3.14}$$

it would have a variance smaller than 1. Consequently, η_ν^* cannot have asymptotically normal distribution with parameters $(0, \mathsf{D}\eta_\nu^*)$.

If $c > 0$, the distribution of (3.13) converges to a distribution concentrated in the points 0 and $\varepsilon(c(1-c))^{-1/2}$. If (3.14) has an asymptotically normal distribution, this distribution could be decomposed in a convolution of two distributions, one of which is not normal. This is, however, not possible, in view of the well-known theorem by H. Cramér.

The theorem is thus completely proved. □

Remark 3.1. In a paper by Hájek (1961) it is proved that, provided we have a fixed double sequence $\{N_\nu, y_{\nu i}\}$, the Lindeberg condition (3.2) is fulfilled for any sequence $\{n_\nu\}$, such that $n_\nu \to \infty$ and $N_\nu - n_\nu \to \infty$, if and only if the relation

$$\lim_{\nu \to \infty} \frac{\sum_{i \in s_{r_\nu}} (y_{\nu i} - \overline{Y}_\nu)^2}{\sum_{i \in S_\nu} (y_{\nu i} - \overline{Y}_\nu)^2} = 0 \tag{3.15}$$

holds for any sequence $\{s_{r_\nu}\}$ such that $s_{r_\nu} \subset S_\nu$ and

$$\lim_{\nu \to \infty} \frac{r_\nu}{N_\nu} = 0 \tag{3.16}$$

where r_ν denotes the number of elements in s_{r_ν}.

Remark 3.2. According to a theorem by Cramér (1937, p. 105) a vector $(\xi_{\nu 1}, \ldots, \xi_{\nu m})$ has an m-dimensional normal limit distribution with parameters $\{\mathsf{E}\xi_{\nu j}, \operatorname{Cov}(\xi_{\nu j}, \xi_{\nu h}),\ h, j = 1, \ldots, m\}$ if any linear combination $\sum_{j=1}^{m} \lambda_j \xi_{\nu j}$ has a one-dimensional normal limit distribution with respective parameters. Let $\xi_{\nu j}$ be given by (1.3) where $y_i = y_{\nu j i}$, where ν labels the experiment and j the variable. Suppose that the sequences $\{y_{\nu j i}, n_\nu, N_\nu\}$, $j = 1, \ldots, m$, fulfil condition (3.2) and that the multiple correlation coefficients $\rho_{\nu j}$ between $\xi_{\nu j}$ and $\{\xi_{\nu j'},\ j' \neq j\}$ are uniformly bounded from 1, i. e. that

$$\lim_{\nu \to \infty} \sup \rho_{\nu j}^2 < 1, \quad j = 1, \ldots, m. \tag{3.17}$$

Then any sequence $\left\{\sum_{j=1}^{m} \lambda_j\, y_{\nu j i},\ n_\nu,\ N_\nu\right\}$, where λ_j are arbitrary constants, fulfils the condition (3.2) and hence the random vector $(\xi_{\nu 1}, \ldots, \xi_{\nu m})$ has an asymptotically normal m-dimensional distribution with respective parameters. Actually, we have

$$\mathsf{D}\left(\sum_{j=1}^{m} \lambda_j\, \xi_{\nu j}\right) \geq \max_{1 \leq j \leq m} (1 - \rho_{\nu j}^2)\, \lambda_j^2\, \mathsf{D}\,\xi_{\nu j} \tag{3.18}$$

and

$$\left|\sum_{j=1}^{m} \lambda_j (y_{\nu j i} - \overline{Y}_{\nu j})\right| \leq m \max_{1 \leq j \leq m} |\lambda_j|\, |y_{\nu j i} - Y_{\nu j}|. \tag{3.19}$$

The rest follows by easy computation.

Remark 3.3. If n_ν / N_ν is bounded from 0 and 1, i. e.

$$0 < \varepsilon < \frac{n_\nu}{N_\nu} < 1 - \varepsilon \quad (\nu \geq \nu_0) \tag{3.20}$$

then the Lindeberg condition (3.2) reduces to the Noether condition (3.5). Really, if (3.5) and (3.20) are satisfied, then the subset $S_{\nu\tau}$ is empty for all sufficiently large ν so that (3.2) clearly holds. If (3.5) is not satisfied, (3.2) does not hold in any case.

4 CONVERGENCE TO THE POISSON DISTRIBUTION

Using the same method as in Section 3, the following theorem will be proved.

Theorem 4.1. *Suppose that* $n_\nu \to \infty$, $n_\nu \le \frac{1}{2} N_\nu$, *and*

$$\lim_{\nu\to\infty} \mathsf{E}\,\xi_\nu = \lim_{\nu\to\infty} \mathsf{D}\,\xi_\nu = \lambda > 0. \tag{4.1}$$

Then the relation (1.7) *is fulfilled if, and only if, first*

$$\lim_{\nu\to\infty} \frac{n_\nu}{N_\nu} = 0 \tag{4.2}$$

and, second,

$$\lim_{\nu\to\infty} \frac{n_\nu}{N_\nu} \sum_{|y_{\nu i}-1|>\tau} y_{\nu i}^2 = 0 \quad \text{for any } \tau > 0. \tag{4.3}$$

Proof. First assume that the infinitesimality condition (3.5) holds. If (4.2) does not hold, then, in view of Remark 3.3, the limiting distribution may be only normal. Consequently, condition (4.2) is necessary. Now η_ν^* has limiting Poisson distribution if and only if

$$\lim_{\nu\to\infty} \sum_{i=1}^{N_\nu} \int_{|x-1|>\tau} x^2 \, \mathrm{d}\,\mathsf{P}\{\zeta_{\nu i} - \mathsf{E}\,\zeta_{\nu i} < x\} = 0. \tag{4.4}$$

We may write, as in (3.6),

$$\sum_{i=1}^{N_\nu} \int_{|x-1|>\tau} x^2 \, \mathrm{d}\,\mathsf{P}\{\zeta_{\nu i} - \mathsf{E}\,\zeta_{\nu i} < x\} \tag{4.5}$$
$$= \frac{n_\nu}{N_\nu}\left(1-\frac{n_\nu}{N_\nu}\right)^2 \sum_{i\in C'_{\nu\tau}} (y_{\nu i} - \overline{Y}_\nu)^2 + \left(1-\frac{n_\nu}{N_\nu}\right)\left(\frac{n_\nu}{N_\nu}\right)^2 \sum_{i\in B'_{\nu\tau}} (y_{\nu i} - \overline{Y}_\nu)^2$$

where $C'_{\nu\tau}$ and $B'_{\nu\tau}$ are subsets of elements of S_ν on which

$$C'_{\nu\tau} : \left|(y_{\nu i} - \overline{Y}_\nu)\left(1 - \frac{n_\nu}{N_\nu} - 1\right)\right| > \tau \tag{4.6}$$

and

$$B'_{\nu\tau} : \left|-(y_{\nu i} - \overline{Y}_\nu)\,\frac{n_\nu}{N_\nu} - 1\right| > \tau. \tag{4.7}$$

In view of (4.1), it holds that

$$\lim_{\nu\to\infty} \frac{n_\nu}{N_\nu} \sum_{i=1}^{N_\nu} y_{\nu i} = \lim_{\nu\to\infty} n_\nu \overline{Y}_\nu = \lambda, \quad \text{i.e.} \quad \lim_{\nu\to\infty} \overline{Y}_\nu = 0 \tag{4.8}$$

and

$$\lim_{\nu\to\infty} \left(1 - \frac{n_\nu}{N_\nu}\right) \frac{n_\nu}{N_\nu} \sum_{i=1}^{n_\nu} (y_{\nu 1} - \overline{Y}_\nu)^2 = \lambda. \tag{4.9}$$

Consequently, in accordance with (4.2),

$$\lim_{\nu\to\infty} \frac{\sum_{i=1}^{N_\nu} \int_{|x-1|>\tau} x^2 \,\mathrm{d}\,\mathsf{P}\{\zeta_{\nu i} - \mathsf{E}\,\zeta_{\nu i} < x\}}{\frac{n_\nu}{N_\nu} \sum_{|y_{\nu i}-1|>\tau} y_{\nu i}^2} = 1,$$

which proves the equivalency of conditions (4.3) and (4.4). □

As for the case when condition (3.5) is not fulfilled, we could prove, as in the proof of Theorem 3.1, that the limiting distribution cannot preserve variance.

Remark 4.1. If the sampling were done with replacement (i. e. as n independent drawings of one element) we would get just the conditions (4.2) and (4.3) for the asymptotic Poisson distribution of the sum of selected values. This coincidence is clearly caused by the fact that the difference between with and without replacement sampling becomes negligible if $n_\nu/N_\nu \to 0$.

5 OTHER CASES

Developing the basic idea further, we obtain the following theorem.

Theorem 5.1. *Suppose that*

$$\lim_{\nu\to\infty} \mathsf{E}\,\xi_\nu = \mu \tag{5.1}$$

and

$$\lim_{\nu\to\infty} \mathsf{D}\,\xi_\nu = \sigma^2 \tag{5.2}$$

and consider an infinitely divisible law – distinct from the normal law – with mean value μ, variance σ^2 and cumulant generating function

$$i\,\mu\,t + \int \left(e^{itu} - 1 - itu\right) \frac{1}{u^2}\,\mathrm{d}\,K(u). \tag{5.3}$$

Then the distribution of ξ_ν converges to the law given by (5.3) if and only if

$$\lim_{\nu\to\infty} \frac{n_\nu}{N_\nu} = 0 \tag{5.4}$$

and

$$\lim_{\nu\to\infty} \frac{n_\nu}{N_\nu} \sum_{y_{\nu i}<u} y_{\nu i}^2 = K(u) \tag{5.5}$$

in all continuity points of $K(u)$.

Proof. The same as for Theorem 4.1. □

6 CONCLUSIONS

If n_ν/N_ν does not converge to 0, normal limiting distribution is possible, namely under conditions established in Theorem 3.1. We can also get a limiting distribution formed as a convolution of a normal distribution and some two-point distributions.

If n_ν/N_ν converges to 0, the variance preserving limiting distribution may be only infinitely divisible. The conditions for this are the same as if the sampling were carried out with replacement.

REFERENCES

Cramér, H. (1937). *Random Variables and Probability Distributions.* Cambridge University Press, London.

Erdős, P. and Rényi, A. (1959). On a central limit theorem for samples from a finite population. *Publ. Math. Inst. Hungar. Acad. Sci. 4*, 49–61.

Gnedenko, B.V. and Kolmogorov, A.N. (1949). *Limit Theorems for Sums of Independent Random Variables.* (In Russian) Gostechizdat, Moscow.

Hájek, J. (1961). Some extensions of the Wald–Wolfowitz–Noether theorem. *Ann. Math. Statist. 32*, 506–523.

Madow, W.G. (1948). On the limiting distributions of estimates based on samples from finite universes. *Ann. Math. Statist. 19*, 535–545.

CHAPTER 16

On Plane Sample and Related Geometrical Problems

Co-authors: T. Dalenius and S. Zubrzycki.
In: Proc. Fourth Berkeley Symp. Math. Statist. Prob.
(L. Le Cam, ed.), 1 (1961), 125–150.
Univ. of California Press.
Reproduced by permission of Lucien Le Cam, editor.

1 SUMMARY

We study the following problem. An isotropic plane stochastic process is observed at points making up a regular pattern. We are interested in finding patterns yielding the least limiting variance of the observed values when the points are situated within a circle with infinitely increasing radius.

In Section 4, a solution is presented for the correlation function (3.5) and some ranges of point densities. The solution is obtained by solving the related geometrical problem of covering a plane by circles in such a way that the circles mutually intersect as little as possible (see Section 3). From the results obtained it follows that, in contradistinction to the linear case, no unique pattern of points is optimum for all convex correlation functions simultaneously. The efficiency of patterns in general use is, however, quite good.

In Section 5, finally, we study a subclass of convex correlation functions of an isotropic plane process, consisting of functions that admit a spectral representation in terms of the simple correlation function (5.2).

Collected Works of Jaroslav Hájek – With Commentary
Edited by M. Hušková, R. Beran and V. Dupač
Published in 1998 by John Wiley & Sons Ltd.

2 INTRODUCTION

In this section, an expository survey of the background of the problem will be presented.

2.1 Applications of Plane Sampling

Many applications of the sampling method may be broadly described as 'plane sampling'. We give some examples.

Forest surveys. In the simplest case, one might want to estimate the area of a certain country or geographical district covered by forest. Similarly, one might want to estimate the proportion of a forest area covered by a certain variety of tree. In another case, one might want to estimate the number of trees or the volume of timber in a forest area. These kinds of applications are discussed in, for example, Matérn (1947).

General agricultural surveys. Illustrative examples of plane sampling are furnished by large-scale surveys carried out in order to estimate the cultivated area of a certain country or geographical district or the total production of a particular crop. We refer here to, for example, Mahalanobis (1944).

Analogous examples are furnished by such small-scale surveys as 'field experiments' restricted to a single field on a farm. There is no need for specific references on this point.

Soil surveys. One specific example concerns the estimation of the proportion of the area of ground of a certain district that is too salty to be cultivated in the usual way; for a discussion, see Sulanke (unpublished).

Another specific example concerns the estimation of the density of worms in soil; see Finney (1946).

Geological surveys proper. The method of plane sampling has been used in order to estimate the extension, volume, and other parameters characterizing such geological deposits as black or brown coal, zinc deposits, and so on. We refer to Zubrzycki (1957).

In this connection, we want to mention briefly applications of plane sampling in connection with the construction of water-power stations. In such situations, there is often a need to estimate the total volume of earth to be removed (for example, as a basis for cost estimates), or the total volume of gravel available for construction purposes.

Some other examples. The examples given so far relate to sampling a geographical area of some sort. The method of plane sampling is, however, applicable to sampling other kinds of areas; some references will be given.

Drápal, Horálek, and Režný (1957) discuss the problem of estimating the proportions of the surface of cast iron composed of crystals of carbon and of iron, respectively. This problem bears considerable resemblance to the 'corpuscle problem' discussed by Wicksell (1925, 1926). Husu (1957) considers the estimation of a parameter that characterizes the 'smoothness' of the surface of a metal plate. Faure (1957) and Savelli (1957) consider the problem of measuring the transparency of photographic film.

2.2 The Sampling Theory

The theoretical task of constructing a (probabilistic) sampling theory to cope with the problems of estimation raised by applications such as those just discussed has long been the subject of considerable research. By and large, the theory of 'field experimentation' is the origin of the theory of plane sampling. However, from a rather early date, somewhat different paths of advancement have been taken.

In field experiments it is often feasible to apply randomization of the experimental units; the use of randomization may be considered as a device for getting around the need for a (realistic) model of the role played by 'topographic variation'. In applications of plane sampling such as those discussed above, it is often desirable, for practical reasons, to use systematic sampling procedures. As a consequence, it is necessary to account for the role played by 'topographic variation' by means of a (realistic) model of this variation.

The theory of linear sampling. To a large extent the theory of plane sampling is a formal extension of the theory of 'linear sampling', this term referring to sampling a one-dimensional stochastic process. Therefore we include some references to the linear case.

Early contributions include such papers as Osborne (1942), Madow and Madow (1944), and Cochran (1946). In the last-mentioned paper it is proved that systematic sampling is, on the average, more precise than stratified sampling, provided that the correlogram is concave upwards. Among recent contributions we may mention Hájek (1955, 1959).

The theory of plane sampling. A classical contribution to the discussion of topographic variation is given by Smith (1938). In the field of plane sampling proper, the work of Mahalanobis (1944) may be considered pioneering. Matérn (1947) presents a most important contribution. In this work, Matérn shows how the theory of stationary stochastic processes, as developed by Khinchin and Cramér, may be used in the construction of a stochastic model of topographic variation. The paper by Quenouille (1949) is another important contribution from the 1940s. Among recent contributions, we may mention Masuyama (1953), Whittle (1954, 1956), Williams (1956), and Zubrzycki (1957, 1958).

3 FORMULATION OF THE PROBLEM OF THE PAPER

Our problem is to find a regular pattern of points which yield the least limiting variance of values associated with an isotropic plane stochastic process observed at these points. In this section we transform this problem into an equivalent geometrical problem, the solution of which is discussed in Section 4.

To begin, we briefly review some previously established results concerning the linear case. The typical problem can be found as follows.

To the points t of a real line there are assigned random variables $\eta(t)$, subject to the following assumptions:

(a) All random variables $\eta(t)$ have common expected value μ and common variance σ^2.

(b) The correlation coefficient between any two random variables $\eta(t')$ and $\eta(t'')$ depends only upon the absolute difference $|t' - t''|$; in symbols,

$$R[\eta(t'), \eta(t'')] = \rho(|t' - t''|). \tag{3.1}$$

(c) The function $\rho(t)$, with $t \geq 0$, called the correlation function of the process, is continuous with $\rho(0) = 1$.

The assumptions (a) to (c) characterize the family of random variables $\eta(t)$ as a continuous stochastic process stationary to the second degree.

Suppose now that we want to estimate the mean $T^{-1}\int_0^T \eta(t)\,\mathrm{d}t$ of the process by the average

$$\overline{\eta} = \frac{1}{n}[\eta(t_1) + \cdots + \eta(t_n)] \tag{3.2}$$

of its values observed at n points $t_1, \ldots, t_n$ selected in a given segment $0 \leq t \leq T$. Consider all probability methods of sampling n points such that the expected number of points selected from any subsegment (α, β), with $0 \leq \alpha < \beta \leq T$, is proportional to $\beta - \alpha$. We ask how these n points should be chosen in order to minimize the quadratic error of estimation, defined as the expected value of $\left[\overline{\eta} - T^{-1}\int_0^T \eta(t)\,\mathrm{d}t\right]^2$. This value extends over both the sampling experiment and nature's experiment in producing the process $\eta(t)$.

Using an argument of Hájek (1959), involving a kind of spectral representation of the correlation function of the process with respect to a one-parameter family of properly chosen simple correlation functions, it follows that if the correlation function is convex, then the best method of choosing the n points $t_1, \ldots, t_n$ is to select them equidistantly.

We now try to generalize the discussion of the linear case to the case of a plane. Thus we consider a family of random variables $\eta(p)$ assigned to the points p of a Euclidean plane, for which the following generalizations (a′) to (c′) of the previously given assumptions (a) to (c) are fulfilled.

(a′) All random variables $\eta(p)$ have common expected value μ and common variance σ^2.

(b′) The coefficient of correlation between any two random variables $\eta(p)$ and $\eta(q)$ depends only on the vector joining the points p and q; in symbols,

$$R[\eta(p), \eta(q)] = \rho(q - p), \tag{3.3}$$

where $q - p$ is the vector difference between q and p.

(c′) The correlation function $\rho(p)$ is a continuous function of p with $\rho(0) = 1$, where 0 is the zero vector.

In the sequel we shall be concerned with processes which are, in addition, isotropic. This means that the correlation function depends only on the length $u = |p - q|$ of the vector $p - q$:

$$\rho(p - q) = \rho(|p - q|) = \rho(u), \tag{3.4}$$

that is, the correlation function is a function of one real variable. In what follows we refer to this function $\rho(u)$ as the correlation function.

In the linear case it turned out that the best method of sampling n points is to select them equidistantly. It is, however, difficult to generalize this result in a straightforward manner to the case of a plane; it is not obvious which domains in the plane can replace the segment $0 \leq t \leq T$. Therefore we have looked for regular allocations that can be dealt with by means of limiting theorems relating to increasing domains. This approach eliminates the troublesome boundary effect from the problem. On the other hand it introduces some problems of a purely geometrical nature. An alternative device to cope with the boundary effect would be to define the stochastic process $\eta(p)$ on the surface of a sphere or a torus.

As shown in Zubrzycki (1957), we can construct a two-dimensional continuous, isotropic stationary stochastic process $\eta(p)$, the correlation function of which is $\rho(u) = r(u/a)$, where

$$r\left(\frac{u}{a}\right) = \begin{cases} \dfrac{2}{\pi}\left\{\arccos\dfrac{u}{a} - \dfrac{u}{a}\left[1 - \left(\dfrac{u}{a}\right)^2\right]^{1/2}\right\}, & 0 \leq u \leq a, \\ 0, & \text{otherwise,} \end{cases} \tag{3.5}$$

where a is a positive constant.

In this analytic form, the correlation function does not reveal its most important feature: the value of $r(u/a)$ for a given u can be computed as

$$r\left(\frac{u}{a}\right) = \frac{|K_1 \cap K_2|}{\frac{\pi a^2}{4}}, \tag{3.6}$$

that is, $r(u/a)$ equals the ratio of the area $|K_1 \cap K_2|$ of the common part of two circles K_1 and K_2 with radius $a/2$ and centers p_1 and p_2 at distance $u = |p_1 - p_2|$ to the area $\pi\, a^2/4$ of such a circle. The geometrical interpretation is illustrated in Figure 1.

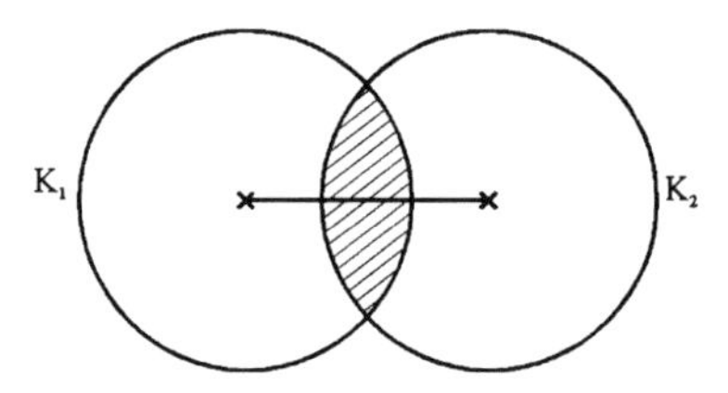

Figure 1 Geometrical interpretation of the value of the correlation function given by (3.6)

The mean value μ of the process may be interpreted as $|D|^{-1} \int_D \eta(p)\, dp$ for a domain D with infinitely large area $|D|$. Suppose now that we have selected n points $p_1, \ldots, p_n$ in the plane and want to estimate μ by the average

$$\overline{\eta} = \frac{1}{n} [\eta(p_1) + \cdots + \eta(p_n)] \tag{3.7}$$

of the values $\eta(p_1), \ldots, \eta(p_n)$ at these points. This average $\overline{\eta}$ is an unbiased estimate of μ. Obviously,

$$\operatorname{var} \overline{\eta} = \frac{\sigma^2}{n^2} \sum_{i=1}^{n} \sum_{j=1}^{n} \rho(|p_i - p_j|). \tag{3.8}$$

The double sum of the right member of (3.8) is, by virtue of the geometrical interpretation of (3.5), equal to the ratio of the sum of the areas of the common parts of all n^2 possible pairs of circles K_i with radius $a/2$ and centers p_i, with $i = 1, \ldots, n$, to the area of such a circle; in symbols,

$$\operatorname{var} \overline{\eta} = \frac{\sigma^2}{n^2} \frac{\sum_{i=1}^{n} \sum_{j=1}^{n} |K_i \cap K_j|}{\frac{\pi a^2}{4}}. \tag{3.9}$$

We now ask which distribution of points in the plane with a given plane density yields (in the limit) the least variance. What we said in connection with the correlation function given by (3.5) shows that this problem is uniquely related to certain questions concerning distributions of circles. Let us consider a sequence of points $p_1, \ldots, p_n$ in the plane and define the density d, if it exists, as

$$d = \lim_{R\to\infty} \frac{1}{\pi R^2} \operatorname{Card}[i : p_i \in K(0, R)], \tag{3.10}$$

that is, as a limit of a ratio of the number of points p_i in a circle $K(0, R)$ of radius R and center 0 to the area πR^2 of this circle, when R tends to infinity. Of course, this limit does not depend on the choice of center 0. Then, given a sequence p_1, $p_2, \ldots$ with density d, let us call a *limiting variance* the limit

$$\sigma^2_\infty = \lim_{R\to\infty} n_R \operatorname{var} \overline{\eta}_R, \tag{3.11}$$

where $n_R = \operatorname{Card}\{i : p_i \in K(0, R)\}$ and $\overline{\eta}_R$ is the average of those variables $\eta(p_i)$ for which p_i is in $K(0, R)$.

Let us introduce two more definitions. Given, in a plane, a sequence of circles K_1, $K_2, \ldots$ with radius $a/2$ and with centers p_1, $p_2, \ldots$ which have a given density d, we define the *mean covering*, for short C', as the limit

$$C' = \lim_{R\to\infty} \frac{1}{\pi R^2} \sum_i |K_i \cap K(0, R)| \tag{3.12}$$

and the *mean double covering*, for short C'', as the limit

$$C'' = \lim_{R\to\infty} \frac{1}{\pi R^2} \sum_i \sum_j |K_i \cap K_j \cap K(0, R)|. \tag{3.13}$$

In this sum the case $i = j$ is not excluded. Of course, the mean covering C' of the circles K_1, $K_2, \ldots$ and density d of their centers p_1, $p_2, \ldots$ are related by the equality $C' = |K_1|\, d$. As a consequence we may use C' as our measure of density of centers in comparisons where $|K_1|$ is kept constant.

Now it is clear that, for stochastic processes $\eta(p)$ with correlation functions given by (3.5), the search for a sequence of points with a prescribed density which yields the minimum limiting variance is equivalent to the search for a corresponding sequence of circles yielding the minimum mean double covering. Moreover, if there exists such a sequence of points that would realize the minimum limiting variance for all positive values of a in the correlation function given by (3.5), then it would be the best sequence also for a process with a correlation function given by

$$\rho(u) = \int_0^\infty r\left(\frac{u}{a}\right) \mathrm{d}F(a), \tag{3.14}$$

where $F(a)$ is a distribution function with $F(0) = 0$. Unfortunately, there do not exist sequences of points that yield minimum mean covering simultaneously for all values of a in (3.5), as will be shown later in this chapter.

4 OPTIMAL NETS OF POINTS

Consider a sequence of congruent circles $K_1, K_2, \ldots$ with centers $p_1, p_2, \ldots$, respectively. Let us denote the indicator function of K_i by $k_i(p)$, that is, let us put

$$k_i(p) = \begin{cases} 1, & p \in K_i, \\ 0, & \text{otherwise.} \end{cases} \tag{4.1}$$

Moreover, we put

$$k(p) = \sum_i k_i(p). \tag{4.2}$$

In other words $k(p)$ is equal to the number of circles covering p. In terms of these functions the definitions of the mean covering C' and the mean double covering C'' given in Section 3 take on the forms

$$\begin{aligned} C' &= \lim_{R\to\infty} \frac{1}{\pi R^2} \sum_i \int_{K(0,R)} k_i(p)\,\mathrm{d}p \\ &= \lim_{R\to\infty} \frac{1}{\pi R^2} \int_{K(0,R)} \sum_i k_i(p)\,\mathrm{d}p \\ &= \lim_{R\to\infty} \frac{1}{\pi R^2} \int_{K(0,R)} k(p)\,\mathrm{d}p \end{aligned} \tag{4.3}$$

and

$$\begin{aligned} C'' &= \lim_{R\to\infty} \frac{1}{\pi R^2} \sum_i \sum_j \int_{K(0,R)} k_i(p)\,k_j(p)\,\mathrm{d}p \\ &= \lim_{R\to\infty} \frac{1}{\pi R^2} \int_{K(0,R)} \sum_i \sum_j k_i(p)\,k_j(p)\,\mathrm{d}p \\ &= \lim_{R\to\infty} \frac{1}{\pi R^2} \int_{K(0,R)} k^2(p)\,\mathrm{d}p. \end{aligned} \tag{4.4}$$

Our problem is to determine sequences of congruent circles with a fixed mean covering for which the mean double covering attains its minimum. We now prove an inequality from which it follows that a sufficient condition for a sequence of circles to have this minimum property is that the set of values of the function $k(p)$ consists of two consecutive integers. This is addressed in the following.

Lemma 4.1. *For any sequence of circles with mean covering C', the following inequality*

$$C'' \geq \{2[C'] + 1\}\, C' - [C']\,\{[C'] + 1\} \tag{4.5}$$

holds, where $[C']$ is the integral part of C'; equality holds if and only if $k(p)$ takes the values $[C']$ and $[C']+1$ only.

Proof. Clearly, we have

$$\begin{aligned}&\frac{1}{\pi R^2}\int_{K(0,R)} k^2(p)\,\mathrm{d}p \\ &= \left\{\frac{1}{\pi R^2}\int_{K(0,R)} k(p)\,\mathrm{d}p\right\}^2 - \left\{\frac{1}{\pi R^2}\int_{K(0,R)} k(p)\,\mathrm{d}p - [C'] - \frac{1}{2}\right\}^2 \\ &\quad + \frac{1}{\pi R^2}\int_{K(0,R)} \left\{k(p) - [C'] - \frac{1}{2}\right\}^2 \mathrm{d}p.\end{aligned} \tag{4.6}$$

Since always

$$\left\{k(p) - [C'] - \frac{1}{2}\right\}^2 \geq \frac{1}{4}, \tag{4.7}$$

we conclude that

$$\begin{aligned}&\frac{1}{\pi R^2}\int_{K(0,R)} k^2(p)\,\mathrm{d}p \\ &\geq \left\{\frac{1}{\pi R^2}\int_{K(0,R)} k(p)\,\mathrm{d}p\right\}^2 - \left\{\frac{1}{\pi R^2}\int_{K(0,R)} k(p)\,\mathrm{d}p - [C'] - \frac{1}{2}\right\}^2 + \frac{1}{4}.\end{aligned} \tag{4.8}$$

For $R \to \infty$, we get

$$C'' \geq C'^2 - \left\{C' - [C'] - \frac{1}{2}\right\}^2 + \frac{1}{4} \tag{4.9}$$

and this is an alternative form of (4.5). Now if $[C']$ and $[C']+1$ are the only values of $k(p)$, then (4.7) and consequently (4.5) become equalities. This proves the lemma. □

We now describe some sequences of circles minimizing the mean double covering. We confine ourselves to the case where the centers of the circles form nets composed of congruent figures such as triangles or squares. This will enable us to compute the mean covering and the mean double covering from a single mesh of a net, and we shall exploit this possibility. The minimum property will follow by our lemma, since the function $k(p)$ will take only two consecutive integers as its values in our examples. We arrange these examples by increasing values of C'.

Example 4.1. If

$$C' \leq \frac{\pi}{2\sqrt{3}} \doteq 0.907, \tag{4.10}$$

then the net of equilateral triangles has the optimal property. This situation is illustrated in Figures 2 and 3. Clearly, the pattern considered here may as well be referred to as a net of rhombuses.

Figure 2 Optimum sampling pattern for $C' < \frac{\pi}{2\sqrt{3}} \doteq 0.907$

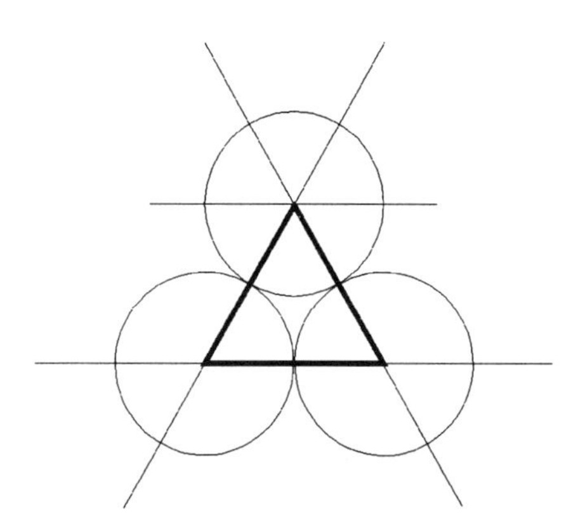

Figure 3 Optimum sampling pattern for $C' = \frac{\pi}{2\sqrt{3}} \doteq 0.907$

For purposes of illustration we present the details of the computation of C' and of C''. In Figure 2 we put the radius of the circles equal to 1, and the side of the triangle equal to $s \geq 2$. Thus the area of the triangle is $A = \left(s^2\sqrt{3}\right)/4$. The circles divide this area into four parts, $A = T_0 + 3T_1 = A_0 + A_1$. The meaning of T_0 and T_1 is shown in Figure 4.

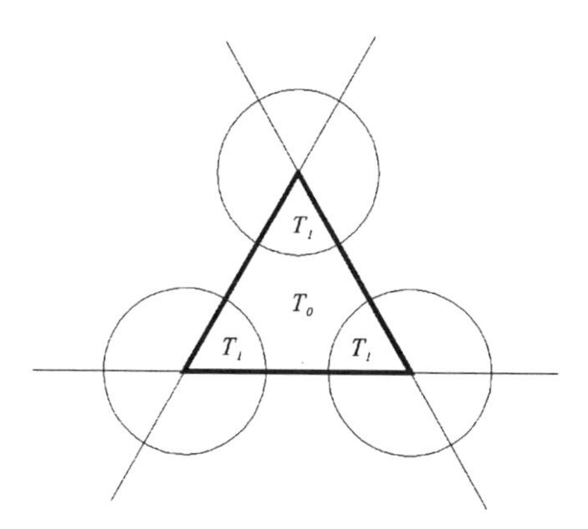

Figure 4 Illustration of the meaning of T_0 and T_1

In T_0 we have $k(p) = 0$, while in T_1 we have $k(p) = 1$. Now

$$C' = \frac{1}{A}\left(\int_{A_0} 0\,\mathrm{d}p + \int_{A_1} 1\,\mathrm{d}p\right) = \frac{1}{A}\int_{A_1} \mathrm{d}p. \tag{4.11}$$

Thus

$$C' = \frac{4}{s^2\sqrt{3}}\left(\frac{3\pi}{6}\right) = \frac{2\pi}{s^2\sqrt{3}}. \tag{4.12}$$

For $s > 2$, we have $C' < \pi/2\sqrt{3} \doteq 0.907$, while for $s = 2$, we have $C' = \pi/2\sqrt{3} \doteq 0.907$. Moreover, $C'' = C'$ since $k^2(p) = k(p)$.

We now compare this value of C'' with the corresponding value of C'' for a net of squares having the same value of C' and, therefore, representing the same point density. We put the side of the square equal to x. Drawing the four circles with radius equal to 1, we obtain the configuration shown in Figure 5.

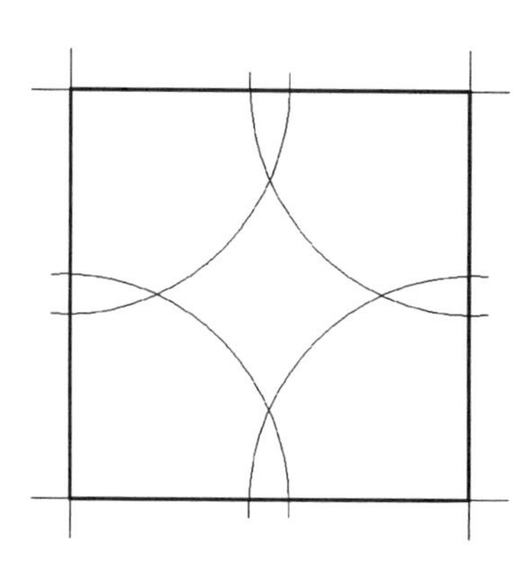

Figure 5 Computation of C'' for a net of squares

Now

$$C' = \frac{1}{A}\left(\int_{A_0} 0\,\mathrm{d}p + \int_{A_1} 1\,\mathrm{d}p + \int_{A_2} 2\,\mathrm{d}p\right) = \frac{\pi}{x^2}. \tag{4.13}$$

From $C' = \pi/2\sqrt{3}$ we get

$$x^2 = \sqrt{12}; \quad x = \sqrt[4]{12} \doteq 1.86, \tag{4.14}$$

that is, the circles intersect as shown in Figure 5.

If A_2 stands for the area of that portion of the square where $k(p) = 2$, we get

$$A_2 = 8\left\{\frac{v\pi}{360} - \frac{y}{2}\frac{\sqrt[4]{12}}{2}\right\} = 8\left\{\frac{v\pi}{360} - \frac{\sqrt[4]{12}}{4}\sin v\right\}, \tag{4.15}$$

with v and y having the meaning indicated in Figure 6. Carrying out the computations gives $A_2 = 0.14$. Clearly $A_1 = \pi - 2A_2 = 2.86$. Thus

$$C' = \frac{1}{\sqrt{12}}\{2.86 + 2(0.14)\} = 0.907 \tag{4.16}$$

as it should be, and

$$C'' = \frac{1}{\sqrt{12}}\{2.86 + 2^2(0.14)\} = 0.988 > 0.907. \tag{4.17}$$

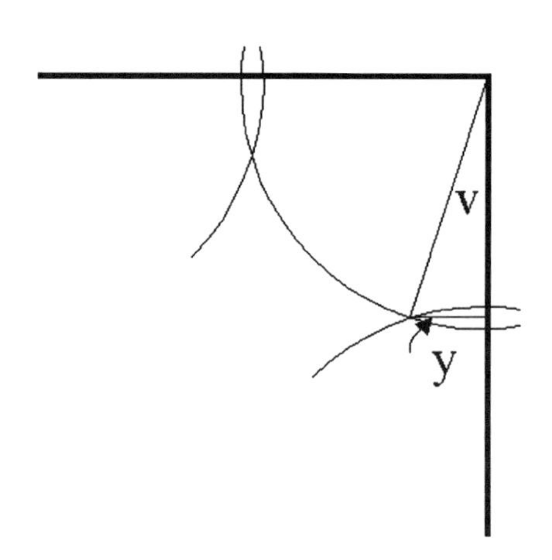

Figure 6 Meaning of v and y

Example 4.2. If

$$0.907 \doteq \frac{\pi}{2\sqrt{3}} \leq C' \leq \frac{2\pi}{3\sqrt{3}} \doteq 1.209, \tag{4.18}$$

then the net of equilateral triangles is still optimal; see Figure 7. However, in this case $k(p)$ has three values: $0, 1$, and 2, so that our Lemma 4.1 does not apply. The optimality of the net in question is a consequence of a known inequality (Fejes Tóth (1953), inequality (3), p. 80), from which it follows that among all convex hexagons of a given area and all circles of a given area, the maximum possible area of a common part of a hexagon and circle is reached when the hexagon is equilateral and the circle is concentric with it.

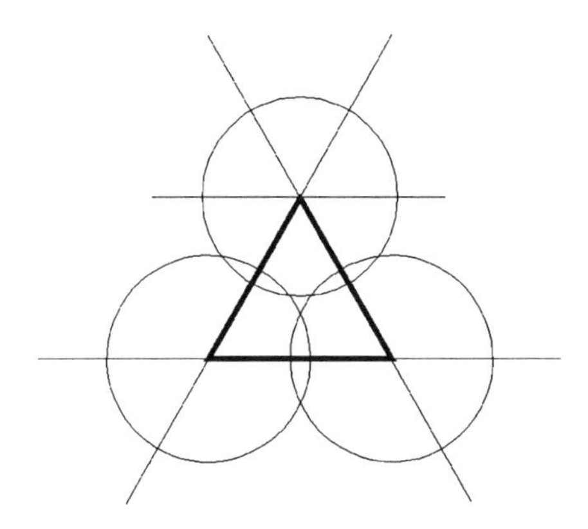

Figure 7 Optimum sampling pattern for $0.907 \doteq \frac{\pi}{2\sqrt{3}} \leq C' \leq \frac{2\pi}{3\sqrt{3}} \doteq 1.209$

The statement of Example 4.2 follows if we apply the quoted inequality to the cells that are formed by attaching each point of a plane to the nearest circle center. The above-mentioned inequality is of its greatest interest when the mean covering is in the range indicated in Example 4.2. Let us note, however, that it also implies the statement in Example 4.1 to the effect that the circles should be disjoint.

Example 4.3. If

$$1.209 \doteq \frac{2\pi}{3\sqrt{3}} \leq C' \leq \frac{2\pi}{2+\sqrt{3}} \doteq 1.684, \tag{4.19}$$

then the net of isosceles triangles is optimal. This is seen as follows. We start with the situation shown in Figure 8. We then increase the mean covering without spoiling the property that $k(p)$ has as values only two consecutive integers, letting the base of the triangle diminish and its height increase, so that the three circles still intersect in one point. We can continue this procedure until the length of the base becomes equal to the radius of our circles, as shown in Figure 9.

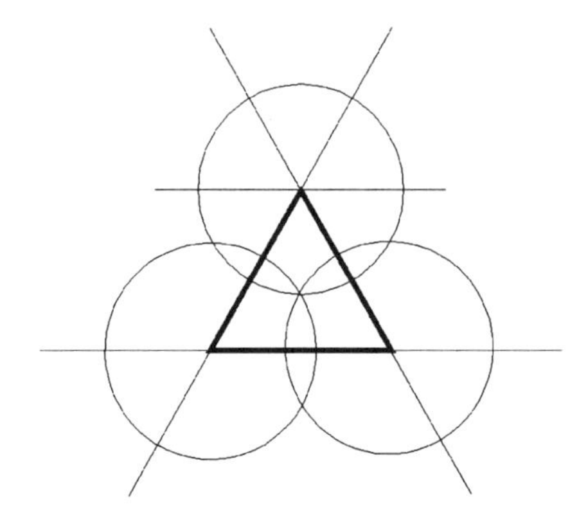

Figure 8 Optimum sampling pattern for $C' = \frac{2\pi}{3\sqrt{3}} \doteq 1.209$

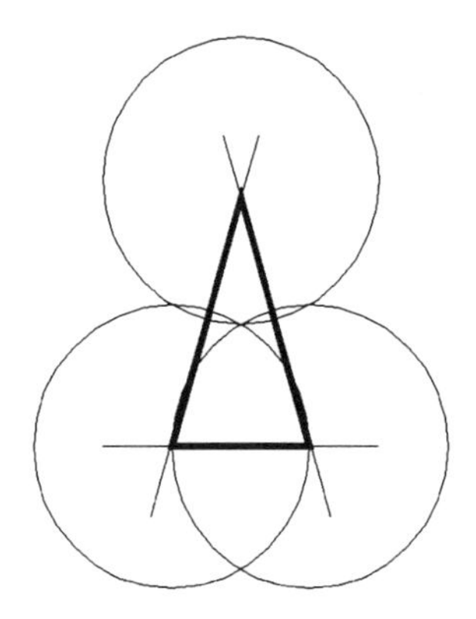

Figure 9 Optimum sampling pattern for $C' = \frac{2\pi}{2+\sqrt{3}} \doteq 1.684$

Example 4.4. If

$$1.571 \doteq \frac{\pi}{2} \leq C' \leq \frac{\pi}{\sqrt{3}} \doteq 1.814, \tag{4.20}$$

a net of rectangles has the optimal property. We start with a net of squares and circles intersecting in the centers of the squares as indicated in Figure 10. This corresponds to $C' = \pi/2 \doteq 1.571$. We then enlarge the mean covering by lengthening two sides of the square and shortening the other two, while the circles still intersect in the middle; the radius of the circles is then equal to the shorter side of the rectangle. This corresponds to $C' = \pi/\sqrt{3} \doteq 1.814$.

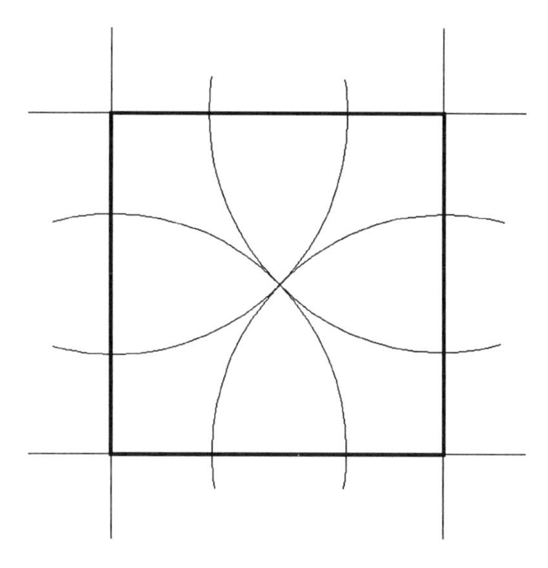

Figure 10 Optimum sampling pattern for $C' = \frac{\pi}{2} \doteq 1.571$

We note that the intervals for C' corresponding to Examples 4.3 and 4.4 respectively overlap. The nature of this situation will be elucidated somewhat. Instead of considering the pattern with which we start in Example 4.3 as made up of a net of equilateral triangles, we think of this pattern in terms of rhombuses, with the base angle $v = 60°$, corresponding to $C' = 2\pi/3\sqrt{3} \doteq 1.209$. If we increase v to $v = 90°$, that is, if we change the rhombuses into squares, C' will increase to $C' = \pi/2 \doteq 1.571$. Thereafter, by 'stretching' the squares into rectangles, we may further increase C' to $C' = \pi/\sqrt{3} \doteq 1.814$.

Example 4.5. If

$$2.418 \doteq \frac{4\pi}{3\sqrt{3}} \leq C' \leq \frac{16\pi}{7\sqrt{7}} \doteq 2.714, \tag{4.21}$$

the optimal property is possessed by a net of hexagons which have two perpendicular axes of symmetry and can be inscribed in a circle; in general, they are not equilateral.

We start with a net of equilateral, congruent hexagons and place the centers of circles at the vertices, the radius of the circles being equal to the side of the hexagons. In this case $C' = 4\pi/3\sqrt{3} \doteq 2.418$. This case is shown in Figure 11. We let C' increase without spoiling the property that $k(p)$ has as values only two consecutive integers, by suitably narrowing our hexagons. We can continue this procedure until the circles corresponding to the vertices of neighboring hexagons touch. Figure 12 shows the extreme situation, which corresponds to $C' = 16\pi/7\sqrt{7} \doteq 2.714$.

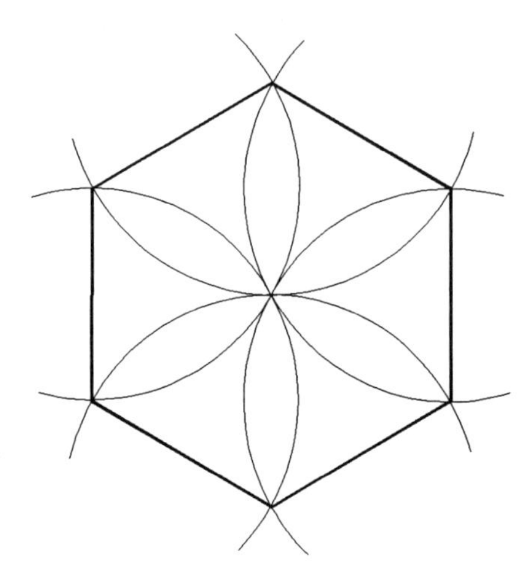

Figure 11 Optimum sampling pattern for $C' = \frac{4\pi}{3\sqrt{3}} \doteq 2.418$

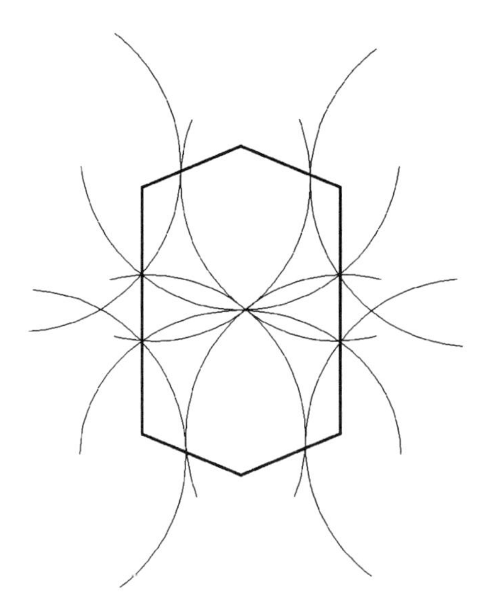

Figure 12 Optimum sampling pattern for $C' = \frac{16\pi}{7\sqrt{7}} \doteq 2.714$

Example 4.6. If

$$C' = \frac{2\pi}{\sqrt{3}} \doteq 3.628, \tag{4.22}$$

the net of equilateral triangles has the optimal property. This pattern is represented in Figure 13.

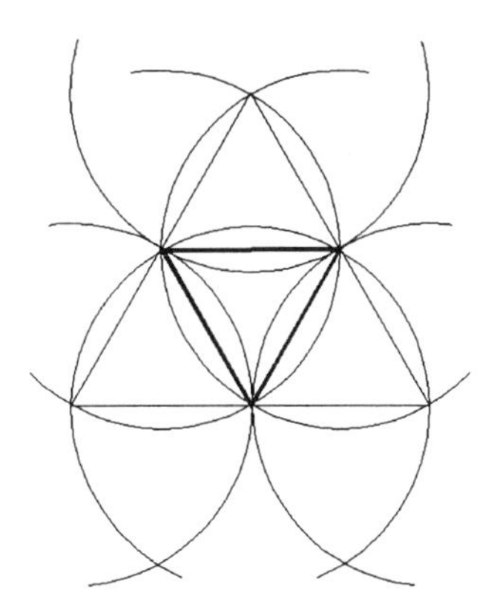

Figure 13 Optimum sampling pattern for $C' = \frac{2\pi}{\sqrt{3}} \doteq 3.628$

We may summarize the previous findings. By means of Lemma 4.1, optimal regular nets of sample points were found for the following values of C' shown in Table 1.

Table 1 Summary of values of C'

Example	
4.1	$C' \leq \frac{\pi}{2\sqrt{3}} \doteq 0.907$
4.2	$0.907 \doteq \frac{\pi}{2\sqrt{3}} \leq C' \leq \frac{2\pi}{3\sqrt{3}} \doteq 1.209$
4.3 and 4.4	$1.209 \doteq \frac{2\pi}{3\sqrt{3}} \leq C' \leq \frac{\pi}{\sqrt{3}} \doteq 1.814$
4.5	$2.418 \doteq \frac{4\pi}{3\sqrt{3}} \leq C' \leq \frac{16\pi}{7\sqrt{7}} \doteq 2.714$
4.6	$C' = \frac{2\pi}{\sqrt{3}} \doteq 3.628$

We observe that Lemma 4.1 does not provide solutions for values of C' close to $1, 2$, and 3 respectively. We conjecture that the same holds true for sufficiently large values of C'. This leads us to formulate the following problem.

Problem 4.1. *Determine the range of values of C' for which*

$$C'' = \{2[C'] + 1\}\, C' - [C']\,\{[C'] + 1\}. \tag{4.23}$$

In the solution of this problem the paper by Heppes (1959) should prove valuable.

It is of some interest to compare, for given values of C', the corresponding values of C'' for different nets of sample points. In Table 2 we present such a comparison.

Table 2 The relation between C'' and the pattern of the net of points for some values of C'

C' Exact value	C' Num. value	$[C']$	C'' when the pattern of the net of points is: Equilateral triangles	Squares	Equilateral hexagons
$\frac{\pi}{2\sqrt{3}}$	0.907	0	0.907	0.988	1.165
$\frac{2\pi}{3\sqrt{3}}$	1.209	1	1.627	1.695	1.870
$\frac{\pi}{2}$	1.571	1	2.861	2.712	2.814
$\frac{4\pi}{3\sqrt{3}}$	2.418	2	6.395	6.389	6.110
$\frac{16\pi}{7\sqrt{7}}$	2.714	2	7.711	7.936	7.736

The examples discussed above and given in Table 2 show that the net of equilateral triangles is not, in general, the optimal net. In other words, this net does not yield the minimum limiting variance for an isotropic stochastic process in a plane with correlation function given by (3.5). The differences are, however, rather small and their sign alternates as C' increases. Therefore we state the following problem.

Problem 4.2. *Is it true that the net of equilateral triangles yields the minimum limiting variance for a stationary isotropic stochastic process with exponential correlation function?*

An affirmative answer to this problem would imply the optimality of the net of equilateral triangles for all isotropic processes with any completely monotone correlation function. Numerical examples contained in Matérn (1960) give some support to such an affirmative answer. Matérn compares the limiting variances for an isotropic process with correlation function $\exp(\mu)$, if triangular, square, and hexagonal nets of some chosen densities are used. The effect is that in all cases the triangular net proved to be somewhat better than the other ones.

Finally, we mention without formal proof the almost obvious relation concerning the limiting behavior of the mean double covering. This is expressed as Lemma 4.2.

Lemma 4.2. *For any regular net of points the ratio of the mean double covering C'' and the square of the mean covering C'^2 tends to unity when the mean covering increases over all bounds; in symbols,*

$$\lim_{C' \to \infty} \frac{C''}{C'^2} = 1. \tag{4.24}$$

5 A CLASS OF PLANAR ISOTROPIC CORRELATION FUNCTIONS

In this section, which is due to Hájek and Zubrzycki, we discuss the class of convex planar correlation functions expressed as

$$\rho(u) = \int_0^\infty r\left(\frac{u}{a}\right) \mathrm{d}F(a), \tag{5.1}$$

where $F(a)$ is a distribution function with $F(0+) = 0$ and

$$r(u) = \begin{cases} \frac{2}{\pi}\left[\arccos u - u(1-u^2)^{1/2}\right], & 0 \le u \le 1, \\ 0, & u > 1. \end{cases} \tag{5.2}$$

For some related results, see Hammersley and Nelder (1955).

Clearly,

$$r'(u) = \begin{cases} -\frac{4}{\pi}(1-u^2)^{1/2}, & 0 \le u < 1, \\ 0, & u > 1, \end{cases} \tag{5.3}$$

and

$$r''(u) = \begin{cases} \frac{4}{\pi}\frac{u}{(1-u^2)^{1/2}}, & 0 \le u < 1, \\ 0, & u > 1. \end{cases} \tag{5.4}$$

Lemma 5.1.

$$\int_0^\infty \frac{1}{u^2} r''\left(\frac{a}{u}\right) r''\left(\frac{u}{\mu}\right) \mathrm{d}u = \begin{cases} \frac{8}{\pi\mu}, & \mu > a > 0, \\ 0, & a > \mu. \end{cases} \tag{5.5}$$

Proof. The case $a > \mu$ is clear. If $\mu > a$, we have, in view of (5.4),

$$\int_0^\infty \frac{1}{u^2} r''\left(\frac{a}{u}\right) r''\left(\frac{u}{\mu}\right) = \left(\frac{4}{\pi}\right)^2 \int_a^\mu \frac{a\,\mathrm{d}u}{u[(u^2-a^2)(\mu^2-u^2)]^{1/2}} = \frac{8}{\pi\mu}. \tag{5.6}$$

□

Theorem 5.1. *The correlation functions $\rho(u)$ of the type given by (5.1) are characterized by the following properties:*

(a) $\rho(u)$ is continuous, convex, and with $\rho(\infty) = 0$.

(b) $\rho'(u)$ is absolutely continuous.

(c) $\int_0^\infty (1/u^2)\, r''(a/u)\, \rho''(u)\, \mathrm{d}u$ is a nonincreasing function of a.

The functions $F(a)$ and $\rho''(u)$ are linked by the inversion formulas

$$\mathrm{d}F(a) = -\frac{\pi}{8} a^3 \,\mathrm{d} \int_0^\infty \frac{1}{u^2} r''\left(\frac{a}{u}\right) \rho''(u)\, \mathrm{d}u, \tag{5.7}$$

and

$$\rho''(u) = \int_0^\infty \frac{1}{a^2} r''\left(\frac{u}{a}\right) \mathrm{d}F(a). \tag{5.8}$$

Condition (c) is fulfilled, for example, if $\rho''(u)/u$ is a nonincreasing function of u, which means, provided that $\rho''(u)$ is absolutely continuous, that

$$\rho''(u) - u\rho'''(u) \geq 0, \quad u \geq 0. \tag{5.9}$$

If (5.9) holds, then $F(a)$ is absolutely continuous and therefore

$$\frac{\mathrm{d}F(a)}{\mathrm{d}a} = -\frac{1}{2} a^2 \int_0^{\pi/2} \rho'''\left(\frac{a}{\sin\theta}\right) \frac{\mathrm{d}\theta}{\sin^2\theta}. \tag{5.10}$$

Proof. Property (a) follows easily from the corresponding property of correlation functions $r(u/a)$. Property (b), and simultaneously the relation given by (5.8), will be proved if we show that the indefinite integral of the right side of (5.8) equals $\rho'(u)$. Now

$$\begin{aligned} &\int_u^\infty \int_0^\infty \frac{1}{a^2} r''\left(\frac{u}{a}\right) \mathrm{d}F(a)\, \mathrm{d}u \\ &= \int_0^\infty \frac{1}{a} \left[\int_u^\infty \frac{1}{a} r''\left(\frac{u}{a}\right) \mathrm{d}u\right] \mathrm{d}F(a) \\ &= -\int_0^\infty \frac{1}{a} r'\left(\frac{u}{a}\right) \mathrm{d}F(a) = -\rho'(u). \end{aligned} \tag{5.11}$$

The change of integration order is justified by Fubini's theorem, since $r''(u) \geq 0$. The last identity follows from (5.1) by differentiation under the integral sign, which is justified, since $(1/a)\, r'(u/a)$ is uniformly bounded for $u > \epsilon > 0$.

In view of (5.8) we have, by using Fubini's theorem again,

$$\begin{aligned} &\int_0^\infty \frac{1}{u^2} r''\left(\frac{a}{u}\right) \rho''(u)\, \mathrm{d}u \\ &= \int_0^\infty \frac{1}{u^2} r''\left(\frac{a}{u}\right) \int_0^\infty \frac{1}{\mu^2} r''\left(\frac{u}{\mu}\right) \mathrm{d}F(\mu)\, \mathrm{d}u \\ &= \int_0^\infty \frac{1}{\mu^2} \left[\int_0^\infty \frac{1}{u^2} r''\left(\frac{a}{u}\right) r''\left(\frac{u}{\mu}\right) \mathrm{d}u\right] \mathrm{d}F(\mu), \end{aligned} \tag{5.12}$$

which gives, in accordance with (5.5),

$$\int_0^\infty \frac{1}{u^2} r''\left(\frac{a}{u}\right) \rho''(u)\,\mathrm{d}u = \frac{8}{\pi}\int_a^\infty \frac{1}{\mu^3}\,\mathrm{d}F(\mu). \tag{5.13}$$

The last relation, however, is equivalent to (5.7).

Now assume that the correlation function $\rho(u)$ fulfills conditions (a), (b) and (c), and consider the function $F(a)$ given by (5.7). In view of property (c), $F(a)$ will be nondecreasing.

The fact that the total variation of $F(a)$ equals 1 follows from the subsequent Lemmas 5.2 and 5.3 and from the following relations:

$$\begin{aligned}\int_0^\infty \mathrm{d}F(a) &= -\frac{\pi}{8}\int_0^\infty a^3\,\mathrm{d}\int_0^\infty \frac{1}{u^2} r''\left(\frac{a}{u}\right)\rho''(u)\,\mathrm{d}u\mathrm{d}a \qquad (5.14)\\ &= \left[-\frac{\pi}{8}a^3\int_0^\infty \frac{1}{u^2} r''\left(\frac{a}{u}\right)\rho''(u)\mathrm{d}u\right]_0^\infty \\ &\quad +\frac{3\pi}{8}\int_0^\infty\left[\int_0^\infty \frac{a^2}{u^2} r''\left(\frac{a}{u}\right)\mathrm{d}a\right]\rho''(u)\,\mathrm{d}u \\ &= \int_0^\infty u\rho''(u)\,\mathrm{d}u = -\int_0^\infty \rho'(u)\,\mathrm{d}u \\ &= \rho(0)-\rho(\infty) = 1.\end{aligned}$$

It remains to show that $\rho'(u)$ is uniquely determined by relation (5.7), that is, that $\rho(u)$ coincides with the correlation function obtained from (5.1). Howcvcr, (5.7) is equivalent to (5.13), if rewritten in the form

$$\frac{8}{\pi}\int_0^\infty \frac{1}{\mu^3}\,\mathrm{d}F(\mu) = \int_0^\infty \frac{1}{u^2} r''\left(\frac{a}{u}\right)\,\mathrm{d}\rho'(u). \tag{5.15}$$

Comparing (5.15) with (5.8) we can see that the $\rho'(u)$ may be determined from $(8/\pi)\int_a^\infty(1/\mu^3)\,\mathrm{d}F(\mu)$ in the same way as $F(a)$ has been determined from $\rho''(u)$. The fact that $\rho'(u)$ may not have finite variation is irrelevant.

Before proceeding to the rest of the proof, we observe that by substituting $u = (a/\sin\theta)$ into (5.7) we get

$$\mathrm{d}F(a) = -\frac{1}{2}a^3\,\mathrm{d}\int_0^{\pi/2}\frac{\rho''\left(\frac{a}{\sin\theta}\right)}{\frac{a}{\sin\theta}}\,\mathrm{d}\theta. \tag{5.16}$$

From this form it is easily seen that condition (c) is fulfilled if $\rho''(u)/u$ is a nonincreasing function of u, or, more especially, if (5.9) holds; notice that $\{[\rho''(u)/u]\}' = -1/u^2[\rho''(u)-u\rho'''(u)]$. Now if (5.9) holds, we can differentiate (5.16) under the integral sign (Fubini's theorem), which gives

$$\frac{\mathrm{d}F(a)}{\mathrm{d}a} = -\frac{1}{2}a^3\int_0^{\pi/2}\left[\rho'''\left(\frac{a}{\sin\theta}\right)\frac{1}{a} - \rho''\left(\frac{a}{\sin\theta}\right)\frac{\sin\theta}{a^2}\right]\mathrm{d}\theta \tag{5.17}$$

$$
\begin{aligned}
&= -\frac{1}{2}a^2\int_0^{\pi/2}\rho'''\left(\frac{a}{\sin\theta}\right)\mathrm{d}\theta + \left[-\frac{1}{2}a\,\rho''\left(\frac{a}{\sin\theta}\right)\cos\theta\right]_0^{\pi/2} \\
&\qquad -\frac{1}{2}\int a^2\,\rho'''\left(\frac{a}{\sin\theta}\right)\frac{\cos^2\theta}{\sin^2\theta}\,\mathrm{d}\theta \\
&= -\frac{1}{2}a^2\int_0^{\pi/2}\rho'''\left(\frac{a}{\sin\theta}\right)\frac{\sin^2\theta+\cos^2\theta}{\sin^2\theta}\,\mathrm{d}\theta \\
&= -\frac{1}{2}a^2\int_0^{\pi/2}\rho'''\left(\frac{a}{\sin\theta}\right)\frac{\mathrm{d}\theta}{\sin^2\theta}.
\end{aligned}
$$

The relation $\rho''(\infty) = 0$ which we have used follows from the subsequent Lemma 5.3. Our theorem is thus completely proved. □

Lemma 5.2.

$$\int_0^\infty u^2\,r''(u)\,\mathrm{d}u = \frac{8}{3\pi}. \tag{5.18}$$

Proof. Integrating by parts, we have

$$
\begin{aligned}
\int_0^\infty u^2\,r''(u)\,\mathrm{d}u &= -\int_0^\infty 2u\,r'(u)\,\mathrm{d}u \qquad (5.19)\\
&= \frac{8}{\pi}\int_0^1 u(1-u^2)^{1/2} = \frac{8}{3\pi}.
\end{aligned}
$$

□

Lemma 5.3. *Any correlation function $\rho(u)$ which fulfills conditions* (a), (b) *and* (c) *of Theorem* 5.1 *has the properties*

$$\lim_{a\to 0} a^3\int_0^\infty \frac{1}{u^2}\,r''\left(\frac{a}{u}\right)\rho''(u)\,\mathrm{d}u = \lim_{a\to\infty} a^3\int_0^\infty \frac{1}{u^2}\,r''\left(\frac{a}{u}\right)\rho''(u)\,\mathrm{d}u = 0 \tag{5.20}$$

and

$$\lim_{a\to 0} a\,\rho'(a) = \lim_{a\to\infty} a\,\rho'(a) = 0. \tag{5.21}$$

If, moreover, $\rho''(u)/u$ is nonincreasing, then

$$\lim_{a\to 0} a^2\,\rho''(a) = \lim_{u\to\infty} a^2\,\rho''(a) = 0. \tag{5.22}$$

Proof. Since $\rho(u)$ is convex, $-\rho'$ is nonnegative and nonincreasing, and we have

$$0 \le -\frac{1}{2}a\,\rho'(a) \le -\int_{a/2}^a \rho'(u)\,\mathrm{d}u = \rho\left(\frac{a}{2}\right) - \rho(a). \tag{5.23}$$

Since $\rho(u)$ is continuous at the points 0 and ∞, (5.21) is clear.

Now, if (5.9) holds true, the function $\rho''(u)/u$ is nonnegative and nonincreasing, so that

$$\begin{aligned} 0 &\leq \frac{1}{3}\, a^2\, \rho''(a) = \frac{1}{3}\, a^3\, \frac{\rho''(a)}{a} \leq a^2 \int_{a/2}^{a} \frac{\rho''(u)}{u}\, \mathrm{d}u \\ &\leq 2a \int_{a/2}^{a} \rho''(u)\, \mathrm{d}u = 2a \left[\rho'(a) - \rho'\left(\frac{1}{2}\, a\right)\right], \end{aligned} \tag{5.24}$$

where the expression converges to 0 if $a \to 0$ or $a \to \infty$.

The same consideration will be used in proving (5.20). In view of condition (c) we have

$$\begin{aligned} 0 &\leq \frac{1}{3}\, a^3 \int_0^\infty \frac{1}{u^2}\, r''\left(\frac{a}{u}\right) \rho''(u)\, \mathrm{d}u \\ &\leq \frac{8}{7} \int_{a/2}^{a} \int_0^\infty \frac{\mu^2}{u^2}\, r''\left(\frac{\mu}{u}\right) \rho''(u)\, \mathrm{d}u \mathrm{d}\mu \\ &= \frac{8}{7} \int_0^\infty \left[\int_{a/2}^{a} \frac{\mu^2}{u^2} r''\left(\frac{\mu}{u}\right) \mathrm{d}\mu\right] \rho''(u)\, \mathrm{d}u \\ &= \frac{8}{7} \int_0^\infty \left[\int_{a/2u}^{a/u} \mu^2\, r''(\mu)\, \mathrm{d}\mu\right] u\, \rho''(u)\, \mathrm{d}u, \end{aligned} \tag{5.25}$$

where, using (5.18),

$$\int_{a/2u}^{a/u} \mu^2\, r''(\mu)\, \mathrm{d}\mu \begin{cases} \leq \dfrac{8}{3\pi}, & \dfrac{a}{u} > 0, \\ < \dfrac{8}{\pi}\, \epsilon, & \dfrac{a}{u} \leq \epsilon, \\ = 0, & \dfrac{a}{2u} > 1. \end{cases} \tag{5.26}$$

Consequently, on the one hand,

$$\int_0^\infty \left[\int_{a/2u}^{a/u} \mu^2\, r''(\mu)\, \mathrm{d}\mu\right] u\, \rho''(u)\, \mathrm{d}u \leq \frac{8}{3\pi} \left[3\epsilon + \int_0^{a/\epsilon} u\, \rho''(u)\, \mathrm{d}u\right] \tag{5.27}$$

and, on the other hand,

$$\int_0^\infty \left[\int_{a/2u}^{a/u} u^2\, r''(u)\, \mathrm{d}u\right] u\, \rho''(u)\, \mathrm{d}u \leq \frac{8}{3\pi} \int_{a/2}^{+\infty} u\, \rho''(u)\, \mathrm{d}u. \tag{5.28}$$

The inequalities (5.25) together with the inequality (5.27) or (5.28) prove the relation (5.20) for $a \to 0$ or $a \to \infty$, respectively. Notice that $u\, \rho''(u)$ is integrable with $\int_0^\infty u\, \rho''(u) = 1$. □

Example 5.1. The convex correlation function $\exp(-cu)$ has a negative third derivative and therefore fulfills condition (5.9). Hence it admits the representation (5.1), where the spectral density is given by (5.10).

Example 5.2. The convex correlation function

$$\rho(u) = \frac{2}{\sqrt{2\pi}} \int_u^\infty e^{-u^2/2}\, \mathrm{d}u \tag{5.29}$$

also admits the representation (5.1), since $\rho''(u)/u = (2/\sqrt{2\pi}) \exp(-u^2/2)$ is a nonincreasing function of u.

Example 5.3. The linear convex correlation function

$$\rho(u) = \begin{cases} 1-u, & u \le 1, \\ 0, & u \ge 1, \end{cases} \tag{5.30}$$

has a discontinuous first derivative, and therefore does not admit the representation (5.1). It may be shown, however, that $\rho(u)$ is not a planar isotropic correlation function. Actually, considering a square net of points with coordinates

$$p_{ij} = \left(\frac{i}{\sqrt{2}}, \frac{j}{\sqrt{2}}\right), \quad 0 \le i, j \le n-1, \tag{5.31}$$

we have

$$\sum_{i,j=0}^{n} \sum_{i',j'=0}^{n} \rho\left(|p_{ij} - p_{i'j'}|\right) (-1)^{i+j+i'+j'} = n^2 - 4\left(1 - \frac{1}{\sqrt{2}}\right) n(n-1), \tag{5.32}$$

which is negative for a sufficiently large n.

Example 5.3 shows that the class of planar isotropic convex correlation functions is smaller than the class of linear convex correlation functions. One might suspect that all planar isotropic correlation functions are expressible in the form (5.1). This is, however, disproved by the following theorem.

Theorem 5.2. *There exist isotropic stationary stochastic processes in a plane with correlation function $g(x,y) = f[(x^2+y^2)^{1/2}]$ such that $f(u)$, where $0 \le u < \infty$, is a convex function which cannot be represented in the form*

$$f(u) = \int_0^\infty r\left(\frac{u}{a}\right) \mathrm{d}F(a), \tag{5.33}$$

where

$$r(u) = \begin{cases} \dfrac{2}{\pi}\left\{\arccos u - u(1-u^2)^{1/2}\right\}, & 0 \le u \le 1, \\ 0, & 1 \le u < \infty, \end{cases} \tag{5.34}$$

and $F(a)$ is a distribution function with $F(0+) = 0$.

This theorem follows from the following two lemmas.

Lemma 5.4. *The function $g(x,y) = f[(x^2+y^2)^{1/2}]$, where*

$$f(u) = \begin{cases} \dfrac{2}{\pi}\left(\dfrac{\pi}{2} - u\right), & 0 \le u \le 1, \\ \dfrac{2}{\pi}\left\{\left(\arcsin \dfrac{1}{u}\right) - [u - (u^2-1)^{1/2}]\right\}, & 1 \le u \le \infty, \end{cases} \tag{5.35}$$

is a correlation function of a stationary isotropic stochastic process (see Figure 14).

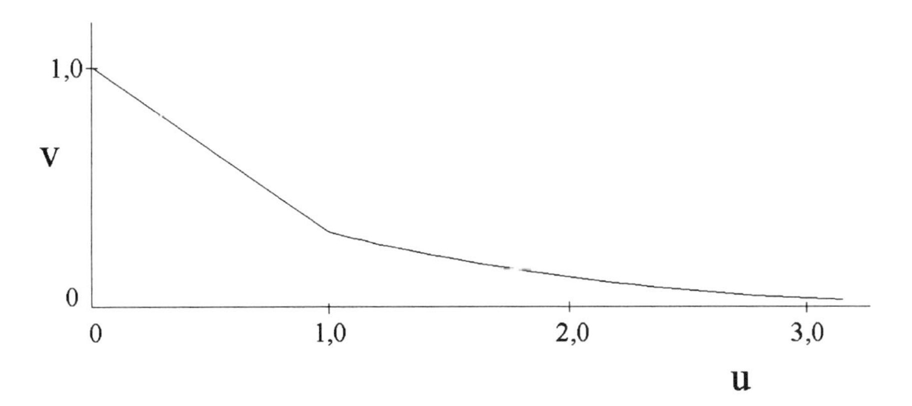

Figure 14 $v = f(u)$, with $f(u)$ given by (5.35)

Lemma 5.5. *If the function $f(u)$ is representable in the form (5.33), then $f''(u) > 0$ for all $u > 0$ with $u < a$, where $F(a) < 1$.*

To prove Lemma 5.4 we consider a linear stationary stochastic process $\eta(t)$ with correlation function $h(t)$ given by

$$h(t) = \begin{cases} 1 - |t|, & |t| \le 1, \\ 0, & \text{otherwise.} \end{cases} \tag{5.36}$$

Define then a plane stochastic process $\xi(x,y)$ by putting

$$\xi(x,y) = \eta(x\cos\alpha + y\sin\alpha), \tag{5.37}$$

where α is a random variable independent of the process $\eta(t)$ with

$$\mathsf{P}\{\alpha < a\} = \begin{cases} 0, & a \le 0, \\ \dfrac{a}{2\pi}, & 0 \le a \le 2\pi, \\ 1, & 2\pi \le a. \end{cases} \tag{5.38}$$

In other words we first define a plane stochastic process which depends only upon one coordinate and has with respect to it correlation function $h(t)$, and then we randomize the direction. It is seen that $\xi(x,y)$ is an isotropic stationary stochastic process with correlation function $g(x,y) = f[(x^2+y^2)^{1/2}]$, where

$$f(u) = \frac{1}{2\pi}\int_0^{2\pi} h(u\sin\psi)\,\mathrm{d}\psi, \tag{5.39}$$

which leads to (5.35).

To prove Lemma 5.5 we note that the second derivative of a function $f(u)$ given by (5.33) is given by

$$f''(u) = \int_0^\infty \frac{1}{a^2} r''\left(\frac{u}{a}\right) \mathrm{d}F(a), \tag{5.40}$$

where

$$r''(u) = \begin{cases} \dfrac{4}{\pi}\,\dfrac{u}{(1-u^2)^{1/2}} > 0, & 0 < u < 1, \\ 0, & 1 < u < \infty. \end{cases} \tag{5.41}$$

This proves Lemma 5.5.

Now the function $f(u)$ given by (5.35) has a second derivative which vanishes for $0 < u < 1$. This contradicts Lemma 5.5 and thus proves our Theorem 5.2.

REFERENCES

Cochran, W. G. (1946). Relative accuracy of systematic and stratified random samples for a certain class of populations. *Ann. Math. Statist. 17*, 164–177.

Drápal, S., Horálek V. and Režný, Z. (1957). Quantitative metallographic lattice analysis. *Hutnické Listy 12*, 485–491.

Faure, P. (1957). Sur quelques résultats relatifs aux fonctions aléatoires stationnaires isotropes introduites dans l'étude expérimentale de certains phénomènes de fluctuations. *C. R. Acad. Sci. Paris 244*, 842–844.

Fejes Tóth, L. (1953). *Lagerungen in der Ebene auf der Kugel und im Raum*. Springer, Berlin, Göttingen, and Heidelberg.

Finney, D. J. (1946). Field sampling for the estimation of wireworm populations. *Biometrics 2*, 1–7.

Hájek, J. (1955). Some Contributions to the Theory of Estimation. Ph. D. Thesis, University of Prague. (In Czech, unpublished.)

Hájek, J. (1959). Optimum strategy and other problems in probability sampling. *Časopis Pěst. Mat. 19*, 387–423.

Hammersley, J. M. and Nelder, J. A. (1955). Sampling from an isotropic Gaussian process. *Proc. Cambridge Philos. Soc. 51*, 652–662.

Heppes, A. (1959). Mehrfache gitterförmige Kreslagerungen in der Ebene. *Acta Math. Acad. Sci. Hungar. 10*, 141–148.

Husu, A. (1957). On some functionals on random fields. *Vestnik Leningrad. Univ. 12*, 37–45. (In Russian, with English summary.)

Madow, W. G. and Madow, L. H. (1944). On the theory of systematic sampling. *Ann. Math. Satistist. 15*, 1–24.

Mahalanobis, P. C. (1944). On large–scale sample surveys. *Philos. Trans. Roy. Soc. London, Ser. B, 231*, 329–451.

Masuyama, M. (1953). A rapid method of estimating basal area in timber survey – an application of integal geometry to areal sampling problems. *Sankhyā 12*, 291–302.

Matérn, B. (1947). Methods of estimating the accuracy of line and sample plot surveys. *Medd. Statens Skogsforskningsinst. 36*, 1–138. (In Swedish)

Matérn, B. (1960). Spatial variation. *Medd. Statens Skogsforskningsinst. 49*, 1–144.

Osborne, J. G. (1942). Sampling errors of systematic and random surveys of cover-type area. *J. Amer. Statist. Assoc. 37*, 256–270.

Quenouille, M. H. (1949). Problems in plane sampling. *Ann. Math. Statist. 20*, 355–375.

Savelli, M. (1957). Etude expérimentale du spectre de la transparence locale d'un film photographique uniformément inpressioné. *C. R. Acad. Sci. Paris 244*, 871–873.

Smith, H. F. (1938). An empirical law describing heterogeneity in the yields of agricultural crops. *J. Agric. Sci. 28*, 1–23.

Sulanke, A. B., unpublished Ph. D. Thesis.

Whittle, P. (1954). On stationary processes in the plane. *Biometrika 41*, 434–449.

Whittle, P. (1956). On the variation of yield variance with plot size. *Biometrika 43*, 337–343.

Wicksell, S. D. (1925). The corpuscle problem. A mathematical study of a biometric problem. *Biometrika 17*, 84–99.

Wicksell, S. D. (1926). The corpuscle problem. Second memoir. Case of ellipsoidal corpuscles. *Biometrika 18*, 151–172.

Williams, R. M. (1956). The variance of the mean of systematic samples. *Biometrika 43*, 137–148.

Zubrzycki, S. (1957). On estimating gangue parameters. *Zastos. Mat. 3*, 105–153. (In Polish)

Zubrzycki, S. (1958). Remarks on random, stratified and systematic sampling in a plane. *Colloq. Math. 6*, 251–264.

CHAPTER 17

Some Extensions of the Wald–Wolfowitz–Noether Theorem

Ann. Math. Statist. *32* (1961), 506–523.
Reproduced by permission of The Institute of Mathematical Statistics.

1 SUMMARY AND INTRODUCTION

Let $(R_{\nu 1}, \ldots, R_{\nu N_\nu})$ be a random vector which takes on the $N_\nu!$ permutations of $(1, \ldots, N_\nu)$ with equal probabilities. Let $\{b_{\nu i},\ 1 \leq i \leq N_\nu,\ \nu \geq 1\}$ and $\{a_{\nu i},\ 1 \leq i \leq N_\nu,\ \nu \geq 1\}$ be double sequences of real numbers. Put

$$S_\nu = \sum_{i=1}^{N_\nu} b_{\nu i}\, a_{\nu R_{\nu i}}. \tag{1.1}$$

We shall prove that the sufficient and necessary condition for asymptotic ($N_\nu \to \infty$) normality of S_ν is of Lindeberg type. This result generalizes previous results by Wald–Wolfowitz (1944), Noether (1949), Hoeffding (1951), Dwass (1953, 1955) and Motoo (1957). In respect to Motoo (1957) we show, in fact, that his condition, applied to our case, is not only sufficient but also necessary.

Cases encountered in rank-test theory are studied in more detail in Section 6 by means of the theory of martingales. The method of this paper consists in proving asymptotic equivalency in the mean of (1.1) to a sum of infinitesimal independent components.

Collected Works of Jaroslav Hájek – With Commentary
Edited by M. Hušková, R. Beran and V. Dupač
Published in 1998 by John Wiley & Sons Ltd.

2 THREE LEMMAS

Consider a sequence $U_1, \ldots, U_N$ of independent random variables each having uniform (rectangular) distribution over the interval $(0,1]$. Let R_i be the rank of U_i, i.e.,

$$U_i = Z_{R_i}, \tag{2.1}$$

where $Z_1 < \cdots < Z_N$ is the sequence $U_1, \ldots, U_N$, reordered in ascending magnitude.

Take a nondecreasing sequence $a_1 \leq \cdots \leq a_N$ of real numbers and put

$$a(\lambda) = a_i \quad \text{for} \quad (i-1)/N < \lambda \leq i/N \quad (1 \leq i \leq N). \tag{2.2}$$

The function $a(\lambda)$ will be called a quantile function of $a_1 \leq \cdots \leq a_N$. As $(i-1)/N < i/(N+1) < i/N$, we have

$$a_i = a(i/N) = a[i/(N+1)]. \tag{2.3}$$

Furthermore,

$$\overline{a} = \frac{1}{N}\sum_{i=1}^{N} a_i = \int_0^1 a(\lambda)\,\mathrm{d}\lambda \tag{2.4}$$

$$\text{and} \quad \sigma^2 = \frac{1}{N}\sum_{i=1}^{N}(a_i - \overline{a})^2 = \int_0^1 [a(\lambda) - \overline{a}]^2\,\mathrm{d}\lambda.$$

Lemma 2.1.

$$\mathsf{E}\left[a(U_1) - a\left(\frac{R_1}{N}\right)\right]^2 \leq 2\max_{1\leq i\leq N}|a_i - \overline{a}|\,\frac{1}{N}\left[2\sum_{i=1}^{N}(a_i - \overline{a})^2\right]^{\frac{1}{2}}, \tag{2.5}$$

where the function $a(\cdot)$ is given by (2.2).

Proof. If $Z_1, \ldots, Z_N$ are fixed, then U_1 takes on each of the values Z_i with probability $1/N$, and $U_1 = Z_i$ is equivalent to $R_1 = i$. Therefore,

$$\begin{aligned} &\mathsf{E}[a(U_1) - a(R_1/N)]^2 \\ &= \mathsf{E}\mathsf{E}\{[a(U_1) - a(R_1/N)]^2 \mid Z_1, \ldots, Z_n\} \\ &= (1/N)\,\mathsf{E}\sum_{i=1}^{N}[a(Z_i) - a(i/N)]^2. \end{aligned} \tag{2.6}$$

Now, first, consider a special quantile function

$$\begin{aligned} \epsilon[\lambda - (k/N)] &= 0 \quad \text{if} \quad \lambda \leq k/N \\ &= 1 \quad \text{if} \quad \lambda > k/N. \end{aligned} \tag{2.7}$$

The quantile function $\epsilon[\lambda - (k/N)]$ corresponds, obviously, to the sequence $a_1 = \cdots = a_k = 0$, $a_{k+1} = \cdots = a_N = 1$. Let K denote the number of the U_is smaller than k/N. Clearly, $Z_K < k/N < Z_{K+1}$. If $K \leq k$, we have

$$\begin{aligned} \epsilon[Z_i - k/N] - \epsilon[(i-k)/N] &= 0 \quad \text{if } i = 1, \ldots, K,\, k+1, \ldots, N \\ &= 1 \quad \text{otherwise} \end{aligned} \tag{2.8}$$

so that

$$\sum_{i=1}^{N} \left[\epsilon\left(Z_i - \frac{k}{N}\right) - \epsilon\left(\frac{i-k}{N}\right)\right]^2 = \mathsf{E}|K - k|. \tag{2.9}$$

We can easily see that (2.9) also holds for $K \geq k$. The result (2.9) together with (2.6) gives

$$\mathsf{E}\left[\epsilon\left(U_1 - \frac{k}{N}\right) - \epsilon\left(\frac{R_1 - k}{N}\right)\right]^2 = \mathsf{E}\frac{|K-k|}{N}. \tag{2.10}$$

The distribution of K is, obviously, binomial with mean value k and variance $k[1 - (k/N)]$, so that

$$\mathsf{E}|K - k| \leq [\mathsf{E}(k-k)^2]^{\frac{1}{2}} = [k\{1 - (k/N)\}]^{\frac{1}{2}},$$

and, therefore,

$$\mathsf{E}\left\{\epsilon[U_1 - k/N] - \epsilon[(R_1 - k)/N]\right\}^2 \leq N^{-1}\left\{k(1-k/N)\right\}^{\frac{1}{2}}. \tag{2.11}$$

Now let us only suppose that $a_1 = 0$, and otherwise the sequence $a_1 \leq \cdots \leq a_N$ can be arbitrary. The quantile function of any such sequence may be expressed in the form

$$a(\lambda) = \sum_{k=1}^{N-1} (a_{k+1} - a_k)\,\epsilon[\lambda - k/N] \quad (a_1 = 0,\ 0 < \lambda \leq 1). \tag{2.12}$$

Actually, e. g., for $\lambda = i/N$, we have

$$a(i/N) = \sum_{k=1}^{N-1} (a_{k+1} - a_k)\,\epsilon[(i-k)/N] = \sum_{k=1}^{i-1} (a_{k+1} - a_k) = a_i \quad (1 \leq i \leq N).$$

Now from (2.12) it follows that, first,

$$\begin{aligned} \sum_{i=1}^{N} a_i^2 &= \sum_{k=1}^{N-1} \sum_{j=1}^{N-1} (a_{k+1} - a_k)(a_{j+1} - a_j) \sum_{i=1}^{N} \epsilon[(i-k)/N]\,\epsilon[(i-j)/N] \\ &= \sum_{k=1}^{N-1} \sum_{j=1}^{N-1} (a_{k+1} - a_k)(a_{j+1} - a_j)(N - \max(k,j)), \end{aligned} \tag{2.13}$$

and, second,

$$\begin{aligned}[a(Z_i) - a\,(i/N)]^2 = & \qquad\qquad (2.14)\\ = \sum_{k=1}^{N-1}\sum_{j=1}^{N-1}(a_{k+1}-a_k)\,(a_{j+1}-a_j)\left[\epsilon\left(Z_i-\frac{k}{N}\right)-\epsilon\left(\frac{i-k}{N}\right)\right] & \\ \times\left[\epsilon\left(Z_i-\frac{j}{N}\right)-\epsilon\left(\frac{i-j}{N}\right)\right]. &\end{aligned}$$

Because $\epsilon[Z_i-(k/N)]-\epsilon[(i-k)/N]$ and $\epsilon[Z_i-(j/N)]-\epsilon[(i-j)/N]$ take on only the values 0 and ± 1, we have

$$\begin{aligned}&\left[\epsilon\left(Z_i-\frac{k}{N}\right)-\epsilon\left(\frac{i-k}{N}\right)\right]\left[\epsilon\left(Z_i-\frac{j}{N}\right)-\epsilon\left(\frac{i-j}{N}\right)\right] \qquad (2.15)\\ &\leq\left[\epsilon\left(Z_i-\frac{\max(k,j)}{N}\right)-\epsilon\left(\frac{i-\max(k,j)}{N}\right)\right]^2.\end{aligned}$$

On combining (2.14), (2.15) and (2.11), we get

$$\begin{aligned}&\mathsf{E}\left[a(U_1)-a\left(\frac{R_1}{N}\right)\right]^2=\frac{1}{N}\mathsf{E}\sum_{i=1}^{N}\left[a(Z_1)-a\left(\frac{i}{N}\right)\right]^2\\ &\leq\sum_{k=1}^{N-1}\sum_{j=1}^{N-1}(a_{k+1}-a_k)\,(a_{j+1}-a_j)\\ &\qquad\times\frac{1}{N}\,\mathsf{E}\sum_{i=1}^{N}\left[\epsilon\left(Z_i-\frac{\max(k,j)}{N}\right)-\epsilon\left(\frac{i-\max(k,j)}{N}\right)\right]^2\\ &=\sum_{k=1}^{N-1}\sum_{j=1}^{N-1}(a_{k+1}-a_k)\,(a_{j+1}-a_j)\\ &\qquad\times\mathsf{E}\left[\epsilon\left(U_i-\frac{\max(k,j)}{N}\right)-\epsilon\left(\frac{R_1-\max(k,j)}{N}\right)\right]^2\\ &\leq\sum_{k=1}^{N-1}\sum_{j=1}^{N-1}(a_{k+1}-a_k)\,(a_{j+1}-a_j)\\ &\qquad\times\frac{1}{N}\left[\max(k,j)\,\frac{N-\max(k,j)}{N}\right]^{\frac{1}{2}}\\ &\leq\frac{1}{N}\sum_{k=1}^{N-1}\sum_{j=1}^{N-1}(a_{k+1}-a_k)\,(a_{j+1}-a_j)\,[N-\max(k,j)]^{\frac{1}{2}}\\ &\leq\frac{1}{N}\left\{\sum_{k=1}^{N-1}\sum_{j=1}^{N-1}(a_{k+1}-a_k)\,(a_{j+1}-a_j)\right.\end{aligned}$$

$$\times \sum_{k=1}^{N-1}\sum_{j=1}^{N-1}(a_{k+1}-a_k)(a_{j+1}-a_j)(N-\max(k,j))\Bigg\}^{\frac{1}{2}}.$$

According to (2.13) and to

$$\sum_{k=1}^{N-1}\sum_{j=1}^{N-1}(a_{k+1}-a_k)(a_{j+1}-a_j)=a_N^2$$

the last expression equals $N^{-1}a_N\left(\sum_{i=1}^N a_i^2\right)^{\frac{1}{2}}$. This means that, for $a_1=0$,

$$\mathsf{E}\left[a(U_1)-a\left(\frac{R_1}{N}\right)\right]^2 \le \frac{1}{N}a_N\left[\sum_{i=1}^N a_i^2\right]^{\frac{1}{2}}. \tag{2.16}$$

Generally, for an arbitrary a_1, we have

$$\mathsf{E}\left[a(U_1)-a\left(\frac{R_1}{N}\right)\right]^2 \le \frac{1}{N}(a_N-a_1)\left[\sum_{i=1}^N (a_i-a_1)^2\right]^{\frac{1}{2}}. \tag{2.17}$$

On making use of the values $-a_N\le\cdots\le -a_1$ instead of $a_1\le\cdots\le a_N$, we also get from (2.17) that

$$\mathsf{E}\left[a(U_1)-a\left(\frac{R_1}{N}\right)\right]^2 \le \frac{1}{N}(a_N-a_1)\left[\sum_{i=1}^N (a_i-a_N)^2\right]^{\frac{1}{2}}. \tag{2.18}$$

It is now easy to derive (2.4). Let us put

$$\begin{aligned} a^+(\lambda) &= \overline{a} \quad \text{if} \quad a(\lambda)\le\overline{a} \\ &= a(\lambda) \quad \text{if} \quad a(\lambda)\ge\overline{a} \end{aligned}$$

and

$$\begin{aligned} a^-(\lambda) &= a(\lambda)-\overline{a} \quad \text{if} \quad a(\lambda)\le\overline{a} \\ &= 0 \quad \text{if} \quad a(\lambda)\ge\overline{a}. \end{aligned}$$

We then have

$$a(\lambda)=a^+(\lambda)+a^-(\lambda)$$

and, in view of (2.17) and (2.18), the following inequalities are clear:

$$\begin{aligned} &\mathsf{E}\left[a(U_1)-a\left(\frac{R_1}{N}\right)\right]^2 \\ &\le 2\mathsf{E}\left[a^+(U_1)-a^+\left(\frac{R_1}{N}\right)\right]^2 + 2\mathsf{E}\left[a^-(U_1)-a^-\left(\frac{R_1}{N}\right)\right]^2 \end{aligned}$$

$$\leq 2(a_N - \overline{a}) \left[\sum_{a_i \geq \overline{a}} (a_i - \overline{a})^2 \right]^{\frac{1}{2}} + 2(\overline{a} - a_1) \left[\sum_{a_i \leq \overline{a}} (a_i - \overline{a})^2 \right]^{\frac{1}{2}}$$

$$\leq 2 \max_{1 \leq i \leq N} |a_i - \overline{a}| \left\{ \left[\sum_{a_i \geq \overline{a}} (a_i - \overline{a})^2 \right]^{\frac{1}{2}} + \left[\sum_{a_i \leq \overline{a}} (a_i - \overline{a})^2 \right]^{\frac{1}{2}} \right\}$$

$$\leq 2 \max_{1 \leq i \leq N} |a_i - \overline{a}| \left[2 \sum_{i=1}^{N} (a_i - \overline{a})^2 \right]^{\frac{1}{2}}.$$

This completes the proof. □

Let us have a nondecreasing squared integrable function $\varphi(\lambda)$, $0 < \lambda < 1$, and put

$$\varphi_N(\lambda) = \varphi[i/(N+1)] \quad \text{if} \quad (i-1)/N < \lambda \leq i/N. \tag{2.19}$$

Lemma 2.2. *The functions $\varphi_N^2(\lambda)$, $0 < \lambda < 1$, are uniformly $(N \geq 1)$ integrable and*

$$\lim_{N \to \infty} \int_0^1 [\varphi_N(\lambda) - \varphi(\lambda)]^2 \, d\lambda = 0. \tag{2.20}$$

Proof. Suppose that $\varphi(0) \geq 0$ so that both $\varphi(\lambda)$ and $\varphi^2(\lambda)$ are nondecreasing. Let A be a subset of $[0,1]$ and $\mu(A)$ its Lebesgue measure. Put

$$I_k = ((k-1)/N, \, k/N)$$

and $J_k = (k/N - \mu(A \cap I_k), \, k/N)$ and note that $\varphi_N(\lambda) = \varphi(k/(N+1))$ for $\lambda \in I_k$ and $\varphi_N(\lambda) \leq \varphi(\lambda)$ for $k/(N+1) \leq \lambda < k/N$. The following obviously holds:

$$\int_{A \cap I_k} \varphi_N^2(\lambda) \, d\lambda = \mu(A \cap I_k) \, \varphi^2(k/(N+1)) \leq (N+1) \, k^{-1} \int_{J_k} \varphi^2(\lambda) \, d\lambda \tag{2.21}$$

so that, on making use of the first right hand expression for $k \leq 3N/4$ and of the second one for $k \geq N/4$, we get

$$\int_A \varphi_N^2(\lambda) \, d\lambda = \sum_{k=1}^{N} \int_{A \cap I_k} \varphi_N^2(\lambda) \, d\lambda \leq \varphi^2(3/4) \, \mu(A) + 4 \int_B \varphi^2(\lambda) \, d\lambda \tag{2.22}$$

where $\mu(B) = \mu(A)$. Inequality (2.22) clearly proves uniform integrability of the functions $\{\varphi_N^2(\lambda)\}$. A general $\varphi(\lambda)$ may be written in the form $\varphi(\lambda) = \varphi_1(\lambda) - \varphi_2(1-\lambda)$, where $\varphi_1(0) \geq 0$, $\varphi_2(0) \geq 0$ and both $\varphi_1(\lambda)$ and $\varphi_2(\lambda)$ are nondecreasing. This completes the proof of uniform integrability.

In order to prove (2.20), let us observe that $\varphi_N(\lambda) \to \varphi(\lambda)$ on the set of continuity points of $\varphi(\lambda)$, which, however, has Lebesgue measure 1. Convergence of $\varphi_N(\lambda)$ to $\varphi(\lambda)$ almost everywhere, together with uniform integrability of the function $\varphi_N^2(\lambda)$, implies (2.20). The proof is completed.

□

Lemma 2.3. *Let $c_1, \ldots, c_N, d_1, \ldots, d_N$ be arbitrary real numbers and put $\bar{c} = N^{-1}\sum_{i=1}^N c_i$, $\bar{d} = N^{-1}\sum_{i=1}^N d_i$. Then*

$$\begin{aligned} \operatorname{var}\left(\sum_{i=1}^N c_i\, d_{R_i}\right) &= \frac{1}{N-1}\sum_{i=1}^N (c_i - \bar{c})^2 \sum_{i=1}^N (d_i - \bar{d})^2 \\ &\le \frac{1}{N-1}\sum_{i=1}^N (c_i - \bar{c})^2 \sum_{i=1}^N d_i^2. \end{aligned} \tag{2.23}$$

The proof is obvious. □

3 ASYMPTOTIC EQUIVALENCY IN THE MEAN

Random variables S_ν and T_ν will be called asymptotically equivalent in the mean (symbolically, $S_\nu \sim T_\nu$):

$$\lim_{\nu\to\infty} \mathsf{E}(S_\nu - T_\nu)^2/(\operatorname{var} T_\nu) = 0. \tag{3.1}$$

The relation is symmetric, and, if $S_\nu \sim T_\nu$ and $T_\nu \sim V_\nu$, then, clearly, $S_\nu \sim V_\nu$.

Let us take a sequence of independent random variables $U_1, U_2, \ldots$ each uniformly distributed over $(0,1]$ and denote the rank of U_i in the partial sequence $U_1, \ldots, U_N$, by $R_{\nu i}$, $1 \le i \le N_\nu$, $\nu \ge 1$. The partial sequence $U_1, \ldots, U_{N_\nu}$, reordered in ascending magnitude, will be denoted by $Z_{\nu 1}, \ldots, Z_{\nu N_\nu}$.

The distribution of S_ν given by (1.1) does not depend on the ordering of the *a*s. So we may suppose that

$$a_{\nu 1} \le \cdots \le a_{\nu N_\nu} \quad (\nu \ge 1). \tag{3.2}$$

We shall assume that the *a*s fulfil the condition

$$\lim_{\nu\to\infty} \frac{\max_{1\le i\le N_\nu}(a_{\nu i} - \bar{a}_\nu)^2}{\sum_{i=1}^{N_\nu}(a_{\nu i} - \bar{a}_\nu)^2} = 0. \tag{3.3}$$

Theorem 3.1. *Under the assumptions (3.2) and (3.3) the statistic S_ν given by (1.1) is asymptotically equivalent in the mean to the statistic*

$$T_\nu = \sum_{i=1}^{N_\nu}(b_{\nu i} - \bar{b}_\nu)\, a_\nu(U_i) + \bar{b}_\nu \sum_{i=1}^{N_\nu} a_{\nu i}, \tag{3.4}$$

where $a_\nu(\cdot)$ denotes the quantile functions of $a_{\nu 1} \leq \cdots \leq a_{\nu N_\nu}$ given by (2.2), and $\bar{b}_\nu = N_\nu^{-1} \sum_{i=1}^{N_\nu} b_{\nu i}$.

Proof. On making use of (2.1) and (2.3), we may write

$$S_\nu - T_\nu = \sum_{i=1}^{N_\nu}(b_{\nu i} - \bar{b}_\nu) \left[a_\nu(Z_{\nu R_{\nu i}}) - a_\nu\left(\frac{R_{\nu i}}{N}\right)\right]. \tag{3.5}$$

As is well known, the distribution of the ranks $(R_{\nu 1}, \ldots, R_{\nu N_\nu})$ is independent of the vector $(Z_{\nu 1}, \ldots, Z_{\nu N_\nu})$. In view of Lemma 2.3, where we put

$$c_i = b_{\nu i} - \bar{b}_\nu$$

and $d_i = a_\nu(Z_{\nu i}) - a_\nu(i/N)$, we can write

$$\begin{aligned}
&\mathsf{E}\left\{(S_\nu - T_\nu)^2 \mid Z_{\nu 1}, \ldots, Z_{\nu N_\nu}\right\} = \mathsf{var}\left\{S_\nu - T_\nu \mid Z_{\nu 1}, \ldots, Z_{\nu N_\nu}\right\} \\
&\leq \frac{1}{N_\nu - 1} \sum_{i=1}^{N_\nu}(b_{\nu i} - \bar{b}_\nu)^2 \sum_{i=1}^{N_\nu} \left[a_\nu(Z_{\nu R_{\nu i}}) - a_\nu\left(\frac{R_{\nu i}}{N_\nu}\right)\right]^2 \\
&= \frac{1}{N_\nu - 1} \sum_{i=1}^{N_\nu}(b_{\nu i} - \bar{b}_\nu)^2 \sum_{i=1}^{N_\nu} \left[a_\nu(U_i) - a_\nu\left(\frac{R_{\nu i}}{N_\nu}\right)\right]^2.
\end{aligned} \tag{3.6}$$

The first equality in (3.6) is ensured by $\mathsf{E}(S_\nu - T_\nu \mid Z_1, \ldots, Z_{\nu N_\nu}) = 0$ which follows from $\sum(b_{\nu i} - \bar{b}_\nu) = 0$. Taking the mean value over $Z_{\nu 1}, \ldots, Z_{\nu N_\nu}$ on both sides of the inequality (3.6), we obtain

$$\mathsf{E}(S_\nu - T_\nu)^2 \leq \frac{1}{N_\nu - 1} \sum_{i=1}^{N_\nu}(b_{\nu i} - \bar{b}_\nu)^2 \sum_{i=1}^{N_\nu} \mathsf{E}\left[a_\nu(U_i) - a_\nu\left(\frac{R_{\nu i}}{N_\nu}\right)\right]^2. \tag{3.7}$$

Clearly,

$$\mathsf{E}\left[a_\nu(U_i) - a_\nu\left(\frac{R_{\nu i}}{N_\nu}\right)\right]^2 = \mathsf{E}\left[a_\nu(U_i) - a_\nu\left(\frac{R_{\nu 1}}{N_\nu}\right)\right]^2 \quad (1 \leq i \leq N_\nu,\ \nu \geq 1), \tag{3.8}$$

so that (3.7) may be put in the form

$$\mathsf{E}(S_\nu - T_\nu)^2 \leq \frac{N_\nu}{N_\nu - 1} \mathsf{E}\left[a_\nu(U_1) - a_\nu\left(\frac{R_{\nu 1}}{N_\nu}\right)\right]^2 \sum_{i=1}^{N_\nu}(b_{\nu i} - \bar{b}_\nu)^2. \tag{3.9}$$

On the other hand, it follows from (3.4) that

$$\operatorname{var} T_\nu = 1/N_\nu \sum_{i=1}^{N_\nu}(a_{\nu i} - \overline{a}_\nu)^2 \sum_{i=1}^{N_\nu}(b_{\nu i} - \overline{b}_\nu)^2. \tag{3.10}$$

Making use of Lemma 2.1, (3.9) and (3.10) yield

$$\frac{\mathsf{E}(S_\nu - T_\nu)^2}{\operatorname{var} T_\nu} \leq \frac{2\sqrt{2}\, N_\nu}{N_\nu - 1} \frac{\max |a_{\nu i} - \overline{a}_\nu|}{\left[\sum_{i=1}^{N_\nu}(a_{\nu i} - \overline{a}_\nu)^2\right]^{\frac{1}{2}}}. \tag{3.11}$$

Consequently, (3.1) follows from (3.11) and (3.3). The proof is complete. □

Theorem 3.2. *Let the function $\varphi(\lambda)$ be nondecreasing, nonconstant and squared integrable. Suppose that*

$$\lim_{\nu\to\infty} N_\nu = \infty. \tag{3.12}$$

Then the statistics

$$S_\nu = \sum_{i=1}^{N_\nu} b_{\nu i}\, \varphi[R_{\nu i}/(N_\nu + 1)] \tag{3.13}$$

and

$$T_\nu = \sum_{i=1}^{N_\nu}(b_{\nu i} - \overline{b}_\nu)\, \varphi(U_i) + \overline{b}_\nu \sum_{i=1}^{N_\nu} \varphi[i/(N_\nu + 1)] \tag{3.14}$$

are asymptotically equivalent in the mean.

Proof. If we put

$$a_{\nu i} = \varphi[i/(N_\nu + 1)], \tag{3.15}$$

the quantile function of $a_{\nu 1} \leq \cdots \leq a_{\nu N_\nu}$ will equal $\varphi_{N_\nu}(\lambda)$ expressed by (2.19). According to Lemma 2.2, the functions $\varphi^2_{N_\nu}(\lambda)$ are uniformly integrable and hence

$$\lim_{\nu\to\infty} N_\nu^{-1} \max_{1\leq i\leq N_\nu} |a_{\nu i}|^2 = \lim_{\nu\to\infty} \max_{1\leq i\leq N_\nu} \int_{(i-1)/N_\nu}^{i/N_\nu} \varphi^2_{N_\nu}(\lambda)\, \mathrm{d}\lambda = 0. \tag{3.16}$$

On the other hand, from (2.20) and from nonconstancy of $\varphi(\lambda)$, it follows that

$$\lim_{\nu\to\infty} \frac{1}{N_\nu} \sum_{i=1}^{N_\nu}(a_{\nu i} - \overline{a}_\nu)^2 = \int_0^1 \left[\varphi(\lambda) - \int_0^1 \varphi(x)\, \mathrm{d}x\right]^2 \mathrm{d}\lambda > 0. \tag{3.17}$$

Relations (3.16) and (3.17) imply that the sequences $a_{\nu 1} \leq \cdots \leq a_{\nu N_\nu}$ fulfil condition (3.3). Therefore, in view of Theorem 3.1, the statistic (3.13) is asymptotically equivalent to the statistic

$$T'_\nu = \sum_{i=1}^{N_\nu} (b_{\nu i} - \bar{b}_\nu)\, \varphi_{N_\nu}(U_i) + \bar{b}_\nu \sum_{i=1}^{N_\nu} \varphi[i/(N_\nu + 1)]. \tag{3.18}$$

It remains to show that (3.18) is equivalent to (3.14). However, as is easy to see,

$$\frac{\mathsf{E}(T_\nu - T'_\nu)^2}{\operatorname{var} T_\nu} = \frac{\int_0^1 [\varphi(\lambda) - \varphi_{N_\nu}(\lambda)]^2 \,\mathrm{d}\lambda}{\int_0^1 \left[\varphi(\lambda) - \int_0^1 \varphi(x)\,\mathrm{d}x\right]^2 \mathrm{d}\lambda}, \tag{3.19}$$

so that $T_\nu \sim T'_\nu$ is a consequence of the assumption (3.12) and of Lemma 2.2. Theorem 3.2 is proved. □

4 NECESSARY AND SUFFICIENT CONDITION FOR ASYMPTOTIC NORMALITY OF S_ν

If $S_\nu \sim T_\nu$, then obviously the asymptotic variance and the asymptotic distribution of S_ν and T_ν exist under the same conditions, and, if they exist, are the same. Thus the problem of the asymptotic distribution of S_ν is reduced to the problem of the asymptotic distribution of T_ν. The statistic T_ν, however, is a sum of independent addends, so that, if these addends are infinitesimal, it suffices to use well-known theory (Gnedenko and Kolmogorov (1954)).

Theorem 4.1. *Let us suppose that*

$$\lim_{\nu\to\infty} \frac{\max_{1\leq i\leq N_\nu}(a_{\nu i} - \bar{a}_\nu)^2}{\sum_{i=1}^{N_\nu}(a_{\nu i} - \bar{a}_\nu)^2} = 0 \tag{4.1}$$

and

$$\lim_{\nu\to\infty} \frac{\max_{1\leq i\leq N_\nu}(b_{\nu i} - \bar{b}_\nu)^2}{\sum_{i=1}^{N_\nu}(b_{\nu i} - \bar{b}_\nu)^2} = 0. \tag{4.2}$$

Then the statistic (1.1) *has an asymptotically normal distribution with mean value* $\mathsf{E}\, S_\nu$ *and variance* $\operatorname{var} S_\nu$ *if, and only if, for any* $\tau > 0$

$$\lim_{\nu\to\infty} 1/N_\nu \sum\sum_{|\delta_{\nu ij}|>\tau} \delta^2_{\nu ij} = 0, \tag{4.3}$$

where

$$\delta_{\nu ij} = \frac{(b_{\nu i} - \bar{b}_\nu)(a_{\nu i} - \bar{a}_\nu)}{\left[\frac{1}{N_\nu}\sum_{i=1}^{N_\nu}(b_{\nu i} - \bar{b}_\nu)^2 \sum_{i=1}^{N_\nu}(a_{\nu i} - \bar{a}_\nu)^2\right]^{\frac{1}{2}}} \quad (1 \le i,\, j \le N_\nu,\ \nu \ge 1). \tag{4.4}$$

Proof. Assuming that $a_{\nu 1} \le \cdots \le a_{\nu N_\nu}$, then from (4.1) and Theorem 3.1 it follows that (1.1) is asymptotically equivalent in the mean to (3.4). Therefore, it suffices to show that, under the additional assumption (4.2), the condition (4.3) is necessary and sufficient for the asymptotic normality of T_ν with mean value $\mathsf{E}\, T_\nu$ and variance $\mathsf{var}\, T_\nu$. The assumption (4.2), however, implies that the addends in (3.4) are infinitesimal, because

$$\begin{aligned} &\frac{\max_{1\le i\le N_\nu} \mathsf{var}[(b_{\nu i} - \bar{b}_\nu)\, a_\nu(U_i)]}{\mathsf{var}\, T_\nu} \\ &= \frac{\max_{1\le i\le N_\nu}(b_{\nu i} - \bar{b}_\nu)^2 \sum_{i=1}^{N_\nu}(a_{\nu i} - \bar{a}_\nu)^2/N_\nu}{\sum_{i=1}^{N_\nu}(b_{\nu i} - \bar{b}_\nu)^2 \sum_{i=1}^{N_\nu}(a_{\nu i} - a_\nu)^2/N_\nu} = \frac{\max_{1\le i\le N_\nu}(b_{\nu i} - \bar{b}_\nu)^2}{\sum_{i=1}^{N_\nu}(b_{\nu i} - \bar{b}_\nu)^2}. \end{aligned} \tag{4.5}$$

Consequently, we have to prove that (4.3) coincides with the Lindeberg condition for T_ν, namely with

$$\lim_{\nu\to\infty} \sum_{i=1}^{N_\nu} \frac{1}{\mathsf{var}\, T_\nu} \int_{|x|>\tau(\mathsf{var}\, T_\nu)^{\frac{1}{2}}} x^2 \,\mathrm{d}\mathsf{P}\left\{(b_{\nu i} - \bar{b}_\nu)\, a_\nu(U_i) < x\right\} = 0. \tag{4.6}$$

Clearly, we have

$$\begin{aligned} &\frac{1}{\mathsf{var}\, T_\nu} \int_{|x|>\tau(\mathsf{var}\, T_\nu)^{\frac{1}{2}}} x^2 \,\mathrm{d}P\left\{(b_{\nu i} - \bar{b}_\nu)\, a_\nu(U_i) < x\right\} \\ &= \frac{\sum_{i=1}^{N_\nu}\left[(b_{\nu i} - \bar{b}_\nu)^2 \sum_{j\in E_{\nu i}}(a_{\nu j} - \bar{a}_\nu)^2\right]/N_\nu}{\sum_{i=1}^{N_\nu}(b_{\nu i} - \bar{b}_\nu)^2 \sum_{i=1}^{N_\nu}(a_{\nu i} - \bar{a}_\nu)^2/N_\nu} \end{aligned} \tag{4.7}$$

where

$$E_{\nu i} = \left\{ j : |(b_{\nu i} - \bar{b}_\nu)(a_{\nu j} - \bar{a}_\nu)| > \tau \left[\frac{1}{N_\nu}\sum_{i=1}^{N_\nu}(b_{\nu i} - \bar{b}_\nu)^2 \sum_{i=1}^{N_\nu}(a_{\nu i} - \bar{a}_\nu)^2\right]^{\frac{1}{2}} \right\}. \tag{4.8}$$

Now, observe that

$$\frac{1}{N_\nu} \sum\sum_{|\delta_{\nu ij}|>\tau} \delta_{\nu ij}^2 = \frac{\sum_{i=1}^{N_\nu}\left[(b_{\nu i} - \bar{b}_\nu)^2 \sum_{j\in E_{\nu i}}(a_{\nu j} - \bar{a}_\nu)^2\right]}{\sum_{i=1}^{N_\nu}(b_{\nu i} - \bar{b}_\nu)^2 \sum_{i=1}^{N_\nu}(a_{\nu j} - \bar{a}_\nu)^2} \tag{4.9}$$

so that (4.3) is actually equivalent to (4.6). The proof is complete. □

The condition (4.3) is symmetrical in the *a*s and the *b*s. In applications, however, the *a*s and the *b*s often play a somewhat different role. In such cases the following theorem is useful.

Theorem 4.2. *A double sequence $\{a_{\nu i}, 1 \leq i \leq N_\nu, \nu \geq 1\}$ satisfying the condition (4.1) fulfils the condition (4.3) for any double sequence $\{b_{\nu i}, 1 \leq i \leq N_\nu, \nu \geq 1\}$ satisfying the condition (4.2) if, and only if,*

$$\left[\lim_{\nu\to\infty} \frac{k_\nu}{N_\nu} = 0\right] \Longrightarrow \lim_{\nu\to\infty} \frac{\max_{1\leq i_1<\cdots<i_{k_\nu}\leq N_\nu} \sum_{\alpha=1}^{k_\nu} (a_{\nu i_\alpha} - \bar{a}_\nu)^2}{\sum_{i=1}^{N_\nu} (a_{\nu i} - \bar{a}_\nu)^2} = 0. \quad (4.10)$$

Proof. First, let us prove that (4.2) and (4.10) implies (4.3). Put, for a fixed $\tau > 0$,

$$E_\nu = \left\{ j : (a_{\nu j} - \bar{a}_\nu)^2 > \frac{\tau^2 \sum_{i=1}^{N_\nu} (b_{\nu i} - \bar{b}_\nu)^2}{\max_{1\leq i\leq N_\nu} (b_{\nu i} - \bar{b}_\nu)^2} \frac{1}{N_\nu} \sum_{i=1}^{N_\nu} (a_{\nu i} - \bar{a}_\nu)^2 \right\} \quad (4.11)$$

and denote the number of elements in E_ν by k_ν. Clearly

$$\frac{\tau^2 \sum_{i=1}^{N_\nu} (b_{\nu i} - \bar{b}_\nu)^2}{\max_{1\leq i\leq N_\nu} (b_{\nu i} - \bar{b}_\nu)^2} \frac{k_\nu}{N_\nu} \sum_{i=1}^{N_\nu} (a_{\nu j} - \bar{a}_\nu)^2 < \sum_{j\in E_\nu} (a_{\nu j} - \bar{a}_\nu)^2 \leq \sum_{j=1}^{N_\nu} (a_{\nu j} - \bar{a}_\nu)^2,$$

i.e.,

$$\frac{k_\nu}{N_\nu} \leq \tau^{-2} \frac{\max_{1\leq i\leq N_\nu} (b_{\nu i} - \bar{b}_\nu)^2}{\sum_{i=1}^{N_\nu} (b_{\nu i} - \bar{b}_\nu)^2},$$

from which, in view of (4.2), it follows that

$$\lim_{\nu\to\infty} (k_\nu / N_\nu) = 0. \quad (4.12)$$

Relation (4.12), according to (4.10), implies that

$$\lim_{\nu\to\infty} \frac{\sum_{j\in E_\nu} (a_{\nu j} - \bar{a}_\nu)^2}{\sum_{j=1}^{N_\nu} (a_{\nu j} - \bar{a}_\nu)^2} = 0. \quad (4.13)$$

Now, from (4.8) and (4.11) it follows that $E_{\nu i} \subset E_\nu$ and, consequently,

$$\frac{\sum_{i=1}^{N_\nu} \left[(b_{\nu i} - \bar{b}_\nu)^2 \sum_{j\in E_{\nu i}} (a_{\nu j} - \bar{a}_\nu)^2 \right]}{\sum_{i=1}^{N_\nu} (b_{\nu i} - \bar{b}_\nu)^2 \sum_{j=1}^{N_\nu} (a_{\nu j} - \bar{a}_\nu)^2} \leq \frac{\sum_{j\in E_\nu} (a_{\nu j} - \bar{a}_\nu)^2}{\sum_{j=1}^{N_\nu} (a_{\nu j} - \bar{a}_\nu)^2}. \quad (4.14)$$

Finally, (4.3) is an obvious consequence of (4.9), (4.14) and (4.3).

Second, let us assume that (4.10) does not hold. Then there exists a sequence B_ν of sets of integers such that, first, $B_\nu \subset \{1, \ldots, N_\nu\}$, second, the numbers of elements in B_ν, say ℓ_ν, satisfy the relation

$$\lim_{\nu\to\infty} (\ell_\nu / N_\nu) = 0, \tag{4.15}$$

and, third,

$$\limsup_{\nu\to\infty} \frac{\sum_{j\in B_\nu}(a_{\nu j} - \overline{a}_\nu)^2}{\sum_{j=1}^{N_\nu}(a_{\nu j} - \overline{a}_\nu)^2} > 0. \tag{4.16}$$

If

$$C_\nu = \left\{ j : (a_{\nu j} - \overline{a}_\nu)^2 > \frac{(N_\nu/\ell_\nu)^{\frac{1}{2}}}{N_\nu} \sum_{i=1}^{N_\nu}(a_{\nu i} - \overline{a}_\nu)^2 \right\}, \tag{4.17}$$

and $C_\nu^* = \{1, 2, \ldots, N_\nu\} - C_\nu$ then, clearly,

$$\frac{\sum_{j\in B_\nu \cap C_\nu^*}(a_{\nu j} - \overline{a}_\nu)^2}{\sum_{i=1}^{N_\nu}(a_{\nu i} - \overline{a}_\nu)^2} \leq \left(\frac{N_\nu}{\ell_\nu}\right)^{\frac{1}{2}} \frac{\ell_\nu}{N_\nu} = \left(\frac{N_\nu}{\ell_\nu}\right)^{\frac{1}{2}}$$

and, therefore, according to (4.15),

$$\lim_{\nu\to\infty} \frac{\sum_{j\in B_\nu \cap C_\nu^*}(a_{\nu j} - \overline{a}_\nu)^2}{\sum_{i=1}^{N_\nu}(a_{\nu i} - \overline{a}_\nu)^2} = 0.$$

Consequently, in view of (4.16),

$$\limsup_{\nu\to\infty} \frac{\sum_{j\in C_\nu}(a_{\nu j} - \overline{a}_\nu)^2}{\sum_{i=1}^{N_\nu}(a_{\nu i} - \overline{a}_\nu)^2} \geq \limsup \frac{\sum_{j\in B_\nu}(a_{\nu j} - \overline{a}_\nu)^2}{\sum_{i=1}^{N_\nu}(a_{\nu i} - \overline{a}_\nu)^2} > 0. \tag{4.18}$$

Now, put

$$b_{\nu 1} = \cdots = b_{\nu n_\nu} = 1, \quad b_{\nu n_\nu + 1} = \cdots = b_{\nu N_\nu} = 0, \tag{4.19}$$

where n_ν is determined by

$$n_\nu \leq (N_\nu/\ell_\nu)^{\frac{1}{2}} < n_\nu + 1 \quad (\nu \geq 1). \tag{4.20}$$

We have, obviously,

$$\lim_{\nu\to\infty} (n_\nu/N_\nu) = 0 \tag{4.21}$$

and, in view of (4.15),

$$\lim_{\nu\to\infty} n_\nu = \infty. \tag{4.22}$$

Furthermore,

$$\overline{b}_\nu = n_\nu / N_\nu \tag{4.23}$$

$$\sum_{i=1}^{N_\nu}(b_{\nu i} - \overline{b}_\nu)^2 = [n_\nu(N_\nu - n_\nu)]/N_\nu, \tag{4.24}$$

and

$$\begin{aligned}
\frac{\max_{1\le i\le N_\nu}(b_{\nu i}-\bar b_\nu)^2}{\sum_{i=1}^{N_\nu}(b_{\nu i}-\bar b_\nu)^2}
&= \frac{\max\left\{(1-n_\nu/N)^2,\,(n_\nu/N_\nu)^2\right\}}{n_\nu(N_\nu-n_\nu)/N_\nu}\\
&\le \max\left\{\frac{1}{n_\nu},\,\frac{1}{N_\nu-n_\nu}\right\}. \qquad (4.25)
\end{aligned}$$

Consequently, the relations (4.21) and (4.22) ensure that the condition (4.2) is satisfied.

The sets $E_{\nu i}$, given generally by (4.8), will be now for $1\le i\le n_\nu$ defined by

$$E_{\nu i}=E_{\nu 1}=\left\{j:(a_{\nu j}-\bar a_\nu)^2>\tau^2\frac{n_\nu}{N_\nu-n_\nu}\sum_{i=1}^{N_\nu}(a_{\nu i}-\bar a_\nu)^2\right\}\quad(1\le i\le n_\nu). \qquad (4.26)$$

From (4.20) it follows that, for ν sufficiently large, the set C_ν given by (4.17) will be included in the set $E_{\nu 1}$ given by (4.26). Therefore

$$\begin{aligned}
\frac{1}{N_\nu}\sum\sum_{|\delta_{\nu ij}|>\tau}\delta^2_{\nu ij}
&= \frac{\sum_{i=1}^{N_\nu}\left[(b_{\nu i}-\bar b_\nu)^2\sum_{j\in E_{\nu i}}(a_{\nu j}-\bar a_\nu)^2\right]}{\sum_{i=1}^{N_\nu}(b_{\nu i}-\bar b_\nu)^2\sum_{i=1}^{N_\nu}(a_{\nu j}-\bar a_\nu)^2}\\
&\ge \frac{\sum_{i=1}^{N_\nu}\left[(b_{\nu i}-\bar b_\nu)^2\sum_{j\in E_{\nu i}}(a_{\nu j}-\bar a_\nu)^2\right]}{\sum_{i=1}^{N_\nu}(b_{\nu i}-\bar b_\nu)^2\sum_{j=1}^{N_\nu}(a_{\nu j}-\bar a_\nu)^2}\\
&= \frac{\left(1-\frac{n_\nu}{N_\nu}\right)^2 n_\nu\sum_{j\in E_{\nu 1}}(a_{\nu j}-\bar a_\nu)^2}{n_\nu\left(1-\frac{n_\nu}{N_\nu}\right)\sum_{j=1}^{N_\nu}(a_{\nu j}-\bar a_\nu)^2} \qquad (4.27)\\
&\ge \left(1-\frac{n_\nu}{N_\nu}\right)\frac{\sum_{j\in C_\nu}(a_{\nu j}-\bar a_\nu)^2}{\sum_{j=1}^{N_\nu}(a_{\nu j}-\bar a_\nu)^2}\quad(\nu\ge\nu_0).
\end{aligned}$$

The relations (4.27), (4.18) and (4.21) imply that (4.3) cannot hold, and the theorem is thereby proved. □

Let $G_\nu(x)$ denote the distribution function of the numbers $a_{\nu 1}\le\cdots\le a_{\nu N_\nu}$, i.e.

$$G_\nu(x)=\frac{\text{number of the } as \text{ smaller than or equal to } x}{N_\nu}. \qquad (4.28)$$

Lemma 4.1. *Condition* (4.10) *may be expressed in any of the following three forms:*

(i) *The functions* $[a_\nu(\lambda) - \overline{a}_\nu]^2 \left\{\int_0^1 [a_\nu(\lambda) - \overline{a}_\nu]^2 \, \mathrm{d}\lambda\right\}^{-1}$ *are uniformly integrable.*

(ii)

$$\left[\lim_{\nu\to\infty} \epsilon_\nu = 0\right] \Longrightarrow \left[\lim_{\nu\to\infty} \frac{\left(\int_0^{\epsilon_\nu} + \int_{1-\epsilon_\nu}^1\right)[a_\nu(\lambda) - \overline{a}_\nu]^2 \, \mathrm{d}\lambda}{\int_0^1 [a_\nu(\lambda) - \overline{a}_\nu]^2 \, \mathrm{d}\lambda} = 0\right]. \tag{4.29}$$

(iii)

$$\left[\lim_{\nu\to\infty} K_\nu = \infty\right] \Longrightarrow \left[\lim_{\nu\to\infty} \frac{1}{\sigma_\nu^2} \int_{|x-\overline{a}_\nu|>K_\nu\sigma_\nu} (x - \overline{a}_\nu)^2 \, \mathrm{d}G_\nu(x) = 0\right], \tag{4.30}$$

where

$$\sigma_\nu^2 = \frac{1}{N_\nu} \sum_{i=1}^{N_\nu} (a_{\nu i} - \overline{a}_\nu)^2 = \int_0^1 [a_\nu(\lambda) - \overline{a}_\nu]^2 \, \mathrm{d}\lambda = \int_{-\infty}^{\infty} (x - \overline{a}_\nu)^2 \, \mathrm{d}G_\nu(x). \tag{4.31}$$

Proof. If $a_{\nu 1} \leq \cdots \leq a_{\nu N_\nu}$, then surely

$$\max_{1 \leq i_1 < \cdots < i_{k_\nu} \leq N_\nu} \sum_{\alpha=1}^{k_\nu} (a_{\nu i_\alpha} - \overline{a}_\nu)^2 = \sum_{i=1}^{k_{\nu 1}} (a_{\nu i} - \overline{a}_\nu)^2 + \sum_{i=N_\nu - k_{\nu 2}}^{N_\nu} (a_{\nu i} - \overline{a}_\nu)^2, \tag{4.32}$$

where $k_{\nu 1} + k_{\nu 2} = k_\nu$. On the other hand, we have

$$\begin{aligned} &\sum_{i=1}^{k_{\nu 1}} (a_{\nu i} - \overline{a}_\nu)^2 + \sum_{i=N_\nu - k_{\nu 2}}^{N_\nu} (a_{\nu i} - \overline{a}_\nu)^2 \\ &= N_\nu \left(\int_0^{k_{\nu 1}/N_\nu} + \int_{1-(k_{\nu 2}/N_\nu)}^1 \right) [a_\nu(\lambda) - \overline{a}_\nu]^2 \, \mathrm{d}\lambda, \end{aligned} \tag{4.33}$$

which proves the equivalence of (4.10) to (ii) and, of course, to (i) as well.

Now we shall prove the equivalence of (4.10) to (4.30). Clearly

$$\frac{1}{\sigma_\nu^2} \int_{|x-\overline{a}_\nu|>K_\nu\sigma_\nu} (x - \overline{a}_\nu)^2 \, \mathrm{d}G_\nu(x) = \frac{1}{\sigma_\nu^2} \frac{1}{N_\nu} \sum_{|a_{\nu i} - \overline{a}_\nu| > K_\nu\sigma_\nu} (a_{\nu i} - \overline{a}_\nu)^2. \tag{4.34}$$

Denoting the number of as such that $|a_{\nu i} - \overline{a}_\nu| > K_\nu\sigma_\nu$ by k_ν, we have the following form of the Chebyshev inequality:

$$k_\nu K_\nu^2 \leq \frac{1}{\sigma_\nu^2} \sum_{|a_{\nu i} - \overline{a}_\nu| > K_\nu\sigma_\nu} (a_{\nu i} - \overline{a}_\nu)^2 \leq N_\nu. \tag{4.35}$$

Assume, first, that (4.10) holds. Then $K_\nu \to \infty$ implies, in view of (4.35), $k_\nu/N_\nu \to 0$, so that, according to (4.10), the right side of (4.34) tends to 0, and, consequently, also the left side of (4.34) tends to 0. Thus (4.10) implies (4.30).

Assume, second, that (4.10) does not hold. On repeating the respective part of the proof of Theorem 4.2, we get again the relation (4.18), which is equivalent to

$$\limsup_{\nu\to\infty} \frac{1}{\sigma_\nu^2} \int_{|x-\overline{a}_\nu|>(N_\nu/\ell_\nu)^{\frac{1}{2}}\sigma_\nu} (x-\overline{a}_\nu)^2 \, \mathrm{d}G_\nu(x) > 0. \tag{4.36}$$

This means that (4.30) is not satisfied for $K_\nu = (N_\nu/\ell_\nu)^{\frac{1}{2}}$. So the negation of (4.10) implies the negation of (4.30), i. e., (4.30) implies (4.10). Lemma 4.1 is proved. □

Corollary. The statistic (3.13) is asymptotically normal with mean value $\mathsf{E}\, S_\nu$ and variance $\mathsf{var}\, S_\nu$ for any double sequence $\{b_{\nu i},\ 1 \le i \le N_\nu,\ \nu \ge 1\}$ satisfying (4.2).

Proof. By Lemma 2.2 and Lemma 4.1 the numbers $a_{\nu i} = \varphi[i/(N_\nu+1)]$ fulfil the condition (4.10). It suffices to apply Theorem 4.2. □

Example 4.1. If the bs are given by (4.19), then the statistic

$$S_\nu = \sum_{i=1}^{N_\nu} b_{\nu i}\, a_{\nu R_{\nu i}} = \sum_{i=1}^{n_\nu} a_{\nu R_{\nu i}} \tag{4.37}$$

represents a sum of n_ν elements selected by simple random sampling from the population $\{a_{\nu 1}, \ldots, a_{\nu N_\nu}\}$. The condition (4.2) is fulfilled, according to (4.25), if, and only if,

$$\lim_{\nu\to\infty} n_\nu = \lim_{\nu\to\infty} (N_\nu - n_\nu) = \infty. \tag{4.38}$$

Hence, provided that (4.1) holds, the distribution of S_ν is asymptotically normal with mean value $\mathsf{E}\, S_\nu$ and variance $\mathsf{var}\, S_\nu$ for all n_ν satisfying (4.38) if, and only if, the as satisfy the condition (4.10).

As we shall see in Section 5, (4.10) is fulfilled, for example, if the populations $\{a_{\nu 1}, \ldots, a_{\nu N_\nu}\}$ have uniformly bounded excesses. See also Erdős and Rényi (1959) and Hájek (1960) for further results concerning sampling from a finite population.

5 COMPARISON OF VARIOUS CONDITIONS

First let us introduce the following notation.

Notation. The condition (4.1), introduced by Noether (1949) and simplified by Hoeffding (1951), will be denoted by N.

The Lindeberg condition (4.3) will be denoted by L.

The condition (4.10) will be denoted by Q.

The Wald–Wolfowitz (1944) condition,

$$\frac{\sum_{i=1}^{N_\nu}(a_{\nu i}-\bar{a}_\nu)^r \Big/ N_\nu}{\left[\sum_{i=1}^{N_\nu}(a_{\nu i}-\bar{a}_\nu)^2\right]^{\frac{r}{2}} \Big/ N_\nu} = O(1) \quad (r=3,\,4,\ldots), \tag{5.1}$$

where $O(1)$ denotes uniform boundedness, will be denoted by W.

The Hoeffding (1951) condition

$$\lim_{\nu\to\infty} N_\nu^{\frac{1}{2}r-1}\frac{\sum_{i=1}^{N_\nu}(a_{\nu i}-\bar{a}_\nu)^r\sum_{i=1}^{N_\nu}(b_{\nu i}-\bar{b}_\nu)^r}{\left[\sum_{i=1}^{N_\nu}(a_{\nu i}-\bar{a}_\nu)^2\sum_{i=1}^{N_\nu}(b_{\nu i}-\bar{b}_\nu)^2\right]^{r/2}} = O \tag{5.2}$$

will be denoted by H.

Observe that the conditions L and H concern $\{a_{\nu i}, b_{\nu i}, 1\le i\le N_\nu, \nu\ge 1\}$ whereas the conditions N, Q, W are applied to each double sequence $\{a_{\nu i},\, 1\le i\le N_\nu,\, \nu\ge 1\}$ and $\{b_{\nu i},\, 1\le i\le N_\nu,\, \nu\ge 1\}$ separately. The fact that $\{b_{\nu i},\, 1\le i\le N_\nu,\, \nu\ge 1\}$ satisfies N and $\{a_{\nu i},\, 1\le i\le N_\nu,\, \nu\ge 1\}$ satisfies Q will be denoted by NQ and the symbols NN, NW and WW will have similar interpretations.

Theorem 5.1. $WW \Longrightarrow NW \Longrightarrow H,\ NQ \Longrightarrow L \Longrightarrow NN.$

Proof. For $WW \Longrightarrow NW \Longrightarrow H$ see Hoeffding (1951), and for $H \Longrightarrow L$ Motoo (1957). $NQ \Longrightarrow L$ follows from Theorem 4.2. Thus it remains to prove $NW \Longrightarrow NQ$, i. e. $W \Longrightarrow Q$ and $L \Longrightarrow NN$.

$W \Longrightarrow Q$. If we take $r=4$ in (5.1) and use the quantile function form, we get

$$\frac{\int_0^1[a_\nu(\lambda)-\bar{a}_\nu]^4\,\mathrm{d}\lambda}{\left(\int_0^1[a_\nu(\lambda)-\bar{a}_\nu]^2\,\mathrm{d}\lambda\right)^2} = O(1). \tag{5.3}$$

As is well known from a theorem due to Vallé–Poussin, (5.3) implies that the functions $[a_\nu(\lambda)-\bar{a}_\nu]^2 \Big/ \int_0^1[a_\nu(\lambda)-\bar{a}_\nu]\,\mathrm{d}\lambda$ are uniformly integrable, which is equivalent to Q (see Lemma 4.1).

$L \Longrightarrow NN$. This fact follows from the inequality

$$\frac{\max_{1\le j\le N_\nu}(a_{\nu j}-\bar{a}_\nu)^2}{\sum_{i=1}^{N_\nu}(a_{\nu i}-\bar{a}_\nu)^2}$$

$$= \sum_{i=1}^{N_\nu} \frac{\max_{1 \le j \le N_\nu} (b_{\nu i} - \bar{b}_\nu)^2 \, (a_{\nu j} - \bar{a}_\nu)^2}{\sum_{i=1}^{N_\nu} (b_{\nu i} - \bar{b}_\nu)^2 \sum_{i=1}^{N_\nu} (a_{\nu i} - \bar{a}_\nu)^2} \tag{5.4}$$

$$= \frac{1}{N_\nu} \sum_{i=1}^{N_\nu} \max_{1 \le j \le N_\nu} \delta_{\nu ij}^2 \;\le\; \epsilon + \frac{1}{N_\nu} \sum\sum_{|\delta_{\nu ij}| > \epsilon} \delta_{\nu ij}^2 \quad (\epsilon > 0),$$

where the *a*s can be replaced by the *b*s. □

Remark 5.1. The condition (5.3) means that the excesses are uniformly bounded. Denoting this condition by W_4, we have $W \Longrightarrow W_4 \Longrightarrow Q$.

Dwass (1953) considers the empirical distribution function $G_\nu(x)$ of the values $a_{\nu 1}, \ldots, a_{\nu N_\nu}$ and supposes that, first, $\lim_{\nu \to \infty} G_\nu(x) = G(x)$ at every continuity point of a distribution function $G(x)$ and, second,

$$\int x \, \mathrm{d}G_\nu(x) \;=\; \int x \, \mathrm{d}G(x) \;=\; 0 \tag{5.5}$$

$$\int x^2 \, \mathrm{d}G_\nu(x) \;=\; \int x^2 \, \mathrm{d}G(x) \;=\; 1. \tag{5.6}$$

These assumptions imply that

$$\left[\lim_{\nu \to \infty} K_\nu = \infty\right] \Longrightarrow \left[\lim_{\nu \to \infty} \int_{|x| > K_\nu} x^2 \, \mathrm{d}G_\nu = 0\right].$$

Consequently, by Lemma 4.1, the *a*s fulfil the condition Q, so that the respective part of the Dwass theorem (1955) is contained in Theorem 4.2.

6 A SPECIAL CASE ENCOUNTERED IN RANK ORDER TEST THEORY

In rank order test theory there are used locally most powerful tests based on statistics of the form

$$S_\nu = \sum_{i=1}^{N_\nu} b_{\nu i} \, \mathsf{E}\{\varphi(U_i) \mid R_{\nu i}\} \tag{6.1}$$

where $\mathsf{E}\{\cdot \mid R_{\nu i}\}$ denotes the conditional mean value under the condition that the rank of U_i among the observations $U_1, \ldots, U_{N_\nu}$ equals $R_{\nu i}$. Let us observe that (6.1) is a special case of (1.1) for

$$a_{\nu i} = \mathsf{E}\{\varphi(U_1) \mid R_{\nu 1} = i\} = \mathsf{E}\{\varphi(Z_{\nu i})\}, \tag{6.2}$$

where, in the middle expression, the index 1 might be replaced by any index $j = 1, \ldots, N_\nu$.

For simplicity we shall suppose that, as in previous sections, the Us have a uniform distribution. This causes no loss of generality, since arbitrarily distributed observations may be expressed as (nondecreasing) functions of uniformly distributed observations. We shall also assume that

$$\int_0^1 \varphi(\lambda)\,\mathrm{d}\lambda = 0 \tag{6.3}$$

and that

$$\int_0^1 \varphi^2(\lambda)\,\mathrm{d}\lambda < \infty. \tag{6.4}$$

The assumption (6.3), however, is not essential. The integers N_ν will be assumed to tend to ∞ monotonically, i.e., $N_\nu \leq N_{\nu+1}$, $\nu \geq 1$.

Lemma 6.1. *Put*

$$Y_\nu = Y_\nu(R_{\nu 1}) = \mathsf{E}\{\varphi(U_1)\,|\,R_{\nu 1}\} \tag{6.5}$$

and assume that (6.3) *and* (6.4) *hold. Then*

(i)

$$\mathsf{P}\left\{\lim_{\nu\to\infty} Y_\nu = \varphi(U_1)\right\} = 1, \tag{6.6}$$

(ii) $\{Y_1, Y_2, \dots, \varphi(U_1)\}$ *is a martingale and* $\{Y_1^2, Y_2^2, \dots, \varphi^2(U_1)\}$ *is a semimartingale,*

iii) *the random variances* Y_ν^2 $(\nu \geq 1)$ *are uniformly integrable,*

iv)

$$\lim_{\nu\to\infty} \mathsf{E}|Y_\nu^2 - \varphi^2(U_1)| = 0 \tag{6.7}$$

and

$$\lim_{\nu\to\infty} \mathsf{E}\,Y_\nu^2 = \mathsf{E}\,\varphi^2(U_1). \tag{6.8}$$

Proof. The Borel fields $\mathcal{F}_\nu$ generated by the random vectors $(R_{\nu 1}, \dots, R_{\nu N_\nu})$ form an increasing sequence of Borel fields. Denote by $\mathcal{F}_\infty$ the smallest Borel field containing $\bigcup_1^\infty \mathcal{F}_\nu$. As is well known, the conditional distribution of U_1 for given $R_{\nu 1} = j$ is the Beta distribution with $p = j$ and $q = N - j + 1$. Hence

$$\mathsf{E}\left(U_1 - \frac{R_{\nu 1}}{N_{\nu+1}}\right)^2 = \frac{1}{N_\nu}\sum_{i=1}^{N_\nu} \frac{j(N_\nu - j + 1)}{(N_\nu+1)^2\,(N_\nu+2)} < \frac{1}{N_\nu}, \tag{6.9}$$

from which it follows that U_1 is equivalent (with probability 1) to a random variable measurable with respect to $\mathcal{F}_\infty$. Consequently, $\varphi(U_1)$ is also equivalent to a random variable measurable with respect to $\mathcal{F}_\infty$.

Since the conditional distribution of U_1 for fixed $R_{\nu 1}$ is independent of $R_{\nu 2}, \ldots, R_{\nu N_\nu}$, we may also write

$$Y_\nu = \mathsf{E}\{\varphi(U_1) \,|\, R_{\nu 1}, \ldots, R_{\nu N_\nu}\} = \mathsf{E}\{\varphi(U_1) \,|\, \mathcal{F}_\nu\}. \tag{6.10}$$

Now we can apply the theory of martingales. The assertions (i) through (iv) are consequences of Doob (1953), Chap. VII, Theorem 4.4, § 1 Example 1, Theorem 1.1, Theorem 3.3 (III), Theorem 4.1 s, respectively. (6.8) is a simple consequence of (6.7). □

Lemma 6.2. *The numbers $a_{\nu i}$ given by* (6.2) *satisfy the condition* Q *given by* (4.10).

Proof. Observe that Y_ν takes on the values $a_{\nu i}$ with equal probabilities so that condition (4.10) coincides, in view of (6.3) and (6.8), with uniform integrability of the random variables Y_ν, $\nu \geq 1$, which has been proved by Lemma 6.1. □

Theorem 6.1. *If $\{b_{\nu i}, 1 \leq i \leq N_\nu, \nu \geq 1\}$ fulfil the condition N given by* (4.2) *and the function $\varphi(\lambda)$ is nonvanishing and squared integrable, then the statistic* (6.1) *has an asymptotically normal distribution with mean value* $\mathsf{E}\, S_\nu$ *and variance* $\operatorname{var} S_\nu$.

Proof. It suffices to apply Theorem 4.2 and Lemma 6.2. □

Theorem 6.2. *Under the assumptions* (6.3) *and* (6.4), *the statistic* (6.1) *is asymptotically equivalent in the mean to the statistic*

$$T_\nu = \sum_{i=1}^{N_\nu} (b_{\nu i} - \bar{b}_\nu)\, \varphi(U_i). \tag{6.11}$$

Proof. By the method used in proving (3.9), we can show that

$$\mathsf{E}(S_\nu - T_\nu)^2 \leq [N_\nu/(N_\nu - 1)]\, \mathsf{E}[\varphi(U_1) - Y_\nu]^2 \sum_{i=1}^{N_\nu} (b_{\nu i} - \bar{b}_\nu)^2. \tag{6.12}$$

In view of (6.5), (6.12) is equivalent to

$$\mathsf{E}(S_\nu - T_\nu)^2 \leq [N_\nu/(N_\nu - 1)]\, [\mathsf{E}\, \varphi^2(U_1) - \mathsf{E}\, Y_\nu^2] \sum_{i=1}^{N_\nu} (b_{\nu i} - \bar{b}_\nu)^2. \tag{6.13}$$

Now it only remains to divide both sides of (6.13) by $\operatorname{var} T_\nu$ and to apply (6.8). □

Remark 6.1. Theorem 6.1 generalizes the Dwass theorem (1953), and the equality (6.8) generalizes the respective part of Theorem 2 in Hoeffding (1953). The condition of convexity of $\varphi(\lambda)$ is removed, which proves the conjecture made by Dwass (1956, p. 358).

7 VECTOR CONSIDERATIONS

We shall briefly touch the question of asymptotical m-dimensional normality of a vector $(S_\nu^1, \ldots, S_\nu^m)$ where

$$S_\nu^g = \sum_{i=1}^{N_\nu} b_{\nu i}\, a_{\nu R_{\nu i}}^g \quad (g = 1, \ldots, m), \tag{7.1}$$

i. e., the bs are fixed and the as depend on $g = 1, \ldots, m$.

Theorem 7.1. *Suppose that for any constants* $\lambda_1, \ldots, \lambda_m$

$$\sum_{i=1}^{N_\nu} \left[\sum_{g=1}^{N_\nu} \lambda_g (a_{\nu i}^g - \bar{a}_\nu^g) \right]^2 \geq \epsilon \max_{1 \leq g \leq m} \left[\lambda_g^2 \sum_{i=1}^{m} (a_{\nu i}^g - \bar{a}_\nu^g)^2 \right] \tag{7.2}$$

where ϵ *is positive and independent of* $\nu \geq 1$. *Assume that the* bs *fulfil the condition* N *given by* (4.2) *and the* a^gs $(g = 1, \ldots, m)$ *the condition* Q *given by* (4.10). *Then the vector* $(S_\nu^1, \ldots, S_\nu^m)$ *has an asymptotically normal distribution with mean values* $\mathsf{E}\, S_\nu^g$ *and covariances* $\mathsf{cov}(S_\nu^g, S_\nu^h)$, $1 \leq g, h \leq N_\nu$, $\nu \geq 1$.

Proof. According to a theorem by Cramér (1946, p. 105), it suffices to prove asymptotic normality for any linear combination

$$\sum_{g=1}^{m} \lambda_g\, S_\nu^g = \sum_{i=1}^{N_\nu} b_{\nu i} \left[\sum_{g=1}^{m} \lambda_g\, a_{\nu R_{\nu i}}^g \right].$$

This will be done if we show that the numbers $c_{\nu i} = \sum_{g=1}^m \lambda_g\, a_{\nu i}^g$ fulfil the condition (4.10) for any $\lambda_1, \ldots, \lambda_m$. However, in view of (7.2), we have

$$\begin{aligned}
\frac{\sum_{\alpha=1}^{k_\nu} (c_{\nu i_\alpha} - \bar{c}_\nu)^2}{\sum_{i=1}^{N_\nu} (c_{\nu i} - \bar{c}_\nu)^2} &= \frac{\sum_{\alpha=1}^{k_\nu} \left[\sum_{g=1}^m \lambda_g (a_{\nu i_\alpha}^g - \bar{a}_\nu^g) \right]^2}{\sum_{i=1}^{N_\nu} \left[\sum_{g=1}^m \lambda_g (a_{\nu i_\alpha}^g - \bar{a}_\nu^g) \right]^2} \\
&\leq \frac{\sum_{\alpha=1}^{k_\nu} m \sum_{g=1}^m \lambda_g^2 (a_{\nu i_\alpha}^g - \bar{a}_\nu^g)^2}{\epsilon \max_{1 \leq g \leq m} \left[\lambda_g^2 \sum_{i=1}^{N_\nu} (a_{\nu i}^g - \bar{a}_\nu^g)^2 \right]} \leq \frac{m}{\epsilon} \sum_{g=1}^{m} \frac{\sum_{\alpha=1}^{k_\nu} (a_{\nu i_\alpha}^g - \bar{a}_\nu^g)^2}{\sum_{i=1}^{N_\nu} (a_{\nu i}^g - \bar{a}_\nu^g)^2}.
\end{aligned}$$

Now it suffices to note that, according to our suppositions, the values $a_{\nu i}^g$ fulfil the condition (4.10). □

Remark 7.1. The condition (7.2) simply means that all multiple correlation coefficients are uniformly bounded from 1.

REFERENCES

Cramér H. (1946). *Mathematical Methods of Statistics.* Princeton University Press, Princeton.

Doob J. L. (1953). *Stochastic Processes.* John Wiley & Sons, New York.

Dwass M. (1953). On the asymptotic normality of certain rank order statistics. *Ann. Math. Stat. 24*, 303–306.

Dwass M. (1955). On the asymptotic normality of some statistics used in non–parametric tests. *Ann. Math. Stat. 26*, 334–339.

Dwass M. (1956). The large sample power of rank order tests in the two–sample problem. *Ann. Math. Stat. 27*, 352–374.

Erdős P. and Rényi A. (1959). On a central limit theorem for samples from a finite population. *Publ. Math. Inst. Hungar. Acad. Sci. 4*, 49–61.

Gnedenko B. V. and Kolmogorov A. N. (1954). *Limit Distributions for Sums of Independent Random Variables.* Addison–Wesley, Cambridge.

Hájek J. (1960). Limiting distributions in simple random sampling from a finite population. *Publ. Math. Inst. Hungar. Acad. Sci. 5*, 361–374.

Hoeffding W. (1951). A combinatorial central limit theorem. *Ann. Math. Stat. 22*, 169–566.

Hoeffding W. (1953). On the distribution of the expected values of the order statistics. *Ann. Math. Stat. 24*, 93–100.

Madow W. G. (1948). On the limiting distributions of estimates based on samples from finite universe. *Ann. Math. Stat. 19*, 535–545.

Motoo M. (1957). On the Hoeffding's combinatorial central limit theorem. *Ann. Inst. Stat. Math. 8*, 145–154.

Noether G. E. (1949). On a theorem by Wald and Wolfowitz. *Ann. Math. Stat. 20*, 455–558.

Wald A. and Wolfowitz J. (1944). Statistical tests based on permutations of the observations. *Ann. Math. Stat. 15*, 358–372.

CHAPTER 18

On Linear Estimation Theory for an Infinite Number of Observations

Teor. Verojatnost. i Primenen. 6 (1961), 181–193.
Reproduced by permission of Hájek family: Al. Hájková, H. Kazárová and Al. Slámová.

1 INTRODUCTION

We shall deal with linear estimation of parameters $\theta = \sum_{\nu=1}^{m} c_\nu \alpha_\nu$, or random variables y with $\mathsf{E}\, y = \sum_{\nu=1}^{m} c_\nu \alpha_\nu$, on the basis of an arbitrary infinite family of observations $\{x_t,\, t \in T\}$. It will be shown that all classical theorems on least-square estimates and Markov estimates remain true for *any* family of observations $\{x_t,\, t \in T\}$ with *any* covariance structure and with *any* functions $\varphi_{\nu t}$ in the linear hypothesis (model (2.1)). Our proofs use elementary facts on linear functionals in a Hilbert space and the concept of sufficient statistics.

Under the assumption of normality non-correlation is equivalent to independence, the best linear unbiased estimates are equivalent to the best unbiased estimates, and wide-sense sufficient statistics (see Section 3) are equivalent to strict-sense sufficient statistics. The Gaussian approach has been applied in papers Grenander (1949) and Hájek (1960 a), which deal with the single parameter case ($m = 1$). In this paper, however, we shall not assume normality. Using the same arguments as in Hájek (1960 a), we could, however, easily derive the strict-sense formulations of all the theorems. The assumption of normality is indispensable in hypothesis-testing problems. Our attention will be focused, however, on estimation problems. Let us only point out that the existence of a sufficient finite set of random variables makes it possible to carry the problem of hypothesis testing back to the classical one.

As an illustration let us look at the problem discussed by L. A. Zadeh and

Collected Works of Jaroslav Hájek – With Commentary
Edited by M. Hušková, R. Beran and V. Dupač
Published in 1998 by John Wiley & Sons Ltd.

J. R. Ragazzini in their paper (1950). They considered a time series

$$x(t) = M(t) + N(t) + \alpha_0 + \alpha_1 t + \cdots + \alpha_n t^n$$

consisting of a stationary random signal $M(t)$, a stationary random noise $N(t)$ and a deterministic signal of polynomial form, where $\alpha_0, \alpha_1, \ldots, \alpha_n$ are unknown constants. They supposed that the degree n of the polynomial were fixed and that $M(t)$ and $N(t)$ had zero mean values. They looked for a best linear unbiased estimate (i. e. prediction, or filtration) of the whole signal $y = M(t_0) + \alpha_0 + \alpha_1 t_0 + \cdots + \alpha_n t_0^n$ at a time point t_0, on the basis of $\{x(t),\, 0 \leq t \leq T\}$. If we replace $M(t) + N(t)$ by an arbitrary stochastic process, the interval $[0, T]$ by an arbitrary set T and the functions t^ν by arbitrary functions $\varphi_{\nu t} = \varphi_\nu(t)$, we arrive at the general problem that is going to be treated.

The Zadeh – Ragazzini model was generalized in Yaglom (1955), where the functions t^ν were replaced by arbitrary analytic functions. Very interesting asymptotic results for stationary processes and $\varphi_t \equiv 1$ are found in Grenander (1949) and Blum (1956). These results were generalized for $\varphi_t = c^{i\lambda t}$ in Chiang Tse-pei (1959).[1]

2 FORMULATION OF THE PROBLEM AND THE CLASS OF LINEAR ESTIMATES

Let $\alpha = (\alpha_1, \ldots, \alpha_m)$ run through an m-dimensional vector space A_m, and let $\varphi_{1t}, \ldots, \varphi_{mt}$ be m known functions of $t \in T$, where T is an arbitrary set. Consider a family of random variables $\{x_t,\, t \in T\}$ and a system of probability distributions $\mathsf{P}_\alpha(\cdot)$, $\alpha \in A_m$, of this family with the following first and second moments:

$$\mathsf{E}_\alpha\, x_t = \sum_{\nu=1}^{m} \alpha_\nu\, \varphi_{\nu t} \quad (\alpha \in A_m,\ t \in T), \tag{2.1}$$

$$\mathsf{E}_\alpha\, x_t\, \overline{x}_s = \mathsf{E}_0\, x_t\, \overline{x}_s + \mathsf{E}_\alpha\, x_t\, \mathsf{E}_\alpha\, \overline{x}_s \quad (\alpha \in A_m;\ t,\ s \in T). \tag{2.2}$$

In the paper we shall denote mean value by indices of corresponding probability distributions, so that $\mathsf{E}_\alpha(\cdot) = \int(\cdot)\, \mathrm{d}\mathsf{P}_\alpha$, etc. The random variables x_t as well as the functions $\varphi_{\nu t}$ and parameters α_ν may either be real-valued or complex-valued; the notation will however refer to the complex case. Note that the covariance does not vary with $\alpha \in A_m$ and equals $\mathsf{E}_0\, x_t\, \overline{x}_s$, where

[1] A compact presentation of the problem with many new results may be found in the monograph Pugachev (1960).

$0 = (0, \ldots, 0)$. Briefly speaking, we are dealing with a linear hypothesis (model) with m parameter $\alpha_1, \ldots, \alpha_m$.

We shall look for best linear unbiased estimates for linear functionals θ defined on $\alpha \in A_m$:

$$\theta = \theta(\alpha) = \sum_{\nu=1}^{m} c_\nu \, \alpha_\nu \tag{2.3}$$

where $c_1, \ldots, c_m$ are either real or complex numbers. All possible linear functionals θ given by all possible $c = (c_1, \ldots, c_m)$ also form an m-dimensional vector space, say A_m^+. The linear functionals θ will also be called *parameters*.

More generally, we will try to estimate a random variable y such that for $\theta \in A_m^+$

$$\mathsf{E}_\alpha \, y = \theta(\alpha), \quad (\alpha \in A_m) \tag{2.4}$$

and that the variance of y and the covariances of y and x_t, $t \in T$, are independent of $\alpha \in A_m$. A random variable v will be called an unbiased estimate of y if

$$\mathsf{E}_\alpha \, v = \theta(\alpha), \quad (\alpha \in A_m) \tag{2.5}$$

and a best unbiased estimate of y if, under the restriction of (2.5), $\mathsf{E}_0|v - y|^2$ will be minimum. Best linear unbiased estimates of y are sometimes called unbiased projections of y or unbiased predictions of y.

We shall choose the estimates in the class of statistics which either are finite linear combinations of random variables x_t, $t \in T$, or are in-mean-limits of such combinations. Of course, all in-mean-limits have to be well defined with respect to any probability distribution $P_\alpha(\cdot)$, $\alpha \in A_m$. Let us denote the set of all finite linear combinations $x = \sum c_j \, x_{t_j}$ by M and try to find a probability distribution $\mathsf{P}_1(\cdot)$ which dominates over all the distributions $\mathsf{P}_\alpha(\cdot)$ in the following sense. For any $\alpha \in A_m$ there exists a finite constant K_α such that

$$\mathsf{E}_\alpha|x|^2 \leq K_\alpha \, E_1|x|^2 \quad (x \in M) \tag{2.6}$$

where K_α is independent of $x \in M$. It is obvious that the in-mean-limits with respect to $\mathsf{P}_1(\cdot)$ will then be well defined via each $\mathsf{P}_\alpha(\cdot)$, $\alpha \in A_m$.

Such a distribution $\mathsf{P}_1(\cdot)$ may be constructed, for example, as follows.

Put

$$\mathsf{P}_{\nu 1} = \mathsf{P}_{(\underbrace{0, \ldots, 0}_{\nu-1}, 1, \underbrace{0, \ldots, 0}_{m-\nu})} \tag{2.7}$$

and

$$\mathsf{P}_1(\cdot) = \frac{1}{m} \sum_{\nu=1}^{m} P_{\nu 1}(\cdot). \tag{2.8}$$

Lemma 2.1. *It is true that*

$$\mathsf{E}_\alpha|x|^2 \leq \left(1 + m\sum_{\nu=1}^{m}|\alpha_\nu|^2\right)\mathsf{E}_1|x|^2 \quad (x \in M) \tag{2.9}$$

where $\mathsf{P}_1(\cdot)$ *is defined by* (2.8).

Proof. The relation (2.2) may be easily extended to M:

$$\mathsf{E}_\alpha\, xy = \mathsf{E}_0\, x\overline{y} + \mathsf{E}_\alpha x\, \mathsf{E}_\alpha \overline{y} \quad (x,\, y \in M). \tag{2.10}$$

From (2.7) and (2.1) it follows that

$$\mathsf{E}_\alpha x = \sum \alpha_\nu\, \mathsf{E}_{\nu 1} x \quad (x \in M). \tag{2.11}$$

Making use of (2.10) and (2.11), we have

$$\mathsf{E}_\alpha|x|^2 = \mathsf{E}_0|x|^2 + \left|\sum_{\nu=1}^{m}\alpha_\nu\, \mathsf{E}_{\nu 1}x\right|^2 \leq \mathsf{E}_0|x|^2 + \sum_{\nu=1}^{m}|\alpha_\nu|^2\sum_{\nu=1}^{m}|\mathsf{E}_{\nu_1}x|^2. \tag{2.12}$$

On the other hand, in view of (2.8), we obtain (2.13)

$$\mathsf{E}_1 x\overline{y} = \mathsf{E}_0 x\overline{y} + \frac{1}{m}\sum_{\nu=1}^{m}\mathsf{E}_{\nu 1}x\, \mathsf{E}_{\nu 1}\overline{y} \quad (x,\, y \in M), \tag{2.13}$$

and especially

$$\mathsf{E}_1|x|^2 = \mathsf{E}_0|x|^2 + \frac{1}{m}\sum_{\nu=1}^{m}|\mathsf{E}_{\nu 1}x|^2 \quad (x \in M). \tag{2.14}$$

Now, (2.9) is a simple consequence of (2.12) and (2.14). □

Let us denote the set of in-mean-limits via $\mathsf{P}_1(\cdot)$ by $\mathcal{M}$. The set $\mathcal{M}$, in which random variables with $\mathsf{P}_1(x = y) = 1$ are considered as identical and an inner product is introduced, $[x, y] = \mathsf{E}_1 xy$ $(x,\, y \in \mathcal{M})$, obviously is a Hilbert space, which in what follows will be denoted by $\mathcal{M}_1$.

Lemma 2.2. *The relations* (2.9), (2.10), (2.11) *and* (2.13) *extend to the whole set* $\mathcal{M}_1$.

Proof. This is obvious. □

The set $\mathcal{M}_1$ will represent the class of admissible linear estimates.

3 THEOREMS ON EXISTENCE OF BEST LINEAR UNBIASED ESTIMATES

Let us first classify the vectors $\alpha \in A_m$ and parameters $\theta \in A_m^+$.

We denote by A_p the p-dimensional linear subspace of A_m consisting of the αs for which

$$\sum_{\nu=1}^{m} \alpha_\nu \varphi_{\nu t} = 0 \quad (t \in T). \tag{3.1}$$

In view of (2.1), (3.1) is equivalent to

$$\mathsf{E}_\alpha x = 0 \quad (x \in \mathcal{M}_1). \tag{3.2}$$

We denote by A_q the q-dimensional subspace of A_m consisting of the αs, for which there exists a finite constant K_α such that

$$\mathsf{E}_\alpha |x|^2 \leq K_\alpha \, \mathsf{E}_0 |x|^2 \quad (x \in \mathcal{M}_1) \tag{3.3}$$

i.e. the square mean value is uniformly dominated by variance. Clearly $0 \leq p \leq q \leq m$, and $A_p \subset A_q$.

A parameter $\theta \in A_m^+$, for which a linear unbiased estimate exists, will be called *estimable*. A parameter $\theta \in A_m^+$, for which a linear unbiased estimate with variance zero exists, will be called *exactly estimable*.

The following theorem shows the close relation of estimability and exact estimability of the θs to the subspaces A_p and A_q.

Theorem 3.1. *A parameter $\theta \in A_m^+$ is estimable if, and only if, $\theta(\alpha) = 0$ for $\alpha \in A_p$, i. e. for all the αs satisfying (3.2). The estimable θs form a linear subspace A_{m-p}^+ of dimension $m - p$.*

A parameter $\theta \in A_{m-p}^+$ is exactly estimable if, and only if, $\theta(\alpha) = 0$ for $\alpha \in A_q$, i. e. for all the αs satisfying (3.3). The exactly estimable θs form a linear subspace A_{m-q}^+ of dimension $m - q$.

Each estimable parameter θ possesses a unique best linear unbiased estimate. The class of best linear unbiased estimates forms an $(m-p)$-dimensional linear subspace of $\mathcal{M}_1$, say W, and this subspace consists of random variables $w_\alpha \in \mathcal{M}_1$, $\alpha \in A_m$, uniquely determined by the relations

$$\mathsf{E}_\alpha x = \mathsf{E}_1 \, x \, \overline{w}_\alpha \quad (x \in \mathcal{M}_1). \tag{3.4}$$

The relation between a parameter $\theta \in A_{m-p}^+$ and its best linear unbiased estimate v^θ is given by the equation

$$\theta = \sum_{\nu=1}^{m} \alpha_\nu \, \mathsf{E}_1 (v^\theta \, \overline{w}_{\nu 1}), \tag{3.5}$$

where $w_{\nu 1}$, $1 \leq \nu \leq m$, are uniquely determined by the relation

$$\mathsf{E}_{\nu 1} x = \mathsf{E}_{1x} \overline{w}_{\nu 1} \quad (x \in \mathcal{M}_1). \tag{3.6}$$

Proof. In view of (2.9), we have

$$|\mathsf{E}_\alpha x|^2 \leq \mathsf{E}_\alpha |x|^2 \leq \left(1 + m \sum_{\nu=1}^{m} |\alpha_\nu|^2\right) \mathsf{E}_1 |x|^2 \tag{3.7}$$

so that $\mathsf{E}_\alpha x$ is a linear functional on $\mathcal{M}_1$. Consequently, as is well known, there exists a uniquely determined random variable $w_\alpha \in \mathcal{M}_1$ such that (3.4) holds true (recall that $\mathsf{E}_1 x\overline{y}$ is the inner product in $\mathcal{M}_1$).

The random variables w_α defined by (3.4) obviously form an $(m-p)$-dimensional subspace W on $\mathcal{M}_1$. Take a random variable $x \in \mathcal{M}_1$ and denote its projection on W by x'. We will show that

$$\mathsf{E}_\alpha x' = \mathsf{E}_\alpha x \quad (\alpha \in A_m) \tag{3.8}$$

and

$$\mathsf{E}_0 |x'|^2 = \mathsf{E}_0 |x|^2 - \mathsf{E}_0 |x - x'|^2. \tag{3.9}$$

In view of (3.4), (3.8) may be rewritten as

$$\mathsf{E}_1 x' \, \overline{w}_\alpha = \mathsf{E}_1 \, x \, \overline{w}_\alpha \tag{3.10}$$

where $w_\alpha \in W$. However, (3.10) is a direct consequence of the fact that x' is a projection of x on W. Further, from (2.13) and (3.8) it follows that

$$\begin{aligned} 0 &= \mathsf{E}_1 \{x'(\overline{x} - \overline{x}')\} = \mathsf{E}_0 \{x'(\overline{x} - \overline{x}')\} + \frac{1}{m} \sum_{\nu=1}^{m} \mathsf{E}_{\nu 1} \, x' \, \mathsf{E}_{\nu 1}(\overline{x} - \overline{x}') \\ &= \mathsf{E}_0 \{x'(\overline{x} - \overline{x}')\}, \end{aligned}$$

which is equivalent to (3.9). The relations (3.8) and (3.9) show that every best linear unbiased estimate must belong to W.

Now (3.5), where $\theta(\alpha) = \mathsf{E}_\alpha \, v^\theta$, is a simple consequence of (2.11) and (3.6). If we show that $v^\theta \in W$ is a unique unbiased estimate of θ given by (3.5), the assertion that W coincides with the class of all best linear unbiased estimates will be proved. It suffices, however, to note that the random variables $w_{11}, \ldots, w_{m1}$ form a complete basis in W so that any element $w \in W$ is uniquely determined by Fourier coefficients $\mathsf{E}_1 \, w \, \overline{w}_{\nu 1}$, $1 \leq \nu \leq m$.

The assertion concerning estimability will be proved as follows. If $\theta(\alpha) \neq 0$ for an α satisfying (3.2), then, clearly, no $x \in \mathcal{M}_1$ is an unbiased estimate of θ. Conversely, if $\theta(\alpha) = 0$ for all αs satisfying (3.2), then $\theta(\alpha) \neq \theta(\beta)$ implies $w_\alpha \neq w_\beta$, so that θ may also be interpreted as a functional defined on W.

This functional is obviously linear, so that there exists an element v^θ such that

$$\theta(w_\alpha) = \mathsf{E}_1\, v^\theta\, \overline{w}_\alpha,$$

which in conjunction with $\theta(\alpha) = \theta(w_\alpha)$ and (3.4) gives

$$\theta(\alpha) = \mathsf{E}_\alpha\, v^\theta,$$

so that θ is estimable.

In order to prove the last assertion concerning exact estimability, let us note that (2.13) and (3.4) imply

$$\begin{aligned} \mathsf{E}_0(x\,\overline{w}_\alpha) &= \mathsf{E}_1(x\,\overline{w}_\alpha) - \frac{1}{m}\sum_{\nu=1} \mathsf{E}_{\nu 1}\, x\, \mathsf{E}_{\nu 1}\, w_\alpha \\ &= \mathsf{E}_\alpha x - \frac{1}{m}\sum_{\nu=1} \mathsf{E}_{\nu 1}\, x\, \mathsf{E}_{\nu 1}\, \overline{w}_\alpha, \quad (\alpha \in A_m). \end{aligned} \tag{3.11}$$

By putting

$$\beta = \alpha - \frac{1}{m}\left(\mathsf{E}_{11}\,\overline{w}_\alpha, \ldots, \mathsf{E}_{m1}\,\overline{w}_\alpha\right), \quad (\alpha \in A_m) \tag{3.12}$$

and making use of (2.11), we can rewrite (3.11) as follows:

$$\mathsf{E}_0\, x\, \overline{w}_\alpha = \mathsf{E}_\beta\, x \quad (\alpha \in A_m). \tag{3.13}$$

Now, if $w_\alpha = v^\theta$ is the best linear unbiased estimate of θ, then (3.13) implies that

$$\mathsf{E}_0|v^\theta|^2 = \mathsf{E}_\beta\, v^\theta = \theta(\beta). \tag{3.14}$$

From (3.13) it follows that β satisfies the condition (3.3):

$$|\mathsf{E}_\beta x|^2 \le \mathsf{E}_0|w_\alpha|^2\, \mathsf{E}_0|x|^2 \quad (x \in \mathcal{M}_1). \tag{3.15}$$

Consequently, if $\theta(\beta) = 0$ for every β satisfying (3.3), then, in view of (3.14), $\mathsf{E}_0|v^\theta|^2 = 0$. Conversely, if $\theta(\beta) = \mathsf{E}_\beta\, v^\theta \neq 0$ and the condition (3.3) is satisfied for β, then $K_\beta\, \mathsf{E}_0|v^\theta|^2 \ge \mathsf{E}_\beta|v^\theta|^2 \ge |\mathsf{E}_\beta\, v^\theta|^2 > 0$, so that θ cannot be exactly estimable.

The assertions concerning the dimension of subspaces of estimable and exactly estimable parameters θ are clear. □

Remark 3.1. Theorem 3.1 implies that any best linear unbiased estimate is a linear combination of a system of random variables $v_1, \ldots, v_\ell$ complete in W. As the dimension of W equals $m - p$, the minimum number of random variables $v_1, \ldots, v_\ell$ is $\ell = m-p$. We can say that any such vector $v_1, \ldots, v_{m-p}$ is sufficient for linear estimation (or sufficient in the wide sense).

Remark 3.2. If the functions $\varphi_{\nu t}$ are linearly independent, then a best linear unbiased estimate exists for any parameter $\theta \in A_m^+$. In particular there exist best linear estimates $\hat{\alpha}_1, \ldots, \hat{\alpha}_m$ for $\alpha_1, \ldots, \alpha_m$. If $v^1, \ldots, v^m$ are other unbiased estimates of $\alpha_1, \ldots, \alpha_m$, then, for any constants $c_1, \ldots, c_m$, we have

$$\mathsf{E}_0 \left| \sum_{\nu=1}^{m} c_\nu \, \hat{\alpha}_\nu \right|^2 \leq \mathsf{E}_0 \left| \sum_{\nu=1}^{m} c_\nu \, v^\nu \right|^2 \tag{3.16}$$

because $\sum c_\nu \, \hat{\alpha}_\nu$ is the best linear estimate of $\sum c_\nu \, \alpha_\nu$. Consequently, the random vector $(\hat{\alpha_1}, \ldots, \hat{\alpha}_m)$ is a Markov estimate for the vector $(\alpha_1, \ldots, \alpha_m)$. Thus, for example, the assertion of Theorem 1 of paper Chiang Tse-pei (1959) remains true even if we drop the assumption that the functions $\varphi_\nu(t) = \varphi_{\nu t}$ are continuous.

The following theorem shows that the case of estimating a random variable needs no substantially new theory.

Theorem 3.2. *Suppose that* $\mathsf{E}_\alpha y = \theta(\alpha)$ *is estimable and denote the projection of* y *on* $\mathcal{M}_1$ *by* y_1. *Then*

$$\theta_1(\alpha) = \mathsf{E}_\alpha (y - y_1) \tag{3.17}$$

is also estimable and the best linear unbiased estimate of y, *say* $\hat{y}$, *exists, is unique, and equals*

$$\hat{y} = y_1 + \hat{\theta}_1, \tag{3.18}$$

where $\hat{\theta}_1$ *is the best linear unbiased estimate of* θ_1. *It holds that*

$$\mathsf{E}_0 |y - \hat{y}|^2 = \mathsf{E}_1 |y - y_1|^2 + \mathsf{E}_1 |\hat{\theta}_1|^2. \tag{3.19}$$

Proof. If v is an unbiased estimate of y, then, since y_1 is a projection of y,

$$\mathsf{E}_0 |y - v| = \mathsf{E}_1 |y - v|^2 = \mathsf{E}_1 |y - y_1|^2 + \mathsf{E}_1 |v - y_1|^2.$$

Now, $v - y_1$ is an unbiased estimate of θ_1, because $\mathsf{E}_\alpha v = \mathsf{E}_\alpha y$. Hence, $\mathsf{E}_1 |v - y_1|^2$ will be minimum if $v - y_1 = \hat{\theta}_1$, where $\hat{\theta}_1$ is the best linear unbiased estimate of θ_1. θ_1 is estimable, since $\mathsf{E}_\alpha y$ is assumed to be estimable and $\mathsf{E}_\alpha y_1$ is unbiasedly estimable simply by $y_1 \in \mathcal{M}_1$. □

4 REGULAR MODELS

If the condition (3.3) is fulfilled for all $\alpha \in A_m$, we shall speak about a regular model. In regular models no parameter $\theta \not\equiv 0$ is exactly estimable. In some important regular linear models explicit solutions are available.

As is seen from (3.3), in a regular model the P_0-distribution may serve as the dominating distribution. The set of in-mean-limits via $\mathsf{P}_0(\cdot)$ coincides with the set $\mathcal{M}$ introduced in Section 2. The set $\mathcal{M}$, in which random variables with $\mathsf{P}_0(x = y) = 1$ are considered as identical, and the following inner product is introduced

$$[x, y] = \mathsf{E}_0\, xy, \tag{4.1}$$

is obviously a Hilbert space. In what follows, the set $\mathcal{M}$ will be denoted by $\mathcal{M}_0$. $\mathcal{M}_0$ consists of the same elements as $\mathcal{M}_1$, but has a different inner product. It is again clear that the mean values $\mathsf{E}_\alpha x$ are linear functionals on $\mathcal{M}_0$, so that for any $\alpha \in A_m$ there exists an element $v_\alpha \in \mathcal{M}_0$ such that

$$\mathsf{E}_\alpha x = \mathsf{E}_0 x\, \overline{v}_\alpha \quad (x \in \mathcal{M}_0) \tag{4.2}$$

and that the class $V = \{v_\alpha,\ \alpha \in A_m\}$ represents the class of all best linear unbiased estimates of estimable parameters $\theta \in A^+_{m-p}$. Clearly $V = W$ but, generally, $v_\alpha \neq w_\alpha$, $\alpha \in A_m$.

Theorem 4.1. *If the model is regular, then a random variable $v^\theta \in V$ determined by (4.2) is the best linear estimate for*

$$\theta = \sum_{\nu=1}^{m} \alpha_\nu\, \mathsf{E}_0(v^\theta\, \overline{v}_{\nu 1}) \tag{4.3}$$

where $v_{\nu 1}$, $1 \leq \nu \leq m$, are determined by

$$\mathsf{E}_{\nu 1} x = \mathsf{E}_0\, x\, \overline{v}_{\nu 1} \quad (x \in \mathcal{M}_0). \tag{4.4}$$

The best linear unbiased estimate of the random variable y, considered in Theorem 3.2, equals

$$\hat{y} = y_0 + \hat{\theta}_0 \tag{4.5}$$

where y_0 is the projection of y on $\mathcal{M}_0$ and $\hat{\theta}_0$ is the best linear estimate of

$$\theta_0(\alpha) = \mathsf{E}_\alpha(y - y_0).$$

The following equation holds:

$$\mathsf{E}_0|y - \hat{y}|^2 = \mathsf{E}_0|y - y_0|^2 + \mathsf{E}_0|\theta_0|^2. \tag{4.6}$$

Proof. This is obvious. □

Remark 4.1. The random variable $v_{\nu 1}$ is also uniquely determined by the relation

$$\varphi_{\nu t} = \mathsf{E}_0\, x_t\, \overline{v_{\nu 1}} \quad (t \in T) \tag{4.7}$$

because $\varphi_{\nu t} = \mathsf{E}_{\nu 1}\, x_t$, and if the relation (4.4) is fulfilled for x_t, $t \in T$, then it is satisfied by any $x \in \mathcal{M}_0$.

Remark 4.2. The formulas (4.5) and (4.6) have been derived by A. M. Yaglom (1955) for $m = 1$, $\varphi_t = 1$, and stationary processes.

Theorem 4.2. *Suppose that no $\alpha \neq (0, \ldots, 0)$ fulfils the relation (3.1) i.e. that the functions $\varphi_{\nu t}$, $1 \leq \nu \leq m$, are linearly independent. Introduce the matrix*

$$D = \|\mathsf{E}_0\, v_{\nu 1}\, \overline{v}_{\mu 1}\|_{\nu,\mu=1}^{m} \tag{4.8}$$

where $v_{\nu 1}$ are determined by (4.4) (or by (4.7)), and denote the best linear estimates of $\alpha_1, \ldots, \alpha_m$ by $\hat{\alpha}_1, \ldots, \hat{\alpha}_m$. Then

$$\begin{bmatrix} \hat{\alpha}_1 \\ \vdots \\ \hat{\alpha}_m \end{bmatrix} = D^{-1} \begin{bmatrix} v_{11} \\ \vdots \\ v_{m1} \end{bmatrix} \tag{4.9}$$

and

$$\|\mathsf{E}_0\, \hat{\alpha}_\nu\, \overline{\hat{\alpha}}_\mu\|_{\nu,\mu=1}^{m} = D^{-1}. \tag{4.10}$$

Best linear estimates of parameter $\theta = \sum c_\nu\, \alpha_\nu$ equal $\hat{\theta} = \sum c_\nu\, \hat{\alpha}_\nu$.

Proof. The matrix D is regular because, in view of (4.7), linear independence of $\varphi_{\nu t}$ is equivalent to linear independence of $v_{\nu 1}$, $1 \leq \nu \leq m$. Thus in order to prove (4.9) it suffices to show that

$$v_{\nu 1} = \sum_{\mu=1}^{m} \hat{\alpha}_\mu\, \mathsf{E}_0(v_{\nu 1}\, \overline{v}_{\mu 1}) \quad (1 \leq \nu \leq m) \tag{4.11}$$

i. e. that the right side of (4.11) fulfils (4.4). When $x = v_{j1}$, first of all,

$$\mathsf{E}_{\nu 1}\, v_{j1} = \mathsf{E}_0(v_{j1}\, \overline{v}_{\nu 1}) \tag{4.12}$$

and, secondly,

$$\mathsf{E}_0 \left(v_{j1} \sum_{\mu=1}^{m} \overline{\hat{\alpha}}_\mu\, \overline{\mathsf{E}_0(v_{\nu 1}\, \overline{v}_{\mu 1})} \right) = \sum_{\mu=1}^{m} \mathsf{E}_0 \left(v_{j1}\, \overline{\hat{\alpha}}_\mu \right) \mathsf{E}_0 \left(\overline{v}_{\nu 1}\, v_{\mu 1} \right) \tag{4.13}$$

$$= \sum_{\mu=1}^{m} \overline{\mathsf{E}_{j1}\, \hat{\alpha}_\mu}\, \mathsf{E}_0(\overline{v}_{\nu 1}\, v_{\mu 1}) = \mathsf{E}_0(\overline{v}_{\nu 1}\, v_{j1}),$$

where we have made use of the obvious relation

$$\mathsf{E}_{j1}\, \hat{\alpha}_\mu = \begin{cases} 1 & \text{if} \quad j = \mu, \\ 0 & \text{if} \quad j \neq \mu. \end{cases} \tag{4.14}$$

By comparing (4.14) and (4.13), we see that (4.4) is fulfilled for $x = v_{j1}$, $1 \leq \nu \leq m$ and hence by an arbitrary $x \in V$. Now, from (4.4) it is easily seen that

$\mathsf{E}_{\nu 1}x = \mathsf{E}_{\nu 1}x'$, where x' is the projection of x on $\mathcal{M}_0$. Hence (4.4) is fulfilled for any $x \in \mathcal{M}_0$, and (4.9) is thus proved.

Finally, (4.10) is a simple consequence of (4.9), and $\sum c_\nu \hat{\alpha}_\nu$ is the best unbiased estimate of $\sum c_\nu \alpha_\nu$, because it is unbiased and belongs to V, where each unbiased estimate is unique. □

Remark 4.3. The vector $v_{11}, \ldots, v_{m1}$ is sufficient in the sense of Remark 3.1.

5 STRICT SENSE (GAUSSIAN) INTERPRETATION

If the random variables x_t, as well as the functions $\varphi_{\nu t}$ and components α_ν are real-valued, then the first and second moments (2.1) and (2.2) uniquely determine a Gaussian distribution system. Using the same reasoning as in the proof of Theorem 3.1 in Hájek (1960 a), we can show that the vector $v_1, \ldots, v_{m-p}$ mentioned in Remark 3.1 is sufficient (in the strict sense) for the Gaussian distribution system considered. The model is regular if, and only if, all Gaussian distributions in the system are equivalent as measures. A parameter θ is exactly estimable if, and only if, $\theta(\alpha) = 0$ for all αs such that $\mathsf{P}_\alpha(\cdot)$ is equivalent to $\mathsf{P}_0(\cdot)$.

6 EXAMPLES

Now we shall present three examples, which illustrate the regularity conditions and show an explicit solution for $v_{11}, \ldots, v_{m1}$ in the case where $\{x_t,\, t \in T\}$ is a stationary process of the autoregressive type.

Example 6.1. Let $\{x_t,\, t = \ldots -1, 0, 1, \ldots\}$ be a discrete stationary process with spectral density $f(\lambda)$ satisfying the following condition:

$$0 < \varepsilon < f(\lambda) < K < \infty \quad (-\pi \le \lambda \le \pi). \tag{6.1}$$

Let the α-mean value of this process be given by (2.1), where T is the set of all integers. We will show that the model is regular if, and only if,

$$\sum_{t=-\infty}^{\infty} |\varphi_{\nu t}|^2 < \infty \quad (1 \le \nu \le m). \tag{6.2}$$

Making use of the well-known isometric correspondence $x_t \leftrightarrow e^{it\lambda}$, and, generally, $x \leftrightarrow g(\lambda)$, where $x \in \mathcal{M}_0$ and $g(\lambda)$ is a quadratically integrable function with weight $f(\lambda)$,

$$\int_{-\pi}^{\pi} |g(\lambda)|^2 f(\lambda)\,\mathrm{d}\lambda < \infty, \tag{6.3}$$

we can write (4.7) as follows:

$$\varphi_{\nu t} = \int_{-\infty}^{\infty} e^{it\lambda}\overline{g_\nu(\lambda)}\, f(\lambda)\,\mathrm{d}\lambda \quad (1 \le \nu \le m). \tag{6.4}$$

If $\varphi_{\nu t}$ can be presented in this way, then the $\varphi_{\nu t}$ represent the Fourier coefficients of a quadratically integrable function, namely, of the function $g_\nu(\lambda)\, f(\lambda)$ (recall that $f(\lambda)$ is bounded and $g_\nu(\lambda)$ fulfils (6.3)). Consequently, (6.2) is satisfied. Conversely, if (6.2) is satisfied, then there exists a quadratically integrable function $h_\nu(\lambda)$ such that

$$\varphi_{\nu t} = \int_{-\pi}^{\pi} e^{it\lambda}\,\overline{h_\nu(\lambda)}\,\mathrm{d}\lambda \quad (\lambda \le \nu \le m). \tag{6.5}$$

By putting

$$g_\nu(\lambda) = \frac{h_\nu(\lambda)}{f(\lambda)} \tag{6.6}$$

we can rewrite (6.5) in the form (6.4), where $g_\nu(\lambda)$ is quadratically integrable, because $f(\lambda)$ is bounded from zero. Thus, we have proved that when (6.2) is satisfied, it is possible to present $\varphi_{\nu t}$ in the form (6.4) with $g_\nu(\lambda)$ fulfilling (6.3) and $f(\lambda)$ fulfilling (6.1).

Now, (6.4) is equivalent to

$$\mathsf{E}_{\nu 1}\, x_t = \mathsf{E}_0\, x_t\, \overline{v}_{\nu 1} \quad (1 \le \nu \le m) \tag{6.7}$$

which easily extends to (4.4). However, in view of (2.11), (4.4) is equivalent to (3.3), so that our assertion is completely proved.

Example 6.2. Let $\{x_t,\ A \le t \le B\}$ be a stationary process with continuous time-parameter t, $-\infty \le A \le \infty$. Suppose that the spectral density is rational, i. e.

$$f(\lambda) = \left| \frac{\sum_{k=1}^{r} b_k (i\lambda)^k}{\sum_{k=1}^{n} a_k (i\lambda)^k} \right|^2, \tag{6.8}$$

where $r < n$. The results derived in Hájek (1960 b) imply that the model is regular if, and only if, the functions $\varphi_{\nu t}$ have derivatives up to the order $m-r$, and then the $(m-r)$th derivative $\varphi_{\nu t}^{(m-r)}$ is quadratically integrable:

$$\int_A^B \left|\varphi_{\nu t}^{(m-r)}\right|^2 \mathrm{d}t < \infty. \tag{6.9}$$

Example 6.3. Let $\{x_t,\ 0 \le t \le T\}$ be a stationary process with continuous time-parameter t. Suppose that the spectral density has the form

$$f(\lambda) = \frac{1}{2\pi} \frac{1}{\left|\sum_{k=0}^{n} a_k (i\lambda)^k\right|^2}, \tag{6.10}$$

where $a_0, \ldots, a_n$ are real and all the roots of the equation $\sum_{k=0}^{n} a_k \lambda^k = 0$ have negative real parts. Introduce the following differential operators L and L^*:

$$(L\varphi)_t = \sum_{k=0}^{n} a_k \varphi_t^{(k)}, \tag{6.11}$$

$$(L^*\varphi)_t = \sum_{k=0}^{n} (-1)^k a_k \varphi_t^{(k)} \tag{6.12}$$

and the following inner product:

$$(\psi, \varphi)_L = \int_0^T (L\psi)_t \overline{(L\varphi)_t} \,\mathrm{d}t + 2 \sum_{j+k \text{ odd},\, k>j}^{n} \sum^{n} a_k a_j \sum_{\nu=0}^{k-j-1} (-1)^\nu \psi_0^{(k-1-\nu)} \overline{\varphi_0}^{(j+\nu)} \tag{6.13}$$

defined on functions φ_t, $0 \le t \le T$, possessing quadratically integrable derivatives up to the order n.

In Hájek (1960) it is shown that the solution $v_{\nu 1}$ of equation (4.4) in our case is

$$v_{\nu 1} = (x, \varphi_\nu)_L \tag{6.14}$$

where $\int_0^T \overline{(L\varphi)_t}\,(Lx)_t \,\mathrm{d}t$ should be interpreted as $\int_0^T \overline{(L\varphi)_t} \,\mathrm{d}\left[L \int_0^t x_s \,\mathrm{d}s\right]$. Note that $L \int_0^t x_s \,\mathrm{d}s$ is a process with uncorrelated increments. Furthermore,

$$\mathsf{E}_0\, v_{\nu 1}\, \overline{v}_{\mu 1} = (\varphi_\mu, \varphi_\nu)_L \quad (1 \le \nu,\ \mu \le m) \tag{6.15}$$

which gives the elements of the matrix D appearing in Theorem 4.2.

Let the functions $\varphi_{\nu t}$ $(1 \le \nu \le m)$ be linearly independent; let y be a random variable such that

$$\mathsf{E}_\alpha y = \sum_{\nu=1}^{m} c_\nu \alpha_\nu \tag{6.16}$$

and that its variance and covariances with x_t are independent of $\alpha \in A_m$. Now, the projection y_0 of y on $\mathcal{M}_0$ is determined by the relation

$$\mathsf{E}_0\, x_t\, \overline{y}_0 = \mathsf{E}_0\, x_t\, \overline{y} \quad (t \in T) \tag{6.17}$$

i. e., in view of (4.7), y_0 equals the random variable v_{01} corresponding to the function

$$\varphi_{0t} = \mathsf{E}_0\, x_t\, \overline{y}. \tag{6.18}$$

Consequently,

$$y_0 = (x, \varphi_0)_L \tag{6.19}$$

and, moreover,

$$\mathsf{E}_{\nu 1}\, y_0 = (\varphi_\nu, \varphi_0)_L \quad (1 \leq \nu \leq m) \tag{6.20}$$

so that

$$\theta_0(\alpha) = \mathsf{E}_\alpha(y - y_0) = \sum_{\nu=1}^{m} [c_\nu - (\varphi_\nu, \varphi_0)_L]\, \alpha_\nu.$$

In view of (4.5) the best linear unbiased estimate of y is

$$\hat{y} = (x, \varphi_0)_L + \sum_{\nu=1}^{m} [c_\nu - (\varphi_\nu, \varphi_0)_L]\, \hat{\alpha}_\nu \tag{6.21}$$

where $\hat{\alpha}_\nu$ are determined by (4.9), and $D = \|(\varphi_\mu \varphi_\nu)_L\|$. Further, in view of (4.6) and (4.10), we have

$$\mathsf{E}_0|y - \hat{y}_0|^2 = \mathsf{E}_0|y|^2 - (\varphi_0, \varphi_0)_L + (k_1, \ldots, k_m)\, D^{-1} \begin{bmatrix} k_1 \\ \vdots \\ k_m \end{bmatrix}, \tag{6.22}$$

where $k_\nu = c_\nu - (\varphi_\nu, \varphi_0)_L$, $1 \leq \nu \leq m$.

If the function φ_t possesses derivatives up to the order $2n$, we can write (6.13) in the following alternative form:

$$\begin{aligned} (\psi, \varphi)_L &= \int_0^T (L^* L\overline{\varphi})_t\, \psi_t\, \mathrm{d}t \qquad (6.23) \\ &\quad + \sum_{h=1}^{n} a_h \sum_{j+k=h-1} \left[(-1)^k\, (L\overline{\varphi})_T^k\, \psi_T^{(j)} + (-1)^j (L^*\overline{\varphi})^{(k)}\, \psi_0^{(j)}\right]. \end{aligned}$$

As it is shown in Hájek (1960), explicit results may be obtained also in some other simple cases.

In conclusion, I wish to thank Prof. A. N. Kolmogorov for very helpful criticism and suggestions.

REFERENCES

Blum, M. (1956). Generalization of the class of nonrandom inputs of the Zadeh–Ragazzini prediction model. *IRE Trans. Inform. Theory IT–2*, 76–81.

Chiang Tse-pei (1959). On the estimation of regression coefficients of a continuous parameter time series with a stationary residual. *Teor. Verojat-nost. i Primenen. IV*, 405–423.

Grenander, U. (1949). Stochastic processes and statistical inference. *Arkiv för Matematik 1*, 195–277.

Grenander, U., and Rosenblatt, M. (1956). *Statistical Analysis of Stationary Time Series*. Stockholm.

Hájek, J. (1960 a). On a simple linear model in Gaussian processes. *Transactions of the 2nd Prague Conference on Information Theory, Statistical Decision Functions, Random Processes*. (ed. J. Kožešník). Liblice, pp. 185–197.

Hájek, J. (1960 b). On linear statistical problems in stochastic processes. *Czechoslovak Math. J. 12*, 404–444.

Pugachev, V. S. (1960). *Theory of Random Functions and its Applications to Problems of Automatic Control.* (In Russian) Fizmatgiz, Moscow.

Yaglom, A. M. (1955). Extrapolation, interpolation and smoothing of stationary random processes with rational spectral density. (In Russian) *Trudy Moskov. Mat. Obšč. 4*, 333–374.

Zadeh, L. A., and Ragazzini, J. R. (1950). An extention of Wiener's theory of prediction. *Journ. Appl. Physics 21*, 645–655.

CHAPTER 19

Concerning Relative Accuracy of Stratified and Systematic Sampling in a Plane

Colloq. Math. 8 (1961), 133–134.
Reproduced by permission of the Institute of Mathematics of the Polish Academy of Sciences.

The conjecture P 254 formulated in Zubrzycki (1958) is disproved by the following theorem. We use Zubrzycki's notation.

Theorem. *Consider a plane stationary stochastic process $y(p)$, $p \in E_2$, with correlation function $f(p,q)$ given by*

$$f(p,q) = R[y(p),\, y(q)] = e^{-a|p-q|} \quad (a > 0), \tag{1.1}$$

where $R[y(p),\, y(q)]$ denotes the coefficient of correlation between $y(p)$ and $y(q)$ and $|p-q|$ denotes the distance of the points p and q, and consider two disjoint domains D and D' congruent by translation. Then for a sufficiently small positive number a the systematic sampling of two points from $D \cup D'$ is less efficient than the stratified sampling, i. e. we have

$$s_{\mathrm{sy}}^2 > s_{\mathrm{st}}^2. \tag{1.2}$$

Proof. In view of theorem 2 of Zubrzycki (1958), it suffices to show that for a sufficiently small a

$$e^{-a|p_0-p_0'|} > \frac{1}{|D|^2} \iint_D \iint_{D'} e^{-a|p-p'|} \mathrm{d}p\, \mathrm{d}p', \tag{1.3}$$

where p_0 and p_0' are centres of gravity of D and D' and $|D|$ denotes the area of D (which is the same as that of D').

Collected Works of Jaroslav Hájek – With Commentary
Edited by M. Hušková, R. Beran and V. Dupač
Published in 1998 by John Wiley & Sons Ltd.

Now

$$e^{-a|p-p'|} = 1 - a|p-p'| + o(a|p-p'|), \tag{1.4}$$

where $o(\cdot)$ denotes a quantity of smaller order; $|p-p'|$ being bounded, the last term on the right side of (1.4) becomes negligible if a approaches zero. Therefore (1.3) will be implied by

$$|p_0 - p_0'| < \frac{1}{|D|^2} \iint_D \iint_{D'} |p-p'| \, \mathrm{d}p \, \mathrm{d}p'. \tag{1.5}$$

Putting $p_0 = (x_0, y_0)$, $p_0' = (x_0', y_0')$, $p = (x, y)$, $p' = (x', y')$, we may rewrite (1.5) as follows:

$$\begin{aligned} &\sqrt{(x_0 - x_0')^2 + (y_0 - y_0')^2} \\ < \; &\frac{1}{|D|^2} \iint_D \iint_{D'} \sqrt{(x-x')^2 + (y-y')^2} \, \mathrm{d}x \mathrm{d}x' \mathrm{d}y \mathrm{d}y'. \end{aligned} \tag{1.6}$$

In view of the Cauchy inequality

$$\begin{aligned} &\sqrt{(x_1 - x_1')^2 + (y_1 - y_1')^2} + \sqrt{(x_2 - x_2')^2 + (y_2 - y_2')^2} \\ \geq \; &\sqrt{(x_1 + x_2 - x_1' - x_2')^2 + (y_1 + y_2 - y_1' - y_2')^2} \\ = \; &2\sqrt{\left(\frac{x_1 + x_2}{2} - \frac{x_1' + x_2'}{2}\right)^2 + \left(\frac{y_1 + y_2}{2} - \frac{y_1' + y_2'}{2}\right)^2} \end{aligned} \tag{1.7}$$

the function $\sqrt{(x-x')^2 + (y-y')^2}$ of arguments x, x', y, y' is convex. Thus (1.6) is a special case of the Jensen inequality (see Hardy, Littlewood and Pólya (1934)). The theorem is thereby proved. □

Comparing this theorem with Theorem 4 of Zubrzycki (1958), we can see that there is no simple relation between the efficiencies of systematic and stratified sampling in a plane, even if we confine ourselves to exponential correlation functions and such regions as squares or regular hexagons.

REFERENCES

Hardy G. H., Littlewood J. E. and Pólya G. (1934). *Inequalities.* Cambridge.

Zubrzycki S. (1958). Remarks on random, stratified and systematic sampling in a plane. *Colloquium Mathematicum 6*, 251–264.

CHAPTER 20

On Linear Statistical Problems in Stochastic Processes

Czechoslovak Math. J. 12 (1962), 404–444.
Reproduced by permission of Hájek family: Al. Hájková, H. Kazárová and Al. Slámová

A unified theoretical basis is developed for the solution of such problems as prediction and filtration (including the unbiased ones), estimation of regression parameters, establishing probability densities for Gaussian processes etc. The results are applied in deriving explicit solutions for stationary processes. The paper is a continuation of Hájek (1958 b), Hájek (1958 c), Hájek (1960) and Hájek (1961).

1 INTRODUCTION AND SUMMARY

Linear statistical problems have been examined in many papers. Most of them are referred to in the extensive monographs Pugachev (1960) and Middleton (1960). No doubt the topic attracted so much attention because of its great practical and theoretical interest. The aim of this chapter is to contribute to developing a unified theory and to provide explicit results for some particular classes of stationary processes.

In Section 2 we introduce closed linear manifolds generated by random variables x_t and covariances R_{ts}, respectively, and study their interplay. We also discuss the abstract feature of a linear problem and enumerate some possible applications. In Section 4 we analyze conditions under which the solution of a linear problem may be interpreted in terms of individual trajectories (i. e. not only as a limit in the mean). Section 5 contains explicit solutions for a finite segment of a stationary process with a rational spectral density. With respect to previous papers Zadeh and Ragazzini (1950), Doplh and Woodbury (1952) and Yaglom (1955) on this topic, our results exhaust all possibilities, are more explicit, and we indicate when they may be interpreted

Collected Works of Jaroslav Hájek – With Commentary
Edited by M. Hušková, R. Beran and V. Dupač
Published in 1998 by John Wiley & Sons Ltd.

in terms of individual trajectories. The last two sections are devoted to Gaussian processes. We define strong equivalency of normal (i. e. Gaussian) distributions of a stochastic process and study the determinant and quadratic form defining the probability density of a normal distribution with respect to another one, strongly equivalent to it. In Section 7 we present explicit probability densities for stationary processes with rational spectral densities.

2 BASIC CONCEPTS AND PRELIMINARY CONSIDERATIONS

Let $\{x_t,\ t \in T\}$ be an arbitrary stochastic process with finite second moments. Suppose that the mean value $\mathsf{E}\, x_t$ vanishes, $t \in T$, and that the covariances of x_t and x_s equal $\mathsf{E}\, x_t \overline{x}_s = R(x_t, x_s) = R_{ts},\ t \in T,\ s \in T$. Let $\mathcal{X}$ be the closed linear manifold of random variables consisting of finite linear combinations $\sum c_\nu x_{t_\nu},\ t_\nu \in T$, and of their limits in the mean. Obviously, the covariance of any two random variables $x,\ y \in \mathcal{X}$, say $R(x, y)$, is uniquely determined by $R_{ts},\ t \in T,\ s \in T$, and the mean value of any random variable $x \in \mathcal{X}$ equals 0. The variance of x will be denoted either by $R(x, x)$ or by $R^2(x)$. The closed linear manifold $\mathcal{X}$ is a Hilbert space with norm $R(x)$ and inner product $R(x, y)$. As usually, random variables $x,\ y$ such that $R(x - y) = 0$ are considered as identical.

Let U be the following mapping of $\mathcal{X}$ in the space of complex-valued functions, $\varphi_t,\ t \in T$:

$$(Uv)_t = R(x_t, v) \quad (v \in \mathcal{X};\ t \in T), \tag{2.1}$$

where $R(x_t, v) = \varphi_t$ is considered as a function of $t \in T$. Let Φ be the set of functions φ such that $\varphi = Uv$ for some $v \in \mathcal{X}$, i. e. such that $\varphi_t = R(x_t, v)$. Obviously $Uv_1 = Uv_2$ implies $v_1 = v_2$, so that there is an inverse operator $U^{-1},\ U^{-1}\varphi = v,\ \varphi \in \Phi$. If we introduce in Φ the norm $\mathcal{Q}(\varphi) = R(U^{-1}\varphi)$ and the inner product

$$\mathcal{Q}(\psi, \varphi) = R(U^{-1}\varphi,\ U^{-1}\psi) \quad (\varphi,\ \psi \in \Phi), \tag{2.2}$$

the mapping U will be unitary (i. e. isometric and one-to-one) and Φ also will be a Hilbert space.

If $R^s = R_{ts}$ is considered as a function of t only, and s is fixed, we have $U^{-1}R_s = x_s$, and

$$\mathcal{Q}(R^s, R^t) = R_{ts}, \quad (t,\ s \in T). \tag{2.3}$$

Obviously

$$\varphi_t = \mathcal{Q}(\varphi, R^t) \quad (t \in T), \tag{2.4}$$

so that $Q(\varphi^n) \to 0$ implies $\varphi_t^n \to 0$ in every point $t \in T$. Clearly, Φ consists of finite linear combinations $\sum c_\nu R^{t_\nu}$ and of their limits in the $\mathcal{Q}$-norm.

Definition 2.1. We shall say that $\mathcal{X}$ and Φ, or, more explicitly, $(\mathcal{X}, R)$ and $(\Phi^x, \mathcal{Q}^x)$, are closed linear manifolds generated by random variables x_t and covariances R_{ts}, respectively.

To any bounded linear operator A defined on $\mathcal{X}$ there corresponds a bounded linear operator $\overline{A}$ on Φ defined by the following relation:

$$(\overline{A}\varphi)_t = R(Ax_t, U^{-1}\varphi) \quad (\varphi \in \Phi), \tag{2.5}$$

where U^{-1} is the inverse of the unitary mapping (2.1) of $\mathcal{X}$ on Φ. We may also write $\overline{A}\varphi = U\, A^*\, U^{-1}\varphi$, where A^* is the adjoint of A. Also conversely, to any linear operator $\overline{A}$ defined on Φ, there corresponds an operator A on $\mathcal{X}$ defined by the relation $Ax = U^{-1}\,\overline{A}^*\, U\, x$. Obviously, $\overline{AB} = \overline{B}\,\overline{A}$. In what follows we shall omit the bar so that any bounded linear operator A will be considered as defined on both spaces Φ and $\mathcal{X}$. We have to bear in mind, of course, that AB in $\mathcal{X}$ must be interpreted as BA in Φ, and vice versa.

Let $\mathcal{X}_0$ be the set of all finite linear combinations $\sum c_\nu\, x_{t\nu}$. Let $\mathcal{X}^+$ be the closure of $\mathcal{X}_0$ with respect to a covariance R^+, and A be an operator in $\mathcal{X}^+$. Let R be another covariance. If

$$R(x, y) = R^+(Ax,\, Ay), \quad (x,\, y \in \mathcal{X}_0) \tag{2.6}$$

then

$$R(x) \leq k\, R^+(x), \tag{2.7}$$

where $k = ||A||$ and $||A||$ is the norm of A. Conversely, (2.7) implies existence of a bounded, positive and symmetric linear operator A in $\mathcal{X}^+$ such that (2.6) holds. Actually, if y is fixed, then $R(y, x)$ represents a linear functional in $\mathcal{X}^+$, and consequently $R(y, x) = R^+(z, x)$, $x \in \mathcal{X}^+$. By setting $z = By$ and $A = B^{1/2}$ we get the required result. Obviously,

$$||B|| = \sup_{x \in \mathcal{X}_0} \frac{R(x, x)}{R^+(x, x)}. \tag{2.8}$$

Definition 2.2. If (2.7) is satisfied, we say that the R-norm is dominated by the R^+-norm.

Lemma 2.1. *Let A be a bounded linear operator in the closed linear manifold $\mathcal{Z}$ generated by random variables z_t, $t \in T$, and let $\mathcal{X}$ be the closed linear manifold generated by random variables $x_t = Az_t$, $t \in T$. Let $(\Phi^z, \mathcal{Q}^z)$ and $(\Phi^x, \mathcal{Q}^x)$ be closed linear manifolds generated by covariances $R^+_{ts} = R(z_t, z_s)$ and $R_{ts} = R(x_t, x_s)$, respectively.*

Then $\Phi^x \subset \Phi^z$. Moreover, Φ^x consists of functions expressible in the form $\varphi = A\chi$, $\chi \in \Phi^z$, and

$$\mathcal{Q}^x(\psi, \varphi) = \mathcal{Q}^z(A_x^{-1}\psi,\, A_x^{-1}\varphi), \tag{2.9}$$

where the function $A_x^{-1}\varphi$, $\varphi \in \Phi^x$, is uniquely determined by the conditions

$$A(A_x^{-1}\varphi)_t = \varphi_t \quad \text{and} \quad (A_x^{-1}\varphi)_t = R(z_t, v), \quad v \in \mathcal{X}. \tag{2.10}$$

Proof. If $\varphi \in \Phi^x$, then $\varphi_t = R(Az_t, v) = AR(z_t, v) = (A\chi)_t$; and conversely $AR(z_t, v) = R(x_t, v) \in \Phi^x$. If we suppose that $v \in \mathcal{X}$, then v is determined by $R(x_t, v)$ uniquely, and $(A_x^{-1}\varphi)_t = R(z_t, v)$ is a unique solution of (2.10). The proof is complete. □

Remark 2.1. From Lemma 2.2 it follows that $\Phi^x = \Phi^z$ if R and R^+ dominate each other.

The values φ_t of every $\varphi \in \Phi$ may be considered as values of a linear functional $f(x) = R(x, U^{-1}\varphi)$ in the points $x = x_t$, $t \in T$. On applying the well-known extension theorem Riesz and Sz.–Nagy (1954, § 35), we get the following result.

Lemma 2.2. *A function φ_t belongs to Φ if and only if there exists a finite constant k such that, for any linear combination,*

$$\left|\sum c_\nu \varphi_{t_\nu}\right| \leq k\, R\left(\sum c_\nu x_{t_\nu}\right). \tag{2.11}$$

Example 2.1. Let the family $\{x_t, t \in T\}$ be finite, $\{x_t, t \in T\} = \{x_1, \ldots, x_n\}$, and let the matrix (R_{ij}), $R_{ij} = R(x_i, x_j)$, $1 \leq i, j \leq n$, be regular. Then Φ consists of all functions $\varphi = \varphi_i$, $1 \leq i \leq n$, and

$$\mathcal{Q}(\psi, \varphi) = \sum\sum \psi_i \overline{\varphi}_j Q_{ij}, \tag{2.12}$$

where (Q_{ij}) is the inverse of (R_{ij}), $(Q_{ij}) = (R_{ij})^{-1}$. Moreover

$$U^{-1}\varphi = \sum\sum x_i \overline{\varphi}_j Q_{ij}. \tag{2.13}$$

Now, consider the x_is as functions of an elementary event $\omega \in \Omega$. Then, for a fixed ω, the sample sequence (trajectory) $x_1(\omega), \ldots, x_n(\omega)$ represents a function of i, $1 \leq i \leq n$. If we denote the latter function by x^ω, then (2.13) may be rewritten in a form dual to (2.1),

$$(U^{-1}\varphi)_\omega = \mathcal{Q}(x^\omega, \varphi). \tag{2.14}$$

Proof. In view of (2.3) and (2.4), the formulas (2.12) and (2.13) are clearly true for $\varphi = R^t$ and $\psi = R^s$, $1 \leq t, s \leq n$, and the general case may be obtained on putting $\varphi = \sum_{\nu=1}^n c_\nu R^\nu$, $\psi = \sum_{\nu=1}^n d_\nu R^\nu$. □

The determination of Φ and $\mathcal{Q}(\psi, \varphi)$ and the solution of the equation $Uv = \varphi$ constitute what will be called a linear problem. If T is a finite set, then, as we have seen in Example 2.1, the linear problem is equivalent to inverting the covariance matrix.

Remark 2.2. If $\mathsf{E}\,x_t$ does not vanish, then $R(x,y) = \mathsf{E}\,xy - \mathsf{E}\,x\,\mathsf{E}\,y$, and, generally, $R(x) = 0$ is compatible with $\mathsf{E}\,x \neq 0$. Consequently, $R(x-y) = 0$ does not imply that $x = y$ with probability 1. This difficulty does not arise if $\varphi_t = \mathsf{E}\,x_t$, $t \in T$, belongs to Φ, because then $\mathsf{E}\,x \neq 0$ only if $R(x) > 0$, in view of

$$\mathsf{E}\,x = R(x, U^{-1}\varphi). \tag{2.15}$$

So, if necessary, the above condition $\mathsf{E}\,x_t \equiv 0$ may be replaced by a more general condition $\mathsf{E}\,x_t \in \Phi$. If, and only if, $\mathsf{E}\,x_t \in \Phi$, $\mathsf{E}\,x$ represents a linear functional on $(\mathcal{X}, R)$.

Now, let us show in what kinds of applications the linear problems appear.

Application 2.1. Let y be a random variable not belonging to $\mathcal{X}$. The projection of y on $\mathcal{X}$, say $\operatorname{Proj} y$, is given by

$$\operatorname{Proj} y = U^{-1}\,R(x_t, y). \tag{2.16}$$

In particular circumstances, the projection is called prediction, interpolation, filtration, or a regression estimate.

Application 2.2. Let $\varphi_1, \ldots, \varphi_m$ be known linearly independent functions belonging to Φ, and let the mean value of the process depend on an unknown vector $\alpha = (\alpha_1, \ldots, \alpha_m)$, where $\alpha_1, \ldots, \alpha_m$ are arbitrary real or complex numbers, so that

$$\mathsf{E}_\alpha\, x_t = \sum_{j=1}^{m} \alpha_j\, \varphi_{jt}. \tag{2.17}$$

Let us choose linear estimates $\widehat{\alpha}_j$ of the α_js, which are best according to an arbitrary criterion with the following property: $\mathsf{E}_\alpha\,\widehat{\Theta} = \mathsf{E}_\alpha\widehat{\Theta}'$ for any α and $R(\widehat{\Theta}) < R(\widehat{\Theta}')$ imply that $\widehat{\Theta}$ is a better estimate than $\widehat{\Theta}'$ (recall that $R^2(\cdot)$ denotes the variance). Then the vector $\widehat{\alpha} = (\widehat{\alpha_1}, \ldots, \widehat{\alpha}_m)$, where the $\widehat{\alpha}_j$s are the best linear estimates of the α_js, has the form

$$\widehat{\alpha} = Cv, \tag{2.18}$$

where $v = (U^{-1}\varphi_1, \ldots, U^{-1}\varphi_m)$ and C is a matrix which depends on other properties of the aforementioned criterion. If we postulate that the estimate should be unbiased and of minimum variance, then $C = B^{-1}$, where $B = (\mathcal{Q}(\varphi_j, \varphi_k))$. If we postulate that the mean value of $\mathsf{E}_\alpha|\widehat{\alpha}_j - \alpha_j|^2$ with respect to an a priori distribution of $\alpha_1, \ldots, \alpha_m$ should be minimum, then $C = (B + D^{-1})^{-1}$, where $D = (\mathsf{E}_{\text{ap}}\alpha_j\overline{\alpha}_k)$ and $\mathsf{E}_{\text{ap}}(\cdot)$ denotes the a priori mean value. In the former case B^{-1} also represents the covariance matrix of the vector $(\widehat{\alpha}_1, \ldots, \widehat{\alpha_m})$, and in the latter case $(B + D^{-1})^{-1}$ also represents the matrix with elements $\mathsf{E}_{\text{ap}}\mathsf{E}_\alpha(\widehat{\alpha}_j - \alpha_j)\,\overline{(\widehat{\alpha}_k - \alpha_k)}$. The main idea of the proof is as follows. From (2.15) it follows that $\mathsf{E}_\alpha x = \sum_{j=1}^m \alpha_j\, R(x,\, U^{-1}\varphi_j)$, so that projection of any estimate $\widehat{\Theta}$ on the subspace $\mathcal{X}_m$ spanned by $U^{-1}\varphi_1, \ldots, U^{-1}\varphi_m$ has the same mean value as $\widehat{\Theta}$ for any α, and, if $\widehat{\Theta}$ does not belong to $\mathcal{X}_m$, has a smaller variance. See Hájek (1961) and Pugachev (1960).

Application 2.3. Suppose that (2.17) still holds and that y is a random variable not belonging to $\mathcal{X}$ such that $\mathsf{E}_\alpha y = \sum_{j=1}^m \alpha_j c_j$ where c_j are known constants. Let us choose an estimate $\widehat{y} \in \mathcal{X}$ of y according to a criterion with the following property: $\mathsf{E}_\alpha \widehat{y} = \mathsf{E}_\alpha \widehat{y}'$ for any α and $R(\widehat{y} - y) < R(\widehat{y}' - y)$ implies that $\widehat{y}$ is a better estimate than $\widehat{y}'$. (Obviously, this situation is a generalization of one considered in Application 2.2.) Then the best linear estimate of y equals

$$\widehat{y} = y_0 + \sum_{j=1}^{m} \widehat{\alpha}_j (c_j - \mathcal{Q}(\varphi_j, \varphi_0)), \tag{2.19}$$

where the $\widehat{\alpha}_j$s have been defined in Application 2.2, and $y_0 = U^{-1}\varphi_0$, where $\varphi_{0t} = R(x_t, y)$, $t \in T$. If we postulate that $\mathsf{E}_\alpha \widehat{y} = \mathsf{E}_\alpha y$ for any α and that $R(\widehat{y} - y)$ should be minimum, then the best unbiased linear estimate (2.19) has the following property:

$$R^2(\widehat{y} - y) = R^2(y_0 - y) + R^2 \left(\sum_{j=1}^{m} \widehat{\alpha}_j (c_j - \mathcal{Q}(\varphi_j, \varphi_0)) \right) \tag{2.20}$$

(see Hájek (1961) and Pugachev (1960)).

Application 2.4. Let P and P^+ be two normal distributions of a real stochastic process $\{x_t,\, t \in T\}$ defined by a common covariance $R_{ts} = R^+_{ts}$ and mean values φ_t and $\varphi^+_t \equiv 0$, respectively. If $\varphi \in \Phi$, then P is absolutely continuous with respect to P^+ and

$$\frac{\mathrm{d}\mathsf{P}}{\mathrm{d}\mathsf{P}^+} = \exp\left\{ U^{-1}\varphi - \frac{1}{2}\, \mathcal{Q}(\varphi, \varphi) \right\}. \tag{2.21}$$

If φ does not belong to Φ, then P and P^+ are perpendicular (mutually singular). The proof follows from the fact that $U^{-1}\varphi$ is a sufficient statistic for the pair $(\mathsf{P}, \mathsf{P}^+)$, as is shown in Hájek (1960). The variance of $U^{-1}\varphi$ equals $\mathcal{Q}(\varphi, \varphi)$ with respect to both P and P^+. In view of (2.16), the mean value of $U^{-1}\varphi$ equals $R(U^{-1}\varphi,\, U^{-1}\varphi) = \mathcal{Q}(\varphi, \varphi)$ if P holds true, and obviously equals 0 if P^+ holds true. Now,

$$-\frac{1}{2}\left[U^{-1}\varphi - \mathcal{Q}(\varphi, \varphi)\right]^2 / \mathcal{Q}(\varphi, \varphi) + \frac{1}{2}\left[U^{-1}\varphi\right]^2 / \mathcal{Q}(\varphi, \varphi) = U^{-1}\varphi - \frac{1}{2}\, \mathcal{Q}(\varphi, \varphi),$$

which proves (2.21). The case of different covariances will be treated in Sections 6 and 7.

Application 2.5. A somewhat different application is given by the following lemma. A difference of two covariances $R^+_{ts} - R_{ts}$ represents again a covariance, if and only if R^+ dominates R and the operator B determined by $R(x, y) = R^+(Bx, y)$ has a norm smaller than or equal to 1.

Proof. If $\|B\| \leq 1$, then $I - B$ is a positive operator, so that $R^+_{ts} - R_{ts} = R((I-B)\,x_t, x_s)$ is a covariance. Conversely, if $R^+_{ts} - R_{ts}$ is a covariance, then $R^+(x, x) \geq R(x, x)$, i.e. R^+ dominates R, and the norm of the operator B defined by $R(x, y) = R(Bx, y)$ is smaller than or equal to 1, in view of (2.8). In particular, $R_{ts} - \varphi_t\,\overline{\varphi}_s$ is a covariance if and only if $\varphi \in \Phi$ and $\mathcal{Q}(\varphi) \leq 1$. This corollary generalizes a result by A. V. Balakrishnan (1959). □

3 STOCHASTIC INTEGRALS

Let $\mathcal{G}$ be a Borel field of measurable subsets Λ of T, and $\mu = \nu(\Lambda)$ be a σ-finite measure on $\mathcal{G}$. Let $Y_\Lambda = Y(\Lambda)$ be an additive random set function defined on subsets of finite measure and such that $\mathsf{E}\,Y_\Lambda = 0$ and

$$R(Y_\Lambda, Y_{\Lambda'}) = \mu(\Lambda \cap \Lambda') \quad (\Lambda,\, \Lambda' \in \mathcal{G}). \tag{3.1}$$

The closed linear manifold $\mathcal{Y}$ generated by random variables Y_Λ consists of random variables expressible in the form

$$v = \int_T h_t\, Y(\mathrm{d}t), \tag{3.2}$$

where h_t is a squared integrable function, $h \in \mathcal{L}^2(\mu)$. The stochastic integral (3.2) is defined as a limit in the mean (see J. L. Doob (1953)).

The closed linear manifold generated by covariances (3.1) consists of all σ-additive set functions $\nu(\Lambda)$ such that $\nu(\Lambda) = \int_\Lambda f\,\mathrm{d}\mu$ and

$$\int_T |f|^2\,\mathrm{d}\mu < \infty. \tag{3.3}$$

We put $f = \mathrm{d}\nu/\mathrm{d}\mu$ and denote $\mu(\mathrm{d}t)$ briefly by $\mathrm{d}\mu$. Moreover, we have

$$\mathcal{Q}(\nu_1, \nu_2) = \int_T \left(\frac{\mathrm{d}\nu_1}{\mathrm{d}\mu}\right) \overline{\left(\frac{\mathrm{d}\nu_2}{\mathrm{d}\mu}\right)} \mathrm{d}\mu, \tag{3.4}$$

and the equation $\nu(\Lambda) = R(Y_\Lambda, v)$ is solved by (3.2), where $h = \mathrm{d}\nu/\mathrm{d}\mu$.

It is often preferable to introduce the formal derivative $y_t = (\mathrm{d}Y/\mathrm{d}\mu)_t$ (so called white noise) and to write $\int h_t\, y_t\, \mu(\mathrm{d}t)$ and $R(y_t, v)$ instead of $\int h_t\, Y(\mathrm{d}t)$ and $[(\mathrm{d}/\mathrm{d}\mu)\, R(Y_\Lambda, v)]$, respectively. Unless the point t has a positive measure, y_t itself has no meaning, but the integrals $\int h\, y\,\mathrm{d}\mu,\ h \in \mathcal{L}^2(\mu)$, and covariances

$R(y_t, v)$, $t \in T$, are well-defined in the above sense. We confine ourselves to this without entering into the theory of random distributions (see Gelfand (1955)).

$\mathcal{L}^2(\mu)$, with the usual inner product $(h, g) = \int h\,\overline{g}\,\mathrm{d}\mu$, may be considered as the closed linear manifold generated by the covariance function of the white noise y_t (formally, $R(y_t, y_s) = 0$ if $t \neq s$, and $R(y_t, y_t) = 1/\mu(\mathrm{d}t)$). The relations

$$R(y_t, v) = t_h \quad \text{and} \quad v = \int \overline{h}_t\, y_t\, \mu(\mathrm{d}t) \tag{3.5}$$

define a unitary transformation $Uv = h$ $(U^{-1}h = v)$ of $\mathcal{Y}$ on $\mathcal{L}^2(\mu)$ (of $\mathcal{L}^2(\mu)$ on $\mathcal{Y}$).

Now consider an operator K in $\mathcal{L}^2(\mu)$, generated by a kernel $K(t, s)$, which is measurable on $\mathcal{G} \times \mathcal{G}$ and squared integrable w.r.t. $\mu \times \mu$. We have

$$(Kh)_t = \int K(t, s)\, h_s\, \mu(\mathrm{d}s). \tag{3.6}$$

This operator may be carried over to $\mathcal{Y}$ according to the formula $Kv = U^{-1} K^* Uv$, where K^* is generated by $K^*(t, s) = \overline{K(s, t)}$. So, if $v = \int \overline{h}\, y\, \mathrm{d}\mu$, then

$$Kv = \int (K^* h)\, y\, \mathrm{d}\mu. \tag{3.7}$$

The domain of definition of the operator K in $\mathcal{Y}$ may be extended to include the white noise y_t by putting

$$x_t = K\, y_t = \int K(t, s)\, y_s\, \mu(\mathrm{d}s). \tag{3.8}$$

Then we also may write

$$Kv = \int \overline{h}(Ky)\, \mathrm{d}\mu = \int \overline{h}_t\, x_t\, \mathrm{d}\mu, \tag{3.9}$$

where the integral is defined in the weak sense (see Karhunen (1947)), i. e. as a random variable ξ belonging to the closed linear manifold $\mathcal{X}$ spanned by random variables $x_t = Ky_t$, $t \in T$, and such that

$$R(y_\tau, \xi) = \int R(y_\tau, x_t)\, h_t\, \mathrm{d}\mu.$$

Actually, we have $R(y_\tau, x_t) = \overline{K(t, \tau)} = K^*(\tau, t)$, so that

$$\int R(y_\tau, x_t)\, h_t\, \mathrm{d}\mu = (K^* h)_\tau = R(y_\tau, Kv),$$

where the last equality follows from (3.7).

The equality

$$\int \overline{(K^*h)}\, y\, \mathrm{d}\mu = \int \overline{h}(Ky)\, \mathrm{d}\mu, \tag{3.10}$$

obtained from (3.7) and (3.9), will now be generalized for arbitrary bounded operators. Let $X(\Lambda) = AY(\Lambda)$, $\Lambda \in \mathcal{G}$, where $Y(\Lambda)$ is an additive random set function satisfying (3.1) and A is a bounded operator defined in the closed linear manifold $\mathcal{Y}$ generated by $\{Y(\Lambda),\ \Lambda \in \mathcal{G}\}$. Put, by definition,

$$\int g(t)\, X(\mathrm{d}t) = A \int g(t)\, Y(\mathrm{d}t) \tag{3.11}$$

for any $g \in \mathcal{L}_2(\mu)$. Then

$$\int g(t)\, AY(\mathrm{d}t) = \int (A^* g)_t\, Y(\mathrm{d}t). \tag{3.12}$$

In fact, (3.12) is a direct consequence of the identity $A = U^{-1} A^* U$, where U and U^{-1} are unitary operators defined by (2.5).

Application 3.1. Let $\{x_t,\ t \in T\}$ be the process expressed by (3.8). Then (i) the closed linear manifold Φ^x generated by covariances $R(x_t, x_s) = R(Ky_t, Ky_s)$, $t \in T$, $s \in T$, consists of functions expressible in the form $\varphi_t = (Kh)_t$, $h \in \mathcal{L}^2(\mu)$, (ii) the inner product in Φ^x, say $\mathcal{Q}^x$, is given by

$$\mathcal{Q}^x(\psi, \varphi) = \int (K_x^{-1}\psi)_t\, \overline{(K_x^{-1}\varphi)_t}\, \mathrm{d}\mu, \tag{3.13}$$

and (iii) the equation $\varphi_t = R(x_t, v)$ is solved by

$$v^\varphi = \int (K_x^{-1}\varphi)_t\, Y(\mathrm{d}t). \tag{3.14}$$

The formula (3.14) is based on Lemma 2.2, where $K_x^{-1}\varphi$ is defined as a function such that $KK_x^{-1}\varphi = \varphi$ and $(K_x^{-1}\varphi)_t = R(y_t, x)$, where x belongs to the closed linear manifold $\mathcal{X}$ generated by random variables x_t, $t \in T$. If $\mathcal{X} = \mathcal{Y}$ then $K_x^{-1} = K^{-1}$ is an ordinary inverse of K. Actually, then $0 \equiv \varphi_t = R(x_t, v)$ implies $v \equiv 0$, and hence $(Kh)_t \equiv 0$ if and only if $h_t \equiv 0$.

Application 3.2. The operator K^* is generated by the kernel $K^*(t,s) = \overline{K(s,t)}$, and the operator KK^* by the kernel

$$R(x_t, x_s) = \int K(t,\tau)\, \overline{K(s,\tau)}\, \mu(\mathrm{d}\tau). \tag{3.15}$$

The equality of both sides in (3.15) follows from (3.8). If the function φ is expressible in the form $\varphi = KK^*h$, $h \in \mathcal{L}^2(\mu)$, i.e.

$$\varphi_t = \int R(x_t, x_s)\, h_s\, \mu(\mathrm{d}s), \tag{3.16}$$

then the equation $\varphi_t = R(x_t, v)$ is solved by

$$v = \int \overline{h}_t \, x_t \, \mathrm{d}\mu = \int \overline{((KK^*)^{-1}\varphi)_t} \, x_t \, \mathrm{d}\mu. \tag{3.17}$$

Actually, $v \in \mathcal{X}$ and $\varphi_t = R(x_t, v)$ follows directly from (3.16) and the weak definition of the integral $\int \overline{h}_t \, x_t \, \mathrm{d}\mu$.

Example 3.1. Let $\{x_t, t \leq t_0\}$ be a semi-infinite segment of a regular stationary process. As is well known (J. L. Doob (1953, Chap. XII, § 5)), there exist a uniquely determined squared integrable function $c(\tau)$ vanishing for $\tau < 0$, and a process with uncorrelated increments Y_t, $E|\mathrm{d}Y_t|^2 = \mathrm{d}t$, such that $x_t = \int_{-\infty}^{t} c(t-s)\,\mathrm{d}Y_s$, and the closed linear manifolds $\mathcal{X}$ and $\mathcal{Y}$ spanned by

$$\{x_t, t \leq t_0\} \quad \text{and} \quad \left\{ y_t = \frac{\mathrm{d}Y_t}{\mathrm{d}t}, t \leq t_0 \right\},$$

respectively, coincide: $\mathcal{X} = \mathcal{Y}$. So we may apply the preceding results with $K(t,s) = c(t-s)$, $\mu(\mathrm{d}t) = \mathrm{d}t$ and $T = (-\infty, t_0]$. In the present case equation (3.16) is called the Wiener–Hopf equation. The functions given by (3.16), however, do not exhaust the whole set Φ^x of functions φ for which the linear problem $\varphi_t = R(x_t, v)$ may be solved.

For a moment, let us consider the whole process x_t, $-\infty < t < \infty$, and denote the usual Fourier–Plancherel transform by F. We know that $F^* = F^{-1}$, which implies, in view of (3.12), that $x_t = \int [Fc(t-\cdot)]_\lambda \, \mathrm{d}(FY)_\lambda$, where, as is well known, $[Fc(t-\cdot)]_\lambda = e^{it\lambda}a(\lambda)$. On putting $\mathrm{d}Z_\lambda = a(\lambda)\,\mathrm{d}(FY)_\lambda$, we get the well-known spectral representation $x_t = \int e^{it\lambda}\mathrm{d}Z_\lambda$, where $\mathsf{E}|\mathrm{d}Z_\lambda|^2 = |a(\lambda)|^2\,\mathrm{d}\lambda$, $|a(\lambda)|^2$ being the spectral density.

Example 3.2. Let Y_t be a process with uncorrelated increments such that $\mathsf{E}|\mathrm{d}Y_t|^2 = \mathrm{d}t$, $-1 \leq t \leq T_0$. We are interested in $x_t = Y_t - Y_{t-1}$, $0 \leq t \leq t_0$, which is a stationary process with correlation function $R(x_t, x_{t-\tau}) = \max(0, 1-|\tau|)$. If we add the random variables $x_t = Y_t - Y_{t-1}$, $-1 \leq t \leq 0$, we obtain $\mathcal{X} = \mathcal{Y}$. The kernel $K(t,s)$ is apparent from relations

$$\begin{aligned} x_t &= \int_{-1}^{t} y_s \, \mathrm{d}s, \quad \text{if} \quad -1 \leq t \leq 0, \\ &= \int_{t-1}^{t} y_s \, \mathrm{d}s, \quad \text{if} \quad 0 \leq t \leq t_0. \end{aligned} \tag{3.18}$$

Let $K_0^{-1}\varphi$ be a function such that $(KK_0^{-1}\varphi)_t = \varphi_t$, $0 \leq t \leq t_0$, and that $(K_0^{-1}\varphi)_t = R(y_t, v)$, where v belongs to the closed linear manifold $\mathcal{X}_0$ generated by random variables $\{x_t, 0 \leq t \leq t_0\}$. We may find, after some computations, that

$$(K_0^{-1}\varphi)_t = \varphi'_t + \varphi'_{t-1} + \varphi'_{t-2} + \dots, \tag{3.19}$$

where the sum stops when the index falls into the interval $[-1,0)$ and φ'_i has been extended so that

$$\varphi'_t = \begin{cases} \frac{1}{N+1}\left(c - N\,\varphi'_{t+1} - (N-1)\,\varphi'_{t+2} - \cdots - \varphi'_{t+N}\right) \\ \qquad \text{for } t_0 - N - 1 < t < 0, \\ \frac{1}{N+2}\left(c - (N+1)\,\varphi'_{t+1} - N\,\varphi'_{t+2} - \cdots - \varphi'_{t+N+1}\right) \\ \qquad \text{for } -1 < t < t_0 - N - 1, \end{cases} \tag{3.20}$$

where N is the greatest integer not exceeding t_0 and

$$c = \frac{(N+1)\,(\varphi_0 + \varphi_{t_0}) + \cdots + (\varphi_N + \varphi_{t_0-N})}{2N - t_0 + 2}. \tag{3.21}$$

For $\varphi_t \equiv 1$ we could obtain the best linear estimate by Hájek (1956) of a constant mean value.

For $\varphi_t = R(t_{x_0+\tau}, x_t) = \max(0, 1 - |t_0 + \tau - t|)$ and $t_0 = N$, we would obtain the best predictor by Hájek (1958 a).

The same method could be applied to processes expressed generally by $x_t = \sum_{k=0}^{m} b_{m-k}\, Y_{t-k}$. In all cases Φ^x consists of absolutely continuous functions with a squared integrable derivative.

4 SOLUTIONS EXPRESSED IN TERMS OF INDIVIDUAL TRAJECTORIES

In what follows we shall assume that the random variables $x_t = x_t(\omega)$ are defined on a probability space $(\Omega, \mathcal{F}, \mathsf{P})$. The covariances and finite-dimensional distributions of random variables $x_t(\omega)$ are not influenced by any adjustment (modification) of every $x_t(\omega)$ on a ω-subset having probability 0. The properties of individual trajectories $x^\omega = x_t(\omega)$, $t \in T$, however, may be changed very substantially.

Theorem 4.1. *The process $x_t = x_t(\omega)$ defined by (3.8) may be adjusted on ω-subsets having probability zero, so that the solution of the equation $\varphi_t = R(x_t, v)$, φ_t expressible as $\varphi_t = \int R(x_t, x_s)\, h_s \,\mathrm{d}\mu$, $h \in \mathcal{L}^2(\mu)$, is given by*

$$v(\omega) = \int_T \overline{h}_t\, x_t(\omega)\,\mathrm{d}\mu, \tag{4.1}$$

where for almost every ω the integral on the right side is to be understood in the usual Lebesgue–Stieltjes sense.

Proof. The compactness of KK^* in $\mathcal{L}^2(\mu)$ follows from (3.6) by standard arguments. Let $\chi_n(t)$ and $\kappa_n (n \geq 1)$ be the eigen-elements and corresponding eigen-values of KK^*, respectively. The equality of some κ_ns is not excluded. We first show that $\sum \kappa_n < \infty$. As the system $\{\chi_n(t),\, n \geq 1\}$ is complete and orthonormal, we have

$$\begin{aligned}\sum_1^\infty \kappa_n &= \sum_1^\infty \int |K^*\chi_n|^2\, \mathrm{d}\mu \\ &= \int \sum_1^\infty \left| \int K^*(t,s)\, \chi_n(s)\, \mu(\mathrm{d}s)\right|^2 \mu(\mathrm{d}t) \\ &= \int\int |K(s,t)|^2\, \mu(\mathrm{d}s)\, \mu(\mathrm{d}t),\end{aligned} \tag{4.2}$$

where the last expression is finite according to our assumptions.

Introduce the unitary transforms U and U^{-1} defined by $(Uv)_t = R(y_t, v)$. Obviously, the random variables $v_n = U^{-1}\chi_n$ form a complete orthonormal system in $\mathcal{Y}$. Moreover,

$$R(Kv_m, Kv_n) = R(v_m,\, K^*Kv_n) = \mathcal{Q}(KK^*\chi_n, \chi_m) = \kappa_n\, Q(\chi_n, \chi_m),$$

so that $\{Kv_n,\, n \geq 1\}$ is an orthogonal system with $R(Kv_n, Kv_n) = \kappa_n$, $n \geq 1$. In view of (3.9), $Kv_n \in \mathcal{X}$, where $\mathcal{X}$ is the closed linear manifold spanned by random variables $x_t = Ky_t$, $t \in T$. If $R(y, Kv_n) = 0$, $n \geq 1$, then $R(K^*y, v_n) = 0$, $n \geq 1$, so that $K^*y \equiv 0$ and $(KUy)_t = (UK^*y)_t \equiv 0$. Consequently,

$$R(x_t, y) = \int K(t,s)\,(Uy)_s\, \mathrm{d}\mu = (KUy)_t = 0,$$

$t \in T$, which shows that $y \perp Kv_n$, $n \geq 1$, only if $y \perp \mathcal{X}$. We conclude that $\{Kv_n,\, n \geq 1\}$, where Kv_n with $\kappa_n = 0$ may be omitted, forms a complete system in $\mathcal{X}$.

Consequently, for every $t \in T$, we have

$$x_t = \sum_1^\infty Kv_n\, \frac{R(x_t, Kv_n)}{\kappa_n} = \sum_1^\infty Kv_n\, \frac{R(y_t, K^*Kv_n)}{\kappa_n} = \sum_1^\infty Kv_n \cdot \chi_n(t), \tag{4.3}$$

where the sum converges in the mean. Further, the sequence

$$\left\{ \sum_1^N \chi_n(t)\, Kv_n(\omega),\ N \geq 1 \right\}$$

converges in the $(\mu \times \mathsf{P})$-mean on $T \times \Omega$ because

$$\int_T \int_\Omega \left| \sum_{N+1}^{N+p} \chi_n(t)\, Kv_n(\omega) \right|^2 \mathrm{d}\mathsf{P}\mathrm{d}\mu = \sum_{N+1}^{N+p} \kappa_n \tag{4.4}$$

and $\sum_1^\infty \kappa_n < \infty$. Consequently, we may draw a subsequence such that

$$\lim_{k\to\infty} \sum_1^{N_k} \chi_n(t)\, K v_n(\omega) = \overline{x}_t(\omega) \tag{4.5}$$

exists for almost all (t,ω) with respect to $\mu \times \mathsf{P}$. Let T_0 be a subset of T such that $\mu(T - T_0) = 0$ and that for $t \in T_0$ the limit (4.5) exists with probability one. On putting

$$\begin{aligned} \tilde{x}_t(\omega) &= \overline{x}_t(\omega), \quad \text{for} \quad t \in T_0, \\ &= x_t(\omega), \quad \text{for} \quad t \in T - T_0, \end{aligned} \tag{4.6}$$

and recalling that the limits (4.3) and (4.5) coincide with probability 1 for $t \in T_0$, we conclude that $\{\tilde{x}_t(\omega),\, t \in T\}$ is an equivalent modification of the process $\{x_t(\omega),\, t \in T\}$.

Now $\int \left(\sum_1^\infty |K v_n|^2\right) \mathrm{d}\mathsf{P} = \sum_1^\infty \kappa_n < \infty$, so that

$$\sum_1^\infty |K v_n(\omega)|^2 < \infty \tag{4.7}$$

with probability one. Without any loss of generality, we may assume that (4.7) holds true for any $\omega \in \Omega$. Then $\sum_1^\infty \chi_n(t)\, K v_n(\omega)$ for every fixed ω represents an orthonormal expansion of $\tilde{x}_t(\omega)$ in $\mathcal{L}^2(\mu)$. So, if $v(\omega)$ is given by (4.1), where $x_t(\omega) \equiv \tilde{x}_t(\omega)$, we get

$$\begin{aligned} v(\omega) &= \sum_1^\infty K v_n(\omega) \int_t \overline{h}_t\, \chi_n(t)\, \mathrm{d}\mu = \sum_1^\infty \kappa_n^{-1}\, K v_n(\omega) \\ &\quad \times \int_T \overline{h}(K K^* \chi_n)_t\, \mathrm{d}\mu = \sum_1^\infty \kappa_n^{-1}\, K v_n(\omega) \int_T \overline{\varphi}_t\, \chi_n(t)\, \mathrm{d}\mu \end{aligned} \tag{4.8}$$

and, in view of (4.3),

$$R(x_t, v) = \sum_1^\infty \chi_n(t) \int_T \overline{\varphi}_t\, \chi_n(t)\, \mathrm{d}\mu = \varphi_t, \tag{4.9}$$

which concludes the proof. □

Remark 4.1. Let x_{t_1}, $t_1 \in T_1$, be arbitrary random variables from the closed linear manifold $\mathcal{X}$ spanned by random variables x_t, $t \in T$. Clearly, the extended process $\{x_t,\, T \in T \cup T_1\}$ spans the same closed linear manifold as the process $\{x_t,\, t \in T\}$. However, the class of random variables v expressible in the form

$$v = \int_{T \cup T_1} \overline{h}_t\, x_t\, \mathrm{d}\mu$$

is never smaller than the class of random variables $v = \int_T \overline{h}_t\, x_t \,\mathrm{d}\mu$, and will usually be larger. By this device, we may enlarge considerably the class of functions φ expressible as $\varphi_t = \int R(x_t, x_s)\, \mu(\mathrm{d}s)$, permitting the solution to be written in the form (4.1). For example, if the process is originally defined on a segment $0 \le t \le T$, the added random variables may be derivatives in the endpoints (see the next section). In the extreme case we could include in the set $\{x_t,\, t \in T \cup T_1\}$ every random variable from $\mathcal{X}$, i. e. put $\{x_t,\, t \in T \cup T_1\} = \mathcal{X}$.

Another way to enlarge the class of functions φ for which the solution is expressible in terms of individual trajectories consists in replacing the white noise y_t by an ordinary random process z_t, $t \in T$. We shall suppose that $\mathcal{Z}$ is the closed linear manifold spanned by the z_ts and K is a compact linear operator in $\mathcal{Z}$. Let Φ^z, Φ^x and Φ^0 be closed linear manifolds generated by covariances $R(z_t, z_s)$, $R(Kz_t, Kz_s)$ and $R(K^*Kz_t,\, K^*Kz_s)$, respectively. From Lemma 2.2 it follows that

$$\Phi^0 \subset \Phi^x \subset \Phi^z. \tag{4.10}$$

Let $\mathcal{Q}^z$ and $\mathcal{Q}^x$ be the inner product in Φ^z and Φ^x, respectively.

Lemma 4.1. *If $\varphi \in \Phi^0$ and $\psi \in \Phi^x$, then*

$$\mathcal{Q}^x(\psi, \varphi) = \mathcal{Q}^z(\psi,\, (KK^*)^{-1}\varphi), \tag{4.11}$$

where $(KK^*)^{-1}\varphi$ is any function such that $KK^*(KK^*)^{-1}\varphi = \varphi$.

Proof. Clearly, $\mathcal{Q}^z(\psi,\, (KK^*)^{-1}\varphi) = \mathcal{Q}(K^{-1}\psi,\, K^*(KK^*)^{-1}\varphi)$, where $K^{-1}\psi$ is any solution if the equation $K\chi = \psi$. In particular, we may take the solution $K_x^{-1}\psi$ considered in Lemma 2.2. Now if $(KK^*)^{-1}\varphi = R(z_t, v)$, then $K^*(KK^*)^{-1}\varphi = R(z_t, Kv)$, where Kv belongs to the closed linear manifold $\mathcal{X}$ spanned by the random variables $x_t = Kz_t$, $t \in T$. Consequently, $K^*(KK^*)^{-1}\varphi = K_x^{-1}\varphi$, where K_x^{-1} is defined by (2.10). So, we have $\mathcal{Q}^z(\psi,\, (KK^*)^{-1}\varphi) = \mathcal{Q}^z(K_x^{-1}\psi,\, K_x^{-1}\varphi)$, which gives (4.11), in accordance with (2.9). The proof is complete. □

Note that the right side of (4.11) is an extension of $\mathcal{Q}^x$, originally defined on $\Phi^x \times \Phi^x$, on $\Phi^z \times \Phi^0$. Also observe that in Lemma 2.2 we need a particular choice of K^{-1}, $K^{-1} = K_x^{-1}$, whereas $(KK^*)^{-1}$ may be chosen in any way, if non-unique. Now we shall extend the formula (4.1) to the case when the white noise y_t is replaced by an ordinary process z_t, $t \in T$.

Theorem 4.2. *Suppose that the above-considered compact operator K in $\mathcal{Z}$ is such that the eigen-values κ_n of K^*K form a convergent series, $\sum_1^\infty \kappa_n < \infty$. Then the process $x_t = Kz_t$, $t \in T$, may be adjusted so that* *almost every trajectory* $x^\omega = x_t(\omega)$ *belongs to* Φ^z, *and the solution of the equation* $\varphi_t = R(x_t, v)$, $\varphi \in \Phi^0$, *equals*

$$v(\omega) = Q^z(x^\omega, (KK^*)^{-1}\varphi). \tag{4.12}$$

Proof. We may proceed as for the proof of Theorem 4.1, with the only exception that the limit

$$\sum_1^\infty Kv_n(\omega)\,\chi_n(t) = \sum_1^\infty Kv_n(\omega)\,R(z_t, v_n) = R\left(z_t, \sum_1^\infty v_n\,Kv_n(\omega)\right) = \overline{x}_t(\omega) \tag{4.13}$$

exists for every t and with each ω satisfying (4.7). The other arguments need no changes.

The formula (4.12) presumes that almost every trajectory $x^\omega = x_t(\omega)$ belongs to Φ^z. The following theorem shows that in the Gaussian case x^ω cannot belong to Φ^z almost surely, unless the operator K has the property assumed in Theorem 4.2, so that this theorem cannot be strengthened. □

Theorem 4.3. *Let* $x_t(\omega)$ *and* $z_t(\omega)$ *be Gaussian processes related by a bounded linear operator* K, $x_t = Kz_t$, $t \in T$, *and* Φ^z *be the closed linear manifold generated by covariances* $R(z_t, z_s)$. *Then, for an equivalent modification of the process* $x_t(\omega)$,

$$\mathsf{P}(x^\omega \in \Phi^z) = 1, \tag{4.14}$$

if and only if the operator K^*K *is compact and the series of its eigen-values* κ_n *is convergent,* $\sum_1^\infty \kappa_n < \infty$.

Proof. Sufficiency follows from Theorem 4.2. For necessity, if (4.14) holds true, then there exists a function $w(\omega, \omega')$ representing a random variable from $\mathcal{Z}$ for ω fixed, and such that

$$x_t(\omega) = \int z_t(\omega')\,w(\omega, \omega')\,\mathsf{P}(d\omega') = Kz_t(\omega). \tag{4.15}$$

If $\{z_n\}$ converges weakly to 0, then

$$\lim_{n\to\infty} x_n(\omega) = \lim_{n\to\infty} \int z_n(\omega')\,w(\omega, \omega')\,\mathsf{P}(d\omega') = 0$$

for almost every ω, and, consequently, $x_n \to 0$ in probability. If the x_ns are Gaussian, then

$$\mathsf{P}(|x_n| \geq \varepsilon) \geq 2\int_{\varepsilon/R(x_n)}^\infty (2\pi)^{-\frac{1}{2}}\,e^{-\frac{1}{2}\tau^2}\,d\tau,$$

so that convergence in probability implies convergence in the mean, $R(x_n) \to 0$. It means that operator K transforms weakly convergent sequences in strongly convergent ones, and consequently is compact (see Riesz and Sz.-Nagy (1954, § 85)). Obviously, the operator K^*K is also compact. Let κ_n be the non-zero eigen-values of K^*K, and let $v_n(\omega)$ be the corresponding eigen-elements. We have

$$\sum_1^\infty \kappa_n = \sum_1^\infty R(Kv_n,\, Kv_n) = \mathsf{E}\left\{\sum_1^\infty |Kv_n(\omega)|^2\right\} \tag{4.16}$$

and, for almost all ω,

$$\sum_1^\infty |Kv_n(\omega)|^2 = \sum_1^\infty |R(v_n, w^\omega)|^2 = R(w^\omega, w^\omega) < \infty, \tag{4.17}$$

where $w^\omega = w(\omega, \cdot)$. The random variables Kv_n are uncorrelated, because $R(Kv_n,\, Kv_m) = R(K^*Kv_n,\, v_m) = \kappa_n\, R(v_n, v_m)$. If we suppose that they are Gaussian, then they are independent. It remains to show that finiteness of (4.17) for almost all ω implies finiteness of (4.16), or, equivalently, that infiniteness of (4.16) would imply infiniteness of (4.17) with positive probability. If $\sum_1^\infty \kappa_n = \infty$, then we may form an infinite sequence of partial sums

$$\xi_k = \sum_{n_k+1}^{n_{k+1}} Kv_n$$

such that each partial sum consists of either (i) a unique element Kv_n such that $\kappa_n \geq 1$, or (ii) several elements such that $\kappa_n \leq 1$ and $\sum_{n_k+1}^{n_k} \kappa_n \geq 4$. The ξ_ks have mean values $\mathsf{E}\,\xi_k = \sum_{n_{k+1}}^{n_k+1} \kappa_n$ and variances $R^2(\xi_k) = 2\sum_{n_k+1}^{n_{k+1}} \kappa_n^2$ so that

$$\mathsf{P}\left(\xi_k > \frac{1}{4}\right) = 2\int_{\frac{1}{2}}^\infty (2\pi)^{-\frac{1}{2}}\, e^{-\frac{1}{2}\tau^2}\,\mathrm{d}\tau > \frac{1}{3}$$

in case (i) and

$$\mathsf{P}\left(\xi_k > \frac{1}{4}\right) \geq 1 - \frac{2\sum \kappa_n^2}{\left|\sum \kappa_n - \frac{1}{4}\right|} \geq 1 - \frac{2\sum \kappa_n}{\left|\sum \kappa_n - \frac{1}{4}\right|^2} \geq \frac{1}{3} \quad (n_k < n \leq n_{k+1}) \tag{4.18}$$

in case (ii). So $\sum_1^\infty \mathsf{P}\left(\xi_k > \frac{1}{4}\right) = \infty$ where the ξ_ks are independent. On applying this to the well-known lemma (Borel–Cantelli), we get that with probability 1 the event $\xi_k > \frac{1}{4}$ takes place for an infinite number of indices k. It means that $\sum \xi_k = \sum Kv_n = \infty$ with probability one. So (4.17) cannot hold unless $\sum \kappa_n < \infty$. The proof is complete. □

5 APPLICATION TO STATIONARY PROCESSES WITH RATIONAL SPECTRAL DENSITY

Let us consider a finite segment $\{x_t, 0 \leq t \leq T\}$ of a second-order stationary process with spectral density of the form

$$f(\lambda) = \frac{1}{2\pi} \left| \sum_{k=0}^{n} a_{n-k}(i\lambda)^k \right|^{-2} \quad (-\infty < \lambda < \infty), \tag{5.1}$$

where the constants a_{n-k} are real and chosen so that all roots of $\sum a_{n-k}\lambda^k = 0$ have negative real parts. Put

$$X_t = \int_0^t x_s \, \mathrm{d}s \quad (0 \leq t \leq T), \tag{5.2}$$

and

$$Y_t = \sum_{k=0}^{n} a_{n-k} \, X_t^{(k)} \quad (0 \leq t \leq T), \tag{5.3}$$

where $X_t^{(k)} = (\mathrm{d}^k/\mathrm{d}t^k)\, X_t$. Of course, $X_t^{(k)} = x_t^{(k-1)}$ for $k > 0$.

Lemma 5.1. *The process Y_t, defined by (5.3), has uncorrelated increments and $\mathsf{E}|\mathrm{d}Y_t|^2 = \mathrm{d}t$. $Y_t - Y_0$ is uncorrelated with random variables $x_0, x_0', \ldots, x_0^{(n-1)}$, and the closed linear manifold $\mathcal{Y}$ generated by $\{Y_t - Y_0,\ 0 \leq t \leq T,\ x_0, \ldots, x_0^{(n-1)}\}$ equals the closed linear manifold $\mathcal{X}$ generated by $\{x_t,\ 0 \leq t \leq T\}$, $\mathcal{Y} = \mathcal{X}$.*

Proof. On making use of the well-known unitary mapping W, defined by $Wx_t = e^{it\lambda}$, and remembering that $R(x, y) = \int Wx\, Wy\, f(\lambda)\, \mathrm{d}\lambda$ we get from (5.2) and (5.3) that

$$W(Y_t - Y_0) = \frac{e^{it\lambda} - 1}{i\lambda} \sum_{k=0}^{n} a_{n-k}(i\lambda)^k \tag{5.4}$$

and $Wx_0^{(j)} = (i\lambda)^j$, $0 \leq j \leq n-1$. Consequently,

$$R(Y_t - Y_0,\, Y_s - Y_0) = \frac{1}{2\pi} \int_{-\infty}^{\infty} \frac{(e^{is\lambda} - 1)\,(e^{is\lambda} - 1)}{\lambda^2}\, \mathrm{d}\lambda = \min(t, s), \tag{5.5}$$

which proves that Y_t has uncorrelated increments and $\mathsf{E}|\mathrm{d}Y_t|^2 = \mathrm{d}t$. Further, we have

$$R(Y_t - Y_0,\, x_0^{(j)}) = \frac{1}{2\pi} \int_{-\infty}^{\infty} \frac{(e^{it\lambda} - 1)\,(-i\lambda)^{j-1}}{\sum a_{n-k}(-i\lambda)^k}\, \mathrm{d}\lambda = 0, \tag{5.6}$$

because, according to our assumption, the function under the integral sign is analytic in the upper hyperplane, including the real line, and is of order $|\lambda|^{j-1-n}$. Finally, if $v \in \mathcal{X}$ and $v \perp y_t$, $0 \le t \le T$, and $v \perp x_0^{(j)}$, $0 \le j \le n-1$, then the function $\varphi_t = R(x_t, v)$ satisfies the relations $(L\varphi)_t = 0$, $0 \le t \le T$, $\varphi_0^{(j)} = 0$, $0 \le j \le n-1$, i.e. $\varphi_t = 0$, $0 \le t \le T$ and $v \equiv 0$. The proof is concluded. □

Let $\mathcal{L}_n^2$ denote the set of complex-value functions of $t \in [0, T]$ processing squared integrable derivatives up to the order n. Introduce in $\mathcal{L}_n^2$ the linear operators

$$(L\varphi)_t = \sum_{k=0}^{n} a_{n-k}\, \varphi_t^{(k)}, \tag{5.7}$$

$$(L^*\varphi)_t = \sum_{k=0}^{n} (-1)^k\, a_{n-k}\, \varphi_t^{(k)}, \tag{5.8}$$

where a_{n-k} are taken from (5.1). Introduce in $\mathcal{L}_{2n}^2$ the operator

$$\begin{aligned}(L^* L\varphi)_t &= \sum_{k=0}^{n}\sum_{j=0}^{n} (-1)^k\, a_{n-k}\, a_{n-j}\, \varphi_t^{(j+k)} \\ &= (-1)^n\, a_0^2\, \varphi_t^{(2n)} + \sum_{k=0}^{n} (-1)^k\, A_{n-k}\, \varphi_t^{(2k)}.\end{aligned} \tag{5.9}$$

Theorem 5.1. *Let $\{x_t,\ t \in T\}$ be a finite segment of the stationary process with spectral density (5.1). Let $(\Phi^x,\ \mathcal{Q}^x)$ be a closed linear manifold generated by covariances $R(x_t, x_s)$, $0 \le t,\ s \le T$.*

Then, in the above notation, $\Phi^x = \mathcal{L}_n^2$ and

$$\mathcal{Q}^x(\psi, \varphi) = \int_0^T (L\psi)_t\, (L\overline{\varphi}_t)\, \mathrm{dt} \tag{5.10}$$

$$+ \sum_{\substack{0 \le j,\, k \le n-1 \\ j+k \text{ even}}}\sum \psi_0^{(j)}\, \overline{\varphi}^{(k)} 2 \sum_{i=\max(0, j+k+1-n)}^{\min(j,k)} (-1)^{j-i}\, a_{n-i}\, a_{n+i-j-k-1},$$

or, equivalently,

$$\mathcal{Q}^x(\psi, \varphi) = a_0^2 \int_0^T \psi_t^{(n)}\, \overline{\varphi}_t^{(n)}\, \mathrm{dt} + \sum_{k=1}^{n} A_{n-k} \int_0^T \psi_t^{(k)}\, \overline{\varphi}_t^{(k)}\, \mathrm{dt} \tag{5.11}$$

$$+ \sum_{0 \le j}\sum_{k \le n-1} \left[\psi_t^{(j)}\, \overline{\varphi}_T^{(k)} + \psi_0^{(j)}\, \overline{\varphi}_0^{(k)}\right] \sum_{i=\max(0, j+k+1-n)}^{\min(j,k)} (-1)^{j-i}\, a_{n-i}\, a_{n+i-j-k-1}.$$

If $\varphi \in \mathcal{L}^2_{2n}$, we may write

$$\begin{aligned}\mathcal{Q}^x(\psi,\varphi) = &\int_0^T \psi_t (L^* L\overline{\varphi})_t \, \mathrm{d}t \\ &+ \sum_{j=0}^{n-1} \psi_T^{(j)} \sum_{k=0}^{n-1-j} (-1)^k \, (L\overline{\varphi})_T^{(k)} \, a_{n-j-k-1} \\ &+ \sum_{j=0}^{n-1} \psi_0^{(j)} \sum_{k=0}^{n-1-j} (-1)^j \, (L^*\overline{\varphi})_0^{(k)} \, a_{n-k-j-1}.\end{aligned} \tag{5.12}$$

The solution $v^\varphi = U^{-1}\varphi$ of the equation $\varphi_t = R(x_t, v)$, $\varphi \in \Phi^x$, is given by (5.10) or (5.11) on substituting $\mathrm{d}X_t^{(k)}$ for $\psi_t^{(k)}\mathrm{d}t$, $0 \le k \le n$, and defining the integrals in the sense of Section 3.1. Random variables $x_t(\omega)$ may be adjusted on zero-probability ω-subsets so that the trajectories $x^\omega = x_t(\omega)$ belong to $\mathcal{L}^2_{n-1}$ with probability 1 and for $\varphi \in \mathcal{L}^2_{n+1}$, $v^\varphi(\omega)$ is given by (5.10) or (5.11) on substituting $x_t^{(k)}(\omega)$ for $\psi_t^{(k)}$, $0 \le k \le n-1$, and

$$x_T^{(n-1)}(\omega)\, \overline{\varphi}_T^{(j)} - x_0^{(n-1)}(\omega)\, \overline{\varphi}_0^{(j)} - \int_0^T x_t^{(n-1)}(\omega)\, \overline{\varphi}_t^{(j+1)} \, \mathrm{d}t$$

$$\text{for} \quad \int_0^T x_t^{(n)} \, \varphi_t^{(j)} \, \mathrm{d}t.$$

If $\varphi \in \mathcal{L}^2_{2n}$, $v^\varphi(\omega)$ is given by

$$\begin{aligned} v^\varphi(\omega) \;=\; &\int_0^T x_t(\omega) \, (L^* L\overline{\varphi})_t \, \mathrm{d}t \\ &+ \sum_{j=0}^{n-1} x_T^{(j)}(\omega) \sum_{k=0}^{n-1-j} (-1)^k \, (L\overline{\varphi})_T^{(k)} \, a_{n-j-k-1} \\ &+ \sum_{j=0}^{n-1} x_0^{(j)}(\omega) \sum_{k=0}^{n-1-j} (-1)^j \, (L\overline{\varphi})_0^{(k)} \, a_{n-j-k-1}.\end{aligned} \tag{5.13}$$

Proof. From Lemma 5.1 it follows that every element $v \in \mathcal{X}$ is of form

$$v = v_1 + v_2 = \sum_{j=0}^{n-1} c_j \, x_0^{(j)} + \int_0^T g(t) \, \mathrm{d}Y_t, \tag{5.14}$$

where $\int |g(t)|^2 \, \mathrm{d}t < \infty$, and the random variable $v_1 = \sum c_j \, x_0^{(j)}$ is uncorrelated with $\mathrm{d}Y_t$, $0 \le t \le T$, and $v_2 = \int g_t \, \mathrm{d}Y_t$ is uncorrelated with $x_0^{(j)}$, $0 \le j \le n-1$. If $\varphi_t = R(x_t, v)$, then

$$\varphi_0^{(j)} = R(x_0^{(j)}, v) = R(x_0^{(j)}, v_1) \quad (0 \le j \le n-1), \tag{5.15}$$

and, in view of (5.3),

$$(L\varphi)_t = \frac{\mathrm{d}}{\mathrm{d}t} R(Y_t, v) = R(y_t, v_2) \quad (0 \le t \le T), \tag{5.15'}$$

where $y_t = \mathrm{d}Y_t/\mathrm{d}t$ is the white noise and $L\varphi \in \mathcal{L}^2$. Consequently, if $\varphi \in \Phi^x$, then $\varphi \in \mathcal{L}_n^2$. Now, let $\varphi \in \mathcal{L}_n^2$, and choose the constants $c_0, \ldots, c_{n-1}$ and the function g_t such that (5.15) and (5.15′) hold true for v given by (5.14). A direct determination of $c_0, \ldots, c_{n-1}$ would involve cumbersome inversion of the matrix $(R(x_0^{(j)}, x_0^{(k)}))_{j,k-1}^n$. For this reason, we first establish g_t and, subsequently, we find $c_0, \ldots, c_{n-1}$ by an indirect method. In accordance with (3.5), (5.15′) is solved by

$$v_2 = \int_0^T (L\overline{\varphi})_t \, \mathrm{d}Y_t = \int_0^T (L\overline{\varphi})_t \, \mathrm{d}(LX)_t. \tag{5.16}$$

The last form is based on (5.3). On inserting (5.16) in (5.14), we obtain

$$v = \sum_{j=0}^{n-1} c_j \, x_0^{(j)} + \int_0^T (L\overline{\varphi})_t \, \mathrm{d}(LX)_t. \tag{5.17}$$

Now consider a unitary and self-adjoint operator V defined in $\mathcal{X}$ by $Vx_t = x_{T-t}$. On carrying over V to Φ^x, we obtain $(V\varphi)_t = R(Vx_t, v^\varphi) = R(x_{T-t}, v^\varphi) = \varphi_{T-t}$. If $\varphi_{T-t} = \varphi_t$, $0 \le t \le T$, we have $V\varphi = \varphi$ and $v = V^*v = Vv$ for $v = U_x^{-1}\varphi$ (i.e. for v such that $\varphi_t = R(x_t, v)$). Consequently, if $\varphi_{T-t} = \varphi_t$, the formula (5.17) must be invariant with respect to the substitution of x_{T-t} for x_t, i.e. of $(L^*X)_{T-t}$ for $(LX)_t$ and $(-1)^j \, x_T^{(j)}$ for $x_0^{(j)}$. This means that, for $\varphi_{T-t} = \varphi_t$, we have

$$\begin{aligned} v &= \int_0^T (L\overline{\varphi})_t \, \mathrm{d}(L^*X)_{T-t} + \sum_{j=0}^{n-1} c_j(-1)^j \, x_T^{(j)} \\ &= \int_0^T (L^*\varphi)_t \, \mathrm{d}(L^*X)_t + \sum_{j=0}^{n-1} c_j(-1)^j \, x_T^{(j)}, \end{aligned} \tag{5.18}$$

where the last expression is obtained from the preceding one on substituting $(L\overline{\varphi})_t = (L^*\overline{\varphi})_{T-t}$, which is justified because $\varphi_{T-t} = \varphi_t$. It is easy to show that

$$\int_0^T (L\overline{\varphi})_t \, \mathrm{d}(LX_t) \tag{5.19}$$

$$= \int_0^T (L^*\overline{\varphi})_t \, \mathrm{d}(L^*X)_t + \sum_{\substack{j+k \text{ odd} \\ k>j}}^{n-1} \sum^{n-1} a_{n-k} \, a_{n-j}$$

$$\times \sum_{\nu=0}^{k-j-1} (-1)^\nu \left[\overline{\varphi}_T^{(j+\nu)} x_T^{(k-j-\nu)} - \overline{\varphi}_0^{(j+\nu)} x_0^{(k-j-\nu)}\right]$$

$$= \int_0^T (L^*\overline{\varphi})_t \,\mathrm{d}(L^*X)_t + \sum_{j+k \text{ even}}^{n-1}\sum^{n-1} (x_T^{(j)}\overline{\varphi}_T^{(k)} - x_0^{(j)}\overline{\varphi}_0^{(k)})$$

$$\times \sum_{i=\max(0,j+k+1-n)}^{\min(j,k)} (-1)^{j-i} a_{n-i}\, a_{n+i-j-k-1}.$$

On comparing (5.17), (5.18) and (5.19), we can easily see that

$$v_1 = \sum_{j=0}^{n-1} c_j\, x_0^{(j)} \tag{5.20}$$

$$= \sum_{j+k \text{ even}}^{n-1}\sum^{n-1} x_0^{(j)}\overline{\varphi}_0^{(k)}\, 2 \sum_{i=\max(0,j+k+1-n)}^{\min(j,k)} (-1)^{j-i} a_{n-i}\, a_{n+i-j-k-1},$$

which inserted in (5.17) gives the required result. Since the condition $\varphi_{T-t} = \varphi_t$ imposes no restrictions on $\varphi_0, \ldots, \varphi_0^{(n-1)}$, (5.20) represents a general solution of (5.15).

As we know, $f(x) = R(x, v^\varphi)$ is a linear functional attaining the value φ_t at the point $x = x_t$. Consequently, $\mathcal{Q}(\psi,\varphi) = R(v^\varphi, v^\psi)$ is obtained from (5.17) simply by replacing x_t by ψ_t. Bearing in mind (5.20), we thus obtain (5.10). Formulas (5.11) and (5.12) are obtained from (5.10) simply by integration by parts, the details of which we shall not reproduce here. In deriving (5.12) we may apply Green's formula and the principle of symmetry, assuming that $\varphi_{T-t} = \varphi_t$.

In order to show that (5.13) may be considered as a special case of (4.1), let us extend the parameter set by adding parameter values $01, \ldots, 0n$ and $T1, \ldots, Tn$ and putting $x_{0i} = x_0^{(i-1)}$ and $x_{Ti} = x_T^{(i-1)}$, $1 \le i \le n$. The white noise y_t will be extended so that y_{0i} are any orthonormal linear combinations of $x_{0i} = x_0^{(i-1)}$, and $y_{T1}, \ldots, y_{Tn}$ are any orthonormal random variables uncorrelated with x_t, $0 \le t \le T$ (or with y_t, $0 \le t \le T$, and $y_{01}, \ldots, y_{0n}$). Defining the linear operator K by $x_t = Ky_t$, $0 \le t \le T$, $t = 01, \ldots, 0n, T1, \ldots, Tn$, we can see that, for $0 \le t \le T$, L coincides with K_x^{-1} defined by (2.10). In our case $K_x^{-1}\varphi$ simply is such a solution of $K(\cdot) = \varphi$ that vanishes for $t = T1, \ldots, Tn$. After this extension of the set of parameter values, (5.13) is equivalent to (3.17), where μ is defined as the Lebesgue measure for $0 \le t \le T$, and $\mu(0i) = \mu(Ti) = 1$, $1 \le i \le n$. Moreover, (5.13) is equivalent (4.1). The only purpose of adding points $T1, \ldots, Tn$ is to enlarge the range of the operator KK^* so that it contains all functions from $\mathcal{L}_{2n}^2$ (see Remark 4.1).

Finally, if $n > 1$ then the derivative $x_t' = z_t$ exists. Considering the operator K defined by

$$x_t = K z_t = x_0 + \int_0^t z_s \, \mathrm{d}s,$$

and applying Theorem 4.2, we readily see that $x_t(\omega)$ may be adjusted so that $x^\omega \in \mathcal{L}_{n-1}^2$ with probability 1, and that, for $\varphi \in \mathcal{L}_{n+1}^2$, the solution of $\varphi_t = R(x_t, v)$ has the form mentioned in the theorem. The proof is complete. □

Remark 5.1. From (5.20) it follows that the inverse of $(R(x_0^{(j)}, x_0^{(k)}))_{j,k=0}^{n-1}$ consists of elements

$$\begin{aligned} D_{jk} &= 2 \sum_{i=\max(0,\, j+k+1-n)}^{\min(j,k)} (-1)^{j-i} a_{n-i}\, a_{n+i-j-k-1} \quad \text{for} \quad j+k \text{ even}, \\ &= 0 \quad \text{for} \quad j+k \text{ odd}. \end{aligned} \tag{5.21}$$

Now, we shall consider a general rational spectral density

$$g(\lambda) = \frac{1}{2\pi} \left| \frac{\sum_{k=0}^m b_{m-k}(i\lambda)^k}{\sum_{k=0}^n a_{n-k}(i\lambda)^k} \right|^2 \quad (-\infty < \lambda < \infty;\ m < n), \tag{5.22}$$

where the as and bs are real and all roots of equations $\sum a_{n-k}\lambda^k = 0$ and $\sum b_{m-k}\lambda^k = 0$ have negative real parts. As is well known, if x_t is a process with spectral density (5.1), then the process

$$z_t = \sum_{k=0}^m b_{m-k}\, x_t^{(k)} \quad (m < n) \tag{5.23}$$

has the spectral density (5.22).

Let us add $2m$ parameter values $01, \dots, 0m, T1, \dots, Tm$ and define $z_{0i} = x_{0i} = x_0^{(i-1)}$, $z_{Ti} = x_{Ti} = x_T^{(i-1)}$, $1 \le i \le m$. Then the operator H defined by $x_t = H z_t$, $0 \le t \le T$, $t = 01, \dots, 0m, T1, \dots, Tm$ is bounded and may be expressed by

$$x_t = \sum_{i=1}^m x_0^{(i-1)} h_j(t) + \int_0^T H(t,s)\, z_s \, \mathrm{d}s \quad (0 \le t \le T), \tag{5.24}$$

where $h_j(t)$ are m linearly independent solutions of the equation $\sum b_{m-k}\, h(t) = 0$, and $H(t,s)$ may be established by well-known methods by Coddington and Levinson (1955) of the theory of linear differential equations. The differential operator involved in (5.23) will be denoted by

$$M = \sum_{k=0}^m b_{m-k} \frac{\mathrm{d}^k}{\mathrm{d}t^k}. \tag{5.25}$$

Further, we put

$$M^* = \sum_{k=1}^{m} b_{m-k}(-1)^k \frac{\mathrm{d}^k}{\mathrm{d}t^k} \quad \text{and} \quad MM^* = \sum_{k=1}^{m}\sum_{j=1}^{m} b_{m-k}\, b_{m-j}(-1)^k \frac{\mathrm{d}^{k+1}}{\mathrm{d}t^{k+j}}.$$

Theorem 5.2. *Let $\{z_t,\ 0 \le t \le T\}$ be a finite segment of the stationary process with spectral density (5.22). Let $(\Phi^z, \mathcal{Q}^z)$ be the closed linear manifold generated by covariances $R(z_t, z_s)$.*

Then $\Phi^z = \mathcal{L}^2_{n-m}$ and the solution $v^\chi = U_z^{-1}\chi$ of the equation $\chi_t = R(z_t, v)$ is the same as the solution of the equation $\psi_t = R(x_t, v)$, where x_t is given by (5.24) and ψ_t is uniquely determined by

$$(M\psi)_t = \chi_t \quad \text{and} \quad \mathcal{Q}^x(\psi, h_j) = 0 \quad (1 \le j \le m), \tag{5.26}$$

where M is given by (5.25), $h_1, \ldots, h_m$ are m independent solutions of $\sum b_{m-k}h(t) = 0$ and $\mathcal{Q}^x$ is given by (5.10) or (5.11). The inner product $\mathcal{Q}^z$ in Φ^z is given by $\mathcal{Q}^z(\nu, \chi) = \mathcal{Q}^x(\xi, \psi)$, where $\mathcal{Q}^x$ is given by (5.10), ψ is determined by (5.26), and ξ is also determined by (5.26) on substituting ν for χ.

Random variables $z_t(\omega)$ may be adjusted on zero-probability ω-subsets so that the trajectories $z^\omega = R_t(\omega)$ belong to $\mathcal{L}^2_{n-m-1}$ with probability 1, and that the involved integral and differential operators may be applied to individual trajectories for $\chi \in \mathcal{L}^2_{n-m+1}$. If $\chi \in \mathcal{L}^2_{2n-2m}$, then the solution is given by

$$\begin{aligned} v^\chi(\omega) \;=\; & \int_0^T z_t(\omega)\,(L^*L\varphi)_t\,\mathrm{d}t \\ & + \sum_{j=0}^{n-m-1} z_T^{(j)}(\omega) \sum_{k=0}^{n-1-j} (-1)^z\,(L\varphi)_T^{(k)}\,a_{n-j-k-1} \\ & + \sum_{j=0}^{n-m-1} z_0^{(j)}(\omega) \sum_{k=0}^{n-1-j} (-1)^j\,(L^*\varphi)_0^{(k)}\,a_{n-j-k-1}, \end{aligned} \tag{5.27}$$

where φ_t is uniquely determined by

$$(MM^*\varphi)_t = \chi_t, \tag{5.28}$$

and

$$\sum_{k=0}^{n-1-j} (-1)^k\,(L\varphi)_T^{(k)}\,a_{n-j-k-1} = 0 \qquad (n-m \le j \le n-1), \tag{5.29}$$

$$\sum_{k=0}^{n-1-j} (-1)^j\,(L^*\varphi)_T^{(k)}\,a_{n-j-k-1} = 0 \qquad (n-m \le j \le n-1). \tag{5.30}$$

Proof. Variances generated by spectral density (5.22) and variances generated by spectral density (5.1), where $n \equiv n-m$, dominate each other, so that the closed linear manifolds generated by respective covariances coincide (cf. Remark 2.1). So, in view of Theorem 5.1, $\Phi^z = \mathcal{L}^2_{n-m}$.

Now, if $R(x_t, v) = \varphi_t$, and φ_t satisfies (5.26), then obviously $R(z_t, v) = \chi_t$. Further, it is easy to verify on the basis of (5.24) that $Q^x(\varphi, h_j) = 0$, $1 \leq j \leq m$, guarantees that the solution v belongs to the closed linear manifold $\mathcal{Z}$ generated by random variables z_t, $0 \leq t \leq T$.

Further, if $\chi \in \mathcal{L}^2_{2n-2m}$, and the conditions (5.29) and (5.30) are satisfied, then the operator M may be applied to the right-hand side of (5.13) term by term under the integral sign, which shows that $Mv^\varphi = v^\chi$, where v^χ is given by (5.27). Consequently $R(z_t, v^\chi) = MM^* R(x_t, v\varphi) = (MM^*\varphi)_t = \chi_t$, $0 \leq t \leq T$, in accordance with (5.27).

The assertions concerning individual trajectories are implied by Theorems 4.1 and 4.2 as in the preceding theorem. □

Remark 5.2. If $h_1, \ldots, h_m$ are the solutions of $Mh = 0$, then the solutions of $MM^*h = 0$ are $h_1, \ldots, h_m\, h_1^*, \ldots, h_m^*$, where $h^*(t) = h(T-t)$. Let φ_0 be a particular solution of $MM^*\varphi = \chi$. Then

$$\varphi = \varphi_0 + \sum_1^m (c_j h_j + d_j h_j^*), \quad \text{where} \quad c_1, \ldots, c_m,\, d_1, \ldots, d_m$$

are determined by (5.29) and (5.30). If $\chi_t = \chi_{T-t}$, then $c_j = d_j$, and if $\chi_t = -\chi_{T-t}$, then $c_j = -d_j$. In both cases the number of unknown constants is m, and not $2m$, and they are determined by (5.29) or (5.30). Denoting $\chi_t^* = \chi_{T-t}$, we have

$$\varphi = \varphi^0 + \frac{1}{2}\sum_{j=1}^m \left\{ c_j(\chi + \chi^*)\,[h_j + h_j^*] + c_j(\chi - \chi^*)\,[h_j - h_j^*] \right\}, \qquad (5.31)$$

where $c_j(\nu)$ denote the constants corresponding to a function ν.

Example 5.1. If $M\varphi = \varphi' + \beta\varphi$, $\beta > 0$, then the solution of (5.28) is given as

$$\varphi_t = \frac{1}{2\beta}\int_0^T e^{-|t-s|\beta}\chi_s \, ds + \frac{1}{2}(c_1 + c_2)\, e^{-t\beta} + \frac{1}{2}(c_1 - c_2)\, e^{-(T-t)\beta}, \quad (5.31')$$

where

$$c_1 = \frac{\sum_{1\leq j\leq n/2} \beta^{2j-2} \sum_{k=0}^{n-2j} a_{n-2j-k}(\chi_0^{(k)} + \chi_T^{(k)}) - L(-\beta)\int_0^T (e^{-t\beta} + e^{-(T-t)\beta})\chi_t dt}{L(\beta) + L(-\beta)\, e^{-T\beta}} \qquad (5.32)$$

and

$$c_2 = \left(\sum_{1 \le j \le n/2} \beta^{2j-2} \sum_{k=0}^{n-2j} a_{n-2j-k} (\chi_0^{(k)} - (-1)^k \chi_T^{(k)}) \right. \tag{5.33}$$
$$\left. - L(-\beta) \int_0^T (e^{-t\beta} - e^{-(T-t)\beta}) \chi_t \mathrm{d}t \right) \left(L(\beta) - L(-\beta)\, e^{-T\beta} \right)^{-1}$$

where

$$L(\beta) = \sum_{k=0}^{n} a_{n-k}\, \beta^k .$$

Moreover,

$$(L^* L \varphi)_t = \varphi_t L^* L(\beta) + \frac{L^* L\left(\frac{\mathrm{d}}{\mathrm{d}t}\right) - L^* L(\beta)}{\beta^2 - \left(\frac{\mathrm{d}}{\mathrm{d}t}\right)^2} \chi_t, \tag{5.34}$$

where $L^* L(\beta) = L(-\beta)\, L(\beta)$ and $\mathrm{d}/\mathrm{d}t$ denotes the differential operator. We omit the details.

Integral-valued parameter t. The spectral density of the nth order Markovian process with integral-valued t is given by

$$f(\lambda) = \frac{1}{2\pi} \left| \sum_{k=0}^{n} a_{n-k}\, e^{i\lambda k} \right|^{-2} \quad (-\pi \le \lambda \le \pi), \tag{5.35}$$

where a_{n-k} are real and such that all roots of $\sum_{k=0}^{n} a_{n-k}\, \lambda^k = 0$ are greater than 1 in absolute value.

It may be easily shown that the random variables

$$y_t = \sum_{k=0}^{n} a_{n-k}\, x_{t-k} \quad (n \le t \le N) \tag{5.36}$$

are uncorrelated mutually as well as with $x_0, \ldots, x_{n-1}$, and have unit variance, $R(y_t) = 1$. If we consider a finite segment of the process $\{x_t,\, 0 \le t \le N\}$, $N \ge 2n$, then Φ^x consists of all complex-valued functions φ_t, $0 \le t \le N$, and it may be shown by the previous method that

$$Q^x(\psi, \varphi) = \sum_{t=0}^{N} \sum_{s=0}^{N} \psi_t\, \overline{\varphi}_s\, Q^x_{ts},$$

where for $|t-s|>n$ $Q^x_{ts}=0$, and for $|t-s|\le n$,

$$\begin{aligned}
Q^x_{ts} &= \sum_{i=0}^{\min[N-t,N-s,n-|t-s|]} a_{n-i}\, a_{n-i-|t-s|}, && \text{if } \max(t,s) > N-n,\\
&= \sum_{i=0}^{n-|t-s|} a_{n-i}\, a_{n-i-|t-s|}, && \text{if } n \le t,\ s \le N-n,\\
&= \sum_{i=0}^{\min[t,s,n-|t-s|]} a_{n-i}\, a_{n-i-|t-s|}, && \text{if } \min(t,s) < n.
\end{aligned} \tag{5.37}$$

The solution of $\varphi_t = R(x_t, v)$ is given by

$$v^\varphi = \sum_{i=0}^{N}\sum_{s=0}^{N} x_t\, \overline{\varphi}_s\, Q^x_{ts}, \tag{5.38}$$

which may be put in a form analogous to (5.13):

$$v^\varphi = \sum_{t=n}^{N-n} x_t (L^* L\varphi)_t + \sum_{t=N-n+1}^{N} x_t \sum_{k=0}^{N-t} (L\varphi)_{t+k}\, a_{n-k} + \sum_{t=0}^{n-1} x_t \sum_{k=0}^{t} (L^*\varphi)_{t-k}\, a_{n-k}, \tag{5.39}$$

where

$$(L\varphi)_t = \sum_{k=0}^{n} a_{n-k}\, \varphi_{t-k} \quad \text{and} \quad (L^*\varphi)_t = \sum_{k=0}^{n} a_{n-k}\, \varphi_{t+k}.$$

We also find that the inverted covariance matrix $(R(x_j, x_k))_{j,k=0}^{n-1}$ consists of the following elements:

$$D_{jk} = \sum_{i=0}^{\min[j,k,n-|j-k|]} a_{n-i}\, a_{n-i-|j-k|} - \sum_{i=n-\max(j,k)}^{n-|j-k|} a_{n-i}\, a_{n-i+|j-k|} \qquad (0 \le j,\ k \le n-1). \tag{5.40}$$

If we have a process $\{z_t,\ 0 \le t \le N\}$ possessing a general rational spectral density

$$g(\lambda) = \frac{1}{2\pi} \left| \frac{\sum_{k=0}^{m} b_{m-k}\, e^{i\lambda k}}{\sum_{k=0}^{n} a_{n-k}\, e^{i\lambda k}} \right|^2 \quad (-\pi \le \lambda \le \pi), \tag{5.41}$$

where both the as and bs are real and such that the roots of $\sum_{k=0}^{n} a_{n-k}\lambda^k = 0$ and $\sum_{k=0}^{m} b_{m-k}\lambda^k = 0$ lie outside the unit circle, we may proceed as we did in the continuous-parameter case.

First, we may consider the process $\{x_t, -m \leq t \leq N\}$ having spectral density (5.35) related to z_t by the difference equation

$$z_t = \sum_{k=0}^{m} b_{m-k}\, x_{t-k} = (Mx)_t \quad (0 \leq t \leq N). \tag{5.42}$$

Then the equation $\chi_t = R(z_t, v)$ is solved by

$$v^{\varphi} = \sum_{t=-m}^{N} \sum_{s=-m}^{N} x_t\, \overline{\psi}_t\, Q_x^{ts},$$

where ψ is uniquely determined by $(M\psi)_t = \sum_{k=0}^{m} b_{m-k}\, \psi_{t-k} = \chi_t$, $0 \leq t \leq N$, and $\mathcal{Q}^x(\psi, h_j) = 0$ for m arbitrary linearly independent solutions of $(Mh)_t = 0$, $0 \leq t \leq N$. Similarly $\mathcal{Q}^z(\nu, \chi) = \mathcal{Q}^x(\xi, \psi)$, where ξ is an arbitrary solution of the equation $M\xi = \nu$.

Second, we may consider the process $\{x_t, -m \leq t \leq N+m\}$ and the adjoint operator

$$(M^*\varphi)_t = \sum_{k=0}^{m} b_{m-k}\, \varphi_{t+k}.$$

Then we obtain the following analogues of equation (5.27):

$$v = \sum_{t=0}^{N} z_t (L^* L\varphi)_t \quad (m \geq n) \tag{5.43}$$

and

$$\begin{aligned} v \quad = \quad & \sum_{t=n-m}^{N-n+m} z_t (L^* L\varphi)_t + \sum_{t=N-n+m}^{N} z_t \sum_{k=0}^{N-t+m} (L\varphi)_{t+k}\, a_{n-k} \\ & + \sum_{t=0}^{n-m-1} z_t \sum_{k=0}^{t+m} (L^*\varphi)_{t-k}\, a_{n-k} \quad (m < n), \end{aligned} \tag{5.44}$$

where φ_t is uniquely determined by $(MM^*\varphi)_t = \chi_t$, $0 \leq t \leq N$, and

$$\sum_{k=0}^{r} (L^*\varphi)_{r-m-k}\, a_{n-k} = 0 \quad (r = 0, \ldots, m-1) \tag{5.45}$$

and

$$\sum_{k=0}^{r} (L\varphi)_{N+m-r+k}\, a_{n-k} = 0 \quad (r = 0, \ldots, m-1), \tag{5.46}$$

cf. (5.29) and (5.30). Remark 5.2 is also valid for the present case. If $n = 0$, we get results for the moving-average scheme.

6 STRONG EQUIVALENCY OF NORMAL DISTRIBUTIONS

Let us first consider two normal distributions P and P^+ of a random sequence $\{v_n, n \geq 1\}$ defined by vanishing mean values $\mathsf{E}v_n = \mathsf{E}^+v_n = 0$, $n \geq 1$, and covariances

$$\begin{aligned} R(v_n, v_m) &= 0, \quad R^+(v_n, v_m) = 0, \qquad \text{if} \quad n \neq m, \\ R(v_n, v_n) &= 1, \quad R^+(v_n, v_n) = \frac{1}{\lambda_n}, \qquad (n \geq 1). \end{aligned} \tag{6.1}$$

The J-divergence of P and P^+ restricted to the vector $\{v_1, \ldots, v_n\}$, say J_n, equals (see Hájek (1958 c))

$$J_n = \frac{1}{2}\sum_{i=1}^{n} \frac{(1-\lambda_i)^2}{\lambda_i}, \tag{6.2}$$

and, consequently, $J_\infty = \lim_{n\to\infty} J_n < \infty$, if and only if

$$\sum_{n=1}^{\infty} (1-\lambda_n)^2 < \infty. \tag{6.3}$$

So, according to Hájek (1958 c), P and P^+ defined on the Borel field generated by $\{v_n, n \geq 1\}$ are equivalent, $\mathsf{P} \sim \mathsf{P}^+$, if and only if (6.3) is true. If $\mathsf{P} \sim \mathsf{P}^+$, then it follows from theory of martingales (Doob (1953, Th. 4.3, Ch. VIII)) that

$$\frac{d\mathsf{P}}{d\mathsf{P}^+} = \exp\left\{-\frac{1}{2}\sum_{1}^{\infty}[v_n^2(1-\lambda_n) + \log\lambda_n]\right\}, \tag{6.4}$$

where the sum converges with probability 1. In general, however, we cannot write

$$\frac{d\mathsf{P}}{d\mathsf{P}^+} = \left(\prod_1^\infty \lambda_n\right)^{-\frac{1}{2}} \exp\left\{-\frac{1}{2}\sum_1^\infty v_n^2(1-\lambda_n)\right\}, \tag{6.5}$$

unless the product $\prod_1^\infty \lambda_n$ is absolutely convergent. The well-known necessary and sufficient condition for absolute convergence of $\prod_1^\infty \lambda_n$ is $\lambda_n \neq 0$ and

$$\sum_1^\infty |1-\lambda_n| < \infty, \tag{6.6}$$

which is stronger then (6.3). If (6.6) is satisfied, then almost surely

$$\sum_1^\infty v_n^2(\omega)\,|1-\lambda_n| < \infty, \tag{6.7}$$

because $\mathsf{E}\left\{\sum_1^\infty v_n^2(\omega)\,|1-\lambda_n|\right\} \leq \sum_1^\infty |1-\lambda_n| < \infty$.

Thus (6.6) is a necessary and sufficient condition for $d\mathsf{P}/d\mathsf{P}^+$ being of the form (6.5), where the product $\prod_1^\infty \lambda_n$ is absolutely convergent and the quadratic form $\sum_1^\infty v_n^2(1-\lambda_n)$ is absolutely convergent with probability 1. This leads us to the notion of strongly equivalent normal distributions introduced in the following.

Definition 6.1. Two covariances R and R^+ will be called strongly equivalent if they dominate each other (Definition 2.2), the operator B defined by $R(x,y) = R^+(Bx,y)$ has a pure point spectrum, and the non-unity eigen-values λ_n of B satisfy the condition (6.6). Two normal distributions P and P^+ defined by strongly equivalent covariances R and R^+ and by vanishing mean values will be called strongly equivalent.

Lemma 6.1. *The probability density of a normal distribution* P *with respect to another normal distribution* P^+, *strongly equivalent to* P, *is given by* (6.5), *where* v_n *and* λ_n *denote the eigen-elements and eigen-values of the operator* B *mentioned in Definition* 6.1. *The* v_n*s are normed by* $R(v_n, v_n) = 1,\ n \geq 1$.

Proof. Obvious. □

Unfortunately, the right-hand side of (6.5) scarcely may be considered as an ultimate expression for $d\mathsf{P}/d\mathsf{P}^+$, because the eigen-values and eigen-elements are difficult to establish even if $d\mathsf{P}/d\mathsf{P}^+$ may be found explicitly (see Section 7). This leads us to derive a theory of the product $\prod_1^\infty \lambda_n$ and of the quadratic form $\sum_1^\infty v_n^2(1-\lambda_n)$.

We begin with the following definition.

Definition 6.2. Let H be a compact symmetric operator, and κ_n be the non-zero eigen-values of H. If $\sum_1^\infty |\kappa_n| < \infty$, we say that H has a finite trace, and put $\operatorname{tr} H = \sum_1^\infty \kappa_n$.

Definition 6.3. Let $B - I$ be a compact symmetric operator and λ_n be the non-unity eigen-values of B. If the product $\prod_1^\infty \lambda_n$ is absolutely convergent, we say that B has a determinant, and put $\det B = \prod_1^\infty \lambda_n$.

Obviously, B has a determinant if and only if the eigen-values of B are different from 0 and $B - I$ has a finite trace. If $\{v_n\}$ is the orthonormal system of eigen-elements of B, corresponding to the non-zero eigen-values, and $\{x_n\}$ is another orthonormal system in $(\mathcal{X}, R^+)$, then

$$\sum_1^\infty |(Hx_n, x_n)| = \sum_{n=1}^\infty \left|\sum_{j=1}^\infty (Hx_n, v_j)(x_n, v_j)\right| = \sum_{n=1}^\infty \left|\sum_{j=1}^\infty \kappa_j |(x_n, v_j)|^2\right|$$

$$\leq \sum_{j=1}^{\infty} |\kappa_j| \sum_{n=1}^{\infty} |(x_n, v_j)|^2,$$

i. e.

$$\sum_{1}^{\infty} |(Hx_n, x_n)| \leq \sum_{1}^{\infty} |\kappa_j|, \tag{6.8}$$

where $(\cdot, \cdot) = R^+(\cdot, \cdot)$ and $H = B - I$.

Denote by $\log B$ the operator which has the same eigen-elements as the operator B but eigen-values $\log \lambda_n$ instead of λ_n. Denoting the eigen-elements of B by v_n, we have

$$(Bx, x) = \sum_{1}^{\infty} \lambda_n |(x, v_n)|^2 \quad \text{and} \quad (\log Bx, x) = \sum_{1}^{\infty} \log \lambda_n |(x, v_n)|^2.$$

On making use of Jensen inequality, we get for $(x, x) = 1$

$$\log(Bx, x) \geq (\log Bx, x) \quad [(x, x) = 1]. \tag{6.9}$$

Let $\mathcal{X}_B$ be the subspace of $\mathcal{X}$ spanned by eigen-elements of B corresponding to non-unity eigen-values. $\mathcal{X}_B$ is separable even if $\mathcal{X}$ is not separable. For any orthonormal system $\{x_n\}$ complete in $\mathcal{X}_B$ we have

$$\log \det B = \operatorname{tr} \log B = \sum_{1}^{\infty} (\log Bx_n, x_n) \leq \sum_{1}^{\infty} \log(Bx_n, x_n),$$

i. e.

$$\det B \leq \prod_{n=1}^{\infty} (Bx_n, x_n), \tag{6.10}$$

where the absolute convergence on the right-hand side is guaranteed by (6.8).

Let $\mathcal{X}_n$ be an n-dimensional subspace of $\mathcal{X}$, and let I_n^+ denote the projection on $\mathcal{X}_n$. The contraction of B on $\mathcal{X}_n$, say B_n, will be defined as follows:

$$B_n = I + I_n^+ (B - I) I_n^+. \tag{6.11}$$

Obviously $(B_n x, x) = (Bx, x)$ if $x \in \mathcal{X}_n$, and $(B_n x, x) = (x, x)$ if $x \perp \mathcal{X}_n$. The operator B_n is well defined on $\mathcal{X}$ as well as on $\mathcal{X}_n$ and in both cases the determinant $\det B_n$ is the same. If $\mathcal{X}_n$ is spanned by linearly independent elements $x_1, \ldots, x_n$, then

$$\det B_n = \frac{|(Bx_i, x_j)|}{|(x_i, x_j)|} \quad (i, j = 1, \ldots, n), \tag{6.12}$$

where $|a_{ij}|$ is the ordinary determinant of a matrix (a_{ij}). Relation (6.12) will be clear if we take for $x_1, \ldots, x_n$ the eigen-elements of B_n.

Theorem 6.1. *Let $\mathcal{X}_1 \subset \mathcal{X}_2 \subset \ldots$ be a sequence of finite-dimensional subspaces of $\mathcal{X}$ such that the smallest subspace containing $\bigcup_1^\infty \mathcal{X}_n$ contains all eigen-elements of B corresponding to non-unity eigen-values. Let B_n be defined by* (6.11). *Then*

$$\det B = \lim_{n\to\infty} \det B_n . \tag{6.13}$$

Proof. Let v_j and $\lambda_j \neq 1$ be the eigen-elements and eigen-values of B, respectively. For any $\varepsilon > 0$ we may choose ℓ so that $\sum_{\ell+1}^\infty |\lambda_j - 1| < \varepsilon$ and then N so that in $\mathcal{X}_N$ there exists an orthonormal system $x_1, \ldots, x_\ell$ such that

$$\sum_1^\ell |((B-I)\, x_j\, , x_j)| \geq \sum_1^\ell |((B-I)\, v_j\, , v_j)| - \varepsilon = \sum_1^\ell |\lambda_j - 1| - \varepsilon . \tag{6.14}$$

In view of (6.8) we have for any orthonormal system $\{z_n\}$, such that $z_1 = x_1, \ldots, z_\ell = x_\ell$,

$$\sum_{\ell+1}^\infty |((B-I)\, z_n\, , z_n)| \leq \sum_1^\infty |\lambda_n - 1| - \sum_1^\ell |((B-I)\, z_n\, , z_n)| < 2\varepsilon . \tag{6.15}$$

Now let $n > N$, and $c^{-1} = \min_{1\leq j\leq \ell}(Bx_j, x_j)$. Let $\{v_{nj}\}$ and $\{v_j\}$ be the eigen-elements of B_n and B, respectively. From (6.10), (6.14) and (6.15) it follows that

$$\det B_n = \prod_{j=1}^n (Bv_{nj}\, , v_{nj}) \geq (1-2\varepsilon) \det B \tag{6.16}$$

and, in view of $\sum_{\ell+1}^\infty |\lambda_j - 1| < \varepsilon$,

$$\begin{aligned} \det B \ &\geq \ (1-\varepsilon) \prod_1^\ell (Bv_j\, , v_j) \geq (1-\varepsilon)\,(1-c\varepsilon) \prod_1^\ell (Bx_j\, , x_j) \\ &\geq \ (1-\varepsilon)\,(1-2\varepsilon)\,(1-c\varepsilon) \prod_1^n (Bx_j\, , x_j) \geq (1-3\varepsilon - c\varepsilon) \det B_n , \end{aligned} \tag{6.17}$$

where $x_1, \ldots, x_n$ is an orthonormal system from $\mathcal{X}_n$ such that $x_1, \ldots, x_\ell$ satisfy (6.14). Now we may let $\varepsilon \to 0$ and $c^{-1} \geq c_0^{-1} > 0$, where c_0^{-1} is independent of ε. The proof is complete. □

Now we shall study the ratio $\det B / \det B_n$, where B_n is given by (6.11), or more generally, $\det B / \det B_c$, where $B_c = I + I_c^+ (B-I)\, I_c^+$, I_c^+ denoting the projection on a subspace $\mathcal{X}_c$ (not necessarily finite-dimensional) of $\mathcal{X}$. More precisely, I_c^+ denotes the projection on $\mathcal{X}_c$ with respect to the covariance R^+. The projection with respect to the covariance $R(x, y) = R^+(Bx, y)$, say

I_c, will be generally different from I_c^+. Now, let us introduce the following 'conditional' covariances

$$R^+(x,y|\mathcal{X}_c) = R^+(x - I_c^+x,\, y - I_c^+y), \tag{6.18}$$
$$R(x,y|\mathcal{X}_c) = R(x - I_cx,\, y - I_cy). \tag{6.19}$$

If $R^+(\cdot|\mathcal{X}_c)$ dominates $R(\cdot|\mathcal{X}_c)$, denote by B^c the operator defined by

$$R(x,y|\mathcal{X}_c) = R^+(B^cx, y|\mathcal{X}_c). \tag{6.20}$$

Theorem 6.2. *If* $\det B$ *exists, then the determinants* $\det B_c$ *and* $\det B^c$ *also exist, and*

$$\det B = \det B_c \det B^c. \tag{6.21}$$

Proof. First suppose that we have $n+m$ random variables $x_1, \ldots, x_n, x_{n+1}, \ldots, x_{n+m}$ which are independent with unit variances if P^+ is true, and have an arbitrary non-singular normal distribution with covariance matrix $R = (R_{ij})$ if P is true. Let $R_n = (R_{ij})_{i,j=1}^n$ and $R^n = (R_{ij}^n)_{i,j=n+1}^{n+m}$, where R_{ij}^n is the conditional (partial) covariance of x_i and x_j $(i,\, j > n)$ for given $x_1, \ldots, x_n$. In this case (6.21) is equivalent to

$$|R| = |R_n|\,|R^n|, \tag{6.22}$$

where $|\cdot|$ denotes the determinant of the corresponding matrix. Equation (6.22) may be proved as follows. We introduce random variables $z_i = x_i$ $(1 \leq i \leq n)$ and $z_i = x_i - I_nx_i$ $(n+1 \leq i \leq n+m)$, where I_n is the projection on the subspace spanned by $x_1, \ldots, x_n$. In matrix notation $z = Ax$, $A = \{a_{ij}\}$, where $|A| = 1$, because $a_{ij} = 0$ $(i < j)$ and $a_{ii} = 1$ $(1 \leq i \leq n+m)$. The covariance matrix of random variables z_i equals

$$ARA^* = \begin{pmatrix} R_n & 0 \\ 0 & R^n \end{pmatrix}, \tag{6.23}$$

from which it follows that $|R| = |A|\,|R|\,|A^*| = |ARA^*| = |R_n|\,|R^n|$.

The general case will be obtained by Theorem 6.1. We take random variables $y_1, \ldots, y_m$ from $\mathcal{X} \ominus \mathcal{X}_c$ and n random variables $x_1, \ldots, x_n$ from $\mathcal{X}_c$ so that the closed linear manifold spanned by $x_1, \ldots, x_n$ contains the projection I_cy_i and $I_c^+y_i$ of y_i on $\mathcal{X}_c$. Then we let $n \to \infty$, and subsequently $m \to \infty$ so that the closed linear manifold spanned by $\{y_1, y_2, \ldots, x_1, x_1, \ldots\}$ contains all eigen-elements of B with non-unity eigen-values. The proof is finished. □

Now we shall study the quadratic form $\sum_1^\infty v_n^2(1-\lambda_n)$ appearing in (6.5).

Theorem 6.3. *Let* P *and* P^+ *be two normal distributions of a stochastic process* $\{x_t, t \in T\}$. *Let* $\int x_t \, d\mathsf{P} = \int x_t \, d\mathsf{P}^+ = 0$, $t \in T$. *Assume that* $x_t = Kz_t$, $t \in T$, *where* z_t *is a stochastic process or a white noise, and* K *is a compact operator such that* K^*K *has a finite trace. Let* $(\Phi, \mathcal{Q})$, $(\Phi^+, \mathcal{Q}^+)$, $(\Phi^z, \mathcal{Q}^z)$ *be closed linear manifolds generated by covariances* $R(x_t, x_s)$, $R^+(x_t, x_s)$ *and* $R(z_t, z_s)$, *respectively. Let* B *be the linear operator defined by* $R(x, y) = R^+(Bx, y)$. *Assume that* $\Phi^+ = \Phi$ *and for some constant* C

$$|\mathcal{Q}((I - B)\,\varphi, \psi)| \leq C\,\mathcal{Q}^z(\varphi)\,\mathcal{Q}^z(\psi) \quad (\varphi, \psi \in \Phi^x). \tag{6.24}$$

Then P *and* P^+ *are strongly equivalent. Moreover, there exists a unique extension* $\widehat{\mathcal{Q} - \mathcal{Q}^+}$ *of* $\mathcal{Q} - \mathcal{Q}^+$ *from* $\Phi \times \Phi$ *on* $\Phi^z \times \Phi^z$, *continuous in the* $\mathcal{Q}^z$*-norm, and we have*

$$\frac{d\mathsf{P}}{d\mathsf{P}^+} = (\det B)^{-\frac{1}{2}} \exp\left\{\frac{1}{2}\widehat{\mathcal{Q} - \mathcal{Q}^+}(x^\omega, x^\omega)\right\}, \tag{6.25}$$

where $x^\omega = x_t(\omega)$ *is the trajectory of the process modified according to Theorem 4.2 or 4.1.*

Proof. $I - B$ has a finite trace because K^*K has a finite trace and because (6.24) holds, where $\mathcal{Q}^z(\varphi) = \mathcal{Q}(K\varphi)$. Since $\Phi^+ = \Phi$, the covariances R and R^+ dominate each other and B has all eigen-values different from 0. Hence B has a determinant, and P and P^+ are strongly equivalent.

In Φ the operator B is defined by $\mathcal{Q}^+(\psi, \varphi) = \mathcal{Q}(B\psi, \varphi)$, in view of Lemma 2.2. Let $\chi_n(t)$ be the eigen-elements of B corresponding to non-unity eigen-values, $\lambda_n \neq 1$. We first show that $\chi_n = KK^*h_n$, where $h_n \in \Phi^z$. In view of (6.24), we have

$$|\mathcal{Q}(\psi, \chi_n)| = \frac{|\mathcal{Q}(\psi, (I - B)\,\chi_n)|}{|1 - \lambda_n|} \leq C\frac{\mathcal{Q}^z(\chi_n)}{|1 - \lambda_n|}\,\mathcal{Q}^z(\psi), \tag{6.26}$$

which shows that $\mathcal{Q}(\psi, \chi_n)$ is a linear functional on $(\Phi^z, \mathcal{Q}^z)$. So $\mathcal{Q}(\psi, \chi_n) = \mathcal{Q}^z(\psi, h_n)$ for some $h_n \in \Phi^z$, which implies, in accordance with (4.11), that $\chi_n = KK^*h_n$ $(\mathcal{Q}^x \equiv \mathcal{Q})$.

Now from (6.24) it follows that $(\mathcal{Q} - \mathcal{Q}^+)(\varphi, \varphi) = \mathcal{Q}^z(A\varphi, \varphi)$, where A is a bounded operator in Φ^z. Consequently, if $\varphi_n \to \varphi$, where $\varphi_n \in \Phi$ and $\varphi \in \Phi^z$, there exists a limit

$$\widehat{\mathcal{Q} - \mathcal{Q}^+}(\varphi, \varphi) = \lim_{n\to\infty} (\mathcal{Q} - \mathcal{Q}^+)(\varphi_n, \varphi_n) = \mathcal{Q}^z(A\varphi, \varphi), \tag{6.27}$$

which represents a unique continuous extension of $\mathcal{Q} - \mathcal{Q}^+$ on $\Phi^z \times \Phi^z$. Because

$$(\mathcal{Q} - \mathcal{Q}^+)(\chi_n, \chi_n) = \mathcal{Q}((1 - \lambda_n)\,\chi_n, \chi_n) = \mathcal{Q}^z((1 - \lambda_n)\,\chi_n, h_n),$$

we conclude that $A\chi_n = (1-\lambda_n)\,h_n$. So, on developing $\mathcal{Q} - \mathcal{Q}^+(\varphi,\varphi)$ into a series, we get

$$\mathcal{Q}\widehat{-}\mathcal{Q}^+(\varphi,\varphi) = \sum_1^\infty |\mathcal{Q}(\varphi,\chi_n)|^2\,(1-\lambda_n) \tag{6.28}$$

$$= \sum_1^\infty \left|\mathcal{Q}\widehat{-}\mathcal{Q}^+\left(\varphi, \frac{\chi_n}{1-\lambda_n}\right)\right|^2 (1-\lambda_n) = \sum_1^\infty |\mathcal{Q}^z(\varphi,h_n)|^2\,(1-\lambda_n).$$

Random variables v_n satisfying $\chi_n(t) = R(x_t, v_n)$ are eigen-elements of B in $\mathcal{X}$. Because $\chi_n = KK^*h_n$, from Theorem 4.2 or 4.1, we have

$$v_n(\omega) = \mathcal{Q}^z(x^\omega, h_n). \tag{6.29}$$

On comparing (6.25) and (6.29), we get $\mathcal{Q}\widehat{-}\mathcal{Q}^+(x^\omega, x^\omega) = \sum_1^\infty v_n^2(\omega)\,(1-\lambda_n)$. Returning to the equation (6.5) and noting that $\det B = \prod_1^\infty \lambda_n$, we see that the proof is complete. □

7 PROBABILITY DENSITIES FOR STATIONARY GAUSSIAN PROCESSES

We begin with the case of an integral-valued parameter $t = 0, 1, \ldots, N$, which is easier. The probability densities may be taken with respect to Lebesgue measure, because the number of random variables is finite.

Theorem 7.1. (i) *The probability density of a finite part $\{x_t,\, 0 \le t \le N\}$ of a Gaussian stationary process with vanishing mean values and covariances generated by the spectral density* (5.35) *is given by*

$$p(x_0,\ldots,x_N) = |Q_{ts}|^{-\frac{1}{2}} \exp\left(-\frac{1}{2}\sum_{t=0}^N\sum_{t=0}^N Q_{ts}x_t x_s\right) \tag{7.1}$$

$$= |D_{jk}|^{\frac{1}{2}} a_0^{N-n+1} \exp\left[-\frac{1}{2}\sum_{j=0}^{n-1}\sum_{k=0}^{n-1} D_{jk}x_j x_k - \frac{1}{2}\sum_{t=n}^{n}\left(\sum_{s=0}^{n} a_k\, x_{t-k}\right)^2\right],$$

where Q_{ts}, $0 \le t,\ s \le N$, and D_{jk}, $0 \le j,\ k \le n-1$ are given by (5.37) *and* (5.40), *respectively, and $|\cdot|$ denotes the determinant.*

(ii) *The probability density of a finite part $\{z_t,\, 0 \le t \le N\}$ of a stationary Gaussian process with vanishing mean values and covariances generated by* *the spectral density* (5.41) *is given by*

$$p(z_0, \ldots, z_N) = |D_{jk}|^{\frac{1}{2}} \, |R(x_i, x_h | z_0, \ldots, z_N)|^{\frac{1}{2}} \, a_0^{N+m+1-n} \, b_0^{-N-1}$$
$$\times \exp\left[-\frac{1}{2} \sum_{j=0}^{n-1} \sum_{k=0}^{n-1} D_{jk} \, \breve{x}_{j-m} \, \breve{x}_{k-m} - \frac{1}{2} \sum_{t=n-m}^{N} \left(\sum_{k=0}^{n} a_k \, \breve{x}_{t-k} \right)^2 \right], \tag{7.2}$$

where $x_{-m}, \ldots, x_N$ *is a process considered in* (i) *and such that* $z_t = \sum b_k \, x_{t-k}$, *and* $R(x_i, x_h | z_0, \ldots, z_N)$, $-m \leq i, \; h \leq -1$, *are conditional (partial) covariances of* x_i, x_h, *when* $z_0, \ldots, z_N$ *are fixed. Moreover,* $\breve{x}_{-m}, \ldots, \breve{x}_N$ *is the solution of equations* $z_t = \sum b_k \, \psi_{t-k}$ *satisfying the conditions*

$$\sum_{-m}^{N} \sum_{-m}^{N} Q_{ts} \, \psi_t \, h_s = 0$$

for some m *linearly independent solutions of* $\sum b_k \, h_{t-k} = 0$.

Proof. (i) As Q_{ts} given by (5.37) represents elements of inverted covariance matrix $R = (R_{ts})$, the first expression for $p(x_0, \ldots, x_N)$ in (7.1) is clear. If we put $R_n = (R(x_i, x_j))_{i,j=0}^{n-1}$ then, in accordance with Theorem 6.2,

$$|Q_{ts}| = |R_n|^{-1} \left[\prod_{t=n}^{N} R^2(x_t - I_{t-1} x_t) \right]^{-1}, \tag{7.3}$$

where $I_{t-1} x_t$ is the projection of x_t on the subspace spanned by the random variables $x_0, \ldots, x_{t-1}$. However, we have $|R_n|^{-1} = |D_{jk}|$ and $R^2(x_t - I_{t-1} x_t) = a_0^{-2}$, as $y_t = a_0(x_t - I_{t-1} x_t)$, where y_t is given by (5.36). Thus we obtain the determinant of the second expression for $p(x_0, \ldots, x_N)$ in (7.1). The quadratic form of the second expression is a mere transcription of $\sum \sum Q_{ts} \, x_t \, x_s$, and is based on the fact that the y_ts given by (5.36) are independent of each other and of $x_0, \ldots, x_{n-1}$.

(ii) First we derive the determinant. Put $z_t = x_t$ if $-m \leq t \leq -1$, and $z_t = \sum b_k \, x_{t-k}$ if $0 \leq t \leq N$. This transformation may be denoted in matrix form as $z = Cx$, where the matrix $C = \{c_{ij}\}$ is such that $c_{ij} = 0$, $i < j$, and $c_{ii} = 1$ if $-m \leq i \leq 1$, and $c_{ii} = b_0$ if $0 \leq i \leq N$. Consequently $|C| = b_0^{N+1}$ and

$$|R(z_t, z_s)| = b_0^{2N+2} \, |R(x_t, x_s)|, \quad -m \leq t, \; s \leq N.$$

Now we know from (i) that

$$|R(x_t, x_s)| = |D_{jk}|^{-1} \, a_0^{-2(N+m+1-n)}$$

which gives

$$|R(z_t, z_s)| = |D_{jk}|^{-1} \, a_0^{-2(N+m+1-n)} \, b_0^{2(N+1)} \quad (-m \leq t, \; s \leq N). \tag{7.4}$$

Finally, according to Theorem 6.2,

$$|R(z_t, z_s)| = |R(z_{t'}, z_{s'})|\,|R(x_i, x_h|z_0, \dots, z_N)| \tag{7.5}$$
$$(-m \le t,\ s \le N;\ 0 \le t',\ s' \le N;\ -m \le i,\ h \le -1),$$

where $x_i = z_i$, if $-m \le i \le -1$. On combining (7.4) and (7.5), we get for $|R(z_{t'}, z_{s'})|$, $0 \le t'$, $s' \le N$, the expression appearing in (7.2). The quadratic form in (7.2) follows from the form of $\mathcal{Q}^z(\nu, \chi)$ described below equation (5.42).
□

Now we proceed to the case of a continuous t, and first consider the nth order Markovian processes. The probability density cannot be taken with respect to the Lebesgue measure, because the system of random variables is infinite. The dominating distribution $\mathsf{P}^+ = \mathsf{P}^+_{n,a}$ of $\{x_t,\ 0 \le t \le T\}$ used in the next theorem will be defined by the following conditions:

the vector $(x_0, x'_0, \dots, x_0^{(n-1)})$ is distributed according to the n-dimensional Lebesgue measure; $x_t^{(n-1)}$ is a Gaussian process with independent increments such that

$$\mathsf{E}|\mathrm{d}\, x_t^{(n-1)}|^2 = a^{-2}\, \mathrm{d}t; \tag{7.6}$$

$x_t^{(n-1)} - x_0^{(n-1)}$ is independent of $(x_0, \dots, x_0^{(n-1)})$; the mean values vanish.

Theorem 7.2. *Let $\{x_t,\ 0 \le t \le T\}$ be a finite segment of a stationary Gaussian process with vanishing mean values and with covariances generated by spectral density* (5.1). *Then the distribution of $\{x_t,\ 0 \le t \le T\}$, say P, is strongly equivalent to $\mathsf{P}^+ = \mathsf{P}^+_{n,a_0}$ defined by* (7.6), *and*

$$\frac{\mathrm{d}\mathsf{P}}{\mathrm{d}\mathsf{P}^+} = |D_{jk}|^{\frac{1}{2}} \exp\left\{\frac{1}{2}\frac{a_1}{a_0}T - \frac{1}{2}\sum_{k=0}^{n-1} A_{n-k}\int_0^T |x_t^{(k)}(\omega)|^2\, \mathrm{d}t \right. \tag{7.7}$$
$$\left. -\frac{1}{4}\sum_{j+k \text{ even}}^{n-1}\sum^{n-1} [x_T^{(j)}\, x_T^{(k)} + x_0^{(j)}\, x_0^{(k)}]\, D_{jk}\right\},$$

where D_{jk}, $0 \le j,\ k \le n-1$, are given by (5.21) and A_{n-k} by (5.9).

Proof. Let $(\Phi^+, \mathcal{Q}^+)$ be the closed linear manifold generated by covariances $R^+(x_t, x_s)$ corresponding to the P^+_{n,a_0}-distribution. It is easy to see that $\Phi^+ = \mathcal{L}_n^2$ and

$$\mathcal{Q}^+(\psi, \varphi) = \mathcal{Q}^+_{n,a_0}(\psi, \varphi) = a_0^2 \int_0^T \psi_t^{(n)}\, \overline{\varphi}_t^{(n)}\, \mathrm{d}t \quad (\psi,\ \varphi \in \mathcal{L}_n^2). \tag{7.8}$$

Consequently, for $\mathcal{Q}$ given by (5.11), where we have made use of (5.21), we get

$$(\mathcal{Q}-\mathcal{Q}^{+})(\psi,\psi) = \sum_{k=0}^{n-1} A_{n-k}\int_0^T |\psi_t^{(k)}|^2\,dt + \frac{1}{2}\sum_{j=0}^{n-1}\sum_{k=0}^{n-1}[\psi_T^{(j)}\overline{\psi}_T^{(k)} + \psi_0^{(j)}\overline{\psi}_0^{(k)}]\,D_{jk}. \tag{7.9}$$

Now let α_1 be a root of $L(\lambda)=\sum a_{n-k}\lambda^k=0$, and let $L_1(\lambda)=L(\lambda)/(\lambda-\alpha_1)$. Then $z_t = x' + \alpha_1 x_t$ is an $(n-1)$th order Markovian process if $n>1$, and a white noise if $n=1$. We have

$$x_t = x_0\,e^{\alpha_1 t} + e^{\alpha_1 t}\int_0^t e^{-\alpha_1 s}\,z_s\,ds = K z_t,$$

where K^*K has a finite trace. Obviously $\mathcal{Q}^z$ satisfies (6.24). Consequently, according to Theorem 6.3, P and $\mathsf{P}^+ = \mathsf{P}^+_{n,a_0}$ are strongly equivalent. Since $\Phi^z = \mathcal{L}^2_{n-1}$, $\mathcal{Q}-\mathcal{Q}^+$ may be extended to $\mathcal{L}^2_{n-1}$. The right-hand side of (7.9) is, however, adjusted so that it directly represents the extension of $\mathcal{Q}-\mathcal{Q}^+$ to $\mathcal{L}^2_{n-1}$. Now, in view of (6.25), we only have to substitute x_t for ψ_t in (7.9), which yields the quadratic form of (7.7).

It is now necessary to find the determinant. The determinant corresponding to the vector $(x_0,\ldots,x_0^{(n-1)})$ equals $|D_{jk}|$, D_{jk} given by (5.21). This determinant is to be multiplied by the determinant of the operator B such that

$$R\left(x_t^{(n-1)}, x_s^{(n-1)}|x_0,\ldots,x_0^{(n-1)}\right) = R^+\left(Bx_t^{(n-1)}, x_s^{(n-1)}|x_0,\ldots,x_0^{(n-1)}\right). \tag{7.10}$$

Let B_N be the restriction of B to the subspace spanned by random variables

$$u_i = x_{iT/N}^{(n-1)}, \quad (i=1,\ldots,N). \tag{7.11}$$

According to Theorem 6.2, we have

$$\det B_N = \prod_{i=1}^{N}\frac{R(u_i-u_i^0,\,u_i-u_i^0)}{R^+(u_i-u_i^+,\,u_i-u_i^+)}, \tag{7.12}$$

where u_i^0 and u_i^+ are the projections of u_i on the subspace spanned by $(x_0,\ldots,x_0^{(n-1)}, u_1,\ldots,u_{i-1})$ with respect to the R-covariance and R^+-covariance, respectively. According to the assumptions (7.6), $u_i^+ = u_{i-1}$ and

$$R^+(u_i-u_i^+,\,u_i-u_i^+) = \frac{T}{N}\,a_0^{-2}. \tag{7.13}$$

If R-covariance holds true, the situation is more complicated. First we find the projection of u_i, say $\overline{u}_i$, on the subspace spanned by $\{x_t,\ 0 \leq t \leq [(i-1)/N]T\}$. Because the process $\{x_{T-t},\ 0 \leq t \leq T\}$ has the same distribution as $\{x_t,\ 0 \leq t \leq T\}$, we may write (5.10) in the following equivalent form:

$$\mathcal{Q}(\psi, \varphi) = \int_0^T (L^* \overline{\varphi})_t\, (L^* \psi)_t \,\mathrm{d}t + \sum_{j=0}^{n-1} \sum_{k=0}^{n-1} \psi_T^{(j)}\, \overline{\varphi}_T^{(k)}\, D_{jk}, \tag{7.14}$$

where D_{jk} is given by (5.21). When looking for $\overline{u}_i$ we have to put $T \equiv [(i-1)/N] \cdot T$ and $\varphi_t = R(x_t, u_i)$. However,

$$(L^* \varphi)_t = L^* R(x_t, u_i) = R(L^* x_t, u_i) = 0, \quad 0 \leq t \leq [(i-1)/N] \cdot T, \tag{7.15}$$

because $L^* x_t$ is a white noise independent of $x_s,\ s > t$, similarly as $L x_t$ was a white noise independent of $x_s,\ s < t$. This means that for $\varphi_t = R(x_t, u_i)$

$$\mathcal{Q}(\psi, \varphi) = \sum_{j=0}^{n-1} \sum_{k=0}^{n-1} \psi_{(i-1)T/N}^{(j)}\, \overline{\varphi}_{(i-1)T/N}^{(k)}\, D_{jk}.$$

If we replace ψ_t by x_t and substitute

$$\varphi_t^{(k)} = \frac{\partial^k}{\partial t^k} R(x_t, x_{iT/N}^{(n-1)}) = (-1)^k\, R_{t-iT/N}^{(n-1+k)},$$

where $R_{t-s} = R_{ts}$, we get

$$\begin{aligned} \overline{u}_i &= \sum_{j=0}^{n-1} x_{(i-1)T/N}^{(j)} \sum_{k=0}^{n-1} (-1)^k\, R_{T/N}^{(n-1+k)}\, D_{jk} \qquad (7.16) \\ &= \sum_{j=0}^{n-1} x_{(i-1)T/N}^{(j)} \sum_{k=0}^{n-1} (-1)^k \left[R_0^{(n-1+k)} + \frac{T}{N} R_{0+}^{(n+k)} + O(N^{-2}) \right] D_{jk}. \end{aligned}$$

Now,

$$\begin{aligned} \textstyle\sum_{k=0}^{n-1} (-1)^k\, R_0^{(n-1+k)}\, D_{jk} &= 1, \quad \text{if} \quad j = n-1, \\ &= 0, \quad \text{if} \quad h < n-1, \end{aligned} \tag{7.17}$$

because $((-1)^k\, R_0^{(j+k)})$ is the covariance matrix of the vector $(x_t, \ldots, x_t^{(n-1)})$, and (D_{jk}) is its inverse. Moreover, in view of $L_t^*\, R_{s-t}^{(k)} = 0,\ t < s$, we have

$$R_{0+}^{(n+k)} = -\frac{1}{a_0} \sum_{h=0}^{n-1} a_{n-k}\, R_{0+}^{(h+k)} \quad (0 \leq k \leq n), \tag{7.18}$$

which, in connection with (7.17) gives

$$\sum_{k=0}^{n-1}(-1)^k R_{0+}^{(n+k)} D_{jk} = -\frac{a_{n-j}}{a_0}, \quad 0 \le j \le n-1. \tag{7.19}$$

If we insert (7.17) and (7.19) into (7.16), we get

$$\overline{u}_i = u_{i-1} - \frac{T}{N}\sum_{j=0}^{n-1}\frac{a_{n-j}}{a_0}x_{(i-1)T/N}^{(j)} + O(N^{-2}). \tag{7.20}$$

From (7.20) it follows that

$$\begin{aligned}
& R(u_i - \overline{u}_i,\, u_i - \overline{u}_i) \qquad\qquad (7.21)\\
= {} & 2\left(1 - (-1)^n R_{T/N}^{(2n-2)}\right) + \frac{2T}{N}\sum_{j=0}^{n-1}\frac{a_{n-j}}{a_0}(-1)^j\left(R_{T/N}^{(n-1+j)} - R_{0+}^{(n-1+j)}\right)\\
& + \frac{T^2}{N^2}\sum_{j=0}^{n-1}\sum_{k=0}^{n-1}\frac{a_{n-j}\,a_{n-k}}{a_0^2}(-1)^j R_{0+}^{(j+k)} + O(N^{-3})\\
= {} & 2(-1)^n\frac{T}{N}R_{0+}^{(2n-1)} + (-1)^n\frac{T^2}{N^2}R_{0+}^{(2n)} + \frac{2T^2}{N^2}\sum_{j=0}^{n-1}\frac{a_{n-j}}{a_0}(-1)^j R_{0+}^{(n+j)}\\
& + \frac{T^2}{N^2}\sum_{j=0}^{n-1}\sum_{k=0}^{n-1}\frac{a_{n-j}\,a_{n-k}}{a_0^2}(-1)^j R_{0+}^{(j+k)} + O(N^{-3})\\
= {} & 2(-1)^n\frac{T}{N}R_{0+}^{(2n-1)} + \frac{T^2}{N^2}\sum_{j=0}^{n-1}\frac{a_{n-j}}{a_0}(-1)^j R_{0+}^{(n+j)} + O(N^{-3})\\
= {} & 2(-1)^n\frac{T}{N}R_{0+}^{(2n-1)} - 2(-1)^n\frac{T^2}{N^2}\frac{a_1}{a_0}R_{0+}^{(2n-1)} + O(N^{-3})\\
= {} & \frac{T}{N}a_0^{-2}\left(1 - \frac{T}{N}\frac{a_1}{a_0}\right) + O(N^{-3}),
\end{aligned}$$

where we have made use of (7.18), and of the adjoint relation

$$(-1)^n R_{0-}^{(n+k)} = -\sum_{h=0}^{n-1}(-1)^h\frac{a_{n-h}}{a_0}R_{0-}^{(h+k)},$$

together with $R_{0\pm}^{2j+1} = 0$, $j < n-1$, $R_{0+}^{(2j)} = R_{0-}^{(2j)} = R_0^{(2j)}$, $0 \le j \le n-1$, and

$$R_{0+}^{(2n-1)} = -R_{0-}^{(2n-1)} = (-1)^n\frac{1}{2}a_0^{-2}.$$

If we choose the projection u_i^0 of u_i on the subspace spanned by $(x_0, \ldots, x_0^{(n-1)},\ u_1, \ldots, u_{i-1})$, we cannot make use of random variables

$x^{(j)}_{(i-1)T/N}$, $0 \leq j \leq n-2$, appearing in (7.20). However, we may approximate them by sums

$$\hat{x}^{(j)}_{(i-1)T/N} = \sum_{j=0}^{n-1} x_0^{(j)} c_j + \sum_{k=1}^{i-1} u_k d_k,$$

so that

$$R^2\left(x^{(j)}_{(i-1)T/N} - \hat{x}^{(j)}_{(i-1)T/N}\right) = O(N^{-1}).$$

For example we may put

$$\hat{x}^{(n-2)}_{(i-1)T/N} = x_0^{(n-2)} + \sum_{k=1}^{i-1} u_k \frac{T}{N}$$

etc. Consequently, we have

$$R(u_i - u_i^0,\, u_i - u_i^0) = \frac{T}{N} a_0^{-2}\left(1 - \frac{T}{N}\frac{a_1}{a_0}\right) + O(N^{-3}). \tag{7.22}$$

So, in accordance with Theorem 6.1 and with (7.13) and (7.22),

$$\det B = \lim_{\substack{n\to\infty \\ N=2^n}} \det B_N = \lim_{\substack{n\to\infty \\ N=2^n}} \prod_{i=1}^{N} \frac{\frac{T}{N} a_0^{-2}\left(1 - \frac{T}{N}\frac{a_1}{a_0}\right) + O(N^{-3})}{\frac{T}{N} a_0^{-2}} = e^{-a_1/a_0 T}, \tag{7.23}$$

which concludes the proof. □

In the case of a general rational spectral density the leading term of $\mathcal{Q}^z(\chi,\chi)$ is

$$a_0^2 b_0^{-2} \int_0^T \left|\chi_t^{(n-m)}\right|^2 \mathrm{d}t = \mathcal{Q}^+(\chi,\chi),$$

$$\mathcal{Q}^z(\chi,\chi) = a_0^2 b_0^{-2} \int_0^T |\chi_t^{(n-m)}|^2 \mathrm{d}t + \dots \tag{7.24}$$

and $\mathcal{Q}^z(\chi,\chi) - \mathcal{Q}^+(\chi,\chi)$ is dominated in the sense of (6.29) by $\overline{\mathcal{Q}}(\chi,\chi)$ corresponding to any rational spectral density with $\overline{n} - \overline{m} = n - m - 1$. Without entering into details let us present the following theorem.

Theorem 7.3. *Let $\{z_t,\, 0 \leq t \leq T\}$ be a finite segment of a stationary Gaussian process with vanishing mean values and with covariances generated by spectral density (5.22). Then the distribution of $\{z_t,\, 0 \leq t \leq T\}$, say P, is strongly equivalent to $\mathsf{P}^+ = \mathsf{P}^+_{n-m,\, a_0/b_0}$ and*

$$\begin{aligned}\frac{\mathrm{d}\mathsf{P}}{\mathrm{d}\mathsf{P}^+} &= |D_{jk}|^{\frac{1}{2}}\, b_0^{m-n}\, |R(x_0^{(i)}, x_0^{(h)} | z_t,\, 0 \leq t \leq T)|^{\frac{1}{2}} \\ &\times \exp\left\{\frac{1}{2}\left(\frac{a_1}{a_0} - \frac{b_1}{b_0}\right)T - \frac{1}{2}\widehat{\mathcal{Q}^z - \mathcal{Q}^+}(x^\omega, x^\omega)\right\},\end{aligned} \tag{7.25}$$

where $\mathcal{Q}^+ = \mathcal{Q}^+_{n-m,\,a_0/b_0}$ *is given by* (7.8), *and the* D_{jk}*s*, $0 \le j,\, k \le n-1$, *are given by* (5.21). *Further,* x_t *is a process considered in Theorem* 7.2 *such that* $z_t = \sum_{k=0}^{m} b_{m-k}\, x_t^{(k)}$, *and* $R(x_0^{(i)}, x_0^{(h)}|z_t,\, 0 \le t \le T)$, $0 \le i,\, h < m-1$, *are conditional (partial) covariances of* $x_0^{(i)}$ *and* $x_0^{(h)}$, *when* $\{z_t,\, 0 \le t \le T\}$ *is fixed.*

Proof. The assertions concerning the quadratic form follow from Theorems 5.2 and 6.3 and from the above discussion. As for the determinant, let us go back to (7.7). If the dominating distribution P^+ were modified so that

$$\sum_{k=0}^{m} b_{m-k}\, x_t^{(m-m-1+k)} = Y_t$$

has independent increments and $\mathsf{E}^+|\mathrm{d}Y_t|^2 = a_0^2\, b_0^{-2}$, then the only change in the determinant would consist in replacing

$$\exp\left\{\frac{1}{2}\frac{a_1}{a_0}T\right\} \quad \text{by} \quad \left\{\frac{1}{2}\left(\frac{a_1}{a_0} - \frac{b_1}{b_0}\right)T\right\}.$$

This could be proved by arguments similar to those used in the proof of Theorem 7.2. Now, if we put $z_t = \sum_{k=0}^{m} b_{m-k}\, x_t^{(k)}$, we can easily see that the P^+-distribution of z_t is the one used in (7.25). Further, the transformation $(z_0^{(-m)}, \ldots, z^{(n-m-1)}) = A(x_0, \ldots, x_0^{(n-1)})$, defined by

$$\begin{aligned} z_0^{(j)} &= \sum_{k=0}^{m} b_{m-k}\, x_0^{(j+k)}, && \text{if } \; 0 \le j \le n-m-1, \\ &= x_0^{(j+m)}, && \text{if } \; -m \le j \le -1, \end{aligned} \tag{7.26}$$

has the determinant $|A| = b_0^{n-m}$. Now replacing x_0 by z_0 amounts to multiplying the determinant by b_0^{n-m}, and excluding random variables $x_0^{(j+m)} = z_0^{(j)}$, $-m \le j \le -1$, amounts to dividing the determinant by the factor $|R(x_0^{(i)}, x_0^{(h)}|\, z_t,\, 0 \le t \le T)|$, which follows from Theorem 6.2. The proof is complete. □

Remark 7.1. Obviously, the following limit exists:

$$\lim_{T\to\infty}\left|R(x_0^{(i)},\, x_0^{(h)}|z_t,\, 0 \le t \le T)\right| = \left|R(x_0^{(i)},\, x_0^{(h)}|z_t,\, 0 \le t < \infty)\right|. \tag{7.27}$$

Remark 7.2. Consider a stationary Gaussian process $\{x_t,\, 0 \le t \le T\}$ with correlation function $R(\tau) = \max(0,\, 1-|\tau|)$. From Example 3.2 it follows, after some computations, that, for $0 < T < 1$,

$$\mathcal{Q}(\varphi, \varphi) = \frac{1}{2}\frac{(\varphi_0 + \varphi_T)^2}{2-T} + \frac{1}{2}\int_0^T |\varphi_t'|^2\, \mathrm{d}t. \tag{7.28}$$

Consequently, the respective distribution, say P, is strongly equivalent to $\mathsf{P}^+ = \mathsf{P}^+_{1,2}$ defined by (7.6), and

$$\frac{d\mathsf{P}}{d\mathsf{P}^+} = \text{const} \exp\left[-\frac{1}{4}\frac{(x_0 + x_T)^2}{2 - T}\right]. \tag{7.29}$$

If, however, $1 < T < 2$, we get

$$\begin{aligned} \mathcal{Q}(\varphi, \varphi) &= \frac{1}{6}\frac{(2\varphi_0 + 2\varphi_T + \varphi_1 + \varphi_{T-1})^2}{4 - T} + \frac{1}{2}\int_{T-1}^{1} |\varphi'_t|^2 \, dt \\ &\quad + \frac{2}{3}\int_1^T \left(|\varphi'_t|^2 + |\varphi'_{t-1}|^2 + \varphi'_t \varphi'_{t-1}\right)^2 dt, \end{aligned} \tag{7.30}$$

which shows that the distribution is not equivalent to any distribution we have met with.

If we introduce two parameters by putting $R(\tau) = d^2 \max(0, 1 - |\tau|/a)$, then for $0 < T < a$ and $\mathsf{P}^+ = \mathsf{P}^+_{1,2d^2/a}$,

$$\frac{d\mathsf{P}}{d\mathsf{P}^+} = \left(\frac{2a - T}{2a}\right)^{\frac{1}{2}} \exp\left[-\frac{1}{4}\frac{(x_0 + x_T)^2}{d^2(2 - T/a)}\right]. \tag{7.31}$$

The determinant was established as follows.

If we put $\varphi_t = R(s + \Delta - t) = d^2(1 - (s + \Delta - t)/a),\ 0 \le t \le s$, then

$$\mathcal{Q}(\varphi, \varphi) = d^2(1 - 2\Delta a^{-1} + 2\Delta^2 a^{-1}(2a - s)^{-1})$$

gives the variance of the projection of $x_{s+\Delta}$ on the subspace spanned by $\{x_t,\ 0 \le t \le s\}$, so that the residual variance equals $\Delta 2d^2 a^{-1}(1 - \Delta(2a - s)^{-1})$. Then we may proceed with a development similar to that used in evaluating (7.23). Note that $x_0 + x_T$ is a sufficient statistic for estimating a. As is well known,

$$2d^2\, a^{-1} = \operatorname*{l.i.m.}_{n\to\infty} \sum_{i=1}^{n} (x_{(i-1)T/n} - X_{iT/n})^2.$$

Remark 7.3. From (7.7) it follows that the vector

$$\left\{\int_0^T |x_t|^2 \, dt, \ldots, \int_0^T |x_t^{(n-1)}|^2 \, dt,\ x_0, \ldots, x_0^{(n-1)},\ x_T, \ldots, x_T^{(n-1)}\right\}$$

represents a sufficient statistic for all nth order Markovian processes with fixed a_0. In the case of a general rational spectral density with $m > 0$ apparently no sufficient statistic exists, which would not be equivalent to the whole process $\{z_t,\ 0 \le t \le 1\}$. See Example 7.3.

Remark 7.4. We have proved, by the way, that distributions P_1 and P_2 corresponding to two rational spectral densities are strongly equivalent if $n_1 - m_1 = n_2 - m_2$ and $a_{01}\, b_{02} = a_{02}\, b_{01}$, and perpendicular in other cases. This result (with equivalence instead of strong equivalence) has been announced by V. F. Pisarenko (1959).

Example 7.1. If $n = 1$, we get

$$R(\tau) = (2a_0 a_1)^{-1}\, e^{-(a_1/a_0)|\tau|}$$

and

$$\frac{\mathrm{dP}}{\mathrm{dP+}} = (2a_0 a_1)^{\frac{1}{2}} \exp\left\{\frac{1}{2}\frac{a_1}{a_0}T - \frac{1}{2}a_1^2 \int_0^T |x_t|^2\,\mathrm{d}t\right\}, \tag{7.32}$$

where $x_t = x_t(\omega)$. See also Striebel (1959).

Example 7.2. If

$$\begin{aligned} f(\lambda) &= \frac{1}{2\pi}\frac{a_0^{-2}}{(\lambda^2+\alpha_1^2)(\lambda^2+\alpha_2^2)} \qquad (7.33) \\ &= \frac{1}{2\pi}\frac{a_0^{-2}}{|(i\lambda)^2 + (i\lambda)(\alpha_1+\alpha_2) + \alpha_1\alpha_2|^2}, \qquad (\alpha_2 > \alpha_1 > 0) \end{aligned}$$

then $a_1/a_0 = \alpha_1 + \alpha_2$, $a_2/a_0 = \alpha_1\,\alpha_2$. Consequently,

$$R(\tau) = \frac{1}{2a_0^2}\left(\frac{1}{\alpha_1}\,e^{-\alpha_1|\tau|} - \frac{1}{\alpha_2}\,e^{-\alpha_2|\tau|}\right), \tag{7.34}$$

$D_{12} = 0$, $D_{00} = 2a_2 a_1 = 2(\alpha_1+\alpha_2)\,\alpha_1\,\alpha_2\,a_0^2$, $D_{11} = 2a_0 a_1 = 2(\alpha_1+\alpha_2)\,a_0^2$, and

$$\begin{aligned} \frac{\mathrm{dP}}{\mathrm{dP+}} &= 2(\alpha_1+\alpha_2)(\alpha_1\,\alpha_2)^{\frac{1}{2}}\,a_0^2 \exp\left[\frac{1}{2}(\alpha_1+\alpha_2)\,T\right] \qquad (7.35) \\ &\quad \times \exp\left\{\frac{1}{2}(\alpha_1^2+\alpha_2^2)\,a_0^2 \int_0^T |x_t'|^2\,\mathrm{d}t - \frac{1}{2}\,\alpha_1^2\,\alpha_2^2\,a_0^2 \int_0^T |x_t|^2\,\mathrm{d}t \right. \\ &\quad \left. + \frac{1}{2}(x_0'^2 + x_T'^2)(\alpha_1+\alpha_2)\,a_0^2 - \frac{1}{2}(x_0^2+x_T^2)(\alpha_1+\alpha_2)\,\alpha_1\,\alpha_2\,a_0^2\right\}. \end{aligned}$$

Example 7.3. Consider a general spectral density with $n = 2$ and $m = 1$,

$$g(\lambda) = \frac{b_0^2(\lambda^2+\beta^2)}{|a_0(i\lambda)^2 + a_1(i\lambda) + a_2|^2} \tag{7.36}$$

and put $L(\beta) = a_0\beta^2 + a_1\beta + a_2$, $L^*(\beta) = a_0\beta^2 - a_1\beta + a_2$ and $LL^*(\beta) = L(\beta)\,L^*(\varepsilon) = a_0^2\beta^4 + (2a_0a_2 - a_1^2)\,\beta^2 + a_2^2$, etc. In order to find $\mathcal{Q}^z(\chi,\chi)$, let us first suppose that $\chi \in \mathcal{L}_2^2$, and use the form of $\mathcal{Q}^z(\chi,\chi)$ resulting from the right-hand side of (5.27) after substituting χ_t for $z_t(\omega)$. On obtaining φ_t from formula (5.31), we get

$$\varphi_t = \frac{1}{2\beta}\int_0^T e^{-|t-s|\beta}\chi_s\,\mathrm{d}s + c_1\,e^{-t\beta} + c_2\,e^{(\tau-t)\beta}, \tag{7.37}$$

where

$$c_1 = \left(a_0 L(\beta)(\chi_0 - L^*(\beta)\frac{1}{2\beta}\int_0^T e^{-t\beta}\chi_t\mathrm{d}t) - L^*(\beta)e^{-T\beta}\right. \tag{7.38}$$
$$\left.\times(\chi_T - L^*(\beta)\frac{1}{2\beta}\int_0^T e^{-(T-t)\beta}\chi_t\mathrm{d}t)\right)\left(LL(\beta) - L^*L^*(\beta)\,e^{-2T\beta}\right)^{-1}$$

and

$$c_2 = \left(a_0 L(\beta)(\chi_T - L^*(\beta)\frac{1}{2\beta}\int_0^T e^{-(T-t)\beta}\chi_t\mathrm{d}t) - L^*(\beta)e^{-T\beta}\right. \tag{7.39}$$
$$\left.\times(\chi_0 - L^*(\beta)\frac{1}{2\beta}\int_0^T e^{-t\beta}\chi_t\mathrm{d}t\right)\left(LL(\beta) - L^*L^*(\beta)\,e^{-2T\beta}\right)^{-1}.$$

Now in view of (5.29) and (5.30) $(L\varphi)_T = (L^*\varphi)_0 = 0$, so that

$$\mathcal{Q}^z(\chi,\chi) = \int_0^T \chi_t(LL^*\varphi)_t\,\mathrm{d}t + \chi_T\,a_0\left[a_0\varphi_T''' + a_1\varphi_T'' + a_2\varphi_T'\right]$$
$$+\chi_0 a_0\left[a_0\varphi_0''' - a_1\varphi_0'' + a_2\varphi_0'\right].$$

On using the relations $\varphi_t'' = -\chi_t + \beta\varphi_t$ and (7.37), we get, after some transformations, that

$$\mathcal{Q}^z(\chi,\chi) = \left(\frac{a_0}{b_0}\right)^2\int_0^T |\chi_t'|^2\,\mathrm{d}t + (a_1^2 - 2a_0a_2 - \beta^2a_0^2)\,b_0^{-2}\int_0^T |\chi_t|^2\,\mathrm{d}t \tag{7.40}$$
$$+\left(\frac{a_0}{b_0}\right)^2 LL^*(\beta)\int_0^T\int_0^T e^{-|t-s|\beta}\chi_t\chi_s\,\mathrm{d}t\mathrm{d}s$$
$$+\left(|\chi_0|^2 + |\chi_T|^2\right)b_0^{-2}\left(a_0a_1 - \beta a_0^2\frac{LL(\beta) + L^*L^*(\beta)\,e^{-2T\beta}}{LL(\beta) - L^*L^*(\beta)\,e^{-2T\beta}}\right)$$
$$-\frac{1}{2}\beta\left(\frac{a_0}{b_0}\right)^2\left(\left|\frac{1}{\beta}\int_0^T e^{-t\beta}\chi_t\,\mathrm{d}t\right|^2 + \left|\frac{1}{\beta}\int_0^T e^{-(T-t)\beta}\chi_t\,\mathrm{d}t\right|^2\right)$$
$$\times\frac{LLL^*L^*(\beta)}{LL(\beta) - L^*L^*(\beta)\,e^{-2T\beta}}$$

$$+2\left(\frac{a_0}{b_0}\right)^2\left(\chi_T\int_0^T e^{-(T-t)\beta}\chi_t\,dt+\chi_0\int_0^T e^{-t\beta}\chi_t\,dt\right)$$
$$\times\frac{LLL^*(\beta)}{LL(\beta)-L^*L^*(\beta)\,e^{-2T\beta}}$$
$$+4\beta\left(\frac{a_0}{b_0}\right)^2\left(\chi_0\chi_T+L^*L^*(\beta)\frac{1}{2\beta}\int_0^T e^{-t\beta}\chi_t\,dt\,\frac{1}{2\beta}\int_0^T e^{-(T-t)\beta}\chi_t\,dt\right)$$
$$\times\frac{LL^*(\beta)\,e^{-T\beta}}{LL(\beta)-L^*L^*(\beta)\,e^{-2T\beta}}$$
$$-2\left(\frac{a_0}{b_0}\right)^2\left(\chi_0\int_0^T e^{-(T-t)\beta}\chi_t\,dt+\chi_T\int_0^T e^{-t\beta}\chi_t\,dt\right)$$
$$\times\frac{LL^*L^*(\beta)\,e^{-T\beta}}{LL(\beta)-L^*L^*(\beta)\,e^{-2T\beta}}.$$

If $T\to\infty$ the terms involving $e^{-T\beta}$ become negligible. The quadratic form (7.40) is obviously well defined for any $\chi\in\mathcal{L}_1^2$. On putting

$$\widehat{\mathcal{Q}^z-\mathcal{Q}^+}(\chi,\chi)=\mathcal{Q}^z(\chi,\chi)-\left(\frac{a_0}{b_0}\right)^2\int_0^T|\chi_t'|^2\,dt, \tag{7.41}$$

$\widehat{\mathcal{Q}^z-\mathcal{Q}^+}$ will be well defined for any $\chi\in\mathcal{L}_0^2$. The probability density will equal

$$\frac{\mathrm{dP}}{\mathrm{dP+}}=2b_0^{-1}a_1(a_0a_2)^{\frac{1}{2}}R(x_0|z_t,\,0\le t\le T)\exp\left[\frac{1}{2}\left(\frac{a_1}{a_0}-\beta\right)T\right] \tag{7.42}$$
$$\times\exp\left\{-\frac{1}{2}\widehat{\mathcal{Q}^z-\mathcal{Q}^+}(x^\omega,x^\omega)\right\},$$

where $\widehat{\mathcal{Q}^z-\mathcal{Q}^+}$ is given by (7.41) and (7.40). The conditional variance of x_0 equals

$$R^2(x_0|z_t,\,0\le t\le T)=R^2(x_0)-\mathcal{Q}^z(\nu,\nu), \tag{7.43}$$

where $R^2(x_0)$ is the absolute variance and $\nu_t=R(x_0,z_t)=b_0R(x_0,x_t'+\beta x_t)\equiv b_0[R'(t)+\beta R(t)]$. In the special case considered in Example 7.2, we have

$$R^2(x_0)=\frac{1}{2}\frac{|\alpha_1-\alpha_2|}{a_0^2\,\alpha_1\,\alpha_2},\quad \nu_t=\frac{1}{2a_0^2}\left(e^{-\alpha_1t}\frac{\beta-\alpha_1}{\alpha_1}-e^{\alpha_2t}\frac{\beta-\alpha_2}{\alpha_2}\right). \tag{7.44}$$

REFERENCES

Balakrishnan A.V. (1959). On a characterization of covariances. *Ann. Math. Stat.* *30*, 670–675.

Coddington E. A. and Levinson N. (1955). *Theory of Ordinary Differential Equations*. New York – Toronto – London.

Dolph C. L. and Woodbury M. A. (1952). On the relations between Green's functions and covariance of certain stochastic processes. *Trans. Amer. Math. Soc. 72*, 519–550.

Doob J. L. (1953). *Stochastic Processes*. New York – Toronto.

Gelfand I. M. (1955). Generalized stochastic processes. (In Russian) *Dokl. Akad. Nauk 100 (5)*, 853–856.

Gelfand I. M. and Yaglom A. M. (1957). On calculating the amount of information about a random function, contained in another one. (In Russian) *Uspekhi Mat. Nauk 12*, 3–52.

Grenander U. (1949). Stochastic processes and statistical inference. *Arkiv för Matematik 1*, 195–277.

Grenander U. (1952). On empirical spectral analysis of stochastic processes. *Arkiv för Matematik 1*, 503–531.

Grenander U. and Rosenblatt M. (1956). *Statistical Analysis of Stationary Time Series*. Stockholm.

Hájek J. (1956). Linear estimation of the mean value of a stationary random process with convex correlation function. *Czechoslovak Math. J. 6*, 94–117 (In Russian) (English translation in *Selected Trans. Statist. Prob. 2* (1962), 41–61).

Hájek J. (1958 a). Predicting a stationary process when the correlation function is convex. *Czechoslovak Math. J. 8*, 150–154.

Hájek J. (1958 b). A property of J-divergences of marginal probability distributions. *Czechoslovak Math. J. 8*, 460–463.

Hájek J. (1958 c). On a property of normal distribution of any stochastic process. *Czechoslovak Math. J. 8*, 610–618. (In Russian. English translation in *Selected Transl. Math. Statist. Prob. 1*, (1961), 245–252.)

Hájek J. (1960). On a simple linear model in Gaussian processes. *Trans. Second Prague Conf. Inf. Th. etc., Praha*, 185–197.

Hájek J. (1961). On linear estimation theory for an infinite number of observations. *Teor. Veryatnost. i Primenen. 6*, 182–193.

Kallianpur G. (1959). A problem in optimum filtering with finite date. *Ann. Math. Stat. 30*, 659–669.

Karhunen K. (1947). Über die linearen Methoden in der Wahrscheinlichkeitsrechnung. *Ann. Acad. Sci. Fennicae, Ser. A, I. Math. Phys. 37*, 3–79.

Kolmogorov A. N. (1941). Stationary sequences in Hilbert space. (In Russian) *Bull. MGU 2 (6)*, 1–40.

Li Heng Won (1959). On the theory of optimal filtering of a single in the presence of internal noise. (In Russian) *Teor. Veroyatnost. i Primenen. 4*, 458–464.

Middleton D. (1960). *Introduction to Statistical Communication Theory*. London.

Pinsker M. S. (1960). Entropy, speed of entropy growth and entropy stability of Gaussian random variables and processes. (In Russian) *Dokl. Akad. Nauk 133 (3)*, 531–533.

Pisarenko V. F. (1959). On absolute continuity of measures corresponding to Gaussian processes with rational spectral density. (In Russian) *Teor. Veroyatnost. i Primenen 6*, 481–481.

Pugachev V. S. (1959). An effective method of founding Bayes solution. (In Russian) *Trans. Second Prague Conf. Inf. Th. etc., Praha, Nakl. ČSAV*, 531–540.

Pugachev V. S. (1960). *Theory of Random Functions and its Applications to Problems of Automatic Control.* (In Russian) Moscow.

Riesz F. and Sz.-Nagy B. (1953). *Lecons d'analyse fonctionnelle.* Budapest. (Russian translation Moskva, 1954.)

Skorokhod A. V. (1957). On differentiation of measures corresponding to random processes. I. Processes with independent increments. (In Russian) *Teor. Veroyatnost. i Primenen. 2*, 417–442.

Skorokhod A. V. (1960). On differentiation of measures corresponding to random processes. II. Markov processes. (In Russian) *Teor. Veroyatnost. i Primenen. 5*, 45–53.

Skorokhod A. V. (1960 a). On a problem in statistics of Gaussian processes. (In Russian) *Dopovidi Akademii nauk USSR 9*, 1167–69.

Striebel C. T. (1959). Densities for stochastic processes. *Ann. Math. Stat. 30*, 549–567.

Wiener N. (1949). *Extrapolation, Interpolation and Smoothing of Stationary Time Series.* Cambridge – New York.

Yaglom A. M. (1955). Extrapolation, interpolation and smoothing of stationary random processes with rational densities. (In Russian) *Trudy Mosk. Mat. Obshch. 4*, 333–374.

Yaglom A. M. (1960). Explicit formulas for extrapolation, smoothing and evaluating the amount of information in the theory of Gaussian stochastic processes. (In Russian) *Trans. Sec. Prague Conf. Inf. Th. etc., Praha*, 251–262.

Zadeh L. A. and Ragazzini R. (1950). Extension of Wiener's theory of prediction. *Journ. Appl. Phys. 21*, 645–655.

CHAPTER 21

An Inequality Concerning Random Linear Functionals on a Linear Space with a Random Norm and Its Statistical Application

Czechoslovak Math. J. 12 (1962), 486–491.

It is shown that the norm of a mixed (averaged) linear functional with respect to a mixed (averaged) norm in a linear space cannot exceed the mean square norm of the random functional with respect to the random norm in the linear space. Applications are given concerning projection (prediction etc.) of random variables and linear estimation of regression coefficients in stochastic processes.

1 THEORY

Let us have a complex linear space $\mathsf{M} = \{x\}$ and a probability space $(\Lambda, \mathcal{B}, F)$, where $\mathcal{B}$ is a Borel field of subsets of Λ, and $F(B)$, $B \in \mathcal{B}$, is a probability measure. Let $s_{\lambda x}$ be a real function on $\Lambda \times \mathsf{M}$, representing for every fixed λ a norm on M and for every fixed x a measurable and quadratically integrable function of $\lambda \in \Lambda$.

Further, let $f_{\lambda x}$ be a function on $\Lambda \times \mathsf{M}$, representing for every fixed λ a linear functional on M, if M is normed by $s_{\lambda x}$, and for every fixed x a measurable function of λ. Let $n_\lambda(f_\lambda)$ be the norm of $f_{\lambda x}$, if M is normed by $s_{\lambda x}$, $\lambda \in \Lambda$. We shall assume that $n_\lambda(f_\lambda)$ is a measurable and quadratically integrable function of $\lambda \in \Lambda$.

Collected Works of Jaroslav Hájek – With Commentary
Edited by M. Hušková, R. Beran and V. Dupač
Published in 1998 by John Wiley & Sons Ltd.

Put

$$\sigma_x = \sqrt{\mathsf{E}\, s^2_{\lambda x}} \quad (x \in \mathsf{M}), \tag{1.1}$$

$$\varphi_x = \mathsf{E}\, f_{\lambda x} \quad (x \in \mathsf{M}), \tag{1.2}$$

where $\mathsf{E}(\cdot) = \int (\cdot)\, \mathrm{d}F(\lambda)$. Since $|f_{\lambda x}| \leq n_\lambda(f_\lambda)\, s_{\lambda x}$, the integrability of $f_{\lambda x}$ follows from quadratic integrability of $n_\lambda(f_\lambda)$ and $s_{\lambda x}$.

Theorem 1. *The function σ_x given by (1.1) is a norm on M and φ_x given by (1.2) is a linear functional on M, if M is normed by σ_x. Also*

$$\nu(\varphi) \leq \sqrt{\mathsf{E}\, n^2_\lambda(f_\lambda)} \tag{1.3}$$

where $\nu(\varphi)$ denotes the norm of φ, if M is normed by σ_x.

If $f_{\lambda x} = \varphi_x$, $\lambda \in \Lambda$, $x \in \mathsf{M}$, we have the following inequality:

$$\nu(\varphi) \leq \frac{1}{\sqrt{\mathsf{E}\left(\frac{1}{n^2_\lambda(\varphi)}\right)}}. \tag{1.4}$$

Proof. We have

$$\begin{aligned}
\sigma^2_{x+y} &= \mathsf{E}\, s^2_{\lambda, x+y} \leq \mathsf{E}\, s^2_{\lambda x} + \mathsf{E}\, s^2_{\lambda y} + 2\mathsf{E}\, s_{\lambda x}\, s_{\lambda y} \\
&\leq \mathsf{E}\, s^2_{\lambda x} + \mathsf{E}\, s^2_{\lambda y} + 2\sqrt{\mathsf{E}\, s^2_{\lambda x}\, \mathsf{E}\, s^2_{\lambda y}} \\
&= \sigma^2_x + \sigma^2_y + 2\sigma_y \sigma_y = (\sigma_x + \sigma_y)^2,
\end{aligned}$$

which shows that σ_x fulfils the triangle inequality. If $\sigma_x = 0$, then $s_{\lambda x} = 0$ for at least one λ, and hence $x = 0$. Finally, $\sigma_{ax} = |a|\, \sigma_x$ follows from (1.1).

Now, additivity and homogeneity of φ_x follows from (1.2) and from the same property of $f_{\lambda x}$. Boundedness of φ as well as (1.3) are implied by the following inequality:

$$|\varphi_x| \leq \mathsf{E}|f_{\lambda x}| \leq \mathsf{E}\, n_\lambda(f_\lambda)\, s_{\lambda x} \leq \sqrt{\mathsf{E}\, n^2_\lambda(f_\lambda)\, \mathsf{E}\, s^2_{\lambda x}} = \sigma_x \sqrt{\mathsf{E}\, n^2_\lambda(f_\lambda)}.$$

If $f_{\lambda x} = \varphi_x$, then

$$\sigma^2_x = \mathsf{E}\, s^2_{\lambda x} \geq \mathsf{E}\, \frac{|\varphi_x|^2}{n^2_\lambda(\varphi)} = |\varphi_x|^2\, \mathsf{E}\, \frac{1}{n^2_\lambda(\varphi)}$$

which is equivalent to (1.4). The proof is complete. □

Now we shall assume that the norms $s_{\lambda x}$ are defined by an inner product $s_{\lambda xy}$, $x, y \in \mathsf{M}$, $\lambda \in \Lambda$, i.e. that

$$s_{\lambda x} = \sqrt{s_{\lambda xx}} \quad (x \in \mathsf{M},\ \lambda \in \Lambda). \tag{1.5}$$

Then, provided that $s_{\lambda x}$ are quadratically integrable, $s_{\lambda xy}$ are integrable and

$$\sigma_{xy} = \mathsf{E}\, s_{\lambda xy} \tag{1.6}$$

is an inner product of M.

Suppose that the linear functionals f_λ and φ admit the representation

$$f_{\lambda x} = s_{\lambda x\, v(\lambda)} \quad (x \in \mathsf{M},\ \lambda \in \Lambda), \tag{1.7}$$

$$\varphi_x = \sigma_{xh}. \tag{1.8}$$

The elements $v(\lambda) \in \mathsf{M}$, $h \in \mathsf{M}$, if they exist, are unique and will be called generating elements of the linear functionals f_λ and φ, respectively. For the existence of such elements it is sufficient that M be complete with norms $s_{\lambda x}$ and σ_x, i. e. that M be a Hilbert space. As is well known,

$$n_\lambda(f_\lambda) = s_{\lambda\, v(\lambda)} \quad (\lambda \in \Lambda), \tag{1.9}$$

$$\nu(\varphi) = \sigma_h. \tag{1.10}$$

In view of (1.7) and (1.8), the relation (1.2) may be written as

$$\sigma_{xh} = \mathsf{E}\, s_{\lambda x\, v(\lambda)} \quad (x \in \mathsf{M}). \tag{1.11}$$

Now we can improve (1.3) as follows.

Theorem 2. *If the norms $s_{\lambda x}$ are defined by inner products $s_{\lambda xy}$, and* (1.7) *and* (1.8) *hold, then*

$$\nu^2(\varphi) = \mathsf{E}\, n_\lambda^2(f_\lambda) - \mathsf{E}\, s^2_{\lambda,\, h-v(\lambda)} \tag{1.12}$$

or, equivalently,

$$\sigma_h^2 = \mathsf{E}\, s^2_{\lambda v(\lambda)} - \mathsf{E}\, s^2_{\lambda,\, h-v(\lambda)}. \tag{1.13}$$

Proof. We have

$$\mathsf{E}\, s^2_{\lambda,\, h-v(\lambda)} = \mathsf{E}\, s^2_{\lambda v(\lambda)} \mathsf{E}\, s^2_{\lambda h} - \mathsf{E}\, s_{\lambda h(\lambda)} - \mathsf{E}\, \overline{s_{\lambda h v(\lambda)}}. \tag{1.14}$$

Now, in view of (1.1) and (1.11), $\mathsf{E}\, s^2_{\lambda h} = \sigma_h^2$ and $\mathsf{E}\, s_{\lambda h v(\lambda)} = \sigma_{hh} = \sigma_h^2$, and therefore, also $\overline{\mathsf{E}\, s_{\lambda h v(\lambda)}} = \sigma_h^2$. So (1.14) is equivalent to (1.13), which is equivalent to (1.12). The theorem is proved. □

2 APPLICATIONS

In what follows, the space M will consist of all finite linear combinations $\sum c_\nu x_{t_\nu}$, $t_\nu \in T$, of values of a stochastic process x_t, $t \in T$. We shall consider a system of second moments

$$s_{\lambda xy} = \int x\, \overline{y}\, \mathrm{d}P_\lambda \quad (\lambda \in \Lambda)$$

generated by a system of probability measures P_λ, $\lambda \in \Lambda$, and assume that $\int x \, dP_\lambda = 0$, $x \in \mathsf{M}$, $\lambda \in \Lambda$. The inner product $s_{\lambda xy}$ will be defined on the set K consisting of all random variables having finite second moments with respect to all measures P_λ, $\lambda \in \Lambda$. We suppose that $\mathsf{M} \subset K$. We shall not complete the space M by adding limit points, because the set of limit points may depend on $\lambda \in \Lambda$, which causes complications unnecessary in this context.

Now we shall apply the derived inequalities in three situations.

2.1 Projections

If $z \in K$, then

$$f_{\lambda x} = s_{\lambda xz} \tag{2.1}$$

is a linear functional on M normed by $s_{\lambda x}$, and $d_{\lambda z}$ given by

$$d_{\lambda z}^2 = s_{\lambda z}^2 - n_\lambda^2(f_\lambda) \tag{2.2}$$

represents the distance of z from M. If M is complete, then $d_{\lambda z}$ is the distance of z from its projection onto M.

Now consider a probability space $(\Lambda, \mathcal{B}, F)$. Suppose that all required measurability and integrability conditions are satisfied, and put

$$\delta_z^2 = \sigma_z^2 - \nu^2(\varphi), \tag{2.3}$$

where σ and φ are given by (1.1) and (1.2). Thus δ_z represents the distance of z from M, when K is normed by σ_x. From (1.3) it follows that

$$\delta_z^2 \geq \mathsf{E}\, d_{\lambda z}^2. \tag{2.4}$$

If the operation of taking means with respect to dF is called mixing, then (2.4) may be expressed thus: the square of z from M normed by mixed second moments is not less than the mixed square distance.

Example 1. If M consists of all linear combinations of past values of a stationary process $\{x_t, t \leq 0\}$ and $z = x_\tau$, $\tau > 0$, then (2.3) determines the minimum square error of prediction. If we have

$$s_{\lambda x_t x_s} = b^2\, r\left(\frac{t-s}{\lambda}\right) \quad (0 < \lambda < \infty) \tag{2.5}$$

where b^2 denotes the variance, and

$$\begin{aligned} r(u) &= 1 - |u|, \quad \text{if } |u| \leq 1, \\ &= 0 \qquad\quad\ \text{otherwise,} \end{aligned} \tag{2.6}$$

then (see Hájek (1958)) we have

$$d^2_{\lambda x_\tau} = b^2(1 - r(\tau/\lambda)). \tag{2.7}$$

To any distribution function $F(\lambda)$ with $F(0) = 0$ there corresponds a mixed correlation function

$$R(u) = \int_0^\infty r\left(\frac{u}{\lambda}\right) \,\mathrm{d}F(\lambda) \tag{2.8}$$

for which the inequality (2.4) takes on the following form:

$$\delta^2_{x_\tau} \geq b^2(1 - R(\tau)). \tag{2.9}$$

As any convex correlation function with $R(\infty) = 0$ may be represented in the form (2.8), the inequality (2.9) remains true for any convex correlation function. This is shown in Hájek (1958).

Example 2. If in (2.5) we take, instead of (2.6),

$$r(u) = e^{-|u|} \tag{2.10}$$

then, as is well-known,

$$d^2_{\lambda x_\tau} = b^2\left(1 - e^{-2\tau/\lambda}\right). \tag{2.11}$$

To any distribution function $F(\lambda)$ with $F(0) = 0$ there corresponds a mixed correlation function

$$R(u) = \int_0^\infty e^{-|u|/\lambda} \,\mathrm{d}F(\lambda) \tag{2.12}$$

for which the inequality (2.4) takes on the form

$$\delta^2_{x_\tau} \geq b^2(1 - R(2\tau)). \tag{2.13}$$

Thus (2.13) holds for any correlation function which is obtained by mixing exponential correlation functions $e^{-|u|/\lambda}$.

2.2 Linear Estimation of Regression Coefficients

As is shown in Hájek (1960), $1/n^2(f)$ is the variance of the best unbiased linear estimate $\hat{\alpha}$ of the parameter α in a linear model, where covariance of x_t and x_s is (independently of α) $s_{x_t x_s}$ and the mean value of x_t is αf_{x_t}. So, adjoining an index λ, we have

$$\mathsf{var}\,(\hat{\alpha}_\lambda) = \frac{1}{n^2_\lambda(f_\lambda)}, \tag{2.14}$$

where $\mathsf{var}(\cdot)$ denotes the variance, $\mathsf{var}(\hat{\alpha}_\lambda) = s^2_{\lambda\hat{\alpha}_\lambda}$. If $f_{\lambda x} = f_x = \varphi_x$, the inequality (1.4) yields

$$\mathsf{var}\,\hat{\alpha} \geq \mathsf{E}\,\mathsf{var}\,\hat{\alpha}_\lambda, \tag{2.15}$$

where $\hat{\alpha}$ corresponds to σ_x given by (1.1). If $f_{\lambda x_t} = \varphi_{x_t} = 1$, $t \in T$, then α denotes the mean value.

Example 3. The inequality (5.10) in Hájek (1956).

Example 4. If x_t, $0 \leq t \leq T_0$, is a stationary process with correlation function $b^2 e^{-|u|/\lambda}$, and α denotes the mean value, then, as is well known,

$$\mathsf{var}\,\hat{\alpha}_\lambda = \frac{2\lambda b^2}{2\lambda + T_0} \quad (0 < \lambda < \infty). \tag{2.16}$$

Thus (2.15) gives

$$\mathsf{var}\,\hat{\alpha} \geq b^2 \int_0^\infty \frac{2\lambda}{2\lambda + T_0}\,\mathrm{d}F(\lambda) \tag{2.17}$$

for any correlation function of form (2.12).

2.3 The General Case

Let us have m linear functionals $\varphi_1, \ldots, \varphi_m$ on M, and denote by M_n the subset of M consisting of those elements for which

$$\varphi_{\nu x} = c_\nu \quad (\nu = 1, \ldots, m) \tag{2.18}$$

where $c_1, \ldots, c_m$ are fixed constants. Now let $\mathsf{M}_m - x_0$ consist of elements $x' = x - x_0$, where x_0 is a fixed element from M_m and x is any element of M_m. Then, clearly, $M_m - x_0$ is a linear space, and the distance of any element $z \in K$ from M_m equals the distance of $z - x_0$ from $\mathsf{M}_m - x_0$. Thus the inequality (2.4) is applicable in this general case also.

The scheme just described has the following statistical interpretation. We have a process $\{x_t, t \in T\}$ with a class of covariance functions $s_{\lambda x_t x_s}$, $\lambda \in \Lambda$, and a class of mean values $\sum_{\nu=1}^m \alpha_\nu \varphi_{\nu x_t}$, where $-\infty < \alpha_1, \ldots, \alpha_m < \infty$ are unknown constants and $\varphi_{\nu t} = \varphi_{\nu x_t}$, $1 \leq \nu \leq m$, are known functions. Observe that covariances do not depend on $\alpha_1, \ldots, \alpha_m$ and mean values do not depend on $\lambda \in \Lambda$. Let us assume that the functions $\varphi_{\nu x_t}$ can be extended to the whole space as linear functionals, which we denote by $\varphi_{\nu x}$, $x \in \mathsf{M}$, $1 \leq \nu \leq m$ (then none of the α_νs can be linearly estimated with zero variance). Now, let us have a random variable z such that its mean value equals $\sum_{\nu=1}^m \alpha_\nu c_\nu$ and that the covariances $s_{\lambda z x_t}$ do not depend on $\alpha_1, \ldots, \alpha_m$. Then M_m defined by (2.17) consists of unbiased linear estimates of z (i. e of elements of M having the same mean value as z for any $\alpha_1, \ldots, \alpha_m$) and the square distance of z from M_m denotes the minimum possible residual variance in linear unbiased estimation of z by elements from M, or from the closure of M. In particular, if z is a constant, $z = \sum_{\nu=1}^m \alpha_\nu c_\nu$, then the square distance denotes the minimum possible variance of linear unbiased estimates of $\sum_{\nu=1}^m \alpha_\nu c_\nu$. In all such cases the residual variance (or simply variance, if z is a constant) satisfies (2.4). For details see Hájek (1961).

REFERENCES

Hájek, J. (1956). Linear estimation of the mean value of a stationary random process with convex correlation function (In Russian) *Czechoslovak Math. J. 6*, 94–117. (English translation in *Selected Transl. Math. Statist. Prob. 2*, (1962), 41–61.)

Hájek, J. (1958). Predicting a stationary process when the correlation function is convex. *Czechoslovak Math. J. 8*, 150–154.

Hájek, J. (1960). On a simple linear model in Gaussian processes. In *Trans. Second Prague Conf. on Inf. Theory, etc.*, Academia, Prague, 185–197.

Hájek, J. (1961). On linear estimation theory for an infinite number of observations. *Teor. Verojatnost. i Primenen. 6*, 182–193.

CHAPTER 22

Asymptotically Most Powerful Rank-Order Tests

Ann. Math. Statist. *33* (1962), 1124–1147.

Reproduced by permission of The Institute of Mathematical Statistics.

SUMMARY

Having observed $X_i = \alpha + \beta c_i + \sigma Y_i$, we test the hypothesis $\beta = 0$ against the alternative $\beta > 0$. We suppose that the square root of the probability density $f(x)$ of the residuals Y_i possesses a square-integrable derivative, and define a class of rank-order tests which are asymptotically most powerful for given f. The main result is exposed in the following succession: theorem, corollaries and examples, comments, preliminaries and proof. The proof is based on results by Hájek (1961) and Le Cam (1960, 196–). Section 6 deals with asymptotic efficiency of rank-order tests, which is shown, on the basis of Mikulski's results (Mikulski (1961)), to be presumably never less than the asymptotic efficiency of corresponding parametric tests of Neyman type Neyman (1958). This would extend the well-known result obtained by Chernoff and Savage (1958) for the Student t-test. Furthermore, it is shown that the efficiency may be negative, i. e. asymptotic power may be less than the asymptotic size. In Section 7 we consider parallel rank-order tests of symmetry for judging paired comparisons. Section 8 is devoted to rank-order tests for densities such that $(f(x))^{\frac{1}{2}}$ does not possess a square-integrable derivative. In Section 9, we construct a test which is asymptotically most powerful simultaneously for all densities $f(x)$ such that $(f(x))^{\frac{1}{2}}$ possesses a square-integrable derivative.

Collected Works of Jaroslav Hájek – With Commentary
Edited by M. Hušková, R. Beran and V. Dupač
Published in 1998 by John Wiley & Sons Ltd.

1 THE MAIN THEOREM

Consider a sequence of random vectors $(X_{\nu 1}, \ldots, X_{\nu N_\nu})$, $1 \leq \nu < \infty$, where the Xs are independent and

$$\mathsf{P}(X_{\nu i} \leq x \,|\, \alpha, \beta\, \sigma) = F((x - \alpha - \beta c_{\nu i})/\sigma), \quad 1 \leq i \leq N_\nu,\ 1 \leq \nu < \infty, \tag{1.1}$$

where F is a known distribution function, $c_{\nu i}$ are known constants, α and σ are nuisance parameters, $-\infty < \alpha < \infty$, $\sigma > 0$, and β is the parameter under test. We test the hypothesis $\beta = 0$ against the alternative $\beta > 0$.

Assume that the density $f(x) = F'(x)$ exists and that $(f(x))^{\frac{1}{2}}$ possesses a square-integrable derivative. As

$$\left((f(x))^{\frac{1}{2}}\right)' = \frac{1}{2}\left[f'(x)/(f(x))^{\frac{1}{2}}\right], \tag{1.2}$$

it means that

$$\int_{-\infty}^{\infty} [f'(x)/f(x)]^2\, f(x)\, \mathrm{d}x < \infty. \tag{1.3}$$

Let the constants $c_{\nu i}$ satisfy the Noether condition

$$\lim_{\nu \to \infty} \left\{ \left[\max_{1 \leq i \leq N_\nu} (c_{\nu i} - \bar{c}_\nu)^2\right] \Big/ \left[\sum_{i=1}^{N_\nu} (c_{\nu i} - \bar{c}_\nu)^2\right] \right\} = 0, \tag{1.4}$$

where $\bar{c}_\nu = N_\nu^{-1} \sum_{i=1}^{N_\nu} c_{\nu i}$, and the boundedness condition

$$\sup_\nu \sum_{i=1}^{N_\nu} (c_{\nu i} - \bar{c}_\nu)^2 < \infty. \tag{1.5}$$

Denoting the inverse of F by F^{-1}, let us introduce the following function:

$$\varphi(u) = -\left[f'(F^{-1}(u))/f(F^{-1}(u))\right], \quad 0 < u < 1. \tag{1.6}$$

From (1.3) it follows that

$$\int_0^1 \varphi(u)\, \mathrm{d}u = \int_{-\infty}^{\infty} f'(x)\, \mathrm{d}x = 0 \tag{1.7}$$

and

$$\int_0^1 \varphi^2(u)\, \mathrm{d}u = \int_{-\infty}^{\infty} [f'(x)/f(x)]^2\, f(x)\, \mathrm{d}x < \infty. \tag{1.8}$$

If to an f there corresponds φ, $f \to \varphi$, then to $(1/\sigma)\, f((x-\alpha)/\sigma)$ there corresponds $(1/\sigma)\, \varphi(u)$:

$$(1/\sigma)\, f((x-\alpha)/\sigma) \longrightarrow (1/\sigma)\, \varphi(u), \tag{1.9}$$

i.e. the map is invariant under translation and a change of scale only introduces a multiplicative constant.

We shall need the following function and constants:

$$\Phi(x) = (2\pi)^{-\frac{1}{2}} \int_{-\infty}^{x} \exp\left(-\frac{1}{2}t^2\right) \mathrm{d}t, \tag{1.10}$$

$$\Phi(K_\epsilon) = 1 - \epsilon, \tag{1.11}$$

$$d_\nu^2 = \sum_{i=1}^{N_\nu} (c_{\nu i} - \bar{c}_\nu)^2 \int_{-\infty}^{\infty} [f'(x)/f(x)]^2 \, f(x) \, \mathrm{d}x. \tag{1.12}$$

Now let $R_{\nu i}$ be the rank of $X_{\nu i}$ in the ordered sample $V_{\nu 1} < \cdots < V_{\nu N_\nu}$, i.e.

$$X_{\nu i} = V_{\nu R_{\nu i}}, \quad 1 \leq i \leq N_\nu, \; 1 \leq \nu < \infty. \tag{1.13}$$

Furthermore, let $\varphi_\nu(u)$ be a function defined on $(0,1)$ and constant over intervals $(i/N_\nu, (i+1)/N_\nu)$, $0 \leq i < N_\nu$, i.e.

$$\varphi_\nu(u) = \varphi_\nu(i/(N_\nu + 1)), \quad (i-1)/N_\nu < u < i/N_\nu. \tag{1.14}$$

Consider rank-order statistics S_ν expressible in the following form:

$$S_\nu = \sum_{i=1}^{N_\nu} (c_{\nu i} - \bar{c}_\nu) \, \varphi_\nu(R_{\nu i}/(N_\nu + 1)), \quad 1 \leq \nu < \infty. \tag{1.15}$$

Let ϵ_ν and $\mathrm{Q}_\nu(\beta, \sigma)$ be the size and power of the test based on the critical region

$$S_\nu > K_\epsilon \, d_\nu, \quad 1 \leq \nu < \infty, \tag{1.16}$$

where K_ϵ and d_ν are given by (1.11) and (1.12).

Definition 1.1. If $\lim_{\nu \to \infty} \epsilon_\nu = \epsilon$ and

$$\liminf_{\nu \to \infty} [\mathrm{Q}_\nu(\beta, \sigma) - \mathrm{Q}_\nu^*(\alpha, \beta, \sigma)] \geq 0, \quad -\infty < \alpha < \infty, \; \beta > 0, \; \sigma > 0, \tag{1.17}$$

where $\mathrm{Q}_\nu^*(\beta, \sigma)$ is the power of any other test of limiting size ϵ, we say that the test based on (1.15) is asymptotically uniformly most powerful (see Neyman (1958)). Of course, uniformity is meant with respect to α, β and σ.

Theorem 1.1. *Consider model* (1.1) *and suppose that conditions* (1.3), (1.4) *and* (1.5) *are satisfied. Let*

$$\lim_{\nu \to \infty} \int_0^1 [\varphi_\nu(u) - \varphi(u)]^2 \, \mathrm{d}u = 0, \tag{1.18}$$

where φ is given by (1.6). Then the critical region (1.16) provides an asymptotically uniformly most powerful test of limiting size ϵ, and the asymptotic power equals

$$1 - \Phi(K_\epsilon - (\beta/\sigma)\, d_\nu), \tag{1.19}$$

where Φ, K_ϵ and d_ν are given by (1.10), (1.11) and (1.12).

2 COROLLARIES AND EXAMPLES

Step-functions φ_ν form an N_ν-dimensional subspace in the space of square-integrable functions on $(0,1)$. The projection of φ on this subspace, say φ_ν^0, equals

$$\varphi_\nu^0(i/(N_\nu+1)) = N_\nu \int_{(i-1)/N_\nu}^{i/N_\nu} \varphi(u)\, du, \quad 1 \le i \le N_\nu. \tag{2.1}$$

If we extend the definition of φ_ν^0 to the whole interval $(0,1)$ according to (1.14), we easily see that (1.18) is satisfied.

Let $V_{\nu 1} < \cdots < V_{\nu N_\nu}$ be the ordered sample referring to F and $\alpha = \beta = 0$ and $\sigma = 1$, so that $Z_{\nu 1} < \cdots < Z_{\nu N_\nu}$, where $Z_{\nu i} = F(V_{\nu i})$, is an ordered sample from the uniform distribution over $(0,1)$. Put

$$\varphi_\nu^+(i/(N_\nu+1)) = \mathsf{E}[-f'(V_{\nu i})/f(V_{\nu i})] = \mathsf{E}[\varphi(Z_{\nu i})], \quad 1 \le i \le N_\nu, \tag{2.2}$$

and complete the definition of φ_ν according to (1.14). Then, on account of Lemma 6.1 of Hájek (1961), φ_ν^+ satisfies the condition (1.18), too. So we get the following corollary.

Corollary 2.1. Let the conditions (1.3), (1.4) and (1.5) be satisfied. Put

$$S_\nu^0 = \sum_{i=1}^{N_\nu} (c_{\nu i} - \bar{c}_\nu)\, \varphi_\nu^0(R_{\nu i}/(N_\nu+1)) \tag{2.3}$$

and

$$S_\nu^+ = \sum_{i=1}^{N_\nu} (c_{\nu i} - \bar{c}_\nu)\, \varphi_\nu^+(R_{\nu i}/(N_\nu+1)), \tag{2.4}$$

where φ_ν^0 and φ_ν^+ are defined by (2.1) and (2.2). Then the critical regions $S_\nu^0 > K_\epsilon\, d_\nu$ and $S_\nu^+ > K_\epsilon\, d_\nu$ both provide an asymptotically most powerful test of limiting size ϵ and of asymptotic power (1.19).

Another step-function of particular interest is defined as follows:

$$\varphi_\nu^*(i/(N_\nu+1)) = \varphi(i/(N_\nu+1)), \quad 1 \le i \le N_\nu. \tag{2.5}$$

Here of course some continuity properties are needed. From Lemma 2.2 of Hájek (1961) it follows that monotoneity of φ is sufficient. Obviously, it also suffices if φ is continuous and bounded on $(0,1)$. So we get the following corollary.

Corollary 2.2. Let the conditions (1.3), (1.4) and (1.5) be satisfied. Put

$$S_\nu^* = \sum_{i=1}^{N_\nu} (c_{\nu i} - \bar{c}_\nu)\,\varphi(R_{\nu i}/(N_\nu + 1)) \tag{2.6}$$

and suppose that $\varphi(u)$ is either monotone or continuous and bounded. Then the critical region $S_\nu^* > K_\epsilon\, d_\nu$ provides an asymptotically most powerful test of limiting size ϵ and of asymptotic power (1.19).

Table 1

No.	$f(x)$	$-\frac{f'(x)}{f(x)}$	$\varphi(u)$
1	$\frac{1}{2}e^{-\lvert x\rvert}$	$\operatorname{sign} x$	$\operatorname{sign}\left(u-\frac{1}{2}\right)$
2	$\sqrt{3}\,e^{-x\sqrt{3}}\,(1+e^{-x\sqrt{3}})^{-2}$	$\sqrt{3}\,(1-e^{-x\sqrt{3}})\,(1+e^{-x\sqrt{3}})^{-1}$	$2\sqrt{3}\left(u-\frac{1}{2}\right)$
3	$(2\pi)^{-\frac{1}{2}}\exp\left(-\frac{1}{2}\,x^2\right)$	x	$\Phi^{-1}(u)$
4	$\exp(x-e^x)$	e^x-1	$-\log(1-u)-1$
5	$\frac{1}{3}\left[\Phi\left(\frac{10}{3}x\right)-\Phi\left(\frac{10}{3}x-10\right)\right]$	$\frac{\exp\left[-\frac{1}{2}\left(\frac{10}{3}x-10\right)^2\right]-\exp\left[-\frac{1}{2}\left(\frac{10}{3}x\right)^2\right]}{\Phi\left(\frac{10}{3}x\right)-\Phi\left(\frac{10}{3}x-10\right)}$ $\times\frac{10}{3}(2\pi)^{-\frac{1}{2}}$	Figure 1

In Table 1 there are listed five various densities together with corresponding logarithmic derivatives and functions $\varphi(u)$. The functions $\varphi(u)$ are drawn on Figure 1.

The first density in Table 1 is called the double exponential. The corresponding S_ν^*-test may be called the median test, because in the two-sample problem (when the $c_{\nu i}$s attain only two different values) it consists in counting the number of observations from the first sample, which are greater than the median of the pooled sample.

The logistic density is listed as the second one. The constant $\sqrt{3}$ has been introduced to get comparable curves in Figure 1. All three tests based on S_ν^0, S_ν^+ and S_ν^*, respectively, reduce to the Wilcoxon test in Wilcoxon (1945).

The fourth density has been obtained by the transformation $y = e^x$ from the exponential density e^{-y}, $y \geq 0$. So testing location changes of the former density may be reinterpreted as testing scale changes of the latter one. The

S_ν^+-test may be called the I. R. Savage (1956) test; he showed that

$$1 + \varphi_\nu^+(i/(N_\nu + 1)) = \sum_{j=N_\nu - i+1}^{N_\nu} (1/j).$$

The last distribution is a convolution of the normal distribution with zero mean and $\sigma = 0,3$ and of the uniform distribution over the interval $(0, 3)$. In this case the rank-order tests emphasize the role of the tails of the ordered sample.

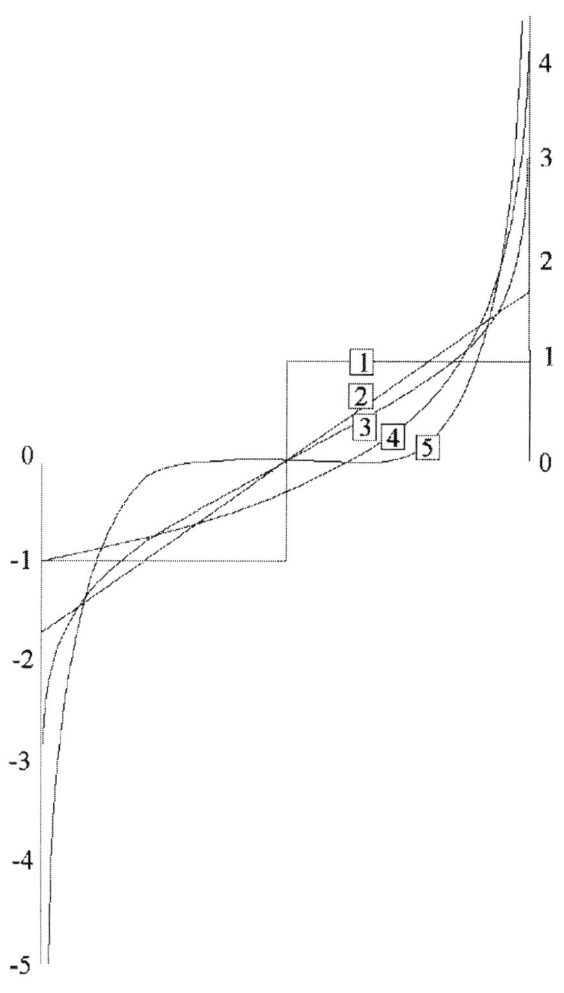

Figure 1 The curves $\varphi(u)$ are numbered according to Table 1

3 SOME COMMENTS

1. The two sample problem. If

$$\begin{aligned} c_{\nu i} &= 0, \qquad 1 \le i \le N_\nu \\ &= \Delta_\nu, \qquad n_\nu < i \le N_\nu, \end{aligned} \tag{3.1}$$

we get the two sample problem as a special case of our regression model. Assumptions (1.4) and (1.5) then reduce to

$$\lim_{\nu \to \infty} \max \left\{ n_\nu^{-1} - N_\nu^{-1}, (N_\nu - n_\nu)^{-1} - N_\nu^{-1} \right\} = 0 \tag{3.2}$$

and

$$\sup_\nu n_\nu \, \Delta_\nu^2 \, (1 - (n_\nu / N_\nu)) < \infty, \tag{3.3}$$

respectively. Obviously, (3.2) is equivalent to the requirement that the sizes of the two samples, n_ν and $N_\nu - n_\nu$, both tend to ∞. Otherwise, however, they may be quite arbitrary. For example, the ratio n_ν / N_ν may tend to 0 or 1.

2. Nuisance parameters. Introducing the nuisance parameters α and σ avoids some artificial assumption concerning the constants $c_{\nu i}$ and, furthermore, makes the knowledge of F more plausible. If the distribution functions of $X_{\nu i}$ were $F(x; \theta_0 + \beta c_{\nu i})$, we would have assumed that $\sum_{i=1}^{N_\nu} c_{\nu i} = 0$, or at least

$$N_\nu \, \bar{c}_\nu^2 \left(\sum_{i=1}^{N_\nu} c_{\nu i}^2 \right)^{-1} \longrightarrow 0,$$

in order to get a situation where asymptotically most powerful tests may be found among the rank-order tests. For this reason we have confined ourselves to a regression model, though the model with $\mathsf{P}(X_{\nu i} < x) = F(x; \theta_0 + \beta c_{\nu i})$ could be discussed with our methods, too.

3. We do not assume that $f(x) > 0$, $-\infty < x < \infty$, so that individual admissible distributions may not be absolutely continuous with respect to each other.

4. The best parametric test statistic of Neyman type. It has the following simple form, see Neyman (1958):

$$M_\nu = -\sum_{i=1}^{N_\nu} (c_{\nu i} - \bar{c}_\nu) \frac{f'((X_{\nu i} - \hat{\alpha}_\nu)/\hat{\sigma}_\nu)}{f((X_{\nu i} - \hat{\alpha}_\nu)/\hat{\sigma}_\nu)} \tag{3.4}$$

where $\hat{\alpha}_\nu$ and $\hat{\sigma}_\nu$ are some proper estimates of α and σ. If the estimates $\hat{\alpha}$ and $\hat{\sigma}$ satisfy the invariance conditions

$$\hat{\alpha}(bX_{\nu 1} + a, \ldots, bX_{\nu N_\nu} + a) = b\hat{\alpha}(X_{\nu 1}, \ldots, X_{\nu N_\nu}) + a \tag{3.5}$$

and

$$\hat{\sigma}\left(bX_{\nu 1}+a,\ldots,bX_{\nu N_\nu}+a\right)=b\hat{\sigma}\left(X_{\nu 1},\ldots,X_{\nu N_\nu}\right), \tag{3.6}$$

then the distribution of M_ν will be invariant with respect to α and σ. If, moreover,

$$\lim_{\nu\to\infty}\mathsf{E}\left[\frac{f'\left((X_{\nu 1}-\hat{\alpha})/\hat{\sigma}\right)}{f\left((X_{\nu 1}-\hat{\alpha})/\hat{\sigma}\right)}-\frac{f'\left((X_{\nu 1}-\alpha)/\sigma\right)}{f\left((X_{\nu 1}-\alpha)/\sigma\right)}\right]^2=0, \tag{3.7}$$

where E refers to $\alpha=\alpha$, $\beta=0$, $\sigma=\sigma$, then it may be shown, by methods of Section 5, that the critical region $M_\nu > K_\epsilon\, d_\nu$ provides an asymptotically uniformly most powerful test of limiting size ϵ and of asymptotic power (1.19). Discussion of the M_ν-test will be continued in Section 6.

5. Transformation of the observations. This may considerably enlarge the scope of the regression model (1.1). For example, if

$$\mathsf{P}(X_{\nu i}\le x)=F\left[x^{1/\sigma}\exp((-\alpha-\beta c_{\nu i})/\sigma)\right], \tag{3.8}$$

where $F(0)=0$, then we have

$$\mathsf{P}(\log X_{\nu i}\le y)=F\left[\exp((y-\alpha-\beta c_{\nu i})/\sigma)\right], \tag{3.9}$$

so that the transformed observations $\log X_{\nu i}$ obey the model (1.1) with $F(y)$ replaced by $F(\exp(y))$. In this way we have obtained density no. 4 in Table 1 from the density e^{-y}, $y>0$.

If (3.8) still holds, and the distribution is symmetrical about zero (i.e. $F(x)+F(-x)=1$), then the absolute values $|X_{\nu i}|,\ldots,|X_{\nu N_\nu}|$ represent a sufficient statistic. Furthermore, we have

$$\mathsf{P}(\log|X_{\nu i}|\le y)=2F\left[\exp((y-\alpha-\beta c_{\nu i})/\sigma)\right]-1, \tag{3.10}$$

so that the transformed observations $\log|X_{\nu i}|$ obey the model (1.1) with $F(y)$ replaced by $2F(\exp(y))-1$.

In this manner our model may be used not only to detect location trends but also to detect scalar trends in one-sided and symmetrical distributions. So, for example, all particular cases considered by Capon (1961) fall within the scope of model (1.1). The paper by Capon (1961) is methodically based on the paper by Chernoff and Savage (1958). From the point of view of our particular problem, however, the assumptions imposed in Chernoff and Savage (1958) are too restrictive.

6. If σ were known, say $\sigma=1$, all assertions would remain valid. If α were known, then we would have to suppose that $N_\nu\,\bar{c}_\nu^2\left(\sum_{i=1}^{N_\nu}c_{\nu i}^2\right)^{-1}\to 0$.

7. In order to remove assumption (1.5), we need to prove that $d_\nu \to \infty$ entails the power of test (1.16) tending to 1. It is very plausible, but we failed to prove it.

8. Given the function $\varphi(u)$, we can construct $f(x)$ as follows. First we establish

$$f(u) = -\int_{-\infty}^{u} \varphi(\lambda)\,\mathrm{d}\lambda, \tag{3.11}$$

and then, provided the integral exists,

$$x(u) = \int_{\frac{1}{2}}^{u} \mathrm{d}\lambda / f(\lambda). \tag{3.12}$$

The integral (3.11) exists, for example, if $f(u)$ is positive for $0 < u < 1$. Then, upon eliminating the parameter u, we get $f(x)$.

9. The distributions with non-decreasing $\varphi(u)$ are called strongly unimodal. Ibragimov (1956) proved that the convolution of $f(x)$ with every unimodal $g(x)$ is a unimodal distribution if and only if $f(x)$ is strongly unimodal.

4 PRELIMINARIES

Contiguity. Let $\{\mathsf{P}_\nu\}$ and $\{\mathsf{Q}_\nu\}$ be two sequences of probability measures defined on a sequence of measurable spaces $\{\mathcal{X}_\nu, \mathcal{U}_\nu\}$, $1 \le \nu < \infty$.

Definition 4.1. If for any sequence of events $\{A_\nu\}$, $A_\nu \in \mathcal{U}_\nu$, $\mathsf{P}_\nu(A_\nu) \to 0$ entails $\mathsf{Q}_\nu(A_\nu) \to 0$, we say that the probability measures Q_ν are contiguous to the probability measures P_ν, $1 \le \nu < \infty$.

Contiguity implies that the singular part of Q_ν tends to zero as $\nu \to \infty$, and that any sequence of random variables $\{Y_\nu\}$ converging to Y in P_ν-probability converges to Y in Q_ν-probability as well.

Let $\{B_\nu\}$ be a sequence of events such that $\mathsf{P}_\nu(B_\nu) = 1$ and Q_ν is absolutely continuous with respect to P_ν on B_ν, $1 \le \nu < \infty$. Put

$$\begin{aligned} r_\nu &= \mathrm{d}\mathsf{Q}_\nu/\mathrm{d}\mathsf{P}_\nu \quad \text{on} \quad B_\nu, \\ &= 0 \qquad\qquad\quad \text{on} \quad \mathcal{X}_\nu - B_\nu, \end{aligned} \tag{4.1}$$

where $\mathrm{d}\mathsf{Q}_\nu/\mathrm{d}\mathsf{P}_\nu$ denotes the usual Radon–Nikodym derivative. The probability measures Q_ν are contiguous to the probability measures P_ν if and only if, first,

$$\lim_{\nu\to\infty} \int r_\nu\,\mathrm{d}\mathsf{P}_\nu = \lim_{\nu\to\infty} \mathsf{Q}_\nu(B_\nu) = 1, \tag{4.2}$$

and, second, for any $\epsilon > 0$ there exists a $\delta > 0$ such that

$$[\mathsf{P}_\nu(A_\nu) < \delta] \Longrightarrow \left[\int_{A_\nu} r_\nu \, \mathrm{d}\mathsf{P}_\nu = \mathsf{Q}_\nu(A_\nu \cap B_\nu) < \epsilon\right], \quad 1 \leq \nu < \infty. \tag{4.3}$$

Condition (4.3) means, however, that the likelihood ratios r_ν are uniformly integrable.

Let us mention two well-known criteria for uniform integrability of a sequence of functions $\{h_\nu\}$. The first, one called the Vallée Poussin theorem, is based on boundedness of integrals

$$\sup_\nu \int |h_\nu| \, \Psi(|h_\nu|) \, \mathrm{d}\mathsf{P}_\nu < \infty, \tag{4.4}$$

where $\Psi(t)$, $t \geq 0$, is any function such that $\Psi(t) \to \infty$ if $t \to \infty$, and $t\Psi(t)$ is bounded from below. For example, we can take $\Psi(t) = t$ or $\log t$.

The second criterion may be formulated as follows. If $\mathsf{P}_\nu = \mathsf{P}$, h_ν are non-negative and $h_\nu \to h$ in P-probability, then the sequence $\{h_\nu\}$ is uniformly P-integrable, if, and only if,

$$\lim_{\nu \to \infty} \int h_\nu \, \mathrm{d}\mathsf{P} = \int h \, \mathrm{d}\mathsf{P}. \tag{4.5}$$

Observe that from the Fatou lemma it follows that

$$\liminf_{\nu \to \infty} \int \int h_\nu \, \mathrm{d}\mathsf{P} \geq \int h \, \mathrm{d}\mathsf{P}.$$

Le Cam (1960) recognized that the condition $\mathsf{P}_\nu = \mathsf{P}$ may be avoided if we replace the convergence in probability by convergence in distribution. He derived a basic result, which will be presented below as Lemma 4.1.

Denote

$$F_\nu(x) = \mathsf{P}_\nu(r_\nu \leq x). \tag{4.6}$$

The sequence $\{r_\nu\}$ converges in distribution if there exists a distribution function $F(x)$ such that

$$\lim_{\nu \to \infty} F_\nu(x) = F(x) \tag{4.7}$$

in all continuity points of $F(x)$. Since

$$1 \geq \mathsf{Q}_\nu(r_\nu > x) \geq x \, \mathsf{P}_\nu(r_\nu > x) = x[1 - F_\nu(x)],$$

it holds that

$$F_\nu(x) \geq 1 - (1/x), \quad x > 0. \tag{4.8}$$

From (4.8) it follows that the sequence $\{F_\nu\}$ is relatively compact, i. e. from every subsequence $\{k\} \subset \{\nu\}$ we can draw a further subsequence $\{j\} \subset \{k\}$ such that (4.7) holds true for ν running through $\{j\}$, and every $F(x)$ such that (4.7) holds true is a distribution function.

Lemma 4.1 (Le Cam). *Keep the above notations. The probability measures Q_ν are contiguous to the probability measures P_ν, if, and only if, for any subsequence $\{j\} \subset \{\nu\}$, $F_j(x) \to F(x)$ in continuity points of $F(x)$ implies*

$$\int_{-\infty}^{\infty} x \, \mathrm{d}F(x) = 1. \tag{4.9}$$

Proof. *Sufficiency.* Suppose that, on the contrary, for some subsequence $\{k\} \subset \{\nu\}$ there exist a sequence of events $\{A_k\}$ and a positive number $\epsilon > 0$ such that $\mathsf{P}_k(A_k) \to 0$ and

$$\mathsf{Q}_k(A_k) \geq \epsilon \quad k \in \{k\}. \tag{4.10}$$

Obviously, we can assume that the events are of the form $A_k = \{r_k \geq x_k\}$ with $x_k \to \infty$. Draw a further subsequence $\{j\} \subset \{k\}$ such that $F_j(x) \to F(x)$ in continuity points of $F(x)$. If x_0 is a continuity point of $F(x)$, then from (4.10) and $A_j = \{r_j \geq x_j\}$ it follows that

$$\begin{aligned} 1 - \epsilon &\geq \limsup_{j\to\infty} \mathsf{Q}_j(r_j < x_j) \\ &\geq \lim_{j\to\infty} \mathsf{Q}_j(r_j < x_0) = \lim_{j\to\infty} \int_{\{r_j < x_0\}} r_j \, \mathrm{d}\mathsf{P}_j \\ &= \lim_{j\to\infty} \int_{-0}^{x_0} x \, \mathrm{d}F_j(x) = \int_{-0}^{x_0} x \, \mathrm{d}F(x), \end{aligned}$$

which contradicts (4.9), since x_0 may be arbitrarily large.

Necessity. Suppose that for some $\epsilon > 0$

$$\int_{-0}^{\infty} x \, \mathrm{d}F(x) < 1 - 2\epsilon,$$

where $F(x)$ is the limit of subsequence $\{F_j\}$. Now, if

$$\mathsf{Q}_j(B_j) = \int_{-0}^{\infty} x \, \mathrm{d}F_j(x) \not\to 1,$$

the proof is finished. In the opposite case, for every continuity point x_0

$$\mathsf{Q}_j(r_j > x_0) = \int_{x_0}^{\infty} x \, \mathrm{d}F_j(x) \longrightarrow 1 - \int_{-0}^{\infty} x \, \mathrm{d}F(x) \geq 2\epsilon,$$

which entails the existence of random events $A_j = \{r_j > x_j\}$, $x_j \to \infty$, such that (4.10) holds true, whereas $\mathsf{P}(r_j > x_j) = 1 - F_j(x) \to 0$, on account of $x_j \to \infty$ and (4.8). The proof is complete. □

Now we shall analyse the most important case where P_ν and Q_ν are two probability distributions of a vector $(X_{\nu 1}, \ldots, X_{\nu N_\nu})$, whose components are independent. In this case, under rather general conditions, the only possible limits of subsequences $\{F_j\}$ are log-normal distributions. Let us have

$$\mathsf{P}_\nu\left(X_{\nu i} \le x_i,\, 1 \le i \le N_\nu\right) = \prod_{i=1}^{N_\nu} \mathsf{P}_{\nu i}\left(X_{\nu i} \le x_i\right) \tag{4.11}$$

$$\mathsf{Q}_\nu\left(X_{\nu i} \le x_i,\, 1 \le i \le N_\nu\right) = \prod_{i=1}^{N_\nu} \mathsf{Q}_{\nu i}\left(X_{\nu i} \le x_i\right) \tag{4.12}$$

and put

$$r_{\nu i} = \mathrm{d}\mathsf{Q}_{\nu i}/\mathrm{d}\mathsf{P}_{\nu i}, \quad 1 \le i \le N_\nu,\ 1 \le \nu < \infty, \tag{4.13}$$

on sets $B_{\nu i}$ such that $\mathsf{P}_{\nu i}(B_{\nu i}) = 1$ and $\mathsf{Q}_{\nu i}$ is absolutely continuous with respect to $\mathsf{P}_{\nu i}$, otherwise put $r_{\nu i} = 0$. If $p_{\nu i}(x)$ and $q_{\nu i}(x)$ are densities corresponding to distribution functions $\mathsf{P}_{\nu i}$ and $\mathsf{Q}_{\nu i}$ respectively, then

$$r_{\nu i}(x) = q_{\nu i}(x)/p_{\nu i}(x), \quad \text{if } p_{\nu i}(x) > 0, \tag{4.14}$$

and $r_{\nu i}(x) = 0$ otherwise.

Introduce the statistics

$$W_\nu = 2\sum_{i=1}^{N_\nu}\left((r_{\nu i})^{\frac{1}{2}} - 1\right) \tag{4.15}$$

and

$$L_\nu = \sum_{i=1}^{N_\nu} \log r_{\nu i}, \tag{4.16}$$

and suppose that for every $\epsilon > 0$

$$\lim_{\nu\to\infty} \max_{1\le i\le N_\nu} \mathsf{P}_{\nu i}\left(|r_{\nu i} - 1| > \epsilon\right) = 0. \tag{4.17}$$

Let $\mathcal{L}[Y_\nu \,|\, \mathsf{P}_\nu] \to N[a_\nu, b_\nu^2]$ denote that the distribution of $b_\nu^{-1}(Y_\nu - a_\nu)$ under P_ν tends to the normal distribution with mean 0 and variance 1. A similar meaning will be ascribed to $\mathcal{L}[Y_\nu \,|\, \mathsf{Q}_\nu] \to N[a_\nu, b_\nu^2]$.

The following lemma has been compiled from Le Cam's results in Le Cam (1960, 196–).

Lemma 4.2 (Le Cam). *Assume that* (4.17) *holds true and that* $\mathcal{L}[W_\nu \,|\, \mathsf{P}_\nu] \to N\left[-\frac{1}{4}\sigma^2, \sigma^2\right]$. *Then*

(i) *the probability distributions* Q_ν *are contiguous to the probability measures* P_ν,

(ii) *$W_\nu \to L_\nu \to \frac{1}{4}\sigma^2$ in P_ν-probability and $\mathcal{L}[Y_\nu, L_\nu \,|\, \mathsf{P}_\nu] \to N[-\sigma^2/2, \sigma^2]$,*
(iii) *if $\mathcal{L}[Y_\nu \,|\, \mathsf{P}_\nu] \to N(a, b^2)$ and $\mathcal{L}[Y_\nu, L_\nu \,|\, \mathsf{P}_\nu]$ tends to a bivariate normal distribution with correlation coefficient ρ, then*

$$\mathcal{L}[Y_\nu \,|\, \mathsf{Q}_\nu] \longrightarrow N[a + \rho b\sigma,\, b^2]. \tag{4.18}$$

Proof. Upon setting $u_{\nu i} = 2\left((r_{\nu i})^{\frac{1}{2}} - 1\right)$, we easily find that

$$L_\nu = W_\nu - \frac{1}{4}\sum_{i=1}^{N_\nu} u_{\nu i}^2 \int_0^1 \frac{2(1-\lambda)\,\mathrm{d}\lambda}{(1 + (1/2)\,\lambda u_{\nu_i})^2}. \tag{4.19}$$

Now, on account of (4.17) and of convergence of W_ν to the normal distribution, we have for any $\epsilon > 0$

$$\lim_{\nu\to\infty} \mathsf{P}\left\{\max_{1\le i\le N_\nu} |u_{\nu_i}| > \epsilon\right\} = 0, \tag{4.20}$$

which implies that the last term in (4.19) has the same limiting distribution as $\frac{1}{4}\sum_{i=1}^{N_\nu} u_{\nu i}^2$, if any. Furthermore, the limiting distribution of $\sum_{i=1}^{N_\nu} u_{\nu i}^2$ will be the same as that of $\sum_{i=1}^{N_\nu} u_{\nu i}^2(\delta)$, where

$$\begin{aligned} u_{\nu i}(\delta) &= u_{\nu i}, \quad \text{if } |u_{\nu i}| < \delta, \\ &= 0, \qquad \text{otherwise.} \end{aligned} \tag{4.21}$$

For any $\delta > 0$ the mean value of $\sum_{i=1}^{N_\nu} u_{\nu i}^2(\delta)$ equals σ^2 asymptotically and the variance is clearly less than $\delta^2\,\sigma^2$. This implies that $\sum_{i=1}^{N_\nu} u_{\nu i}^2 \to \sigma^2$ in P_ν-probability, which concludes the proof of assertion (ii).

Since $r_\nu = \exp(L_\nu)$, (ii) entails $F_\nu(x) \to F(x)$, where $F(e^y)$ is the normal distribution function with parameters $\left[-\frac{1}{2}\sigma^2,\, \sigma^2\right]$. This implies that (4.9) holds true, and hence, according to Lemma 4.1, assertion (i) holds true.

It remains to prove (iii). Let $F_\nu(u,v)$ be the distribution function of (Y_ν, L_ν) and $F(u,v)$ be the limiting bivariate normal distribution. Obviously,

$$\begin{aligned} \mathsf{Q}_\nu(Y_\nu < x) &= \int_{\{Y_\nu < x\}} r_\nu\,\mathrm{d}\mathsf{P}_\nu + \mathsf{Q}_\nu(A_\nu) \\ &= \int_{-\infty}^{x}\int_{-\infty}^{\infty} e^v\,\mathrm{d}F_\nu(u,v) + \mathsf{Q}_\nu(A_\nu), \end{aligned}$$

where $\mathsf{Q}_\nu(A_\nu) \to 0$ because of contiguity. For the same reason, the r_νs are uniformly integrable, and hence

$$\lim_{\nu\to\infty}\int_{-\infty}^{x}\int_{-\infty}^{\infty} e^v\,\mathrm{d}F_\nu(u,v) = \int_{-\infty}^{x}\int_{-\infty}^{\infty} e^v\,\mathrm{d}F(u,v).$$

Now, after easy computations,

$$\begin{aligned}
&\int_{-\infty}^{x}\int_{-\infty}^{\infty} e^{v}\,\mathrm{d}F(u,v)\\
&= \int_{-\infty}^{x}\int_{-\infty}^{\infty} b^{-1}\sigma^{-1}[2\pi(1-\rho^2)]^{-\frac{1}{2}}\exp\left\{v-[2(1-\rho^2)]^{-1}\left\{[(u-a)^2/b^2]\right.\right.\\
&\qquad \left.\left.-\left[2\rho(u-a)\left(v+\frac{1}{2}\sigma^2\right)/b\sigma\right]+\left[\left(v+\frac{1}{2}\sigma^2\right)^2\Big/\sigma^2\right]\right\}\right\}\mathrm{d}v\mathrm{d}u\\
&= b^{-1}(2\pi)^{-\frac{1}{2}}\int_{-\infty}^{x}\exp\left[-\frac{1}{2}((u-a-\rho b\sigma)^2/b^2)\right]\,\mathrm{d}u,
\end{aligned}$$

which concludes the proof. □

Remark 4.1. The notion of contiguity has been developed independently by Le Cam and the present author. In the original version of this paper the less efficient criterion (4.4) had been used, which led to much weaker results. Then, under the influence of Le Cam's ideas, the paper was revised and put in the present form.

In conclusion, we formulate a lemma concerning functions possessing a square-integrable derivative.

Lemma 4.3. *Let $s(x)$, $-\infty < x < \infty$, be an absolutely continuous function possessing a square-integrable derivative $s'(x)$. Then*

$$\lim_{h\to 0}\int_{-\infty}^{\infty}\{([s(x-h)-s(x)]/h)-s'(x)\}^2\,\mathrm{d}x = 0. \tag{4.22}$$

Proof. From

$$\{[s(x-h)-s(x)]/h\}^2 = \left[(1/h)\int_{x-h}^{x} s'(y)\,\mathrm{d}y\right]^2 \le (1/h)\int_{x-h}^{x}[s'(y)]^2\,\mathrm{d}y$$

it follows that

$$\int_{-\infty}^{\infty}\{[s(x-h)-s(x)]/h\}^2\,\mathrm{d}x \le \int_{-\infty}^{\infty}[s'(x)]^2\,\mathrm{d}x. \tag{4.23}$$

Furthermore, $h^{-1}[s(x) = s(x-h)] \to s'(x)$ almost surely, so that, due to (4.23) and Fatou's lemma,

$$\lim_{h\to 0}\int_{-\infty}^{\infty}\{[s(x-h)-s(x)]/h\}^2\,\mathrm{d}x = \int_{-\infty}^{\infty}[s'(x)]^2\,\mathrm{d}x. \tag{4.24}$$

However, (4.24) implies, according to the above-mentioned criterion (4.5), that the functions $h^{-2}[s(x)-s(x-h)]^2$ are uniformly integrable, which entails (4.22). The proof is complete. □

5 PROOF OF THE MAIN THEOREM

Put

$$s(x) = (f(x))^{\frac{1}{2}}, \tag{5.1}$$

so that

$$s'(x) = \frac{1}{2}\left[f'(x)/(f(x))^{\frac{1}{2}}\right]. \tag{5.2}$$

Consequently, if (1.3) is satisfied, $s(x)$ has a square-integrable derivative.

Now take a particular alternative, say $[\alpha = \alpha_0, \beta = \beta_0, \sigma = \sigma_0]$, and associate with it the particular hypothesis $[\alpha = \alpha_0 + \beta_0 \bar{c}_\nu, \beta = 0, \sigma = \sigma_0]$. Denoting the distribution under the alternative and the hypothesis by Q_ν and P_ν, respectively, we can express the statistic (4.15) in the form

$$W_\nu = 2\sum_{i=1}^{N_\nu} \{([s(Y_{\nu_i} - \gamma\, c_{\nu i} + \gamma\, \bar{c}_\nu)]/s(Y_{\nu i})) - 1\}, \tag{5.3}$$

where

$$Y_{\nu i} = [X_{\nu i} - \alpha_0 - \beta_0\, \bar{c}_\nu]\,/\sigma_0, \tag{5.4}$$

and

$$\gamma = \beta_0/\sigma_0. \tag{5.5}$$

Let E denote the mean value referring to P_ν. We easily see that, because of Lemma 4.3,

$$\begin{aligned} 2\mathsf{E}\left[\frac{s(Y_{\nu i} - h)}{s(Y_{\nu i})} - 1\right] &= -h^2 \int_{-\infty}^{\infty} \left[\frac{s(x-h) - s(x)}{h}\right]^2 \mathrm{d}x \\ &\sim -h^2 \int_{-\infty}^{\infty} [s'(x)]^2\, \mathrm{d}x = -h^2 \int_{-\infty}^{\infty} [f'(x)/f(x)]^2\, f(x)\, \mathrm{d}x, \end{aligned} \tag{5.6}$$

where the sign $\sim$ denotes that the ratio of both sides tends to 1 as $h \to 0$. Consequently, in view of (5.3),

$$\mathsf{E}W_\nu \sim -\frac{1}{4}\gamma^2\, d_\nu^2, \tag{5.7}$$

where γ and d_ν^2 are given by (5.5) and (1.12).

Now we shall approximate W_ν by

$$T_\nu = \sum_{i=1}^{N_\nu} (c_{\nu i} - \bar{c}_\nu)\, [f'(Y_{\nu i})/f(Y_{\nu i})]. \tag{5.8}$$

Under the hypothesis, obviously,

$$\operatorname{var} T_\nu = d_\nu^2. \tag{5.9}$$

Furthermore, upon setting $h_{\nu i} - \gamma c_{\nu i} - \gamma \bar{c}_\nu$,

$$\begin{aligned} &\mathsf{E}(W_\nu - \mathsf{E}W_\nu - \gamma T_\nu)^2 \\ &\leq 4\sum_{i=1}^{N_\nu} \mathsf{E}\left[\frac{s(Y_{\nu i} - h_{\nu i})}{S(Y_{\nu i})} - 1 - \frac{1}{2}h_{\nu i}\frac{f'(Y_{\nu i})}{f(Y_{\nu i})}\right]^2 \\ &\leq 4\gamma^2 \sum_{i=1}^{N_\nu}(c_{\nu i} - \bar{c}_\nu)^2 \int \left[\frac{s(x - h_{\nu i}) - s(x)}{h_{\nu i}} - s'(x)\right]^2 \mathrm{d}x. \end{aligned} \tag{5.10}$$

From (1.5), (5.9), (5.10) and Lemma 4.3 it follows that

$$\lim_{\nu\to\infty} \mathsf{E}(W_\nu - \mathsf{E}W_\nu - \gamma T_\nu)^2 = 0. \tag{5.11}$$

Furthermore, introducing the statistic

$$T_\nu^* = \sum_{i=1}^{N_\nu}(c_{\nu i} - \bar{c}_\nu)\,\varphi_\nu(F(Y_{\nu i})), \tag{5.12}$$

and consulting (1.18), we can see that

$$\lim_{\nu\to\infty} \mathsf{E}(T_\nu - T_\nu^*)^2 \leq \lim_{\nu\to\infty} \sum_{\nu=1}^{N_\nu}(c_{\nu i} - \bar{c}_\nu)^2 \int_0^1 [\varphi_\nu(u) - \varphi(u)]^2\,\mathrm{d}u = 0. \tag{5.13}$$

Now, (5.7), (5.11) and (5.13) imply

$$\mathsf{P}_\nu \lim_{\nu\to\infty}\left(W_\nu - \frac{1}{4}\gamma^2 d_\nu^2 - \gamma T_\nu\right) = 0 \tag{5.14}$$

and

$$\mathsf{P}_\nu \lim_{\nu\to\infty}(T_\nu^* - T_\nu) = 0, \tag{5.15}$$

where $\mathsf{P}_\nu \lim$ denotes the limit in P_ν-probability. Furthermore, according to Theorem 3.1 of Hájek (1961),

$$\lim_{\nu\to\infty} \mathsf{E}(S_\nu - T_\nu^*)^2 = 0, \tag{5.16}$$

which in conjunction with (5.13) implies

$$\mathsf{P}_\nu \lim_{\nu\to\infty}(S_\nu - T_\nu) = 0. \tag{5.17}$$

However, it is easy to see (Gnedenko and Kolmogorov (1954), Theorem 4, p. 103), that

$$\mathcal{L}[T_\nu \,|\, \mathsf{P}_\nu] \longrightarrow N[0, d_\nu^2], \tag{5.18}$$

and, consequently, on account of (5.14),

$$\mathcal{L}(W_\nu \,|\, \mathsf{P}_\nu) \longrightarrow N\left[-\frac{1}{4}\gamma^2\, d_\nu^2,\ \gamma^2\, d_\nu^2\right]. \tag{5.19}$$

Now observe that if the assertion of Theorem 1.1 were not true, there would exist a subsequence $\{j\} \subset \{\nu\}$ such that $d_j^2 \to d^2$, and that the assertion would still be false for the problem reduced to this subsequence $\{j\}$. So we may assume, without any loss of generality, that

$$\lim_{\nu\to\infty} d_\nu^2 = d^2. \tag{5.20}$$

Upon making use of Lemma 4.1, we conclude that the distributions Q_ν are contiguous to the distributions P_ν, and that

$$\mathsf{P}_\nu \lim_{\nu\to\infty}\left(L_\nu - \frac{1}{2}\gamma^2\, d^2 - \gamma\, T_\nu\right) = 0. \tag{5.21}$$

Now (5.21), (5.17) and contiguity imply that

$$\mathsf{Q}_\nu \lim_{\nu\to\infty}\left(L_\nu - \frac{1}{2}\gamma^2\, d^2 - \gamma\, S_\nu\right) = 0. \tag{5.22}$$

Consequently, the best critical region for distinguishing the particular alternative Q_ν from the particular hypothesis P_ν, known to be based on the statistic L_ν (Neyman–Pearson lemma), may be approximated asymptotically by the test (1.16). Since the statistic S_ν is distribution free, the test has the correct limiting size for all possible particular hypotheses, and, consequently, it is really asymptotically uniformly most powerful.

It remains to find the limit of $\mathcal{L}[S_\nu \,|\, \mathsf{Q}_\nu]$. From contiguity and (5.17) it follows that, if one of the limits exists,

$$\lim_{\nu\to\infty} \mathcal{L}[S_\nu \,|\, \mathsf{Q}_\nu] = \lim_{\nu\to\infty} \mathcal{L}[T_\nu \,|\, \mathsf{Q}_\nu]. \tag{5.23}$$

However, as we have seen, $\mathcal{L}[T_\nu \,|\, \mathsf{P}_\nu] \to N[0, d^2]$, and, further, on account of (5.21) $\mathcal{L}[L_\nu,\, T_\nu \,|\, \mathsf{P}_\nu]$ tends to a degenerated bivariate normal distribution with correlation coefficient $\rho = 1$ and variances $\gamma^2\, d^2$ and d^2 referring to L_ν and T_ν, respectively. Consequently, according to assertion 3 of Lemma 4.1,

$$\lim_{\nu\to\infty} \mathcal{L}[T_\nu \,|\, \mathsf{Q}_\nu] = N[\gamma\, d^2,\, d^2]. \tag{5.24}$$

Now, from (5.23) and (5.24) we easily conclude (1.19), and the proof is complete. □

6 ASYMPTOTIC EFFICIENCY

If the true density, say $g(x)$, differs from the supposed one, say $f(x)$, then the tests based on S_ν will no longer be asymptotically most powerful. Denote

$$\psi(u) = -\left[g'(G^{-1}(u))/g(G^{-1}(u))\right], \tag{6.1}$$

where $G(x) = \int_{-\infty}^{x} g(y)\,\mathrm{d}y$, so that $\psi(u)$ parallels $\varphi(u)$, given by (1.6). Assume that

$$\int_0^1 \psi^2(u)\,\mathrm{d}u = \int_{-\infty}^{\infty} \left[g'(x)/g(x)\right]^2 g(x)\,\mathrm{d}x < \infty, \tag{6.2}$$

and introduce the correlation coefficient

$$\rho = \left[\int_0^1 \varphi(u)\,\psi(u)\,\mathrm{d}u\right] \Big/ \left[\int_0^1 \varphi^2(u)\,\mathrm{d}u \int_0^1 \psi^2(u)\,\mathrm{d}u\right]^{\frac{1}{2}}. \tag{6.3}$$

Theorem 6.1. *Let the conditions* (1.3), (1.4), (1.5), (1.18) *and* (6.2) *be satisfied. Then the asymptotic power of the test* $S_\nu > K_\epsilon\, d_\nu$ *equals*

$$1 - \Phi\left(K_\epsilon - \rho(\beta/\sigma)\, d_\nu\right) \tag{6.4}$$

where ρ *is given by* (6.3).

Proof. Let P_ν correspond to G and to $\alpha = \alpha_0 + \beta_0\,\bar{c}_\nu,\ \beta = 0,\ \sigma = \sigma_0$, and let Q_ν correspond to G and $\alpha = \alpha_0,\ \beta = \beta_0,\ \sigma = \sigma_0$. Introduce statistics

$$L_\nu = \sum_{i=1}^{N_\nu} \log\left[g(Y_{\nu i} - \gamma\, c_{\nu i} + \gamma\,\bar{c}_\nu)/g(Y_{\nu i})\right] \tag{6.5}$$

and

$$T_\nu = \sum_{i=1}^{N_\nu} (c_{\nu i} - \bar{c}_\nu)\,\varphi(G(Y_{\nu i})). \tag{6.6}$$

Now, upon following literally the pattern used in proving Theorem 1.1, we prove that (5.23) still holds with our redefined Q_ν and T_ν, and that $\mathcal{L}[L_\nu, T_\nu \,|\, \mathsf{P}_\nu]$ tends to the bivariate normal distribution with correlation coefficient ρ given by (6.3). Hence the result. □

The number ρ^2 is usually called the efficiency. In the two-sample problem where $n_\nu/N_\nu \to \lambda,\ 0 < \lambda < 1$, the efficiency may be interpreted as the ratio of sample sizes needed to attain the same power. In the general regression problem the sample size N_ν should be replaced by the sum $\sum_{i=1}^{N_\nu}(c_{\nu i} - \bar{c}_\nu)^2$. However, even then the interpretation fails if $\rho < 0$. This case may arise only if φ and ψ are both non-monotonous, for example, if

$$\begin{aligned}\varphi(u) &= -u, & 0 < u < \frac{1}{4},\\ &= u - \frac{1}{2}, & \frac{1}{4} < u \leq \frac{3}{4},\\ &= 1 - u, & \frac{3}{4} < u < 1,\end{aligned}$$

and

$$\begin{aligned}\psi(u) &= -u, & 0 < u < \frac{1}{8},\\ &= \frac{1}{64}, & \frac{1}{8} < u \leq \frac{1}{2},\\ &= -\frac{1}{64}, & \frac{1}{2} < u \leq \frac{7}{8},\\ &= u, & \frac{7}{8} < u < 1.\end{aligned}$$

The functions $f(u) = -\int_0^u \varphi(\lambda)\,\mathrm{d}\lambda$ and $g(u) = -\int_0^u \psi(\lambda)\,\mathrm{d}\lambda$ are positive for $0 < u < 1$, so that, according to Subsection 3.8, the corresponding densities exist. However, the correlation coefficient equals

$$\rho = -(10/16\sqrt{73}),$$

so that the asymptotic power is less than the limiting size, provided that $\lim_{\nu\to\infty} \inf d_\nu > 0$. So, in a wider model, where F is not known, the $S_\nu(\varphi)$-test with non-monotonous φ may not be unbiased.

Now we shall compare the efficiency of the rank-order tests with the efficiency of the test based on Neyman's statistic M_ν given by (3.4). Suppose that there exist numbers $\alpha(g)$ and $\sigma(g)$ such that

$$\lim_{\nu\to\infty} \mathsf{E}\left[\frac{f'((X_{\nu 1} - \hat{\alpha})/\hat{\sigma})}{f((X_{\nu 1} - \hat{\alpha})/\hat{\sigma})} - \frac{f'((X_{\nu 1} - \alpha(g))/\sigma(g))}{f'((X_{\nu 1} - \alpha(g))/\sigma(g))}\right]^2 = 0, \tag{6.7}$$

where E refers to the hypothesis that $\mathsf{P}(X_{\nu i} < x) = G(x)$, $1 \leq i \leq N_\nu$, $1 \leq \nu < \infty$. Moreover, in accordance with (3.5) and (3.6), let us assume that

$$\alpha(g_{a,b}) = b\alpha(g) + a \tag{6.8}$$

and

$$\sigma(g_{a,b}) = b\sigma(g), \tag{6.9}$$

where

$$g_{a,b}(x) = (1/b)\,g((x-a)/b), \quad -\infty < a < \infty,\ b > 0. \tag{6.10}$$

Put $g_0 = g_{a,b}$ with a and b chosen so that

$$\alpha(g_0) = 0, \quad \sigma(g_0) = 1. \tag{6.11}$$

Introduce another correlation coefficient

$$\rho^* = C/(AB)^{\frac{1}{2}}, \tag{6.12}$$

where

$$A = \int_{-\infty}^{\infty} [g_0'(x)/g_0(x)]^2 \, g_0(x) \, \mathrm{d}x, \tag{6.13}$$

$$B = \int_{-\infty}^{\infty} [f'(x)/f(x)]^2 \, g_0(x) \, \mathrm{d}x - \left\{ \int_{-\infty}^{\infty} [f'(x)/f(x)] \, g_0(x) \, \mathrm{d}x \right\}^2 \tag{6.14}$$

and

$$C = \int_{-\infty}^{\infty} [f'(x)/f(x)] \, [g_0'(x)/g_0(x)] \, g_0(x) \, \mathrm{d}x. \tag{6.15}$$

Under the above assumption, the asymptotic power of Neyman's M_ν-test equals

$$1 - \Phi(K_\epsilon - \rho^*(\beta/\sigma) \, d_\nu). \tag{6.16}$$

An interesting question arises: what is the relation of ρ and ρ^*? If f is the normal density, then Chernoff and Savage (1958) proved that

$$\rho^* \leq \rho, \tag{6.17}$$

with equality only if $g \equiv f$ up to a change of location and scale. Directly from the definition of ρ^* it follows that it does not depend on changes of location and scale (the same is true of ρ, because of (1.9)). If f is normal and $\hat{\alpha} = \overline{x}$ and $\hat{\sigma} = s$, where s is the sample standard deviation, then (6.11) means that in computing ρ^* according to (6.12) through (6.15) we restrict ourselves to densities with zero means and unity variances. The thesis of Mikulski (1961, remarks on p. 17) strongly supports the conjecture that (6.17) holds true generally. To avoid misunderstanding, let us point out that Mikulski's paper aims at disproving the conjecture and succeeds in doing it in the two-sample problem, where, however, nuisance parameters are not involved. Upon introducing the nuisance parameters α and σ and using invariant estimates $\hat{\alpha}$ and $\hat{\sigma}$, the distribution of the statistic M_ν becomes independent of changes of location and scale, which brings it closer to the statistic S_ν. The solution of the corresponding variational problem is then to be chosen in the class of densities satisfying the side conditions (6.11). The author hopes to explore the question in a subsequent paper.

7 TESTS OF SYMMETRY

Let $f(x)$ be symmetric density, i. e.

$$f(x) = f(-x), \quad -\infty < x < \infty. \tag{7.1}$$

Consider the hypothesis that the $X_{\nu i}$s are independent and have a common probability density $f(x/\sigma)$, where σ is a nuisance parameter ($\sigma > 0$), against the alternative that the densities equal $f((x - \beta\, c_{\nu i})/\sigma)$, where

$$\lim_{\nu\to\infty} \left\{ \max_{1\le i\le N_\nu} c_{\nu i}^2 \Big/ \sum_{i=1}^{N_\nu} c_{\nu i}^2 \right\} = 0 \tag{7.2}$$

and

$$\sup_\nu \sum_{i=1}^{N_\nu} c_{\nu i}^2 < \infty. \tag{7.3}$$

Now let $R_{\nu i}$ be the rank of $|X_{\nu i}|$ in the sequence $|X_{\nu i}|, \dots, |X_{\nu N_\nu}|$ rearranged according to ascending magnitude. Put

$$S_\nu = \sum_{i=1}^{N_\nu} c_{\nu i}\, \varphi_\nu(R_{\nu i}/(N_\nu + 1)) \operatorname{sign} X_{\nu i}, \tag{7.4}$$

and extend the definition of $\varphi_\nu(u)$ to the whole interval (0,1) according to (1.14). Furthermore, set

$$\varphi(u) = -\left\{ f'\left(F^{-1}\left(\frac{1}{2} + \frac{1}{2}u\right)\right) \Big/ f\left(F^{-1}\left(\frac{1}{2} + \frac{1}{2}u\right)\right)\right\} \tag{7.5}$$

and

$$d_\nu^2 = \sum_{i=1}^{N_\nu} c_{\nu i}^2 \int_{-\infty}^{\infty} \left[f'(x)/f(x)\right]^2 f(x)\, \mathrm{d}x. \tag{7.6}$$

Theorem 7.1. *Let the conditions* (1.3), (7.1), (7.2) *and* (7.3) *be satisfied. Let* (1.18) *hold true for* φ_ν *used in (7.4) and* φ *defined by* (7.5).

Then the critical region $S_\nu > K_\epsilon\, d_\nu$, *where* S_ν, K_ϵ *and* d_ν *are given by* (7.4), (1.11) *and* (7.6), *respectively, provides an asymptotically uniformly most powerful test of* $\beta = 0$ *against* $\beta > 0$ *with limiting size* ϵ *and asymptotic power* (1.19), *where* d_ν *is given by* (7.6).

Proof. The proof would be just the same as in Theorem 1.1. Usually the *c*s are constant, so that $S_\nu = 2S'_\nu - \text{const}$, where

$$S'_\nu = \sum_{X_{\nu i}>0} \varphi_\nu(R_{\nu i}/(N_\nu + 1)). \tag{7.7}$$

Special forms of φ_ν are obtained by setting $\varphi_\nu(u) = \varphi_\nu^0\left(\frac{1}{2}+\frac{1}{2}u\right)$, $\varphi_\nu(u) = \varphi_\nu^+\left(\frac{1}{2}+\frac{1}{2}u\right)$ and $\varphi_\nu(u) = \varphi_\nu^*\left(\frac{1}{2}+\frac{1}{2}u\right)$, where φ_ν^0, φ_ν^+ and φ_ν^* are given by (2.1), (2.2) and (2.5), respectively. Upon taking $\varphi(u) = \operatorname{sign}\left(u - \frac{1}{2}\right)$, appearing in the first row of Table 1, and putting $\varphi_\nu = \varphi^*_\nu\left(\frac{1}{2} + \frac{1}{2}u\right)$, we get the sign test, based on the statistic

$$S'_\nu = \sum_{X_{\nu i}>0} 1. \tag{7.8}$$

Upon taking $\varphi(u) = 2u - 1$, we get the Wilcoxon test in Wilcoxon (1945) for paired comparisons, based on the statistic

$$(N_\nu + 1)\, S'_\nu = \sum_{X_{\nu i}>0} R_{\nu i}. \tag{7.9}$$

□

8 A TEST FOR THE UNIFORM DISTRIBUTION

Let $f(x)$ be a density and set

$$f_\sigma(x) = \sigma^{-1}(2\pi)^{-\frac{1}{2}} \int_{-\infty}^{\infty} f(x-y)\exp\left(-\frac{1}{2}(y^2/\sigma^2)\right) dy. \tag{8.1}$$

It may be easily seen that $f_\sigma(x)$ satisfies condition (1.3) for any f and $\sigma > 0$. So we can take, for every $\sigma > 0$, one of the statistics S^0_ν, S^+_ν and S^*_ν, defined in Section 2, and observe what happens with the critical region when $\sigma \to 0$. At least in some cases, we obtain a limiting critical region, and may hope that it has some good asymptotic properties.

Let us illustrate this idea in the case of the uniform distribution over $(0,1)$. The probability density of its convolution with the normal probability density equals

$$\begin{aligned} f_\sigma(x) &= \sigma^{-1}(2\pi)^{-\frac{1}{2}} \int_{x-1}^{x} \exp\left(-\frac{1}{2}t^2/\sigma^2\right) dt \\ &= \Phi\left(\frac{x}{\sigma}\right) - \Phi\left(\frac{x-1}{\sigma}\right). \end{aligned}$$

Put $\varphi_\sigma(u) = -f'_\sigma(F^{-1}_\sigma(u))/f_\sigma(F^{-1}_\sigma(u))$. Then

$$\frac{\varphi'_\sigma(u)}{\varphi_\sigma(u)} = -\frac{f''_\sigma\, f_\sigma - f'^2_\sigma}{f^2_\sigma\, f'_\sigma} = \left[f'_\sigma - \frac{f''_\sigma\, f_\sigma}{f'_\sigma}\right] f_\sigma^{-2}$$

where, for fixed u, $0 < u < 1$, $f_\sigma(x) \to 1$ and $f'_\sigma \to 0$, if $\sigma \to 0$. So

$$\frac{\varphi'_\sigma(u)}{\varphi_\sigma(u)} \sim -\frac{f''_\sigma}{f'_\sigma} = \frac{\frac{1}{\sigma^3}\left[x\, e^{-\frac{1}{2}x^2/\sigma^2} - (x-1)\, e^{-\frac{1}{2}(x-1)^2/\sigma^2}\right]}{\frac{1}{\sigma}\left[e^{-\frac{1}{2}x^2/\sigma^2} - e^{-\frac{1}{2}(x-1)^2/\sigma^2}\right]}.$$

This in conjunction with $u \sim x$ shows that

$$\lim_{\sigma\to 0} |\varphi'_\sigma(u)/\varphi_\sigma(u)| = \infty, \quad 0 < u < 1,\ u \neq \frac{1}{2}. \tag{8.2}$$

Now consider the statistic (2.6), where we drop the index ν. From (8.2) it follows that the critical region generated by the statistic

$$S^*_\sigma = \sum_{i-1}^{N} (c_i - \bar{c})\, \varphi_\sigma(R_i/(N+1)) \tag{8.3}$$

converges to one generated by the statistic

$$W = 0.I_1\, I_2 \cdots I_{[\frac{1}{2}N]}, \tag{8.4}$$

where the right side represents the development of W in the triadic system, $[\frac{1}{2}N] = \frac{1}{2}N$ for N even and $\frac{1}{2}(N-1)$ for N odd, and $I_1, \cdots, I_{[\frac{1}{2}N]}$ are random variables defined as follows:

$$\begin{aligned} I_k &= 2, &\quad &\text{if} \quad c_{D_{N-k}} > c_{D_k}, \\ &= 1, &\quad &\text{if} \quad c_{D_{N-k}} = c_{D_k}, \\ &= 0, &\quad &\text{if} \quad c_{D_{N-k}} < c_{D_k}, \end{aligned} \tag{8.5}$$

where $D_1, \ldots, D_N$ is a random permutation of $(1, \ldots, N)$ defined by

$$R_{D_k} = k, \tag{8.6}$$

i.e. D_k is the ordinal number of the observation whose rank (according to magnitude) is k. For example, if we have the following cs and xs

i	1	2	3	4	5	6
x_i	56	28	−1	96	103	8
c_i	1	1	0	0	1	1

(8.7)

then to the largest and smallest observation there corresponds $c_5 = 1$ and $c_3 = 0$, respectively, so that $c_{D_6} - c_{D_1} = c_5 - c_3 = 1$, and $I_1 = 2$, etc. So we get $W = 0.201$.

The test may be used for testing the hypothesis that the Xs are uniformly distributed over (a, b), where both a and b are unknown, against the alternative that X_i is uniformly distributed over $(a - \beta\, c_j,\, b - \beta\, c_i)$, where a, b, are unknown and the cs are known.

Obviously, the limiting distribution of W is not normal even under the hypothesis. Nevertheless the corresponding test might be asymptotically most powerful. The discussion of this problem would go, however, beyond the scope of the present paper.

9 A UNIVERSAL ASYMPTOTICALLY MOST POWERFUL TEST

Now we shall construct a test which is asymptotically most powerful for all densities $f(x)$ satisfying (1.3). However interesting this construction may be from the point of view of theory, it is of little use in practice because of slow convergence. First we generalize Theorem 1.1.

Lemma 9.1. *If the values of the functions φ_ν are random and under the hypothesis independent of the vectors $(X_{\nu 1}, \ldots, X_{\nu N_\nu})$, $1 \leq \nu < \infty$, and if condition* (1.18) *is replaced by*

$$\lim_{\nu\to\infty} \mathsf{P}_\nu \left\{ \int_0^1 [\varphi_\nu(u) - (1/\sigma)\,\varphi(u)]^2 \,\mathrm{d}u > \epsilon \,|\, \alpha = \alpha,\ \beta = 0,\ \sigma = \sigma \right\} = 0 \quad (9.1)$$
$$\text{for every } \epsilon > 0,\ -\infty < \alpha < \infty,\ \sigma > 0,$$

then Theorem 1.1 *still holds.*

Proof. Let $\mathsf{P}_\nu\{\cdot\}$ and the conditional mean value $\mathsf{E}_\nu[\cdot \,|\, \varphi_\nu]$ given φ_ν refer to some particular hypothesis $\alpha = \alpha$, $\beta = 0$, $\sigma = \sigma$. Let T_ν and T_ν^* be given by (5.8) and (5.12). Then from (9.1) and from Theorem 3.1 of Hájek (1961) it follows that for every $\epsilon > 0$

$$\lim_{\nu\to\infty} \mathsf{P}_\nu \left\{ \mathsf{E}_\nu \left[(S_\nu - T_\nu^*)^2 \,|\, \varphi_\nu \right] > \epsilon \right\} = 0 \quad (9.2)$$

and

$$\lim_{\nu\to\infty} \mathsf{P}_\nu \left\{ \mathsf{E}_\nu \left[(T_\nu^* - T_\nu)^2 \,|\, \varphi_\nu \right] > \epsilon \right\} = 0. \quad (9.3)$$

Now, (9.2) and (9.3) imply (5.17). All other parts of the proof of Theorem 1.1 need no modification. The proof is finished. □

Remark 9.1. The function $(1/\sigma)\,\varphi(u)$ corresponds to the density $f((x - \alpha)/\sigma)$ according to (1.9).

Now we shall split the sample $(X_{\nu 1}, \ldots, X_{\nu N_\nu})$ into two subsamples $(X_{\nu 1}, \ldots, X_{\nu N_\nu})$ and $(X_{\nu K_\nu + 1}, \ldots, X_{\nu N_\nu})$, where K_ν is chosen so that

$$\lim_{\nu\to\infty} K_\nu = \infty, \quad \lim_{\nu\to\infty} [K_\nu/(N_\nu - K_\nu)] = 0 \quad (9.4)$$

and

$$\lim_{\nu\to\infty} \left\{ \sum_{i=K_\nu+1}^{N_\nu} (c_{\nu i} - \bar{\bar{c}}_\nu)^2 \Big/ \sum_{i=1}^{N_\nu} (c_{\nu i} - \bar{\bar{c}}_\nu)^2 \right\} = 1, \quad (9.5)$$

where

$$\bar{\bar{c}}_\nu = (N_\nu - K_\nu)^{-1} \sum_{i=K_\nu+1}^{N_\nu} c_{\nu i}. \tag{9.6}$$

This is always possible in view of (1.4). Condition $K_\nu \to \infty$ will enable us to find a φ_ν satisfying (9.1) and depending on $(X_{\nu 1}, \dots, X_{\nu K_\nu})$ only. Condition (9.5) shows, on the other hand, that the S_ν-test based on the subsample $(X_{\nu K_\nu+1}, \dots, X_{\nu N_\nu})$ will have the same asymptotic power as the same test based on the whole sample $(X_{\nu 1}, \dots, X_{\nu N_\nu})$. So our problem will be solved by application of the S_ν-test, with φ_ν determined from $(X_{\nu 1}, \dots, X_{\nu K_\nu})$ to $(X_{\nu K_\nu+1}, \dots, X_{\nu N_\nu})$. The first subsample $(X_{\nu 1}, \dots, X_{\nu K_\nu})$ will be small enough not to decrease the efficiency and large enough to get an estimator of $\varphi(u)$ which is consistent in the sense of (9.1).

Lemma 9.2. *Introduce numbers* m_ν,

$$K_\nu^{\frac{4}{5}} < m_\nu \le K_\nu^{\frac{4}{5}} + 1 \tag{9.7}$$

and sequences $0 = h_{\nu 0} < h_{\nu 1} < \cdots < h_{\nu n_\nu} < h_{\nu, n_\nu+1} = K_\nu$ *such that*

$$\lim_{\nu\to\infty} \max_{0\le j\le n_\nu} \frac{|h_{\nu,j+1} - h_{\nu j}|}{K_\nu^{\frac{5}{6}}} = \lim_{\nu\to\infty} \min_{0\le j\le n_\nu} \frac{|h_{\nu,j+1} - h_{\nu j}|}{K_\nu^{\frac{5}{6}}} = 1, \tag{9.8}$$

and set, for $1 \le j \le n_\nu$,

$$k_{\nu j} = h_{\nu j} + m_\nu, \quad \ell_{\nu j} = h_{\nu j} - m_\nu. \tag{9.9}$$

Let $V_{\nu 1} < \cdots < V_{\nu K_\nu}$ *be the observations* $X_{\nu 1}, \dots, X_{\nu K_\nu}$ *rearranged according to ascending magnitude. Suppose that* (9.4) *is satisfied. Finally, for* $h_{\nu j}/K_\nu < i/(N_\nu - K_\nu + 1) \le h_{\nu j+1}/K_\nu$, $1 \le j < n_\nu$, *put*

$$\varphi_\nu[i/(N_\nu - K_\nu + 1)] = \frac{1}{2} K_\nu^{-\frac{1}{30}} \left[(V_{k_{\nu j}} - V_{\ell_{\nu j}})^{-1} - (V_{k_{\nu,j+1}} - V_{\ell_{\nu,j+1}})^{-1} \right], \tag{9.10}$$

and set $\varphi_\nu[i/(N_\nu - K_\nu + 1)] = 0$ *otherwise. Complete the definition of* $\varphi(u)$ *so as to be constant over intervals* $[i/(N_\nu - K_\nu), (i+1)/(N_\nu - K_\nu)]$, $0 \le i < N_\nu - K_\nu$. *Then* (9.1) *holds true for any density satisfying* (1.3).

Proof. Obviously, it suffices to prove (9.1) for $\alpha = 0$, $\sigma = 1$. Then the random variables $U_{\nu i} = F(X_{\nu i})$, $1 \le i \le K_\nu$, will be uniformly distributed over $(0, 1)$ and $Z_{\nu \ell} < \cdots < Z_{\nu k_\nu}$, where $Z_{\nu i} = F(V_{\nu i})$, will be the ordered sample from the uniform distribution. We know that

$$\mathsf{E} Z_{\nu i} = i/(K_\nu + 1) \tag{9.11}$$

and

$$\operatorname{var} Z_{\nu i} = i(K_\nu - i + 1)/(K_\nu + 1)^2 (K_\nu + 2) < 1/K_\nu. \tag{9.12}$$

Consequently, by the Tchebyshev inequality, the probability of the event A_ν

$$A_\nu = \bigcup_{j=1}^{n_\nu} \left\{ |Z_{\nu k_{\nu j}} - [k_{\nu j}/(K_\nu + 1)]| > (m_\nu/K_\nu)\, K_\nu^{-\frac{1}{5}} \right\} \tag{9.13}$$
$$\cup \bigcup_{j=1}^{n_\nu} \left\{ |Z_{\nu \ell_{\nu j}} - [\ell_{\nu j}/(K_\nu + 1)]| > (m_\nu/K_\nu)\, K_\nu^{-\frac{1}{5}} \right\}$$

is of order $(n_\nu/K_\nu)\,(K_\nu^{-\frac{1}{5}} m_\nu/K_\nu)^{-2} \sim K_\nu^{-\frac{1}{30}}$, so that $\mathsf{P}_\nu(A_\nu) \to 0$. This implies that for every $\epsilon > 0$

$$\lim_{\nu\to\infty} \mathsf{P}_\nu \left\{ \max_{i \le j \le n_\nu} K_\nu^{-\frac{1}{30}} \left| [Z_{\nu k_{\nu j}} - Z_{\nu \ell_{\nu j}}]^{-1} - \frac{1}{2}(K_\nu/m_\nu) \right| > \epsilon \right\} = 0. \tag{9.14}$$

On the other hand, from $Z_{\nu i} = F(V_{\nu i})$ it follows that

$$Z_{\nu k_{\nu j}} - Z_{\nu \ell_{\nu j}} = \left(V_{\nu k_{\nu j}} - V_{\nu \ell_{\nu j}} \right) f(B_{\nu j}), \tag{9.15}$$

where

$$V_{\nu \ell_{\nu j}} \le B_{\nu j} \le V_{\nu k_{\nu j}}, \quad 1 \le j \le n_\nu. \tag{9.16}$$

Hence (9.10) may be rewritten as follows:

$$\varphi_\nu \left(\frac{1}{N_\nu - K_\nu + 1} \right) = \frac{1}{2} K_\nu^{-\frac{1}{30}} \left[\frac{f(B_{\nu j})}{Z_{\nu k_{\nu j}} - Z_{\nu \ell_{\nu j}}} - \frac{f(B_{\nu j+1})}{Z_{\nu k_{\nu j+1}} - Z_{\nu \ell_{\nu j+1}}} \right]. \tag{9.17}$$

Now from boundedness of $f(x)$ (implied by (1.3)) and from (9.14) it follows that for every $\epsilon > 0$

$$\lim_{\nu\to\infty} \mathsf{P}_\nu \left\{ \max_{h_{\nu 1}/K_\nu < i/(N_\nu - K_\nu + 1) \le h_{\nu n_\nu}/K_\nu} \left| \varphi[i/(N_\nu - K_\nu + 1)] \right. \right. \tag{9.18}$$
$$\left. \left. - K_\nu^{-\frac{1}{6}} [f(B_{\nu j}) - f(B_{\nu, j+1})] \right| > \epsilon \right\} = 0.$$

Moreover, from (9.7), (9.8), (9.9), (9.16) and $\mathsf{P}_\nu(A_\nu) \to 0$, A_ν given by (9.13), it follows that for every $\epsilon > 0$

$$\lim_{\nu\to\infty} \mathsf{P}_\nu \left\{ \max_{\ell \le j \le n_\nu} \left| \left[K_\nu^{\frac{1}{6}} / (F(B_{\nu,j+1}) - F(B_{\nu j})) \right] - 1 \right| > \epsilon \right\} = 0. \tag{9.19}$$

The relations (9.18) and (9.19) imply that it suffices to prove (9.1) for the step function defined by

$$\begin{aligned} \overline{\varphi}(i/(N_\nu - k_\nu + 1)) = {} & \{[f(B_{\nu j}) - f(B_{\nu,j+1})]/[F(B_{\nu,j+1}) - F(B_{\nu j})]\} \\ & \text{if } (h_{\nu j}/K_\nu) < (i/(N_\nu - K_\nu + 1)) \le (h_{\nu,j+1}/K_\nu), \\ & 1 \le j < n_\nu, \\ = {} & 0 \quad \text{otherwise} \end{aligned} \tag{9.20}$$

and by the requirement to be constant over intervals $(i/(N_\nu - K_\nu), (i+1)/(N_\nu - K_\nu))$, $0 \le i < N_\nu - K_\nu$. Put $C_{\nu j} = F(B_{\nu j})$, $1 \le j \le n_\nu$, and observe that

$$\frac{f(B_{\nu j}) - f(B_{\nu,j+1})}{F(B_{\nu,j+1}) - F(B_{\nu j})} = \frac{\int_{C_{\nu j}}^{C_{\nu,j+1}} \varphi(u)\,du}{C_{\nu,j+1} - C_{\nu j}}, \tag{9.21}$$

and, on account of (9.16),

$$Z_{\nu \ell_{\nu j}} \le C_{\nu j} \le Z_{\nu k_{\nu j}}, \quad 1 \le j \le n_\nu. \tag{9.22}$$

In view of $\mathsf{P}_\nu(A_\nu) \to 0$, A_ν given by (9.13), we have

$$\lim_{\nu\to\infty} \mathsf{P}_\nu \left\{ \max_{1\le j\le n_\nu} |C_{\nu j} - (h_{\nu j}/K_\nu)| > 4(m_\nu/K_\nu) \right\} = 0, \tag{9.23}$$

and, on account of $n_\nu\, m_\nu/K_\nu \to 0$, for every $\epsilon > 0$,

$$\lim_{\nu\to\infty} \mathsf{P}_\nu \left\{ \sum_{j=1}^{n_\nu} |C_{\nu j} - (h_{\nu j}/K_\nu)| > \epsilon \right\} = 0. \tag{9.24}$$

Introduce another step function $\overline{\overline{\varphi}}_\nu$:

$$\begin{aligned} \overline{\overline{\varphi}}_\nu(u) &= \frac{f(B_{\nu j}) - f(B_{\nu,j+1})}{F(B_{\nu,j+1}) - F(B_{\nu j})} \quad \text{if} \quad C_{\nu j} < u \le C_{\nu,j+1},\ 1 \le j < n_\nu, \\ &= 0 \qquad \text{otherwise.} \end{aligned} \tag{9.25}$$

Now, because $\varphi(u)$ is square-integrable, to any $\epsilon > 0$ there exists a $\delta > 0$ such that

$$\max_{0\le j\le n_\nu} |C_{\nu j} - C_{\nu,j+1}| < \delta, \quad C_{\nu 0} = 0,\ C_{\nu,n_\nu+1} = 1 \tag{9.26}$$

entails

$$\int_0^1 [\overline{\overline{\varphi}}_\nu(u) - \varphi(u)]^2\,du < \epsilon. \tag{9.27}$$

(We remind the reader of relation (9.21).) Now (9.26) is, obviously, fulfilled for every $\delta > 0$ with limiting probability 1, and, consequently, for every $\epsilon > 0$,

$$\lim_{\nu\to\infty} \mathsf{P}_\nu \left\{ \int_0^1 [\overline{\overline{\varphi}}_\nu(u) - \varphi(u)]^2\,du > \epsilon \right\} = 0. \tag{9.28}$$

On the other hand, from the very definitions of $\overline{\varphi}_\nu$ and $\overline{\overline{\varphi}}_\nu$, it follows that

$$\int_0^1 [\overline{\varphi}_\nu(u) - \overline{\overline{\varphi}}_\nu(u)]^2\,du < 2\int_{\Lambda_\nu} \overline{\overline{\varphi}}{}_\nu^2(u)\,du, \tag{9.29}$$

where the Lebesgue measure of Λ_ν tends to 0 in P_ν-probability in view of $K_\nu/(N_\nu - K_\nu) \to 0$ and (9.24). This, in connection with uniform integrability of functions $\overline{\overline{\varphi}}_\nu^2$ corresponding to all possible divisions $[C_{\nu 1}, \ldots, C_{\nu n_\nu}]$, and on account of (9.28), implies, for every $\epsilon > 0$,

$$\lim_{\nu \to \infty} \mathsf{P}_\nu \left\{ \int_0^1 [\overline{\varphi}_\nu(u) - \varphi(u)]^2 \, \mathrm{d}u > \epsilon \right\} = 0. \tag{9.30}$$

The proof is concluded. □

Theorem 9.1. *Let assumptions* (1.3), (1.4) *and* (1.5) *be satisfied. Let* K_ν *satisfy the conditions* (9.4) *and* (9.5). *Let* $V_{\nu 1} < \cdots < V_{\nu K_\nu}$ *be the ordered subsample* $(X_{\nu 1}, \ldots, X_{\nu K_\nu})$ *and let* $R_{\nu i}$ *denote the rank of* $X_{\nu i}$ *in the subsample* $(X_{\nu, K_\nu+1}, \ldots, X_{\nu N_\nu})$, $K_\nu < i \leq N$, $1 \leq \nu < \infty$. *Let* φ_ν *be defined by* (9.10) *and* $\overline{\overline{c}}_\nu$ *by* (9.6). *Then the critical region*

$$\sum_{i=K_\nu+1}^{N_\nu} (c_{\nu i} - \overline{\overline{c}}_\nu) \varphi_\nu(R_{\nu i}/(N_\nu - K_\nu + 1)) \tag{9.31}$$

$$> K_\epsilon \left[\sum_{i=K_\nu+1}^{N_\nu} (c_{\nu i} - \overline{\overline{c}}_\nu)^2 \int_0^1 \varphi_\nu^2(u) \, \mathrm{d}u \right]^{\frac{1}{2}}$$

provides an asymptotically uniformly most powerful test of $\beta = 0$ *against* $\beta > 0$ *with limiting size* ϵ *and asymptotic power* (1.19) *for all densities in question.*

Proof. The proof follows immediately from Lemmas 9.1 and 9.2. □

Acknowledgement. This work was carried out with the partial support of the National Science Foundation, Grant G–14648.

REFERENCES

Capon J. (1961). Asymptotic efficiency of certain locally most powerful rank tests. *Ann. Math. Statist. 32*, 88–100.

Chernoff H. and Savage I. R. (1958). Asymptotic normality and efficiency of certain nonparametric tests statistics. *Ann. Math. Statist. 29*, 972–994.

Dwass M. (1956). The large sample power of rank order tests in the two sample case. *Ann. Math. Statist. 27*, 352–374.

Fraser D. A. S. (1956). *Nonparametric Methods in Statistics.* John Wiley and Sons, New York.

Gnedenko B. V. and Kolmogorov A. N. (1954). *Limit Distributions for Sums of Independent Random Variables.* Addison–Wesley, Cambridge.

Hájek J. (1961). Some extensions of the Wald–Wolfowitz–Noether theorem. *Ann. Math. Statist. 32*, 506–523.

Ibragimov, J. A. (1956). On the composition of unimodal distributions. *Teor. Veroyatnost. i Primenen. 1*, 283–288.

Le Cam L. (1960). Locally asymptotically normal families of distributions. *Univ. of Calif. Pubs. in Stat. 3*, 37–98.

Le Cam L. (196–). Asymptotic normality. Independent identically distributed case. Unpublished manuscript.

Mikulski P. W. (1961). Some Problems in the Asymptotic Theory of Testing Statistical Hypotheses (Efficiency of Non–parametric Procedures). Ph.D. Thesis, University of California, Graduate Division, Northern Section.

Neyman J. (1958). Optimal asymptotic tests of composite statistical hypotheses. In *The H. Cramér Jubilee Volume*, Almquist and Wiksell, Uppsala.

Savage I. R. (1956). Contributions to the theory of rank order statistics – the two sample case. *Ann. Math. Statist. 27*, 590–615.

Stein C. (1956). Efficient nonparametric testing and estimation. In *Third Berkeley Symp. Math. Statist. Prob.*, pp. 187–615.

van der Waerden B. L. (1957). *Mathematische Statistik*. Springer–Verlag, Berlin – Göttingen – Heidelberg.

Wilcoxon F. (1945). Individual comparisons by ranking methods. *Biometrics 1*, 80–83.

CHAPTER 23

Cost Minimization in Multiparameter Estimation

The methodological part of the paper published in Apl. Math. 7 (1962), 405–425 (in Czech).
Translated by J. Dupačová.
Reproduced by permission of Hájek family: Al. Hájková, H. Kazárová and Al. Slámová.

Assuming that the total cost C is a linear function of some parameters $r_1, \ldots, r_H$ and that the variances of the considered estimates are linear in the reciprocal values of these parameters, the problem is to choose parameter values which minimize C under prescribed upper bounds on the variances. The main field of applications comes from sample surveys.

1 INTRODUCTION AND SUMMARY

Consider a random experiment providing estimates $t_1, \ldots, t_J$ of parameters $\theta_1, \ldots, \theta_J$. Both the experiment and the estimation procedure are fixed except for H parameters $r_1, \ldots, r_H$ which are nonnegative and such that the total cost $C = C(r_1, \ldots, r_H)$ of the experiment and of evaluating the estimates is increasing in each parameter $r_h, 1 \leq h \leq H$. We assume further that the estimates t_j, $1 \leq j \leq J$, are unbiased and their variances $D_j = D_j(r_1, \ldots, r_H)$ are decreasing in parameters $r_h, 1 \leq h \leq H$. The problem is to choose parameters $r_1, \ldots, r_H$ which guarantee that the conditions

$$D_j = D_j(r_1, \ldots, r_H) \leq b_j \quad (1 \leq j \leq J) \tag{1.1}$$

on the precision of all estimates are satisfied for prescribed constants b_j and that the cost C is minimal.

The solution of the above problem will be given for the case when the cost is linear in parameters $r_1, \ldots, r_H$ and the variances D_j are linear in the

Collected Works of Jaroslav Hájek – With Commentary
Edited by M. Hušková, R. Beran and V. Dupač
Published in 1998 by John Wiley & Sons Ltd.

reciprocal values of these parameters:

$$C = \sum_{h=1}^{H} c_h r_h + c_0 \tag{1.2}$$

$$D_j = \sum_{h=1}^{H} \frac{d_{jh}}{r_h} - d_{j0} \quad (1 \leq j \leq J). \tag{1.3}$$

All constants which enter the above formulas are nonnegative. In the theory of probability sampling (sample surveys) there is a host of examples for which the linearity assumptions are fulfilled; see for instance the book Hájek (1960), pp. 117, 158, 163, 169, 262. Parameters $r_1, \ldots, r_H$ denote the sample sizes in individual strata or characterize the size and the structure of a two-stage or double sampling, etc. Sample surveys, however, are not the only opportunity for application of the subsequent results.

The problem was formulated in the context of stratified sampling by T. Dalenius in his book Dalenius (1957) on page 200, including a geometric solution for $H = 2$. For $H > 2$, exploitation of nonlinear programming techniques is proposed (Dalenius (1957), p. 206); however, neither details nor a hint on how to realize the ideas in practice are given. The problem can also be formulated as a two-person zero-sum game with $r_1, \ldots, r_H$ strategies of the first player and $t_1, \ldots, t_J$ strategies of the second player. This idea is delineated in § 8 of the original Hájek paper, which also contains a numerical example (§ 7) with $J = 2$ and $H = 3$ related to Kozák (1961), and treats the frequent case of independent estimates $t_1, \ldots, t_H$ and $t_0 = t_1 + \ldots + t_H$ (§ 6) and suggests a method of solution for a cost function C depending on the r_h in a nonlinear way (§ 9).

2 THE SIZE AND THE STRUCTURE OF THE EXPERIMENT

Based on already determined parameters $r_1, \ldots, r_H$, our evaluation of the experiment is related to $J + 1$ numbers $C, D_1, \ldots, D_J$ each of which depends on these parameters. To be able to compare different choices of parameters $r_1, \ldots, r_H$ in a simple way an evaluation only by one number would be desirable. Such a number should depend on $C, D_1, \ldots, D_J$, and various aggregating functions have been proposed. None of them, however, have been (nor could apparently be) fully satisfactory. Let us see how this problem can be overcome.

Assume for simplicity that there is only one variance, D, so that our attitude is quantified by the couple of numbers C, D. Observe first that the experiment can be equivalently determined by numbers $R, \rho_1, \ldots, \rho_H$, where

$R = r_1 + \ldots + r_H$ and $\rho_h = r_h/R, 1 \le h \le H$. Further notice that

$$(C - c_0)(D + d_0) = \sum_{h=1}^{H} c_h r_h \sum_{h=1}^{H} d_h r_h^{-1} = \sum_{h=1}^{H} c_h \rho_h \sum_{h=1}^{H} d_h \rho_h^{-1} \tag{2.1}$$

so that $(C - c_0)(D + d_0)$ depends only on $\rho_1, \ldots, \rho_H$. Let us call the number R *the size of the experiment* and numbers $\rho_1, \ldots, \rho_H$ *the structure of the experiment*. The number $(C - c_0)(D + d_0)$, can be taken as our evaluation of the structure of the experiment: the smaller $(C - c_0)(D + d_0)$ is, the better the structure of the experiment. If the optimal structure is determined, it remains to fix the size R of the experiment so that the prescribed value of the variance D is reached or the prescribed value of the cost C is obtained. Evidently, this approach solves at the same time both the problem to minimize the cost C under a prescribed variance D and that of minimization of the variance D under a prescribed value of C. The main idea of this paragraph can be summarized as follows: using one number one cannot value the experiment as a whole but only one aspect of the experiment, for instance, its structure.

Let us return now to the general case of J variances $D_1, \ldots, D_J$. Assume that our evaluation of the experiment depends only on a function $\varphi = \varphi(D_1, \ldots, D_J)$. Consider two special choices of function φ:

$$\varphi = \sum_{j=1}^{J} v_j (D_j + d_{j0}) \tag{2.2}$$

and

$$\varphi = \max_{1 \le j \le J} \frac{D_j + d_{j0}}{b_j + d_{j0}}, \tag{2.3}$$

where v_j and b_j are certain nonnegative constants. Accordingly, our evaluation of the structure of the experiment is determined by the product $(C - c_0)\varphi$, where φ is given by equation (2.2) or (2.3).

If φ is given by equation (2.2) and the variances D_j by equations (1.3), then

$$\varphi = \sum_{h=1}^{H} d_h r_h^{-1} - d_0 \tag{2.4}$$

where

$$d_h = \sum_{j=1}^{J} v_j d_{jh} \quad (0 \le h \le H). \tag{2.5}$$

It means that in this case the optimal structure is the same as for the case of only one variance which is equal to the right-hand side of equation (2.4).

An essentially different situation arises when φ is defined by equation (2.3); this will be addressed in this chapter. At this point we shall only clarify the relationship between the problem to minimize the product $(C-c_0)\max_j(D_j+d_{j0})(b_j+d_{j0})^{-1}$ and the problem of minimization of C subject to constraints (1.1). This relationship is given by the identity

$$\min_{r_1,\dots,r_H}\left[(C-c_0)\max_j\frac{D_j+d_{j0}}{b_j+d_{j0}}\right]=\min_{r_1,\dots,r_H;D_j\le b_j}(C-c_0), \tag{2.6}$$

where $\min_{r_1,\dots,r_H}$ denotes the minimum over all nonnegative values of the parameters $r_1,\dots,r_H$ and the minimum on the right-hand side is carried over those nonnegative values of parameters $r_1,\dots,r_H$ for which $D_j\le b_j, 1\le j\le J$.

Proof of identity (2.6). Let the minimum on the right-hand side of (2.6) be attained for $r_h=r_h^0, 1\le h\le H$. Put $D_j^0=D_j(r_1^0,\dots,r_H^0)$; then evidently

$$\max_j\frac{D_j+d_{j0}}{b_j+d_{j0}}=1. \tag{2.7}$$

Indeed, otherwise it would be possible to decrease parameters r_h^0 and thus to decrease C without violating constraints $D_j\le b_j$, which contradicts the assumed optimality of $r_h=r_h^0$. Hence for $C^0:=C(r_1^0,\dots,r_H^0)$

$$\begin{aligned}\min_{r_1,\dots,r_H;D_j\le b_j}(C-c_0)&=(C^0-c_0)\max_j\frac{D_j^0+d_{j0}}{b_j+d_{j0}}\\ &\ge\min_{r_1,\dots,r_H}\left[(C-c_0)\max_j\frac{D_j+d_{j0}}{b_j+d_{j0}}\right].\end{aligned}$$

As the opposite inequality is a trivial one, the identity (2.6) follows. □

We have proved at the same time that the same values r_h^0 which minimize C subject to conditions $D_j\le b_j$ do also minimize the product

$$(C-c_0)\max_j\frac{D_j+d_{j0}}{b_j+d_{j0}}=\max_j\left[\sum_{h=1}^H c_hr_h\sum_{h=1}^H\frac{d_{jh}}{b_j+d_{j0}}r_h^{-1}\right]. \tag{2.8}$$

The product (2.8) is also minimized by arbitrary values of the parameters of the form $r_h=\lambda r_h^0$, $\lambda>0$, as the value of (2.8) does not depend on the positive multiplier λ. Let us summarize: by solving the problem of minimization of C subject to conditions $D_j\le b_j$ the problem of optimal structure is also solved provided that the structure is valued (indirectly proportional) by the product (2.8). It is obvious, too, that the same solution, as concerns the structure, also solves the third possible problem – minimization of (2.3) subject to a fixed cost.

To conclude, let us note that one could define the *size of the experiment* by means of an arbitrary function $R=R(r_1,\dots,r_H)$ for which

$$R(\lambda r_1,\dots,\lambda r_H)=\lambda R(r_1,\dots,r_H). \tag{2.9}$$

This condition is satisfied for instance for $R = C - c_0$, $R = r_h$ or $R = (D_j + d_{j0})^{-1}$, where j, h are arbitrary and C and D_j are given by equations (1.2) and (1.3). Similarly, the structure need not be defined by ratios $\rho_h = r_h/R$, $1 \leq h \leq H$, but can be simply defined by the property that the vectors $(r_1, \ldots, r_H)$ and $(r'_1, \ldots, r'_H)$ have the same structure if and only if $r'_h = \lambda r_h$, $1 \leq h \leq H$, for a constant multiplier $\lambda > 0$.

3 THE GENERAL THEOREM

The following theorem describes the general properties of the solution of our problem.

Theorem 1. *The values $r_1^0, \ldots, r_H^0$ that minimize the cost C subject to conditions $D_j \leq b_j$, $1 \leq j \leq J$, equal*

$$r_h^0 = \frac{1}{c_h}\left(\sum_{j=1}^{J} a_{jh} p_j^0\right)^{1/2} \sum_{k=1}^{H}\left(\sum_{j=1}^{J} a_{jk} p_j^0\right)^{1/2} \qquad (1 \leq h \leq H), \tag{3.1}$$

where $p_1^0, \ldots, p_J^0$ are nonnegative numbers whose sum equals 1 and such that

$$\sum_{h=1}^{H}\left(\sum_{j=1}^{J} a_{jh} p_j^0\right)^{1/2} = \max_{p_1,\ldots,p_J, p_j \geq 0 \sum p_j = 1} \sum_{h=1}^{H}\left(\sum_{j=1}^{J} a_{jh} p_j\right)^{1/2}, \tag{3.2}$$

where

$$a_{jh} = \frac{c_h d_{jh}}{b_j + d_{j0}} \quad (1 \leq j \leq J, 1 \leq h \leq H). \tag{3.3}$$

The characteristic property of numbers p_j^0 is

$$\begin{aligned} \sum_{h=1}^{H} a_{jh}\left(\sum_{i=1}^{J} a_{ih} p_i^0\right)^{-1/2} &= \sum_{h=1}^{H}\left(\sum_{i=1}^{J} a_{ih} p_i^0\right)^{1/2} \quad \text{if } p_j^0 > 0 \\ &\leq \sum_{h=1}^{H}\left(\sum_{i=1}^{J} a_{ih} p_i^0\right)^{1/2} \quad \text{if } p_j^0 = 0. \end{aligned} \tag{3.4}$$

If $p_j^0 > 0$ then $D_j(r_1^0, \ldots, r_h^0) = b_j$.

Proof. Conditions $D_j \leq b_j$, $1 \leq j \leq J$, imply that

$$\max_{1 \leq j \leq J} \frac{D_j + d_{j0}}{b_j + d_{j0}} \leq 1, \tag{3.5}$$

so that also

$$C-c_0 \geq \max_{1\leq j\leq J}\left[(C-c_0)\frac{D_j+d_{j0}}{b_j+d_{j0}}\right] = \max_{1\leq j\leq J}\left[\sum_{h=1}^{H} c_h r_h \sum_{h=1}^{H} d_{jh}(b_j+d_{j0})^{-1} r_h^{-1}\right]. \tag{3.6}$$

As no convex combination (i. e. with nonnegative coefficients whose sum equals 1) of a set of numbers can be greater than their maximum we can write

$$\max_{1\leq j\leq J}\left[\sum_{h=1}^{H} c_h r_h \sum_{h=1}^{H} d_{jh}(b_j+d_{j0})^{-1} r_h^{-1}\right]$$
$$= \max_{p_j\geq 0,\sum p_j=1}\left\{\sum_{h=1}^{H} c_h r_h \sum_{h=1}^{H}\left[r_h^{-1}\sum_{j=1}^{J} d_{jh}(b_j+d_{j0})^{-1} p_j\right]\right\}. \tag{3.7}$$

Using the well-known Cauchy inequality we get

$$\sum_{h=1}^{H} c_h r_h \sum_{h=1}^{H}\left[r_h^{-1}\sum_{j=1}^{J} d_{jh}(b_j+d_{j0})^{-1} p_j\right]$$
$$\geq \left\{\sum_{h=1}^{H}\left[c_h \sum_{j=1}^{J} d_{jh}(b_j+d_{j0})^{-1} p_j\right]^{1/2}\right\}^2$$
$$= \left[\sum_{h=1}^{H}\left(\sum_{j=1}^{J} a_{jh} p_j\right)^{1/2}\right]^2, \tag{3.8}$$

where the constants are defined by (3.3). Relations (3.6) – (3.8) imply that

$$C-c_0 \geq \max_{p_j\geq 0,\sum p_j=1}\left[\sum_{h=1}^{H}\left(\sum_{j=1}^{J} a_{jh} p_j\right)^{1/2}\right]^2 = \left[\sum_{h=1}^{H}\left(\sum_{j=1}^{J} a_{jh} p_j^0\right)^{1/2}\right]^2. \tag{3.9}$$

We have thus obtained a lower bound for C. Before proving that this bound is attained for $r_1^0,\ldots,r_H^0$ given by (3.1) and with the conditions $D_j \leq b_j$ satisfied let us derive (3.4) from (3.2).

The expression $\sum_{h=1}^{H}\left(\sum_{j=1}^{J} a_{jh} p_j\right)^{1/2}$ is evidently a concave function of variables $(p_1,\ldots,p_J)$, $p_j \geq 0, \sum_j p_j = 1$; hence, each local maximum is the global one. Points of global maxima are characterized by the fact that increments of function values in all directions which are feasible with respect to the set defined by inequalities $p_j \geq 0$ and equation $\sum_j p_j = 1$ are nonpositive.

In other words, $(p_1^0, \ldots, p_J^0)$ is a point of global maximum if and only if the total differential of the function

$$F(p_1, \ldots, p_J) = \sum_{h=1}^{H} \left(\sum_{j=1}^{J} a_{jh} p_j \right)^{1/2} \tag{3.10}$$

is nonpositive at the point $(p_1^0, \ldots, p_J^0)$ in all feasible directions. As

$$\frac{\partial F}{\partial p_j} = \frac{1}{2} \sum_{h=1}^{H} a_{jh} \left(\sum_{i=1}^{J} a_{ih} p_i \right)^{-1/2} \qquad (1 \leq j \leq J)$$

the total differential of F at the point $(p_1^0, \ldots, p_J^0)$ equals

$$\Delta F = \frac{1}{2} \sum_{j=1}^{J} \left[\Delta p_j^0 \sum_{h=1}^{H} a_{jh} \left(\sum_{i=1}^{J} a_{ih} p_i^0 \right)^{-1/2} \right] \tag{3.11}$$

and the increments are limited by conditions

$$\sum_{j=1}^{J} \Delta p_j^0 = 0 \quad \text{and} \quad \Delta p_j^0 \geq 0 \tag{3.12}$$

if

$$p_j^0 = 0, \; (1 \leq j \leq J)$$

which follow from the constraints $\sum p_j^0 = \sum (p_j^0 + \Delta p_j^0)$ and $p_j^0 + \Delta p_j^0 \geq 0$. It is easy to realize that the condition of nonpositive total differential

$$\Delta F \leq 0 \tag{3.13}$$

for all Δp_j^0 which fulfil (3.12) is identical to the condition that

$$\begin{aligned} \sum_{h=1}^{H} a_{jh} \left(\sum_{i=1}^{J} a_{ih} p_i^0 \right)^{-1/2} &= \alpha \quad \text{if} \quad p_j^0 > 0 \\ &\leq \alpha \quad \text{if} \quad p_j^0 = 0 \end{aligned} \tag{3.14}$$

for a constant α. Let us show that (3.14) is sufficient, leaving the proof of its necessity to the reader. If (3.14) holds true then according to (3.11)

$$2\Delta F = \alpha \sum_{j=1}^{J} \Delta p_j^0 - \sum_{p_j^0 = 0} \left[\alpha - \sum_{h=1}^{H} a_{jh} \left(\sum_{i=1}^{J} a_{ih} p_i^0 \right)^{-1/2} \right] \Delta p_j^0, \tag{3.15}$$

where the summation in the second term is carried only over those j for which $p_j^0 = 0$. According to (3.12), the first sum on the right-hand side of (3.15) is 0 and the second one is nonpositive as for $p_j^0 = 0$, $\Delta p_j^0 \geq 0$ according to (3.12) and $\alpha - \sum a_{jh} \left(\sum a_{ih} p_i^0\right)^{-1/2} \geq 0$ according to (3.14). Multiplication of equations and inequalities (3.4) by p_j^0 and summation over j implies that α in (3.14) equals the right-hand side of (3.4), and hence (3.4) is sufficient.

Now it is easy to finish the proof. Substituting $r_1^0, \ldots, r_H^0$ given by (3.1) into formula (1.2) for C one gets

$$C - c_0 = \left[\sum_{h=1}^{H} \left(\sum_{j=1}^{J} a_{jh} p_j^0\right)^{1/2}\right]^2 \tag{3.16}$$

so that the lower bound (3.9) is attained. Moreover, substituting these $r_1^0, \ldots, r_H^0$ into formula (1.3) for D_j one gets

$$\frac{D_j + d_{j0}}{b_j + d_{j0}} = \sum_{h=1}^{H} a_{jh} c_h^{-1} (r_h^0)^{-1}$$

$$= \sum_{h=1}^{H} a_{jh} \left(\sum_{i=1}^{J} a_{ih} p_i^0\right)^{-1/2} \left[\sum_{h=1}^{H} (\sum_{j=1}^{J} a_{jh} p_j^0)^{1/2}\right]^{-1}.$$

Considering (3.4) this implies that

$$\frac{D_j + d_{j0}}{b_j + d_{j0}} \leq 1 \quad (1 \leq j \leq J) \tag{3.17}$$

for all j for which $p_j^0 > 0$. Inequalities (3.17) are evidently equivalent to inequalities $D_j \leq b_j$, which completes the proof. □

4 DELETION OF NONESSENTIAL ESTIMATES

To apply Theorem 1 it is necessary to compute p_j^0; we shall call them weights (p_j^0 is the weight of the estimate t_j). Firstly, it is necessary to check in each single case which weights equal zero. If $p_j^0 = 0$ then the solution is the same as if the estimate t_j were not taken into account. This is evident from (3.1). To check cases $p_j^0 = 0$ means to find out which estimates t_j can be exempt from consideration. For such estimates, the inequalities $D_j \leq b_j$ will be a consequence of the satisfied inequalities for variances of the remaining estimates. The estimates t_j for which $p_j^0 = 0$ will be called *nonessential* estimates, the remaining ones being called *essential* estimates. It may happen that just one estimate, say t_q, is essential, in which case

$$r_h^0 = c_h^{-1} a_{qh}^{1/2} \sum_{k=1}^{H} a_{qk}^{1/2} = \left(\frac{d_{qh}}{c_h}\right)^{1/2} (b_q + d_{q0})^{-1} \sum_{k=1}^{H} \sqrt{(c_k d_{qk})} \quad (1 \leq h \leq H). \tag{4.1}$$

Theorem 2. *If there are nonnegative numbers λ_i, $i \neq k$, $\sum_{i \neq k} \lambda_i = 1$ and such that for all $h = 1, \ldots, H$*

$$a_{kh} \leq \sum_{i \neq k} \lambda_i a_{ih} \quad (1 \leq h \leq H), \tag{4.2}$$

then the maximum in (3.9) can be attained at a point for which $p_k^0 = 0$.

Let A be a subset of $\{1, \ldots, J\}$ and B its complement, $B = \{1, \ldots, J\} - A$. If

$$\sum_{h=1}^{H} a_{ih}^{1/2} \geq \sum_{h=1}^{H} a_{jh} a_{ih}^{-1/2} \quad (i \in A, j \in B) \tag{4.3}$$

holds true for each $i \in A$ and each $j \in B$ then the maximum can be attained at a point for which $p_j^0 = 0$ for all $j \in B$.

Proof. Inequalities (4.2) imply immediately that for each h

$$\begin{aligned} \sum_{j=1}^{J} a_{jh} p_j &= a_{kh} p_k + \sum_{i \neq k} a_{ih} p_i \leq \sum_{i \neq k} \lambda_i a_{ih} p_k + \sum_{i \neq k} a_{ih} p_i \\ &= \sum_{i \neq k} a_{ih}(p_i + \lambda_i p_k) = \sum_{j=1}^{J} a_{jh} p'_j, \end{aligned}$$

where $p'_k = 0$. It means that the maximum of $\sum_{h=1}^{H} (\sum_{j=1}^{J} a_{jh} p_j)^{1/2}$ can be attained when $p_k^0 = 0$. Let us now prove the second part of the theorem. Put

$$\alpha = \sum_{i \in A} p_i$$

$$q_i = \frac{p_i}{\alpha} \quad \text{for } i \in A$$

$$\ell_j = \frac{p_j}{1 - \alpha} \quad \text{for } j \in B.$$

This notation allows to write

$$\sum_{j=1}^{J} a_{jh} p_j = \alpha \sum_{i \in A} a_{ih} q_i + (1 - \alpha) \sum_{j \in B} a_{jh} \ell_j,$$

where $\sum_{i \in A} q_i = \sum_{j \in B} \ell_j = 1$. Further,

$$\sum_{h=1}^{H} \left(\sum_{j=1}^{J} a_{jh} p_j^0 \right)^{1/2} = \sum_{h=1}^{H} \left[\alpha \sum_{i \in A} a_{ih} q_i + (1-\alpha) \sum_{j \in B} a_{jh} \ell_j \right]^{1/2} := G(\alpha). \tag{4.4}$$

Differentiation leads to the conclusion that, as a function of α, the right-hand side is concave for $0 \leq \alpha \leq 1$ (the second derivative is negative). We want to prove that

$$G(1) \geq G(\alpha) \; (0 \leq \alpha \leq 1)$$

i. e. that annulation of weights p_j, $j \in B$ does not lead to a decrease of (4.4). Thanks to the concavity of $G(\alpha)$ it is enough to prove that the derivative at $\alpha = 1$ is nonnegative. We have

$$\left. \frac{\partial G}{\partial \alpha} \right|_{\alpha=1} = \frac{1}{2} \sum_{h=1}^{H} \left(\sum_{i \in A} a_{ih} q_i \right)^{1/2} - \frac{1}{2} \sum_{h=1}^{H} \left(\sum_{j \in B} a_{jh} \ell_j \right) \left(\sum_{i \in A} a_{ih} q_i \right)^{-1/2}.$$

It remains to prove that for arbitrary q_i, ℓ_j (4.3) implies

$$\sum_{h=1}^{H} \left(\sum_{i \in A} a_{ih} q_i \right)^{1/2} \geq \sum_{h=1}^{H} \left(\sum_{j \in B} a_{jh} \ell_j \right) \left(\sum_{i \in A} a_{ih} q_i \right)^{-1/2}. \tag{4.5}$$

Recall that the function $f(x_1, \ldots, x_H) = \sum_{h=1}^{H} x_h^{1/2}$, $x_h \geq 0$, is concave. For $x_h = a_{ih}$ by Jensen's inequality (see Hardy et al. (1934))

$$\sum_{h=1}^{H} \left(\sum_{i \in A} a_{ih} q_i \right)^{1/2} \geq \sum_{i \in A} q_i \sum_{h=1}^{H} a_{ih}^{1/2}. \tag{4.6}$$

According to assumption (4.3) we also have

$$\sum_{i \in A} q_i \sum_{h=1}^{H} a_{ih}^{1/2} = \sum_{i \in A} q_i \sum_{h=1}^{H} \sum_{j \in B} \ell_j a_{ih}^{1/2} \geq \sum_{i \in A} q_i \sum_{h=1}^{H} a_{ih}^{-1/2} \left(\sum_{j \in B} a_{jh} \ell_j \right). \tag{4.7}$$

The function $g(x_1, \ldots, x_H) = \sum_{h=1}^{H} x_h^{-1/2} (\sum_{j \in B} \ell_j a_{jh})$, $x_h \geq 0$, is convex. If we put $x_h = a_{ih}$ in (4.7) and apply Jensen's inequality once more we get

$$\sum_{i \in A} q_i \sum_{h=1}^{H} a_{ih}^{-1/2} \left(\sum_{j \in B} a_{jh} \ell_j \right) \geq \sum_{h=1}^{H} \left(\sum_{i \in A} q_i a_{ih} \right)^{-1/2} \left(\sum_{j \in B} \ell_j a_{jh} \right). \tag{4.8}$$

Now, apparently, (4.5) follows from (4.6), (4.7) and (4.8), which completes the proof. □

To apply the Theorem we write first the matrix of numbers

$$\begin{pmatrix} a_{11}, a_{12}, \ldots, a_{1H} \\ a_{21}, a_{22}, \ldots, a_{2H} \\ \vdots \\ a_{J1}, a_{J2}, \ldots, a_{JH} \end{pmatrix} \tag{4.9}$$

and check all couples of rows for the dominance property in the sense that $a_{ih} \geq a_{kh}$ for all h. If a row is dominated by another row, it is deleted from the matrix and in the choice of experiment the corresponding estimate is not taken into account. By the method of trials and errors we check then whether some row is dominated by a convex combination of other rows in the sense of condition (4.2). Finally, we use the matrix consisting of the remaining rows to construct the square matrix

$$\begin{pmatrix} M_{11}, \ldots, M_{1I} \\ \vdots \\ M_{I1}, \ldots, M_{II} \end{pmatrix}, \tag{4.10}$$

where

$$M_{ij} = \sum_{h=1}^{H} a_{jh} a_{ih}^{-1/2} \quad (1 \leq i, j \leq I) \tag{4.11}$$

and $I \leq J$ denotes the number of rows in the reduced matrix, these rows being renumbered by indices from 1 to I. Using matrix (4.10) we check if it is possible to construct sets of indices A and B so that condition (4.3) is fulfilled; recall that the union of sets A and B is the whole set $\{1, \ldots, I\}$. If such decomposition is found, all rows indexed by indices belonging to B can be deleted and only rows indexed by indices from A are kept. This is the last step of deleting nonessential estimates according to Theorem 2.

Remark. Classification of estimates into essential and nonessential ones is unique only when the maximum of $\sum(\sum a_{jh} p_j)^{1/2}$ is attained at a unique point, which can be considered as the usual case.

REFERENCES

Blackwell, D. and Girshick, M. A. (1954). *Theory of Games and Statistical Decisions.* Wiley, New York.

Dalenius, T. (1957). *Sampling in Sweden, Contributions to the Methods and Theories of Sample Survey Practice.* Almquist and Wiksell, Stockholm.

Hájek, J. (1960). *Theory of Probability Sampling with Applications to Sample Surveys.* (In Czech) Publ. House of the Czechoslovak Academy of Sciences, Prague.

Hardy, G. H., Littelwood, J. L. and Pólya, G. (1934). *Inequalities.* Cambridge University Press, Cambridge, 1934.

Kozák, J. (1961). On the Use of Multipurpose Sampling in the Frame of Socialist State Statistics (with Emphasis on Agricultural Statistics) (in Czech). PhD thesis, ÚÚSKS, Prague.

Nordbotten, S. (1956). Allocation in stratified sampling by means of linear programming. *Skandinavisk Aktuarietidskrift 39*, 1–6.

CHAPTER 24

Asymptotic Theory of Rejective Sampling with Varying Probabilities from a Finite Population

Ann. Math. Statist. *35* (1964), 1491–1523.
Reproduced by permission of The Institute of Mathematical Statistics.

SUMMARY

In Hájek (1960) the author established necessary and sufficient conditions for asymptotic normality of estimates based on simple random sampling without replacement from a finite population, and thus solved a comparatively old problem initiated by W. G. Madow (1948). The solution was obtained by approximating simple random sampling by so called Poisson sampling, which may be decomposed into independent subexperiments, each associated with a single unit in the population. In the present chapter the same method is used for deriving asymptotic normality conditions for a special kind of sampling with varying probabilities called here rejective sampling. Rejective sampling may be realized by n independent draws of one unit with fixed probabilities, generally varying from unit to unit, given the condition that samples in which all units are not distinct are rejected. If the drawing probabilities are constant, rejective sampling coincides with simple random sampling without replacement, and so the present chapter is a generalization of Hájek (1960).

Basic facts about rejective sampling are given in Section 2. To obtain more refined results, Poisson sampling is introduced and analyzed (Section 3) and then related to rejective sampling (Section 4). The next three sections deal with probabilities of inclusion, variance formulas and asymptotic normality of

Collected Works of Jaroslav Hájek – With Commentary
Edited by M. Hušková, R. Beran and V. Dupač
Published in 1998 by John Wiley & Sons Ltd.

estimators for rejective sampling. In Section 8 asymptotic formulas are tested numerically and applications to sample surveys are indicated. The chapter is concluded by short-cuts in the practical performance of rejective sampling.

Readers who are interested in applications only may concentrate upon Sections 1, 8 and 9. Those interested in the theory of mean values and variances only may omit Lemma 4.3 and Section 7.

1 INTRODUCTION

Consider a population U consisting of N identifiable units of arbitrary nature, so that they may be represented by integers $1, 2, \ldots, N$, $U = \{1, 2, \ldots, N\}$. There are many different formal definitions of a sample from the population U, no one satisfactory from all points of view. Here we shall adhere to the simplest possible one: *a sample is a subset of* U. Thus there are 2^N possible samples from U. A typical sample of this kind will be denoted by s, $s \subset U$.

The above definition does not reflect some features of common sampling designs, such as ordering of units (due to carrying out sampling by successive draws of the unit), repetition of units (due to replacement or interpenetration), and hierarchy of units (due to stages). Consequently, within the theory of s-samples some common terms such as 'sampling with or without replacement' or 'two-stage sampling' lose their meaning. However, this is justified, at least in theory, by a theorem proving that in cases where the sample is defined so as to distinguish ordering, repetitions, stages, random strata, etc., the sample s together with values ascertained thereon represent a sufficient statistic (see Hájek (1959) or Roy and Chakraborti (1960), for example). In multistage sampling, s should be understood as a subset of the population of ultimate units.

Some sampling procedures define directly a probability distribution on the space of all s-samples. This is true about systematic sampling, rejective sampling and Poisson sampling, for example. In other cases the outcome of the sampling experiment, say ω, is described in more detail, and the sample s is merely an abstract function of it, $s = s(\omega)$. For example, if ω denotes a sequence of n units from U with possible repetitions (there are N^n such sequences), then $s = s(\omega)$ will denote the subset of U consisting of all distinct units appearing at least once in ω. Outcomes of this kind are assumed in successive sampling and in sampling with replacement, for example. In all cases, however, we can project the probability distribution $\mathsf{P}_1(\cdot)$ defined on the outcomes ω into the space of probability distributions defined on the samples s. In doing that we put

$$\mathsf{P}(s) = \mathsf{P}_1(\omega \in \Omega_s), \tag{1.1}$$

where Ω_s consists of all ωs which yield the given subset of distinct units s.

$\mathsf{P}(s)$, $s \in U$, given by (1.1), will be called a *projection* of $\mathsf{P}_1(\omega)$.

Thus any sampling experiment may be described by a probability distribution $\mathsf{P}(s)$ of the samples s, and by a set of conditional distributions $\mathsf{P}_1(\omega \mid \Omega_s)$, $s \subset U$. The conditional distributions $\mathsf{P}_1(\omega \mid \Omega_s)$, however, are irrelevant for estimation purposes, and may be neglected in theoretical considerations. For this reason, any probability distribution $\mathsf{P}(s)$ defined on the space of the samples s will be called a *probability sampling*, and probability sampling so defined will be considered as a mathematical model of what is usually called 'sampling plan', 'sampling design', 'sampling procedure', 'sampling experiment' or 'sampling method'.

The probability of selecting a sample s which contains the unit i, say π_i, equals

$$\pi_i = \sum_{s \ni i} \mathsf{P}(s), \quad (i \in U), \tag{1.2}$$

where $\sum_{s \ni i}$ denotes summation extended over all samples containing the unit i. Similarly,

$$\pi_{ij} = \sum_{s \ni i,j} \mathsf{P}(s), \quad (i,\, j \in U) \tag{1.3}$$

is the probability of including both units i and j in the sample. The probabilities π_i and π_{ij} will be called *probabilities of inclusion*.

For current practice it is not necessary to know the probabilities $\mathsf{P}(s)$ for every s. All we need is to know the probabilities of inclusion π_i and π_{ij}, at least approximately, and be certain that the estimates are approximately normal.

A particular kind of probability sampling is given by a formula for $\mathsf{P}(s)$, $s \subset U$, or for $\mathsf{P}_1(\omega)$ to be projected into $\mathsf{P}(s)$, containing some free parameter, which may be controlled by the statistician. Most frequently these parameters are some non-negative numbers $\alpha_1, \ldots, \alpha_N$ related in some way to the 'size' of the units, and we may assume that

$$\sum_{i=1}^{N} \alpha_i = 1. \tag{1.4}$$

For example, *rejective sampling* of size n corresponding to $(\alpha_1, \ldots, \alpha_N)$, say $\mathsf{R}(s \mid n,\, \alpha_1, \ldots, \alpha_N)$, is defined by

$$\begin{aligned} \mathsf{R}(s \mid n,\, \alpha_1, \ldots, \alpha_N) &= c(n,\, \alpha_1, \ldots, \alpha_N) \prod_{i \in s} \alpha_i, \quad \text{if } s \text{ contains } n \text{ units} \\ &= 0, \quad \text{otherwise,} \end{aligned} \tag{1.5}$$

where the constant $c(n,\, \alpha_1, \ldots, \alpha_N)$ is chosen so that $\sum \mathsf{R}(s \mid n,\, \alpha_1, \ldots, \alpha_N) = 1$, s running through all subsets of size n.

In accordance with (1.5) rejective sampling may be regarded as projected sampling with replacement of size n with probabilities α_i given the condition that the number of distinct units in the sample equals n.

Successive sampling is originally defined on ordered sequences $\omega = (r_1, \ldots, r_n)$ of distinct units, and depends on n and numbers $\alpha_1, \ldots, \alpha_N$ as follows:

$$\begin{aligned} \mathsf{P}_1(\omega \mid n, \alpha_1, \ldots, \alpha_N) &= \frac{\alpha_{r_1} \cdots \alpha_{r_n}}{(1-\alpha_{r_1}) \cdots (1-\alpha_{r_1} - \cdots - \alpha_{r_{n-1}})}, \\ \omega &= (r_1, \ldots, r_N). \end{aligned} \tag{1.6}$$

Sampling with replacement is originally defined on ordered sequences $\omega' = (r_1, \ldots, r_n)$ of not necessarily distinct units by the formula

$$\begin{aligned} \mathsf{P}'_1(\omega' \mid n, \alpha_1, \ldots, \alpha_N) &= \alpha_{r_1} \cdots \alpha_{r_n}, \\ \omega' &= (r_1, \ldots, r_n), \end{aligned} \qquad 1 \leq r_1, \ldots, r_n \leq N. \tag{1.7}$$

For all three kinds of probability sampling just introduced, the correspondence between the parameters $(n, \alpha_1, \ldots, \alpha_N)$ and probabilities of inclusion π_i, π_{ij}, $1 \leq i, j \leq N$, is rather complicated, and even approximations are not too simple, unless n is of lower order than N.

Next, two kinds of probability sampling are distinguished by a simple relation between the parameters $(n, \alpha_1, \ldots, \alpha_N)$ and the probabilities of inclusion $\pi_1, \ldots, \pi_N$. First, *Poisson sampling* is defined by

$$\mathsf{P}(s \mid n, \alpha_1, \ldots, \alpha_N) = \prod_{i \in s} (n\alpha_i) \prod_{j \in U - s} (1 - n\alpha_j), \quad s \subset U, \tag{1.8}$$

where $\alpha_i < n^{-1}$, $1 \leq i \leq N$. Second, *randomized systematic sampling* is defined by the following experiment: permute randomly the units in the population and denote the result by $U^+ = \{R_1, \ldots, R_N\}$; then take an observation ξ from the uniform distribution over $[0, 1]$ and include in s each unit i such that for some $k = 0, 1, \ldots, n-1$ and $m = 1, \ldots, N$ we have $R_m = i$ and

$$n\left(\alpha_{R_1} + \cdots + \alpha_{R_{m-1}}\right) \leq \xi + k < n\left(\alpha_{R_1} + \cdots + \alpha_{R_m}\right). \tag{1.9}$$

For both Poisson and randomized systematic sampling

$$\pi_i = n\alpha_i, \quad i = 1, \ldots, N, \tag{1.10}$$

provided that the right side does not exceed 1 for any $i = 1, \ldots, N$. On the other hand, Poisson sampling is not very suitable in practice, because it makes the sample size a random variable, and even an empty sample may occur. Also randomized systematic sampling has some disadvantages: (i) the procedure of randomly permuting the population may be very tedious if N is very large; (ii) the dependence of the π_{ij}s on the π_is is very complicated, and the recent asymptotic results by Hartley and Rao (1962) are applicable only if n is of smaller order than N.

For illustration, let us consider a population consisting of four units, $U = \{1, 2, 3, 4\}$, and compute the probabilities $\mathsf{P}(s)$ for all five kinds of sampling with $n = 2$ and $\alpha_1 = .1$, $\alpha_2 = .2$, $\alpha_3 = .3$ and $\alpha_4 = .4$. The results are given in Table 1. The successive sampling, randomized systematic sampling and sampling without replacement have been projected according to (1.1).

Table 1 $n = 2,\ \alpha_1 = .1,\ \alpha_2 = .2,\ \alpha_3 = .3,\ \alpha_4 = .4$

The Sample s	Sampling type				
	Rejective	Poisson	Projected successive	Projected with replacement	Projected randomized systematic
0		.038			
1		.010		.010	
2		.026		.040	
3		.058		.090	
4		.154		.160	
12	.057	.006	.047	.040	.067
13	.086	.014	.076	.060	.067
14	.114	.038	.111	.080	.067
23	.171	.038	.161	.120	.067
24	.229	.102	.233	.160	.266
34	.343	.230	.371	.240	.466
123		.010			
124		.026			
134		.058			
234		.154			
1234		.038			
	1.000	1.000	.999	1.000	1.000

The rest of this section will be devoted to the question of motivation and integration with the previous literature.

Sampling with unequal probabilities presents no especial difficulties, if it amounts, within individual strata, either to sampling with equal probabilities, as in Neyman (1934), or to sampling just one unit, as in Hansen and Hurwitz (1943). The general case, however, raises very difficult mathematical problems, which has stimulated attempts to establish a unified and modernized theory of probability sampling (another name for sampling from finite populations). One aspect of this theory should be an appropriate definition of a sample, and of the class of admissible probability samplings and estimators of, say, the population total

$$Y = \sum_{i=1}^{N} y_i. \tag{1.11}$$

Horwitz and Thompson (1952) showed that

$$\hat{Y} = \sum_{i \in s} (y_i / \pi_i) \tag{1.12}$$

is an unbiased estimator of Y, if $\pi_i > 0$, $i = 1, \ldots, N$, for any sampling design, and established the variance of $\hat{Y}$ in terms of the probabilities π_{ij}, $1 \leq i, j \leq N$. Yates and Grundy (1953) noted that for sampling design where the sample is of fixed size, the variance of $\hat{Y}$ may be rewritten in the following form:

$$\mathrm{var}(\hat{Y}) = \sum\sum_{i<j} \left[(y_i/\pi_i) - (y_j/\pi_j)\right]^2 (\pi_i \pi_j - \pi_{ij}). \tag{1.13}$$

However, in order to be able to utilize the general formulas, we have to study in more detail some particular kinds of probability sampling, and solve for them, at least approximately, the following three problems:

P1. To establish relation between controlled parameters $(n, \alpha_1, \ldots, \alpha_N)$ and the probabilities of inclusion $\pi_1, \ldots, \pi_N$.

P2. To establish relation between the probabilities $\pi_1, \ldots, \pi_N$ and the probabilities π_{ij}, $1 \leq i < j \leq N$.

P3. To find conditions for asymptotic normality of estimators such as $\hat{Y}$.

Without solving these problems, we cannot control the π_is in (1.12), simplify and estimate the variance (1.13), and justify the confidence intervals based on the assumption of normality.

All three of the above problems are solved below for rejective sampling, and the author hopes to do the same for successive sampling in some subsequent work. As for the previous literature, there is just one paper in this line, namely by Hartley and Rao (1962). They presented a solution of Problem P2 for randomized systematic sampling by a quite different method than the one used in the present chapter (recall that P1 is trivial for this kind of sampling in view of (1.10)). Moreover, their approach is asymptotic in a quite different sense, because, roughly speaking, they are interested in the case when n is fixed and $N \to \infty$, while the present chapter assumes that $n \to \infty$ and $(N - n) \to \infty$. As a matter of fact, under the conditions assumed by Hartley and Rao,

$$n(\pi_i \pi_j - \pi_{ij})/(\pi_i \pi_j) \longrightarrow 1 \tag{1.14}$$

for randomized systematic sampling as well as for rejective and successive sampling. Formula (1.14), however, is false for all three kinds of sampling, if $n \to \infty$ and $(N-n) \to \infty$, or more precisely if

$$\sum_{i=1}^{N} \pi_i(1-\pi_i) \longrightarrow \infty. \tag{1.15}$$

For rejective sampling under condition (1.15),

$$\left[\sum_{i=1}^{N} \pi_i(1-\pi_i)\right] (\pi_i\pi_j - \pi_{ij}) / \left[\pi_i(1-\pi_i)\,\pi_j(1-\pi_n)\right] \longrightarrow 1 \tag{1.16}$$

uniformly in i, j, as shown in Section 5. Hartley and Rao leave open the problem of what happens under condition (1.15), and center their attention upon deriving expansions for π_{ij} in terms of π_i accurate to higher orders in N than that given by (1.14). Consequently, their formulas are applicable, if N is much larger than n only, and it is not reasonable to compare them with those obtained here. To make a fruitful comparison possible, one should solve the problem P2 for rejective sampling under the Hartley–Rao conditions, and for randomized systematic sampling under condition (1.15).

In conclusion, let us mention the paper (1962) by Rao, Hartley and Cochran (1962), where they present a clever attempt to circumvent the above problems by introducing random strata and selecting just one unit from each stratum with unequal probabilities. However the simplicity of their approach is somewhat invalidated by the fact that the estimator they proposed is not admissible, i. e. there exists another estimator with uniformly smaller variance. Actually, the outcome of the sampling experiment suggested in Rao, Hartley and Cochran (1962) may be described as $\omega = (S_1, \ldots, S_L, s)$, where S_h are random strata and s the selected subset of distinct units. Now s, together with the values $y_i, i \in s$, represents sufficient statistics, and the conditional mean values of the Rao–Hartley–Cochran estimator with respect to these sufficient statistics has the above mentioned superior property. Unfortunately this conditional mean value is very difficult to compute and thus only indicates a possibility of improvement. If the Rao–Hartley–Cochran design were used with the estimator (1.12), one should first solve the problems P1 and P2. Thus also in this line further development is necessary.

Finally, as noted by the referee, the sampling referred to here as *Poisson sampling* is but a generalization of *binomial sampling*, introduced by L. A. Goodman (1949), for the case of unequal probabilities.

2 REJECTIVE SAMPLING

We shall begin with a definition of rejective sampling equivalent to (1.4).

Definition 2.1. Let $p_1, \ldots, p_N$ be some fixed numbers such that $0 < p_i < 1$, $i = 1, \ldots, N$. A probability sampling $\mathsf{R}(s)$ will be called *rejective sampling of size* n *with probabilities* $p_1, \ldots, p_N$ if

$$\begin{aligned}\mathsf{R}(s) &= C \prod_{i \in s} p_i \prod_{i \in U-s} (1 - p_i), \quad \text{if } s \text{ contains } n \text{ units}, \\ &= 0, \quad \text{otherwise.}\end{aligned} \tag{2.1}$$

The representation (2.1) will be called *canonical* if

$$\sum_{i=1}^{N} p_i = n. \tag{2.2}$$

Of course, (2.1) is equivalent to (1.5) if

$$\alpha_i = p_i/(1 - p_i) \left[\sum_{j=1}^{N} p_j/(1 - p_j) \right]^{-1}, \quad i = 1, \ldots, N. \tag{2.3}$$

There exist infinitely many other probabilities $p_1^*, \ldots, p_N^*$ and constants C^* such that (2.1) may be rewritten as

$$\begin{aligned}\mathsf{R}^*(s) &= C^* \prod_{i \in s} p_i^* \prod_{i \in U-s} (1 - p_i^*) \quad \text{if } s \text{ contains } n \text{ units}, \\ &= 0 \quad \text{otherwise.}\end{aligned} \tag{2.4}$$

Obviously $\mathsf{R}^*(s) = \mathsf{R}(s)$ for all s, if, and only if,

$$p_i^*/(1 - p_i^*) = c[p_i/(1 - p_i)], \quad 1 \le i \le N, \tag{2.5}$$

for some constant c.

Lemma 2.1. Let the numbers $p_1, \ldots, p_N$ and $p_1^*, \ldots, p_N^*$ satisfy (2.5). Put

$$h = \sum_{i=1}^{N} p_i, \tag{2.6}$$

$$h^* = \sum_{i=1}^{N} p_i^* \tag{2.7}$$

and

$$d = \sum_{i=1}^{N} p_i(1 - p_i). \tag{2.8}$$

Then

$$p_i^*(1-p_i) = p_i(1-p_i^*)\,\{1+(h^*-h)/d + o[(h^*-h)/d]\}, \tag{2.9}$$

where $o(x)$ means that $o(x)\,x^{-1} \to 0$ for $x \to 0$.

Proof. From (2.5) it easily follows that

$$p_i - p_i^* = (1-c)\,p_i(1-p_i^*) \tag{2.10}$$

and

$$p_i - p_i^* = (c^{-1}-1)\,p_i^*(1-p_i). \tag{2.11}$$

Assume that $h^* > h$, so that $p_i^* > p_i$ and $1-p_i^* < 1-p_i$. Carrying out a summation in (2.10) and (2.11) with respect to i we get

$$h - h^* = (1-c)\sum_{i=1}^{N} p_i(1-p_i^*) < (1-c)\,d, \tag{2.12}$$

$$h - h^* = (c^{-1}-1)\sum_{i=1}^{N} p_i^*(1-p_i) > (c^{-1}-1)\,d, \tag{2.13}$$

from which we immediately obtain

$$\{1-[(h^*-h)/d]\}^{-1} < c < 1+[(h^*-h)/d]. \tag{2.14}$$

Now, (2.9) is an easy consequence of (2.14) and (2.5). The case when $h^* < h$ may be treated quite similarly. The proof is complete. □

Now, let $p_1, \ldots, p_N$ be any numbers such that $0 < p_i < 1,\ 1 \le i \le N$.

Lemma 2.2. If $\sum_{i=1}^{N} p_i = n$ and s_n is a sample of size n, then

$$\sum_{i \in U-s_n} p_i = \sum_{i \in s_n} (1-p_i) > \frac{1}{2}\,d. \tag{2.15}$$

If $\sum_{i=1}^{N} p_i = h$ and s_k is a sample of size k, then

$$h - k + \sum_{i \in s_k}(1-p_i) = \sum_{i \in U-s_k} p_i. \tag{2.16}$$

Proof. The proof is immediate. A property of rejective sampling is that conditional probabilities $\mathsf{P}(s \mid s \ni i)$, $\mathsf{P}(s \mid s \ni i, j)$ etc. and 'complementary' probabilities $\mathsf{P}^*(s) = \mathsf{P}(U-s)$ again represent rejective sampling with properly modified parameters. □

3 POISSON SAMPLING

Here, also, we need to rewrite definition (1.8).

Definition 3.1. A probability sampling will be called *Poisson sampling with probabilities* $p_1, \ldots, p_N$ if

$$\mathsf{P}(s) = \prod_{i \in s} p_i \prod_{i \in U - s} (1 - p_i), \quad (s \subset U). \tag{3.1}$$

Let $K(s)$ denote the number of units contained in s.

From formula (3.1) it follows that Poisson sampling may be carried out by N independent subexperiments, the ith of them deciding with probabilities p_i or $1 - p_i$ whether the ith unit will be included in s or not, respectively.

Obviously $K = K(s)$ is a random variable in Poisson sampling, and it ranges from 0 to N. Now, introduce zero–one random variables $\xi_i = \xi_i(s)$ characterizing the presence or the absence of the ith unit in s:

$$\begin{aligned} \xi_i &= 1, \quad \text{if } s \ni i, \\ &= 0, \quad \text{otherwise.} \end{aligned} \qquad (1 \le i \le N) \tag{3.2}$$

Then

$$K = \sum_{i=1}^{N} \xi_i, \tag{3.3}$$

and the distribution of K is the one considered in Poisson's generalization of Bernoulli's scheme, since the random variables ξ_i are independent, as can be easily seen from (3.1) . Hence our label 'Poisson sampling'.

From (3.3) it follows that

$$\mathsf{E}K = h = \sum_{i=1}^{N} p_i \tag{3.4}$$

and

$$\operatorname{var} K = d = \sum_{i=1}^{N} p_i(1 - p_i). \tag{3.5}$$

The following Lemma follows from asymptotic normality of K for $d \to \infty$ (Cramér (1946), pp. 216–217) and the unimodality of the distribution of K which can be demonstrated.

Lemma 3.1. For any constant C it holds that

$$\max_{h - Cd^{\frac{1}{2}} < k < h + Cd^{\frac{1}{2}}} \left| \frac{(2\pi d)^{\frac{1}{2}} \, \mathsf{P}(K = k)}{\exp\left[-\frac{1}{2}(k - h)^2 \, d^{-1}\right]} - 1 \right| \longrightarrow 0 \tag{3.6}$$

if $d = \sum_{i=1}^{N} p_i(1-p_i) \to \infty$.

Now, let $y_1, \ldots, y_N$ be some real values associated with units $1, \ldots, N$. In Poisson sampling with probabilities $p_1, \ldots, p_N$ the variance of the estimator

$$T' = \sum_{i \in s} y_i / p_i \tag{3.7}$$

will be due to (a) the variability of the ratios y_i/p_i and (b) the variability of the sample size $K = K(s)$. The latter source of variability is not present in rejective sampling and, hence, is undesirable in our consideration. To eliminate it, let us shift the values y_i/p_i by a constant c chosen so that we may obtain minimum variance. Using the random variables ξ_i defined by (3.3), we easily see that

$$\begin{aligned}
\mathsf{var}\left\{\sum_{i \in s}[(y_i/p_i) - c]\right\} &= \mathsf{var}\left\{\sum_{i=1}^{N}[(y_i/p_i) - c]\,\xi_i\right\} \\
= \sum_{i=1}^{N}[(y_i/p_i) - c]^2\,\mathsf{var}\,\xi_i &= \sum_{i=1}^{N}[(y_i/p_i) - c]^2\,p_i(1-p_i) \\
\geq \sum_{i=1}^{N}[(y_i/p_i) - R]^2\,p_i(1-p_i) &= \sum_{i=1}^{N}(y_i - R\,p_i)^2\,(p_i^{-1} - 1),
\end{aligned}$$

where

$$R = \sum_{i=1}^{N} y_i(1-p_i) \Big/ \sum_{i=1}^{N} p_i(1-p_i) \tag{3.8}$$

and the minimum is obtained for $c = R$. This invites us to replace (3.7) by

$$T = \sum_{i \in s}[(y_i/p_i) - R] + R\sum_{i=1}^{N} p_i. \tag{3.9}$$

The quantity T will have an auxiliary but important role in our considerations. Note that T is not an estimator, since it depends on y_i, $i \in U - s$, unless the size of s is $\sum_{i=1}^{N} p_i$. However, for samples of size $\sum_{i=1}^{N} p_i$ it coincides with T'.

Throughout the chapter we formulate the asymptotic relations for the whole class of Poisson or rejective samplings corresponding to all possible $N = 1, 2, \ldots$ and $p_1, \ldots, p_N$, $0 < p_i < 1$, $i = 1, \ldots, N$. This approach has been applied in Lemmas 2.1 and 3.2. To be able to state the condition of asymptotic normality for T in the same way, we need a modification of the Lindeberg condition implicit in the following Lemma.

Lemma 3.2. With the above notation, we have

$$\mathsf{E}T = \sum_{i=1}^{N} y_i \tag{3.10}$$

and

$$\mathsf{var}\,T = \sum_{i=1}^{N} (y_i - R\,p_i)^2\,(p_i^{-1} - 1). \tag{3.11}$$

Moreover,

$$\sup_x \left| \mathsf{P}(T < x) - \Phi\left(\frac{x - \mathsf{E}T}{(\mathsf{var}\,T)^{\frac{1}{2}}}\right) \right| \longrightarrow 0, \tag{3.12}$$

where Φ denotes the normed normal distribution function if

$$e = e(y_1, \ldots, y_N, p_1, \ldots, p_N) \longrightarrow 0, \tag{3.13}$$

where

$$e = \inf\{\epsilon : L_\epsilon^* \le \epsilon\} \tag{3.14}$$

and

$$L_\epsilon^* = (\mathsf{var}\,T)^{-1} \sum_{i \in A_\epsilon} (y_i - R\,p_i)^2\,(p_i^{-1} - 1), \tag{3.15}$$

where

$$A_\epsilon = \left\{i : (y_i - R\,p_i)^2 > \epsilon^2\,p_i^2\,\mathsf{var}\,T\right\}. \tag{3.16}$$

Proof. Putting $T = \sum_{i=1}^{N} [(y_i/p_i) - R]\,\xi_i + R\sum_{i=1}^{N} p_i$ we readily see that the Lindeberg condition for T amounts to $L_\epsilon \to 0$ for every $\epsilon > 0$, where

$$\begin{aligned} L_\epsilon \;=\; & (\mathsf{var}\,T)^{-1} \sum_{i \in C_\epsilon} [(y_i/p_i) - R]^2\,(1 - p_i)^2\,p_i \\ & + \sum_{i \in B_\epsilon} [(y_i/p_i) - R]^2\,(1 - p_i)\,p_i^2, \end{aligned} \tag{3.17}$$

where

$$C_\epsilon = \left\{i : |(y_i/p_i) - R|\,(1 - p_i) > \epsilon(\mathsf{var}\,T)^{\frac{1}{2}}\right\}$$

and

$$B_\epsilon = \left\{i : |(y_i/p_i) - R|\,p_i > \epsilon(\mathsf{var}\,T)^{\frac{1}{2}}\right\}.$$

Now, it may be easily verified that $L_\epsilon \le L_\epsilon^* \le 2L_{\frac{1}{2}\epsilon}$. Consequently $L_\epsilon \to 0$ for every $\epsilon > 0$ is equivalent to $L_\epsilon^* \to 0$ for every $\epsilon > 0$. Finally, as $\left[e < \frac{1}{2}\epsilon\right] \Rightarrow [L_e^* \le L_{2\epsilon} \le 2e]$, condition (3.13) entails $L_\epsilon^* \to 0$ for every $\epsilon \to 0$ for any sequence of Poisson samplings. The proof is complete. □

Remark. Following the pattern of the proof of Theorem 3.1 in Hájek (1960), we could show that (3.13) is also necessary for (3.12).

4 LEMMAS CONNECTING REJECTIVE SAMPLING AND POISSON SAMPLING

A comparison of (2.1) and (3.1) shows that rejective sampling of size n may be regarded as Poisson sampling given the condition that the sample size equals n. If the probabilities $p_1, \dots, p_N$ are chosen so that $n = \sum_{i=1}^{N} p_i$ the sample size of the rejective sampling will equal the average sample size of the Poisson sampling.

We have also seen that the rejective sampling may be regarded as projected sampling with replacement of size n given the condition that the number of distinct units equals n.

Poisson sampling and sampling with replacement may both be decomposed into independent subexperiments, and, for this reason, they both possess simple variance formulas and simple conditions for asymptotic normality. However, whereas all the relevant properties of Poisson sampling may be carried over to rejective sampling, at least asymptotically, sampling with replacement gives formulas diverging from those for rejective sampling even in the limit, unless the probability of rejection tends to zero. This somewhat surprising result is probably due to the fact that rejective sampling may represent an 'average result' of Poisson sampling, whereas in sampling with replacement rejective sampling always represents an extreme case (all units must be distinct). Thus the parallelism of rejective sampling and sampling with replacement, though at first sight very inviting, is misleading.

First we show that convergence in probability in Poisson sampling implies the same in rejective sampling for an important class of 'hereditary' events.

Lemma 4.1. Consider Poisson sampling $\mathsf{P}(\cdot)$ and rejective sampling $\mathsf{R}(\cdot)$ of size n, both with probabilities $p_1, \dots, p_N$ such that $\sum_{i=1}^{N} p_i = n$. Let A be an event such that either

$$[s \in A,\, s^* \subset s] \Longrightarrow (s^* \in A) \quad \text{for all } s,\, s^* \in U \tag{4.1}$$

or

$$[s \in A,\, s^* \supset s] \Longrightarrow (s^* \in A) \quad \text{for all } s,\, s^* \in U. \tag{4.2}$$

Then there is a d_0 such that for $d > d_0$

$$\mathsf{P}(A) \geq \frac{1}{20}\,\mathsf{R}(A). \tag{4.3}$$

Proof. Assume that (4.1) holds, for example. Then, in view of (2.16),

$$\begin{aligned} \mathsf{P}(A,\, K=k) &= \sum_{s_k \in A} \mathsf{P}(s_k) \\ &\geq \sum_{s_{k+1}\in A} \mathsf{P}(s_{k+1}) \sum_{j \in s_{k+1}} \frac{1-p_j}{n-k-1+p_j+\sum_{i\in s_{k+1}}(1-p_i)}. \end{aligned}$$

Furthermore, from (2.15) it follows that

$$\sum_{j \in s_{k+1}} \frac{1-p_j}{n-k-1+p_j+\sum_{i\in s_{k+1}}(1-p_i)} \geq 1-4(n-k)/d, \quad (k<n). \quad (4.4)$$

On combining (4.3) and (4.4), we get

$$\begin{aligned} \mathsf{P}(A,\, K=k) &\geq (1-4(k-n)/d)\, \mathsf{P}(A,\, K=k+1) \quad (4.5) \\ &\geq (1-4(k-n)^2/d)\, \mathsf{P}(A,\, K=n), \quad (k<n). \end{aligned}$$

After easy computations, (4.5) yields

$$\begin{aligned} \mathsf{P}(A) &= \sum_{k=0}^{N} \mathsf{P}(A,\, K=k) \geq \sum_{n-\frac{1}{3}d^{\frac{1}{2}}<k<n} \mathsf{P}(A,\, K=k) \quad (4.6) \\ &\geq \frac{5}{9}\left(\frac{1}{3}d^{\frac{1}{2}}-1\right) \mathsf{P}(A,\, K=n). \end{aligned}$$

Now, noting that

$$\mathsf{R}(A) = \mathsf{P}(A \,|\, K=n) = [\mathsf{P}(A,\, K=n)]/[\mathsf{P}(K=n)], \quad (4.7)$$

and recalling (3.6), we see that (4.6) implies (4.3) immediately.

The case where (4.2) holds may be carried over to the previous one by considering the complementary rejective sampling $\mathsf{P}^*(s) = \mathsf{P}(U-s)$ of size $N-n$ with probabilities $1-p_1, \ldots, 1-p_N$. The proof is complete. □

From Lemma 4.1 we obtain another Lemma almost immediately.

Lemma 4.2. Consider rejective sampling $\mathsf{R}(\cdot)$ of size n with probabilities $p_1, \ldots, p_N$, $\sum_{i=1}^N p_i = n$, and some numbers $b_1, \ldots, b_N$ such that

$$0 \leq b_i \leq 1 \quad (1 \leq i \leq N). \quad (4.8)$$

Denote $d = \sum_{i=1}^N p_i(1-p_i)$.

Then

$$\left(d^{-1}\sum_{i\in s} b_i - d^{-1}\sum_{i=1}^{N} b_i p_i\right) \longrightarrow 0 \quad (4.9)$$

in R-probability if $d \to \infty$, uniformly in $b_1, \ldots, b_N$.

Proof. The convergence in probability follows for the corresponding Poisson sampling $\mathsf{P}(\cdot)$ by the Chebyshev inequality, since

$$\mathsf{E}^{\mathsf{P}}\left(d^{-1}\sum_{i\in s} b_i\right) = d^{-1}\sum_{i=1}^{N} b_i p_i$$

and, for $b_i^2 \leq 1$,

$$\mathsf{var}^{\mathsf{P}}\left(d^{-1}\sum_{i\in s} b_i\right) = d^{-2}\sum_{i=1}^{N} b_i^2 p_i(1-p_i) \leq d^{-1},$$

where E^{P} and $\mathsf{var}^{\mathsf{P}}$ refer to $\mathsf{P}(\cdot)$. Now the events

$$d^{-1}\sum_{i\in s} b_i > d^{-1}\sum_{i=1}^{N} b_i p_i + \epsilon \tag{4.10}$$

and

$$d^{-1}\sum_{i\in s} b_i < d^{-1}\sum_{i=1}^{N} b_i p_i - \epsilon \tag{4.11}$$

are of the hereditary type considered in Lemma 4.1, for $b_i \geq 0$. Consequently the probability of (4.10) and (4.11) converges to 0 also for rejective sampling if $d \to \infty$, in view of (4.3). □

Examples. Lemma 4.2 entails, for example, that

$$d^{-1}\sum_{i\in s}(1-p_i) \longrightarrow 1$$

and

$$\left[d^{-1}\sum_{i\in s} p_i(1-p_i) - d^{-1}\sum_{i=1}^{N} p_i^2(1-p_i)\right] \longrightarrow 0$$

in probability for rejective sampling in canonical form, if $d \to \infty$.

Now we shall establish an important result, which may be regarded as the corner-stone of the present chapter. Our aim is to construct an experiment yielding jointly a rejective sample s_n and Poisson sample s_K in such a way that

$$\text{either} \quad s_K \subset s_n \quad \text{or} \quad s_K \supset s_n \tag{4.12}$$

with probability 1. The experiment will consist of two successive steps.

Experiment. First, we select a rejective sample s_n in accordance with its distribution, and, independently, ascertain the size $K = k$ of the Poisson sample. Second, if $k = n$, we put $s_K = s_n$; if $k > n$, we select a rejective sample s_{k-n} from $U - s_n$, considered as a population, with probabilities $(k-n)\,D_n^{-1}\,p_i$, $i \in U - s_n$, where

$$D_n = D(s_n) = \sum_{i\in s_n}(1-p_i) = \sum_{i\in U-s_n} p_i, \tag{4.13}$$

and put $s_K = s_n \cup s_{k-n}$; if $k < n$, we select a rejective sample s_{n-k} from s_n, considered as a population, with probabilities $(n-k)\,D_n^{-1}(1-p_i)$, and put $s_K = s_n - s_{n-k}$.

In the above experiment a pair (s_k, s_n), such that either $s_k \subset s_n$ or $s_k \supset s_n$, will have probability

$$\begin{aligned} \mathsf{Q}(s_k, s_n) &= \mathsf{R}(s_n)\,\mathsf{P}(K=k)\,\mathsf{R}^*(s_k - s_n \mid U - s_n,\, k-n), && \text{if } k > n,\\ &= \mathsf{R}(s_n)\,\mathsf{P}(K=n), && \text{if } k = n,\\ &= \mathsf{R}(s_n)\,\mathsf{P}(K=n)\,\mathsf{R}^+(s_n - s_k \mid s_n,\, n-k), && \text{if } k < n, \end{aligned} \tag{4.14}$$

where

$$\begin{aligned} &\mathsf{R}^*(s_k - s_n \mid U - s_n,\, k-n) \\ &= C^*(U - s_n,\, k-n) \prod_{i\in s_k - s_n} (k-n)\,D_n^{-1}\,p_i \prod_{i\in U-s_k} \left[1-(k-n)\,D_n^{-1}\,p_i\right], \end{aligned} \tag{4.15}$$

and

$$\begin{aligned} &\mathsf{R}^+(s_n - s_k \mid s_n,\, n-k) \\ &= C^+(s_n,\, n-k) \prod_{i\in s_n - s_k} (n-k)\,D_n^{-1}(1-p_i) \prod_{i\in s_k} \left[1-(n-k)\,D_n^{-1}(1-p_i)\right]. \end{aligned} \tag{4.16}$$

Denoting the Poisson samplings corresponding to R^* and R^+ by P^* and P^+, respectively, and by K^* and K^+ the sizes of the respective Poisson samples, we can write

$$C^*(U - s_n,\, k-n) = [\mathsf{P}^*(K^* = k-n)]^{-1} \tag{4.17}$$

and

$$C^+(s_n,\, n-k) = [\mathsf{P}^+(K^+ = n-k)]^{-1}.$$

Furthermore,

$$\begin{aligned} \mathsf{E}K^* &= \sum_{i\in U-s_n} (k-n)\,D_n^{-1}\,p_i = k-n \\ \operatorname{var} K^* &= \sum_{i\in U-s_n}{}'(k-n)\,D_n^{-1}\,p_i[1-(n-k)\,D_n^{-1}p_i] \end{aligned}$$

so that, in view of (2.15),

$$(k-n)\,[1-2(k-n)/d] < \mathsf{var}\, K^* < (k-n). \tag{4.18}$$

Combining (4.17) and (4.18) and applying Lemma 3.1, we get

$$[d \to \infty,\ |k-n|/d \to 0] \Longrightarrow \left[C^*(2\pi(k-n))^{-\frac{1}{2}} \to 1\right]. \tag{4.19}$$

The same result holds for C^+.

Now, return to the joint distribution (4.14) and ask whether the marginal distribution of s_k, say $\mathsf{P}_0(s_k)$, corresponds to Poisson sampling $\mathsf{P}(\cdot)$ with probabilities $p_1, \ldots, p_N$, as we wished. Unfortunately, this is not so, unless $p_1 = p_2 = \cdots = p_N$, as in Hájek (1960). On the other hand, we shall be able to prove that the difference $\mathsf{P}_0 - \mathsf{P}$ tends to zero in variation as $d \to \infty$. Consequently any random variable T has the same limiting distribution, if any, for P_0 as for P. We shall denote the variation of $\mathsf{P}_0 - \mathsf{P}$ by $||\mathsf{P}_0 - \mathsf{P}||$ and define it as follows:

$$||\mathsf{P}_0 - \mathsf{P}|| = 2 \sum_{\mathsf{P}_0(s)<\mathsf{P}(s)} [1 - (\mathsf{P}_0(s)/\mathsf{P}(s))]\, \mathsf{P}(s) \tag{4.20}$$

where the sum extends over all samples s satisfying the condition under the sign $\sum$.

Lemma 4.3. Let $\mathsf{P}(\cdot)$ denote Poisson sampling with probabilities $p_1, \ldots, p_N$, $\sum p_i = n$, and define $\mathsf{P}_0(\cdot)$ by

$$\begin{aligned} \mathsf{P}_0(s_k) &= \sum_{s_n \subset s_k} \mathsf{Q}(s_k, s_n) \quad \text{if } k > n, \\ &= \mathsf{P}(s_k) \qquad\qquad\quad \text{if } k = n, \\ &= \sum_{s_n \supset s_k} \mathsf{Q}(s_k, s_n) \quad \text{if } k < n, \end{aligned} \tag{4.21}$$

where Q is given by (4.14) and the sums extend over all s_n satisfying the indicated conditions.

Then $d = \sum_{i=1}^{N} p_i(1-p_i) \to \infty$ implies

$$||\mathsf{P}_0 - \mathsf{P}|| \to 0. \tag{4.22}$$

Proof. Let $a \approx b$ and $a \sim b$ denote that $a/b \to 1$ and $(a-b) \to 0$ in P-probability, respectively, if $d \to \infty$. Furthermore, let $a \gtrsim b$ denote that for every $\epsilon > 0$, $a > (1-\epsilon)\,b$ holds with P-probability tending to 1 if $d \to \infty$. If a and b also depend on s_n, then the convergences are assumed uniform with respect to s_n. For example, Lemma 3.1 implies that

$$[\mathsf{P}(K=k)/\mathsf{P}(K=n)] \approx \exp\left[-\frac{1}{2}\,d^{-1}(k-n)^2\right] \tag{4.23}$$

and

$$d^{-2}(k-n)^3 \sim 0. \tag{4.24}$$

In both the above relations the left side depends on s_k only, namely on the size k of s_k. However, (4.24) and (2.15) also imply that

$$D_n^{-2}(k-n)^3 \sim 0, \tag{4.25}$$

uniformly in s_n (recall (4.13)).

In view of (4.20), our task is to prove that

$$\mathsf{P}_0(s_k)/\mathsf{P}(s_k) \;\gtrsim\; 1. \tag{4.26}$$

From now on we shall assume that $k > n$, so that probability and the signs $\approx$, $\sim$, $\gtrsim$ should be understood as conditional given $k > n$. Note that

$$\begin{aligned} \mathsf{R}(s_n) &= \mathsf{P}(s_n)\,[\mathsf{P}(K=n)]^{-1} \\ &= \mathsf{P}(s_k)\,[\mathsf{P}(K=n)]^{-1} \prod_{i \in s_k - s_n} (1-p_i)/p_i \end{aligned} \tag{4.27}$$

and put Q in the following form:

$$\begin{aligned} \mathsf{Q}(s_k, s_n) &= \mathsf{P}(s_k)\,[\mathsf{P}(K=k)/\mathsf{P}(K=n)]\,M(s_k, s_n) \\ &\quad \times C^*(U - s_n,\, k-n) \prod_{i \in s_k - s_n} (k-n)\,D_k^{-1}(1-p_i) \\ &\quad \times \prod_{i \in s_n} [1 - (k-n)\,D_k^{-1}(1-p_i)], \end{aligned} \tag{4.28}$$

where

$$D_k = \sum_{i \in s_k} (1-p_i) \tag{4.29}$$

and

$$\begin{aligned} M(s_k, s_n) &= (D_k/D_n)^{k-n} \prod_{i \in U - s_k} [1 - (k-n)\,D_n^{-1} p_i] \\ &\quad \times \prod_{i \in s_n} [1 - (k-n)\,D_k^{-1}(1-p_i)]^{-1}. \end{aligned} \tag{4.30}$$

Now using the usual development $1 - x = \exp\left[-x - \frac{1}{2}x^2 + O(x^3)\right]$, we can write

$$M(s_k, s_n) \approx \exp\Bigg[(k-n)\left[(D_k - D_n)/D_n\right] - (k-n)\,D_n^{-1}\sum_{i\in U-s_k} p_i + (k-n)\,D_k^{-1}\sum_{i\in s_n}(1-p_i) - \frac{1}{2}(k-n)^2\,D_n^{-2}\sum_{i\in U-s_k} p_i^2 + \frac{1}{2}(k-n)^2\,D_k^{-2}\sum_{i\in s_n}(1-p_i)^2\Bigg], \tag{4.31}$$

where we have left out terms dominated (uniformly with respect to s_n) by a multiple of $d^{-2}(k-n)^3$. Now, by (2.16),

$$(k-n)\,(D_k - D_n)/D_n - (k-n)\,D_n^{-1}\sum_{i\in U-s_k} p_i + (k-n)\,D_k^{-1}\sum_{i\in s_n}(1-p_i) = D_n^{-1}(k-n)^2 - (k-n)\,D_k^{-1}\sum_{i\in s_k - s_n}(1-p_i) \tag{4.32}$$

and, further,

$$|D_n^{-1} - D_k^{-1}| = (D_k - D_n)/(D_n\,D_k) \le 4d^{-2}(k-n). \tag{4.33}$$

Consequently, (4.31) may be continued as follows:

$$M(s_k, s_n) \approx \exp\Bigg[-(k-n)\,D_k^{-1}\sum_{i\in s_k - s_n}(1-p_i) + (k-n)^2\,D_k^{-2}\sum_{i\in s_k}(1-p_i)^2 + D_k^{-1}(k-n)^2 - \frac{1}{2}(k-n)^2\,D_k^{-2}\left(\sum_{i\in U-s_k} p_i^2 + \sum_{i\in s_k}(1-p_i)^2\right)\Bigg]. \tag{4.34}$$

Now, by the Chebyshev inequality, $D_k \approx d$, $D_k^{-1}(k-n)^2$ is bounded in probability and

$$d^{-1}\left[\sum_{i\in U-s_k} p_i^2 + \sum_{i\in s_k}(1-p_i)^2\right] \sim d^{-1}\sum_{i=1}^{N}\left[p_i^2(1-p_i) + (1-p_i)^2\,p_i\right] = 1.$$

Thus, finally,

$$M(s_k, s_n) \approx \exp\left[\frac{1}{2}\,d^{-1}(k-n)^2\right] \times \exp\left[-(k-n)\,D_k^{-1}\sum_{i\in s_k - s_n}(1-p_i) + (k-n)^2\,D_k^{-2}\sum_{i\in s_k}(1-p_i)^2\right]. \tag{4.35}$$

As for C^*, it follows from (4.19) that

$$C^*(U - s_n, k - n) \approx [2\pi(k-n)]^{\frac{1}{2}}. \tag{4.36}$$

On combining (4.23), (4.28), (4.35) and (4.36), we get

$$\begin{aligned} \mathsf{Q}(s_k, s_n) &\approx \mathsf{P}(s_k)\,[2\pi(k-n)]^{\frac{1}{2}} \\ &\times \exp\left[(k-n)^2 D_k^{-2} \sum_{i\in s_k}(1-p_i)\right]^2 \prod_{i\in s_k - s_n}(k-n)\,D_k^{-1}(1-p_i) \\ &\times \prod_{i\in s_n}[1-(k-n)\,D_k^{-1}(1-p_i)] \exp\left[-(k-n)\,D_k^{-1}\prod_{i\in s_k-s_n}(1-p_i)\right]. \end{aligned} \tag{4.37}$$

Let $\mathsf{R}^+(\cdot \mid s_k, k-n)$ denote rejective sampling of size $k-n$ with probabilities $(k-n)\,D_k^{-1}(1-p_i)$ from the population s_k, i. e.

$$\begin{aligned} \mathsf{R}^+(s_k - s_n \mid s_k, k-n) &= C^+(s_k, k-n) \prod_{i\in s_k-s_n}(k-n) \\ &\times D_k^{-1}(1-p_i) \prod_{i\in s_n}[1-(k-n)\,D_k^{-1}(1-p_i)], \quad (s_n \subset s_k). \end{aligned} \tag{4.38}$$

By the same considerations as we used in the proof of (4.19), we could show that

$$C^+(s_k, k-n) \approx [2\pi(k-n)]^{\frac{1}{2}}. \tag{4.39}$$

On comparing (4.21), (4.37), (4.38) and (4.39), we easily see that

$$\begin{aligned} \mathsf{P}_0(s_k) \approx\ & \mathsf{P}(s_k) \exp\left[(k-n)^2 D_k^{-2} \sum_{i\in s_k}(1-p_i)^2\right] \\ & \times \mathsf{E}^+\left\{\exp\left[-(k-n)\,D_k^{-1}\sum_{i\in s_k-s_n}(1-p_i)\right]\Bigg|\, s_k, k-n\right\} \end{aligned} \tag{4.40}$$

where $\mathsf{E}^+(\cdot \mid s_k, k-n)$ denotes the mean value via $\mathsf{R}^+(\cdot \mid s_k, k-n)$.

To conclude the proof for $k > n$, it remains to show that

$$\begin{aligned} & \mathsf{E}^+\left\{\exp\left[-(k-n)\,D_k^{-1}\sum_{i\in s_k-s_n}(1-p_i)\right]\Bigg|\, s_k, k-n\right\} \\ \gtrsim\ & \exp\left[-(k-n)^2 D_k^{-2}\sum_{i\in s_k}(1-p_i)^2\right]. \end{aligned} \tag{4.41}$$

This is the most intricate part of the whole proof. First, introduce d_+ corresponding to R^+:

$$d_+ = d_+(s_k) = \sum_{i\in s_k}(k-n)\,D_k^{-1}(1-p_i)\,[1-(k-n)\,D_k^{-1}(1-p_i)] \tag{4.42}$$

$$= (k-n) - (k-n)^2 D_k^{-2} \sum_{i \in s_k} (1-p_i)^2.$$

Obviously,

$$d^+ \approx (k-n), \tag{4.43}$$

and $d^+ \to \infty$ in P-probability. In view of Lemma 4.2 with U and p_i replaced by s_k and $(k-n)\, D_k^{-1}(1-p_i)$, respectively, for any $\varepsilon > 0$ there exists an $a(\varepsilon)$ such that for $d_+ > a(\varepsilon)$

$$\left| d_+^{-1} \sum_{i \in s_k - s_n} (1-p_i) - d_+^{-1} \sum_{i \in s_k} (k-n)\, D_k^{-1}(1-p_i)^2 \right| < \varepsilon, \tag{4.44}$$

with R^+-probability greater than $1-\varepsilon$. Further, there exists a $b(\varepsilon)$ such that for $d > b(\varepsilon)$

$$d_+(s_k) > a(\varepsilon) \tag{4.45}$$

and

$$d_k(k-n)\, D_k^{-1} < \varepsilon^{-\frac{1}{2}} \tag{4.46}$$

hold with P-probability greater than $1-\varepsilon$. Now, the relations (4.44) through (4.46) imply that

$$(1-\varepsilon) \exp\left[-\varepsilon^{\frac{1}{2}} - (k-n)^2 D_k^{-2} \sum_{i \in s_k} (1-p_i)^2 \right] \tag{4.47}$$
$$\leq \mathsf{E}^+ \left\{ \exp\left[-(k-n)\, D_k^{-1} \sum_{i \in s_k - s_n} (1-p_i) \right] \middle| s_k,\, k-n \right\}$$

with P-probability greater than $1-\varepsilon$ if $d > b(\varepsilon)$. However, (4.47) is equivalent to (4.41). Thus the proof is finished for $k > n$. The proof for $k < n$ would be the same in principle, and is left to the reader. □

Remark 4.1. By a similar method we could obtain a sample s_n with approximately rejective distribution by starting from a Poisson sample s_k, and then extracting (if $k > n$) or adding (if $k < n$) $|k-n|$ units by rejective sampling of size $|k-n|$ from s_k or $U - s_k$ with probabilities $(k-n)\, D_k^{-1}(1-p_i)$ or $(k-n)\, D_k^{-1} p_i$, respectively. This possibility will be considered in Section 9.

5 PROBABILITIES OF INCLUSION π_i AND π_{ij}

We begin with a Theorem.

Theorem 5.1. *Consider rejective sampling of size n with probabilities p_i, $1 \leq i \leq N$, such that $\sum_{i=1}^N p_i = n$, and denote by π_i the probability of including the ith unit in the sample, $i = 1, \ldots, N$ (see (1.2)).*

Then

$$\pi_i(1-p_i) = p_i(1-\pi_i)\left[1 - [(\overline{\overline{\pi}} - \pi_i)/d_0] + o(d_0^{-1})\right], \quad 1 \leq i \leq N, \quad (5.1)$$

$$\pi_i(1-p_i) = p_i(1-\pi_i)\left[1 - [(\overline{\overline{p}} - p_i)/d] + o(d^{-1})\right], \quad 1 \leq i \leq N, \quad (5.2)$$

where

$$d_0 = \sum_{i=1}^N \pi_i(1-\pi_i), \quad (5.3)$$

$$d = \sum_{i=1}^N p_i(1-p_i), \quad (5.4)$$

$$\overline{\overline{\pi}} = d_0^{-1}\sum_{i=1}^N \pi_i^2(1-\pi_i) \quad (5.5)$$

and

$$\overline{\overline{p}} = d^{-1}\sum_{i=1}^N p_i^2(1-p_i), \quad (5.6)$$

and $d \cdot o(d^{-1}) \to 0$ and $d_0 \cdot o(d_0^{-1}) \to 0$ if $d \to \infty$ and $d_0 \to \infty$, respectively, uniformly in $i = 1, \ldots, N$.

Furthermore,

$$\max_{1\leq i\leq N} |(\pi_i/p_i) - 1| \to 0, \quad (5.7)$$

$$\max_{1\leq i\leq N} |[(1-\pi_i)/(1-p_i)] - 1| \to 0, \quad (5.8)$$

$$\max_{1\leq i\leq N} |[(\pi_i(1-\pi_i))/(p_i(1-p_i))] - 1| \to 0 \quad (5.9)$$

and

$$|(d_0/d) - 1| \to 0, \quad (5.10)$$

if either $d \to \infty$ or $d_0 \to \infty$.

Proof. Consider $\mathsf{P}(s)$ given by (3.1) and recall the fact that for samples of size n, $K(s) = n$,

$$\sum_{h\in U-s}\left[p_h \Big/ \sum_{j\in s}(1-p_j)\right] = \left[\sum_{h\in U-s} p_h \Big/ \sum_{j\in s}(1-p_j)\right] = 1, \quad (K(s) = n),$$

(see (2.15)). Thus we have

$$\begin{aligned}\mathsf{P}(s) &= \mathsf{P}(s) \sum_{h \in U-s} \left[p_h \Big/ \sum_{j \in s} (1-p_j) \right] \qquad (5.11)\\ &= p_i/(1-p_i) \sum_{h \in U-s} \mathsf{P}(s \cup \{h\} - \{i\}) \left[(1-p_h) \Big/ \sum_{j \in s} (1-p_j) \right], \\ &\qquad (K(s)=n),\end{aligned}$$

and, consequently,

$$\sum_{s \ni i} \mathsf{P}(s) = p_i/(1-p_i) \sum_{U-s \ni i} \mathsf{P}(s) \sum_{h \in s} \left[(1-p_h) \Big/ \left(p_h - p_i + \sum_{j \in s}(1-p_j) \right) \right], \qquad (5.12)$$

where s on the right side is the subset denoted by $s \cup \{h\} - \{i\}$ in (5.11). From the definition (1.2) of π_i it follows that

$$(1-\pi_i) \sum_{s \ni i} \mathsf{P}(s) = \pi_i \sum_{U-s \ni i} \mathsf{P}(s), \quad K(s) = n \qquad (5.13)$$

since in rejective sampling the probability of s, $\mathsf{R}(s)$ is proportional to $\mathsf{P}(s)$ (see (2.1)). On combining (5.12) and (5.13) we get

$$\begin{aligned}\pi_i(1-p_i) &= p_i(1-\pi_i) \left[\sum_{U-s \ni i} \mathsf{P}(s) \right]^{-1} \qquad (5.14)\\ &\times \sum_{U-s \ni i} \mathsf{P}(s) \left\{ \sum_{h \in s} \frac{1-p_h}{p_h - p_i + \sum_{j \in s}(1-p_j)} \right\}, \quad 1 \le i \le N.\end{aligned}$$

Now denote by $\mathsf{R}_i(\cdot)$ the rejective sampling of size n with probabilities p_j, $j \neq i$, from $U - \{i\}$. Then (5.14) may be rewritten as

$$\pi_i(1-p_i) = p_i(1-\pi_i)\, \mathsf{E}_i \left\{ \sum_{h \in s} \frac{1-p_h}{p_h - p_i + \sum_{j \in s}(1-p_j)} \right\}, \quad 1 \le i \le N, \qquad (5.15)$$

where E_i denotes the mean value via $\mathsf{R}_i(\cdot)$.

The next step is to analyze the rejective sampling $\mathsf{R}_i(\cdot)$. To put it into canonical form let us introduce probabilities p_j^*, $j \neq i$, such that (2.5) holds and $\sum_{j \neq i} p_j^* = n$. According to (2.9) we have, denoting $d_i = \sum_{j \neq i} p_j(1-p_j)$,

$$p_j^*(1-p_j) = p_j(1-p_j^*) \left[1 + (p_i/d_i) + o(d_i^{-1}) \right], \quad j \neq i, \qquad (5.16)$$

since $\sum_{j\neq i} p_j^* - \sum_{j\neq i} p_j = n - (n - p_i) = p_i$. Obviously, d_i may be replaced by d in (5.16), since $d/d_i = 1 + o(1)$ if $d \to \infty$. Thus we can modify (5.16) as follows:

$$p_j^*(1-p_j) = p_j(1-p_j^*)\left[1 + (p_i/d) + o(d^{-1})\right], \quad j \neq i. \tag{5.17}$$

Since $p_j^* \geq p_j$, $j \neq i$, (5.17) implies

$$1 \leq p_j^*/p_j \leq 1 + (p_i/d) + o(d^{-1}) \tag{5.18}$$

and

$$1 \leq (1-p_j)/(1-p_j^*) \leq 1 + (p_i/d) + o(d^{-1}). \tag{5.19}$$

Consequently,

$$\max_{j\neq i}\left|[p_j^*(1-p_j^*)/p_j(1-p_j)] - 1\right| \to 0 \quad \text{if } d \to \infty. \tag{5.20}$$

This means that $d \to \infty$ implies

$$d_{0i}^* = \sum_{j\neq i} p_j^*(1-p_j^*) \to \infty \tag{5.21}$$

uniformly in i (for brevity, the dependence on i is not indicated in the symbol p_j^*).

Now, on account of (2.15),

$$\sum_{h\in s} \frac{1-p_h}{p_h - p_i + \sum_{j\in s}(1-p_j)} = 1 - d^{-1}B_i + o(d^{-1}) \tag{5.22}$$

where

$$B_i = d\,\frac{\sum_{j\in s}(1-p_j)(p_j - p_i)}{\left[\sum_{j\in s}(1-p_j)\right]^2} \tag{5.23}$$

with

$$|B_i| \leq 4, \quad 1 \leq i \leq N. \tag{5.24}$$

Now, we infer from (5.18) through (5.21), (5.23) and Lemma 4.2, that

$$(B_i - \overline{\overline{p}} + p_i) \to 0 \tag{5.25}$$

in R_i-probability as $d \to \infty$, uniformly with respect to i. From (5.24) and (5.25) it follows that $\mathsf{E}_i(B_i) = \overline{\overline{p}} - p_i + o(1)$, if $d \to \infty$, which inserted into (5.15) gives (5.2).

Now from (5.2) it follows that

$$p_i - \pi_i = \left[(\overline{\overline{p}} - p_i)/d + o(d^{-1})\right] p_i(1-\pi_i)$$

and, consequently,

$$|(\pi_i/p_i) - 1| = (1-\pi_i)\,|(\overline{\overline{p}} - p_i)/d + o(d^{-1})|, \quad 1 \le i \le N. \tag{5.26}$$

However, (5.26) is equivalent to (5.7). Furthermore, (5.8) may be proved in a similar way, and (5.9) is an easy consequence of (5.7) and (5.8). Finally (5.10) follows from (5.9), and (5.1) follows from (5.2) and (5.7), (5.9) and (5.10). The proof is complete. □

The following theorem is basic for applications.

Theorem 5.2. *Let π_{ij} be the probability of including both the ith and the jth unit in the sample for rejective sampling of size n with probabilities p_i, $1 \le i \le N$, such that $\sum_{i=1}^N p_i = n$.*

Then

$$\pi_i\pi_j - \pi_{ij} = d_0^{-1}\,\pi_i(1-\pi_i)\,\pi_j(1-\pi_j)\,[1+o(1)], \tag{5.27}$$

where $d_0 = \sum_{i=1}^N \pi_i(1-\pi_i)$ and $o(1) \to 0$ uniformly for $d_0 \to \infty$.

Proof. First observe that π_{ij}/π_i may be regarded as the probability of inclusion of the jth unit in rejective sampling of size $n-1$ with probabilities p_j, $j \ne i$, from the population $U - \{i\}$. Corresponding canonical probabilities p_j satisfy, in accordance with (2.9),

$$p_j^+(1-p_j) = p_j(1-p_j^+)\left[1 + [(p_i - 1)/d] + o(d^{-1})\right], \quad j \ne i, \tag{5.28}$$

since $\sum_{j\ne i} p_j^+ - \sum_{j \ne i} p_j = n-1-(n-p_i) = p_i - 1$, and $\sum_{j\ne i} p_j(1-p_j)$ may be replaced by $d = \sum_{j=1}^N p_j(1-p_j)$. On the other hand we may apply (5.2) to the present sampling (i. e. put π_{ij}/π_i for π_i and p_j^+ for p_i), which yields

$$(\pi_{ij}/\pi_i)(1-p_j^+) = p_j^+[1-(\pi_{ij}/\pi_i)]\left[1 - [(\overline{\overline{p}}^+ - p_j^+)/d_+] + o(d_+^{-1})\right], \tag{5.29}$$

where

$$d_+ = \sum_{j \ne i} p_j^+(1-p_j^+).$$

Utilizing (5.28) in the same way as we did (5.2) in proving (5.7) through (5.10), we may show that $\overline{\overline{p}}^+$, p_j^+ and d_+ in the brackets in (5.29) may be replaced by $\overline{\overline{p}}$, p_j and d, which, in turn, may be replaced by $\overline{\overline{\pi}}$, π_j and d_0, in view of (5.7) through (5.10). Thus, (5.28) and (5.29) may be rewritten as follows:

$$p_j^+(1-p_j) = p_j(1-p_j^+)\left[1 + [(\pi_i - 1)/d_0] + o(d_0^{-1})\right], \quad j \ne i, \tag{5.30}$$

$$(\pi_{ij}/\pi_i)(1-p_j^+) = p_j^+[1-(\pi_{ij}/\pi_i)]\left[1-[(\overline{\overline{\pi}}-\pi_j)/d_0]+o(d_0^{-1})\right], \quad j \ne i. \tag{5.31}$$

Eliminating p_j^+ and p_j from (5.1), (5.30) and (5.31), we get

$$(\pi_{ij}/\pi_i)(1-\pi_j) = \pi_j[1-(\pi_{ij}/\pi_i)]\left[1-[(1-\pi_i)/d_0]+o(d_0^{-1})\right], \quad j \neq i. \tag{5.32}$$

However, carrying out the multiplications, (5.32) becomes

$$\pi_i\pi_j - \pi_{ij} = d_0^{-1}\pi_i\pi_j[1-(\pi_{ij}/\pi_i)]\,[1-\pi_i+o(1)], \quad j \neq i. \tag{5.33}$$

Since (5.32) also implies that

$$|\{[1-(\pi_{ij}/\pi_i)]/[1-\pi_j]\}-1| \to 0 \quad \text{if } d_0 \to \infty,$$

uniformly in $j \neq i$, (5.33) is equivalent to (5.27) provided that $\pi_i \leq \frac{1}{2}$. If $\pi_i \geq \frac{1}{2}$, it suffices to consider the complementary rejective sampling with $\mathsf{P}'(s) = \mathsf{P}(U-s)$, which has the same d_0 and probabilities of inclusion $\pi_i' = 1-\pi_i$ and $\pi_{ij}' = 1+\pi_{ij}-\pi_i-\pi_j$, so that $\pi_i(1-\pi_i) = \pi_i'(1-\pi_i')$ and $\pi_i'\pi_j' - \pi_{ij}' = \pi_i\pi_j - \pi_{ij}$. The proof is complete. □

6 THE VARIANCE OF THE SIMPLE LINEAR ESTIMATOR

We shall first assume that we know the exact values of the π_is.

Theorem 6.1. *Consider a rejective sample s of size n selected with probabilities of inclusion $\pi_1, \ldots, \pi_N$, and a population of values $y_i, \ldots, y_N$. Put*

$$\hat{Y}_0 = \sum_{i \in s} y_i/\pi_i. \tag{6.1}$$

Then

$$\mathsf{E}(\hat{Y}_0) = \sum_{i=1}^{N} y_i \tag{6.2}$$

and

$$\mathsf{var}(\hat{Y}_0) = [1+o(1)]\sum_{i=1}^{N}(y_i - R\pi_i)^2\,[(1/\pi_i)-1] \tag{6.3}$$

where

$$R = \sum_{i=1}^{N} y_i(1-\pi_i) \Bigg/ \sum_{i=1}^{N} \pi_i(1-\pi_i) \tag{6.4}$$

and $o(1) \to 0$ if $\sum_{i=1}^{N} \pi_i(1-\pi_i) \to \infty$.

Proof. The assertion (6.2) is obvious and (6.3) follows immediately from (5.27) and (1.13). □

Now assume, as in practice, that we want the π_is to be proportional to some non-negative numbers x_i such that

$$0 < nx_i < X = \sum_{i=1}^{N} x_i, \quad 1 \leq i \leq N. \tag{6.5}$$

Since $\sum_{i=1}^{N} \pi_i = n$, we should have

$$\pi_i = nx_i/X, \quad 1 \leq i \leq N. \tag{6.6}$$

However, we are only able to control the probabilities p_i, of which the π_is are rather complicated functions. None the less we have established the relations (5.9) and (5.27), from which it follows that

$$\pi_i \pi_j - \pi_{ij} = d^{-1} p_i(1-p_i)\, p_j(1-p_j)\, [1 + o(1)]. \tag{6.7}$$

So, replacing (6.6) by

$$p_i = nx_i/X, \quad 1 \leq i \leq N, \tag{6.8}$$

we still have the following modification of Theorem 6.1.

Theorem 6.2. *Consider rejective sampling of size n from a population of values $y_1, \ldots, y_N$ with probabilities $p_i, \ldots, p_N$ given by (6.8), and the estimator*

$$\hat{Y} = (X/n) \sum_{i \in s} y_i / x_i. \tag{6.9}$$

Then

$$\mathsf{F}(\hat{Y} - Y)^2 - [1 + o(1)] \sum_{i=1}^{N} (y_i - Rx_i)^2 \, [(X/nx_i) - 1], \tag{6.10}$$

where $o(1) \to 0$ if $d \to \infty$,

$$d = \sum_{i=1}^{N} (nx_i/X)\, [1 - (nx_i/X)] \tag{6.11}$$

and

$$R = \sum_{i=1}^{N} y_i [1 - (nx_i/X)] \Bigg/ \sum_{i=1}^{N} x_i [1 - (n\, x_i/X)]\,. \tag{6.12}$$

Proof. Obviously,

$$\mathsf{E}(\hat{Y} - Y)^2 = \mathsf{var}(\hat{Y}) + (\mathsf{E}\hat{Y} - Y)^2. \tag{6.13}$$

Now, in accordance with (6.7) and (6.8),

$$\operatorname{var}\hat{Y} = \frac{1}{2}\sum\sum_{i\neq j} [(y_i/x_i) - (y_j/x_j)]^2 (X^2/n^2)(\pi_i\pi_j - \pi_{ij}) \tag{6.14}$$

$$= \frac{1}{2}\sum\sum_{i\neq j} [(y_i/x_i) - (y_j/x_j)]^2 x_i x_j [1-(nx_i/X)][1-(nx_j/X)]\, d^{-1}[1+o(1)]$$

$$= [1+o(1)]\sum_{i=1}^{N}(y_i - Rx_i)^2 [(X/nx_i) - 1].$$

Furthermore, in view of (6.12) and (5.26),

$$\begin{aligned}
(\mathsf{E}\hat{Y} - Y)^2 &= \left[\sum_{i=1}^{N} y_i((\pi_i X/nx_i) - 1)\right]^2 \\
&= \left[\sum_{i=1}^{N}(y_i - Rx_i)\frac{\pi_i - nx_i/X}{nx_i/X}\right]^2 \\
&= \left[\sum_{i=1}^{N}(y_i - Rx_i)(1-\pi_i)\left(\frac{\overline{\overline{p}} - p_i}{d} + o(d^{-1})\right)\right]^2 \\
&\leq \sum_{i=1}^{N}(y_i - Rx_i)^2(\pi_i^{-1} - 1)\sum_{i=1}^{n}\pi_i(1-\pi_i)\left[\frac{\overline{\overline{p}} - p_i}{d} + o(d^{-1})\right]^2 \\
&= \sum_{i=1}^{N}(y_i - Rx_i)^2 [(X/nx_i) - 1][o(1)]
\end{aligned}$$

which shows that

$$(\mathsf{E}\hat{Y} - Y)^2/\operatorname{var}\hat{Y} \to 0.$$

Consequently, the asymptotic expression for $\operatorname{var}\hat{Y}$ (6.14) is correct for $\mathsf{E}(\hat{Y} - Y)^2$ as well. The proof is complete. □

Remark 6.1. In Section 8 we shall suggest instead of (6.8) another relation between the p_is and the x_is yielding a better approximation of (6.6). Generally, it suffices that

$$p_i[1-(nx_i/X)] = (nx_i/X)(1-p_i)[1+o(1)], \quad 1 \leq i \leq N, \tag{6.15}$$

where $o(1) \to 0$, uniformly in i, if $d \to \infty$.

7 NECESSARY AND SUFFICIENT CONDITIONS FOR ASYMPTOTIC NORMALITY

Under the conditions of Theorem 6.2 consider the estimator $\hat{Y}$ given by (6.9) and simultaneously the quantity

$$T = (X/n)\sum_{i\in s}[(y_i/x_i) - R] + XR, \tag{7.1}$$

where R is given by (6.12) and s denotes a Poisson sample with the same probabilities (6.8). From (3.11) and (6.10) it easily follows that

$$\mathsf{var}^{\mathsf{P}}(T) = \mathsf{E}^R(\hat{Y} - Y)^2\,[1 + o(1)] \quad \text{if } d \to \infty, \tag{7.2}$$

where $\mathsf{var}^{\mathsf{P}}$ and E^R refer to Poisson and rejective sampling, respectively.

There is, however, a still deeper relation between T and $\hat{Y}$, which may be explained heuristically as follows. Let s_n and s_k denote a typical rejective and Poisson sample, respectively. If, for example, $s_n \subset s_k$, then

$$T - \hat{Y} = (X/n) \sum_{i\in s_k - s_n} [(y_i/x_i) - R]. \tag{7.3}$$

Consequently, if the difference $s_k - s_n$ contains a relatively small number of units, the difference $T - \hat{Y}$ will be small, too, and may be asymptotically negligible.

Theorem 7.1. *Consider rejective sampling of size n with probabilities $p_i = nx_i/X$ from a population of values $y_1, \ldots, y_N$. Define $\hat{Y}$ by (6.9) and put*

$$B^2 = \sum_{i=1}^{N}(y_i - Rx_i)^2\,[(X/nx_i) - 1], \tag{7.4}$$

where R is given by (6.12). Moreover put

$$e = e(y_1, \ldots, y_N, x_1, \ldots, x_N) = \inf\{\varepsilon : L_\varepsilon^* \le \varepsilon\}, \tag{7.5}$$

where

$$L_\varepsilon^* = B^{-2} \sum_{i\in A_\varepsilon} (y_i - Rx_i)^2\,[(x/nx_i) - 1] \tag{7.6}$$

with

$$A_\varepsilon = \{i : |y_i - Rx_i| > \varepsilon(nx_i/X)\,B\}. \tag{7.7}$$

Then

$$e \to 0 \tag{7.8}$$

implies

$$\sup_x \left| \mathsf{P}(\hat{Y} < x) - \Phi[(x - Y)/B] \right| \to 0, \tag{7.9}$$

where $\mathsf{P}(\cdot)$ *denotes the probability referring to the rejective sampling and* Φ *denotes the normed normal distribution function.*

Proof. We know from Lemma 3.3 that the theorem holds for corresponding Poisson sampling and the statistic (7.1). Now, according to Lemma 4.3 the asymptotic distribution of T will be the same also for the distribution $\mathsf{P}_0(\cdot)$ considered there. We also may consider the distribution $\mathsf{Q}(s_k, s_n)$ from Lemma 4.2, since P_0 is its marginal distribution for s_k. However, under Q we always have either $s_k \supset s_n$ or $s_k \subset s_n$, and $s_k - s_n$ (or $s_n - s_k$) represents a rejective sample of size $k - n$ (or $n - k$) from $U - s_n$ (or s_n). Thus it suffices to show that the difference (7.3) is asymptotically negligible in the following sense:

$$B^{-2}(T - \hat{Y})^2 \sim 0, \tag{7.10}$$

which means that the left side converges to the right side in Q-probability. Furthermore, for $k > n$ (7.10) is equivalent to

$$B^{-2} \operatorname{var}^*(T - \hat{Y} \mid U - s_n, k - n) \sim 0 \tag{7.11}$$

and

$$B^{-1} \mathsf{E}^*(T - \hat{Y} \mid U - s_n, k - n) \sim 0, \tag{7.12}$$

where var^* and E^* refer to the rejective sampling R^* of size $k - n$ from $U - s_n$ given by (4.15). Leaving the case $k < n$ to the reader, we shall conclude the proof by establishing (7.11) and (7.12).

According to (1.13) and (7.3)

$$\operatorname{var}^*(T - \hat{Y} \mid U - s_n, k - n) \tag{7.13}$$

$$= \frac{1}{2}(X/n)^2 \sum_{i,j \in U - s_n}\sum [(y_i/x_i) - (y_j/x_j)]^2 (\pi_i^* \pi_j^* - \pi_{ij}^*)$$

and, furthermore,

$$\mathsf{E}^*(T - y \mid U - s_n, k - n) = X/n \sum_{i \in U - s_n} [(y_i/x_i) - R]\, \pi_i^*. \tag{7.14}$$

Denote by $\approx$ that the ratio of both sides tends to 1 in Q-probability if $d \to \infty$ and note that (4.18) entails

$$d^* = \sum_{i \in U - s_n} \pi_i^*(1 - \pi_i^*) = \operatorname{var}^* K^* \approx (k - n).$$

Since $|K-n| \to \infty$ in Q-probability, and in view of (5.1) and (5.7), we have

$$\pi_i^* \approx (k-n)\, D_n^{-1} p_i \approx (k-n)\, d^{-1}(nx_i/X) \quad \text{uniformly in } i = 1, \ldots, N, \tag{7.15}$$

where $d = \sum_{i=1}^{N}(nx_i/X)\,(1-nx_i/X)$. Furthermore, in view of (5.27),

$$\pi_i^*\, \pi_j^* - \pi_{ij}^* \approx (k-n)^{-1}\, \pi_i^*\, \pi_j^* \approx (k-n)\, d^{-2}(n/X)^2\, x_i\, x_j. \tag{7.16}$$

Now, (7.13) in connection with (7.15) and (7.16) yields

$$\begin{aligned} &\mathsf{var}^*(T-Y \mid U-s_n, k-n) \qquad (7.17)\\ &\approx \frac{1}{2}(X/n)^2\,(k-n)^{-1} \sum\sum_{i,j \in U-s_n} [(y_i/x_i)-(y_j/x_j)]^2\, \pi_i^* \pi_j^* \\ &\le (X/n)^2 \sum_{i\in U-s_n} [(y_i/x_i)-R]^2\, \pi_i^* \\ &\approx (k-n)\, d^{-1} \sum_{i \in U-s_n} (y_i - Rx_i)^2\, (X/nx_i). \end{aligned}$$

Consequently,

$$\begin{aligned} &B^{-2}\, \mathsf{var}^*(T-\hat{Y} \mid U-s_n,\, k-n) \qquad (7.18)\\ &\approx (k-n)\, d^{-1}\, B^{-2} \sum_{i\in U-s_n} (y_i - Rx_i)^2\, (X/nx_i), \end{aligned}$$

where

$$(n-k)\, d^{-1} \sim 0 \tag{7.19}$$

and the remaining factor is bounded in Q-probability, since

$$\begin{aligned} &\mathsf{E}\left\{B^{-2} \sum_{i\in U-s_n} (y_i - Rx_i)^2\, (X/nx_i)\right\} \qquad (7.20)\\ &= B^{-2} \sum_{i=1}^{N} (y_i - Rx_i)^2\, (X/nx_i)\,(1-\pi_i) \;\approx\; 1. \end{aligned}$$

Obviously (7.18) through (7.20) entails (7.11).

It remains to prove (7.12), which is somewhat more intricate. In view of (7.14) and (7.15), we have

$$\begin{aligned} &\mathsf{E}^*(T-\hat{Y} \mid U-s_n,\, k-n) \qquad (7.21)\\ &= (X/n) \sum_{i\in U-s_n} [(y_i/x_i)-R]\,(k-n)\, d^{-1}(nx_i/X)\,[1+o_i(1)] \\ &= (k-n)\, d^{-1} \sum_{i\in U-s_n} (y_i - Rx_i)\,[1+o_i(1)], \end{aligned}$$

where $o_i(1) \to 0$ uniformly in i if $d \to \infty$. Thus, denoting by E_0 the mean value referring to s_n, we have

$$\begin{aligned}
&\mathsf{E}_0 \left[\mathsf{E}^*(T - \hat{Y} \mid U - s_n,\, k-n)\right] \qquad (7.22)\\
&= (k-n)\, d^{-1} \sum_{i=1}^{N} (y_i - Rx_i)\, [1 + o_i(1)]\, (1 - \pi_i)\\
&= (k-n)\, d^{-1} \sum_{i=1}^{N} (y_i - Rx_i)\, [1 - (nx_i/X)]\, [1 + o_i(1)]\\
&= (k-n)\, d^{-1} \sum_{i=1}^{N} (y_i - Rx_i)\, [1 - (nx_i/X)]\, [o_i(1)].
\end{aligned}$$

Consequently, by the Cauchy inequality,

$$\left\{\mathsf{E}_0[\mathsf{E}^*(T - \hat{Y} \mid U - s_n,\, k-n)]\right\}^2 = (k-n)^2\, d^{-1}\, B^2[o(1)]. \qquad (7.23)$$

Since $(k-n)^2\, d^{-1}$ is bounded in Q-probability, we have

$$B^{-2} \left\{\mathsf{E}_0[\mathsf{E}(T - \hat{Y} \mid U - s_n,\, k-n)]\right\}^2 \sim 0 \qquad (7.24)$$

if $d \to \infty$.

Further, denoting the variance referring to s_n by var_0, we obtain from (7.21)

$$\begin{aligned}
&\mathsf{var}_0 \left[\mathsf{E}^*(T - \hat{Y} \mid U - s_n,\, k-n)\right] \qquad (7.25)\\
\approx\;& \frac{1}{2}(k-n)^2\, d^{-3} \sum_{i=1}^{N} \sum_{j=1}^{N} [(y_i - Rx_i)\,(1 + o_i(1))\\
&-(y_j - Rx_j)\,(1 + o_j(1))]^2 \cdot [(nx_i\, nx_j)/XX]\, [1 - (nx_i/X)]\, [1 - (nx_i/X)]\\
\gtrsim\;& (k-n)^2\, d^{-2} \sum_{i=1}^{N} (y_i - Rx_i)^2\, (nx_i/X)\, [1 - (nx_i/X)]\\
\leq\;& (k-n)^2\, d^{-2}\, B^2.
\end{aligned}$$

Consequently, also

$$B^{-2}\, \mathsf{var}_0 \left[\mathsf{E}(T - \hat{Y} \mid U - s_n,\, k-n)\right] \sim 0. \qquad (7.26)$$

However, the relations (7.26) and (7.24) are sufficient for (7.12) in the same way as (7.12) and (7.11) were sufficient for (7.10). The proof is complete. □

Remark. Condition (7.8) is also necessary.

8 NUMERICAL INVESTIGATIONS AND APPLICATION

Given some numbers x_i, $1 \le i \le N$, we generally do not know any method of performing rejective sampling such that $\pi_i = nx_i/X$ would hold exactly. However, there are satisfactory approximations. We have seen that rejective sampling with probabilities $p_i = nx_i/X$ will yield probabilities of inclusion such that

$$\pi_i/(nx_i/X) \to 1, \quad 1 \le i \le N, \tag{8.1}$$

uniformly in i, if $d \to \infty$, $d = \sum_{i=1}^N (nx_i/X)[1-(nx_i/X)]$. Convergence of $\pi_i/(nx_i/X)$ to 1 may be accelerated by replacing $p_i = nx_i/X$ by some more intricate relationship obtained from (5.1).

Before doing that, let us recall that definition (2.1) of rejective sampling is associated with the possibility of performing it as Poisson sampling with probabilities $p_1, \dots, p_N$ given the condition that the sample size is n. The alternative definition (1.5) is associated with the possibility of performing rejective sampling as sampling with replacement in which the probability of selecting the ith unit in a single draw equals α_i, $1 \le i \le N$. The parameters p_i and α_i may both be controlled by the statistician and are related by (2.3). Thus, for example, $p_i = nx_i/X$ is equivalent to

$$\alpha_i = (\lambda_1 x_i)/[1-(nx_i/X)], \quad 1 \le i \le N, \tag{8.2}$$

where λ_1 is to be chosen so that $\sum_{i=1}^N \alpha_i = 1$.

Now, from (5.1) it follows that

$$\begin{aligned} \alpha_i &= \lambda_2 p_i/(1-p_i) \doteq [(\lambda_2\pi_i)/(1-\pi_i)]\left\{1-[(\overline{\overline{\pi}}-\pi_i)/d_0]\right\}^{-1} \\ &\doteq [(\lambda_2\pi_i)/(1-\pi_i)]\left\{1+[(\overline{\overline{\pi}}-\pi_i)/d_0]\right\} \\ &= [(\lambda_3\pi_i)/(1-\pi_i)]\left\{1-[\pi_i/(d_0+\overline{\overline{\pi}})]\right\}. \end{aligned} \tag{8.3}$$

So, if we put $\pi_i = nx_i/X$ and recall (5.3) and (5.5), we get for the α_is the following approximation:

$$\alpha_i = \{\lambda x_i/[1-(nx_i/X)]\}[1-(x_i/X^*)], \quad 1 \le i \le N, \tag{8.4}$$

where

$$X^* = \sum_{i=1}^N x_i[1-(nx_i/X)] + \frac{\sum_{i=1}^N x_i^2[1-(nx_i/X)]}{\sum_{i=1}^N x_i[1-(nx_i/X)]} \tag{8.5}$$

and λ is determined from $\sum_{i=1}^N \alpha_i = 1$. The α_is obtained from (8.4) bring $\pi_i/(nx_i/X)$ close to 1 even for small n and N (see example below). The relation (8.4) may be simplified in various ways. For example, we may take

$$\alpha_i = \lambda x_i/\{1-[(n-1)x_i/X]\}, \quad 1 \le i \le N \tag{8.6}$$

or

$$\alpha_i = \lambda x_i[1-(x_i/X)]/[1-(nx_i/X)], \quad 1 \le i \le N. \tag{8.7}$$

Table 2 A comparison of the α_is with the probabilities of inclusion π_i and with approximation $\alpha_i^{\mathrm{I}}, \alpha_i^{\mathrm{II}}, \alpha_i^{\mathrm{III}}, \alpha_i^{\mathrm{IV}}$ based on formulas (8.4), (8.7), (8.6), (8.2) with $x_i = \pi_i$, respectively, for rejective sampling ($N = 10$, $n = 4$, $\alpha_i = i/55$)

	Formula for α_i^g		(8.4)	(8.7)	(8.6)	(8.2)
i	$\alpha_i = \dfrac{i}{55}$	$\dfrac{\frac{1}{4}\pi_i}{\alpha_i}$	$\dfrac{\alpha_i^{\mathrm{I}}}{\alpha_i}$	$\dfrac{\alpha_i^{\mathrm{II}}}{\alpha_i}$	$\dfrac{\alpha_i^{\mathrm{III}}}{\alpha_i}$	$\dfrac{\alpha_i^{\mathrm{IV}}}{\alpha_i}$
1	.0182	1.47	.999	.94	1.03	.85
2	.0364	1.37	.999	.95	1.03	.88
3	.0545	1.28	1.000	.96	1.04	.91
4	.0727	1.19	1.000	.98	1.03	.94
5	.0909	1.12	1.001	.99	1.03	.96
6	.1091	1.05	1.001	1.00	1.03	.99
7	.1273	.98	1.001	1.00	1.01	1.01
8	.1455	.92	1.001	1.01	.99	1.03
9	.1636	.87	.999	1.02	.98	1.04
10	.1818	.82	.998	1.02	.96	1.06

In Table 2 there are numerical values of ratios of the true α_is and the approximate α_is computed from the above formulas for $N = 10$, $n = 4$ and $x_i = \pi_i$, $1 \le i \le 10$. Denoting the probabilities of inclusion referring to α_i, $\alpha_i^{\mathrm{I}}, \alpha_i^{\mathrm{II}}$, α_i^{III}, α_i^{IV} by π_i, $\pi_i^{\mathrm{I}}, \pi_i^{\mathrm{II}}$, π_i^{III}, π_i^{IV}, respectively, we may assume that π_i^g/π_i would be even less variable than α_i^g/α_i, $g = \mathrm{I}, \mathrm{II}, \mathrm{III}, \mathrm{IV}$. This assumption is justified by (8.4).

Hence we can conclude that formula (8.4) and its simplification solve the problem P1 formulated in Section 1 for any situation which may in practice occur.

Now, as for the problem P2, let us calculate the values of $\pi_i\pi_j - \pi_{ij}$ for the same example as in Table 2 and compare them with the approximate values given by (5.27):

$$\pi_i\pi_j - \pi_{ij} \doteq \left[\sum_{i=1}^{N} \pi_i(1-\pi_i)\right]^{-1} \pi_i(1-\pi_i)\,\pi_j(1-\pi_j). \tag{8.8}$$

The results are presented in Tables 3 and 4. In Table 4 we can observe that the ratios tend to be smaller than 1, which is due to the fact that, denoting

$d = \sum \pi_i(1 - \pi_i)$,

$$\sum_{i<j}\sum (\pi_i \pi_j - \pi_{ij}) = \frac{1}{2} d$$

whereas

$$\sum_{i<j}\sum \pi_i(1 - \pi_i)\, \pi_j(1 - \pi_j)\, d^{-1} = \frac{1}{2} d - \frac{1}{2} d^{-1} \sum \pi_i^2 (1 - \pi_i)^2 < \frac{1}{2} d.$$

The relative difference is, however, of order $1/N$ and is not worth correcting for the Ns encountered in practice.

Table 3 $\pi_i \pi_j - \pi_{ij}$ for rejective sampling ($N = 10$, $n = 4$, $\alpha_i = i/55$)

	j								
i	2	3	4	5	6	7	8	9	10
1	.0066	.0087	.0101	.0111	.0116	.0119	.0120	.0119	.0117
2		.0151	.0175	.0191	.0200	.0204	.0205	.0203	.0200
3			.0228	.0247	.0258	.0263	.0263	.0260	.0255
4				.0284	.0296	.0300	.0299	.0295	.0288
5					.0318	.0321	.0319	.0314	.0306
6						.0330	.0328	.0321	.0313
7							.0328	.0321	.0313
8								.0317	.0308
9									.0301

Consequently, we may hold that problem P2 is also satisfactorily solved by formula (8.8).

Now in practice we may proceed as follows. Wanting the π_is to be proportional to some numbers x_i, we carry out rejective sampling with the α_is given (8.6), since this formula is both simple and satisfactory in normal cases. Then we use the estimate

$$\hat{Y} = (X/n) \sum_{i \in s} y_i / x_i \tag{8.9}$$

whose mean square deviation equals, approximately,

$$\mathsf{E}(\hat{Y} - Y)^2 = \sum_{i=1}^{N} (y_i - R x_i)^2 \left[(X/n x_i) - 1\right] \tag{8.10}$$

with

$$R = \frac{\sum_{i=1}^{N} y_i [1 - (n x_i / X)]}{\sum_{i=1}^{N} x_i [1 - (n x_i / X)]} \doteq Y/X. \tag{8.11}$$

Table 4 $\dfrac{\pi_i(1-\pi_i)\,\pi_j(1-\pi_j)}{(\pi_i\pi_j-\pi_{ij})\sum_{i=1}^{10}\pi_i(1-\pi_i)}$

for rejective sampling ($N = 10,\ n = 4,\ \alpha_i = i/55$)

	j								
i	2	3	4	5	6	7	8	9	10
1	1.07	1.03	.99	.97	.95	.93	.92	.91	.91
2		.99	.96	.93	.92	.91	.90	.89	.89
3			.93	.91	.90	.89	.88	.88	.88
4				.89	.88	.88	.87	.87	.88
5					.87	.87	.87	.87	.88
6						.87	.87	.88	.88
7							.88	.88	.89
8								.89	.90
9									.91

Furthermore, (8.10) may be estimated by

$$e(\hat{Y}-Y)^2 = X^2/[n(n-1)]\sum_{i\in s}[(y_i/x_i)-r]^2\,[1-(nx_i/X)], \tag{8.12}$$

where

$$r = \frac{\sum_{i\in s}(y_i/x_i)[1-(nx_i/X)]}{\sum_{i\in s}[1-(nx_i/X)]}; \tag{8.13}$$

with a slight overestimation we may put

$$r = n^{-1}\sum_{i\in s} y_i/x_i. \tag{8.14}$$

Formula (8.12) is a simplified version of the Yates–Grundy estimator

$$\begin{aligned} e(\hat{Y}-Y)^2 &= \frac{1}{2}\sum\sum_{i,j\in s}[(y_i/\pi_i)-(y_j/\pi_j)]^2\,(\pi_i\pi_j-\pi_{ij})/\pi_{ij} \\ &\doteq \frac{1}{2}(X^2/n^2)\sum\sum_{i,j\in s}[(y_i/x_i)-(y_j/x_j)]^2\,[1-(nx_i/X)] \\ &\qquad \times[1-(nx_j/X)]\,[d_0-(1-nx_i/X)\,(1-nx_j/X)]^{-1}, \end{aligned}$$

where we have used $\pi_i \doteq nx_i/X$ and the formula (8.8) for π_{ij}. Now, further approximations

$$[d_0-(1-nx_i/X)\,(1-nx_j/X)]^{-1} \doteq d_0^{-1}[n/(n-1)]$$

and (see examples following Lemma 4.2) $d_0^{-1}\sum_{i\in s}[1-(nx_i/X)] \doteq 1$ lead to formula (8.12).

If we decide to utilize the numbers x_i by ratio estimation, we still may use unequal probabilities in order to make use of the variability of expected values of $(y_i - Rx_i)^2$. Then the π_is should be proportional to some numbers z_i (usually $z_i = x_i^g$, $1 \le g \le 2$), which may be accomplished by using again (8.6) with the x_is replaced by the z_is. The estimator becomes

$$\breve{Y} = X\frac{\sum_{i\in s} y_i/z_i}{\sum_{i\in s} x_i/z_i} \tag{8.15}$$

and its mean square deviation may be approximated by

$$\begin{aligned}
\mathsf{E}(\breve{Y}-Y)^2 &= \mathsf{E}\left\{\left(\frac{Xn}{Z\sum_{i\in s} x_i/z_i}\right)^2\left((Z/n)\sum_{i\in s}\frac{y_i-(Y/X)\,x_i}{z_i}\right)^2\right\} \qquad (8.16)\\
&\doteq \mathsf{E}\left((Z/n)\sum_{i\in s}\frac{y_i-(Y/X)\,x_i}{z_i}\right)^2\\
&= \sum_{i=1}^{N}[y_i-(Y/X)\,x_i-\varphi z_i]^2\,[(Z/nz_i)-1],
\end{aligned}$$

where

$$\varphi = \frac{\sum_{i=1}^{N}(y_i-(Y/X)\,x_i)\,[1-(nz_i/Z_i)]}{\sum_{i=1} z_i[1-(nz_i/Z)]} \tag{8.17}$$

and $Z = \sum_{i=1}^{N} z_i$. Obviously, we have used (8.10) with y_i and x_i replaced by $y_i - (Y/X)\,x_i$ and z_i, respectively. Usually the approximation with $\varphi = 0$ will do as well.

For estimated mean square error we can take, admitting a slight upward bias,

$$\begin{aligned}
&e(\breve{Y}-Y)^2 \qquad (8.18)\\
&= (X/\hat{Y})^2\,[Z^2/n(n-1)]\sum_{i\in S}[(y_i/z_i)-(\hat{Y}/\hat{X})\,(x_i/z_i)]^2\,[1-(nz_i/Z)],
\end{aligned}$$

where

$$\hat{Y} = Zn^{-1}\sum_{i\in s} y_i/z_i, \quad \hat{X} = Zn^{-1}\sum_{i\in s} x_i/z_i. \tag{8.19}$$

9 SHORT–CUTS IN PERFORMING REJECTIVE SAMPLING

Most frequently rejective sampling would be performed by n successive independent (i.e. with replacement) draws of one unit with single-draw probabilities α_i given by, say, (8.6), rejecting the whole partly built up sample if any two units in it coincide (in contradistinction with successive sampling, where only the last unit is rejected, if it coincides with some previously drawn).

In carrying out the single draws, the following short-cuts may prove useful.

Method 1. We choose a number Q such that

$$Q = \max_{1 \leq i \leq N} x_i / \{1 - [(n-1)\, x_i / X]\}$$

and select a random number i in the range from 1 to N and, independently, a random number q in the range from 1 to Q. If

$$q \leq x_i / \{1 - [(n-1)\, x_i / X]\} \,, \tag{9.1}$$

we draw the unit i but reiterate the whole process otherwise.

When applying Method 1 we do not need to compute the numbers $x_i/(1-(n-1)\, x_i/X)$ except for the units selected in the course of the performance, and even for a great portion of them we are able to check (9.1) by eye.

If the sample is rejected with a high probability, then another method based on Remark 4.1 is worth considering.

Method 2. We select N independent random numbers $q_1, \ldots, q_N$ in the range from 1 to $n^{-1}X$ and include in a provisory sample every unit i such that $q_i \leq x_i$. Denote the provisory sample by s_k and suppose it contains k units. Now, if $k = n$, we accept s_k for the definite sample; if $k < n$ we select from the population $U - s_k$ a rejective sample s^0_{n-k} of size $n-k$ with single-draw probabilities

$$\alpha_i^0 = \lambda x_i \left[1 - (n-k-1)\, x_i\, d^{-1}\right]^{-1}, \quad i \in U - s_k$$

where $d = \sum_{i=1}^{N} (nx_i/X)\,[1 - (nx_i/X)]$, and put $s = s_k \cup s^0_{n-k}$; if $k > n$ we select from s_k a rejective sample s^0_{k-n} of size $k-n$ with single-draw probabilities

$$\alpha_i^0 = \lambda\,[1 - (nx_i/X)]\,\left\{1 - (k-n-1)\, d^{-1}[1-(nx_i/X)]\right\}^{-1}$$

and put $s = s_k - s^0_{k-n}$.

If N is large, the first step in Method 2 may be facilitated as follows. We select a number Q such that $Q \geq \max_{1 \leq i \leq N} x_i$ and then draw a simple random sample s_m of size m, where m is a realization of a binomial random variable with number of trials N and probability of success nQ/X (the binomial distribution may be replaced appropriately by a normal or Poisson distribution). Then we proceed with s_m in the same way as with U except that the numbers $q_1, \ldots, q_N$ are selected in the range from 1 to Q.

10 ACKNOWLEDGEMENT

I wish to thank Professor Douglas G. Chapman for his valuable suggestions concerning the structure and intelligibility of the chapter.

The work was carried out with the partial support of the National Science Foundation, Grant G–14648.

REFERENCES

Cramér H. (1946). *Mathematical Methods of Statistics.* Princeton Univ. Press, Princeton, N. J.

Goodman L. A. (1949). On the estimation of the number of classes in a population. *Ann. Math. Statist. 20*, 572–579.

Hájek J. (1959). Optimum strategy and other problems in probability sampling. *Časopis Pěst. Mat. 84*, 387–423.

Hájek J. (1960). Limiting distributions in simple random sampling from a finite population. *Publ. Math. Inst. Hungar. Acad. Sci. 5*, 361–374.

Hansen M. H. and Hurwitz W. N. (1943). On the theory of sampling from finite populations. *Ann. Math. Statist. 20*, 332–362.

Hartley H. O. and Rao J. N. K. (1962). Sampling with unequal probabilities and without replacement. *Ann. Math. Statist. 33*, 350–374.

Horwitz D. G. and Thompson D. J. (1952). A generalization of sampling without replacement from a finite universe. *J. Amer. Statist. Assoc. 47*, 663–685.

Loève M. (1960). *Probability Theory.* Van Nostrand, Princeton.

Madow W. G. (1948). On the limiting distributions of estimates based on samples from finite universes. *Ann. Math. Statist. 19*, 535–545.

Neyman J. (1934). On two different aspects of the representative method: The method of stratified sampling and the method of purposive selection. *J. Roy. Statist. Soc. 97*, 558–606.

Rao J. N. K., Hartley H. O. and Cochran W. G. (1962). On a simple procedure of unequal probability sampling without replacement. *J. Roy. Statist. Soc. Ser. B 24*, 482–491.

Roy J. and Chakraborti I. M. (1960). Estimating the mean of a finite population. *Ann. Math. Statist. 31*, 392–398.

Yates F. and Grundy P. M. (1953). Selection without replacement from within strata with probability proportional to size. *J. Roy. Statist. Soc. Ser. B 15*, 253–261.

CHAPTER 25

Extension of the Kolmogorov–Smirnov Test to Regression Alternatives

In: Bernoulli–Bayes–Laplace; Proceedings of an International Research Seminar, 1963 (J. Neyman and L. Le Cam, eds.), 45–60 Springer–Verlag, Berlin.

SUMMARY

The Kolmogorov–Smirnov test for regression alternatives is defined in Section 3, where an artificial numerical example is also given. The limiting distribution is derived in Section 4 by means of techniques developed in Section 1 (conditions for convergence in distribution in $C[0,1]$) and Section 2 (extension of a Kolmogorov inequality to the case of sampling without replacement from a finite population). The auxiliary results just mentioned also offer some interest in their own right.

1 CONVERGENCE IN DISTRIBUTION IN $C[0,1]$ OF STOCHASTIC PROCESSES

Let us have a sequence of stochastic processes $\{X_\nu(t,\omega_\nu), 0 \leq t \leq 1\}$, $1 \leq \nu < \infty$, defined on some measurable spaces $(\Omega_\nu, \mathcal{F}_\nu)$ and assume that all sample paths $X_\nu(\cdot,\omega_\nu)$, $1 \leq \nu < \infty$, $\omega_\nu \in \Omega_\nu$, are continuous functions on $[0,1]$. Moreover, we shall assume that the subvectors $[X_\nu(t_1),\ldots,X_\nu(t_n)]$, $1 \leq n < \infty$, $0 \leq t_1,\ldots,t_n \leq 1$, converge in distribution to the corresponding

---

Collected Works of Jaroslav Hájek – With Commentary
Edited by M. Hušková, R. Beran and V. Dupač
Published in 1998 by John Wiley & Sons Ltd.

subvectors $[X(t_1), \ldots, X(t_n)]$ of some process $\{X(t),\, 0 \le t \le 1\}$

$$\mathcal{L}\,[X_\nu(t_1), \ldots, X_\nu(t_n)] \longrightarrow \mathcal{L}\,[X(t_1), \ldots, X(t_n)]\,, \qquad (1.1)$$
$$1 \le n < \infty,\ 0 \le t_1, \ldots, t_n \le 1,\ \nu \to \infty.$$

We suppress here the dependence of random variables on elementary events ω_ν or ω because our interest is centered on respective n-dimensional distributions.

The implications of relation (1.1) are rather poor. We cannot even infer that there is a version of the process $X(t)$, say $X(t, \omega_0)$, such that every path $X(\cdot, \omega_0)$ again is a continuous function on $[0, 1]$ for every $\omega_0 \in \Omega_0$. To see this it suffices to take $X_\nu(t, \omega_\nu) = f_\nu(t),\ \omega_\nu \in \Omega_\nu$, where $\{f_\nu(t)\}$ is a sequence of continuous functions converging pointwise to a discontinuous function.

The existence of a version of $X(t)$ with continuous paths is not very consequential either, but it is a prerequisite for the formulation of a stronger kind of convergence of processes $X_\nu(t)$ to the process $X(t)$. Let h be a continuous functional defined on the usual Banach space $C[0, 1]$ of continuous functions $x(t)$ on $[0, 1]$. This means that for every $\varepsilon > 0$ and $x(\cdot)$ there is a $\delta > 0$ such that

$$\left\{\max_{0 \le t \le 1} |x(t) - y(t)| < \delta\right\} \Longrightarrow \{|h(x(\cdot)) - h(y(\cdot))| < \varepsilon\} \qquad (1.2)$$

for any function $y(t)$ from $C[0, 1]$. We give some examples of continuous functionals:

$$h_M[x(\cdot)] = \max_{0 \le t \le 1} x(t), \qquad (1.3)$$
$$h_\delta[x(\cdot)] = \max_{|t-s| < \delta} |x(t) - x(s)|, \quad \delta > 0, \qquad (1.4)$$
$$h_G[x(\cdot)] = \int_0^1 x(t)\, \mathrm{d}G(t), \qquad (1.5)$$

where $G(t)$ is any function of finite variation. Obviously, h_G is a linear functional, while h_M and h_δ are nonlinear.

Definition 1.1. A sequence of processes $\{X_\nu(t),\, 0 \le t \le 1\}$, $1 \le \nu < \infty$, with continuous paths everywhere, will be said to converge in distribution in $C[0, 1]$ to a process $\{X(t),\, 0 \le t \le 1\}$, if there is a version of $X(t)$, say $X(t, \omega_0)$, such that the paths $X(\cdot, \omega_0)$ are continuous everywhere, and if for every continuous functional $h = h[x(\cdot)]$ on $C[0, 1]$, $h[X_\nu(\cdot)]$ converges in distribution to $h[X(\cdot)]$:

$$\mathcal{L}\,\{h[X_\nu(\cdot)]\} \longrightarrow \mathcal{L}\,\{h[X(\cdot)]\}\,, \quad \nu \to \infty. \qquad (1.6)$$

Remark 1.1. If the paths $X(\cdot)$ are continuous everywhere and the functional h is continuous, then $h[X(\cdot)]$ is a random variable (measurable function).

The question of convergence in distribution of stochastic processes in connection with statistical problems was first raised by Doob (1949). His conjecture was then proved by Donsker (1952). First general results in terms of weak convergence of measures were obtained by Prohorov (1956) who utilized earlier results by Alexandrov (1941 – 1943). The matter was further developed by Le Cam (1957), Varadarajan (1958), Driml (1959), Prohorov (1961), Bartoszynski (1962) and others.

In this paper a somewhat simpler formulation of necessary and sufficient conditions for convergence in distribution in $C[0,1]$ are given, and the proof is deliberately self-sufficient (and therefore rather long), because I believe that some reader with a predominantly statistical background might find it convenient.

Theorem 1.1. *Assume that* (1.1) *holds true. Then* $\{X_\nu(t),\, 0 \le t \le 1\}$ *converge to* $\{X(t),\, 0 \le t \le 1\}$ *in distribution in* $C[0,1]$ *if and only if*

$$\lim_{\delta\to 0} \limsup_{\nu\to\infty} \mathsf{P}\left(\max_{|t-s|<\delta} |X_\nu(t) - X_\nu(s)| > \varepsilon\right) = 0 \quad \text{for every } \varepsilon > 0. \tag{1.7}$$

Condition (1.7) *is satisfied if*

$$\lim_{\delta\to 0} \frac{1}{\delta} \limsup_{\nu\to\infty} \max_{0\le s\le 1-\delta} \mathsf{P}\left(\max_{s\le t\le s+\delta} |X_\nu(t) - X_\nu(s)| \ge \varepsilon\right) = 0, \quad \varepsilon > 0, \tag{1.8}$$

where the max within the parentheses refers to t only, while s is assumed fixed.

If the paths of the process $X_\nu(t)$ are linear on intervals $[k/N_\nu, (k+1)/N_\nu]$, $k = 0, \ldots, N_{\nu-1}$, $N_\nu \to \infty$ for $\nu \to \infty$, then (1.7) *is implied by*

$$\lim_{\delta\to 0} \frac{1}{\delta} \limsup_{\nu\to\infty} \max_{0\le k\le N_\nu - m_{\nu\delta}} \mathsf{P}\left(\max_{k\le j<k+m_{\nu\delta}} \left| X_\nu\left(\frac{k}{N_\nu}\right) - X_\nu\left(\frac{j}{N_\nu}\right)\right| > \varepsilon\right) = 0, \quad \varepsilon > 0, \tag{1.9}$$

where the $m_{\nu\delta}$s are integers such that

$$\lim_{\nu\to\infty} \frac{m_{\nu\delta}}{N_\nu} = \delta, \quad \delta > 0. \tag{1.10}$$

Proof. *Necessity of* (1.7). Relation (1.6) must hold for every functional h_δ defined by (1.4). Thus, for every $\varepsilon > 0$

$$\limsup_{\nu\to\infty} \mathsf{P}\left\{h_\delta[X_\nu(\cdot)] > \varepsilon\right\} \le \mathsf{P}\left\{h_\delta[X(\cdot)] > \frac{1}{2}\varepsilon\right\}. \tag{1.11}$$

On the other hand, as all continuous functions on $[0,1]$ are equicontinuous, $h_\delta[X(\cdot)] \to 0$ for $\delta \to 0$ everywhere in $C[0,1]$, so that

$$\lim_{\delta \to 0} \mathsf{P}\left\{h_\delta[X(\cdot)] > \frac{1}{2}\varepsilon\right\} = 0, \quad \varepsilon > 0. \tag{1.12}$$

Now (1.11) and (1.12), in view of (1.4), yield (1.7) immediately.

Sufficiency of (1.7). We shall first prove the existence of a continuous path version of the limiting process $X(t)$. In view of (1.1), we know the finite-dimensional distributions of $X(t)$. These are obviously consistent, and hence we may construct the process in the space of all functions on $[0,1]$, say $R^{[0,1]}$, by the Kolmogorov procedure. We obtain a probability distribution, say, $\mathsf{P}(\cdot)$, on Borel field $\mathcal{B}$ of subsets of $R^{[0,1]}$ generated by subsets $A_{tc} = \{x(\cdot) : x(t) < c\}$, $t \in [0,1]$, $-\infty < c < \infty$, such that the distribution law of coordinate random variables $\{X(t_1), \ldots, X(t_n)\}$ will coincide with the law on the right side of (1.1). Now denote by C, $C \subset R^{[0,1]}$, the subset of continuous functions. If for every two subsets B_1 and B_2 from $\mathcal{B}$

$$[C \cap B_1 = C \cap B_2] \Longrightarrow [\mathsf{P}(B_1) = \mathsf{P}(B_2)], \tag{1.13}$$

then the probability measure $\mathsf{P}(\cdot)$ defined on $\mathcal{B}$ may be carried over to the Borel field $\mathcal{A}$, consisting of subsets $C \cap B$, $B \in \mathcal{B}$. Denoting the corresponding measure on $\mathcal{A}$ by $\mathsf{P}_C(\cdot)$, we have

$$\mathsf{P}_C(A) = \mathsf{P}(B), \quad A \in \mathcal{A},\ B \in \mathcal{B},\ A = C \cap B, \tag{1.14}$$

where B is any subset such that $A = C \cap B$. In view of (1.13) the value of $\mathsf{P}_C(A)$ does not depend on the choice of B. Obviously the coordinate random variables $\{X(t_1), \ldots, X(t_n)\}$ would have the same distribution in $(C, \mathcal{A})$ under $\mathsf{P}_C(\cdot)$ as in $(R^{[0,1]}, \mathcal{B})$ under $\mathsf{P}(\cdot)$. So, if we prove (1.13) we can construct a version of the process $X(t)$ possessing continuous paths everywhere.

Doob (1937) showed that (1.13) holds true if and only if (the proof is quite simple)

$$[B \supset C,\ B \in \mathcal{B}] \Longrightarrow [\mathsf{P}(B) = 1]. \tag{1.15}$$

(Recall that C does not belong to $\mathcal{B}$.) We know that to each event B, $B \in \mathcal{B}$, there corresponds a countable subset of $[0,1]$, say I_B, $I_B \subset [0,1]$, such that

$$[x(\cdot) \in B,\ x(t) = y(t),\ t \in I_B] \Longrightarrow [y(\cdot) \in B]. \tag{1.16}$$

[See, for example, Špaček (1955).] Without loss of generality, we may assume that I_B is dense in $[0,1]$. Now if B contains C, then it also contains all functions $x(\cdot)$ which are uniformly continuous on the set I_B, i. e. for each $\varepsilon > 0$ there is a $\delta > 0$ such that

$$[|t-s| < \delta,\ t \in I_B,\ s \in I_B] \Longrightarrow [|x(t) - x(s)| < \varepsilon]. \tag{1.17}$$

Denote the latter set by $C(I_B)$. So

$$[B \supset C] \Longrightarrow [B \supset C(I_B)]. \tag{1.18}$$

However, the set $C(I_B)$ is measurable, so that it suffices to show that

$$\mathsf{P}[C(I_B)] = 1. \tag{1.19}$$

Let us take an increasing sequence of finite subsets $I_1 \subset I_2 \subset \ldots$ of $[0, 1]$ such that

$$I_B = \bigcup_1^\infty I_n. \tag{1.20}$$

Now, in view of (1.1),

$$\begin{aligned}
&\mathsf{P}\left[\max_{\substack{|t-s|<\delta \\ t,\, s \in I_n}} |X(t) - X(s)| > \varepsilon\right] \qquad (1.21)\\
\leq \quad &\limsup_{\nu\to\infty} \mathsf{P}\left[\max_{\substack{|t-s|<\delta \\ t,\, s \in I_n}} |X_\nu(t) - X_\nu(s)| > \frac{1}{2}\varepsilon\right]\\
\leq \quad &\limsup_{\nu\to\infty} \mathsf{P}\left[\max_{|t-s|<\delta} |X_\nu(t) - X_\nu(s)| > \frac{1}{2}\varepsilon\right].
\end{aligned}$$

Consequently,

$$\begin{aligned}
&\mathsf{P}\left[\max_{\substack{|t-s|<\delta \\ t,\, s \in I_B}} |X(t) - X(s)| > \varepsilon\right] \qquad (1.22)\\
= \quad &\lim_{n\to\infty} \mathsf{P}\left[\max_{\substack{|t-s|<\delta \\ t,\, s \in I_B}} |X(t) - X(s)| > \varepsilon\right]\\
\leq \quad &\limsup_{\nu\to\infty} \mathsf{P}\left[\max_{|t-s|<\delta} |X_\nu(t) - X_\nu(s)| > \frac{1}{2}\varepsilon\right].
\end{aligned}$$

Now, combining (1.22) and (1.7), we obtain

$$\lim_{\delta\to\infty} \mathsf{P}\left[\max_{\substack{|t-s|<\delta \\ t,\, s \in I_B}} |X(t) - X(s)| > \varepsilon\right] = 0, \quad \varepsilon > 0. \tag{1.23}$$

This is, however, equivalent to (1.19). So we have proved the existence of a continuous-path version of the limiting process $X(t)$.

Now, we have to show that (1.6) holds true for any continuous functional h, provided the paths of $X(t)$ are continuous. Let $X^{(k)}(t)$ be a process derived from the process $X(t)$ as follows:

$$X^{(k)}(t) = X\left(\frac{i}{2^k}\right) + 2^k\left(t - \frac{i}{2^k}\right)\left[X\left(\frac{i+1}{2^k}\right) - X\left(\frac{i}{2^k}\right)\right] \quad \text{for } \frac{i}{2^k} \le t \le \frac{i+1}{2^k}. \tag{1.24}$$

Obviously, within the intervals $[i/2^k, (i+1)/2^k]$, we have that $X^{(k)}(t)$ is linear.

Let the processes $X_\nu^{(k)}(t)$ be derived from the processes $X_\nu(t)$ in the very same way. Now, from (1.24) and (1.7) it follows that

$$\lim_{k\to\infty} \mathsf{P}\left[\max_{0\le t\le 1} |X^{(k)}(t) - X(t)| > \varepsilon\right] = 0, \quad \varepsilon > 0, \tag{1.25}$$

$$\lim_{k\to\infty} \sup_\nu \mathsf{P}\left[\max_{0\le t\le 1} |X_\nu^{(k)}(t) - X_\nu(t)| > \varepsilon\right] = 0, \quad \varepsilon > 0. \tag{1.26}$$

Now, since the functional h is continuous, (1.25) and (1.26) imply

$$\lim_{k\to\infty} \mathsf{P}\left\{|h[X^{(k)}(\cdot)] - h[X(\cdot)]| > \varepsilon\right\} = 0, \quad \varepsilon > 0, \tag{1.27}$$

$$\lim_{k\to\infty} \sup_\nu \mathsf{P}\left\{|h[X_\nu^{(k)}(\cdot)] - h[X_\nu(\cdot)]| > \varepsilon\right\} = 0, \quad \varepsilon > 0. \tag{1.28}$$

Now, note that h applied to $X^{(k)}(t)$ and $X_\nu^{(k)}(t)$ may be considered as a continuous functional of the vectors $\{X(i/2^k), i = 0, 1, \ldots, 2^k\}$ and $\{X_\nu(i/2^k), i = 0, 1, \ldots, 2^k\}$. Bearing this in mind, we conclude from (1.1) that

$$\mathcal{L}\left\{h[X_\nu^{(k)}(\cdot)]\right\} \longrightarrow \mathcal{L}\left\{h[X^{(k)}(\cdot)]\right\}, \quad \nu \to \infty. \tag{1.29}$$

Suppose that y is a continuity point of the distribution of $h[X(\cdot)]$. Then for every $\varepsilon > 0$ there exists a $\delta > 0$ such that

$$\begin{aligned} \mathsf{P}\{h[X(\cdot)] < y + \delta\} - \tfrac{1}{4}\varepsilon &\le \mathsf{P}\{h[X(\cdot)] \le y\} \\ &\le \mathsf{P}\{h[X(\cdot)] < y - \delta\} + \tfrac{1}{4}\varepsilon. \end{aligned} \tag{1.30}$$

Furthermore, in view of (1.27) and (1.28), there exists a k_0 such that

$$\mathsf{P}\{h[X(\cdot)] < y + \delta\} \ge \mathsf{P}\left\{h[X^{(k_0)}(\cdot)] < y + \tfrac{2}{3}\delta\right\} - \tfrac{1}{4}\varepsilon, \tag{1.31}$$

$$\mathsf{P}\{h[X(\cdot)] < y - \delta\} \le \mathsf{P}\left\{h[X^{(k_0)}(\cdot)] < y - \tfrac{2}{3}\delta\right\} + \tfrac{1}{4}\varepsilon, \tag{1.32}$$

and, uniformly in ν,

$$\mathsf{P}\left\{h[X_\nu^{(k_0)}(\cdot)] < y + \tfrac{1}{3}\delta\right\} \ge \mathsf{P}\{h[X_\nu(\cdot)] \le y\} - \tfrac{1}{4}\varepsilon, \tag{1.33}$$

$$\mathsf{P}\left\{h[X_\nu^{(k_0)}(\cdot)] < y - \tfrac{1}{3}\delta\right\} \le \mathsf{P}\{h[X_\nu(\cdot)] \le y\} + \tfrac{1}{4}\varepsilon. \tag{1.34}$$

Now, (1.29) implies that there exists ν_0 such that

$$\mathsf{P}\left\{h[X_\nu^{(k_0)}(\cdot)] < y + \frac{1}{3}\delta\right\} \leq \mathsf{P}\left\{h[X^{(k_0)}(\cdot)] < y + \frac{2}{3}\delta\right\} + \frac{1}{4}\varepsilon, \quad (1.35)$$

$$\mathsf{P}\left\{h[X_\nu^{(k_0)}(\cdot)] < y - \frac{1}{3}\delta\right\} \geq \mathsf{P}\left\{h[X^{(k_0)}(\cdot)] < y - \frac{2}{3}\delta\right\} - \frac{1}{4}\varepsilon, \quad (1.36)$$

$$\nu \geq \nu_0.$$

Upon combining (1.30) through (1.36) we can see that

$$\begin{aligned} \mathsf{P}\{h[X_\nu(\cdot)] \leq y\} - \varepsilon &\leq \mathsf{P}\{h[X(\cdot)] \leq y\} \\ &\leq \mathsf{P}\{h[X_\nu(\cdot)] \leq y\} + \varepsilon, \quad \nu \geq \nu_0. \end{aligned} \quad (1.37)$$

However, (1.37) is equivalent to

$$\lim_{\nu \to \infty} \mathsf{P}\{h[X_\nu(\cdot)] \leq y\} = \mathsf{P}\{h[X(\cdot)] \leq y\} \quad (1.38)$$

at each continuity point y, which, in turn, is equivalent to (1.6).

Sufficiency of (1.8) *and* (1.9). For a given $\delta > 0$, divide the interval into n intervals $[s_i, s_{i+1}]$, $1/\delta \leq n < 1/\delta + 1$, $s_i = i\delta$, $i = 0, 1, \ldots, n-1$, $s_n = 1$. We can easily see that for any function $x(t)$

$$\begin{aligned} &\left[\max_{|t-s|<\delta} |x(t) - x(s)| > \varepsilon\right] \\ \Longrightarrow &\left[\max_{s_i \leq s \leq s_{i+1}} |x(s) - x(s_i)| > \frac{1}{3}\varepsilon \text{ for at least one } i = 0, \ldots, n-1\right]. \end{aligned} \quad (1.39)$$

Consequently,

$$\begin{aligned} &\mathsf{P}\left[\max_{|t-s|<\delta} |X_\nu(t) - X_\nu(s)| > \varepsilon\right] \\ &\leq \sum_{i=0}^{n-1} \mathsf{P}\left[\max_{s_i \leq s \leq s_{i+1}} |X_\nu(s) - X_\nu(s_i)| > \frac{1}{3}\varepsilon\right] \\ &\leq \left(\frac{1}{\delta} + 1\right) \max_{0 \leq s \leq 1-\delta} \mathsf{P}\left[\max_{s \leq t \leq s+\delta} |X_\nu(t) - X_\nu(s)| > \frac{1}{3}\varepsilon\right]. \end{aligned} \quad (1.40)$$

Now it is obvious that (1.8) implies (1.7). The proof of sufficiency of (1.9) is then immediate.

The proof is complete. □

Remark 1.2. Consider the sequence of probability measures $\{\mathsf{P}_\nu(\cdot)\}$ induced in $(C, \mathcal{A})$ by processes $\{X_\nu(t), 0 \leq t \leq 1\}$. It is easy to show that $\{\mathsf{P}_\nu(\cdot)\}$ is relatively compact (every subsequence contains weakly convergent subsubsequences) under conditions (1.1) and (1.7). According to Prohorov (1956) it amounts to showing that for each $\varepsilon > 0$ there is a compact subset K_ε of $C[0,1]$ such that

$$\inf_\nu P_\nu(K_\varepsilon) > 1 - \varepsilon. \tag{1.41}$$

In fact, as $X_\nu(0)$ converges in distribution to $x(0)$, there is a constant M_ε such that

$$\inf_\nu \mathsf{P}\left[|X_\nu(0)| < M_\varepsilon\right] > 1 - \frac{1}{2}\varepsilon. \tag{1.42}$$

Furthermore, (1.7) is equivalent to

$$\lim_{\delta\to 0}\sup \sup_\nu \mathsf{P}\left[\max_{|t-s|<\delta} |X_\nu(t) - X_\nu(s)| > \varepsilon\right] = 0. \tag{1.43}$$

Thus, we may choose a sequence $\delta_n \to 0$ such that

$$\sup_\nu \mathsf{P}\left[\max_{|t-s|<\delta_n} |X_\nu(t) - X_\nu(s)| > \frac{1}{n}\right] < \left(\frac{1}{2}\right)^{n+1} \varepsilon. \tag{1.44}$$

Now, it suffices to put

$$K_\varepsilon = \{x(\cdot) : |x(0)| < M_\varepsilon\} \cap \bigcap_{n=0}^{\infty} \left\{x(\cdot) : \max_{|t-s|<\delta_n} |x(t) - x(s)| \le \frac{1}{n}\right\}. \tag{1.45}$$

In this way, utilizing the results of Prohorov (1956), we could shorten considerably the 'sufficiency' part of the proof.

Remark 1.3. Bartoszynski (1962) proved that a process $\{X(t),\ 0 \le t \le 1\}$ may be realized in $(C, \mathcal{A})$ if and only if for every $\varepsilon > 0$ there is a function $\Psi_\varepsilon(\delta)$ such that $\Psi_\varepsilon(\delta) \downarrow 0$ as $\delta \downarrow 0$ and for every finite subset $I = \{t_1, \dots, t_n\}$ of $[0,1]$

$$\mathsf{P}\left[\max_{\substack{|t_i-t_j|<\delta \\ t_i, t_j \in I}} |X(t_i) - X(t_j)| > \varepsilon\right] \le \Psi_\varepsilon(\delta). \tag{1.46}$$

In our case we may simply put

$$\Psi_\varepsilon(\delta) = \limsup_{\nu\to\infty} \mathsf{P}\left[\max_{|t-s|<\delta} |X_\nu(t) - X_\nu(s)| > \frac{1}{2}\varepsilon\right]. \tag{1.47}$$

Then $\Psi_\varepsilon(\delta) \downarrow 0$ as $\delta \downarrow 0$ is ensured by (1.7), and by (1.1):

$$\mathsf{P}\left[\max_{\substack{|t_i-t_j|<\delta \\ t_i, t_j \in I}} |X(t_i) - X(t_j)| > \varepsilon\right] \tag{1.48}$$

$$\leq \limsup_{\nu} \mathsf{P}\left[\max_{\substack{|t_i - t_j| < \delta \\ t_i, t_j \in I}} |X_\nu(t_i) - X_\nu(t_j)| > \frac{1}{2}\varepsilon\right]$$
$$\leq \limsup_{\nu} \mathsf{P}\left[\max_{|t-s|<\delta} |X_\nu(t) - X_\nu(s)| > \frac{1}{2}\varepsilon\right]$$
$$= \Psi_\varepsilon(\delta).$$

Thus we could shorten the proof of existence of the $C[0,1]$ version of $X(t)$ this way also.

Remark 1.4. If we guess that the limiting process $x(t)$ is Gaussian with a certain mean value function $m(t)$ and covariance kernel $K(t,s)$, then (1.1) is implied by

$$\mathcal{L}\left[\sum_{i=1}^{n} \lambda_i\, x_\nu(t_i)\right] \longrightarrow \mathcal{L}\left[\sum_{i=1}^{n} \lambda_i\, x(t_i)\right] \tag{1.49}$$
$$= N\left(\sum_{i=1}^{n} \lambda_i\, m(t_i), \sum_{i=1}^{n}\sum_{j=1}^{n} \lambda_i\, \lambda_j\, K(t_i, t_j)\right),$$
$$1 \leq n < \infty,\ -\infty < \lambda_1, \ldots, \lambda_n < \infty,$$

where $N(a, \sigma^2)$ denotes the normal distribution with mean value a and variance σ^2.

Observe that $\sum_{i=1}^{n} \lambda_i\, x(t_i)$ represents a special kind of linear functional in $C[0,1]$. If the convergence in distribution for these linear functionals is supplemented by the convergence of the functional $h_\delta[x(\cdot)]$ (see (1.4)), which amounts to the condition (1.7), then the convergence for all other continuous functionals follows.

2 AN EXTENSION OF KOLMOGOROV'S INEQUALITY

Let us have a finite population consisting of N values $\{c_1, c_2, \ldots, c_N\}$. Assume that

$$\sum_{i=1}^{N} c_i = 0, \tag{2.1}$$

and draw a sequence of n $(n < N)$ values $Y_1, \ldots, Y_n$ by simple random sampling without replacement. From (2.1) it follows that

$$\mathsf{E}Y_i = 0, \quad i = 1, \ldots, n. \tag{2.2}$$

Now, if random variables $(Y_1, \ldots, Y_k)$, $k < n$, are fixed, then $(Y_{k+1}, \ldots, Y_n)$ represents again a simple random sample of size $n - k$ from the population $\{c'_1, \ldots, c'_{N-k}\}$, where c'_i are identical with values c_i excepting those selected as $Y_1, \ldots, Y_k$, $Y_i = c_{r_i}$, $i = 1, \ldots, k$. Consequently,

$$\begin{aligned} &\mathsf{E}\,(Y_{k+1} + \cdots + Y_n | Y_1, \ldots, Y_n) \\ &= \frac{n-k}{N-k} \sum_{i=1}^{N-k} c'_i = \frac{n-k}{N-k} \left[\sum_{i=1}^{N} c_i - \sum_{j=1}^{k} Y_j \right] = -\frac{n-k}{N-k}(Y_1 + \cdots + Y_k), \end{aligned} \tag{2.3}$$

in view of (2.1).

Lemma 2.1. *For any $r \geq 1$ and $\varepsilon > 0$ it follows that*

$$\mathsf{P}\left(\max_{1 \leq k \leq n} |Y_1 + \cdots + Y_k| > \varepsilon \right) < \frac{\mathsf{E}|Y_1 + \cdots + Y_n|^r}{\varepsilon^r \left(1 - \frac{n}{N}\right)^r}. \tag{2.4}$$

Proof. Introduce disjoint events $A_1, \ldots, A_n$ defined as follows:

$$A_k = \left\{ (Y_1, \ldots, Y_n) : \max_{1 \leq j \leq k-1} |Y_1 + \cdots + Y_j| \leq \varepsilon,\ |Y_1 + \cdots + Y_k| > \varepsilon \right\}. \tag{2.5}$$

Then

$$\mathsf{E}|Y_1 + \cdots + Y_n|^r = \int |Y_1 + \cdots + Y_n|^r \,\mathrm{d}P \geq \sum_{k=1}^{n} \int_{A_k} |Y_1 + \cdots + Y_n|^r \,\mathrm{d}P. \tag{2.6}$$

Now, since A_k depends only on $Y_1, \ldots, Y_k$,

$$\int_{A_k} |Y_1 + \cdots + Y_n|^r \,\mathrm{d}P = \int_{A_k} \mathsf{E}\left\{ |Y_1 + \cdots + Y_n|^r \,|Y_1, \ldots, Y_k \right\} \mathrm{d}P,\ \ 1 \leq k \leq n. \tag{2.7}$$

As for $r \geq 1$ the function $|x|^r$ is convex in x, upon applying the Jensen inequality we get

$$\begin{aligned} &\mathsf{E}\left\{ |Y_1 + \cdots + Y_n|^r \,|Y_1, \ldots, Y_k \right\} \\ &\geq |\mathsf{E}\left\{ Y_1 + \cdots + Y_n | Y_1, \ldots, Y_k \right\}|^r \\ &= |Y_1 + \cdots + Y_k + \mathsf{E}\{Y_{k+1} + \cdots + Y_n | Y_1, \ldots, Y_k\}|^r . \end{aligned}$$

Recalling (2.3), we conclude that

$$\begin{aligned} &\mathsf{E}\left\{ |Y_1 + \cdots + Y_n|^r \,|Y_1, \ldots, Y_k \right\} \\ &\geq \left| Y_1 + \cdots + Y_k - \frac{n-k}{N-k}(Y_1 + \cdots + Y_k) \right|^r \\ &\geq \left(1 - \frac{n}{N}\right)^r |Y_1 + \cdots + Y_k|^r, \quad 1 \leq k \leq n. \end{aligned} \tag{2.8}$$

Now, combining (2.6), (2.7) and (2.8), we obtain

$$\mathsf{E}|Y_1+\cdots+Y_n|^r \geq \left(1-\frac{n}{N}\right)^r \sum_{i=1}^{n} \int_{A_k} |Y_1+\cdots+Y_k|^r \,\mathrm{d}P. \qquad (2.9)$$

However, according to (2.5), $|Y_1+\cdots+Y_k|>\varepsilon$ on A_k, so that (2.9) implies

$$\begin{aligned}\mathsf{E}|Y_1+\cdots+Y_n|^r &\geq \left(1-\frac{n}{N}\right)^r \varepsilon^r \sum_{i=1}^{n} \int_{A_k} \mathrm{d}P \qquad (2.10)\\ &= \left(1-\frac{n}{N}\right)^r \varepsilon^r\, \mathsf{P}\left(\bigcup_{i=1}^{n} A_k\right)\\ &= \left(1-\frac{n}{N}\right)^r \varepsilon^r\, \mathsf{P}\left(\max_{1\leq k\leq n} |Y_1+\cdots+Y_k|>\varepsilon\right).\end{aligned}$$

Noting that (2.10) is equivalent to (2.4), our proof is complete. □

Case $r=2$. Since for simple random sampling without replacement

$$\mathsf{E}(Y_1+\cdots+Y_n)^2 = n\left(1-\frac{n}{N}\right)\frac{1}{N-1}\sum_{i=1}^{N} c_i^2, \qquad (2.11)$$

the inequality (2.4) for $r=2$ yields

$$\mathsf{P}\left(\max|Y_1+\cdots+Y_k|>\varepsilon\right) < \frac{n\sum_{i=1}^{N} c_i^2}{\varepsilon^2(N-1)\left(1-(n/N)\right)} \qquad (2.12)$$

provided (2.1) holds true.

Case $r=4$. We have (see Isserlis (1931))

$$\begin{aligned}&\mathsf{E}(Y_1+\cdots+Y_n)^4 \qquad (2.13)\\ &= \frac{n(N-n)}{N(N-1)(N-2)(N-3)}\left\{[N(N+1)-6n(N-n)]\sum_{i=1}^{N} c_i^4\right.\\ &\qquad \left. +3(N-n-1)(n-1)\left[\sum_{i=1}^{N} c_i^2\right]^2\right\}.\end{aligned}$$

Applying (2.13) to (2.4) we obtain

$$\begin{aligned}&\mathsf{P}\left(\max_{1\leq k\leq n}|Y_1+\cdots+Y_k|>\varepsilon\right) \qquad (2.14)\\ &\leq \left\{\frac{n}{N}\left(1-\frac{n}{N}\right)\sum_{i=1}^{N} c_i^4 + 3\left[\frac{n}{N}\left(1-\frac{n}{N}\right)\sum_{i=1}^{N} c_i^2\right]^2\right\}\frac{1+o(1)}{\varepsilon^4\left(1-(n/N)\right)^4},\end{aligned}$$

($o(1)$ refers to $n\to\infty$, $N-n\to\infty$).

Formula (2.14) will be useful in the sequel.

3 KOLMOGOROV–SMIRNOV TYPE TEST FOR REGRESSION ALTERNATIVES

Let the observations $X_1, \ldots, X_N$ be decomposed as

$$X_i = \alpha + \beta\, c_i + E_i, \quad i = 1, \ldots, N, \tag{3.1}$$

where α and β are unknown parameters, $c_1, \ldots, c_N$ are some known (or approximately known) constants and $E_1, \ldots, E_N$ are independent random variables with common but unknown continuous distribution.

We suggest the following statistic for testing the hypothesis $\beta = 0$ against the alternative $\beta > 0$.

First, arrange the observations according to ascending magnitude,

$$X_{D_1} < X_{D_2} < \cdots < X_{D_N}. \tag{3.2}$$

Then form the corresponding sequence of the c_is,

$$c_{D_1}, c_{D_2}, \ldots, c_{D_N} \tag{3.3}$$

and compute the statistic

$$K = \frac{\max_{1 \le k \le N}(k\bar{c} - c_{D_1} - \cdots - c_{D_k})}{\left[\sum_{i=1}^{N}(c_i - \bar{c})^2\right]^{\frac{1}{2}}}, \tag{3.4}$$

where

$$\bar{c} = \frac{1}{N}\sum_{i=1}^{N} c_i. \tag{3.5}$$

In Section 4 we shall show that the limiting distribution of the statistic K coincides with the limiting distribution of the well-known Kolmogorov–Smirnov statistic for the two-sample problem, i.e. that

$$\mathsf{P}(K > \lambda) \to e^{-2\lambda^2} \tag{3.6}$$

in large samples under conditions of Corollary 4.1.

Example. Let us have model

$$X_i = \alpha + \beta i + E_i, \quad i = 1, \ldots, 9, \tag{3.7}$$

yielding the following observations:

i	1	2	3	4	5	6	7	8	9
x_i	124	131	134	127	128	140	136	149	137

Rearranging the observations, we get

x_{D_i}	124	127	128	131	134	136	137	140	149
c_{D_i}	1	4	5	2	3	7	9	6	8

Now $\overline{c} = 5$ and the successive sum of c_{D_i}s are as follows:

k	1	2	3	4	5	6	7	8	9
$5k - c_{D_1} - \cdots - c_{D_k}$	4	5	5	8	10	8	4	3	0

On the other hand,

$$\sum_{i=1}^{9}(c_i - \overline{c})^2 = 2\left(1^2 + 2^2 + 3^2 + 4^2\right) = 60$$

so that

$$K = \frac{10}{\sqrt{60}} = 1.29.$$

If we correct for discontinuity, then

$$K^* = \frac{9.5}{\sqrt{60}} = 1.23.$$

The critical values for $\alpha = 0.05$ is $K_{0.05} = 1.22$, so that the result is significant on this level.

In conclusion let us show that for

$$\begin{aligned} c_i &= 0 \quad \text{for } i = 1, \ldots, n, \\ &= 1 \quad \text{for } i = n+1, \ldots, N \end{aligned} \tag{3.8}$$

the K-test coincides with the usual Kolmogorov–Smirnov tests for two samples of sizes n and $N - n$, respectively. Actually for the c_is given by (3.8), we have

$$\overline{c} = \frac{N-n}{N} \tag{3.9}$$

and

$$\sum_{i=1}^{N}(c_i - \overline{c})^2 = \frac{n(N-n)}{N}. \tag{3.10}$$

Moreover,

$$k\bar{c} - c_{D_1} + \cdots + c_{D_k} = (N-n)\left[S_N(X_{D_k}) - S_{N-n}(X_{D_k})\right],$$

where $S_N(x)$ and $S_{N-n}(x)$ denote the empirical distribution functions corresponding to the total sample and the second sample (of size $N-n$), respectively. As, furthermore,

$$NS_N(x) = nS_n(x) + (N-n)\,S_{N-n}(x),$$

where $S_n(x)$ corresponds to the first sample, we also have

$$k\bar{c} - (c_{D_1} - \cdots - c_{D_k}) = \frac{n(N-n)}{N}\left[S_n(X_{D_k}) - S_{N-n}(X_{D_k})\right].$$

Consequently, in view of (3.4),

$$\begin{aligned} K &= \left|\frac{n(N-n)}{N}\right|^{\frac{1}{2}} \max_{1\le k\le N} |S_n(X_{D_k}) - S_{N-n}(X_{D_k})| \\ &= \left|\frac{n(N-n)}{N}\right|^{\frac{1}{2}} \max_{-\infty<x<\infty} |S_n(x) - S_{N-n}(x)| \end{aligned}$$

because the maximum is attained at some of the points $X_{D_1}, \ldots, X_{D_n}$.

4 LIMITING DISTRIBUTION

Let us consider a sequence of testing problems with N_ν observations and regression constants c_{ν_i}, $i = 1, \ldots, N_\nu$, $\nu = 1, 2, \ldots$. For simplicity assume that

$$\sum_{i=1}^{N_\nu} c_{\nu i} = 0, \quad 1 \le \nu < \infty, \tag{4.1}$$

$$\sum_{i=1}^{N_\nu} c_{\nu i}^2 = 1, \quad 1 \le \nu < \infty, \tag{4.2}$$

and, furthermore,

$$\lim_{\nu\to\infty} \max_{1\le i\le N_\nu} c_{\nu i}^2 = 0. \tag{4.3}$$

The last condition is usually called the Noether condition.

Now let us recall briefly some results of Hájek (1962). Denote

$$V_{\nu i} = X_{\nu D_{\nu i}}, \quad 1 \le i \le N_\nu, \tag{4.4}$$

and, inversely,

$$X_{\nu i} = V_{\nu R_{\nu i}}, \quad 1 \le i \le N_\nu. \tag{4.5}$$

The vector $(V_{\nu 1}, \ldots, V_{\nu N_\nu})$, $V_{\nu 1} < \cdots < V_{\nu N_\nu}$ is usually called the ordered sample, and the vector $(R_{\nu 1}, \ldots, R_{\nu N_\nu})$ is the vector of ranks. Given a square integrable function $\varphi(u)$, $0 \le u \le 1$, and an N_ν, let us define

$$\varphi_\nu^0\left(\frac{i}{N_\nu + 1}\right) = N_\nu \int_{(i-1)/N_\nu}^{i/N_\nu} \varphi(u)\, \mathrm{d}u, \quad 1 \le i \le N_\nu, \tag{4.6}$$

and introduce statistics

$$S_\nu^0(\varphi) = \sum_{i=1}^{N_\nu} c_{\nu i}\, \varphi_\nu^0\left(\frac{R_{\nu i}}{N_\nu + 1}\right), \quad 1 \le \nu < \infty. \tag{4.7}$$

From Corollary 2.1 of Hájek (1962) we have the following lemma.

Lemma 4.1. *Let conditions* (4.1), (4.2) *and* (4.3) *be satisfied, and let* $\varphi(u)$, $0 \le u \le 1$, *be a square integrable function. Then the distribution of the statistic* $S_\nu^0(\varphi)$ *given by* (4.7) *with* φ_ν^0 *defined by* (4.6) *is asymptotically normal with zero mean and variance*

$$\sigma^2(\varphi) = \int_0^1 \varphi^2(u)\, \mathrm{d}u - \left[\int_0^1 \varphi(u)\, \mathrm{d}u\right]^2. \tag{4.8}$$

In this chapter we shall be interested in functions $\varphi_t(u)$ defined as follows:

$$\begin{aligned} \varphi_t(u) &= 0 \quad \text{if } 0 \le u \le t, \\ &= 1 \quad \text{if } t < u \le 1. \end{aligned} \tag{4.9}$$

Obviously,

$$\begin{aligned} S_\nu^0(\varphi_t) &= \sum_{i=1}^{N_\nu} c_{\nu i}\, \varphi_{t\nu}^0\left(\frac{R_{\nu i}}{N_\nu + 1}\right) \\ &= \sum_{i=1}^{N_\nu} c_{\nu D_{\nu_i}}\, \varphi_{t\nu}^0\left(\frac{i}{N_\nu + 1}\right) \\ &= \sum_{i=k+1}^{N_\nu} c_{\nu D_{\nu i}} + N_\nu\left(\frac{k}{N_\nu} - t\right) c_{\nu D_{\nu k}}, \quad \frac{k-1}{N_\nu} \le t \le \frac{k}{N_\nu}. \end{aligned} \tag{4.10}$$

Or, equivalently, in view of (4.1),

$$S_\nu^0(\varphi_t) = -\sum_{i=1}^{k-1} c_{\nu D_{\nu i}} - N_\nu\left(t - \frac{k-1}{N_\nu}\right) c_{\nu D_{\nu k}}, \quad \frac{k-1}{N_\nu} \le t \le \frac{k}{N_\nu}. \tag{4.11}$$

Consider now a sequence of stochastic processes $\{X_\nu(t),\ 0 \le t \le 1\}$ defined as follows:

$$X_\nu(t) = S_\nu^0(\varphi_t), \quad 0 \le t \le 1, 1 \le \nu < \infty, \tag{4.12}$$

where $S_\nu^0(\varphi_t)$ is given by (4.11). Obviously the process $\{X_\nu(t),\ 0 \le t \le 1\}$ is a mapping of the vector of ranks $(R_{\nu 1}, \dots, R_{\nu N_\nu})$, or, equivalently, of 'antiranks' $(D_{\nu 1}, \dots, D_{\nu N_\nu})$, into the space $C[0,1]$ of continuous functions, and the paths $X_\nu(\cdot)$ are linear within the intervals $[k/N_\nu,\ (k+1)/N_\nu],\ k = 0, \dots, N_\nu - 1$.

Next we shall show that the process $\{X_\nu(t),\ 0 \le t \le 1\}$ converges in distribution in $C[0,1]$ to a process we shall call a Brownian bridge.

Definition 4.1. A stochastic process $\{X(t,\omega),\ 0 \le t \le 1\}$ will be called a Brownian bridge if it is Gaussian with parameters

$$\mathsf{E}[X(t)] = 0, \qquad 0 \le t \le 1, \tag{4.13}$$

$$\mathsf{E}[X(t)\,X(s)] = t(1-s), \qquad 0 \le t \le s \le 1, \tag{4.14}$$

and if all paths $X(\cdot,\omega)$ are continuous functions on $[0,1]$.

In terms of the function $\varphi_t(u)$ introduced by (4.9) the covariance (4.14) may be expressed as follows:

$$\mathsf{E}[X(t)\,X(s)] = \int_0^1 \varphi_t(u)\,\varphi_s(u)\,\mathrm{d}u - \left[\int_0^1 \varphi_t(u)\,\mathrm{d}u\right]\left[\int_0^1 \varphi_s(u)\,\mathrm{d}u\right]. \tag{4.15}$$

Theorem 4.1. *Under conditions (4.1) through (4.3) the sequence of processes $\{X_\nu(t),\ 0 \le t \le 1\}$ defined by (4.12) and (4.11), converge in distribution in $C[0,1]$ to a Brownian bridge defined above.*

Proof. According to Theorem 1.1, we have first to prove that (1.1) holds true. In view of Remark 1.4, (1.1) will be proved if we show that for any n and $t_1, \dots, t_n \in [0,1]$, $-\infty < \lambda_1, \dots, \lambda_n < \infty$,

$$\mathcal{L}\left[\sum_{i=1}^n \lambda_i\, X_\nu(t_i)\right] \longrightarrow \mathcal{L}\left[\sum_{i=1}^n \lambda_i\, X(t_i)\right], \quad \text{as } \nu \to \infty. \tag{4.16}$$

However, in view of (4.13) through (4.15), $\sum_{i=1}^n \lambda_i\, X(t_i)$ is normal with zero mean and variance

$$\sigma^2(\psi) = \int_0^1 \psi^2(u)\,\mathrm{d}u - \left[\int_0^1 \psi(u)\,\mathrm{d}u\right]^2, \tag{4.17}$$

where

$$\psi(u) = \sum_{i=1}^n \lambda_i\, \varphi_{t_i}(u), \quad 0 \le u \le 1. \tag{4.18}$$

On the other hand, from (4.12) and (4.11) it follows that

$$\sum_{i=1}^{n} \lambda_i X_\nu(t_i) = S_\nu^0(\psi), \tag{4.19}$$

where ψ is defined by (4.18). Thus (4.16) follows simply by Lemma 4.1.

Now we need to show that condition (1.9) is satisfied. In view of (4.11)

$$X_\nu\left(\frac{k-1}{N_\nu}\right) - X_\nu\left(\frac{j}{N_\nu}\right) = \sum_{i=k}^{j} c_{\nu D_{\nu i}} = \sum_{i=1}^{j-k+1} Y_{\nu i}, \tag{4.20}$$

where we have put, for simplicity,

$$Y_{\nu i} = c_{\nu D_{\nu k+i-1}}, \quad 1 \le i \le j-k+1. \tag{4.21}$$

Since $c_{\nu D_{\nu 1}}, \ldots, c_{\nu D_{\nu N_\nu}}$ represents a random permutation of $\{c_{\nu 1}, \ldots, c_{\nu N_\nu}\}$, $Y_{\nu 1}, \ldots, Y_{\nu j-k+1}$ represents a simple random sample without replacement from the population $\{c_{\nu 1}, \ldots, c_{\nu N_\nu}\}$. Consequently, by Lemma 2.1 and (2.14), we have

$$\begin{aligned} &\mathsf{P}\left[\max_{k \le j \le k+m_{\nu\delta}} \left| X_\nu\left(\frac{k-1}{N_\nu}\right) - X_\nu\left(\frac{j}{N_\nu}\right)\right| > \varepsilon\right] \\ &= \mathsf{P}\left(\max_{1 \le i \le m_{\nu\delta}} |Y_{\nu 1} + \cdots + Y_{\nu i}| > \varepsilon\right) \\ &\le \left\{\frac{m_{\nu\delta}}{N_\nu}\left(1 - \frac{m_{\nu\delta}}{N_\nu}\right)\sum_{i=1}^{N_\nu} c_{\nu i}^4 + 3\left|\frac{m_{\nu\delta}}{N_\nu}\left(1 - \frac{m_{\nu\delta}}{N_\nu}\right)\sum_{i=1}^{N_\nu} c_{\nu i}^2\right|^2\right\} \\ &\qquad \times \frac{1+o(1)}{\varepsilon^4\left(1 - m_{\nu\delta}/N_\nu\right)^4}, \end{aligned} \tag{4.22}$$

where $o(1)$ refers to

$$m_{\nu\delta} \to \infty, \quad N_\nu - m_{\nu\delta} \to 0, \quad \text{as } \nu \to \infty, \tag{4.23}$$

which is satisfied for every $0 < \delta < 1$, in view of (1.10). Now, utilizing the condition (4.2), we get

$$\begin{aligned} &\mathsf{P}\left[\max_{k \le j < k+m_{\nu\delta}} \left| X_\nu\left(\frac{k-1}{N_\nu}\right) - X_\nu\left(\frac{j}{N_\nu}\right)\right| > \varepsilon\right] \\ &< \frac{\delta \max_{1 \le i \le N_\nu} c_{\nu i}^2 + 3\delta^2(1-\delta)}{\varepsilon^4(1-\delta)^2}[1 + o_\nu(1)], \end{aligned} \tag{4.24}$$

where $o_\nu(1) \to 0$ if $\nu \to \infty$. The right side being independent of k, we have, in view of (4.3),

$$\begin{aligned} &\frac{1}{\delta}\limsup_{\nu\to\infty} \max_{1 \le k \le N_\nu - m_{\nu\delta}+1} \mathsf{P}\left[\max_{k \le j < k+m_{\nu\delta}} \left| X_\nu\left(\frac{k-1}{N_\nu}\right) - X_\nu\left(\frac{j}{N_\nu}\right)\right| > \varepsilon\right] \\ &= \frac{\delta}{\varepsilon^4(1-\delta)^2}. \end{aligned} \tag{4.25}$$

Consequently the condition (1.9) is satisfied and the proof is thereby complete. □

Corollary 4.1. Assume that

$$\lim_{\nu\to\infty} \frac{\max_{1\le i\le N_\nu}(c_{\nu i}-\bar{c}_\nu)^2}{\sum_{i=1}^{N_\nu}(c_{\nu i}-\bar{c}_\nu)^2} \tag{4.26}$$

and consider the statistics K_ν defined by (3.4). Then

$$\lim_{\nu\to\infty} \mathsf{P}(K_\nu < \lambda) = 1 - e^{-2\lambda^2}, \quad \lambda > 0, \tag{4.27}$$

and

$$\lim_{\nu\to\infty} \mathsf{P}(|K_\nu| < \lambda) = \sum_{r=-\infty}^{\infty} (-1)^k\, e^{2k^2\lambda^2}, \quad \lambda > 0. \tag{4.28}$$

Proof. In view of (4.12) and (4.11)

$$K_\nu = \max_{0\le t\le 1} X_\nu(t) = h_M[X_\nu(\cdot)] \quad [h_M \text{ if given by (1.3)}] \tag{4.29}$$

and

$$|K_\nu| = \max_{0\le t\le 1} |X_\nu(t)| \tag{4.30}$$

which are continuous functionals in $C[0,1]$ applied to sample paths $X_\nu(\cdot)$. The rest follows from the definition of convergence in distribution in $C[0,1]$ and the well-known formulas for the Brownian bridge $X(t)$ [see Doob (1949)]:

$$\mathsf{P}\left[\max_{0\le t\le 1} X(t) < \lambda\right] = 1 - e^{-2\lambda^2} \tag{4.31}$$

and

$$\mathsf{P}\left[\max_{0\le t\le 1} |X(t)| < \lambda\right] = \sum_{k=-\infty}^{\infty} (-1)^k\, e^{-2k^2\lambda^2}. \tag{4.32}$$

□

Remark 4.1. (4.16) is sufficient for (1.1) because it implies the convergence of corresponding characteristic functions:

$$\varphi_\nu(\lambda_1,\dots,\lambda_n) = \mathsf{E}\left\{\exp\left[i\sum_{j=1}^{n}\lambda_j\, X_\nu(t_j)\right]\right\} \tag{4.33}$$

$$\longrightarrow \quad \varphi(\lambda_1,\dots,\lambda_n) = \mathsf{E}\left\{\exp\left[i\sum_{j=1}^{n}\lambda_j\, X(t_j)\right]\right\}.$$

REFERENCES

Bartoszynski L. (1962). On a certain characterization of measures in $C[0,1]$. *Bull. Acad. Pol. Sci. Cl. 3 10 (8)*, 445.

Donsker M. D. (1952). Justification and extension of Doob's heuristic approach to the Kolmogorov–Smirnov theorems. *Ann. Math. Statist. 23*, 277.

Doob J. L. (1949). Heuristic approach to the Kolmogorov–Smirnov theorems. *Ann. Math. Statist. 20*, 393.

Driml M. (1959). Convergence of compact measures on metric spaces. In *Transactions of the 2nd Prague Conf. Inf. Th.*, p. 71.

Hájek J. (1962). Asymptotically most powerful rank–order tests. *Ann. Math. Statist. 33*, 1124.

Isserlis L. (1931). On the moment distributions of moments in the case of samples drawn from a limited universe. *Proc. Roy. Soc. 132*, 586.

Le Cam L. (1957). Convergence in distribution of stochastic processes. *Univ. Calif. Publ. Statistics 2 (11)*, 207.

Prohorov Ju. V. (1956). Convergence of stochastic processes and limiting theorems of probability theory. *Probability Theory 1*, 177.

Prohorov Ju. V. (1961). The method of characteristic functionals. In *Proc. Fourth Berkeley Symposium on Mathematical Statistics and Probability*, p. 403.

Smirnov N. V. (1939). Estimation of the difference of two empirical distribution functions from two independent samples (In Russian) *Bull. Mosk. Univ.* 2, 3.

Špaček A. (1955). Regularity properties of random transforms. *Czechoslovak Math. J. 5*, 143.

Varadarajan V. S. (1958). *Convergence of Stochastic Processes.* Thesis. Indian Stat. Institute.

CHAPTER 26

On Basic Concepts of Statistics

In: Proc. Fifth Berkeley Symp. Math. Statist. Prob.
(L. Le Cam, ed.), 1 (1967), 139–162,
Univ. of California Press.
Reproduced by permission of L. Le Cam, editor.

1 INTRODUCTION

This paper is a contribution to current discussions on fundamental concepts, principles, and postulates of statistics. In order to exhibit the basic ideas and attitudes, mathematical niceties are suppressed as much as possible. The heart of the paper lies in definitions, simple theorems, and nontrivial examples. The main issues under analysis are *sufficiency, invariance, similarity, conditionality, likelihood*, and their mutual relations.

Section 2 contains a definition of sufficiency for a subparameter (or sufficiency in the presence of a nuisance parameter), and a criticism of an alternative definition due to A. N. Kolmogorov (1942). In that section, a comparison of the principles of sufficiency in the sense of Blackwell–Girschick (1954) and in the sense of A. Birnbaum (1962) is added. In Theorem 3.5 it is shown that for nuisance parameters introduced by a group of transformations, the sub-σ-field of invariant events is sufficient for the respective subparameter.

Section 4 deals with the notion of similarity in the x-space as well as in the (x, θ)-space, and with related notions such as ancillary and exhaustive statistics. Confidence intervals and fiducial probabilities are shown to involve a postulate of 'independence under ignorance'.

Sections 5 and 6 are devoted to the principles of conditionality and of likelihood, as formulated by A. Birnbaum (1962). Their equivalence is proved and their strict form is criticized. The two principles deny gains obtainable by mixing strategies, disregarding that, in non-Bayesian conditions, the expected maximum conditional risk is generally larger than the maximum overall risk. Therefore, the notion of 'correct' conditioning is introduced, in a general

Collected Works of Jaroslav Hájek – With Commentary
Edited by M. Hušková, R. Beran and V. Dupač
Published in 1998 by John Wiley & Sons Ltd.

enough way to include the examples given in the literature to support the conditionality principle. It is shown that in correct conditioning the maximum risk equals the expected maximum conditional risk, and that in invariant problems the sub-σ-field of invariant events yields the deepest correct conditioning.

A proper field of application of the likelihood principle is shown to consist of families of experiments, in which the likelihood functions, possibly after a common transformation of the parameter, have approximately normal form with constant variance. Then each observed likelihood function allows computation of the risk without reference to the particular experiment.

In Section 7, some forms of the Bayesian approach are touched upon, such as those based on diffuse prior densities, or on a family of prior densities.

In Section 8, some instructive examples are given with comments.

References are confined to papers quoted only.

2 SUFFICIENCY

The notion of sufficiency does not promote many disputes. Nonetheless, there are two points, namely sufficiency for a subparameter and the principle of sufficiency, which deserve a critical examination.

2.1 Sufficiency for a Subparameter

Let us consider an experiment (X, θ), where X denotes the observations and θ denotes a parameter. The random element X takes its values in an x-space, and θ takes its values in a θ-space. A function τ of θ will be called a subparameter. If θ were replaced by a σ-field, then τ would be replaced by a sub-σ-field. Under what conditions may we say that a statistic is sufficient for τ? Sufficiency for τ may also be viewed as sufficiency in the presence of a nuisance parameter. The present author is aware of only one attempt in this direction, due to Kolmogorov (1942).

Definition 2.1. A statistic $T = t(X)$ is called sufficient for a subparameter τ, if the posterior distribution of τ, given $X = x$, depends only on $T = t$ and on the prior distribution of θ.

Unfortunately, the following theorem shows that the Kolmogorov definition is void.

Theorem 2.1. *If τ is a nonconstant subparameter, and if T is sufficient for τ in the sense of Definition 2.1, then T is sufficient for θ as well.*

Proof. For simplicity, let us consider the discrete case only. Let T be not sufficient for θ, and let us try to show that it cannot be sufficient for τ. Since T is not sufficient for θ, there exist two pairs, (θ_1, θ_2) and (x_1, x_2), such that

$$T(x_1) = T(x_2) \tag{2.1}$$

and

$$\frac{\mathsf{P}_{\theta_1}(X = x_1)}{\mathsf{P}_{\theta_2}(X = x_1)} \neq \frac{\mathsf{P}_{\theta_1}(X = x_2)}{\mathsf{P}_{\theta_2}(X = x_2)}. \tag{2.2}$$

If $\tau(\theta_1) \neq \tau(\theta_2)$, let us consider the following prior distribution: $\nu(\theta = \theta_1) = \nu(\theta = \theta_2) = \frac{1}{2}$. Then, for $\tau_1 = \tau(\theta_1)$ and $\tau_2 = \tau(\theta_2)$,

$$\mathsf{P}(\tau = \tau_1 \,|\, X = x_1) = \mathsf{P}_{\theta_1}(X = x_1) / [\mathsf{P}_{\theta_1}(X = x_1) + \mathsf{P}_{\theta_2}(X = x_1)] \tag{2.3}$$

and

$$\mathsf{P}(\tau = \tau_1 \,|\, X = x_2) = \mathsf{P}_{\theta_1}(X = x_2) / [\mathsf{P}_{\theta_1}(X = x_2) + \mathsf{P}_{\theta_2}(X = x_2)]. \tag{2.4}$$

Obviously, (2.2) entails $\mathsf{P}(\tau = \tau_1 \,|\, X = x_1) \neq \mathsf{P}(\tau = \tau_1 \,|\, X = x_2)$ which implies, in view of (2.1), that T is not sufficient for τ.

If $\tau(\theta_1) = \tau(\theta_2)$ held, we would choose θ_3 such that $\tau(\theta_3) \neq \tau(\theta_1) = \tau(\theta_2)$. Note that the equations

$$\frac{\mathsf{P}_{\theta_1}(X = x_1)}{\mathsf{P}_{\theta_3}(X = x_1)} = \frac{\mathsf{P}_{\theta_1}(X = x_2)}{\mathsf{P}_{\theta_3}(X = x_2)} \tag{2.5}$$

and

$$\frac{\mathsf{P}_{\theta_2}(X = x_1)}{\mathsf{P}_{\theta_3}(X = x_1)} = \frac{\mathsf{P}_{\theta_2}(X = x_2)}{\mathsf{P}_{\theta_3}(X = x_2)} \tag{2.6}$$

are not compatible with (2.2). Thus either (2.5) or (2.6) does not hold, and the above reasoning may be accomplished either with (θ_1, θ_3) or (θ_2, θ_3), both pair satisfying the condition $\tau(\theta_1) \neq \tau(\theta_3)$, $\tau(\theta_2) \neq \tau(\theta_3)$. □

Thus we have to try to define sufficiency in the presence of nuisance parameters in some less stringent way.

Definition 2.2. Let $\mathcal{P}_\tau$ be the convex hull of the distributions $\{\mathsf{P}_\theta, \tau(\theta) = \tau\}$ for all possible τ-values. We shall say that T is sufficient for τ if

(i) the distribution of T will depend on τ only, that is,

$$\mathsf{P}_\theta(\mathrm{d}t) = \mathsf{P}_\tau(\mathrm{d}t), \tag{2.7}$$

and

(ii) there exist distributions $\mathsf{Q}_\tau \in \mathcal{P}_\tau$ such that T is sufficient for the family $\{\mathsf{Q}_\tau\}$.

In the same manner we define a sufficient sub-σ-field for τ.

Now we shall prove an analogue of the well-known Rao–Blackwell theorem. For this purpose, let us consider a decision problem with a *convex* set D of decisions d, and with a loss function $L(\tau, d)$, which is *convex* in d for each τ, and depends on θ only through τ, $L(\theta, d) = L(\tau(\theta), d)$. Applying the minimax principle to eliminate the nuisance parameter, we associate with each decision function $\delta(x)$ the following risk:

$$R(\tau, \delta) = \sup_{\tau(\theta)=\tau} \int L[\tau(\theta),\, \delta(x)]\, \mathsf{P}_\theta(\mathrm{d}x), \tag{2.8}$$

where the supremum is taken over all θ-values such that $\tau(\theta) = \tau$. Now, if T is sufficient for τ in the sense of Definition 2.2, we can associate with each decision function $\delta(x)$ another decision function $\overline{\delta}(t)$ defined as follows:

$$\overline{\delta}(t) = \int \delta(x)\, \mathsf{Q}_\tau(\mathrm{d}x \,|\, T = t) \tag{2.9}$$

provided that $\delta(x)$ is integrable. Note that the right side of (2.9) does not depend on τ, since T is sufficient for $\{\mathsf{Q}_\tau\}$, according to Definition 2.2, and that $\overline{\delta}(t) \in D$ for every t in view of the convexity of D. Finally put

$$\delta^*(x) = \overline{\delta}(t(x)). \tag{2.10}$$

Theorem 2.2. *Under the above assumptions,*

$$R(\tau,\, \delta^*) \leq R(\tau, \delta) \tag{2.11}$$

holds for all τ.

Proof. Since the distribution of T depends on τ only, we have

$$\begin{aligned} R(\tau, \delta^*) &= \int L\,[\tau(\theta),\, \delta^*(x)]\, \mathsf{P}_\theta(\mathrm{d}x) \\ &= \int L\,[\tau(\theta),\, \delta^*(x)]\, \mathsf{Q}_\tau(\mathrm{d}x) \end{aligned} \tag{2.12}$$

for all θ such that $\tau(\theta) = \tau$. Furthermore, since $L(\tau, \mathrm{d})$ is convex in d,

$$\begin{aligned} \int L[\tau(\theta),\, \delta^*(x)]\, \mathsf{Q}_\tau(\mathrm{d}x) &\leq \int L[\tau(\theta),\, \delta(x)]\, \mathsf{Q}_\tau(\mathrm{d}x) \\ &\leq \sup_{\tau(\theta)=\tau} \int L[\tau(\theta),\, \delta(x)]\, \mathsf{P}_\theta(\mathrm{d}x) = R(\tau, \delta). \end{aligned} \tag{2.13}$$

This concludes the proof. □

Thus, adopting the minimax principle for dealing with nuisance parameters, and under the due convexity assumptions, one may restrict oneself to decision procedures depending on the sufficient statistic only. This finds its application in points estimation and in hypothesis testing, for example.

Remark 2.1. Le Cam (1964) presented three parallel ways of defining sufficiency. All of them could probably be used in giving equivalent definitions of sufficiency for a subparameter. For example, Kolmogorov's Definition 2.1 could be reformulated as follows: T is called sufficient for τ if it is sufficient for each system $\{\mathsf{P}_{\theta'}\}$ such that $\{\theta'\} \subset \{\theta\}$ and $\theta'_1 \neq \theta'_2 \Rightarrow \tau(\theta'_1) \neq \tau(\theta'_2)$.

Remark 2.2. We also could extend the notion of sufficiency by adding some 'limiting points'. For example, we could introduce ϵ-sufficiency, as in Le Cam (1964), and then say that T is sufficient if it is ϵ-sufficient for every $\epsilon > 0$, or if it is sufficient for every compact subset of the τ-space.

Remark 2.3. (Added in proof.) Definition 2.2 does not satisfy the natural requirement that T should be sufficient for τ if it is sufficient for some finer subparameter τ', $\tau = \tau(\tau')$. To illustrate this point, let us consider a sample from $N(\mu, \sigma^2)$ and put $T = (\overline{x}, s^2)$, $\tau = \mu$, $\tau' = (\mu, \sigma^2)$. Then T is not sufficient for τ in the sense of Definition 2.2, since its distribution fails to be dependent on μ only. On the other hand, T is sufficient for τ', as is well known. The Definition 2.2 should be corrected as follows.

Definition 2.2*. A statistic $T = t(X)$ is called sufficient for a subparameter τ if it is sufficient in the sense of Definition 2.2 for some subparameter τ' such that $\tau = \tau(\tau')$.

Remark 2.4. (Added in proof.) A more stringent (and, therefore, more consequential) definition of sufficiency for a parameter is provided by Lehmann (1959) in Problem 31 of Chapter III: T is sufficient for τ if, first, $\theta = (\tau, \eta)$, second $\mathsf{P}_\theta(\mathrm{d}t) = \mathsf{P}_\tau(\mathrm{d}t)$, third, $\mathsf{P}_\theta(\mathrm{d}x \,|\, T = t) = \mathsf{P}_\eta(\mathrm{d}x \,|\, T = t)$. If T is sufficient in this sense, it is also sufficient in the sense of Definition 2.2 where we may take $\mathsf{Q}_\tau = \mathsf{P}_{\tau,\eta_1}$ for any particular η_1.

2.2 The Principle of Sufficiency

If comparing the formulations of this principle as given in Blackwell–Girshick (1954) and in A. Birnbaum (1962), one feels that there is an apparent discrepancy. According to Birnbaum's sufficiency principle, we are not allowed to use randomized tests, for example, while no such implication follows from the Blackwell–Girshick sufficiency principle. The difference is serious but easy to explain: Blackwell and Girshich consider only convex situations (that is, convex D and convex $L(\theta, d)$ for each θ), where the Rao–Blackwell theorem can be proved, while A. Birnbaum has in mind any possible situation. However, what may be supported in convex situations by a theorem is a rather stringent postulate in a general condition. (If, in Blackwell–Girshick, the situation is not convex, they make it convex by allowing randomized decisions.)

In estimating a real parameter with $L(\theta, d) = (\theta - d)^2$, the convexity conditions are satisfied and no randomization is useful. If, however, θ would run through a discrete subset of real numbers, randomization might bring the same gains from the minimax point of view as in testing a simple hypothesis against a simple alternative. And even in convex situations the principle may not exclude any decision procedure as inadmissible. To this end it is necessary for $L(\tau, d)$ to be strictly convex in d, and not, for example, linear, as in randomized extensions of nonconvex problems.

3 INVARIANCE

Most frequently, the nuisance parameter is introduced by a group of transformations of the x-space on itself. Then, if we have a finite invariant measure on the group, we can easily show that the sub-σ-field of invariant events is sufficient in the presence of the corresponding nuisance parameter.

Let $G = \{g\}$ be a group of one-to-one transformations of the x-space on itself. Take a probability distribution P of X and put

$$\mathsf{P}_g(X \in A) = \mathsf{P}(gX \in A). \tag{3.1}$$

Then, obviously, $\mathsf{P}_h(X \in g^{-1}A) = \mathsf{P}_{gh}(X \in A)$. Let μ be a σ-finite measure and denote by μg the measure such that $\mu g(A) = \mu(gA)$.

Theorem 3.1. *Let $\mathsf{P} \ll \mu$ and $\mu g \ll \mu$ for all $g \in G$. Then $\mathsf{P}_g \ll \mu$. Denoting $p(x, g) = \mathrm{d}\mathsf{P}_g/\mathrm{d}\mu$ and $p(x) = \mathrm{d}\mathsf{P}/\mathrm{d}\mu$, then*

$$p(x, g) = p(g^{-1}x)\, \frac{\mathrm{d}\mu g^{-1}}{\mathrm{d}\mu}(x) \tag{3.2}$$

holds. More generally,

$$p(x, h^{-1}g) = p(h(x), g)\, \frac{\mathrm{d}\mu h}{\mathrm{d}\mu}(x). \tag{3.3}$$

Proof. According to the definition, one can write

$$\mathsf{P}_g(X \in A) = \mathsf{P}(X \in g^{-1}A) = \int_{g^{-1}A} p(x)\, \mathrm{d}\mu \tag{3.4}$$

$$= \int_A p(g^{-1}(y))\, \mathrm{d}\mu g^{-1}(y) = \int_A p(g^{-1}(y))\, \frac{\mathrm{d}\mu g^{-1}}{\mathrm{d}\mu}\, \mathrm{d}\mu(y).$$

□

Condition 3.1. Let $\mathcal{G}$ be a σ-field of subsets of G, and let $\mathcal{A}$ be the σ-field of subsets of the x-space. Assume that

(i) $\mu g \ll \mu$ for all $g \in G$,
(ii) $p(x, g)$ is $\mathcal{A} \times \mathcal{G}$-measurable,
(iii) functions $\phi_h(g) = hg$ and $\psi_h(g) = gh$ are $\mathcal{G}$-measurable,
(iv) there is an invariant probability measure ν on G, that is, $\nu(Bg) = \nu(gB) = \nu(B)$ for all $g \in G$ and $B \in \mathcal{G}$.

Theorem 3.2. *Under Condition* 3.1, *let us put*

$$\overline{p}(x) = \int p(x, g)\, \mathrm{d}\nu(g). \tag{3.5}$$

Then, for each $h \in G$,

$$\overline{p}(h(x)) = \left[\frac{\mathrm{d}\mu h}{\mathrm{d}\mu}(x)\right]^{-1} \overline{p}(x). \tag{3.6}$$

Remark 3.1. Note that the first factor on the right-hand side does not depend on p.

Proof. In view of (3.3), we have

$$\begin{aligned}
\overline{p}(h(x)) &= \int p(h(x),\, g)\, \mathrm{d}\nu(g) = \left[\frac{\mathrm{d}\mu h}{\mathrm{d}\mu}(x)\right]^{-1} \int p(x,\, h^{-1}g)\, \mathrm{d}\nu \qquad (3.7)\\
&= \left[\frac{\mathrm{d}\mu h}{\mathrm{d}\mu}(x)\right]^{-1} \int p(x, f)\, \mathrm{d}\nu h(f)\\
&= \left[\frac{\mathrm{d}\mu h}{\mathrm{d}\mu}(x)\right]^{-1} \int p(x, f)\, \mathrm{d}\nu(f)\\
&= \left[\frac{\mathrm{d}\mu h}{\mathrm{d}\mu}(x)\right]^{-1} \overline{p}(x).
\end{aligned}$$

□

We shall say that an event is G-invariant if $gA = A$ for all $g \in G$. Obviously, the set of G-invariant events is a sub-σ-field $\mathcal{B}$, and a measurable function f is $\mathcal{B}$-measurable if and only if $f(g(x)) = f(x)$ for all $g \in G$. Consider now two distributions P and P_0 and seek the derivative of P relative to P_0 on $\mathcal{B}$, say $[\mathrm{d}\mathsf{P}/\mathrm{d}\mathsf{P}_0]^{\mathcal{B}}$. Assume that $\mathsf{P} \ll \mu$ and $\mathsf{P}_0 \ll \mu$, denote $p = \mathrm{d}\mathsf{P}/\mathrm{d}\mu$, $p_0 = \mathrm{d}\mathsf{P}_0/\mathrm{d}\mu$, and introduce $\overline{p}(x)$ and $\overline{p}_0(x)$ by (3.5).

Theorem 3.3. *Under Condition* 3.1 *and under the assumption that* $\overline{p}_0(x) = 0$ *entails* $\overline{p}(x) = 0$ *almost* μ*-everywhere, we have* $\mathsf{P} \ll \mathsf{P}_0$ *on* $\mathcal{B}$ *and*

$$[\mathrm{d}\mathsf{P}/\mathrm{d}\mathsf{P}_0]^{\mathcal{B}} = \overline{p}(x)/\overline{p}_0(x). \tag{3.8}$$

Proof. Put $\ell(x) = \overline{p}(x)/\overline{p}_0(x)$. Theorem 3.2 entails $\ell(g(x)) = \ell(x)$ for all $g \in G$, namely, $\ell(x)$ is $\mathcal{B}$-measurable. Further, for any $B \in \mathcal{B}$, in view of (3.2) and of $[p = 0] \Rightarrow [p_0 = 0]$,

$$\begin{aligned} \int_B \ell(x)\,\mathrm{d}\mathsf{P}_0 &= \int_B \ell(x)\,p_0(x)\,\mathrm{d}\mu = \int_B \ell(x)\,p_0(g^{-1}(x))\,\mathrm{d}\mu g^{-1} \\ &= \int_B \ell(x)\,p_0(x,g)\,\mathrm{d}\mu = \int_B \ell(x)\,\overline{p}_0(x)\,\mathrm{d}\mu = \int_B \overline{p}(x)\,\mathrm{d}\mu \end{aligned} \tag{3.9}$$

holds. On the other hand,

$$\mathsf{P}(B) = \int_B p(x)\,\mathrm{d}\mu = \int_B p(g^{-1}(x))\,\mathrm{d}\mu g^{-1} = \int_B p(x,g)\,\mathrm{d}\mu = \int_B \overline{p}(x)\,\mathrm{d}\mu. \tag{3.10}$$

Thus $\mathsf{P}(B) = \int_B \ell(x)\,\mathrm{d}\mathsf{P}_0$, $B \in \mathcal{B}$, which concludes the proof. □

Theorem 3.4. *Let the statistic* $T = t(X)$ *have an expectation under* P_h, $h \in G$. *Let Condition* 3.1 *be satisfied. Then the conditional expectation of* T *relative to the sub-*σ*-field* $\mathcal{B}$ *of* G*-invariant events and under* P_h *equals*

$$\mathsf{E}_h(,\,|\,\mathcal{B}, x) = \int t(g^{-1}(x))\,p(x,\,gh)\,\mathrm{d}\nu(g)/\overline{p}(x). \tag{3.11}$$

Proof. The proof would follow the lines of the proofs of Theorems 3.2 and 3.3. □

Consider now a dominated family of probability distributions $\{\mathsf{P}_\tau\}$ and define $\mathsf{P}_{\tau,g}$ by (3.1) for each τ. Putting $\theta = (\tau, g)$, we can say that τ is a subparameter of θ.

Theorem 3.5. *Under Condition* 3.1 *the sub-*σ*-field* $\mathcal{B}$ *of* G*-invariant events is sufficient for* τ *in the sense of Definition* 2.2.

Proof. First, the G-invariant events have a probability depending on τ only, that is, $\mathsf{P}_{\tau,g}(B) = \mathsf{P}_\tau(B)$. Second, for

$$\mathsf{Q}_\tau(A) = \int \mathsf{P}_{\tau,g}(A)\,\mathrm{d}\nu(g) \tag{3.12}$$

we have

$$\begin{aligned} \mathsf{Q}_\tau(A) &= \int \left[\int_A p_\tau(x,g)\,\mathrm{d}\mu\right] \mathrm{d}\nu(g) \\ &= \int_A \int p_\tau(x,g)\,\mathrm{d}\nu(g)\,\mathrm{d}\mu \\ &= \int_A \overline{p}_\tau(x)\,\mathrm{d}\mu. \end{aligned} \tag{3.13}$$

Now let P_0 be some probability measure such that $\mathsf{P}_\tau \ll \mathsf{P}_0 \ll \mu$, and introduce $\overline{p}_0(x)$ by (3.5). Then, according to Theorem 3.3, $\overline{p}_\tau(x)/\overline{p}_0(x)$ is $\mathcal{B}$-measurable for all τ. Thus $\overline{p}(x) = [\overline{p}_\tau(x)/\overline{p}_{\tau 0}(x)]\,\overline{p}_0(x)$, and $\mathcal{B}$ is sufficient for $\{\mathsf{Q}_\tau\}$, according to the factorization criterion. □

Remark 3.2. Considering right-invariant and left-invariant *probability* measures on $\mathcal{G}$ does not provide any generalization. Actually, if ν is right-invariant, then $\overline{\nu}(B) = \int_A \nu(gB)\,\mathrm{d}\nu(g)$ is invariant, that is, both right-invariant and left-invariant.

Remark 3.3. Theorem 3.3 sometimes remains valid even if there exists only a right-invariant countably finite measure ν, as in the case of the groups of location and/or scale shifts (see Hájek and Šidák (1967, Chapter II)).

4 SIMILARITY

The concept of similarity plays a very important role in classical statistics, namely in contributions by J. Neyman and R. A. Fisher. It may be applied in the x-space as well as in the (x, θ)-space.

4.1 Similarity in the x-Space

Consider a family of distributions $\{\mathsf{P}_\theta\}$ on measurable subsets of the x-space. We say that an *event* A is *similar* if its probability is independent of θ:

$$\mathsf{P}_\theta(A) = \mathsf{P}(A) \quad \text{for all } \theta. \tag{4.1}$$

Obviously the events 'whole x-space' and 'the empty set' are always similar.

The class of similar events is closed under complementation and formation of countable *disjoint* unions. Such classes are called λ-field by Dynkin (1959). A λ-field is a broader concept than a σ-field. The class of similar events is usually not a σ-field, which causes ambiguity in applications of the conditionality principle, as we shall see. Generally, if both A and B are similar and are not disjoint, then $A \cap B$ and $A \cup B$ may not be similar. Consequently, the system of σ-fields contained in a λ-field may not include a largest σ-field.

Dynkin (1959) calls a system of subsets a π-field, if it is closed under intersection, and shows that a λ-field containing a π-field contains the smallest σ-field over the π-field. Thus, for example, if for a vector statistic $T = t(X)$ the events $\{x : t(x) < c\}$ are similar for every vector c, then $\{x : T \in A\}$ is similar for every Borel set A.

More generally, a *statistic* $V = v(X)$ is called *similar* if $\mathsf{E}_\theta V = \int v(x)\,\mathsf{P}_\theta(\mathrm{d}x)$ exists and is independent of θ. We also may define similarity with respect to a nuisance parameter only. The notion of similarity forms a basis for the definition of several other important notions.

Ancillary statistics. A statistic $U = u(x)$ is called ancillary if its distribution is independent of θ, that is, if the events generated by U are similar. (We have seen that it suffices that the events $\{U < c\}$ be similar.)

Correspondingly, a sub-σ-field will be called ancillary if all its events are similar.

Exhaustive statistics. A statistic $T = t(x)$ is called exhaustive if (T, U), with U an ancillary statistic, is a minimal sufficient statistic.

Complete families of distributions. A family $\{\mathsf{P}_\theta\}$ is called complete if the only similar statistics are constants.

Our definition of an exhaustive statistic follows Fisher's explanation (ii) in Fisher (1956, p. 49) and the examples given by him.

Denote by I_θ^X, I_θ^T, $I_\theta^T(u)$ Fisher's information for the families $\{\mathsf{P}_\theta(\mathrm{d}x)\}$, $\{\mathsf{P}_\theta(\mathrm{d}t)\}$, $\{\mathsf{P}_\theta(\mathrm{d}t) \mid U = u\}$, respectively. Then, since (T, U) is sufficient and U is ancillary,

$$\mathsf{E} I_\theta^T(U) = I_\theta^X. \tag{4.2}$$

If $I_\theta^T < I_\theta^X$, Fisher calls (4.2) 'recovering the information lost'. What is the real content of this phrase?

The first interpretation would be that $T = t$ contains all information supplied by $X = x$, provided that we know $U = u$. But knowing both $T = t$ and $U = u$, we know $(T, U) = (t, u)$, that is, we know the value of the sufficient statistics. Thus this interpretation is void, and, moreover, it holds even if U is not ancillary.

A more appropriate interpretation seems to be as follows. Knowing $\{\mathsf{P}_\theta(\mathrm{d}t \mid U = u)\}$, we may dismiss the knowledge of $\mathsf{P}(\mathrm{d}u)$ as well as of $\{\mathsf{P}_\theta(\mathrm{d}t \mid U = u')\}$ for $u' \neq u$. This, however, expresses nothing other than the conditionality principle formulated in Section 5.

Fisher makes a two-field use of exhaustive statistics: first, for extending the scope of fiducial distributions to cases where no appropriate sufficient statistic exists, and, second, to a (rather unconvincing) eulogy of maximum likelihood estimates.

The present author is not sure about the appropriateness of restricting ourselves to 'minimal' sufficient statistics in the definition of exhaustive statistics. Without this restriction, there would rarely exist a minimal exhaustive statistic, and with this restriction we have to check, in particular cases, whether the employed sufficient statistic is really minimal.

4.2 Similarity in the (x, θ)-Space

Consider a family of distributions $\{\mathsf{P}_\theta\}$ on the x-space, the family of all possible prior distributions $\{\nu\}$ on the θ-space, and the family of distributions $\{R_\nu\}$ on the (x, θ)-space given by

$$R_\nu(\mathrm{d}x\,\mathrm{d}\theta) = \nu(\mathrm{d}\theta)\,\mathsf{P}_\theta(\mathrm{d}x). \tag{4.3}$$

Then we may introduce the notion of similarity in the (x, θ)-space in the same manner as before in the x-space, with $\{\mathsf{P}_\theta\}$ replaced by $\{R_\nu\}$. Consequently, the prior distribution will play the role of an unknown parameter. Thus an event Λ in the (x, θ)-space will be called similar if

$$R_\nu(\Lambda) = R(\Lambda) \tag{4.4}$$

for all prior distributions ν.

To avoid confusion, we shall call measurable functions of (x, θ) *quantities* and not statistics. A quantity $H = h(X, \Theta)$ will be called similar if its expectation $\mathsf{E}H = \int h(x, \theta)\,\nu(d\theta)\,\mathsf{P}_\theta(\mathrm{d}x)$ is independent of ν. Ancillary statistics in the (x, θ)-space are called *pivotal* quantities of *distribution-free* quantities. Since only the first component of (x, θ) is observable, the applications of similarity in the (x, θ)-space are quite different from those in the x-space.

Confidence regions. This method of region estimation dwells on the following idea. Having a similar event Λ, whose probability equals $1 - \alpha$, with α very small, and knowing that $X = x$, we can feel confidence that $(x, \theta) \in \Lambda$, namely, that the unknown θ lies within the region $S(x) = \{\theta : (x, \theta) \in \Lambda\}$. Here, our confidence that the event Λ has occurred is based on its high probability, and this confidence is assumed to be unaffected by the knowledge of $X = x$. Thus we assume a sort of independence between Λ and X, though their joint distribution is indeterminate. The fact that this intrinsic assumption may be dubious is most appropriately manifested in cases when $S(x)$ is either empty, so that we know that Λ did not occur, or equals the whole θ-space, so we are sure that Λ has occurred. Such a situation arises in the following.

Example 1. For this example, $\theta \in [0, 1]$, x is real, $\mathsf{P}_\theta(\mathrm{d}x)$ is uniform over $[0, \theta + 2]$, and

$$\Lambda = \{(x, \theta) : \theta + \alpha < x < \theta + 2 - \alpha\}. \tag{4.5}$$

Then $S(x) = \emptyset$ for $3-\alpha < x < 3$ or $0 < x < \alpha$, and $S(x) = [0,1]$ for $1+\alpha < x < 2-\alpha$. Although this example is somewhat artificial, the difficulty involved seems to be real.

Fiducial distribution. Let $F(t,\theta)$ be the distribution function of T under θ, and assume that F is continuous in t for every θ. Then $H = F(T,\Theta)$ is a pivotal quantity with uniform distribution over $[0,1]$. Thus $R(F(T,\Theta) \leq x) = x$. Now, if $F(t,\theta)$ is strictly decreasing in θ for each t, and if $F^{-1}(t,\theta)$ denotes its inverse for fixed t, then $F(T,\Theta) \leq x$ is equivalent to $\Theta \geq F^{-1}(T,x)$. Now, again, if we know that $T = t$, and if we feel that it does not affect the probabilities concerning $F(T,\Theta)$, we may write

$$\begin{aligned} x &= R(F(T,\Theta) \leq x) = R(\Theta \geq F^{-1}(T,x)) \\ &= R(\Theta \geq F^{-1}(T,x) \,|\, T = t) = R(\Theta \geq F^{-1}(t,x) \,|\, T = t) \end{aligned} \tag{4.6}$$

namely, for $\theta = F^{-1}(t,x)$,

$$R(\Theta < \theta \,|\, T = t) = 1 - F(t,\theta). \tag{4.7}$$

As we have seen, in confidence regions as well as in fiducial probabilities, a peculiar postulate of independence is involved. The postulate may be generally formulated as follows.

Postulate of independence under ignorance. Having a pair (Z, W) such that the marginal distribution of Z is known, but the joint distribution of (Z, W), as well as the marginal distribution of W are unknown, and observing $W = w$, we assume that

$$\mathsf{P}(Z \in \Lambda \,|\, W = w) = \mathsf{P}(Z \in \Lambda). \tag{4.8}$$

J. Neyman gives a different justification of the postulate than R. A. Fisher. Let us make an attempt to formulate the attitudes of both these authors.

Neyman's justification. Assume that we perform a long series of independent replications of the given experiments, and denote the results by $(Z_1, w_1), \ldots, (Z_N, w_N)$, where the w_is are observed numbers and the Z_is are not observable. Let us decide to accept the hypothesis $Z_i \in \Lambda$ at each replication. Then our decisions will be correct approximately in $100\,P\%$ of cases with $P = \mathsf{P}(Z \in \Lambda)$. Thus, in a long series, our mistakes in determining $\mathsf{P}(Z \in \Lambda \,|\, W = w)$ by (4.8) if any, will compensate each other.

Fisher's justification. Suppose that the only statistics $V = v(W)$, such that the probabilities $\mathsf{P}(Z \in \Lambda \,|\, V = v)$ are well-determined, are constants. Then we are allowed to take $\mathsf{P}(Z \in \Lambda \,|\, W = w) = \mathsf{P}(Z \in \Lambda)$, since our absence of knowledge prevents us from doing anything better.

The above interpretation of Fisher's view is based on the following passage from his book Fisher (1956, pp. 54–55):

The particular pair of values of θ and T appropriate to a particular experimenter certainly belongs to this enlarged set, and within this set the proportion of cases satisfying the inequality

$$\theta > \frac{T}{2n} \chi^2_{2n}(P) \tag{4.9}$$

is certainly equal to the chosen probability P. It might, however, have been true ... that in some recognizable subset, to which his case belongs, the proportion of cases in which the inequality was satisfied should have some value other than P. It is the stipulated absence of knowledge *a priori* of the distribution of θ, together with the exhaustive character of the statistic T, that makes the recognition of any such subset impossible, and so guarantees that in his particular case ... the general probability is applicable.

To apply our general scheme to the case considered by R. A. Fisher, we should put $Z = 1$ if (4.9) is satisfied and $Z = 0$ otherwise, and $W = T$.

Remark 4.1. We can see that Fisher's argumentation applies to a more specific situation than described in the above postulate. He requires that conditional probabilities are not known for any statistics $V = v(W)$ except for $V =$ const. Thus the notion of fiducial probabilities is a more special notion than the notion of confidence regions, since Neyman's justification does not need any such restrictions.

Fisher's additional requirement is in accord with his requirement that fiducial probabilities should be based on the minimal sufficient statistics, and in such a case does not lead to difficulties.

However, if the fiducial probabilities are allowed to be based on exhaustive statistics, then his requirement is contradictory, since no minimal exhaustive statistics may exist. Nonetheless, we have the following.

Theorem 4.1. *If a minimal sufficient statistic $S = s(X)$ is complete, then it is also a minimal exhaustive statistic.*

Proof. If there existed an exhaustive statistic $T = t(S)$ different from S, there would exist a nonconstant ancillary statistic $U = u(S)$, which contradicts the assumed completeness of S. □

If S is not complete, then there may exist ancillary statistics $U = u(S)$, and the family corresponding to exhaustive statistics contains a minimal member if and only if the family of Us contains a maximal member.

Remark 4.2. Although the two above justification are unconvincing, they correspond to habits of human thinking. For example, if one knows that an individual comes from a subpopulation of a population where the proportion of individuals with a property A equals P, and if one knows nothing else about the subpopulation, one applies P to the given individual without hesitation. This is true even if we know the 'name' of the individual, that is, if the subpopulation consists of a single individual. Fisher is right when he claims that we would not use P if the given subpopulation would be a part of a larger one, in which the proportion is known, too. He does not say, however, what to do if there is no minimal such larger subpopulation.

In confidence regions the subpopulation consists of pairs (X, Θ) such that $X = x$. It is true that we know the 'proportion' of elements of that subpopulation for which the event Λ occurs, but we do not know how much of the probability mass each element carries. Thus we can utilize the knowledge of this proportion only if it equals 0 or 1, which is exemplified by Example 1 but occurs rather rarely in practice. The problem becomes still more puzzling if the knowledge of the proportion is based on estimates only, and if these estimates become less reliable as the subpopulation from which they are derived becomes smaller. Is there any reasonable recommendation as to what to do if we know $\mathsf{P}(X \in A \mid U_1 = u_1)$ and $\mathsf{P}(X \in A \mid U_2 = u_2)$ but we do not know $\mathsf{P}(X \in A \mid U_1 = u_1, U_2 = u_2)$?

Remark 4.3. Fisher violently attacked the Bayes postulate as an adequate form of expressing our ignorance mathematically. He reinforced this attitude in Fisher (1956, p. 20), where we may read: 'It is evidently easier for the practitioner of natural science to recognize the difference between knowing and not knowing than it seems to be for the more abstract mathematician.' On the other hand, as we have seen, he admits that the absence of knowledge of proportions in a subpopulation allows us to act in the same manner as if we knew that the proportion is the same as in the whole population. The present author suspects that this new postulate, if not stronger than the Bayes postulate, is by no means weaker. Actually, if a proportion P in a population is interpreted as probability for an individual taken from this population, we tacitly assume that the individual has been selected according to the uniform distribution.

One cannot escape the feeling that all attempts to avoid expressing in a mathematical form the absence of our prior knowledge about the parameter have eventually been a failure. Of course, the mathematical formalization of ignorance should be understood broadly enough, including not only prior distributions, but also the minimax principle, and so on.

4.3 Estimation

Similarity in the (x, θ)-space is widely utilized in estimation. For example, an estimate $\hat{\theta}$ is unbiased if and only if the quantity $H = \hat{\theta} - \theta$ is similar with zero expectation. Further, similarity provides the following class of estimation methods: starting with a similar quantity $H = h(X, \Theta)$ with $\mathsf{E}H = c$, and observing $X = x$, we take for θ the solution of equation

$$h(x, \theta) = c. \tag{4.10}$$

This class includes the method of maximum likelihood for

$$h(x,\theta) = (\partial/\partial\theta)\log p_\theta(x). \tag{4.11}$$

Another method is offered by the pivotal quantity $H = F(T,\Theta)$ considered in connection with fiducial probabilities. Since $\mathsf{E}H = \frac{1}{2}$, we may take for $\hat{\theta}$, given $T = t$, the solution if any of

$$F(t,\theta) = \frac{1}{2}, \tag{4.12}$$

that is, the parameter value for which the observed value is the median.

5 CONDITIONALITY

If the prior distribution ν of θ is known, all statisticians agree that the decisions given $X = x$ should be made on the basis of the conditional distribution $R_\nu(\mathrm{d}\theta \mid X = x)$. This exceptional agreement is caused by the fact that then there is only one distribution in the (x,θ)-space, so that the problems are transferred from the ground of statistics to the ground of pure probability theory.

The problems arise in situations where conditioning is applied to a family of distributions. Here two basically different situations must be distinguished according to whether the conditioning statistic is ancillary or not. Conditioning with respect to an ancillary statistic is considered in the conditionality principle as formulated by A. Birnbaum (1962). On the other hand, for example, conditioning with respect to a sufficient statistic for a nuisance parameter, successfully used in constructing most powerful similar tests (see Lehmann (1959)), is a quite different problem and will not be considered here.

Our discussion will concentrate on the conditionality principle, which may be formulated as follows.

The principle of conditionality. Given an ancillary statistic U, and knowing $U = u$, statistical inference should be based on conditional probabilities $\mathsf{P}_\theta(\mathrm{d}x \mid U = u)$ only; that is, the probabilities $\mathsf{P}_\theta(\mathrm{d}x \mid U = u')$ for $u' \neq u$ and $\mathsf{P}(\mathrm{d}u)$ should be disregarded.

This principle, if properly illustrated, looks very appealing. Moreover, all *proper* Bayesian procedures are concordant with it. We say that a Bayesian procedure is proper if it is based on a prior distribution ν established independently of the experiment. A Bayesian procedure which is not proper is exemplified by taking for the prior distribution the measure ν given by

$$\nu(\mathrm{d}\theta) = \sqrt{I_\theta}\,\mathrm{d}\theta, \tag{5.1}$$

where I_θ denotes the Fisher information associated with the particular experiment. Obviously, such a Bayesian procedure is not compatible with the principle of conditionality (cf. de Finetti and Savage (1962)).

The term 'statistical inference' used in the above definition is very vague. Birnbaum (1962) makes use of an equally vague term, 'evidential meaning'. The present author sees two possible interpretations within the framework of decision theory.

Risk interpretation. A decision procedure δ defined on the original experiment should be associated with the same risk as its restriction associated with the partial experiment given $U = u$. Of course this transfer of risk is possible only after the experiment has been performed and u is known.

Decision rules interpretation. A decision function δ should be interpreted as a function of two arguments, $\delta = \delta(E, x)$, with E denoting an experiment from a class $\mathcal{E}$ of experiments and x denoting one of its outcomes. The principle of conditionality then restricts the class of 'reasonable' decision functions to those for which $\delta(E, x) = \delta(F, x)$ as soon as F is a subexperiment of E, and x belongs to the set of outcomes of F (F being a subexperiment of E means that F equals 'E given $U = u$', where U is some ancillary statistic and u some of its particular values).

In this section we shall analyze the former interpretation, returning to the latter in the next section.

The principle of conditionality, as it stands, is not acceptable in non-Bayesian conditions, since it denies possible gains obtainable by randomization (mixed strategies). On the other hand, some examples exhibited to support this principle (see Birnbaum (1962, p. 280)) seem to be very convincing. Before attempting to delimit the area of proper application of the principle, let us observe that the principle is generally ambiguous. Actually, since generally no maximal ancillary statistic exists, it is not clear which ancillary statistic should be chosen for conditioning.

Let us denote the conditional distribution given $U = u$ by $\mathsf{P}_\theta(\mathrm{d}x \mid U = u)$ and the respective conditional risk by

$$R(\theta, U, u) = \int L(\theta,\, \delta(x))\, \mathsf{P}_\theta(\mathrm{d}x \mid U = u). \tag{5.2}$$

In terms of a sub-σ-field $\mathcal{B}$, the same will be denoted by

$$R(\theta, \mathcal{B}, x) = \int L(\theta,\, \delta(x))\, \mathsf{P}_\theta(\mathrm{d}x \mid \mathcal{B}, x), \tag{5.3}$$

where $R(\theta, \mathcal{B}, x)$, as a function x, is $\mathcal{B}$-measurable. Since the decision function δ will be fixed in our considerations, we have deleted it in the symbol for the risk.

Definition 5.1. A conditioning relative to an ancillary statistic U or an ancillary sub-σ-field $\mathcal{B}$ will be called *correct* if

$$R(\theta, U, u) = R(\theta)\, b(u) \tag{5.4}$$

or

$$R(\theta, \mathcal{B}, x) = R(\theta)\, b(x), \tag{5.5}$$

respectively. In (5.4) and (5.5), $R(\theta)$ denotes the overall risk, and the function $b(x)$ is $\mathcal{B}$-measurable. Obviously, $\mathsf{E}b(U) = \mathsf{E}b(X) = 1$.

The above definition is somewhat too strict, but it covers all examples exhibited in the literature to support the conditionality principle.

Theorem 5.1. *If the conditioning relative to an ancillary sub-σ-field $\mathcal{B}$ is correct, then*

$$\mathsf{E}\left[\sup_\theta R(\theta, \mathcal{B}, x)\right] = \sup_\theta R(\theta). \tag{5.6}$$

The proof follows immediately from (5.5). Obviously, for a noncorrect conditioning we may obtain $\mathsf{E}\,[\sup_\theta R(\theta, \mathcal{B}, X)] > \sup_\theta R(\theta)$, so that, from the minimax point of view, conditional reasoning generally disregards possible gains obtained by mixing (randomization), and, therefore, is hardly acceptable under non-Bayesian conditions. □

Theorem 5.2. *As in Section 3, consider the family $\{\mathsf{P}_g\}$ generated by a probability distribution P and a group G of transformations g. Further, assume the loss function L to be invariant, namely, such that $L(hg,\, \delta(h(x))) = L(g,\, \delta(x))$ holds for all h, g, and x. Then the conditioning relative to the ancillary sub-σ-field $\mathcal{B}$ of G-invariant events is correct. That is, $R(g) = R$ and*

$$R(g, \mathcal{B}, x) = R(\mathcal{B}, x) = \int L\left[g,\, \delta(h^{-1}(x))\right] p(x, hg)\, \mathrm{d}\nu(h)/\overline{p}(x). \tag{5.7}$$

Proof. For $B \in \mathcal{B}$,

$$\begin{aligned}\int_B L(g,\, \delta(x))\, p(x, g)\, \mathrm{d}\mu(x) &= \int_{h^{-1}B} L[g,\, \delta(h(y))]\, p(h(y),\, g)\, \mathrm{d}\mu h(y) \quad (5.8)\\ &= \int_B L(h^{-1}g,\, \delta(y))\, p(y,\, h^{-1}g)\, \mathrm{d}\mu(y).\end{aligned}$$

The theorem easily follows from Theorem 3.4 and from the above relations. □

The following theorem describes an important family of cases of ineffective conditioning.

Theorem 5.3. *If there exists a complete sufficient statistic $T = t(X)$, then $R(\theta, U, u) = R(\theta)$ holds for every ancillary statistic U and every δ which is a function of T.*

Proof. The theorem follows from the well-known fact (see Lehmann (1959), p. 162) that all ancillary statistics are independent of T if T is complete and sufficient. □

Definition 5.2. We shall say that an ancillary sub-σ-field $\mathcal{B}$ yields the *deepest correct conditioning* if, for every nonnegative convex function ψ and every other ancillary sub-σ-field $\overline{\mathcal{B}}$ yielding a correct conditioning,

$$\mathsf{E}\psi[b(X)] \geq \mathsf{E}\psi[\bar{b}(X)] \tag{5.9}$$

holds, with b and $\bar{b}$ corresponding to $\mathcal{B}$ and $\overline{\mathcal{B}}$ by (5.5), respectively.

Theorem 5.4. *Under Condition 3.1 and under the conditions of Theorem 5.2, the ancillary sub-σ-field $\mathcal{B}$ of G-invariant events yields the deepest correct conditioning. Further, if any other ancillary sub-σ-field possesses this property, then $R(\overline{\mathcal{B}}, x) = R(\mathcal{B}, x)$ almost μ-everywhere.*

Proof. Let ν denote the invariant measure on $\mathcal{G}$. Take a $C \in \overline{\mathcal{B}}$ and denote by $\chi_C(x)$ its indicator. Then

$$\phi_C(x) = \int \chi_C(gx)\,\mathrm{d}\nu(g) \tag{5.10}$$

is $\mathcal{B}$-measurable and

$$\mathsf{P}(C) = \int \phi_C(x)\,p(x)\,\mathrm{d}\mu. \tag{5.11}$$

Further, since the loss function is invariant,

$$\int_C L(g,\,\delta(x))\,p(x,g)\,\mathrm{d}\mu = \int \chi_C(g(y))\,L(g_1,\,\delta(y))\,p(y)\,\mathrm{d}\mu(y) \tag{5.12}$$

where g_1 denotes the identity transformation. Consequently, denoting by $R(g, C)$ the conditional risk, given C, we have from (5.10) and (5.12),

$$\mathsf{P}(C)\int R(g, C)\,\mathrm{d}\nu(g) = \int \phi_C(x)\,L(g_1,\,\delta(x))\,p(x)\,\mathrm{d}\mu. \tag{5.13}$$

Now, since $R(g) = R$, according to Theorem 5.2, and since the correctness of $\overline{\mathcal{B}}$ entails $R(g, C) = R\bar{b}_C$, we have

$$\int R(g, C)\,\mathrm{d}\nu(g) = R\bar{b}_C. \tag{5.14}$$

Note that

$$\mathsf{P}(C)\,\bar{b}_C = \int_C \bar{b}(x)\,p(x)\,\mathrm{d}\mu. \tag{5.15}$$

Further, since $\phi_C(x)$ is $\mathcal{B}$-measurable,

$$\begin{aligned}\int \phi_C(x)\,L(g_1,\,\delta(x))\,p(x)\,\mathrm{d}\mu &= \int \phi_C(x)\,R(\mathcal{B},x)\,p(x)\,\mathrm{d}\mu \qquad (5.16)\\ &= \int \phi_C(x)\,Rb(x)\,p(x)\,\mathrm{d}\mu.\end{aligned}$$

By combining together (5.13) through (5.16), we obtain

$$\mathsf{P}(C)\,\bar{b}_C = \int \phi_C(x)\,b(x)\,p(x)\,\mathrm{d}\mu. \tag{5.17}$$

Now, let us assume $\mathsf{E}\psi(\bar{b}(X)) < \infty$. Then, given an $\epsilon > 0$, we choose a finite partition $\{C_k\}$, $C_k \in \overline{\mathcal{B}}$, such that

$$\mathsf{E}\psi(\bar{b}(X)) < \sum_k \mathsf{P}(C_k)\,\psi(\bar{b}_{C_k}) + \epsilon, \tag{5.18}$$

and we note that, in view of (5.11), (5.17), and of convexity of ψ,

$$\mathsf{P}(C_k)\,\psi(\bar{b}_{C_k}) \le \int \phi_{C_k}(x)\,\psi(b(x))\,p(x)\,\mathrm{d}\mu, \tag{5.19}$$

and, in view of (5.10),

$$\sum_k \phi_{C_k}(x) = 1 \tag{5.20}$$

for every x. Consequently, (5.18) through (5.20) entail

$$\mathsf{E}\psi(\bar{b}(X)) < \mathsf{E}\psi(b(X)) + \epsilon. \tag{5.21}$$

Since $\epsilon > 0$ is arbitrary, (5.9) follows. The case $\mathsf{E}\psi(\bar{b}(X)) = \infty$ could be treated similarly.

The second assertion of the theorem follows from the course of proving the first assertion. Actually, we obtain $\mathsf{E}\psi(\bar{b}(X)) < \mathsf{E}\psi(b(X))$ for some ψ, unless $\bar{b}$ is a function of b a. e. Also, conversely, b must be a function of $\bar{b}$, because the two conditionings are equally deep. □

A *restricted conditionality principle.* If all ancillary statistics (sub-σ-fields) yielding the deepest correct conditioning give the same conditional risk, and if there exists at least one such ancillary statistic (sub-σ-field), then the use of the conditional risk is obligatory.

6 LIKELIHOOD

Still more attractive than the principle of conditionality appears the principle of likelihood, which may be formulated along the lines of A. Birnbaum (1962), as follows.

The principle of likelihood. Statistical inference should be based on the likelihood functions only, disregarding the other structure of the particular experiments. Here, again, various interpretations are possible. We suggest the following.

Interpretation. For a given θ-space, the particular decision procedures should be regarded as functionals on a space $\mathcal{L}$ of likelihood functions, where all functions differing in a positive multiplicator are regarded as equivalent. Given an experiment E such that for all outcomes x the likelihood functions $\ell_x(\theta)$ belong to $\mathcal{L}$, and a decision procedure δ in the above sense, we should put

$$\delta(x) = \delta[\ell_x(\cdot)], \tag{6.1}$$

where $\ell_x(\theta) = p_\theta(x)$.

If $\mathcal{L}$ contains only unimodal functions, an estimation procedure of the above kind is the maximum likelihood estimation method.

A. Birnbaum (1962) proved that the principle of conditionality joined with the principle of sufficiency is equivalent to the principle of likelihood. He also conjectured that the principle of sufficiency may be left out in his theorem, and gave a hint, which was not quite clear, for proving it. But his conjecture is really true if we assume that *statistical inferences should be invariant under one-to-one transformations of the x-space to some other homeomorph space.* In the following theorem we shall give the 'decision interpretation' to the principle of conditionality, and the class $\mathcal{E}$ of experiments will be regarded as the class of all experiments such that all their possible likelihood functions belong to some space $\mathcal{L}$.

Theorem 6.1. *Under the above stipulations, the principle of conditionality is equivalent to the principle of likelihood.*

Proof. If F is a subexperiment of E, and if x belongs to the space of possible outcomes of F, the likelihood functions $\ell_x(\theta \mid E)$ and $\ell_x(\theta \mid F)$ differ by a positive multiplicator only, i.e. they are identical. Thus the likelihood principle entails the conditionality principle.

Further, assume that the likelihood principle is violated for a decision procedure δ, that is there exist in $\mathcal{E}$ two experiments $E_1 = (\mathcal{X}_1, \mathsf{P}_{1\theta})$ and $E_2 = (\mathcal{X}_2, \mathsf{P}_{2\theta})$ such that, for some points x_1 and x_2,

$$\delta(E_1, x_1) \neq \delta(E_2, x_2), \tag{6.2}$$

whereas

$$\mathsf{P}_{1\theta}(X_1 = x_1) = c\mathsf{P}_{2\theta}(X_2 = x_2) \quad \text{for all } \theta, \tag{6.3}$$

with $c = c(x_1, x_2)$, but independent of θ. We here assume the spaces $\mathcal{X}_1$ and $\mathcal{X}_2$ finite and disjoint, and the σ-fields to consist of all subsets. Then let us choose a number λ such that

$$0 < \lambda < \frac{1}{c+1} \tag{6.4}$$

and consider the experiment $E = (\mathcal{X}_1 \cup \mathcal{X}_2, \mathsf{P}_\theta)$, where

$$\begin{aligned} \mathsf{P}_\theta(X = x) &= \lambda \mathsf{P}_{1\theta}(x) && \text{if } x \in \mathcal{X}_1, \\ &= \lambda c \mathsf{P}_{2\theta}(x) && \text{if } x \in \mathcal{X}_2 - \{x_2\}, \\ &= 1 - \lambda(c + 1 - \mathsf{P}_{1\theta}(x_1)) && \text{if } x = x_2. \end{aligned} \tag{6.5}$$

Then E conditioned by $x \in \mathcal{X}_1$ coincides with E_1 and E conditioned by $x \in (\mathcal{X}_2 - \{x_2\}) \cup \{x_1\}$ coincides with $\tilde{E}_2$ which is equivalent to E_2 up to a one-to-one transformation $\phi(x) = x$, if $x \in \mathcal{X}_2 - \{x_2\}$ and $\phi(x_2) = x_1$. Thus we should have $\delta(E_1, x_1) = \delta(E, x_1) = \delta(\tilde{E}_2, x_1) = \delta(E_2, x_2)$, according to the conditionality principle. In view of (6.2) this is not true, so that the conditionality principle is violated for δ, too. □

Given a fixed space $\mathcal{L}$, the likelihood principle says what decision procedures of wide scope (covering many experimental situations) are permissible. Having any two such procedures, and wanting to compare them, we must resort either to some particular experiment and compute the risk, or to assume some a priori distribution $\nu(\mathrm{d}\theta)$ and to compute the conditional risk. In both cases we leave the proper ground of the likelihood principle.

However, for some special spaces $\mathcal{L}$, all conceivable experiments with likelihood functions in $\mathcal{L}$ give us the same risk, so that the risk may be regarded as independent of the particular experiment. The most important situation of this kind is treated in the following.

Theorem 6.2. *Let θ be real and $\mathcal{L}_\sigma$ consist of the following functions:*

$$\ell(\theta) = c \exp\left[-\frac{1}{2}\frac{(\theta - t)^2}{\sigma^2}\right], \quad -\infty < t < \infty \tag{6.6}$$

with σ fixed and positive. Then for each experiment with likelihood functions in $\mathcal{L}_\sigma$ there exists a complete sufficient statistic $T = t(X)$, which is normally distributed with expectation θ and variance σ^2.

Proof. Put $\sigma = 1$ for simplicity. We have a measurable space $(\mathcal{X}, t)$ with a σ-finite measure $\mu(\mathrm{d}x)$ such that the densities with respect to μ allow the representation

$$p_\theta(x) = c(x) \exp\left[-\frac{1}{2}(\theta - t(x))^2\right]. \tag{6.7}$$

This relation shows that $T = t(X)$ is sufficient, by the factorization criterion. Further,

$$1 = \int p_\theta(x)\,\mu(\mathrm{d}x) = \int e^{-\frac{1}{2}\theta^2 + \theta t}\,\overline{\mu}(\mathrm{d}t), \tag{6.8}$$

where $\overline{\mu} = \mu^* t^{-1}$ and $\mu^*(\mathrm{d}x) = c(x)\exp\left[-\frac{1}{2}t^2(x)\right]\mu(\mathrm{d}x)$. Now, assuming that there exist two different measures $\overline{\mu}$ satisfying (6.8), we easily derive a contradiction with the completeness of exponential families of distributions (see Lehmann (1959), Theorem 1, p. 132). Thus $\overline{\mu}(\mathrm{d}t) = e^{-t^2/2}\,\mathrm{d}t$ and the theorem is proved. □

The whole work by Fisher suggests that he associated the likelihood principle with the above family of likelihoods, and, asymptotically, with families, which in the limit shrink to $\mathcal{L}_\sigma$ for some $\sigma > 0$ (see Example 8.7). If so, the present author cannot see any serious objections against the principle, and particularly against the method of maximum likelihood. Outside of this area, however, the likelihood principle is misleading, because the information about the kind of experiment does not lose its value even if we know the likelihood.

7 VAGUELY KNOWN PRIOR DISTRIBUTIONS

There are several ways of utilizing a vague knowledge of the prior distribution $\nu(\mathrm{d}\theta)$. Let us examine three of them.

7.1 Diffuse Prior Distributions

Assume that $\nu(\mathrm{d}\theta) = \sigma^{-1} g(\theta/\sigma)\,\mathrm{d}\theta$, where $g(x)$ is continuous and bounded. Then, under general conditions the posterior density $p(\theta \mid X = x, \sigma)$ will tend to a limit for $\sigma \to \infty$, and the limits will, independently of g, correspond to $\nu(\mathrm{d}\theta) = \mathrm{d}\theta$. This will be true especially if the likelihood functions $\ell(\theta \mid X = x)$ are strongly unimodal, that is, if $-\log \ell(\theta \mid X = x)$ is convex in θ for every x, in which case also all moments of finite order will converge. There are, however, at least two difficulties connected with this approach.

Difficulty 1. However large a fixed σ is, for a nonnegligible portion of θ-values placed at the tails of the prior distribution, the experiment will lead to such results x that the posterior distribution with $\nu(\mathrm{d}\theta) = \sigma^{-1} g(\theta/\sigma)\,\mathrm{d}\theta$ will differ significantly from that with $\nu(\theta) = \mathrm{d}\theta$. Thus, independently of the rate of diffusion, our results will be biased in a nonnegligible portion of cases.

Difficulty 2. If interested in rapidly increasing functions of θ, say e^{θ}, and trying to estimate them by the posterior expectation, the expectation will diverge for $\sigma \to \infty$ under the usual conditions (see Example 8.5).

7.2 A Family of Possible Prior Distributions

Given a family of prior distributions $\{\nu_\alpha(\mathrm{d}\theta)\}$, where α denotes some 'meta-parameter', we may either

(i) estimate α by $\hat{\alpha} = \hat{\alpha}(X)$, or
(ii) draw conclusions which are independent of α.

7.3 Making Use of Non-Bayesian Risks

If we know the prior distribution $\nu(\mathrm{d}\theta)$ exactly, we could make use of the Bayesian decision function δ_ν, associating with every outcome x the decision $d = d(x)$ that minimizes

$$R(\nu, x, d) = \int L(\tau(\theta),\, d)\, \nu(\mathrm{d}\theta \mid X = x). \tag{7.1}$$

Simultaneously, the respective minimum

$$R(\nu, x) - \min_{d' \in D} R(\nu, x, d') \tag{7.2}$$

could characterize the risk, given $X = x$.

If the knowledge of ν is vague, we can still utilize δ_ν as a more or less good solution, the quality of which depends on our luck with the choice of ν. Obviously, in such a situation, the use of $R(\nu, x)$ would be too optimistic. Consequently, we had better replace it by an estimate $\hat{R}(\theta, \delta_\nu) = r(X)$ of the usual risk

$$R(\theta, \delta_\nu) = \int L(\theta,\, \delta_\nu(x))\, \mathsf{P}_\theta(\mathrm{d}x). \tag{7.3}$$

Then our notion of risk will remain realistic even if our assumptions concerning ν are not. Furthermore, comparing $R(\nu, x)$ with $\hat{R}(\theta, \delta_\nu) = r(x)$, we may obtain information about the appropriateness of the chosen prior distribution.

8 EXAMPLES

Example 8.1. Let us assume that $\theta_i = 0$ or 1, $\theta = (\theta_1, \ldots, \theta_N)$ and $\tau = \theta_1 + \cdots + \theta_N$. Further, put

$$p_\theta(x_1, \ldots, x_N) = \prod_{i=1}^{N} [f(x_i)]^{\theta_i} [g(x_i)]^{1-\theta_i}, \tag{8.1}$$

where f and g are some one-dimensional densities. In other words, $X = (X_1, \ldots, X_N)$ is a random sample of size N, each member X_i of which comes from a distribution with density either f or g. We are interested in estimating the number of X_i's associated with the density f. Let $T = (T_1, \ldots, T_N)$ be the order statistic, namely $T_1 \leq \cdots \leq T_N$ are the observations $X_1, \ldots, X_N$ rearranged in ascending magnitude.

Proposition 8.1. The vector T is a sufficient statistic for τ in the sense of Definition 2.2, and

$$p_\tau(t) = p_\tau(t_1, \ldots, t_N) = \tau!(N-\tau)! \sum_{s_\tau \in S_\tau} \prod_{i \in s_\tau} f(t_i) \prod_{j \notin s_\tau} g(t_j), \tag{8.2}$$

where S_τ denotes the system of all subsets s_τ of size τ from $\{1, \ldots, N\}$.

Proof. A simple application of Theorem 3.5 to the permutation group. □

Proposition 8.2. For every t

$$p_{\tau+1}(t)\, p_{\tau-1}(t) = p_\tau^2(t) \tag{8.3}$$

holds.

Proof. See Samuels (1965). Relation (8.3) means that $-\log p_\tau(t)$ is convex in τ; that is, the likelihoods are (strongly) unimodal. Thus we could try to estimate τ by the method of maximum likelihood, or still better by the posterior expectation for the uniform prior distribution; if $X_i = 0$ or 1, T is equivalent to $T' = X_1 + \cdots + X_N$. □

This kind of problem occurs in compound decision making. See H. Robbins (1950).

Example 8.2. Let $0 < \sigma^2 < K$, μ real, $\theta = (\mu, \sigma^2)$, and

$$p_\theta(x_1, \ldots, x_n) = \sigma^{-n} (2\pi)^{-n/2} \exp\left\{ -\frac{1}{2} \sum_{i=1}^{n} (x_i - \mu)^2 \sigma^{-2} \right\}. \tag{8.4}$$

Proposition 8.3. The statistic

$$s^2 = (n-1)^{-1} \sum_{i=1}^{n} (x_i - \overline{x})^2$$

is sufficient for σ^2.

Proof. For given $\sigma^2 < K$ we take for the mixing distribution of μ the normal distribution with zero expectation and the variance $(K - \sigma^2)/n$. Then we obtain the mixed density

$$q_\sigma(x_1, \ldots, x_N) = \sigma^{-n+1} (2\pi)^{-n/2} \exp\left\{ -\frac{1}{2} \frac{n\overline{x}^2}{K} - \frac{1}{2} \sum_{i=1}^{n} (x_i - \overline{x})^2 \sigma^{-2} \right\} \tag{8.5}$$

and apply Definition 2.2. □

Remark 1. If $K \to \infty$, then the mixing distribution tends to the uniform distribution over the real line, which is not finite. For practical purposes the bound K means no restriction, since such a K always exists.

Remark 2. A collection of examples appropriate for illustrating the notion of sufficiency for a subparameter could be found in J. Neyman and E. Scott (1948).

Example 8.4. If τ attains only two values, say τ_0 and τ_1 and if it admits a sufficient statistic in the sense of Definition 2.2, then the respective distributions Q_{τ_0} and Q_{τ_1} are *least favorable* for testing the composite hypothesis $\tau = \tau_0$ against the composite alternative $\tau = \tau_1$ (see Lehmann (1959)). In particular, if T is ancillary under $\tau(\theta) = \tau_0$, and if

$$p_\theta(x) = r_\tau(t(x))\, p_{\overline{b}}(x), \quad \theta = (\tau, \overline{b}), \tag{8.6}$$

where the range of $\overline{b}$ is independent of the range of τ, then T is sufficient for τ in the sense of Definition 2.2, and the respective family $\{\mathsf{Q}_\tau\}$ may be defined so that we choose an arbitrary b, say b_0, and then put $\mathsf{Q}_\tau(\mathrm{d}x) = r_\tau(t(x))\, p_{b_0}(x)\, \mu(\mathrm{d}x)$. In this way asymptotic sufficiency of the vector of ranks for a class of testing problems is proved in Hájek and Šidák (1967). (See also Remark 2.4.)

Example 8.5. The standard theory of probability sampling from finite populations is linear and nonparametric. If we have any information about the *type* of distribution in the population, and if this type could make nonlinear estimates preferable, we may proceed as follows.

Consider the population values $Y_1, \ldots, Y_N$ as a sample from a distribution with two parameters. Particularly, assume that the random variables $X_i = h(Y_i)$, where h is a known strictly increasing function, are normal (μ, σ^2). For example, if $h(y) = \log y$, then the Y_is are log-normally distributed. Now a simple random sample may be identified with the partial sequence $Y_1, \ldots, Y_n$, $2 < n < N$. Our task is to estimate $Y = Y_1 + \cdots + Y_N$, or, equivalently, as we know $Y_1 + \cdots + Y_n$, to estimate

$$Z = Y_{n+1} + \cdots + Y_N. \tag{8.7}$$

Now the minimum variance unbiased estimate of Z equals

$$\hat{Z} = (N-n)\left[B\left(\frac{1}{2}n-1, \frac{1}{2}n-1\right)\right]^{-1} \times \int_0^1 h^{-1}\left[\overline{x} + (2v-1)\, s(n-1)\, n^{-1/2}\right] [v(1-v)]^{n/2-2}\, \mathrm{d}y, \tag{8.8}$$

where

$$\overline{x} = \frac{1}{n}\sum_{i=1}^{n} h(y_i), \quad s^2 = \frac{1}{n-1}\sum_{i=1}^{n}(h(y_i) - \overline{x})^2. \tag{8.9}$$

On the other hand, choosing the usual 'diffuse' prior distribution $\nu\,(\mathrm{d}\mu\,\mathrm{d}\sigma) = \sigma^{-1}\,\mathrm{d}\mu\,\mathrm{d}\sigma$, we obtain for the conditional expectation of Z, given $Y_i = y_i$, $i = 1, \ldots, n$, the following result:

$$\tilde{Z} = \mathsf{E}(Z \mid Y_i = y_i,\, 1 \le i \le n, \nu) = (N-n)\,(n-1)^{-1/2} \times \left[B\left(\frac{1}{2}n - \frac{1}{2}, \frac{1}{2}\right)\right]^{-1} \int_{-\infty}^{\infty} h^{-1}\left(\overline{x} + vs(1+1/n)^{1/2}\right)\left(1 + \frac{v^2}{n-1}\right)^{-n/2} \mathrm{d}v. \tag{8.10}$$

However, for most important functions h, for example for $h(y) = \log y$, that is $h^{-1}(x) = e^x$, we obtain $\tilde{Z} = \infty$. Thus the Bayesian approach should be based on some other prior distribution. In any case, however, the Bayesian solution would be too sensitive with respect to the choice of $\nu(\mathrm{d}\mu\,\mathrm{d}\sigma)$.

Exactly the same unpleasant result (8.10) obtains by the method of fiducial prediction recommended by R. A. Fisher (1956, p. 116).

Example 8.6. Consider the following method of sampling from a finite population of size N: each unit is selected by an independent experiment, and the probability of its being included in the sample equals n/N. Then the sample size K is a binomial random variable with expectation n. To avoid empty samples, let us reject samples of size $< k_0$, where k_0 is a positive integer, so that K will have a truncated binomial distribution. Further, let us assume that we are estimating the population total $Y = y_1 + \cdots + y_N$ by the estimator

$$\breve{Y} = \frac{N}{K} \sum_{i \in s_K} y_i, \tag{8.11}$$

where s_K denotes a sample of size K. Then

$$\mathsf{E}\left((\breve{Y} - Y)^2 \mid K = k\right) = \frac{N(N-k)}{k}\,\sigma^2 \tag{8.12}$$

holds, where σ^2 is the population variance $\sigma^2 = (N-1)^{-1}\sum_{i=1}^{N}(y_i - \overline{Y})^2$. Thus K yields a correct conditioning. Further, the deepest correct conditioning is that relative to K.

On the other hand, simple random sampling of fixed size does not allow effective correct conditioning, unless we restrict somehow the set of possible $(y_1, \ldots, y_N)$-values.

Example 8.7. Let $f(x)$ be a continuous one-dimensional density such that $-\log f(x)$ is convex. Assume that

$$I = \int_{-\infty}^{\infty}\left[\frac{f'(x)}{f(x)}\right]^2 f(x)\,\mathrm{d}x < \infty. \tag{8.13}$$

Now for every integer N put

$$p_\theta(x_1, \ldots, x_N) = \prod_{i=1}^{N} f(x_i - \theta) \tag{8.14}$$

and

$$\ell_x(\theta) = p_\theta(x_1, \ldots, x_N), \quad -\infty < \theta < \infty. \tag{8.15}$$

Let $t(x)$ be the mode (or the mid-mode) of the likelihood $\ell_x(\theta)$, that is,

$$\ell_x(t(x)) \geq \ell_x(\theta), \quad -\infty < \theta < \infty. \tag{8.16}$$

Since $f(x)$ is strictly unimodal, $t(x)$ is uniquely defined. Then put

$$\ell_x^*(\theta) = c_N\, I^{1/2}(2\pi)^{-1/2}\exp\left[-\frac{1}{2}(\theta - t(x))^2\, I\right], \tag{8.17}$$

where c_N is chosen so that

$$\int_{-\infty}^{\infty} \ell_x^*(\theta)\,\mathrm{d}x_1 \cdots \mathrm{d}x_N = 1. \tag{8.18}$$

Then $c_N \to 1$ and

$$\lim_{N\to\infty}\int |\ell_x(\theta) - \ell_x^*(\theta)|\,\mathrm{d}x_1 \cdots \mathrm{d}x_N = 0, \tag{8.19}$$

the integrals being independent of θ. Thus, for large N, the experiment with likelihoods ℓ may be approximated by an experiment with normal likelihoods ℓ^*. Similar results may be found in Le Cam (1958) and Huber (1964).

Example 8.8. Let $\theta = (\theta_1, \ldots, \theta_N)$,

$$p_\theta(x_1, \ldots, x_N) = (2\pi)^{-N/2} \exp\left(-\frac{1}{2}\sum_{i=1}^{N}(x_i - \theta_i)^2\right), \tag{8.20}$$

and let $(\theta_1, \ldots, \theta_N)$ be regarded as a sample from a normal distribution (μ, σ^2). If μ and σ^2 were known, we would obtain the following joint density of (x, θ):

$$\begin{aligned} & r_\nu(x_1, \ldots, x_N, \theta_1, \ldots, \theta_N) \qquad (8.21) \\ = \; & \sigma^{-N}(2\pi)^{-N} \exp\left(-\frac{1}{2}\sum_{i=1}^{N}(x_i - \theta_i)^2 - \frac{1}{2}\sigma^{-2}\sum_{i=1}^{N}(\theta_i - \mu)^2\right). \end{aligned}$$

Consequently, the best estimator of $(\theta_1, \ldots, \theta_N)$ would be $(\hat\theta_1, \ldots, \hat\theta_N)$ defined by

$$\hat\theta_i = \frac{\mu + \sigma^2 x_i}{1 + \sigma^2}. \tag{8.22}$$

Now, in the prior experiment, (μ, σ^2) can be estimated by $(\theta, \sigma_\theta^2)$, where

$$\overline{\theta} = \frac{1}{N}\sum_{i=1}^{N}\theta_i, \quad s_\theta^2 = \frac{1}{N-1}\sum_{i=1}^{N}(\theta_i - \overline{\theta})^2. \tag{8.23}$$

In the x-experiment, in turn, a sufficient pair of statistics for $(\overline{\theta}, s_\theta^2)$ is $(\overline{x}, s^2)$, where

$$\overline{x} = \frac{1}{N}\sum_{i=1}^{N}x_i, \quad s^2 = \frac{1}{N-1}\sum_{i=1}^{N}(x_i - \overline{x})^2. \tag{8.24}$$

Now, while $\overline{\theta}$ may be estimated by $\overline{x}$, the estimation of s_θ^2 must be accomplished by some more complicated function of s^2. We know that the distribution of $(N-1)\,s^2$ is noncentral χ^2 with $(N-1)$ degrees of freedom and the parameter of noncentrality $(N-1)\,s_\theta^2$. Denoting the distribution function of that distribution by $F_{N-1}(x, \delta)$, where δ denotes the parameter of noncentrality, we could estimate s_θ^2 by $\hat s_\theta^2 = h(s^2)$, where $h(s^2)$ denotes the solution of

$$F_{N-1}((N-1)\,s^2, (N-1)\,s_\theta^2) = \frac{1}{2} \tag{8.25}$$

(see Section 4.3). On substituting the estimate in (8.21), one obtains modified estimators

$$\tilde\theta_i = \frac{\overline{x} + \hat s_\theta^2 x_i}{1 + \hat s_\theta^2}. \tag{8.26}$$

The estimators (8.26) will be for large N nearly as good as the estimators (8.22) (cf. Stein (1956)).

The estimators (8.26) could be successfully applied to estimating the averages in individual strata in sample surveys. They represent a compromise between estimates based on observations from the same stratum only, which are unbiased but have large variance, and the overall estimates which are biased but have small variance.

The same method could be used for $\theta_i = 1$ or 0, and $(\theta_1, \ldots, \theta_N)$ regarded as a sample from an alternative distribution with unknown p. The parameter p could then be estimated along the lines of Example 8.1 (cf. H. Robbins (1950)).

9 CONCLUDING REMARKS

The genius of classical statistics is based on skillful manipulations with the notions of sufficiency, similarity, and conditionality. The importance of similarity increases after introducing the notion of completeness. An adequate imbedding of similarity and conditionality into the framework of the general theory of decision making is not straightforward. Classical statistics provided richness of various methods and did not care too much about criteria. The decision theory added a great deal of criteria, but stimulated very few new methods. From the point of view of decision making, risk is important only before we make a decision. After the decision has been irreversibly done, the risk is irrelevant, and, for instance, its estimation or speculating about the conditional risk makes no sense. Such an attitude seems to be strange to the spirit of classical statistics. If regarding statistical problems as games against Nature, one must keep in mind that besides randomizations introduced by the statistician, there are randomizations (ancillary statistics) involved intrinsically in the structure of the experiment. Such randomization may make the transfer to conditional risks facultative.

REFERENCES

Birnbaum A. (1962). On the foundations of statistical inference. *J. Amer. Statist. Assoc. 57*, 269–326.

Blackwell D. and Girschick M. A. (1954). *Theory of Games and Statistical Decisions*. New York, Wiley.

de Finetti B. and Savage L. J. (1962). Sul modo di scegliere le probabilita iniziali. *Sui Fondamenti della Statistica, Biblioteca del Metron, Series C, I*, 81–147.

Dynkin E. B. (1959). *The Foundations of the Theory of Markovian Processes*. Moscow, Fizmatgiz. (In Russian)

Fisher R. A. (1956). *Statistical Methods and Scientific Inference*. London, Oliver and Boyd.

Hájek J. and Šidák Z. (1967). *The Theory of Rank Tests*. Prague, Publishing House of the Czech Academy of Sciences.

Huber P. J. (1964). Robust estimation of a location parameter. *Ann. Math. Statist.* *35*, 73–101.

Kolmogorov A. N. (1942). Sur l'estimation statistique des parameters de la loi de Gauss. *Izv. Akad. Nauk SSSR Ser. Mat.* *6*, 3–32.

Le Cam L. (1958). Les propriétés asymptotiques des solutions de Bayes. *Publ. Inst. Statist. Univ. Paris* *7*, 3–4.

Le Cam. L. (1964). Sufficiency and approximate sufficiency. *Ann. Math. Statist.* *35*, 1419–1455.

Lehmann E. L. (1959). *Testing Statistical Hypotheses*. New York, Wiley.

Neyman J. (1962). Two breakthroughs in the theory of decision making. *Rev. Inst. Internat. Statist.* *30*, 11–27.

Neyman J. and Scott E. L. (1948). Consistent estimates based on partially consistent observations. *Econometrica* *16*, 1–32.

Robbins H. (1950). Asymptotically subminimax solutions of compound decision problems. In *Proceedings of the Second Berkeley Symposium on Mathematical Statistics and Probability*, Berkeley and Los Angeles, University of California Press, pp. 131–148.

Samuels S. M. (1965). On the number of successes in independent trials. *Ann. Math. Statist.* *36*, 1272–1278.

Stein Ch. (1956). Inadmissibility of the usual estimator for the mean of multivariate normal distribution. In *Proceedings of the Third Berkeley Symposium on Mathematical Statistics and Probability*, Berkeley and Los Angeles, University of California Press, Vol. I, pp. 197–206.

CHAPTER 27

Locally Most Powerful Rank Tests of Independence

In: Studies in Math. Statist., 1968 (K. Sarkadi and I. Vincze, eds.), 45–51 Akadémiai Kiadó, Budapest.
Reproduced by permission of Akadémiai Kiadó, Budapest.

1 THEORY

Let $(X_1, Y_1), \ldots, (X_N, Y_N)$ denote a random sample from a bivariate distribution. Under a hypothesis, say H, we shall assume that the random variables X_i and Y_i are independent and distributed according to same unknown densities $f(x)$ and $g(x)$, respectively. Thus H is composite and may be regarded as a family of $2N$-dimensional densities

$$p(x_1, y_1, \ldots, x_N, y_N) = \prod_{i=1}^{N} f(x_i)\, g(y_i),$$

where f and g both run through the set of all one-dimensional densities. The hypothesis H will be tested against a simple alternative

$$q_\Delta(x_1, y_1, \ldots, x_N, y_N) = \prod_{i=1}^{N} h_\Delta(x_i, y_i),$$

where

$$h_\Delta(x, y) = \int_{-\infty}^{\infty} f_0(x - \Delta z)\, g_0(x - \Delta z)\, \mathrm{d}M(z)$$

with f_0 and g_0 being some particular densities, and $M(z)$ being a distribution function with positive and finite variance

$$0 < \sigma^2 = \int z^2\, \mathrm{d}M(z) - \left(\int z\, \mathrm{d}M(z)\right)^2 < \infty.$$

Collected Works of Jaroslav Hájek – With Commentary
Edited by M. Hušková, R. Beran and V. Dupač
Published in 1998 by John Wiley & Sons Ltd.

Thus, under the alternative, we assume that

$$X_i = X_i^0 + \Delta Z_i,$$

$$Y_i = Y_i^0 + \Delta Z_i,$$

where X_i^0, Y_i^0, Z_i are independent variables with respective distributions determined by f_0, g_0 and $M(z)$, and Δ denotes a real number.

Since the hypothesis is very wide (non-parametric) we restrict ourselves to rank tests. Moreover, since it is difficult to find explicitly the most powerful rank test, we shall seek the locally most powerful rank test. Recall that a rank test is called locally most powerful for testing H against q_Δ if it is most powerful for all Δ in some neighbourhood of 0. Further recall that a test is called a rank test for testing H if its critical region (or its critical function) is determined by $(R_1, \ldots, R_N, Q_1, \ldots, Q_N)$, where R_i denotes the rank of X_i within the sample $(X_1, \ldots, X_N)$ and Q_i denotes the rank of Y_i within the sample $(Y_1, \ldots, Y_N)$. Denoting the ith order statistic for $(X_1, \ldots, X_N)$ by $X^{(i)}$, and the ith order statistic in $(Y_1, \ldots, Y_N)$ by $Y^{(i)}$, we have

$$X_i = X^{(R_i)}, \quad Y_i = Y^{(Q_i)}, \quad 1 \leq i \leq N.$$

Now we must introduce some regularity conditions on f_0 and g_0. We shall assume that f_0 and g_0 are both absolutely continuous on finite intervals, i.e. that there exist functions f_0' and g_0' such that

$$f(x) - f(x_0) = \int_{x_0}^{x} f_0'(t)\,\mathrm{d}t$$

and

$$g(x) - g(x_0) = \int_{x_0}^{x} g_0'(t)\,\mathrm{d}t.$$

Furthermore, we shall suppose that

$$\int_{-\infty}^{\infty} |f_0'(t)|\,\mathrm{d}t < \infty \tag{1.1}$$

and

$$\int_{-\infty}^{\infty} |g_0'(t)|\,\mathrm{d}t < \infty \tag{1.2}$$

and that f_0' and g_0' are both *continuous almost everywhere*.

Finally, introduce the following quantities (called *scores*):

$$a_N(i, f_0) = \mathsf{E}_0\left\{f_0'(X^{(i)})/f_0(X^{(i)})\right\}, \quad 1 \leq i \leq N,$$

where E_0 refers to a sample $X_1, \ldots, X_N$ from a distribution with density f_0.

Theorem. Under the above assumptions and notations, the test with the critical region

$$\sum_{i=1}^{N} a_N(R_i, f_0)\, a_N(Q_i, g_0) \geq k \tag{1.3}$$

is the locally most powerful rank test for H against q_Δ on level $\alpha = \alpha(k)$, with α equalling the probability of (1.3) under H.

Proof. Let $r = (r_1, \dots, r_N)$, $r' = (r'_1, \dots, r'_N)$ be some permutations of $(1, \dots, N)$, and let us put $R = (R_1, \dots, R_N)$, $Q = (Q_1, \dots, Q_N)$. Further, let us denote by P_Δ the probability distribution corresponding to the density q_Δ, $\mathrm{d}\mathsf{P}_\Delta = q_\Delta\, \mathrm{d}x$. We shall show that

$$\lim_{\Delta \to 0} \frac{1}{\Delta^2} \left[(N!)^2\, \mathsf{P}_\Delta(R = r,\, Q = r') - 1\right] = \sigma^2 \sum_{i=1}^{N} a_N(r_i, f_0)\, a_N(r'_i, g_0),$$

from which our assertion follows immediately. Actually, then each pair (r, r') such that $\Sigma\, a_N(r_i, f_0)\, a_N(r'_i, g_0) \geq k$ is more probable under P_Δ than each pair (s, s') such that $\Sigma\, a_N(s_i, f_0)\, a_N(s'_i, g_0) < k$, for sufficiently small Δ.

Introduce densities

$$f_\Delta(x) = \int f(x - \Delta z)\, \mathrm{d}M(z)$$

$$g_\Delta(x) = \int g(x - \Delta z)\, \mathrm{d}M(z)$$

and note that

$$(N!)^2 \int \underset{\substack{R=r,\\ Q=r'}}{\cdots} \int \prod_{i=1}^{N} f_\Delta(x_i)\, g_\Delta(y_i)\, \mathrm{d}x_1 \cdots \mathrm{d}x_N\, \mathrm{d}y_1 \cdots \mathrm{d}y_N = 1.$$

Consequently,

$$\begin{aligned}
&\frac{1}{\Delta^2} \left[(N!)^2\, \mathsf{P}_\Delta(R = r,\, Q = r') - 1\right] \\
&= (N!)^2 \int \underset{\substack{R=r,\\ Q=r'}}{\cdots} \int \Delta^{-2} \left[\prod_{i=1}^{N} h_\Delta(x_i, y_i) - \prod_{i=1}^{N} f_\Delta(x_i)\, g_\Delta(y_i)\right] \\
&\qquad \times \mathrm{d}x_1 \cdots \mathrm{d}x_N\, \mathrm{d}y_1 \cdots \mathrm{d}y_N \\
&= \sum_{k=1}^{N} (N!)^2 \int \underset{\substack{R=r,\\ Q=r'}}{\cdots} \int \left[\frac{h_\Delta(x_k, y_k) - f_\Delta(x_k)\, g_\Delta(y_k)}{\Delta^2}\right] \\
&\quad \times \left[\prod_{i=k+1}^{N} f_\Delta(x_i)\, g_\Delta(x_i) \prod_{j=1}^{k-1} h_\Delta(x_j, y_j)\right] \mathrm{d}x_1 \cdots \mathrm{d}x_N\, \mathrm{d}y_1 \cdots \mathrm{d}y_N. \qquad (1.4)
\end{aligned}$$

Now, obviously,

$$\begin{aligned}\lim_{\Delta\to 0} f_\Delta(x)\, g_\Delta(y) &= f(x)\, g(y)\\ \lim_{\Delta\to 0} h_\Delta(x,y) &= f(x)\, g(y)\end{aligned}$$

in every point (x, y). If we further show that

$$\lim_{\Delta\to 0} \frac{h_\Delta(x,y) - f_\Delta(x)\, g_\Delta(y)}{\Delta^2} = \sigma^2 f'(x)\, g'(y) \tag{1.5}$$

almost everywhere, and that the limit may be carried out under the integral sign, (1.4) will entail

$$\begin{aligned}&\lim_{\Delta\to 0} \frac{1}{\Delta^2}\left[(N!)^2\, \mathsf{P}_\Delta(R=r,\, Q=r') - 1\right]\\
&= \sum_{k=1}^N (N!)^2 \int \underset{\substack{R=r,\\ Q=r'}}{\cdots} \int \sigma^2 \frac{f_0'(x_k)}{f_0(x_k)} \frac{g_0'(y_k)}{g_0(y_k)} \prod_{i=1}^N f_0(x_i)\, g_0(y_i)\\
&\qquad \times \mathrm{d}x_1 \cdots \mathrm{d}x_N\, \mathrm{d}y_1 \cdots \mathrm{d}y_N\\
&= \sigma^2 \sum_{k=1}^N \left[N! \int \underset{R=r}{\cdots} \int \frac{f_0'(x_k)}{f_0(x_k)} \prod_{i=1}^N f_0(x_i)\, \mathrm{d}x_1 \cdots \mathrm{d}x_N\right]\\
&\qquad \times \left[N! \int \underset{Q=r'}{\cdots} \int \frac{g_0'(y_k)}{g_0(y_k)} \prod_{i=1}^N g_0(y_i)\, \mathrm{d}y_1 \cdots \mathrm{d}y_N\right]\\
&= \sigma^2 \sum_{k=1}^N \mathsf{E}\left\{\frac{f_0'(X^{(r_k)})}{f_0(X^{(r_k)})}\right\} \mathsf{E}\left\{\frac{g_0'(Y^{(r_k')})}{g_0(Y^{(r_k')})}\right\} = \sigma^2 \sum_{k=1}^N a_N(r_k, f_0)\, a_N(r_k', g_0).\end{aligned}$$

In the above relations we have used the fact that $f(x) = 0$ implies $f'(x) = 0$ in every point, where the derivative exists.

Thus it remains to show that (1.5) holds almost everywhere and that passing to the limit under the integral sign is justified.

We begin with the proof of (1.5). We can write

$$\begin{aligned}&h_\Delta(x,y) - f_\Delta(x)\, g_\Delta(y) \qquad (1.6)\\
&= \iint_{z<z'} [f_0(x-\Delta z) - f_0(x-\Delta z')]\, [g_0(y-\Delta z) - g_0(y-\Delta z')]\\
&\qquad \times \mathrm{d}M(z)\, \mathrm{d}M(z').\end{aligned}$$

If x is a continuity point of f_0', and y is a continuity point of g_0', then

$$\begin{aligned}&[f_0(x-\Delta z) - f_0(x-\Delta z')]\, [g_0(y-\Delta z) - g_0(y-\Delta z')] \qquad (1.7)\\
&= \Delta^2 (z-z')^2\, [f_0'(x)\, g_0'(y) + \varepsilon],\end{aligned}$$

where $\varepsilon \to 0$ for $\max\{|\Delta z|, |\Delta z'|\} \to 0$.

Furthermore, in view of (1.1) and (1.2), the integrand in (1.6) is uniformly bounded, and

$$\iint_{\substack{\max\{|\Delta z|, |\Delta z'|\} > \delta \\ z < z'}} \mathrm{d}M(z)\,\mathrm{d}M(z') \leq (\Delta/\delta)^2 \int_{|z|>\delta/\Delta} z^2\,\mathrm{d}M(z). \tag{1.8}$$

Now, (1.5) almost everywhere obviously follows from (1.6), (1.7) and (1.8), recalling that f_0' and g_0' are continuous almost everywhere and that $\int z^2\,\mathrm{d}M(z) < \infty$.

In conclusion we show that

$$\limsup_{\Delta\to 0} \int_{-\infty}^{\infty} \cdots \int_{-\infty}^{\infty} \Delta^{-2} |h_\Delta(x_k, y_k) - f_\Delta(x_k)\, g_\Delta(y_k)| \tag{1.9}$$
$$\times \prod_{1=k+1}^{N} f_\Delta(x_i)\, g_\Delta(x_i) \prod_{j=1}^{k-1} h_\Delta(x_j, y_j)\,\mathrm{d}x_1 \cdots \mathrm{d}x_N\,\mathrm{d}y_1 \cdots \mathrm{d}y_N$$
$$\leq \int_{-\infty}^{\infty} \cdots \int_{-\infty}^{\infty} \sigma^2 |f_0'(x_k)\, g_0'(y_k)| \prod_{i\neq k}^{N} f_0(x_i)\, g_0(x_i)\,\mathrm{d}x_1 \cdots \mathrm{d}x_N\,\mathrm{d}y_1 \cdots \mathrm{d}y_N$$

holds. It may be shown that (1.9) is a sufficient condition for passing to the limit under the integral sign, with the integral being extended over any Borel subset, and a fortiori over the subset $\{R = r, Q = r'\}$. First observe that

$$\int_{-\infty}^{\infty} \cdots \int_{-\infty}^{\infty} \Delta^{-2} |h_\Delta(x_k, y_k) - f_\Delta(x_k)\, g_\Delta(y_k)| \tag{1.10}$$
$$\times \left[\prod_{i=k+1}^{N} f_\Delta(x_i)\, g_\Delta(x_i) \prod_{j=1}^{k-1} h_\Delta(x_j, y_j) \right] \mathrm{d}x_1 \cdots \mathrm{d}x_N\,\mathrm{d}y_1 \cdots \mathrm{d}y_N$$
$$= \int_{-\infty}^{\infty} \int_{-\infty}^{\infty} \Delta^{-2} |h_\Delta(x_k, y_k) - f_\Delta(x_k)\; g_\Delta(y_k)|\,\mathrm{d}x_k\,\mathrm{d}y_k$$

and, similarly,

$$\int_{-\infty}^{\infty} \cdots \int_{-\infty}^{\infty} \sigma^2 |f_0'(x_k)\, g_0'(y_k)| \prod_{i=k}^{N} f_0(x_i)\, g_0(y_i) \tag{1.11}$$
$$\times \mathrm{d}x_1 \cdots \mathrm{d}x_N\,\mathrm{d}y_1 \cdots \mathrm{d}y_N$$
$$= \sigma^2 \int_{-\infty}^{\infty} \int_{-\infty}^{\infty} |f_0'(x_k)\, g_0'(y_k)|\,\mathrm{d}x_k\,\mathrm{d}y_k.$$

Further, note that (1.6) entails

$$|h_\Delta(x_k, y_k) - f_\Delta(x_k)\, g_\Delta(y_k)|$$

$$\leq \iint_{z<z'} |f_0(x-\Delta z) - f_0(x-\Delta z')|\,|g_0(y-\Delta z) - g_0(y-\Delta z')| \times \mathrm{d}M(z)\,\mathrm{d}M(z')$$

$$\leq \iint_{z<z'} \int_{\Delta z}^{\Delta z'} \int_{\Delta z}^{\Delta z'} |f_0'(x-t)\,g_0'(y-s)|\,\mathrm{d}t\,\mathrm{d}s\,\mathrm{d}M(z)\,\mathrm{d}M(z').$$

Consequently, interchanging the integration order (which is justified with non-negative integrands), we obtain

$$\int_{-\infty}^{\infty}\int_{-\infty}^{\infty} |h_\Delta(x_k, y_k) - f_\Delta(x_k)\,g_\Delta(y_k)|\,\mathrm{d}x_k\,\mathrm{d}y_k \tag{1.12}$$

$$\leq \iint_{z<z'} \int_{\Delta z}^{\Delta z'}\int_{\Delta z}^{\Delta z'}\int_{-\infty}^{\infty}\int_{-\infty}^{\infty} |f_0'(x_k - t) \times g_0'(y_k - t)\,|\mathrm{d}x_k\,\mathrm{d}y_k\,\mathrm{d}t\,\mathrm{d}s\,\mathrm{d}M(z)\,\mathrm{d}M(z')$$

$$= \int_{-\infty}^{\infty}\int_{-\infty}^{\infty} |f_0'(x_k)\,f_0'(y_k)|\,\mathrm{d}x_k\,\mathrm{d}y_k \times \iint_{z<z'} \int_{\Delta z}^{\Delta z'}\int_{\Delta z}^{\Delta z'} \mathrm{d}t\,\mathrm{d}s\,\mathrm{d}M(z)\,\mathrm{d}M(z')$$

$$= \int_{-\infty}^{\infty}\int_{-\infty}^{\infty} |f_0'(x_k)\,g_0'(y_k)|\,\mathrm{d}x_k\,\mathrm{d}y_k \iint_{z<z'} \Delta^2(z-z')\,\mathrm{d}M(z)\,\mathrm{d}M(z')$$

$$= \Delta^2\sigma^2 \int_{-\infty}^{\infty}\int_{-\infty}^{\infty} |f_0'(x_k)\,g_0'(y_k)|\,\mathrm{d}x_k\,\mathrm{d}y_k.$$

Now (1.9) follows from (1.10), (1.11) and (1.12). □

2 EXAMPLES

If f_0 and g_0 are both logistic densities,

$$f_0 = g_0 = e^{-x}(1+e^{-x})^{-2},$$

then

$$a_N(i, f_0) = b_N(i, g_0) = \frac{2i}{N+1} - 1$$

and the locally most powerful test coincides with one based on the Spearman statistic (see Kendall (1955))

$$\rho = \frac{12}{N^3 - N} \sum_{i=1}^{N} \left(R_i - \frac{N+1}{2} \right) \left(Q_i - \frac{N+1}{2} \right).$$

If f_0 and g_0 are both normal densities, then the locally most powerful test is based on

$$S = \sum_{i=1}^{N} a_N(R_i)\, a_N(Q_i), \tag{2.1}$$

where

$$a_N(i) = \mathsf{E} Z_N^{(i)},$$

with $Z_N^{(i)}$ being the ith order statistic in a normal sample of size N. Thus S may be called a *rank correlation coefficient with normal scores.*

Very often approximate scores are preferred to exact ones. We put, for example,

$$a_N(i, f_0) \doteq \varphi\left(\frac{i}{N+1}, f_0 \right), \tag{2.2}$$

where

$$\varphi(u, f_0) = -\frac{f'(F^{-1}(u))}{f(F^{-1}(u))}.$$

Thus, for f_0 and g_0 both normal, (2.1) may be used with

$$a_N(i) = \Phi^{-1}\left(\frac{i}{N+1} \right),$$

with Φ^{-1} denoting the inverse of the standardized normal distribution function. In this case we could speak about a rank correlation coefficient of van der Waerden type.

Finally, if

$$f_0(x) = g_0(x) = \frac{1}{2} e^{-|x|},$$

then (2.2) yields

$$a_N(i, f_0) = b_N(i, g_0) \doteq \text{sign}\left(i - \frac{N+1}{2} \right)$$

and the test comes to be the *quadrant* test (see Elandt (1962)).

3 MOMENTS AND DISTRIBUTION

Consider a general statistic

$$S = \sum_{i=1}^{N} a(R_i)\, b(Q_i).$$

Then under the hypothesis,

$$\mathsf{E}\, S = N\, \overline{a}\, \overline{b}$$

and

$$\operatorname{var} S = (N-1)^{-1}\, \sigma_a^2\, \sigma_b^2,$$

where

$$\begin{aligned} \overline{a} &= \frac{1}{N}\sum_{i=1}^{N} a(i), \\ \sigma_a^2 &= (N-1)^{-1}\sum_{i=1}^{N}\,[a(i)-\overline{a}]^2 \end{aligned}$$

and $\overline{b}$ and σ_b^2 are defined similarly.

If

$$\begin{aligned} a(i) &= a_N(i, f_0), \\ b(i) &= b_N(i, g_0), \end{aligned}$$

where

$$\begin{aligned} \int_{-\infty}^{\infty} [f_0'(x)/f_0(x)]^2\, f_0(x)\,\mathrm{d}x &< \infty \\ \int_{-\infty}^{\infty} [g_0'(x)/g_0(x)]^2\, g_0(x)\,\mathrm{d}x &< \infty, \end{aligned}$$

then S is asymptotically normal ($\mathsf{E}\, S$, $\operatorname{var} S$) for $N \to \infty$. This may be derived from the results of Hájek (1961).

REFERENCES

Elandt R. C. (1962). Exact and approximate power of the non–parametric test of tendency. *Ann. Math. Stat. 33*, 471–481.

Hájek J. (1961). Some extensions of the Wald–Wolfowitz–Noether theorem. *Ann. Math. Stat. 32*, 506–523.

Kendall M. G. (1955). *Rank Correlation Method.* London, Charles Griffin and Co. 2nd ed.

CHAPTER 28

Asymptotic Normality of Simple Linear Rank Statistics Under Alternatives

Ann. Math. Statist. 39 (1968), 325–346.
Reproduced by permission of The Institute of Mathematical Statistics.

SUMMARY

Results of Chernoff–Savage (1958) and Govindarajulu–Le Cam–Raghavachari (1966) are extended from the two-sample case to the general regression case and, simultaneously, the conditions on the score-generating function are relaxed. The main results are stated in Section 2 and their proofs are given in Section 5. Sections 3 and 4 contain auxiliary propositions, on which the methods of the present paper are based. Section 6 includes a counterexample, showing that the theorem cannot be extended to discontinuous score-generating functions.

1 INTRODUCTION

Simple linear rank statistics, given by (2.3) below, besides being essential in non-parameteric theory also provide a key to solving some problems of general theory such as establishment of asymptotically most powerful tests for some problems. Under the alternative, the distribution of a simple linear rank statistic is determined by the following three entities: first, regression

Collected Works of Jaroslav Hájek – With Commentary
Edited by M. Hušková, R. Beran and V. Dupač
Published in 1998 by John Wiley & Sons Ltd.

constants $c_1, \ldots, c_N$, second, distribution functions of individual observations $F_1, \ldots, F_N$, third, scores $a(1), \ldots, a(N)$. The scores are usually assumed to be generated by a function φ, e. g. by (2.4) below. The same might be assumed concerning the regression constants $c_1, \ldots, c_N$. The distribution functions may be derived from a parametric family $F(x, \theta)$, $\theta \in \Omega$. If $c_i = 1$ or 0, we have the so-called *two-sample* problem.

The central problem concerning simple linear rank statistics is their asymptotic normality either with 'natural' parameters ($\mathsf{E}S$, $\operatorname{var} S$) or with some other parameters (μ, σ^2). Under some regularity conditions the answer is positive. And, as may be expected, less regularity in one entity may be counterbalanced by more regularity in the other. The most regular $(c_1, \ldots, c_N)$ are those generated by a linear function, i. e. $c_i = a + ib$, $1 \leq i \leq N$; the next condition, in descending restrictivity, is boundedness of $N \max_{1 \leq i \leq N}(c_i - \bar{c})^2 \big/ \sum_{i=1}^N (c_i - \bar{c})^2$; the mildest condition yet used in the literature is the Noether condition $\max_{1 \leq i \leq N}(c_i - \bar{c})^2 \cdot \left[\sum_{i=1}^N (c_i - \bar{c})^2\right]^{-1} \to 0$ (see Hájek (1961, 1962)).

The most regular $(F_1, \ldots, F_N)$ correspond to the null hypothesis $F_1 = \cdots = F_N$; next comes the condition of contiguity (see Hájek (1962)); finally, we may only assume that the variance of S under $(F_1, \ldots, F_N)$ is of the same order as under $F_1 = \cdots = F_N$; or allow for some rate of degeneration of the variance.

The regularity conditions concerning the scores, generated by a function $\varphi(t)$, $0 < t < 1$, are expressed in terms of smoothness and boundedness of φ. An ideal φ is linear; Dwass (1956) succeeded in treating polynomials; Chernoff–Savage (1958) assumed two derivatives, the ith derivative being bounded by $[t(1-t)]^{-i-\frac{1}{2}+\delta}$, $i = 0, 1, 2$, $\delta > 0$, $0 < t < 1$; Govindarajulu –Le Cam–Raghavachari (1966) considered a larger class, removing the assumption of existence of a second derivative and relaxing the boundedness.

The first general attempt to treat the asymptotic distribution of simple linear rank statistics in the two-sample case was made by Dwass (1956). As we have already mentioned, using the method of U-statistics, he was successful with polynomial score-generating functions only. Then Chernoff–Savage published their pioneering paper (1958), which was a basis for all further developments excepting those based on the contiguity approach. The Chernoff–Savage conditions have been improved as recent as that by Govindarajulu–Le Cam–Raghavachari (1966). In all papers mentioned above, the regularity condition concerning the regression constants (boundedness) and the score-generating function are rather stringent, whereas the assumption concerning $(F_1, \ldots, F_N)$ is broad enough (non-degeneration of variance only).

In contradistinction to it, the papers by Hájek (1961, 1962) are concerned with situations, where $(F_1, \ldots, F_N)$ is very regular (either null hypothesis or contiguous alternatives) whereas the cs satisfy the Noether condition only and φ is square integrable only. Thus the existing literature has dealt with somewhat extreme cases, leaving open intermediate ones. For example, one would need a theorem concerning discrete φ for $(F_1, \ldots, F_N)$ satisfying milder conditions than contiguity.

Govindarajulu–Le Cam–Raghavachari (1966) improved the Chernoff–Savage (1958) method by considering the sample distribution functions as processes and proving some refined limiting properties of these. The methods of the present paper are quite elementary, although rather involved. Two new methodical tools are used: first, the inequality of Theorem 3.1; second, the method of projection. The inequality may be used as follows. Supposing Theorem 2.3 has been proved for a certain class $\Phi_0 = \{\varphi_0\}$ of score-generating functions, it then holds for the class $\Phi = \{\varphi\}$ consisting of functions φ possessing the following property: for every $\epsilon > 0$ there is a $\varphi_0 \in \Phi_0$ and non-decreasing functions φ_1 and φ_2 such that $\varphi = \varphi_0 + \varphi_1 - \varphi_2$ and $\int_0^1 (\varphi_1^2 + \varphi_2^2)\, dt < \epsilon$. For example, if Φ_0 consists of polynomials (as in Dwass (1956)), Φ encompasses all functions considered in Govindarajulu–Le Cam–Raghavachari (1966), and a *fortiori*, in Chernoff–Savage (1958).

The method of projection was born by observing that many successful methods of deriving asymptotic normality of statistics, which do not have the standard form of a sum of independent random variables, consisted in finding such a standard approximation. This was the case with U-statistics, with the method of Chernoff–Savage (1958) and also with Hájek (1961). The method of projection is based on Lemma 4.1 giving to any statistic $S = s(X_1, \ldots, X_n)$, where the Xs are independent, the best approximation of the form $\hat{S} = \sum_{i=1}^N \ell_i(X_i)$ where the functions ℓ_i may be chosen arbitrarily except for $\mathsf{E}\ell_i^2(X_i) < \infty$. In this way the due approximation may be found quite mechanically. This is important namely in more complicated situations, where intuition and good luck may easily fail. Moreover, since, for best $\hat{S}$, we have $\mathsf{E}(S - \hat{S})^2 = \operatorname{var} S - \operatorname{var} \hat{S}$ the remainder $S - \hat{S}$ may be treated in terms of mean square convergence, which is usually easier to handle than the convergence in probability.

The problem of a single simple linear rank statistic may be generalized to a finite set of such statistics (needed in the c-sample problem) or to a continuous parameter set of such statistics (needed in dealing with statistics of Kolmogorov–Smirnov types). The former generalization is touched on in Remark 2.4; the latter is out of the scope of this chapter. Also the cases of simple linear statistics for the independence problem (see Bhuchonghul (1964)) as well as the symmetry problem are not considered here.

To keep in touch with Hájek (1961, 1962), we use a different notation than that of Chernoff–Savage: in particular, we use $\varphi(t)$, $a_N(j)$, S instead of $J(u)$, E_{Nj} and T_N, respectively. They also use the indicators Z_{Ni} of the ith order statistics being included in the first sample instead of ranks. Their formula $\sum_{i=1}^N \mathsf{E}_{Ni} Z_{Ni}$ reads $\sum_{i=1}^m a(R_i)$ in our notation, m denoting the number of observations in the first sample and $\mathsf{E}_{Ni} = a(i)$, $1 \leq i \leq N$. To avoid cumbersome subscripts connected with sequences we give to the limiting assertions the ϵ-form.

2 THE MAIN RESULTS

Let $X_1, \ldots, X_N$ be independent random variables with continuous distribution functions $F_1, \ldots, F_N$, and let $R_1, \ldots, R_N$ denote the corresponding ranks. Introducing the function

$$\begin{aligned} u(x) &= 1, \quad x \geq 0, \\ &= 0, \quad x < 0, \end{aligned} \tag{2.1}$$

we may write

$$R_i = \sum_{j=1}^{N} u(X_i - X_j), \quad 1 \leq i \leq N. \tag{2.2}$$

We are interested in asymptotic normality of simple linear rank statistics

$$S = \sum_{i=1}^{N} c_i \, a_N(R_i), \tag{2.3}$$

where $c_1, \ldots, c_N$ are arbitrary 'regression constants', and $a_N(1), \ldots, a_N(N)$ are 'scores' generated by a function $\varphi(t)$, $0 < t < 1$, in either of the following two ways:

$$a_N(i) = \varphi(i/(N+1)), \quad 1 \leq i \leq N, \tag{2.4}$$

$$a_N(i) = \mathsf{E}\varphi\left(U_N^{(i)}\right), \quad 1 \leq i \leq N, \tag{2.5}$$

where $U_N^{(i)}$ denotes the ith order statistic in a sample of size N from the uniform distribution on $(0,1)$. Scores given by (2.5) occur in statistics yielding locally most powerful rank tests. Scores (2.4) are distinguished by simplicity. In what follows φ and the scores are assumed fixed, whereas N, $(c_1, \ldots, c_N)$ and $(F_1, \ldots, F_N)$ are considered variable. Let

$$\bar{c} = N^{-1} \sum_{i=1}^{N} c_i, \quad \bar{\varphi} = \int_0^1 \varphi(t) \, \mathrm{d}t, \quad H(x) = N^{-1} \sum_{i=1}^{N} F_i(x). \tag{2.6}$$

Theorem 2.1. *Consider the statistic S of (2.3), where the scores are given either by (2.4) or (2.5). Assume that φ has a bounded second derivative.*

Then for every $\epsilon > 0$ there exists K_ϵ such that

$$\operatorname{var} S > K_\epsilon \max_{1 \leq i \leq N} (c_i - \bar{c})^2 \tag{2.7}$$

entails

$$\max_{-\infty<x<\infty}\left|\mathsf{P}(S-\mathsf{E}S<x(\mathsf{var}\,S)^{\frac{1}{2}})-(2\pi)^{-\frac{1}{2}}\int_{-\infty}^{x}\exp\left(-\frac{1}{2}y^2\right)\,\mathrm{d}y\right|<\epsilon. \tag{2.8}$$

The assertion remains true if we replace var S *in* (2.7) *and* (2.8) *by*

$$\sigma^2=\sum_{i=1}^{N}\mathsf{var}[\ell_i(X_i)], \tag{2.9}$$

where

$$\ell_i(x)=N^{-1}\sum_{j=1}^{N}(c_j-c_i)\int[u(y-x)-F_i(y)]\,\varphi'(H(y))\,\mathrm{d}F_j(y), \tag{2.10}$$

with φ' *denoting the derivative of* φ *and* $u(\cdot)$ *given by* (2.1).

In applications we usually assume that the distribution functions differ only slightly. Then the following variation of Theorem 2.1 is useful.

Theorem 2.2. *Under assumptions of Theorem* 2.1, *for every* $\epsilon>0$ *there exists a* $\delta_\epsilon>0$ *such that the joint satisfaction of*

$$\sum_{i=1}^{N}(c_i-\overline{c})^2>\delta_\epsilon^{-1}\max_{1\le i\le N}(c_i-\overline{c})^2 \tag{2.11}$$

and

$$\max_{i,j,x}|F_i(x)-F_j(x)|<\delta_\epsilon \tag{2.12}$$

entails

$$\sup_x\left|\mathsf{P}(S-\mathsf{E}S<xd)-(2\pi)^{-\frac{1}{2}}\int_{-\infty}^{x}\exp\left(-\frac{1}{2}y^2\right)\,\mathrm{d}y\right|<\epsilon, \tag{2.13}$$

where

$$d^2=\sum_{i=1}^{N}(c_i-\overline{c})^2\int_0^1[\varphi(t)-\overline{\varphi}]^2\,\mathrm{d}t. \tag{2.14}$$

Definition. We shall say that $\varphi(t)$, $0<t<1$, is *absolutely continuous inside* $(0,1)$ if for every $\epsilon>0$ and $0<\alpha<\frac{1}{2}$ there exists a $\delta=\delta(\epsilon,\alpha)$ such that for each finite set of disjoint intervals (a_k,b_k) the relation

$$\sum_{k=1}^{n}|b_k-a_k|<\delta\quad\text{and}\quad\alpha<a_k,\ b_k<1-\alpha$$

implies

$$\sum_{k=1}^{n} |\varphi(b_k) - \varphi(a_k)| < \epsilon.$$

Thus φ is absolutely continuous inside $(0,1)$ if it is absolutely continuous on $(\alpha,\, 1-\alpha)$ for every $\alpha \in \left(0,\, \frac{1}{2}\right)$.

It is a well-known theorem of calculus that $\varphi(t)$ is absolutely continuous inside $(0,1)$ if and only if there exists a function $\varphi'(x)$ integrable on every interval $(\alpha,\, 1-\alpha)$, $0 < \alpha < \frac{1}{2}$, and such that

$$\varphi(b) - \varphi(a) = \int_a^b \varphi'(t)\, dt, \quad 0 < a < b < 1. \tag{2.15}$$

We further know that $\varphi'(t)$ represents the derivative of $\varphi(t)$ almost everywhere. However, the converse is not true, i.e. the existence of the derivative a.s. does not generally entail (2.15). Of course if the derivative exists *everywhere* and is integrable on every interval $(\alpha,\, 1-\alpha)$, $0 < \alpha < \frac{1}{2}$, then φ is absolutely continuous inside $(0,1)$. There also may be a finite number of points possessing only the left-hand and right-hand derivatives as, e.g., in $\varphi(t) = \left|t - \frac{1}{2}\right|$. Let us emphasize that for $t \to 0$ or 1 the limit of a $\varphi(t)$, which is absolutely continuous inside $(0,1)$, may not exist or be infinite. Consequently (2.15) may even not make sense for $a = 0$ or $b = 1$.

Consider again the statistic (2.3) where the scores are given either by (2.4) or (2.5).

Theorem 2.3. *Let $\varphi(t) = \varphi_1(t) - \varphi_2(t)$, $0 < t < 1$, where the $\varphi_i(t)$ are both non-decreasing, square integrable, and absolutely continuous inside $(0,1)$, $i = 1, 2$.*

Then for every $\epsilon > 0$ and $\eta > 0$ there exists $N_{\epsilon\eta}$ such that

$$N > N_{\epsilon\eta}, \quad \operatorname{var} S > \eta\, N \max_{1 \le i \le N} (c_i - \bar{c})^2 \tag{2.16}$$

entails (2.8).

As in Theorem 2.1, $\operatorname{var} S$ may be replaced by σ^2 of (2.9) in both (2.16) and (2.8).

Theorem 2.4. *Under conditions of Theorem 2.3, for every $\epsilon > 0$ and $\eta > 0$ there exist $N_{\epsilon\eta}$ and $\delta_{\epsilon\eta}$ such that joint satisfaction of*

$$\sum_{i=1}^{N} (c_i - \bar{c})^2 > \eta\, N \max_{1 \le i \le N} (c_i - \bar{c})^2, \tag{2.17}$$

$$N < N_{\epsilon\eta}, \tag{2.18}$$

$$\max_{i,j,x} |F_i(x) - F_j(x)| < \delta_{\epsilon\eta} \tag{2.19}$$

entails (2.13).

Remark 2.1. The theorem also holds if the scores are given by

$$a_N(i) = N \int_{(i-1)/N}^{i/N} \varphi(t)\,\mathrm{d}t, \quad 1 \le i \le N. \tag{2.20}$$

Remark 2.2. The variance d^2, given by (2.14), is asymptotically equivalent (in the ratio sense) to the variance of S under the null hypothesis $F_i = F$, $1 \le i \le N$, and, as a matter of fact, could be replaced by it as well. Theorem 2.1 is as powerful, as far as the cs are concerned, as the corresponding theorem under the null hypothesis. In particular, for the two-sample case it provides asymptotic normality even for cases where $m/n \to 0$ or ∞, and also for cases where $\mathsf{var}\, S$ under the alternative is of smaller order than d^2.

Remark 2.3. (Comparison with Chernoff–Savage and Govindarajulu–Le Cam–Raghavachari). If $c_i = 1$, $1 \le i \le m$, and $c_i = 0$, if $m < i \le N$, and if $F_i = F$, $1 \le i \le m$, and $F_i = G$, $m < i \le N$, we have the problem considered by Chernoff–Savage (1958) and by Govindarajulu–Le Cam–Raghavachari (1966). In the present paper the condition concerning the function $\varphi(t)$, denoted in the above papers by $J(u)$, is relaxed. Actually, taking a $\varphi(t)$ satisfying the condition $|\varphi'(t)| \le K[t(1-t)]^{-3/2+\delta}$, $\delta > 0$, and putting

$$\varphi_1(t) = \varphi\left(\frac{1}{2}\right) + \int_{\frac{1}{2}}^{t} \max[0, \varphi'(s)]\,\mathrm{d}s, \quad \varphi_2(t) = \int_{\frac{1}{2}}^{t} \max[0, -\varphi'(s)]\,\mathrm{d}s \tag{2.21}$$

we have $\varphi(t) = \varphi_1(t) - \varphi_2(t)$, where φ_1 and φ_2 are non-decreasing, square integrable and absolutely continuous on intervals $(\epsilon, 1-\epsilon)$, $0 < \epsilon < \frac{1}{2}$. Also if, according to the assumptions of Govindarajulu–Le Cam–Raghavachari, $|\varphi'(t)| \le f_0 + fg$, where f_0 is integrable, f is integrable and U-shaped, and g is square integrable and U-shaped, the functions (2.21) have the desired properties (see Lemma 2 of Govindarajulu–Le Cam–Raghavachari (1966)).

The present theorem surpasses the corresponding results of Chernoff and Savage (1958) and Govindarajulu–Le Cam–Raghavachari (1966) in the following further respects: first, the scores given by (2.5) are considered throughout, second, a simplified variance is given under (2.12) or (2.19), and, finally, it is shown that the asymptotic normality holds with natural parameters.

On the other hand, both papers Chernoff and Savage (1958) and Govindarajulu–Le Cam–Raghavachari (1966) show that S is asymptotically normal (μ, σ^2) with

$$\mu = \sum_{i=1}^{N} c_i\, \mathsf{E}[\varphi(H(X_i))]. \tag{2.22}$$

The present author did not succeed in showing that this is still true under conditions of Theorems 2.3 and 2.4. From Theorem 2.4 it follows that $\mathsf{E}S$ is replaceable by μ in Theorems 2.1 and 2.2 provided $\sum_{i=1}^N c_i^2$ is bounded by a multiple of $\sum_{i=1}^N (c_i - \bar{c})^2$.

In Govindarajulu–Le Cam–Raghavachari (1966), Theorem 1, uniformity is also shown with respect to a class of φ-functions. This is missing in the present paper, though a similar extension is easily possible. The corresponding class would consist of functions $\varphi = \varphi_1 - \varphi_2$ such that $\varphi_1, \varphi_2 \in \Phi^*$, Φ^* is a class of non-decreasing functions, which are uniformly square integrable, absolutely continuous on intervals $(\epsilon, 1-\epsilon)$, $0 < \epsilon < \frac{1}{2}$, and compact with respect to convergence such that $\varphi_k \to \varphi$ is equivalent to $\int_\epsilon^{1-\epsilon} |\varphi_k' - \varphi'| \, dt \to 0$ for every $\epsilon > 0$.

Remark 2.4. (The c-sample problem.) Theorem 2.1 contains all of the essentials for providing the c-sample limit theorem considered in Govindarajulu–Le Cam–Raghavachari (1966) and Puri (1964). Let $s_1, \ldots, s_c$ be a decomposition of the set $\{1, \ldots, N\}$. Let $n_j = \operatorname{card} s_j$, $1 \leq j \leq c$, and consider statistics

$$S_j = \sum_{i \in s_j} a_N(R_i), \tag{2.23}$$

where the scores are given either by (2.4) or (2.5). Then any linear combination of the statistics S_j, say

$$S = \sum_{j=1}^{c} \lambda_j \, S_j \tag{2.24}$$

is of the form (2.3) with

$$c_i = \lambda_j, \quad \text{if } i \in s_j, \quad 1 \leq j \leq c. \tag{2.25}$$

Remark 2.5. The constants $K_\epsilon \, \delta_\epsilon$, $N_{\epsilon\eta}$, $\delta_{\epsilon\eta}$ appearing in the above theorems depend on φ.

Remark 2.6. By integration by parts we have

$$\begin{aligned} &\int [u(y-x) - F_i(y)] \, \varphi'(H(y)) \, \mathrm{d}F_j(y) \\ &= -\int_{x_0}^{x} \varphi'(H(y)) \, \mathrm{d}F_j(y) + \text{const.} \end{aligned} \tag{2.26}$$

3 VARIANCE BOUND FOR MONOTONE SCORES

We shall start with two elementary lemmas.

Lemma 3.1. *Let $\alpha(x_1, \ldots, x_h)$ be a real function, non-decreasing in each argument. Let $X_1, \ldots, X_h$ be arbitrary independent random variables such that $\mathsf{E}|\alpha(X_1, \ldots, X_h)| < \infty$.*

Then, for $k < h$, the function

$$\overline{\alpha}(x_1, \ldots, x_k) = \mathsf{E}\left[\alpha(X_1, \ldots, X_h) \,|\, X_1 = x_1, \ldots, X_k = x_k\right] \tag{3.1}$$

is non-decreasing in each argument.

Moreover, if $\beta(x_1, \ldots, x_h)$ is another real function, non-decreasing in each argument, then

$$\mathsf{cov}\left[\alpha(X_1, \ldots, X_h),\, \beta(X_1, \ldots, X_h)\right] \geq 0, \tag{3.2}$$

provided the covariance is well-defined.

Proof. The first part is trivial. For the second part, if $h = 1$, (3.2) follows from the identity

$$\mathsf{cov}[\alpha(X_1),\, \beta(X_1)] = \frac{1}{2}\,\mathsf{E}\left\{[\alpha(X_1) - \alpha(Y_1)]\,[\beta(X_1) - \beta(Y_1)]\right\},$$

where Y_1 is an independent copy of X_1. Generally,

$$\begin{aligned}
&\mathsf{cov}\left[\alpha(X_1, \ldots, X_h),\, \beta(X_1, \ldots, X_h)\right] \\
&= \mathsf{cov}\left[\overline{\alpha}(X_1),\, \overline{\beta}(X_1)\right] + \int \mathsf{cov}\left[\alpha(x, X_2, \ldots, X_h),\, \beta(x, X_2, \ldots, X_h)\right] \mathrm{d}F_1(x)
\end{aligned}$$

holds, where $\overline{\alpha}$ (and also $\overline{\beta}$) is defined by (3.1) with $k = 1$. The functions $\overline{\alpha}$ and $\overline{\beta}$ and also the functions $\alpha(x, \ldots)$ and $\beta(x, \ldots)$ are non-decreasing, so that the assertion holds for h if holding for $h - 1$. The proof is terminated. (Another proof follows from Theorem 2 of Lehmann (1966)). □

Lemma 3.2. *Let $R_1, \ldots, R_N$ be ranks of a sequence of independent observations $X_1, \ldots, X_N$ possessing arbitrary continuous distribution functions $F_1, \ldots, F_N$. Let $a(1), \ldots, a(N)$ be arbitrary scores and $u(x)$ be given by (2.1). Then*

$$\begin{aligned}
&\mathsf{E}[a(R_i) \,|\, X_i = x,\, X_j = y] - \mathsf{E}[a(R_i) \,|\, X_i = x] \qquad (3.3)\\
&= [u(x - y) - F_j(x)] \sum_{k=2}^{N} (a(k) - a(k-1)) \\
&\quad \times \mathsf{P}(R_i = k \,|\, X_i = x,\, X_j = x - 1), \quad i \neq j.
\end{aligned}$$

Proof. Specify $i = 1$, $j = 2$ in order to simplify the notation. Denote by $B(k \mid p_1, \ldots, p_N)$ the probability of k successes in N independent trials with respective probabilities of success $p_1, \ldots, p_N$. Thus $B(\cdot \mid \cdot)$ denotes the probabilities of the Poisson binomial distribution. We obviously have, in view of (2.2),

$$\mathsf{P}(R_1 = k \mid X_1 = x,\, X_2 = y) = B(k \mid 1,\, u(x-y),\, F_3(x), \ldots, F_N(x)) \tag{3.4}$$

and

$$\mathsf{P}(R_1 = k \mid X_1 = x) = B(k \mid 1,\, F_2(x),\, F_3(x), \ldots, F_N(x)). \tag{3.5}$$

We easily see that

$$\begin{aligned} &B(k \mid 1, F_2(x), \ldots, F_N(x)) \\ &= F_2(x) B(k \mid 1, 1, F_3(x), \ldots, F_N(x)) \\ &\qquad + (1 - F_2(x)) B(k \mid 1, 0, F_3(x), \ldots, F_N(x)) \\ &= F_2(x) B(k \mid 1, 1, F_3(x), \ldots, F_N(x)) \\ &\qquad + (1 - F_2(x)) B(k+1 \mid 1, 1, F_3(x), \ldots, F_N(x)). \end{aligned} \tag{3.6}$$

Consequently, in accordance with (3.4) through (3.6),

$$\begin{aligned} &\mathsf{P}(R_1 = k \mid X_1 = x,\, X_2 = y) - \mathsf{P}(R_1 = k \mid X_1 = x) \\ &= [u(x-y) - F_2(x)]\, [B(k \mid 1, 1, F_3(x), \ldots, F_N(x)) \\ &\qquad - B(k+1 \mid 1, 1, F_3(x), \ldots, F_N(x))] \\ &= [u(x-y) - F_2(x)]\, [\mathsf{P}(R_1 = k \mid X_1 = x,\, X_2 = x-1) \\ &\qquad - \mathsf{P}(R_1 = k+1 \mid X_1 = x,\, X_2 = x-1)], \end{aligned}$$

where $x - 1$ could be replaced by any number smaller than x. Now the last relation entails

$$\begin{aligned} &\mathsf{E}[a(R_1) \mid X_1 = x,\, X_2 = y] - \mathsf{E}[a(R_1) \mid X_1 = x] \\ &= \sum_{k=1}^{N} a(k)\, [\mathsf{P}(R_1 = k \mid X_1 = x,\, X_2 = y) - \mathsf{P}(R_1 = k \mid X_1 = x)] \\ &= [u(x-y) - F_2(x)] \sum_{k=1}^{N} a(k)\, [\mathsf{P}(R_1 = k \mid X_1 = x,\, X_2 = x-1) \\ &\qquad - \mathsf{P}(R_1 = k+1 \mid X_1 = x,\, X_2 = x-1)] \\ &= [u(x-y) - F_2(x)] \sum_{k=2}^{N} [a(k) - a(k-1)]\, \mathsf{P}(R_1 = k \mid X_1 = x,\, X_2 = x-1). \end{aligned}$$

□

The inequality derived in the following theorem will remove problems caused by the fact that the score-generating function φ may be unbounded in the vicinities of 0 and 1, and may have no bounded second derivative in intervals $(\epsilon, 1-\epsilon)$, $\frac{1}{2} > \epsilon > 0$.

Theorem 3.1. (Variance inequality.) *Let $X_1, \ldots, X_N$ be independent random variables with arbitrary continuous distribution functions $F_1, \ldots, F_N$. Let $c_1, \ldots, c_N$ be arbitrary constants and $a_1 \leq \cdots \leq a_N$ be non-decreasing constants. Let R_i denote the rank of X_i, $1 \leq i \leq N$.*

Then, writing a_i and $a(i)$ interchangeably,

$$\operatorname{var}\left[\sum_{i=1}^{N} c_i\, a(R_i)\right] \leq 21 \max_{1\leq i\leq N}(c_i-\bar{c})^2 \sum_{i=1}^{N}(a_i-\bar{a})^2, \tag{3.7}$$

where

$$\bar{c} = N^{-1}\sum_{i=1}^{N} c_i, \quad \bar{a} = N^{-1}\sum_{i=1}^{N} a_i.$$

Proof. Without losing generality, we may assume $\bar{c} = \bar{a} = 0$. Obviously

$$\operatorname{var}\left[\sum_{i=1}^{N} c_i\, a(R_i)\right] \tag{3.8}$$

$$= \sum_{i=1}^{N} c_i^2 \operatorname{var}[a(R_i)] + \sum\ \sum_{i=1,\, j=1,\, i\neq j}^{N} c_i\, c_j \operatorname{cov}[a(R_i),\, a(R_j)]$$

and

$$\sum_{i=1}^{N} c_i^2 \operatorname{var}[a(R_i)] \leq \mathsf{E}\left\{\sum_{i=1}^{N} c_i^2\, a^2(R_i)\right\} \tag{3.9}$$

$$\leq \max_{1\leq i\leq N} c_i^2\, \mathsf{E}\left\{\sum_{i=1}^{N} a^2(R_i)\right\} = \max_{1\leq i\leq N} c_i^2 \sum_{i=1}^{N} a_i^2,$$

since $\sum_{i=1}^{N} a^2(R_i) = \sum_{i=1}^{N} a_i^2$ constantly.

The covariances are more tedious. (2.2) entails that R_i is a non-decreasing function of $-X_j$ (note the minus sign), $j \neq i$, and of X_i. If $X_i = x$ and $X_j = y$ are fixed, both R_i and R_j are non-decreasing functions of $-X_k$, $k \neq i, j$, and the same is true about $a(R_i)$ and $a(R_j)$, since $a_1 \leq \cdots \leq a_N$. Consequently, by Lemma 3.1,

$$\operatorname{cov}\left[a(R_i),\, a(R_j) \,|\, X_i = x,\, X_j = y\right] \geq 0, \quad 1 \leq i \neq j \leq N. \tag{3.10}$$

Also $\mathsf{E}[a(R_i)\,|\,X_i = x,\, X_j = y]$ as well as $-\mathsf{E}[a(R_j)\,|\,X_i = x,\, X_j = y]$ are both non-decreasing in x and $-y$, $i \neq j$, according to the same lemma. Thus

$$-\mathsf{cov}\,\{\mathsf{E}[a(R_i)\,|\,X_i,\, X_j],\, \mathsf{E}[a(R_j)\,|\,X_i,\, X_j]\} \geq 0, \quad 1 \leq i \neq j \leq N. \tag{3.11}$$

Notice that

$$\begin{aligned}&\mathsf{cov}[a(R_i),\, a(R_j)] \\ &= \mathsf{E}\,\mathsf{cov}[a(R_i),\, a(R_j)\,|\,X_i,\, X_j] + \mathsf{cov}\,\{\mathsf{E}[a(R_i)\,|\,X_i, X_j],\, \mathsf{E}[a(R_j)\,|\,X_i, X_j]\}\end{aligned}$$

and

$$\sum\sum_{i \neq j} \mathsf{cov}[a(R_i),\, a(R_j)] = -\sum_{i=1}^{N} \mathsf{var}[a(R_i)] \leq 0,$$

which entails

$$\begin{aligned}&\sum\sum_{i \neq j} \mathsf{E}\,\mathsf{cov}[a(R_i),\, a(R_j)\,|\,X_i,\, X_j] \\ &\leq -\sum\sum_{i \neq j} \mathsf{cov}\,\{\mathsf{E}[a(R_i)\,|\,X_i,\, X_j],\, \mathsf{E}[a(R_j)\,|\,X_i,\, X_j]\}.\end{aligned} \tag{3.12}$$

Furthermore, in view of (3.10), (3.11) and (3.12),

$$\begin{aligned}&\sum\sum_{i \neq j} c_i\, c_j\, \mathsf{cov}[a(R_i),\, a(R_j)] \\ &\leq \max_{1 \leq i \leq N} c_i^2 \Big\{ \sum\sum_{i \neq j} \mathsf{E}\,\mathsf{cov}[a(R_i),\, a(R_j)\,|\,X_i,\, X_j] \\ &\qquad - \sum\sum_{i \neq j} \mathsf{cov}\{\mathsf{E}[a(R_i)\,|\,X_i,\, X_j],\, \mathsf{E}[a(R_j)\,|\,X_i,\, X_j]\} \Big\} \\ &\leq -2 \max_{1 \leq i \leq N} c_i^2 \sum\sum_{i \neq j} \mathsf{cov}\{\mathsf{E}[a(R_i)\,|\,X_i,\, X_j],\, \mathsf{E}[a(R_j)\,|\,X_i,\, X_j]\}.\end{aligned} \tag{3.13}$$

The last inequality together with (3.8) and (3.9) implies that (3.7) will be proved, if we show that

$$-\sum\sum_{i \neq j} \mathsf{cov}\,\{\mathsf{E}[a(R_i)\,|\,X_i,\, X_j],\, \mathsf{E}[a(R_j)\,|\,X_i,\, X_j]\} \leq 10 \sum_{i=1}^{N} a_i^2. \tag{3.14}$$

Note the identity

$$\begin{aligned}&\mathsf{cov}\,\{\mathsf{E}[a(R_i)\,|\,X_i,\, X_j],\, \mathsf{E}[a(R_j)\,|\,X_i,\, X_j]\} \\ &= \mathsf{cov}\,\{\mathsf{E}[a(R_i)\,|\,X_i],\, \mathsf{E}[a(R_j)\,|\,X_i]\} + \mathsf{cov}\,\{\mathsf{E}[a(R_i)\,|\,X_j],\, \mathsf{E}[a(R_j)\,|\,X_j]\} \\ &\quad + \mathsf{E}\,\{(\mathsf{E}[a(R_i)\,|\,X_i,\, X_j] - \mathsf{E}[a(R_i)\,|\,X_i])\,(\mathsf{E}[a(R_j)\,|\,X_i,\, X_j] \\ &\quad - \mathsf{E}[a(R_j)\,|\,X_j])\}.\end{aligned} \tag{3.15}$$

We shall now try to obtain bounds for the sums of the right-hand side members of (3.15) over all $j \neq i$. The first two members give the same sum, which may be bounded as follows:

$$
\begin{aligned}
&- \sum\sum_{i \neq j} \operatorname{cov}\{\mathsf{E}[a(R_i)\,|X_i],\, \mathsf{E}[a(R_j)|\,X_i]\} \\
&= \sum_{i=1}^{N} \operatorname{cov}\left\{\mathsf{E}[a(R_i)\,|\,X_i],\, \mathsf{E}\left[-\sum_{j\neq i} a(R_j)\middle|\, X_i\right]\right\} \\
&= \sum_{i=1}^{N} \operatorname{cov}\left\{\mathsf{E}[a(R_i)\,|\,X_i],\, \mathsf{E}\left[a(R_i) - \sum_{j=1}^{N} a_j \middle|\, X_i\right]\right\} \\
&= \sum_{i=1}^{N} \operatorname{var}\{\mathsf{E}[a(R_i)\,|\,X_i]\} \le \sum_{i=1}^{N} \operatorname{var}[a(R_i)] \le \sum_{i=1}^{N} a_i^2.
\end{aligned} \tag{3.16}
$$

In the following sum we shall utilize the relation (3.3), the self-evident fact

$$|[u(X_i - X_j) - F_j(X_i)]\,[u(X_j - X_i) - F_i(X_j)]| \le 1,$$

and the inequality

$$
\begin{aligned}
&\mathsf{P}(R_i = k\,|\,X_i = x,\, X_j = x-1) \\
&\le \mathsf{P}(R_i = k\,|\,X_i = x) + \mathsf{P}(R_i = k-1\,|\,X_i = x),
\end{aligned}
$$

entailed by (3.4) through (3.6). We thus obtain

$$
\begin{aligned}
&- \sum\sum_{i\neq j} \mathsf{E}\{(\mathsf{E}[a(R_i)\,|\,X_i,\, X_j] - \mathsf{E}[a(R_i)\,|\,X_i])\,(\mathsf{E}[a(R_j)\,|\,X_i,\, X_j] \\
&\quad - \mathsf{E}[a(R_j)\,|\,X_j])\} \\
&\le \sum\sum_{i\neq j} \sum_k\sum_h (a_k - a_{k-1})\,(a_h - a_{h-1})\,\mathsf{E}\{[\mathsf{P}(R_i = k\,|\,X_i) \\
&\quad + \mathsf{P}(R_i = k-1\,|\,X_i)]\,[\mathsf{P}(R_j = h\,|\,X_j) + \mathsf{P}(R_j = h-1\,|\,X_j)]\} \\
&= \sum_k\sum_h (a_k - a_{k-1})\,(a_h - a_{h-1}) \sum\sum_{i\neq j} [\mathsf{P}(R_i = k)\,\mathsf{P}(R_j = h) \\
&\quad + \mathsf{P}(R_i = k-1)\,\mathsf{P}(R_j = h-1) + \mathsf{P}(R_i = k-1)\,\mathsf{P}(R_j = h) \\
&\quad + \mathsf{P}(R_i = k)\,\mathsf{P}(R_j = h-1)] \\
&\le 4 \sum_k\sum_h (a_k - a_{k-1})\,(a_h - a_{h-1}) = 4(a_N - a_1)^2 \\
&\le 8 \sum_{i=1}^{N} a_i^2.
\end{aligned} \tag{3.17}
$$

Here we have used the fact that the events $R_i = k$, $1 \leq i \leq N$, are disjoint so that $\sum_{i=1}^{N} \mathsf{P}(R_i = k) = 1$. Now (3.14) follows from (3.15), (3.16) used twice, and (3.17). □

4 PROJECTION APPROXIMATION FOR S WITH SMOOTH AND BOUNDED SCORES

The central limit theorem is concerned with sequences of sums of independent random variables. Its scope may however be extended to statistics that are asymptotically (as the number of observations increases) equivalent to such sums. Given a statistic $S = s(X_1, \ldots, X_N)$, defined on a sequence of independent observations $X_1, \ldots, X_N$, we shall try to approximate it by a statistic

$$L = \sum_{i=1}^{N} \ell_i(X_i), \tag{4.1}$$

where the functions ℓ_i may be chosen arbitrarily. Obviously, L is a sum of independent random variables $Y_i = \ell_i(X_i)$, $1 \leq i \leq N$. The set $\mathcal{L}$ of statistics L such that $\mathsf{E}\ell_i^2(X_i) < \infty$, $1 \leq i \leq N$, availed with the usual inner product $(L, L') = \mathsf{E}\, L\, L'$, is a closed subspace of the Hilbert space of square integrable statistics. The best approximation in the mean square is given by a projection on the subspace $\mathcal{L}$. It is worth noting that the possibility of a successful approximation of a statistic by an L-statistic is usually better than one would intuitively expect. The projection may be obtained explicitly in terms of conditional expectations as is shown in the following.

Projection Lemma 4.1. *Let $X_1, \ldots, X_N$ be independent random variables and $S = s(X_1, \ldots, X_N)$ be a statistic such that $\mathsf{E}S^2 < \infty$. Let*

$$\hat{S} = \sum_{i=1}^{N} \mathsf{E}(S \mid X_i) - (N-1)\,\mathsf{E}S. \tag{4.2}$$

Then

$$\mathsf{E}\hat{S} = \mathsf{E}S \tag{4.3}$$

and

$$\mathsf{E}(S - \hat{S})^2 = \operatorname{var} S - \operatorname{var} \hat{S}. \tag{4.4}$$

Moreover, if L is given by (4.1) with $\mathsf{E}\ell_i^2(X_i) < \infty$, $1 \leq i \leq N$, then

$$\mathsf{E}(S - L)^2 = \mathsf{E}(S - \hat{S})^2 + \mathsf{E}(\hat{S} - L)^2. \tag{4.5}$$

Proof. (4.3) is obvious, and (4.4) follows from (4.5), if we take for L the constant $\mathsf{E}S = \mathsf{E}\hat{S}$ and rearrange the formula. Thus it remains to prove (4.5). Without loss of generality, we may assume that $\mathsf{E}S = \mathsf{E}\hat{S} = 0$. We obviously have

$$\mathsf{E}[(S-\hat{S})(\hat{S}-L)] = \sum_{i=1}^{N} \mathsf{E}\left[\mathsf{E}(S-\hat{S}\,|\,X_i)(\mathsf{E}(S\,|\,X_i) - \ell_i(X_i))\right]. \tag{4.6}$$

Now, since $X_1, \ldots, X_N$ are independent,

$$\begin{aligned} \mathsf{E}[\mathsf{E}(S\,|\,X_j)\,|\,X_i] &= \mathsf{E}S, && \text{if} \quad j \neq i \\ &= \mathsf{E}(S\,|\,X_i), && \text{if} \quad j = i. \end{aligned}$$

Consequently, since $\mathsf{E}S = 0$, $\mathsf{E}(\hat{S}\,|\,X_i) = \sum_{j=1}^{N} \mathsf{E}[\mathsf{E}(S\,|\,X_j)\,|\,X_i] = \mathsf{E}(S\,|\,X_i)$, i. e.

$$\mathsf{E}(S-\hat{S}\,|\,X_i) = 0, \quad 1 \leq i \leq N. \tag{4.7}$$

By inserting (4.7) into (4.6), we obtain $\mathsf{E}[(S-\hat{S})(\hat{S}-L)] = 0$, which is equivalent to (4.5). □

If S is a sum of simpler statistics, we may obtain the upper bound for the residual variance $\mathsf{E}(S-\hat{S})^2$ in the following lemma.

Residual variance Lemma 4.2. *Let* $S = \sum_{i=1}^{N} S_i$ *and* $\hat{S} = \sum_{i=1}^{N} \mathsf{E}(S\,|\,X_i) - (N-1)\,\mathsf{E}S$. *Then*

$$\begin{aligned} \mathsf{E}(S-\hat{S})^2 \leq & \sum_{i=1}^{N} \mathsf{E}[S_i - \mathsf{E}(S_i\,|\,X_i)]^2 \\ & + \sum\sum_{i\neq j} \left\{ \mathsf{E}([S_i - \mathsf{E}(S_i\,|\,X_i)][S_j - \mathsf{E}(S_j\,|\,X_j)]) \right. \\ & \left. - \sum_{k\neq i,j} \mathsf{cov}[\mathsf{E}(S_i\,|\,X_k),\, \mathsf{E}(S_j\,|\,X_k)] \right\}. \end{aligned} \tag{4.8}$$

Proof. Utilizing (4.4) we obtain

$$\begin{aligned} \mathsf{E}(S-\hat{S})^2 &= \mathsf{var}\,S - \mathsf{var}\,\hat{S} = \sum_{i=1}^{N} \left[\mathsf{var}\,S_i - \mathsf{var} \sum_{k=1}^{N} \mathsf{E}(S_i\,|\,X_k)\right] \\ & + \sum_{i\neq j} \left\{ \mathsf{cov}(S_i, S_j) - \sum_{k=1}^{N} \mathsf{cov}[\mathsf{E}(S_i\,|\,X_k),\, \mathsf{E}(S_j\,|\,X_k)] \right\}. \end{aligned} \tag{4.9}$$

Next

$$\begin{aligned}\operatorname{var} S_i - \operatorname{var}\sum_{k=1}^{N} \mathsf{E}(S_i \mid X_k) &= \operatorname{var} S_i - \sum_{k=1}^{N} \operatorname{var} \mathsf{E}(S_i \mid X_k) \\ &\leq \operatorname{var} S_i - \operatorname{var}\mathsf{E}(S_i \mid X_i) \\ &= \mathsf{E}[S_i - \mathsf{E}(S_i \mid X_i)]^2.\end{aligned} \tag{4.10}$$

Further, obviously,

$$\operatorname{cov}(S_i, S_j) = \operatorname{cov}[\mathsf{E}(S_i \mid X_i),\, \mathsf{E}(S_j \mid X_i)] \tag{4.11}$$

$$+\operatorname{cov}[\mathsf{E}(S_i \mid X_j),\, \mathsf{E}(S_j \mid X_j)] + \mathsf{E}\{[S_i - \mathsf{E}(S_i \mid X_i)]\,[S_i - \mathsf{E}(S_j \mid X_j)]\}.$$

Now, (4.8) is an easy consequence of (4.9) through (4.11). □

Theorem 4.1. *Let the score-generating function φ possess a bounded second derivative. Consider the statistic $S = \sum_{i=1}^{N} c_i\, \varphi(R_i/(N+1))$ and put $\hat{S} = \sum_{i=1}^{N} \mathsf{E}(S \mid X_i) - (N-1)\, \mathsf{E}S$. Then there exists a constant $K = K(\varphi)$ such that for any N, $(c_1, \ldots, c_N)$ and $(F_1, \ldots, F_N)$*

$$\mathsf{E}(S - \hat{S})^2 \leq K\, N^{-1} \sum_{i=1}^{N} (c_i - \overline{c})^2. \tag{4.12}$$

Proof. Without loss of generality we may assume $\overline{c} = 0$. Put

$$\rho_i = R_i/(N+1), \quad 1 \leq i \leq N, \tag{4.13}$$

so that $S = \sum_{i=1}^{N} c_i\, \varphi(\rho_i)$. By Taylor expansion

$$\begin{aligned}\varphi(\rho_i) &= \varphi\,[\mathsf{E}(\rho_i \mid X_i)] + (\rho_i - \mathsf{E}(\rho_i \mid X_i))\, \varphi'[\mathsf{E}(\rho_i \mid X_i)] \\ &\quad + (\rho_i - \mathsf{E}(\rho_i \mid X_i))^2\, k_i(X_i),\end{aligned} \tag{4.14}$$

where $k_i(x)$ is bounded, say $k_i^2(x) < K_2$, $-\infty < x < \infty$, $1 \leq i \leq N$. Thus $S = S_1 + S_2 + S_3$, where

$$\begin{aligned}S_1 &= \sum_{i=1}^{N} c_i\, \rho_i\, \varphi'\,[\mathsf{E}(\rho_i \mid X_i)], \\ S_2 &= \sum_{i=1}^{N} c_i\, \{\varphi[\mathsf{E}(\rho_i \mid X_i)] - \mathsf{E}(\rho_i \mid X_i)\, \varphi'[\mathsf{E}(\rho_i \mid X_i)]\}, \\ S_3 &= \sum_{i=1}^{N} c_i [\rho_i - \mathsf{E}(\rho_i \mid X_i)]^2\, k_i(X_i).\end{aligned}$$

Now, denoting the projection of S_i by $\hat{S}_i$, we have $\hat{S} = \hat{S}_1 + \hat{S}_2 + \hat{S}_3$. Now, obviously $\hat{S}_2 = S_2$, so that

$$\begin{aligned} \mathsf{E}(S - \hat{S})^2 &\leq 2\mathsf{E}(S_1 - \hat{S}_1)^2 + 2\mathsf{E}(S_3 - \hat{S}_2)^2 \\ &\leq 2\mathsf{E}(S_1 - \hat{S}_1)^2 + 2\mathsf{E}S_3^2. \end{aligned} \tag{4.15}$$

Next, by the Schwartz inequality

$$\begin{aligned} \mathsf{E}S_3^2 &\leq \sum_{i=1}^{N} c_i^2 \sum_{i=1}^{N} \mathsf{E}\left\{[\rho_i - \mathsf{E}(\rho_i \,|\, X_i)]^4 \, k_i^2(X_i)\right\} \\ &\leq K_2 \sum_{i=1}^{N} c_i^2 \sum_{i=1}^{N} \mathsf{E}[\rho_i - \mathsf{E}(\rho_i \,|\, X_i)]^4. \end{aligned}$$

As is seen from (2.2), for $X_i = x$, $R_i = \rho_i(N+1)$ is a sum of N independent zero–one random variables. Consequently

$$\mathsf{E}[\rho_i - \mathsf{E}(\rho_i \,|\, X_i)]^4 \leq \frac{1}{4}(N+1)^{-2}, \quad 1 \leq i \leq N,$$

and

$$\mathsf{E}S_3^2 \leq K_2(N+1)^{-1} \sum_{i=1}^{N} c_i^2. \tag{4.16}$$

Further we shall apply Lemma 4.2 to $S_1 = \sum_{i=1}^{N} S_{1i}$, $S_{1i} = c_i \, \rho_i \, \varphi'[\mathsf{E}(\rho_i \,|\, X_i)]$. Let K_1 be an upper bound for $[\varphi'(t)]^2$, $0 < t < 1$. Then, obviously,

$$\mathsf{E}(S_{1i} - \mathsf{E}(S_{1i} \,|\, X_i))^2 \leq c_i^2(N+1)^{-2} K_1 \, \mathsf{E}\,\mathsf{var}(R_i \,|\, X_i) \leq \frac{1}{4} \, c_i^2(N+1)^{-1} K_1. \tag{4.17}$$

The covariance term in (4.8) is somewhat more complicated. We have

$$\begin{aligned} &\mathsf{E}([S_i - \mathsf{E}(S_i \,|\, X_i)] \, [S_j - \mathsf{E}(S_j \,|\, X_j)]) \\ &= \mathsf{E}\,\mathsf{cov}(S_i, \, S_j \,|\, X_i, \, X_j) \\ &\quad + \mathsf{E}\left\{[\mathsf{E}(S_i \,|\, X_i, \, X_j) - \mathsf{E}(S_i \,|\, X_i)] \, [\mathsf{E}(S_j \,|\, X_i, \, X_j) - \mathsf{E}(S_j \,|\, X_j)]\right\}. \end{aligned} \tag{4.18}$$

Now (2.2) entails, $i \neq j$,

$$\begin{aligned} &\mathsf{cov}(S_{1i}, \, S_{1j} \,|\, X_i, \, X_j) \\ &= c_i \, c_j (N+1)^{-2} \sum_{k \neq i, j} \{\min[F_k(X_i), \, F_k(X_j)] - F_k(X_i) \, F_k(X_j)\} \\ &\qquad \times \varphi'(\mathsf{E}(\rho_i \,|\, X_i)) \, \varphi'(\mathsf{E}(\rho_j \,|\, X_j)) \end{aligned}$$

and, consequently,

$$\begin{aligned} &\mathsf{E}\,\mathsf{cov}(S_{1i},\, S_{1j} \mid X_i,\, X_j) \\ = \;& c_i\, c_j (N+1)^{-2} \sum_{k \neq i,\, j} \int\!\!\int \{\min[F_k(x),\, F_k(y)] - F_k(x)\, F_k(y)\} \\ & \times \varphi'(\mathsf{E}(\rho_i \mid X_i = x))\, \varphi'(\mathsf{E}(\rho_i \mid X_j = y))\, \mathrm{d}F_i(x)\, \mathrm{d}F_j(y). \end{aligned}$$

The right side may be transformed as follows. Introducing

$$\ell_{ik}(s) = \int [u(x-s) - F_k(x)]\, \varphi'(\mathsf{E}(\rho_i \mid X_i = x))\, \mathrm{d}F_i(x), \tag{4.19}$$

we easily see that

$$\begin{aligned} \int\!\!\int \{\min[F_k(x),\, F_k(y)] - F_k(x)\, F_k(y)\}\, \varphi'(\mathsf{E}(\rho_i \mid X_i = x))\, \varphi'(\mathsf{E}(\rho_j \mid X_j = y)) \\ \times \mathrm{d}F_i(x)\, \mathrm{d}F_j(y) \;=\; \mathsf{cov}[\ell_{ik}(X_k),\, \ell_{jk}(X_k)]. \end{aligned}$$

Thus

$$\mathsf{E}\,\mathsf{cov}(S_{1i},\, S_{1j} \mid X_i,\, X_j) = c_i\, c_j (N+1)^{-2} \sum_{k \neq i,\, j} \mathsf{cov}[\ell_{ik}(X_k),\, \ell_{jk}(X_k)]. \tag{4.20}$$

On the other hand, (3.3) yields

$$\begin{aligned} &\mathsf{E}(S_{1i} \mid X_k) - \mathsf{E}(S_{1i}) \\ &= c_i \int [\mathsf{E}(\rho_i \mid X_k,\, X_i = x) - \mathsf{E}(\rho_i \mid X_i = x)]\, \varphi'(\mathsf{E}(\rho_i \mid X_i = x))\, \mathrm{d}F_i(x) \\ &= c_i (N+1)^{-1} \int [u(x - X_k) - F_k(x)]\, \varphi'(\mathsf{E}(\rho_i \mid X_i = x))\, \mathrm{d}F_i(x) \\ &= c_i (N+1)^{-1}\, \ell_{ik}(X_k). \end{aligned}$$

Consequently, $i \neq j,\ k \neq i,\ k \neq j$,

$$\mathsf{cov}[\mathsf{E}(S_{1i} \mid X_k),\, \mathsf{E}(S_{1j} \mid X_k)] = c_i\, c_j (N+1)^{-2}\, \mathsf{cov}[\ell_{ik}(X_k),\, \ell_{jk}(X_k)]. \tag{4.21}$$

Again by (3.3),

$$\mathsf{E}(\rho_i \mid X_i,\, X_j) - \mathsf{E}(\rho_i \mid X_i) = (N+1)^{-1}\, [u(X_i - X_j) - F_j(X_i)]. \tag{4.22}$$

Consequently,

$$\begin{aligned} &|\mathsf{E}\, \{[\mathsf{E}(S_{1i} \mid X_i,\, X_j) - \mathsf{E}(S_{1i} \mid X_i)]\, [\mathsf{E}(S_{1j} \mid X_i,\, X_j) - \mathsf{E}(S_{1j} \mid X_j)]\}| \qquad (4.23) \\ &\leq K_1\, |c_i\, c_j|\, (N+1)^{-2}\, |\mathsf{E}\, \{[u(X_i - X_j) - F_j(X_i)]\, [u(X_j - X_i) - F_i(X_j)]\}| \\ &\leq K_1\, |c_i\, c_j|\, (N+1)^{-2}\, \{\mathsf{E}[u(X_i - X_j) - F_j(X_i)]^2\, \mathsf{E}[u(X_j - X_i) - F_i(X_j)]^2\}^{\frac{1}{2}} \\ &\leq \frac{1}{4} K_1\, |c_i\, c_j|\, (N+1)^{-2}. \end{aligned}$$

Now, combining (4.18) through (4.22), we obtain

$$\Big| \mathsf{E}\,\{[S_{1i} - \mathsf{E}(S_{1i}\,|\,X_i)]\,[S_{1j} - \mathsf{E}(S_{1j}\,|\,X_j)]\} - \sum_{k\neq i,\,j} \mathsf{cov}[\mathsf{E}(S_{1i}\,|\,X_k),\,\mathsf{E}(S_{1j}\,|\,X_k)] \Big| \le \frac{1}{4} K_1\,|\,c_i\,c_j\,|\,(N+1)^{-2}. \tag{4.24}$$

Applying (4.17) and (4.24) to (4.8), we obtain

$$\begin{aligned} &\mathsf{E}(S_1 - \hat{S}_1)^2 \\ &\le \frac{1}{4} K_1 (N+1)^{-1} \sum_{i=1}^{N} c_i^2 + \frac{1}{4} K_1 (N+1)^{-2} \sum\sum_{i\neq j} |\,c_i\,c_j\,| \\ &\le \frac{1}{2} K_1 (N+1)^{-1} \sum_{i=1}^{N} c_i^2. \end{aligned} \tag{4.25}$$

Finally, (4.15), (4.16) and (4.25) yield

$$\mathsf{E}(S - \hat{S})^2 \le \left(K_1 + \frac{1}{2} K_2\right)(N+1)^{-1} \sum_{i=1}^{N} c_i^2.$$

Thus (4.12) is proved with $K = K_1 + \frac{1}{2}K_2$, K_1 and K_2 being upper bounds for the squared first and second derivatives, respectively. □

Acknowledgement. In the original paper inequality (4.12) was proved for φ linear only, whereas for φ'' bounded a weaker version of (4.12) with N^{-1} replaced by $N^{-\frac{1}{2}}$ was proved. The present stronger result is based on a suggestion kindly made to the author by Mr. Peter J. Huber.

Theorem 4.2. *Under conditions of Theorem 4.1 there exists a constant $M = M(\varphi)$ such that for any N, $(c_1, \ldots, c_N)$ and $(F_1, \ldots, F_N)$*

$$\mathsf{E}\left(S - \mathsf{E}S - \sum_{i=1}^{N} Z_i\right)^2 \le M\,N^{-1} \sum_{i=1}^{N} (c_i - \bar{c})^2 \tag{4.26}$$

and

$$(\mathsf{E}S - \mu)^2 \le M\,N^{-1} \sum_{i=1}^{N} c_i^2, \tag{4.27}$$

where

$$Z_i = \ell_i(X_i) \tag{4.28}$$
$$= N^{-1}\sum_{j=1}^{N}(c_j - c_i)\int[u(x - X_i) - F_i(x)]\,\varphi'(H(x))\,\mathrm{d}F_j(x), \quad 1 \le i \le N,$$

and

$$\mu = \sum_{i=1}^{N} c_i \int \varphi(H(x))\,\mathrm{d}F_i(x),$$

with $H(x) = N^{-1}\sum_{j=1}^{N} F_j(x)$.

Proof. We have

$$\hat{S} - \mathsf{E}\hat{S} = \sum_{i=1}^{N}\sum_{j=1}^{N} c_j[\mathsf{E}(\varphi(\rho_j)\,|\,X_i) - \mathsf{E}\,\varphi(\rho_j)].$$

Since, obviously,

$$\sum_{j=1}^{N}\varphi(\rho_j) = \sum_{j=1}^{N}\mathsf{E}(\varphi(\rho_j)\,|\,X_i) = \sum_{j=1}^{N}\mathsf{E}\,\varphi(\rho_j) = \sum_{i=1}^{N}\varphi(i/(N+1)),$$

we also have

$$\hat{S} - \mathsf{E}\hat{S} = \sum_{i=1}^{N}\sum_{j=1}^{N}(c_j - c_i)\,[\mathsf{E}(\varphi(\rho_j)\,|\,X_i) - \mathsf{E}\,\varphi(\rho_j)]. \tag{4.29}$$

Now (3.3) yields

$$\mathsf{E}(\varphi(\rho_j)\,|\,X_i) - \mathsf{E}\,\varphi(\rho_j) \tag{4.30}$$
$$= \int[u(x - X_i) - F_i(x)]\sum_{k=2}^{N}[\varphi(k/(N+1)) - \varphi((k-1)/(N+1))]$$
$$\times\mathsf{P}(R_j = k\,|\,X_j = x,\, X_i = x - 1)\mathrm{d}F_j(x)$$
$$= (N+1)^{-1}\int[u(x - X_i) - F_i(x)]\,\mathsf{E}[\varphi'(\rho_j)\,|\,X_j = x,\, X_i = x - 1]$$
$$\times\mathrm{d}F_j(x) + (N+1)^{-2}\,K_2^{\frac{1}{2}}\,\alpha_i, \quad |\,\alpha_i\,| \le 1.$$

Further,

$$\mathsf{E}[\varphi'(\rho_j)\,|\,X_j = x,\, X_i = x - 1] = \varphi'(H(x)) + 2N^{-\frac{1}{2}}\,K_2^{\frac{1}{2}}\,\beta_i, \quad |\,\beta_i\,| \le 1,$$

so that

$$\mathsf{E}(\varphi(\rho_j)\,|\,X_i) - \mathsf{E}\,\varphi(\rho_j)$$
$$= (N+1)^{-1}\int [u(x-X_i)-F_i(x)]\,\varphi'(H(x))\,\mathrm{d}F_j(x) + 3N^{-\frac{3}{2}}\,K_2^{\frac{1}{2}}\,\gamma_i, \qquad |\,\gamma_i\,| \le 1.$$

Consequently, since $\mathsf{E}\,Z_i = 0$,

$$\begin{aligned}
\mathsf{E}\left(\hat{S} - \mathsf{E}\hat{S} - \sum_{i=1}^{N} Z_i\right)^2 &= \sum_{i=1}^{N}\mathsf{E}\left(\sum_{j=1}^{N}(c_j - c_i)\,3N^{-\frac{3}{2}}\,K_2^{\frac{1}{2}}\gamma_i\right)^2 \\
&\le 9N^{-3}\,K_2\sum_{i=1}^{N}\left(\sum_{j=1}^{N}|\,c_j - c_i\,|\right)^2 \\
&\le 18N^{-1}\,K_2\sum_{i=1}^{N}(c_i - \overline{c})^2.
\end{aligned}$$

This together with (4.12) yields (4.26) with $M = 2K + 36K_2^2$.
Finally,

$$\varphi(\rho_i) = \varphi(H(X_i)) + (\rho_i - H(X_i))\,\varphi'(H(X_i)) + \frac{1}{2}(\rho_i - H(X_i))^2\,k_i(X_i)$$

yields, for proper K_4,

$$|\,\mathsf{E}\,\varphi(\rho_i) - \mathsf{E}\,\varphi(H(X_i))\,| \le K_4\,N^{-1}.$$

Therefore,

$$|\,\mathsf{E}S - \mu\,|^2 = \left(\sum_{i=1}^{N} c_i[\mathsf{E}\,\varphi(\rho_i) - \mathsf{E}\,\varphi(H(X_i))]\right)^2 \le K_4^2\,N^{-1}\sum_{i=1}^{N} c_i^2. \tag{4.31}$$

Obviously, (4.31) provides (4.27) with $M = K_4^2$. The proof is complete. □

5 THE PROOFS TO SECTION 2

Proof of Theorem 2.1. Consider the random variables Z_i given by (4.28) and σ^2 given by (2.9). Obviously, $\mathsf{E}Z_i = 0$ and $\sigma^2 = \sum_{i=1}^{N}\mathsf{var}\,Z_i$. Choose a $\epsilon > 0$. Then by the Lindeberg theorem, there exists a $\delta > 0$ such that the relation

$$\sigma^{-2}\sum_{i=1}^{N}\int_{|\,x\,|>\delta\sigma} x^2\,\mathrm{d}P(Z_i \le x) < \delta \tag{5.1}$$

entails

$$\sup_x \left| \mathsf{P}\left(\sum_{i=1}^N Z_i < x\sigma\right) - \Phi(x)\right| < \frac{1}{4}\,\epsilon, \tag{5.2}$$

where $\Phi(x) = (2\pi)^{-\frac{1}{2}} \int_{-\infty}^{x} \exp\left(-\frac{1}{2}\,y^2\right) dy$. Furthermore, there exists a $\beta > 0$ such that

$$|\,\Phi(x) - \Phi(x \pm \beta)\,| < \frac{1}{4}\,\epsilon, \quad -\infty < x < \infty, \tag{5.3}$$

and, in turn,

$$\sup_x \left| \mathsf{P}\left(\sum_{i=1}^N Z_i < x\sigma \pm \beta\sigma\right) - \Phi(x)\right| < \frac{1}{2}\,\epsilon. \tag{5.4}$$

Now we shall show that (2.8) is entailed by

$$\mathsf{var}\, S \geq \left[2\delta^{-1} \sup_{0<t<1} |\,\varphi'(t)\,| + \left(2\epsilon^{-\frac{1}{2}}\,\beta^{-1} + 1\right) M^{\frac{1}{2}}\right]^2 \max_{1\leq i\leq N}(c_i - \overline{c})^2, \tag{5.5}$$

where M is the constant appearing in (4.26). If the scores are generated from φ by (2.4), Theorem 4.2 entails

$$\left|\sigma - (\mathsf{var}\, S)^{\frac{1}{2}}\right| \leq M^{\frac{1}{2}} \max_{1\leq i\leq N} |\,c_i - \overline{c}\,|, \tag{5.6}$$

and hence (5.5) entails

$$\delta\sigma \geq 2 \max |\,c_i - \overline{c}\,| \sup |\,\varphi'(t)\,|. \tag{5.7}$$

On the other hand (4.28) implies that always

$$|\,Z_i\,| \leq 2 \max |\,c_i - \overline{c}\,| \sup |\,\varphi'(t)\,|. \tag{5.8}$$

Putting together (5.7) and (5.8) we see that the left side of (5.1) is zero for σ satisfying (5.7). Thus (5.5) entails (5.1).

Further, by (4.26) and (5.4)

$$\begin{aligned}
\mathsf{P}(S - \mathsf{E}S < x\sigma) &\leq \mathsf{P}\left(\sum_{i=1}^N Z_i < x\sigma + \beta\sigma\right) + \mathsf{P}\left(\left|S - \mathsf{E}S - \sum_{i=1}^N Z_i\right| > \beta\sigma\right) \\
&\leq \Phi(x) + \frac{1}{2}\,\epsilon + \mathsf{E}\left(S - \mathsf{E}S - \sum_{i=1}^N Z_i\right)^2 / \beta^2\sigma^2 \\
&\leq \Phi(x) + \frac{1}{2}\,\epsilon + M \max_{1\leq i\leq N}(c_i - \overline{c})^2 / \beta^2\sigma^2.
\end{aligned}$$

Now (5.5) and (5.6) entail

$$\beta^2\sigma^2 \geq 4\epsilon^{-1}\, M \max_{1\leq i\leq N}(c_i - \overline{c})^2.$$

Altogether we have

$$\mathsf{P}(S - \mathsf{E}S < x\sigma) \leq \Phi(x) + \frac{1}{2}\epsilon + \frac{1}{4}\epsilon = \Phi(x) + \frac{3}{4}\epsilon. \tag{5.9}$$

Similarly we would prove the opposite inequality leading to

$$\sup_x |\,\mathsf{P}(S - \mathsf{E}S < x\sigma) - \Phi(x)\,| < \frac{3}{4}\epsilon. \tag{5.10}$$

Finally, if $\epsilon < 1$, then (5.5) and (5.6) entail

$$\left|\sigma - (\mathsf{var}\, S)^{\frac{1}{2}}\right| < \beta\sigma,$$

which combined with (5.3) and (5.10) yields

$$\sup_x \left|\mathsf{P}\left(S - \mathsf{E}S < x(\mathsf{var}\, S)^{\frac{1}{2}}\right) - \Phi(x)\right| < \epsilon. \tag{5.11}$$

In the course of the proof we have also proved that (5.10) is satisfied, if $\mathsf{var}\, S$ is replaced by σ^2 in (5.5).

If the scores were generated by (2.5), it suffices to note that, by the Taylor expansion,

$$\mathsf{E}\varphi(U_N^{(i)}) = \varphi(i/(N+1)) + \kappa_{iN},$$

where $|\,\kappa_{iN}\,| \leq k\,N^{-1}$, where k does not depend on i or N. Thus, denoting the statistic $\sum_{i=1}^{N} c_i\, a_N(R_i)$ by S, if (2.4) holds, and by S', if (2.5) holds, we have

$$\mathsf{E}(S - \mathsf{E}S - S' + \mathsf{E}S')^2 \leq \left(k\,N^{-1}\sum_{i=1}^{N}|\,c_i - \bar{c}\,|\right)^2 \leq k^2\,N^{-1}\sum_{i=1}^{N}(c_i - \bar{c})^2.$$

Consequently $S - \mathsf{E}S$ is equivalent to $S' - \mathsf{E}S'$ in asymptotic considerations. □

Proof of Theorem 2.2. Obviously

$$\begin{aligned} \int_{x_0}^{y} \varphi'(H(x))\,\mathrm{d}F_j(x) &= \int_{x_0}^{y} \varphi'(F_j(x))\,\mathrm{d}F_j(x) + R_j(y) \qquad (5.12)\\ &= \varphi(F_j(y)) + \text{const} + R_j(y)\\ &= \varphi(F_i(y)) + \text{const} + R_{ij}(y), \end{aligned}$$

where

$$\begin{aligned} |\,R_{ij}(y)\,| &\leq \left[\sup_t |\,\varphi''(t)\,| + \sup_t |\,\varphi'(t)\,|\right] \max_{i,j,x} |\,F_i(x) - F_j(x)\,| \qquad (5.13)\\ &= L \max_{i,j,x} |\,F_i(x) - F_j(x)\,|, \quad 1 \leq i,\, j \leq N, \end{aligned}$$

where $L = L(\varphi)$ is defined by the last equation. Let

$$V_i = (\bar{c} - c_i)\,\varphi(F_i(X_i)) \tag{5.14}$$

and note, by inspection of (2.14), that

$$d^2 = \sum_{i=1}^{N} \operatorname{var} V_i. \tag{5.15}$$

Now, (2.26), (4.28), (5.12) and (5.14) entail

$$Z_i = V_i + \text{const} + R_i^*(X_i)\,N^{-1} \sum_{i=1}^{N} |\,c_j - c_i\,|, \tag{5.16}$$

where R_i^* is bounded by the right side of (5.13). Consequently, by (5.15),

$$\begin{aligned} |\,\sigma - d\,| &= \left| \left(\operatorname{var} \sum_{i=1}^{N} Z_i\right)^{\frac{1}{2}} - \left(\operatorname{var} \sum_{i=1}^{N} V_i\right)^{\frac{1}{2}} \right| \\ &\le \left(\sum_{i=1}^{N} \operatorname{var}(Z_i - V_i) \right)^{\frac{1}{2}} \\ &\le \left[\sum_{i=1}^{N} \left(N^{-1} \sum_{j=1}^{N} |\,c_j - c_i\,| \right)^2 \right]^{\frac{1}{2}} L \max_{i,j,x} |\,F_i(x) - F_j(x)\,| \\ &\le L \left(2 \sum_{i=1}^{N} (c_i - \bar{c})^2 \right)^{\frac{1}{2}} \max_{i,j,x} |\,F_i(x) - F_j(x)\,|. \end{aligned} \tag{5.17}$$

Now we find δ_ϵ as follows. In accordance with Theorem 2.1, we choose $K_{\frac{1}{2}\epsilon}$ and $\alpha > 0$ such that $\sigma^2 > K_{\frac{1}{2}\epsilon} \max_{1 \le i \le N} (c_i - \bar{c})^2$ entails

$$\sup_x |\mathsf{P}(S - \mathsf{E}S < x\sigma(1 \pm \alpha)) - \Phi(x)| < \epsilon.$$

Then we choose δ_ϵ so that, first, $\delta_\epsilon^{-1} > K_{\frac{1}{2}\epsilon}$, and, second, that (2.12) entails $|\,\sigma/d - 1\,| < \alpha$. The last implication is for sufficiently small δ_ϵ guaranteed by (5.17) and (2.14). The rest easily follows. □

Proof of Theorem 2.3. We shall start with the following lemma.

Lemma 5.1. *If φ satisfies the conditions of Theorem 2.3, then for any $\alpha > 0$ there exists a decomposition*

$$\varphi(t) = \psi(t) + \varphi_1(t) - \varphi_2(t), \quad 0 < t < 1, \tag{5.18}$$

such that ψ is a polynomial, φ_1 and φ_2 are non-decreasing, and

$$\int_0^1 \varphi_1^2(t)\,\mathrm{d}t + \int_0^1 \varphi_2^2(t)\,\mathrm{d}t < \alpha. \tag{5.19}$$

Proof. Without losing generality, we may assume that $\varphi(t)$ itself is nondecreasing. Take an $\epsilon > 0$ and put

$$\begin{aligned} \varphi_0(t) &= \varphi(\epsilon), \qquad 0 < t < \epsilon, &(5.20)\\ &= \varphi(t), \qquad \epsilon \leq t \leq 1-\epsilon, \\ &= \varphi(1-\epsilon), \quad 1-\epsilon < t < 1; \\ \varphi_3(t) &= \min[0,\, \varphi(t) - \varphi(\epsilon)]; &(5.21)\\ \varphi_4(t) &= \max[0,\, \varphi(t) - \varphi(1-\epsilon)]. &(5.22)\end{aligned}$$

Obviously,

$$\varphi(t) = [\varphi_0(t) + \varphi_3(t) + \varphi_4(t)], \tag{5.23}$$

where φ_3 and φ_4 are non-decreasing. As $\varphi(t)$ is absolutely continuous on $(\epsilon,\, 1-\epsilon)$, $\varphi_0(t)$ is absolutely continuous on the whole interval $(0,1)$. Consequently, there exists a derivative $\varphi_0'(t)$ such that

$$\varphi_0(t) = \varphi(\epsilon) + \int_0^t \varphi_0'(s)\,\mathrm{d}s, \quad 0 \leq t \leq 1. \tag{5.24}$$

Further to every $\beta > 0$ there exists a polynomial $q(s)$ such that

$$\int_0^1 |\,\varphi_0'(s) - q(s)\,|\;\mathrm{d}s < \beta, \tag{5.25}$$

because the set of polynomials is a dense subset of the L_1-space of integrable functions.

Putting

$$\begin{aligned} \psi(t) &= \varphi(\epsilon) + \int_0^t q(s)\,\mathrm{d}t, &(5.26)\\ \varphi_1(t) &= \varphi_3(t) + \varphi_4(t) + \int_0^t \max[0,\, \varphi_0'(s) - q(s)]\,\mathrm{d}s, &(5.27)\\ \varphi_2(t) &= \int_0^t \max[0,\, q(s) - \varphi_0'(s)]\mathrm{d}s, &(5.28)\end{aligned}$$

we easily see that

$$\varphi(t) = \psi(t) + \varphi_1(t) - \varphi_2(t), \tag{5.29}$$

where $\psi(t)$ is a polynomial, $\varphi_1(t)$ and $\varphi_2(t)$ are non-decreasing and (5.20) may be satisfied by taking ϵ and β in the above construction sufficiently small.

Now take $\epsilon > 0$ and $\eta > 0$. Then choose $\beta > 0$ and $\gamma > 0$ such that

$$\left|\Phi(x) - \Phi[(x \pm \beta)(1 \pm \gamma)^{-1}]\right| < \frac{1}{4}\epsilon \tag{5.30}$$

and $\alpha > 0$ such that

$$\alpha < \eta \min\left(\gamma^2, \frac{1}{4}\beta^2\epsilon\right)/84. \tag{5.31}$$

Subsequently, we decompose φ according to (5.18) with α satisfying (5.19) and (5.31). We denote

$$S_\psi = \sum_{i=1}^{N} c_i\, \psi(R_i/(N+1)), \tag{5.32}$$

$$S_h = \sum_{i=1}^{N} c_i\, \varphi_h(R_i/(N+1)), \quad h = 1, 2.$$

Obviously $S = S_\psi + S_1 - S_2$. As the functions φ_h are non-decreasing, we have, in view of Theorem 3.1,

$$\begin{aligned} &\left|(\mathsf{var}\, S)^{\frac{1}{2}} - (\mathsf{var}\, S_\psi)^{\frac{1}{2}}\right| \\ &\leq (\mathsf{var}(S - S_\psi))^{\frac{1}{2}} \leq (\mathsf{var}\, S_1)^{\frac{1}{2}} + (\mathsf{var}\, S_2)^{\frac{1}{2}} \\ &\leq (42)^{\frac{1}{2}} \max_{1\leq i\leq N} |c_i - \bar{c}| \left[\sum_{i=1}^{N}(\varphi_{1\alpha}^2(i/(N+1)) + \varphi_{2\alpha}^2(i/(N+1)))\right]^{\frac{1}{2}}. \end{aligned} \tag{5.33}$$

Now, it may be easily shown that for a non-decreasing φ

$$\sum_{i=1}^{N} \varphi^2(i/(N+1)) \leq (N+1)\int_0^1 \varphi^2(t)\, \mathrm{d}t \leq 2N \int_0^1 \varphi^2(t)\, \mathrm{d}t. \tag{5.34}$$

Combining (2.16) and (5.31) through (5.34), we obtain

$$\left|(\mathsf{var}\, S)^{\frac{1}{2}} - (\mathsf{var}\, S_\psi)^{\frac{1}{2}}\right| \leq [\mathsf{var}(S - S_\psi)]^{\frac{1}{2}} \leq (\mathsf{var}\, S)^{\frac{1}{2}} \min\left(\gamma, \frac{1}{2}\beta\, \epsilon^{\frac{1}{2}}\right). \tag{5.35}$$

Finally, let $K_{\frac{1}{2}\epsilon} = K_{\frac{1}{2}\epsilon}(\psi)$ be the constant, the existence of which was established in Theorem 2.1. (Note that ψ is a polynomial, and hence has a bounded second derivative.) Then put $N_{\epsilon\eta} = (1-\gamma)^{-2}\,\eta^{-1}\, K_{\frac{1}{2}\epsilon}$ so that (2.16) in conjunction with (5.35) entails $\mathsf{var}\, S_\psi > K_{\frac{1}{2}\epsilon} \max_{1\leq i\leq N}(c_i - \bar{c})^2$. Consequently, by Theorem 2.1, (5.30) and (5.35),

$$\mathsf{P}\left(S - \mathsf{E}S < x(\mathsf{var}\, S)^{\frac{1}{2}}\right)$$

$$\le \mathsf{P}\left(S_\psi - \mathsf{E}S_\psi < (x+\beta)\,(\mathsf{var}\,S)^{\frac{1}{2}}\right)$$
$$+\mathsf{P}\left(\,|\,S - \mathsf{E}S - S_\psi + \mathsf{E}S_\psi\,| > \beta(\mathsf{var}\,S)^{\frac{1}{2}}\right)$$
$$\le \mathsf{P}\left(S_\psi - \mathsf{E}S_\psi < (x+\beta)\,(\mathsf{var}\,S)^{\frac{1}{2}}\right) + \mathsf{var}(S - S_\psi)/\beta^2\,\mathsf{var}(S)$$
$$\le \mathsf{P}\left(S_\psi - \mathsf{E}S_\psi < (x+\beta)\,(1-\gamma)^{-1}(\mathsf{var}\,S_\psi)^{\frac{1}{2}}\right) + \frac{1}{4}\epsilon$$
$$\le \Phi\left((x+\beta)\,(1-\gamma)^{-1}\right) + \frac{3}{4}\epsilon \;\le\; \Phi(x) + \epsilon.$$

Similarly we would prove the opposite inequality needed in (2.8).

The version of the theorem, in which $\mathsf{var}\,S$ is replaced by σ^2, may again be shown by a decomposition of σ^2, corresponding to the decomposition $\varphi = \psi + \varphi_1 - \varphi_2$. If we define σ_ψ^2, σ_1^2 and σ_2^2 by (2.9) and (2.10), where φ is replaced by ψ, φ_1 and φ_2, respectively, we obtain

$$|\,\sigma - \sigma_\psi\,| \le \sigma_1 + \sigma_2. \tag{5.36}$$

Now, if φ in (2.10) is non-decreasing then the random variables

$$\int [u(x - X_i) - F_i(x)]\,\varphi'(H(x))\,\mathrm{d}F_j(x), \quad 1 \le j \le N,$$

are (see Lemma 3.1) non-negatively correlated. Consequently,

$$\begin{aligned}
\sigma_h^2 \;&\le\; 4 \max_{1\le i\le N}(c_i - \overline{c})^2 \qquad (5.37)\\
&\qquad \times \sum_{i=1}^N \mathsf{var}\left[N^{-1}\sum_{j=1}^N \int [u(x - X_i) - F_i(x)]\,\varphi_h'(H(x))\,\mathrm{d}F_j(x)\right]\\
&\le\; 4 \max_{1\le i\le N}(c_i - \overline{c})^2 \sum_{i=1}^N \mathsf{var}\left(\int [u(x - X_i) - F_i(x)]\,\varphi_h'(H(x))\,\mathrm{d}H(x)\right)\\
&=\; 4 \max_{1\le i\le N}(c_i - \overline{c})^2 \sum_{i=1}^N \mathsf{var}\,\varphi_h(H(X_i))\\
&\le\; 4N \max_{1\le i\le N}(c_i - \overline{c})^2 \int_0^1 \varphi_h^2(t)\,\mathrm{d}t, \quad h = 1, 2.
\end{aligned}$$

Further, in view of (5.36), (5.37) and (5.19),

$$|\,\sigma - \sigma_\psi\,|^2 \le 8N \max_{1\le i\le N}(c_i - \overline{c})^2\,\alpha.$$

Thus, for α sufficiently small, $|\sigma - \sigma_\psi|$ may be made negligible in comparison with $\eta N \max_{1\le i\le N}(c_i - \overline{c})^2$. Since, as we see from (5.6) and (5.35),

$\left|(\text{var}\, S_\psi)^{\frac{1}{2}} - \sigma_\psi\right|$ as well as $\left|(\text{var}\, S)^{\frac{1}{2}} - (\text{var}\, S_\psi)^{\frac{1}{2}}\right|$ are under control, too, the difference $\left|\sigma - (\text{var}\, S)^{\frac{1}{2}}\right|$ must be negligible with respect to σ if

$$\sigma^2 > \eta N \max_{1 \le i \le N} (c_i - \bar{c})^2$$

and N is sufficiently large.

The case of the scores given by (2.5) could be treated similarly as in the proof of Theorem 2.1, with the help of the inequality

$$\sum_{i=1}^{N} [\mathsf{E}\varphi(U_n^{(i)})]^2 \le N \int_0^1 \varphi^2(t)\, \mathrm{d}t$$

holding for any φ. □

Proof of Theorem 2.4. We omit this proof because it consists in the combination of methods used in the proof of Theorems 2.2 and 2.3. □

6 A COUNTEREXAMPLE

The conditions concerning φ in Theorems 2.3 and 2.4 seem to be close to what is necessary, if the other conditions are unchanged. If φ is discontinuous, as it occurs with the median test and in the first stage of the study of the Kolmogorov–Smirnov test, the assertion of the theorems is not valid, as we now show. In particular, we shall assume that

$$\begin{aligned} \varphi(t) &= 0, \quad 0 < t < \frac{1}{2} \\ &= 1, \quad \frac{1}{2} \le t < 1, \end{aligned} \tag{6.1}$$

and that the scores are given by (2.4). Further, let $c_i = 1,\ 1 \le i \le \frac{1}{4}N$, and $= 0,\ \frac{1}{4}N < i \le N$; $F_i = F,\ 1 \le i \le \frac{1}{4}N$, where F is a uniform distribution of $\left(\frac{1}{4}, \frac{1}{2}\right)$ and $F_i = G,\ \frac{1}{4}N < i \le N$, where G is uniform on $\left(0, \frac{1}{4}\right) \cup \left(\frac{1}{2}, 1\right)$. Then, if N is a multiple of 4,

$$S = \sum_{i=1}^{N} c_i\, \varphi(R_i/(N+1)) = \max\left\{0, \sum_{i > \frac{1}{4}N} \left[u\left(\frac{1}{4} - X_i\right) - \frac{1}{3}\right]\right\} \tag{6.2}$$

with u given by (2.1). Since S is a truncated binomial random variable, the order of $\text{var}\, S$ is N, say $\text{var}\, S > bN,\ N \ge 1$. Further, $\max_{1 \le i \le N}(c_i - \bar{c})^2 = 9/16$, so that $\text{var}\, S > \eta N \max_{1 \le i \le N}(c_i - \bar{c})^2$ if $\eta \le (16/9)\, b$. On the other hand $\mathsf{P}(S = 0) \to \frac{1}{2}$, so that (2.8) is not implied by (2.16) for any $N_{\epsilon\eta}$, provided that $\epsilon < \frac{1}{2}$ and $\eta \le b$.

ACKNOWLEDGEMENT

This work has been partially supported by the National Science Foundation, Grant GP–5059.

REFERENCES

Bhuchonghul S. (1964). A class of non-parametric tests for independence in bivariate populations. *Ann. Math. Statist. 35*. 138–149.

Chernoff H. and Savage I. E. (1958). Asymptotic normality and efficiency of certain nonparametric test statistics. *Ann. Math. Statist. 29*, 972–994.

Dwass M. (1956). The large-sample power of rank order tests in the two–sample problem. *Ann. Math. Statist. 27*, 352–374.

Govindarajulu Z., Le Cam L. and Raghavachari M. (1966). Generalizations of theorems of Chernoff and Savage on the asymptotic normality of test statistics. In *Proc. Fifth Berkeley Symp. Math. Statist. Prob.*, Vol. 1, pp. 609–638.

Hoeffding W. (1948). A class of statistics with asymptotically normal distribution. *Ann. Math. Statist. 19*, 293–325.

Hájek J. (1961). Some extensions of the Wald–Wolfowitz–Noether theorem. *Ann. Math. Statist. 32*, 506–523.

Hájek J. (1962). Asymptotically most powerful rank–order tests. *Ann. Math. Statist. 33*, 1124–1147.

Lehmann E. L. (1966). Some concepts of dependence. *Ann. Math. Statist. 37*, 1137–1153.

Puri M. L. (1964). Asymptotic efficiency of a class of *c*–sample tests. *Ann. Math. Statist. 35*, 102–121.

CHAPTER 29

Asymptotic Normality of Simple Linear Rank Statistics Under Alternatives II

Co-author: Václav Dupač.
Ann. Math. Statist. 40 (1969), 1992–2017.
Reproduced by permission of The Institute of Mathematical Statistics.

SUMMARY

This is a straightforward continuation of Hájek (1968). We provide a further extension of the Chernoff–Savage (1958) limit theorem. The requirements concerning the score-generating function are relaxed to a minimum: we assume that this function is a difference of two nondecreasing and square integrable functions. Thus, in contradistinction to Hájek (1968), we have dropped the assumption of absolute continuity. The main results are accumulated in Section 2 without proofs. The proofs are given in Sections 4 through 7. Section 3 contains auxiliary results.

1 INTRODUCTION

The basic tools used in this paper are the same as in Hájek (1968), namely the variance inequality (Theorem 3.1) and the Projection Lemma 4.1. The main problem to overcome was the treatment of the score-generating function, which has just one jump and is constant otherwise (Theorem 1). The solution of this seemingly simple problem took three Sections – 3, 4 and 5. The remaining theorems were then obtained relatively easily by combining

Collected Works of Jaroslav Hájek – With Commentary
Edited by M. Hušková, R. Beran and V. Dupač
Published in 1998 by John Wiley & Sons Ltd.

Theorem 1 with the results of Hájek (1968).

Since we have relaxed the conditions concerning the score-generating function, we had to introduce additional smoothness requirements concerning the distributions F_{Ni} of individual observations. Without it the theorems of Hájek (1968) do not hold for discrete score-generating functions, as is illustrated by a counterexample.

In framing the theorems we had to balance two requirements: the generality of the conditions and the readiness for applications. To satisfy both we presented six variants of the conditions under which the conclusion of Theorem 1 holds.

If the score-generating function is discrete, the assumptions become more complex. For this reason we had to abandon the ϵ-form used in Hájek (1968), and to switch to limiting theorems concerning sequences. To simplify the notation, the sequences are indexed by the sample size, though, in principle, the sample size could be made a function of a new index as well.

Pyke and Shorack (1968) obtained results similar to ours for the two-sample and c-sample problems and for a slightly less general score-generating function, and they provide a simplified expression for the asymptotic expectation, which is not attempted in the present chapter. Their approach, completely different from ours, is based on convergence properties of certain stochastic processes and has some common points with an earlier paper by Govindarajulu–Le Cam–Raghavachari (1966).

2 MAIN RESULTS

For every $N \geq 1$, let $R_{N1}, \ldots, R_{NN}$ denote the ranks of independent random variables $X_{N1}, \ldots, X_{NN}$. Choose some scores $a_N(1), \ldots, a_N(N)$ and regression constants $c_{N1}, \ldots, c_{NN}$, and put

$$S_N = \sum_{i=1}^{N} c_{Ni}\, a_N(R_{Ni}). \tag{2.1}$$

We inquire under what conditions the linear rank statistics S_N are asymptotically normal.

We shall assume that the c_{Ni}s satisfy either the *Noether* condition

$$\lim_{N\to\infty} \max_{1\leq i\leq N} (c_{Ni} - \bar{c}_N)^2 \Big/ \sum_{i=1}^{N} (c_{Ni} - \bar{c}_N)^2 = 0, \tag{2.2}$$

or, more stringently, the *boundedness* condition

$$\limsup_{N\to\infty} N \max_{1\leq i\leq N} (c_{Ni} - \bar{c}_N)^2 \Big/ \sum_{i=1}^{N} (c_{Ni} - \bar{c}_N)^2 < \infty. \tag{2.3}$$

The scores will be generated by a function $\varphi(t)$, $0 < t < 1$, which is representable as a difference of two nondecreasing, square integrable functions:

$$\varphi(t) = \varphi_1(t) - \varphi_2(t), \tag{2.4}$$

φ_i nondecreasing, $\int_0^1 \varphi_i^2 \, \mathrm{d}t < \infty$. The scores are obtained either by interpolation,

$$a_N(i) = \varphi(i(N+1)^{-1}), \quad 1 \leq i \leq N, \tag{2.5}$$

or by a procedure satisfying

$$\sum_{i=1}^{N} \left| a_N(i) - \varphi(i(N+1)^{-1}) \right| = O(1), \tag{2.6}$$

or, more loosely,

$$\sum_{i=1}^{N} \left| a_N(i) - \varphi(i(N+1)^{-1}) \right| = o\left(N^{\frac{1}{2}}\right). \tag{2.7}$$

The distribution functions of $X_{N1}, \ldots, X_{NN}$, say $F_{N1}, \ldots, F_{NN}$, will be assumed *continuous*. We shall put

$$H_N(x) = N^{-1} \sum_{i=1}^{N} F_{Ni}(x), \quad -\infty < x < \infty, \tag{2.8}$$

$$H_N^{-1}(t) = \inf\{x : H_N(x) > t\}, \quad 0 < t < 1, \tag{2.9}$$

and

$$L_{Ni}(t) = F_{Ni}(H_N^{-1}(t)), \quad 0 < t < 1. \tag{2.10}$$

It is easy to show that

$$N^{-1} \sum_{i=1}^{N} L_{Ni}(t) = t, \quad 0 < t < 1, \ N \geq 1. \tag{2.11}$$

In the sequel, v will denote a jump point of the score-generating function φ or, more generally, a point of some set containing the singular set of φ. We shall need a sort of uniform differentiability of the functions L_{Ni} at this point (these points), which will appear as a combination of two requirements. First, for such a $v \in (0,1)$ and for every $K > 0$, $K' > 0$

$$\max_{1 \leq i \leq N,\, KN^{-\frac{1}{2}} \leq |t-v| \leq K' N^{-\frac{1}{2}} \log^{\frac{1}{2}} N} \left| (L_{Ni}(t) - L_{Ni}(v))/(t-v) \right| = O(1). \tag{2.12}$$

Second, we shall assume that the L_{Ni}s are uniformly approximately linear in the vicinity of the point $v \in (0,1)$ in the following sense: there exist numbers $\ell_{Ni}(v)$ such that for every $K > 0$

$$\max_{1 \leq i \leq N,\, |t-v| \leq K N^{-\frac{1}{2}}} \left| L_{Ni}(t) - L_{Ni}(v) - (t-v)\, \ell_{Ni}(v) \right| = o\left(N^{-\frac{1}{2}}\right). \tag{2.13}$$

If (2.13) is satisfied for some ℓ_{ni}s, then setting $t = N^{-\frac{1}{2}}+v$ in (2.13) we see that it is also satisfied for $\ell^*_{Ni}(v) = N^{\frac{1}{2}}\left[L_{Ni}\left(v+N^{-\frac{1}{2}}\right) - L_{Ni}(v)\right]$. Consequently, in view of (2.11), we may assume

$$N^{-1}\sum_{i=1}^{N}\ell_{Ni}(v) = 1 \tag{2.14}$$

in (2.13) without any loss of generality. For the same reason, we may assume that $\ell_{Ni}(v)$ as a function of v is measurable. Further it is important to note that (2.13) neither implies nor is implied by the existence of the derivative $L'_{Ni}(v)$ at the point v. Consequently we generally cannot put $\ell_{Ni}(v) = L'_{Ni}(v)$, even if the latter number exists. Of course, if the derivative exists *uniformly* with respect to i and N, then (2.12) and (2.13) are satisfied with $\ell_{Ni}(v) = L'_{Ni}(v)$. The somewhat artificial looking conditions (2.12) and (2.13) result from our method of proving Theorem 1 and from our recognition that the assumption of uniform differentiability would be too restrictive (see Theorem 4, for example).

Other conditions concerning the L_{Ni}s used in the sequel are as follows:

$$\liminf_{N\to\infty} N^{-1}\sum_{i=1}^{N} L_{Ni}(v)[1 - L_{Ni}(v)] > 0, \tag{2.15}$$

$$\liminf_{N\to\infty}\min_{1\le i\le N} L_{Ni}(v)[1 - L_{Ni}(v)] > 0. \tag{2.16}$$

Obviously, (2.16) entails (2.15).

For applications, it is necessary to replace (2.12), (2.13) and (2.15) by some feasible conditions. For example, we may assume that the distribution functions F_{Ni} possess densities f_{Ni} such that for some open interval $(\alpha,\beta) \subset (-\infty,\infty)$ the following hold:

(a) $f_{Ni}(x) = 0$ for $x < \alpha(x > \beta)$, if $\alpha(\beta)$ is finite; (2.17)

(b) $f_{Ni}(x)$ are continuous on every compact subinterval of (α,β) uniformly in (x, N, i);

(c) for every compact interval $C \subset (\alpha,\beta)$ there exists an $\epsilon > 0$ such that for all $N \ge 1$, $N^{-1}\operatorname{card}\left\{i : \inf_{x\in C} f_{Ni}(x) > \epsilon\right\} > \epsilon$;

(d) $\alpha < \liminf_{N\to\infty} H_N^{-1}(t) \le \limsup_{N\to\infty} H_N^{-1}(t) < \beta$ for all $t \in (0,1)$;

where card A stands for the number of elements of the set A.

In particular, (2.17) is satisfied for

(a) $f_{Ni}(x) = f(x - d_{Ni})$; (2.18)

(b) $f(x)$ is uniformly continuous and positive on $(-\infty, \infty)$;

(c) for every $\epsilon > 0$ there exists a compact interval C such that for all $N \geq 1$, $N^{-1}\,\mathrm{card}\{i : |d_{Ni}| \notin C\} < \epsilon$.

Obviously, (2.18) is entailed by

$$\text{(a)}\quad f_{Ni}(x) = f(x - d_{Ni}); \tag{2.19}$$

(b) $f(x)$ is uniformly continuous and positive on $(-\infty, \infty)$;

(c) $\sup_{i,N} |d_{Ni}| < \infty$.

The last group of conditions concerns the nondegeneration of $\mathsf{var}\, S_N$. The milder form is

$$\liminf_{N\to\infty} \mathsf{var}\, S_N \Big/ \sum_{i=1}^{N} (c_{Ni} - \bar{c}_N)^2 > 0. \tag{2.20}$$

The stricter form is

$$\liminf_{N\to\infty} \mathsf{var}\, S_N \Big/ N \max_{1\leq i\leq N} (c_{Ni} - \bar{c}_N)^2 > 0. \tag{2.21}$$

Obviously (2.21) entails (2.20). Note that

$$\mathsf{var}\, S_N = (N-1)^{-1} \sum_{i=1}^{N} (a_N(i) - \bar{a}_N)^2 \sum_{i=1}^{N} (c_{Ni} - \bar{c}_N)^2 \quad \text{if } F_{N1} = F_{N2} = \cdots = F_{NN},$$

which entails $\mathsf{var}\, S_N = O\left(\sum_{i=1}^{N} (c_{Ni} - \bar{c}_N)^2\right)$ under (2.4) and (2.5) or (2.6) or (2.7). If the regression constants c_{Ni} are bounded in the sense of (2.3), then, obviously, (2.20) and (2.21) are equivalent.

Alternatively we may assume that (2.20) and (2.21) hold with $\mathsf{var}\, S_N$ replaced by some approximative variance σ_N^2:

$$\liminf_{N\to\infty} \sigma_N^2 \Big/ \sum_{i=1}^{N} (c_{Ni} - \bar{c}_N)^2 > 0, \tag{2.22}$$

$$\liminf_{N\to\infty} \sigma_N^2 \Big/ N \max_{1\leq i\leq N} (c_{Ni} - \bar{c}_N)^2 > 0. \tag{2.23}$$

Theorem 1. *Consider statistics* (2.1) *with scores satisfying* (2.6), *where*

$$\begin{aligned} \varphi(t) &= 0 \qquad 0 < t < v, \\ &= 1 \qquad v \leq t < 1. \end{aligned} \tag{2.24}$$

Put

$$\tilde{c}_N(v) = N^{-1} \sum_{i=1}^{N} c_{Ni}\, \ell_{Ni}(v) \tag{2.25}$$

and

$$\sigma_N^2 = \sum_{i=1}^{N} (c_{Ni} - \tilde{c}_N(v))^2\, L_{Ni}(v)\,[1 - L_{Ni}(v)]. \tag{2.26}$$

Then S_N is asymptotically normal with parameters ($\mathsf{E}\, S_N$, $\operatorname{var} S_N$) *and also* ($\mathsf{E}\, S_N$, σ_N^2), *if any of the following sets of conditions is satisfied:*

C$_1$: (2.2), (2.12), (2.13), (2.15), (2.20) *or* (2.22)
C$_2$: (2.2), (2.12), (2.13), (2.16)
C$_3$: (2.2), (2.17), (2.20) *or* (2.22)
C$_4$: (2.2), (2.18), (2.20) *or* (2.22)
C$_5$: (2.2), (2.19)
C$_6$: (2.3), (2.12), (2.13), (2.20) *or* (2.22).

Proof. Asymptotic normality of S_N will be shown by first showing that S_N is asymptotically equivalent to its projection $\hat{S}_N$ onto a space of linear statistics (see Sections 3 and 4) and then showing that $\hat{S}_N$ is asymptotically equivalent to a sum of independent random variables $\sum_{i=1}^{N} Z_{Ni}$ (see Section 5) to which the Lindeberg central limit theorem applies. The components of σ_N^2 in (2.26) are the variances of the Z_{Ni}s. Moreover, for this choice of σ_N^2, conditions (2.20) and (2.22) will be shown to be equivalent. □

Example. If we employ conditions (2.2) and (2.19) we can see that in the two-sample location case the median test statistic is asymptotically normal if (a) the underlying density is uniformly continuous and positive on $(-\infty, \infty)$, (b) the difference of the location parameters remains bounded, and (c) both the sample sizes converge to infinity.

On combining Theorem 1 with Theorem 2.3 of Hájek (1968), we obtain a powerful result for the case of bounded regression constants (see (2.3)). Before formulating it, let us recall that every nondecreasing function may be decomposed into an absolutely continuous part and a singular part. (The singular part includes both the jump and singular continuous components.) We shall have $\varphi = \varphi_1 - \varphi_2$, where each φ_i is nondecreasing. Let us denote their absolutely continuous and singular parts by φ_i^{ac} and φ_i^{s}, respectively:

$$\varphi_i(t) = \varphi_i^{ac}(t) + \varphi_i^{s}(t), \quad i = 1,\, 2,\; 0 < t < 1. \tag{2.27}$$

Then put

$$\varphi_{ac} = \varphi_1^{ac} - \varphi_2^{ac}, \quad \varphi_s = \varphi_1^{s} - \varphi_2^{s}. \tag{2.28}$$

To every nondecreasing φ there corresponds uniquely a measure ν such that $\nu\{(a,b)\} = \varphi(b) - \varphi(a)$, if a and b are continuity points of φ. We shall understand by $\int_A \mathrm{d}\varphi$ and $\int h\,\mathrm{d}\varphi$ the expressions $\nu(A)$ and $\int h\,\mathrm{d}\nu$, respectively.

Theorem 2. *Consider statistics* (2.1) *with scores satisfying* (2.7), *where φ fulfils* (2.4). *Denote by A the set of values $v \in (0,1)$ for which at least one of the conditions* (2.12), (2.13) *and* (2.15) *is not satisfied, and assume that there is a measurable set $B \supset A$ such that*

$$\int_B (\mathrm{d}\varphi_1^s(v) + \mathrm{d}\varphi_2^s(s)) = 0. \tag{2.29}$$

Put

$$\sigma_N^2 = \sum_{i=1}^N \mathsf{var}\left[\int_{\frac{1}{2}}^{H_N(X_{Ni})} (\tilde{c}_N(v) - c_{Ni})\,\mathrm{d}\varphi(v)\right] \tag{2.30}$$

where

$$\tilde{c}_N(v) = N^{-1}\sum_{j=1}^N c_{Nj}\,\ell_{Nj}(v), \tag{2.31}$$

if $v \notin B$, with the ℓ_{Nj}s chosen so as to be measurable and satisfy (2.14), *and*

$$\tilde{c}_N = N^{-1}\sum_{j=1}^N c_{Nj}\,L'_{Nj}(v), \tag{2.32}$$

if $v \in B$ but the derivative $L'_{Nj}(v)$ exists.

Then S_N is asymptotically normal with parameters ($\mathsf{E}\,S_N$, $\mathsf{var}\,S_N$) and also ($\mathsf{E}\,S_N$, σ_N^2) provided either (2.21) *or* (2.23) *holds, with σ_N^2 given by* (2.30).

Proof. See Section 6. □

Remark 1. Since the L_{Nj} are absolutely continuous, and since (2.29) holds, $\tilde{c}_N(v)$ is defined almost everywhere with respect to the measure induced by $\varphi_1 + \varphi_2$. Consequently, the integral (2.30) is well-defined.

Remark 2. Now (2.11), entailing $N^{-1}\sum_{i=1}^N L'_{Ni}(t) = 1$, and (2.14) imply that $\tilde{c}_N(v)$ is always a weighted average of $c_{N1}, \ldots, c_{NN}$. Consequently,

$$|\tilde{c}_N(v) - c_{Ni}| \leq 2 \max_{1\leq j\leq N} |c_{Nj} - \bar{c}_N|, \tag{2.33}$$

where $\bar{c}_N = \sum_{j=1}^N c_{Nj}$.

Remark 3. If (2.17) or (2.18) or (2.19) holds, then we may put

$$\ell_{Ni}(v) = L'_{Ni}(v) = f_{Ni}(H_N^{-1}(v))/h_N(H_N^{-1}(v));$$

(see the part $(C_3) \Rightarrow (C_1)$ of the proof of Theorem 1 in Section 5). Consequently,

$$\tilde{c}_N(v) = \sum_{j=1}^{N} c_{Nj}\, f_{Nj}(H_N^{-1}(v)) \Big/ \sum_{j=1}^{N} f_{Nj}(H_N^{-1}(v)) \,, \quad 0 < v < 1. \tag{2.34}$$

Remark 4. As is shown at the end of Section 6, the conditions of Theorem 2 mean that the regression constants c_{Ni} are bounded in the sense of (2.3).

The following theorem, based on Theorem 1 and on Theorem 2.1 of Hájek (1968), combines unbounded c_{Ni} with a class of bounded score-generating functions.

Theorem 3. *Consider statistics* (2.1) *with scores satisfying* (2.6), *and assume that* $\varphi = \varphi_1 + \varphi_2$, *where* φ_1 *is constant but for a finite number of jumps and* φ_2 *has a bounded second derivative. Assume that* (2.2) *holds. Further assume that* (2.17) *or* (2.18) *or* (2.19) *is true.*

Then S_N *is asymptotically normal with parameters* $(\mathsf{E}\, S_N, \operatorname{var} S_N)$ *and also* $(\mathsf{E}\, S_N, \sigma_N^2)$, *provided* (2.20) *or* (2.22) *holds, with* σ_N^2 *given by* (2.30) *and* $\tilde{c}_N$ *given by* (2.34).

Proof. The proof follows easily from the proofs of our Theorem 1 and Theorem 2.1 of Hájek (1968), since S_N may be represented as a sum of a statistic considered in the latter theorem and a linear combination of statistics considered in the former one. We approximate each component statistic by a sum of independent random variables (denoted by Z_{Ni} in Hájek (1968) and in our Section 5) and then we bound the variance of the difference by a multiple of the sum of the bounds for residual variances of individual terms. It will be clear that (2.30) is the right formula for σ_N^2 if you note that (2.30) reduces to (2.26) when φ is the unit step function. □

In conclusion, we formulate two theorems concerning the important two-sample case, in which

$$\begin{aligned} F_{Ni} &= F(x) \qquad 1 \le i \le m_N, \\ &= G(x) \qquad m_N < i \le N \end{aligned} \tag{2.35}$$

and

$$\begin{aligned} c_{Ni} &= 1 \qquad 1 \le i \le m_N, \\ &= 0 \qquad m_N < i \le N. \end{aligned} \tag{2.36}$$

Then, obviously,

$$S_N = \sum_{i=1}^{m_N} a_N(R_{Ni}). \tag{2.37}$$

□

Theorem 4. *Consider the statistics S_N under the two-sample case (2.35) and (2.36). Assume that φ satisfies (2.4), the scores $a_N(i)$ satisfy (2.7) and that for some fixed λ_0*

$$m_N/N = \lambda_0 + O\left(N^{-\frac{1}{2}}\right) \quad (0 < \lambda_0 < 1) \tag{2.38}$$

Put $H_0 = \lambda_0 F + (1-\lambda_0)\, G$, $L_0 = F\, H_0^{-1}$ and $M = G\, H_0^{-1}$, where $F\, H_0^{-1}$ denotes the composition of functions F and H_0^{-1} and $G\, H_0^{-1}$ has a similar meaning. Denote by B the set of values $t \in (0,1)$ for which the derivatives $L_0'(t)$, $M_0'(t)$ do not exist. Assume that

$$\int_B (\mathrm{d}\varphi_1^s + \mathrm{d}\varphi_2^s) = 0, \tag{2.39}$$

where φ_i^s denotes the singular part of φ_i. Put

$$\begin{aligned}\tau_0^2 = 2\lambda_0(1-\lambda_0) \int_0^1 \int_0^w \{&(1-\lambda_0)\, M_0'(v)\, M_0'(w)\, L_0(v)\, [1-L_0(w)] \\ &+\lambda_0\, L_0'(v)\, L_0'(w)\, M_0(v)\, [1-M_0(w)]\}\ \mathrm{d}\varphi(v)\, \mathrm{d}\varphi(w),\end{aligned} \tag{2.40}$$

and postulate that $\tau_0^2 > 0$.

Then S_N is asymptotically normal with parameters $(\mathsf{E}\, S_N,\ \mathsf{var}\, S_N)$ and also $(\mathsf{E}\, S_N,\ N\, \tau_0^2)$.

Proof. See Section 7. □

Remark 5. The expression τ_0^2 is a λ_0^2-multiple of expression (4.4) in Pyke and Shorack (1968). The statistic T^* considered there is linearly connected with $S_N : S_N = \lambda_0\, N^{\frac{1}{2}}\, T^*$+constant.

Remark 6. Since L_0 and M_0 are both absolutely continuous, their derivative exists almost everywhere with respect to the Lebesgue measure, and, in turn, with respect to the measure generated by $\varphi_1^{ac} + \varphi_2^{ac}$. Consequently, (2.39) entails

$$\int_B (\mathrm{d}\varphi_1 + \mathrm{d}\varphi_2) = 0 \tag{2.41}$$

and (2.40) is well-defined.

Theorem 5. *Consider the two-sample case (2.35) and (2.36). Assume that φ satisfies (2.4) and $a_N(i)$ satisfy (2.7). Furthermore assume that*

$$0 < \liminf_{N\to\infty} m_N/N \le \limsup_{N\to\infty} m_N/N < 1 \tag{2.42}$$

and that the densities $f = F'$ and $g = G'$ exist and $f(x) + g(x) > 0$ for $x \in (\alpha, \beta)$, whereas $f(x) + g(x) = 0$ for $x \notin [\alpha, \beta]$. Let f and g be continuous on (α, β). Put $L_N = F\,H_N^{-1}$ and $M_N = G\,H_N^{-1}$, and

$$\tau_N^2 = 2\lambda_N(1-\lambda_N)\int_0^1\int_0^w \{(1-\lambda_N)\,M_N'(v)\,M_N'(w)\,L_N(v)\,[1-L_N(w)] \tag{2.43}$$
$$+\lambda_N\,L_N'(v)\,L_N'(w)\,M_N(v)\,[1-M_N(w)]\}\,\mathrm{d}\varphi(v)\,\mathrm{d}\varphi(w),$$

where $\lambda_N = m_N/N$. Assume that

$$\liminf \tau_N^2 > 0. \tag{2.44}$$

Then S_N is asymptotically normal with parameters $(\mathsf{E}\,S_N, \mathsf{var}\,S_N)$ and also $(\mathsf{E}\,S_N, N\,\tau_N^2)$.

Proof. See Section 7. □

Remark 7. There are many further possible variants of the above theorems. For example, if we assume that $\varphi(t)$ is continuous at $t = \frac{1}{2}$ and the densities f_{Ni} are *symmetric* with respect to a fixed point x_0, then (2.17) may be relaxed as follows. There exists a $0 < \alpha \le +\infty$ such that

(a) $f_{Ni} = 0$ for $x < x_0 - \alpha$, if α is finite; (2.45)

(b) $f_{Ni}(x)$ are continuous on every compact subinterval of $(x_0 - \alpha,\ x_0)$ uniformly in (x, N, i);

(c) for any compact interval $C \subset (x_0 - \alpha,\ x_0)$ there exists an $\epsilon > 0$ such that for all $N \ge 1$, $N^{-1}\,\mathrm{card}\{i : \inf_{x\in C} f_{Ni}(x) > \epsilon\} > \epsilon$;

(d) $x_0 - \alpha < \liminf_{N\to\infty} H_N^{-1}(t) \le \limsup_{N\to\infty} H_N^{-1}(t) < x_0$ for every $t \in \left(0,\ \frac{1}{2}\right)$.

It may be shown that (2.45) implies the satisfaction of (2.12) and (2.13) with $\ell_{Ni}(v) = L_{Ni}'(v) = f_{Ni}(H_N^{-1}(v))/h_N(H_N^{-1}(v))$ for all $v \ne \frac{1}{2}$, $0 < v < 1$ (see Section 5, the proof of $C_3 \Rightarrow C_1$). If φ is continuous at $v = \frac{1}{2}$, then the singular parts of φ_1 and φ_2 give measure 0 to this single point, and, consequently, Theorems 1, 2, 3 are applicable with (2.12), (2.13) and (2.15) replaced by (2.45). Condition (2.45) is satisfied, for example, if $f_{Ni}(x) = e^{-d_{Ni}}\,f(xe^{-d_{Ni}})$, where $f(x)$ is continuous and positive on $(-\infty, \infty)$, unimodal, symmetric with respect to the origin and the d_{Ni}s satisfy part (c) of (2.18).

Remark 8. The (first) subscript N of the symbols introduced in this Section as well as below will be omitted in the following. Further, K_1, $K_2, \dots$ will denote positive constants, independent of N and numbered in order of appearance.

3 LEMMAS CONCERNING THE POISSON BINOMIAL DISTRIBUTION

Denote by

$$B(k; p_1, \dots, p_N)$$

the probability of k successes in N independent trials with respective probabilities of success $p_1, p_2, \dots, p_N$. Let $B_{i_1,\dots,s_i}^{i'_1,\dots,i'_r}(k; p_1, \dots, p_N)$ denote the above probability, where the parameters $p_{i_1}, \dots, p_{i_s}$ were replaced by zeros and $p_{i'_1}, \dots, p_{i'_r}$ by ones. Let B^* stand for either of the symbols B, B^i, B_i, B^{ij}, B_j^i, B_{ij}, $1 \le i, j \le N$, $i \ne j$. Let us put $q_i = 1 - p_i$. Further, let us abbreviate $B(k; p_1, \dots, p_N)$ as $B(k)$ if $p_1, \dots, p_N$ are fixed.

Lemma 1. *For $A > 4$ we have*

$$\sum B^*(k; p_1, \dots, p_N) < \max \left[\exp(-A^2 / 16 \sum_{i=1}^{N} p_i q_i), \exp(-A/8) \right], \quad (3.1)$$

where the summation on the left extends either over $k > \sum_{i=1}^N p_i + A$ or over $k < \sum_{i=1}^N p_i - A$.

Proof. Let us first prove that

$$\sum_{k > k_0 + 2} B^*(k) \le \sum_{k > k_0} B(k) \le \sum_{k > k_0 - 2} B^*(k). \quad (3.2)$$

The above definitions entail

$$\begin{aligned} B(k) &= p_i p_j B^{ij}(k) + p_i q_j B_j^i(k) + q_i p_j B_i^j(k) + q_i q_j B_{ij}(k) \quad (3.3) \\ &= p_i p_j B^{ij}(k) + p_i q_j B^{ij}(k+1) + q_i p_j B^{ij}(k+1) + q_i q_j B^{ij}(k+2). \end{aligned}$$

Obviously (3.3) yields (3.2) for $B^* = B^{ij}$. The proof of (3.2) for the remaining variants of B^* is similar.

Now, we shall use Theorem 18.1.A, proposition (i) in Loève (1963). A random variable Y with distribution $B(\cdot; p_1, \dots, p_N)$, centered by its mean $\sum_{i=1}^N p_i$, satisfies the conditions of the quoted theorem with $s = \left(\sum_{i=1}^N p_i q_i \right)^{\frac{1}{2}}$

and $c = s^{-1}$, in Loève's notation. If we put, moreover, $\epsilon s = A$, we get

$$\begin{aligned} 0 < A &\leq \sum_{i=1}^{N} p_i\, q_i \Rightarrow \sum_{k > \sum_{i=1}^{N} p_i + A} B(k) < \exp\left(-\frac{1}{4} A^2 \sum_{i=1}^{N} p_i\, q_i\right) \quad (3.4) \\ A &\geq \sum_{i=1}^{N} p_i\, q_i \Rightarrow \sum_{k > \sum_{i=1}^{N} p_i + A} B(k) < \exp\left(-\frac{1}{4} A\right). \end{aligned}$$

As the conditions of the theorem remain satisfied also if the random variable Y is multiplied by -1, the relation (3.4) holds true also when the summation extends over $k < \sum_{i=1}^{N} p_i - A$. Finally, suppose that $A > 4$. Then $A - 2 > A/2$ and (3.2) entails

$$\sum_{k > \sum_{i=1}^{N} p_i + A} B^*(k) \leq \sum_{k > \sum_{i=1}^{N} p_i + \frac{1}{2} A} B(k).$$

Combining this with (3.4), we obtain (3.1) for $k > \sum_{i=1}^{N} p_i + A$. The case $k < \sum_{i=1}^{N} p_i - A$ may be treated similarly. □

In what follows, $\phi(x; \mu, \sigma^2)$ and $\Phi(x; \mu, \sigma^2)$ will denote the normal density and the normal distribution function, respectively, with parameters (μ, σ^2). Further, we shall write $\phi(x)$ and $\Phi(x)$ instead of $\phi(x; 0, 1)$ and $\Phi(x; 0, 1)$.

Lemma 2. *For all x, μ, h_1 and all sufficiently small $|h_2|/\sigma^2$ we have*

$$\begin{aligned} |\phi(x; \mu + h_1,\, \sigma^2 + h_2) - \phi(x; \mu, \sigma^2)| &\leq |h_1|/4\sigma^2 + |h_2|/5\sigma^3, \quad (3.5) \\ |\Phi(x; \mu + h_1,\, \sigma^2 + h_2) - \Phi(x; \mu, \sigma^2)| &\leq 2|h_1|/5\sigma + |h_2|/8\sigma^2. \quad (3.6) \end{aligned}$$

Proof. The proof follows easily from Taylor's formula. □

In the following lemma, k_0, k_1 may depend on N.

Lemma 3. *Suppose that*

$$\lim_{N \to \infty} \sum_{i=1}^{N} p_i\, q_i / \log N = +\infty, \quad (3.7)$$

$$|k_0 - k_1| \leq K_1 \quad \textit{for all } N \geq 1. \quad (3.8)$$

Then we have, for N sufficiently large,

$$\left| B^*(k_0; p_1, \ldots, p_N) - \phi\left(k_1; \sum_{i=1}^N p_i, \sum_{i=1}^N p_i\, q_i\right)\right| \le K_2 \left(\sum_{i=1}^N p_i\, q_i\right)^{-1}, \quad (3.9)$$

$$\left| \sum_{k \le k_0} B^*(k; p_1, \ldots, p_N) - \Phi\left(k_1; \sum_{i=1}^N p_i, \sum_{i=1}^N p_i\, q_i\right)\right| \le K_3 \left(\sum_{i=1}^N p_i q_i\right)^{-\frac{1}{2}}. \quad (3.10)$$

Proof. Specializing Theorem 1 in Petrov (1957), we get, under assumption (3.7),

$$\left| B(k) - \phi\left(k; \sum_{i=1}^N p_i, \sum_{i=1}^N p_i\, q_i\right)\right| \le K_4\, \Lambda \left(\sum_{i=1}^N p_i\, q_i\right)^{-\frac{1}{2}}, \quad (3.11)$$

where Λ is the Lyapunov ratio, i. e.

$$\Lambda = \sum_{i=1}^N (p_i^3 q_i + p_i\, q_i^3) \left(\sum_{i=1}^N p_i\, q_i\right)^{-\frac{3}{2}} < \left(\sum_{i=1}^N p_i\, q_i\right)^{-\frac{1}{2}}. \quad (3.12)$$

Recalling the definition of $B^*(k)$, let $\phi^*(k)$ denote either $\phi\left(k; \sum_{i=1}^N p_i, \sum_{i=1}^N p_i\, q_i\right)$ or the value obtained from ϕ by replacement of one or two numbers p_i (and, correspondingly, q_i) by 0 or 1. Thus we have

$$\phi^*(k) = \phi\left(k; \sum_{i=1}^N p_i + h_1, \sum_{i=1}^N p_i\, q_i + h_2\right), \quad (3.13)$$

where $|h_1| \le 2$, $-\frac{1}{2} \le h_2 \le 0$. From (3.11) we get for N sufficiently large

$$|B^*(k) - \phi^*(k)| < K_5 \left(\sum_{i=1}^N p_i\, q_i - \frac{1}{2}\right)^{-1} \le 2K_5 \left(\sum_{i=1}^N p_i\, q_i\right)^{-1}; \quad (3.14)$$

and (3.13), (3.5) and (3.8) entail

$$\left| \phi^*(k_0) - \phi\left(k_1; \sum_{i=1}^N p_i, \sum_{i=1}^N p_i\, q_i\right)\right| \quad (3.15)$$

$$\le \frac{1}{4}(K_1 + 2) \left(\sum_{i=1}^N p_i\, q_i\right)^{-1} + (1/10) \left(\sum_{i=1}^N p_i\, q_i\right)^{-\frac{3}{2}} < K_6 \left(\sum_{i=1}^N p_i\, q_i\right)^{-1}.$$

Setting $k = k_0$ in (3.14) and combining it with (3.15) we obtain (3.9).

As for the distribution functions, we make use of the Berry–Esseen theorem (Theorem 20.3.B in Loève (1963)) instead of Petrov's theorem. So we get

$$\left|\sum_{k\le k_0} B^*(k) - \Phi\left(k_0; \sum_{i=1}^N p_i + h_1, \sum_{i=1}^N p_i\, q_i + h_2\right)\right| \le K_7 \left(\sum_{i=1}^N p_i\, q_i\right)^{-\frac{1}{2}}; \tag{3.16}$$

and, in view of (3.6),

$$\left|\Phi\left(k_0; \sum_{i=1}^N p_i + h_1, \sum_{i=1}^N p_i\, q_i + h_2\right) - \Phi\left(k_1; \sum_{i=1}^N p_i, \sum_{i=1}^N p_i\, q_i\right)\right| \tag{3.17}$$

$$< K_8 \left(\sum_{i=1}^N p_i\, q_i\right)^{-\frac{1}{2}}.$$

Inequalities (3.16) and (3.17) yield (3.10). □

4 PROOF OF THEOREM 1: UPPER BOUND FOR $E(S - \hat{S})^2$

Lemma 4. *The functions $L_i(t)$, defined by (2.10), satisfy the relation*

$$|L_i(t) - L_i(s)| \le N|t - s|, \quad 0 < s,\ t < 1,\ 1 \le i \le N,\ N \ge 1; \tag{4.1}$$

hence they are absolutely continuous.

Proof. It follows easily from the definitions (2.8) – (2.10) and from the continuity of the F_is. □

Throughout this section, we shall assume that the scores $a(i)$ are defined by (2.5) with the function φ given by (2.24), and that the conditions (2.12) (2.13) and (2.15) are fulfilled.

First we shall prove two auxiliary propositions.

Lemma 5. *Let x, y be real numbers; let us denote as V the integer part of $(N+1)\,v$. Then to each $K_9 > 2$ there exists a $K_{10} > 1$ such that*

$$v - H(x) > K_9\, N^{-\frac{1}{2}} \lg^{\frac{1}{2}} N \Longrightarrow \mathsf{P}(R_i > V \mid X_i = x,\ X_j = y) < N^{-K_{10}} \tag{4.2}$$

and

$$v - H(x) < -K_9\, N^{-\frac{1}{2}} \lg^{\frac{1}{2}} N \Longrightarrow \mathsf{P}(R_i \le V \mid X_i = x,\ X_j = y) < N^{-K_{10}} \tag{4.3}$$

hold for $N > N_0(K_9)$.

The relations (4.2), (4.3) remain true even when the condition $X_j = y$ is omitted.

Proof. If x satisfies the left side of (4.2), then

$$\begin{aligned} V &= \sum_{m=1}^{N} F_m(x) + (V - Nv) + N(v - H(x)) \\ &> \sum_{m=1}^{N} F_m(x) - 1 + K_9 N^{\frac{1}{2}} \lg^{\frac{1}{2}} N > \sum_{m=1}^{N} F_m(x) + K_{11} N^{\frac{1}{2}} \lg^{\frac{1}{2}} N \end{aligned}$$

for each $2 < K_{11} < K_9$ and $N > N_0(K_{11})$.

From Lemma 1 we get (with B^* being equal to B^{ij} or B^i_j for $x \geq y$ and $x < y$, respectively)

$$\mathsf{P}(R_i > V \mid X_i = x,\, X_j = y) = \sum_{k>V} B^*(k; F_1(x), \dots, F_N(x)) \tag{4.4}$$

$$\begin{aligned} &\leq \sum_{k>\sum_{m=1}^{N} F_m(x) + K_{11} N^{\frac{1}{2}} \lg^{\frac{1}{2}} N} B^*(k; F_1(x), \dots, F_N(x)) \\ &< \max\left[\exp\left(-K_{11}^2 N \lg N \Big/ 16 \sum_{m=1}^{N} F_m(x)\,(1 - F_m(x))\right),\right. \\ &\qquad \left.\exp\left(-K_{11} N^{\frac{1}{2}} \lg^{\frac{1}{2}} N/8\right)\right] \\ &< \exp\left(-\frac{1}{4} K_{11}^2 \lg N\right) = N^{-K_{10}}, \quad \text{where } K_{10} = \frac{1}{4} K_{11}^2 \text{ and } N > N_0(K_{11}). \end{aligned}$$

The proof of (4.3) is quite similar. The last assertion of the lemma follows by integration with respect to $\mathrm{d}F_j(y)$. □

In the sequel, we shall denote $D^2 = N^{-1} \sum_{m=1}^{N} L_m(v)\,(1 - L_m(v))$ and, again, $V = [(N+1)\,v]$.

Lemma 6. *Suppose that $|v - H(x)| \leq K_{12} N^{-\frac{1}{2}} \lg^{\frac{1}{2}} N$. Then, for sufficiently large N, we have*

$$\left| \sum_{m=1}^{N} F_m(x)\,(1 - F_m(x)) - ND^2 \right| \leq K_{13} N^{\frac{1}{2}} \lg^{\frac{1}{2}} N, \tag{4.5}$$

$$\left| \phi\left(V; \sum_{m=1}^{N} F_m(x), \sum_{m=1}^{N} F_m(x)\,(1 - F_m(x))\right) \right. \tag{4.6}$$

$$-\phi\left(Nv;\sum_{m=1}^{N}F_m(x),\,ND^2\right)\Bigg| < K_{14}\,N^{-1}\lg^{\frac{1}{2}}N,$$

$$\left|\Phi\left(V;\sum_{m=1}^{N}F_m(x),\sum_{m=1}^{N}F_m(x)\,(1-F_m(x))\right)\right. \tag{4.7}$$

$$-\Phi\left(Nv;\sum_{m=1}^{N}F_m(x),\,ND^2\right)\Bigg| < K_{15}\,N^{-1}\lg^{\frac{1}{2}}N.$$

Proof. Let us put $H(x) = t$. For $K_{16}\,N^{-\frac{1}{2}} \leq |t-v| \leq K_{12}\,N^{-\frac{1}{2}}\lg^{\frac{1}{2}}N$ we have, according to (2.12),

$$\begin{aligned}\sum_{m=1}^{N}F_m(x)\,(1-F_m(x)) &= \sum_{m=1}^{N}L_m(t)\,(1-L_m(t))\\ = \quad \sum_{m=1}^{N}L_m(v)\,(1-L_m(v)) &+ O(N|t-v|) = ND^2 + O\left(N^{\frac{1}{2}}\lg^{\frac{1}{2}}N\right),\end{aligned}$$

and, according to (2.13), we also have the same result for $|t-v| \leq K_{16}\,N^{-\frac{1}{2}}$. Thus we get (4.5). Now, putting

$$\begin{aligned}|h_1| &= |V-Nv| \leq 1,\\ |h_2| &= \left|\sum_{m=1}^{N}F_m(x)\,(1-F_m(x)) - ND^2\right| \leq K_{13}\,N^{\frac{1}{2}}\,\lg^{\frac{1}{2}}N,\\ &\qquad \sigma^2 = ND^2 \geq K_{17}\,N\end{aligned}$$

(according to (2.15)), we obtain (4.6) and (4.7) from Lemma 2. □

The purpose of this section is to derive an upper bound for the residual variance $\mathsf{E}(S-\hat{S})^2$, where

$$\hat{S} = \sum_{i=1}^{N}\mathsf{E}(S\,|\,X_i) - (N-1)\,\mathsf{E}S. \tag{4.8}$$

Combining the formulas 4.8 and 4.18 in Hájek (1968), we get

$$\begin{aligned}\mathsf{E}(S-\hat{S})^2 \;\leq\; &\sum_{i=1}^{N}(c_i-\overline{c})^2\,\mathsf{E}[a(R_i)-\mathsf{E}(a(R_i)\,|\,X_i)]^2 \qquad (4.9)\\ &+\sum\sum_{i\neq j}(c_i-\overline{c})\,(c_j-\overline{c})\left\{\mathsf{E}[\mathrm{cov}(a(R_i),a(R_j)\,|\,X_i,X_j)]\right.\end{aligned}$$

$$+\mathsf{E}[(\mathsf{E}(a(R_i)\,|\,X_i,X_j) - \mathsf{E}(a(R_i)\,|\,X_i))\,(\mathsf{E}(a(R_j)\,|\,X_i,X_j)$$
$$\left. -\mathsf{E}(a(R_j)\,|\,X_j))] - \sum_{m\neq i,j} \mathsf{cov}(\mathsf{E}(a(R_i)\,|\,X_m),\, \mathsf{E}(a(R_j)\,|\,X_m)) \right\}.$$

Let us investigate each term on the right-hand side separately.

Lemma 7. *For $N \to \infty$ we have*

$$\mathsf{E}[\mathsf{cov}(a(R_i),\, a(R_j)\,|\,X_i, X_j)] = N^{-1}\, D^2\, \ell_i(v)\, \ell_j(v) + o(N^{-1})$$

uniformly with respect to $1 \le i,\, j \le N$.

Proof. We have

$$\begin{aligned} & \mathsf{cov}(a(R_i),\, a(R_j)\,|\,X_i = x,\, X_j = y) && (4.10)\\ = \; & \mathsf{E}(a(R_i)\, a(R_j)\,|\,X_i = x,\, X_j = y) - \mathsf{E}(a(R_i)\,|\,X_i = x,\, X_j = y)\\ & \qquad \times \mathsf{E}(a(R_j)\,|\,X_i = x,\, X_j = y)\\ = \; & \mathsf{P}(R_i > V\,|\,X_i = x,\, X_j = y)\, \mathsf{P}(R_j \le V\,|\,X_j = y,\, X_i = x)\\ & \qquad \text{for } x < y,\\ = \; & \mathsf{P}(R_j > V\,|\,X_j = y,\, X_i = x)\, \mathsf{P}(R_i \le V\,|\,X_i = x,\, X_j = y)\\ & \qquad \text{for } x > y. \end{aligned}$$

Let $K_9 > 2$. Denote

$$I = \left\{(x,y) : |H(x) - v| \le K_9\, N^{-\frac{1}{2}} \lg^{\frac{1}{2}} N,\, |H(y) - v| \le K_9\, N^{-\frac{1}{2}} \lg^{\frac{1}{2}} N\right\}.$$

In view of (4.10) and Lemma 5, there exists a $K_{10} > 1$ such that

$$0 \le \mathsf{cov}(a(R_i),\, a(R_j)\,|\,X_i = x,\, X_j = y) < N^{-K_{10}} \tag{4.11}$$

for all $N > N_0(K_{10})$ and all $(x,y) \notin I$. Further, in view of Lemma 3, (2.15) and (4.5), for $(x,y) \in I$ and $x < y$, expression (4.10) may be continued as follows:

$$\begin{aligned} & \mathsf{cov}(a(R_i),\, a(R_j)\,|\,X_i = x,\, X_j = y)\\ = \; & \sum_{k>V} B^i_j(k; F_1(x), \ldots, F_N(x)) \sum_{\ell \le V} B^{ij}(\ell; F_1(y), \ldots, F_N(y))\\ = \; & \left[1 - \Phi\left(V; \sum_{m=1}^N F_m(x),\, \sum_{m=1}^N F_m(x)\,(1 - F_m(x))\right)\right]\\ & \qquad \times \Phi\left(V; \sum_{m=1}^N F_m(y),\, \sum_{m=1}^N F_m(y)\,(1 - F_m(y))\right) + \vartheta_1\, N^{-\frac{1}{2}}, \end{aligned}$$

with $|\vartheta_1| \le K_{18}$. According to (4.7), we may further write

$$\operatorname{cov}[a(R_i), a(R_j) \mid X_i = x,\, X_j = y] \tag{4.12}$$
$$= \Phi\left((H(x)-v)/\left(DN^{-\frac{1}{2}}\right)\right)\left(1-\Phi\left((H(y)-v)/\left(DN^{-\frac{1}{2}}\right)\right)\right)+\vartheta_2\, N^{-\frac{1}{2}} \lg^{\frac{1}{2}} N$$

with $|\vartheta_2| \le K_{19}$. For $(x, y) \in I$, $x > y$, the positions of x and y on the right-hand side of (4.12) should be interchanged.

Result (4.12) for the conditional covariance remains true even when we enlarge the square I to the square

$$I' = \left\{(x,y) : \max(|H(x) - v|,\, |H(y) - v|) \le K_9\, K_{20}^{-1}\, D\, N^{-\frac{1}{2}} \lg^{\frac{1}{2}} N\right\},$$

where K_{20} is such that $D \ge K_{20}$ for all N. Writing now $K_9\, K_{20}^{-1} = K_{21}$, $(H(x) - v)/DN^{-\frac{1}{2}} = p$, $(H(y) - v)/DN^{-\frac{1}{2}} = q$, and

$$I'' = \left\{(p,q) : \max(|p|,\, |q|) \le K_{21} \lg^{\frac{1}{2}} N\right\},$$

we have

$$\mathsf{E}\left[\operatorname{cov}(a(R_i),\, a(R_j) \mid X_i, X_j)\right] \tag{4.13}$$
$$= \int\!\!\int_{I''\cap\{p<q\}} \Phi(p)\,(1-\Phi(q))\, \mathrm{d}_p L_i\left(v + DN^{-\frac{1}{2}}p\right) \mathrm{d}_q L_j\left(v + DN^{-\frac{1}{2}}q\right)$$
$$+ \int\!\!\int_{I''\cap\{p>q\}} \Phi(q)\,(1-\Phi(p))\, \mathrm{d}_p L_i\left(v + DN^{-\frac{1}{2}}p\right) \mathrm{d}_q L_j\left(v + DN^{-\frac{1}{2}}q\right)$$
$$+ N^{-\frac{1}{2}} \lg^{\frac{1}{2}} N \int\!\!\int_{I''} \vartheta_3\, \mathrm{d}_p L_i\left(v+DN^{-\frac{1}{2}}p\right) \mathrm{d}_q L_j\left(v+DN^{-\frac{1}{2}}q\right) + \vartheta_4\, N^{-K_{10}},$$

with $|\vartheta_3| \le K_{22}$, $|\vartheta_4| \le 1$. The last two terms are $o(N^{-1})$ uniformly in i, j as is easily seen from (2.12) and $K_{10} > 1$. The evaluation of the remaining two integrals is essentially the same, so let us calculate only the first of them in detail. We shall divide the domain I'' into two parts, J and $I'' - J$, where $J = \{(p, q) : \max(|p|,\, |q|) \le K_{23}\}$. In K we shall make use of the expansion

$$L_i\left(v + D\, N^{-\frac{1}{2}}p\right) = L_i(v) + \ell_i(v)\, D\, N^{-\frac{1}{2}}p + \Lambda_i(p), \quad 1 \le i \le N, \tag{4.14}$$

where the functions $\Lambda_i(p)$ are absolutely continuous and are of order $o(N^{-\frac{1}{2}})$ uniformly in the interval $[-K_{23},\, K_{23}]$ and with respect to $1 \le i \le N$. This follows from (2.13). Let us first treat the following integral:

$$\int\!\!\int_{J\cap\{p<q\}} \Phi(p)\,(1-\Phi(q))\, \mathrm{d}_p L_i \mathrm{d}_q L_j \tag{4.15}$$
$$- N^{-1} D^2 \ell_i(v)\, \ell_j(v) \int\!\!\int_{J\cap\{p<q\}} \Phi(p)\,(1-\Phi(q))\, \mathrm{d}p\, \mathrm{d}q$$

$$+N^{-\frac{1}{2}} D\ell_i(v) \int\int_{J\cap\{p<q\}} \Phi(p)\,(1-\Phi(q))\,\mathrm{d}p\,\mathrm{d}\Lambda_j(q)$$
$$+N^{-\frac{1}{2}} D\ell_j(v) \int\int_{J\cap\{p<q\}} \Phi(p)\,(1-\Phi(q))\,\mathrm{d}\Lambda_i(p)\,\mathrm{d}q$$
$$+\int\int_{J\cap\{p<q\}} \Phi(p)\,(1-\Phi(q))\,\mathrm{d}\Lambda_j(p)\,\mathrm{d}\Lambda_j(q).$$

If K_{23} is sufficiently large, the integral $\int\int_{J\cap\{p<q\}} \Phi(p)\,(1-\Phi(q))\,\mathrm{d}p\,\mathrm{d}q$ is arbitrarily close to

$$\begin{aligned}
&\int\int_{\{p<q\}} \Phi(p)\,(1-\Phi(q))\,\mathrm{d}p\,\mathrm{d}q \qquad (4.16)\\
&= \int_{-\infty}^{+\infty} (1-\Phi(q)) \left(\int_{-\infty}^{q} \Phi(p)\,\mathrm{d}p\right) \mathrm{d}q\\
&= \int_{-\infty}^{+\infty} (1-\Phi(q))\,(q\Phi(q)+\phi(q))\,\mathrm{d}q = \frac{1}{2}.
\end{aligned}$$

If K_{23} is fixed, then the remaining terms on the right-hand side of (4.15) are $o(N^{-1})$, as may be shown by integration by parts, and making use of (4.14). Moreover, the numbers $\ell_i(v)$ considered as functions of (i, N), $1 \le i \le N < \infty$, are bounded, as appears by setting $t = v + N^{-\frac{1}{2}}$ in (2.12) and (2.13). Consequently,

$$\int\int_{J\cap\{p<q\}} \Phi(p)[1-\Phi(q)]\,\mathrm{d}_pL_i\,\mathrm{d}_qL_j = \frac{1}{2}N^{-1}D^2\ell_i(v)\,\ell_j(v) + o(N^{-1}). \qquad (4.17)$$

The same holds for the integral extending over the region $J \cap \{q < p\}$.

The proof of Lemma 7 will be completed if we show that

$$\begin{aligned}
&\int\int_{(I''-J)\cap\{p<q\}} \Phi(p)\,(1-\Phi(q))\,\mathrm{d}_pL_i\,\mathrm{d}_qL_j \qquad (4.18)\\
&\quad + \int\int_{(I''-J)\cap\{p>q\}} \Phi(q)\,(1-\Phi(p))\,\mathrm{d}_pL_i\,\mathrm{d}_qL_j = o(N^{-1}).
\end{aligned}$$

The left-hand side of (4.18) may be decomposed into four integrals extending over areas similar to the one which is shaded on Figure 1. The integral corresponding to the shaded area equals

$$\int_{-K_{21}\lg^{\frac{1}{2}} N}^{-K_{23}} \Phi(p) \left(\int_{p}^{-p} (1-\Phi(q))\,\mathrm{d}_qL_j\right) \mathrm{d}_pL_i. \qquad (4.19)$$

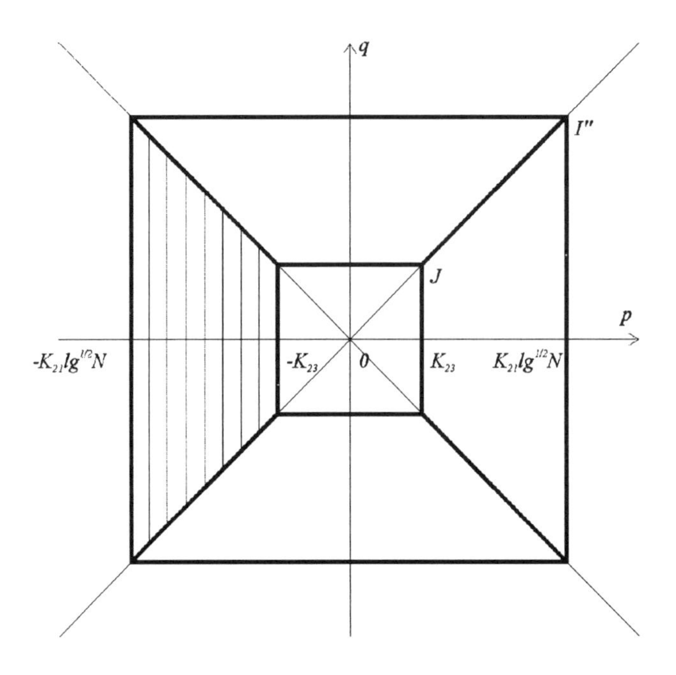

Figure 1

In (4.19), where $p < 0$, we have

$$\int_p^{-p} (1 - \Phi(q))\, \mathrm{d}_q L_j \tag{4.20}$$
$$\leq \; L_j\left(v + DN^{-\frac{1}{2}}|p|\right) - L_j\left(v - DN^{-\frac{1}{2}}|p|\right) \leq K_{24}\, N^{-\frac{1}{2}}|p|,$$

owing to (2.12) and to the inequality $K_{23} \leq |p| \leq K_{21} \lg^{\frac{1}{2}} N$. Thus the expression (4.19) is less than or equal to

$$K_{23}\, N^{-\frac{1}{2}} \int_{-K_{21}\lg^{\frac{1}{2}} N}^{-K_{23}} |p|\, \Phi(p)\, \mathrm{d}_p L_i$$
$$\leq \; K_{23}\, N^{-\frac{1}{2}} \left\{ - \left[(L_i(v + DN^{-\frac{1}{2}}p) - L_i(v))\, p\Phi(p) \right]_{-K_{21}\lg^{\frac{1}{2}} N}^{-K_{23}} \right.$$
$$\left. + \int_{-K_{21}\lg^{\frac{1}{2}} N}^{-K_{23}} \left(L_i(v + DN^{-\frac{1}{2}}p) - L_i(v) \right) (\Phi(p) + p\phi(p))\, \mathrm{d}p \right\}$$
$$< \; K_{25}\, N^{-1} \int_{-\infty}^{-K_{23}} p^2 \phi(p)\, \mathrm{d}p < \epsilon\, N^{-1},$$

with $\epsilon > 0$ arbitrarily small for K_{23} sufficiently large. As the same bound is valid for the three other integrals, too, the proof of Lemma 7 is finished. □

Lemma 8. *For $N \to \infty$ we have*

$$\begin{aligned} &\sum_{1 \le m \le N,\, m \ne i,\, j} \mathrm{cov}(\mathsf{E}(a(R_i) \,|\, X_m),\, \mathsf{E}(a(R_j) \,|\, X_m)) \\ = &\; N^{-1} D^2 \,\ell_i(v)\, \ell_j(v) + o(N^{-1}), \end{aligned}$$

uniformly with respect to $1 \le i,\, j \le N$.

Proof. According to Lemma 3.2 in Hájek (1968) we have

$$\begin{aligned} &\mathsf{E}(a(R_i) \,|\, X_i = x,\, X_m = z) - \mathsf{E}(a(R_i) \,|\, X_i = x) \qquad (4.21) \\ = &\; [u(x - z) - F_m(x)]\, \mathsf{P}(R_i = V + 1 \,|\, X_i = x,\, X_m = x - 1), \quad i \ne m. \end{aligned}$$

(We have used the notation introduced in Lemma 5 and the unit step function $u(x) = 1,\ x \ge 0;\ = 0,\ x < 0$.) Lemma 5 shows that

$$\mathsf{P}(R_i = V + 1 \,|\, X_i = x,\, X_m = x - 1) < N^{-K_{10}}$$

for some $K_{10} > 1$ and for all $|H(x) - v| \ge K_9\, N^{-\frac{1}{2}} \lg^{\frac{1}{2}} N,\ N > N_0(K_{10})$. On the other hand Lemmas 3 and 6 imply

$$\begin{aligned} &\mathsf{P}(R_i = V + 1 \,|\, X_i = x,\, X_m = x - 1) \qquad (4.22) \\ = &\; B^{ij}(V + 1 \,|\, F_i(x), \ldots, F_N(x)) \\ = &\; \phi\left(V; \sum_{i=1}^{N} F_n(x), \sum_{i=1}^{N} F_n(x)\,(1 - F_n(x))\right) \\ &\quad + \vartheta_5 \left(\sum_{i=1}^{N} F_n(x)\,(1 - F_n(x))\right)^{-1} \\ = &\; \phi(Nv; NH(x), ND^2) + \vartheta_5\, N^{-1} \lg^{\frac{1}{2}} N \end{aligned}$$

for some $|\vartheta_5| \le K_{26}$, $|\vartheta_6| \le K_{27}$ and all $|H(x) - v| \le K_9\, N^{-\frac{1}{2}} \lg^{\frac{1}{2}} N$.

Now, we proceed similarly as in the proof of the preceding lemma. The result (4.22) remains true even in the larger interval

$$|H(x) - v| \le K_9\, K_{20}^{-1}\, D\, N^{-\frac{1}{2}} \lg^{\frac{1}{2}} N \quad (\text{where } K_{20} \le D,\ N \ge 1).$$

Writing again $K_{21} = K_9\, K_{20}^{-1}$, $(H(x) - v)/DN^{-\frac{1}{2}} = p$, $(H(z) - v)/DN^{-\frac{1}{2}} = q$, and integrating (4.21) with respect to $\mathrm{d}F_i(x)$, we get after some calculations,

$$\begin{aligned} &\mathsf{E}(a(R_i) \,|\, X_m = z) - \mathsf{E}(a(R_i)) \qquad (4.23) \\ = &\; D^{-1} N^{-\frac{1}{2}} \int_{|p| \le K_{21} \lg^{\frac{1}{2}} N} [u(p - q) - L_m(v)]\, \phi(p)\, \mathrm{d}_p L_i\left(v + DN^{-\frac{1}{2}} p\right) \\ &\quad + o(N^{-1}) \end{aligned}$$

uniformly with respect to $-\infty < z < +\infty$.

Let us again divide the domain of integration into $\{|p| \leq K_{23}\}$ and $\left\{K_{23} \leq |p| \leq K_{21} \lg^{\frac{1}{2}} N\right\}$ and in the first domain let us use the expansion (4.14). We get

$$\begin{aligned} &\int_{-K_{23}}^{K_{23}} [u(p-q) - L_m(v)]\, \phi(p)\, \mathrm{d}_p L_i \\ = \quad & D\, N^{-\frac{1}{2}} \ell_i(v) \int_{-K_{23}}^{K_{23}} [u(p-q) - L_m(v)]\, \phi(p)\, \mathrm{d}p \\ & + \int_{-K_{23}}^{K_{23}} [u(p-q)] - L_m(v)\, \phi(p)\, \mathrm{d}\Lambda_i(p), \end{aligned}$$

where the integral $\int_{-K_{23}}^{K_{23}} [\cdots]\, \phi(p)\, \mathrm{d}p$ is arbitrarily close (uniformly with respect to z) to the expression $1 - \Phi(q) - L_m(v)$ for sufficiently large K_{23}, whereas the integral $\int_{-K_{23}}^{K_{23}} [\cdots]\, \phi(p)\, \mathrm{d}\Lambda_i(p)$ is of the order of magnitude $o\left(N^{-\frac{1}{2}}\right)$, as can be easily shown by integration by parts. Further we have

$$\begin{aligned} &\left| \int_{-K_{23}\sqrt{\log N}}^{-K_{23}} [\cdots]\, \phi(p)\, \mathrm{d}pL_i \right| \\ \leq \quad & \left[\left(L_i\left(v + D\, N^{-\frac{1}{2}} p\right) - L_i(v)\right) \phi(p)\right]_{-K_{21}\sqrt{\log N}}^{-K_{23}} \\ & - \int_{-K_{21}\sqrt{\log N}}^{-K_{23}} \left(L_i\left(v + D\, N^{-\frac{1}{2}} p\right) - L_i(v)\right) \phi'(p)\, \mathrm{d}p \\ \leq \quad & K_{28}\, N^{-\frac{1}{2}} \int_{-\infty}^{-K_{23}} p^2\, \phi(p)\, \mathrm{d}p \ < \ \epsilon\, N^{-\frac{1}{2}} \quad \text{for } K_{23} \text{ large}, \end{aligned}$$

and the same for $\int_{K_{23}}^{K_{21}\sqrt{\log N}} [\cdots]\, \phi(p)\, \mathrm{d}_p L_i$. Altogether, we have

$$\begin{aligned} & \mathsf{E}(a(R_i) \mid X_m = z) - \mathsf{E}(a(R_i)) \\ = \quad & N^{-1} \ell_i(v) \{1 - \Phi(q) - L_m(v)\} + o(N^{-1}), \end{aligned} \tag{4.24}$$

uniformly in $-\infty < z < +\infty$. Hence,

$$\begin{aligned} & \mathsf{cov}(\mathsf{E}(a(R_i) \mid X_m),\, \mathsf{E}(a(R_j) \mid X_m)) \\ = \quad & \int_{-\infty}^{+\infty} (\mathsf{E}(a(R_i) \mid X_m = z) - \mathsf{E}(a(R_i))) \\ & \times (\mathsf{E}(a(R_j) \mid X_m = z) - \mathsf{E}(a(R_j)))\, \mathrm{d}F_m(z) \\ = \quad & N^{-2} \ell_i(v)\, \ell_j(v) \int_{-\infty}^{+\infty} \{1 - \Phi(q) - L_m(v)\}^2\, \mathrm{d}_q L_m\left(v + D\, N^{-\frac{1}{2}} q\right) \\ & + o(N^{-2}). \end{aligned} \tag{4.25}$$

As each of the integrals $\int_{-\infty}^{+\infty}(1-\Phi(q))\,\mathrm{d}_q L_m$ and $\int_{-\infty}^{+\infty}(1-\Phi(q))^2\,\mathrm{d}_q L_m$ evidently equals $L_m(v)+o(1)$, we have

$$\int_{-\infty}^{+\infty}(1-\Phi(q)-L_m(v))^2\,\mathrm{d}_q L_m = L_m(v)\,(1-L_m(v))+o(1).$$

Inserting this into (4.25) and summing over $1\le m\le N,\ m\ne i,\,j$, we finally get the assertion of the lemma. (Observe that deleting the ith and jth terms can diminish the sum $N\,D^2=\sum_{m=1}^{N}L_m(v)\,(1-L_m(v))$ at most by $\frac{1}{2}$.) □

Lemma 9. *For $N\to\infty$ we have*

$$\begin{aligned}&\mathsf{E}\,[(\mathsf{E}(a(R_i)\,|\,X_i,\,X_j)-\mathsf{E}(a(R_i)\,|\,X_i))\,(\mathsf{E}(a(R_j)\,|\,X_i,\,X_j)-\mathsf{E}(a(R_j)\,|\,X_j))]\\&=o(N^{-1}),\end{aligned}$$

uniformly in $1\le i,\,j\le N$.

Lemma 10. *For $N\to\infty$ we have*

$$\mathsf{E}(a(R_i)-\mathsf{E}(a(R_i)\,|\,X_i))^2=o(1),$$

uniformly in $1\le i\le N$.

Proofs of Lemmas 9 and 10 will be omitted; they are similar to the proofs of the two preceding lemmas, but substantially simpler. □

Lemma 11. *For $N\to\infty$ we have*

$$\mathsf{E}(S-\hat{S})^2=o\left(\sum_{i=1}^{N}(c_i-\bar{c})^2\right).$$

Proof. It follows by inserting the results of Lemmas 7 – 10 into the inequality (4.9) and making use of the inequality

$$\sum_{i=1}^{N}\sum_{j=1}^{N}|c_i-\bar{c}|\;|c_j-\bar{c}|\le N\sum_{i=1}^{N}(c_i-\bar{c})^2.$$

□

5 PROOF OF THEOREM 1: COMPLETION

In this section, we shall at first assume that the scores $a(i)$ are defined by (2.5) with the function φ given by (2.24), and that the assumptions $(2.2), (2.12), (2.13), (2.15)$ and (2.20) or (2.22) (with σ^2 given by (2.26)) are satisfied.

From the definitions of S and $\hat{S}$ and from the obvious relation

$$\sum_{j=1}^{N} [\mathsf{E}(a(R_j) \,|\, X_i) - \mathsf{E}(a(R_j))] = 0$$

it follows $\hat{S} - \mathsf{E}\hat{S} = \sum_{i=1}^{N} Y_i$, where

$$Y_i = \sum_{j=1}^{N} (c_j - c_i) \, [\mathsf{E}(a(R_j) \,|\, X_i) - \mathsf{E}(a(R_j))], \tag{5.1}$$

Y_i are independent, $\mathsf{E}Y_i = 0,\ 1 \le i \le N,\ \mathsf{var}\,\hat{S} = \sum_{i=1}^{N} \mathsf{var}\,Y_i$.

Further denote

$$Z_i = (\tilde{c}(v) - c_i)\, [u(v - H(X_i)) - L_i(v)], \quad 1 \le i \le N, \tag{5.2}$$

where $\tilde{c}(v)$ is given by (2.25). Random variables Z_i are independent, with zero expectations, and such that $\sum_{i=1}^{N} \mathsf{var}\,Z_i = \sigma^2$ (defined by (2.26)).

According to (4.24) we can write, using $(2.12), (2.13)$ and (2.15),

$$Y_i = (\tilde{c}(v) - c_i) \left[\Phi\left((v - H(X_i))/DN^{-\frac{1}{2}}\right) - L_i(v)\right] + N^{-1} \sum_{j=1}^{N} (c_j - c_i)\, \eta_j \tag{5.3}$$

where η_j are random variables such that $|\eta_j| \le \epsilon_N,\ 1 \le j \le N$, for some sequence of constants $\epsilon_N \to 0$. Consequently, as in the close of the proof of Lemma 8, we get

$$\mathsf{E}(Y_i - Z_i)^2 = o\left(N^{-1} \sum_{j=1}^{N} (c_j - \overline{c})^2\right), \tag{5.4}$$

uniformly in $1 \le i \le N$.

Now, the inequality $\mathsf{var}\left(S - \sum_{i=1}^{N} Z_i\right) \le 2\sum_{i=1}^{N} \mathsf{E}(Y_i - Z_i)^2 + 2\mathsf{E}(S - \hat{S})^2$ together with (5.4) and with Lemma 11 yield the following.

Lemma 12.

$$\mathsf{var}\left(S - \sum_{i=1}^{N} Z_i\right) = o\left(\sum_{i=1}^{N} (c_i - \overline{c})^2\right).$$

Lemma 13. (2.20) *holds if and only if* (2.22) *holds with* σ^2 *given by* (2.26). *In this case* $\lim_{n\to\infty} \mathsf{var}\, S/\sigma^2 = 1$.

Proof. From Minkowski's inequality we obtain

$$\left((\mathsf{var}\, U_1/\mathsf{var}\, U_2)^{\frac{1}{2}} - 1\right)^2 \leq \mathsf{var}(U_1 - U_2)/\mathsf{var}\, U_2. \tag{5.5}$$

If (2.20) is satisfied, we put $U_1 = \sum_{i=1}^N Z_i$, $U_2 = S$ in (5.5); if (2.22) is satisfied, we put $U_1 = S$, $U_2 = \sum_{i=1}^N Z_i$. In both cases, Lemma 12 entails $\mathsf{var}\, S/\sigma^2 \to 1$ and, consequently, the validity of the entire Lemma 13. □

Lemma 14. *The random variables* $\sum_{i=1}^N Z_i$ *are asymptotically normal with parameters* $(0, \sigma^2)$.

Proof. From (5.2) it follows that $|Z_i| \leq 2 \max_{1\leq j\leq N} |c_j - \bar{c}|$, $1 \leq i \leq N$; (2.2) and (2.22) then imply that $\max_{1\leq i\leq N} |Z_i|/\sigma = o(1)$, but this means that the Lindeberg condition for asymptotic normality is trivially satisfied. □

Now, the assertion of Theorem 1, under conditions listed at the beginning of this Section, is a consequence of the following (already proved) propositions:

$$\mathcal{L}\left(\sum_{i=1}^N Z_i/\sigma\right) \to \mathcal{N}(0,1), \quad \left[\mathsf{var}\left(S - \sum_{i=1}^N Z_i\right)\right]/\sigma^2 \to 0, \quad (\mathsf{var}\, S)^{\frac{1}{2}}/\sigma \to 1.$$

(Namely, we can write

$$(S - \mathsf{E}S)/(\mathsf{var}\, S)^{\frac{1}{2}} = \left(\sum_{i=1}^N Z_i/\sigma + \left(S - \mathsf{E}S - \sum_{i=1}^N Z_i\right)/\sigma\right)\sigma/(\mathsf{var}\, S)^{\frac{1}{2}} \tag{5.6}$$

and apply the Theorem 20.6 in Cramér (1946).)

Let us relax the condition (2.5) to (2.6) and denote the corresponding statistics as S and S^*. Then we have from (2.6) and (2.2)

$$\begin{aligned}\mathsf{var}(S - S^*) &\leq \max_{1\leq i\leq N} (c_i - \bar{c})^2 \cdot \left(\sum_{j=1}^N |a(i) - \varphi(i/(N+1))|\right)^2 \\ &= o\left(\sum_{i=1}^N (c_i - \bar{c})^2\right).\end{aligned}$$

Hence, with the help of (2.22), (5.5), (5.6), the asymptotic normality of S^* easily follows.

Thus we have proved Theorem 1 under conditions C_1. It remains to show that this set of conditions is entailed by the other sets.

$C_2 \Rightarrow C_1$ will follow from (2.16) $\Rightarrow$ [(2.15), (2.22)]. However, (2.16) $\Rightarrow$ (2.15) is obvious, and (2.16) $\Rightarrow$ (2.22) follows from (2.26) and from

$$\sum_{i=1}^{N}(c_i - \tilde{c}(v))^2 \geq \sum_{i=1}^{N}(c_i - \overline{c})^2.$$

$C_3 \Rightarrow C_1$ will follow from (2.17) $\Rightarrow$ [(2.12), (2.13), (2.15)]. If we put $h(x) = N^{-1}\sum_{i=1}^{N} f_i(x)$, we obtain from (c) of (2.17) that on every compact $C \subset (\alpha, \beta)$

$$\inf_{x \in C} h(x) > \epsilon^2. \tag{5.7}$$

Further, from (d) of (2.17) it follows that for every $0 < v_1 < v < v_2 < 1$ there exists a compact $C \subset (\alpha, \beta)$ such that

$$H^{-1}(t) \in C \quad \text{for all } v_1 < t < v_2. \tag{5.8}$$

Putting (5.7) and (5.8) together, and noting that $[h(H^{-1}(t))]^{-1}$ represents the derivative of $H^{-1}(t)$, we obtain for all $N \geq 1$

$$\sup_{v_1 < t < v_2} |\mathrm{d}H^{-1}(t)/\mathrm{d}t| < \epsilon^{-2}.$$

Consequently, $H^{-1}(t)$ is uniformly continuous in a neighborhood of v. The same is true about

$$\mathrm{d}L_i(t)/\mathrm{d}t = f_i(H^{-1}(t))/h(H^{-1}(t))$$

since all three involved functions, f_i, H^{-1} and the reciprocal of h, are uniformly continuous in a neighborhood of v; for f_i, consult (2.17b) and for the reciprocal of h, see (2.17b) and (5.7). From this fact (2.12) and (2.13), with $\ell_i = L_i'$, easily follow.

Now it remains to prove (2.17) $\Rightarrow$ (2.15). Part (d) of (2.17) entails the existence of numbers x_1 and x_2 such that $\alpha < x_1 < x_2 < \beta$ and such that for all sufficiently large N, $x_1 \leq H^{-1}(v) \leq x_2$. Consequently

$$L_i(v)\,[1 - L_i(v)] \geq F_i(x_1)\,[1 - F_i(x_2)]. \tag{5.9}$$

Further, (c) of (2.17) entails that for fixed y_1, y_2 such that $\alpha < y_1 < x_1 < x_2 < y_2 < \beta$ there exists an $\epsilon > 0$ such that

$$N^{-1}\text{card}\left\{i : \inf_{y_1 \leq x \leq y_2} f_i(x) > \epsilon\right\} > \epsilon,$$

which implies

$$N^{-1}\,\text{card}\,\{i : F_i(x_1) > \epsilon(x_1 - y_1),\ 1 - F_i(x_2) > \epsilon(y_2 - x_1)\} > \epsilon.$$

This combined with (5.9) gives

$$N^{-1}\sum_{i=1}^{N} L_i(v)\,[1-L_i(v)] > \epsilon^3 (x_1-y_1)\,(y_2-x_2).$$

Note that C_1 concerns one point v, whereas C_3 ensures the satisfaction of C_1 for any $0<v<1$.

$C_4 \Rightarrow C_2$ follows from (2.18) $\Rightarrow$ (2.17), which is easily shown.

$C_5 \Rightarrow C_4$ follows from (2.19) $\Rightarrow$ (2.18) and (2.19) $\Rightarrow$ (2.16) $\Rightarrow$ (2.22). The first implication is obvious, and (2.16) $\Rightarrow$ (2.22) has been proved by proving $C_2 \Rightarrow C_1$; (2.19) $\Rightarrow$ (2.16) follows from

$$L_i(v)\,[1-L_i(v)] \geq F(F^{-1}(v)-2K)\,[1-F(F^{-1}(v)+2K)] > 0,$$

where F corresponds to f and $K \geq \sup_{i,N} |\mathrm{d}_i|$.

$C_6 \Rightarrow C_1$ is entailed by (2.3) $\Rightarrow$ (2.2) and $[(2.3),(2.22)] \Rightarrow (2.15)$. The former implication is obvious; in the other we utilize the fact that

$$\sigma^2 \leq 2 \max_{1\leq i\leq N} (c_i-\overline{c})^2 \sum_{i=1}^{N} L_i(v)\,[1-L_i(v)].$$

This conclude the proof of Theorem 1. □

Let us show that, under conditions of Theorem 1, σ^2 and $\operatorname{var} S$ are of order $\sum_{i=1}^{N}(c_i-\overline{c})^2$. From (2.26) it follows that

$$\sigma^2 \leq \sum_{i=1}^{N}(c_i-\tilde{c}(v))^2 = \sum_{i=1}^{N}(c_i-\overline{c})^2 + N(\overline{c}-\tilde{c}(v))^2.$$

Now, since the numbers $\ell_i(v)$ are for fixed v uniformly bounded in (N,i) (see below (4.16)), we have

$$N(\overline{c}-\tilde{c}(v))^2 = N^{-1}\left(\sum_{i=1}^{N}(\overline{c}-c_i)\,\ell_i(v)\right)^2 \leq K_v^2 \sum_{i=1}^{N}(\overline{c}-c_i)^2$$

for some K_v. Thus $\sigma^2 = O\left(\sum_{i=1}^{N}(c_i-\overline{c})^2\right)$ and hence, in view of Theorem 1 (see also Lemma 12),

$$\operatorname{var} S = O\left(\sum_{i=1}^{N}(c_i-\overline{c}^2)\right). \tag{5.10}$$

6 PROOF OF THEOREM 2

Assuming first that (2.5) is satisfied, put

$$S_\varphi = \sum_{i=1}^{N} c_i\, \varphi(R_i/(N+1)).$$

Consulting Hájek (1968), namely (3.7) and (5.34), we can see that, for φ *nondecreasing*,

$$\operatorname{var} S_\varphi \leq 42\, N \max_{1\leq i\leq N} (c_i - \overline{c})^2 \int_0^1 (\varphi(t) - \overline{\varphi})^2 \,\mathrm{d}t. \tag{6.1}$$

Further put

$$T_v = \sum_{i=1}^{N} [u(v - H(X_i)) - L_i(v)]\,(\tilde{c}(v) - c_i) \tag{6.2}$$

where $\tilde{c}(v)$ is defined by (2.31) if $v \notin B$ and by (2.32) if $v \in B$. We have, obviously, in view of (2.11)

$$\begin{aligned} \operatorname{var} T_v &= \sum_{i=1}^{N} (\tilde{c}(v) - c_i)^2\, L_i(v)\,[1 - L_i(v)] \\ &\leq 4\,N \max_{1\leq i\leq N} (c_i - \overline{c})^2\, v(1-v). \end{aligned} \tag{6.3}$$

Note that T_v equals $\sum_{i=1}^N Z_i$ from Section 5. Consequently, by Lemma 12,

$$\mathsf{E}(S_\nu - \mathsf{E}S_\nu - T_v)^2 = o\left(\sum_{i=1}^{N} (c_i - \overline{c})^2\right). \tag{6.4}$$

If φ is nondecreasing and square integrable, then $\int_0^1 v(1-v)\,\mathrm{d}\varphi < +\infty$ and

$$T_\varphi = \int_0^T T_v \,\mathrm{d}\varphi(v) \tag{6.5}$$

is well-defined. Similarly as in proving (5.37) in Hájek (1968), we can show that

$$\operatorname{var} T_\varphi \leq 4\,N \max_{1\leq i\leq N} (c_i - \overline{c})^2 \int_0^1 \varphi^2(t)\,\mathrm{d}t. \tag{6.6}$$

If φ is absolutely continuous and the φ-measure vanishes on the complement of B, then T_φ of (6.5) may be equivalently expressed by

$$T_\varphi = \sum_{i=1}^{N} N^{-1} \sum_{j=1}^{N} (c_j - c_i) \int [u(x - X_i) - F_i(x)]\,\varphi'(H(x))\,\mathrm{d}F_j(x). \tag{6.7}$$

This follows from the obvious formula

$$\begin{aligned}&\int [u(x - X_i) - F_i(x)]\, \varphi'(H(x))\, \mathrm{d}F_j(x)\\ = &\int_0^1 [u(v - H(x_i)) - L_i(v)]\, L_j'(v)\, \mathrm{d}\varphi(v).\end{aligned}$$

Now for every $\alpha > 0$, φ may be decomposed as follows:

$$\varphi(t) = \psi(t) - \lambda(t) + \gamma(t) - \eta(t) + g(t) - h(t), \tag{6.8}$$

where all functions on the right side are nondecreasing and square integrable, and

(a) ψ and λ are absolutely continuous such that

$$\int_B (\mathrm{d}\psi + \mathrm{d}\lambda) = \int_0^1 (\mathrm{d}\psi + \mathrm{d}\lambda);$$

(b) $\gamma(t)$ and $\eta(t)$ are bounded such that

$$\int_B (\mathrm{d}\gamma + \mathrm{d}\eta) = 0;$$

(c) $\int_0^1 (g^2(t) + h^2(t))\, \mathrm{d}t < \alpha$, and $\int_B (\mathrm{d}g + \mathrm{d}h) = 0$.

Now, if ψ has a bounded second derivative, Theorem 4.2 of Hájek (1968) entails

$$\lim_{N\to\infty} \left[\mathsf{E}(S_\psi - \mathsf{E}S_\psi - T_\psi)^2\right] / \left[N \max_{1\le i\le N} (c_i - \overline{c})^2\right] = 0. \tag{6.9}$$

However, the bounds (5.33) and (5.37) of Hájek (1968) show that (6.9) holds for an arbitrary nondecreasing, absolutely continuous and square integrable ψ.

Next, we can easily see that

$$S_\gamma - \mathsf{E}S_\gamma - T_\gamma = \int_0^1 (S_v - \mathsf{E}S_v - T_v)\, \mathrm{d}\gamma(v). \tag{6.10}$$

Consequently,

$$\mathsf{E}(S_\gamma - \mathsf{E}S_\gamma - T_\gamma)^2 \le [\gamma(1-) - \gamma(0+)] \int_0^1 \mathsf{E}(S_v - \mathsf{E}S_v - T_v)^2\, \mathrm{d}\gamma. \tag{6.11}$$

Taking into account (6.4), (6.3) and (6.1), which entails $\mathsf{var}\, S \le 42\, N \times \max_{1\le i\le N} (c_i - \overline{c})^2\, v(1 - v)$, we easily see that (6.9) holds also for ψ replaced by γ or η. Finally (6.1) and (6.6) entail

$$\begin{aligned}\mathsf{E}(S_g - \mathsf{E}S_g - T_g)^2 &\le 2\,\mathsf{var}\, S_g + 2\,\mathsf{var}\, T_g \\ &\le 92\, N \max_{1\le i\le N} (c_i - \overline{c})^2 \int_0^1 g^2(t)\, \mathrm{d}t.\end{aligned} \tag{6.12}$$

Thus

$$\mathsf{E}(S_g - S_h - \mathsf{E}S_g + \mathsf{E}S_h - T_g + T_h)^2 \leq \alpha\, 184\, N \max_{1\leq i\leq N}(c_i - \bar{c})^2. \tag{6.13}$$

Now knowing that (6.9) holds for ψ replaced by λ, γ, η and that α in (6.13) may be arbitrarily small, we conclude that

$$[\mathsf{E}(S_\varphi - \mathsf{E}S_\varphi - T_\varphi)^2] \Big/ \left[N \max_{1\leq i\leq N}(c_i - \bar{c})^2\right] \to 0. \tag{6.14}$$

By (6.14) the problem of asymptotic normality of S_φ with natural parameters is reduced to the same problem for T_φ (see (5.6) and the text that follows.) Assumption (2.23) entails

$$\liminf_{N\to\infty}(\mathsf{var}\, T_\varphi) \Big/ \left[N \max_{1\leq i\leq N}(c_i - \bar{c})^2\right] > 0 \tag{6.15}$$

since $\sigma_N^2 = \mathsf{var}\, T_\varphi$. Also (2.21) in connection with (6.14) entails (6.15). Now $T_\varphi = \sum_{i=1}^N Y_i$, where

$$Y_i = \int_0^1 [u(t - H(X_i)) - L_i(t)]\,(\tilde{c}(t) - c_i)\,\mathrm{d}\varphi(t). \tag{6.16}$$

If φ were bounded, we would have

$$|Y_i| \leq 2 \max_{1\leq j\leq N}|c_j - \bar{c}| \quad \text{(a variation of } \varphi\text{)}. \tag{6.17}$$

Therefore, in view of (6.15), $\max|Z_i|/(\mathsf{var}\, T_\varphi) \to 0$, and the central limit theorem trivially applies.

If φ is not bounded, we write $\varphi(t) = b(t) + c(t)$, where $b(t)$ is bounded, and $\int c^2(t)\,\mathrm{d}t < \epsilon$. Then, by (6.6)

$$\begin{aligned} &[\mathsf{E}(T_\varphi - T_b)^2] \Big/ \left[N \max_{1\leq i\leq N}(c_i - \bar{c})^2\right] \\ = \;& [\mathsf{var}\, T_c] \Big/ \left[N \max_{1\leq i\leq N}(c_i - \bar{c})^2\right] \leq 4\epsilon \end{aligned}$$

and

$$\begin{aligned} \left((\mathsf{var}\, T_b/\mathsf{var}\, T_\varphi)^{\frac{1}{2}} - 1\right)^2 &\leq \mathsf{var}\, T_c/\mathsf{var}\, T_\varphi \\ &\leq \left[4\epsilon\, N \max_{1\leq i\leq N}(c_i - \bar{c})^2\right] \Big/ \mathsf{var}\, T_\varphi. \end{aligned}$$

Now, for large N, if ϵ is sufficiently small, $\mathsf{var}\, T_b/\mathsf{var}\, T_\varphi$ will be as close to 1 as we want (see (6.15)) and the difference between $\mathcal{L}\left(T_b(\mathsf{var}\, T_b)^{-\frac{1}{2}}\right)$ and

$\mathcal{L}\left(T_\varphi(\mathsf{var}\,T_b)^{-\frac{1}{2}}\right)$ will also be small. Consequently, since $\mathcal{L}\left(T_b(\mathsf{var}\,T_b)^{-\frac{1}{2}}\right) \to \mathcal{N}(0,1)$, we must also have $\mathcal{L}(T_\varphi)(\mathsf{var}T_\varphi)^{-\frac{1}{2}}) \to \mathcal{N}(0,1)$. From (6.14) and (6.15) we also have $\mathcal{L}\left(S_\varphi - \mathsf{E}S_\varphi)\,(\mathsf{var}\,S_\varphi)^{-\frac{1}{2}}\right) \to \mathcal{N}(0,1)$.

The proof will be complete if we show that (2.5) may be relaxed to (2.7). Putting $S_\varphi = \sum_{i=1}^N c_i\,\varphi(R_i/(N+1))$ and $S_\varphi^* = \sum_{i=1}^N c_i\,a_N(R_i)$, we have from (2.7)

$$\begin{aligned} & \mathsf{E}(S_\varphi - S_\varphi^* - \mathsf{E}S_\varphi + \mathsf{E}S\varphi^*)^2 \\ \leq \; & \max_{1\leq i\leq N}(c_i-\overline{c})^2\left[\sum_{i=1}^N |a_N(i) - \varphi(i/(N+1))|\right]^2 \\ = \; & o\left(N\max_{1\leq i\leq N}(c_i-\overline{c})^2\right). \end{aligned}$$

This concludes the proof. □

In addition we shall show that under conditions of Theorem 2 the regression constants have to be bounded in the sense of (2.3). We shall show that

$$\lim_{N\to\infty}\left[\sum_{i=1}^N (c_i-\overline{c})^2\right]\Big/\left[N\max_{1\leq i\leq N}(c_i-\overline{c})^2\right] = 0 \tag{6.18}$$

entails

$$\lim_{N\to\infty}[\mathsf{var}\,S_\varphi]\Big/\left[N\max_{1\leq i\leq N}(c_i-\overline{c})^2\right] = 0 \tag{6.19}$$

contradicting thus our assumption (2.21). Since $\mathsf{var}\,S_\varphi/\sigma_N^2 \to 1$, (6.19) also contradicts (2.23).

If γ is nondecreasing and bounded, then

$$\mathsf{var}\,S_\gamma \leq [\gamma(1-) - \gamma(0+)]\int_0^1 \mathsf{var}\,S_v\,\mathrm{d}\gamma(v),$$

where

$$\mathsf{var}\,S_v \leq 42\,N\max_{1\leq i\leq N}(c_i-\overline{c})^2\,v(1-v).$$

If $\int_B \mathrm{d}\gamma = 0$, then we can see from (5.10) that (6.19) is satisfied for φ replaced by γ.

If ψ has a bounded second derivative then we may compile from results of Hájek (1968), namely from Theorem 4.2 and (5.17) and (2.14), that $\mathsf{var}\,S_\psi = O\left(\sum_{i=1}^N (c_i-\overline{c})^2\right)$, so that (6.18) entails (6.19), if we replace φ by ψ. Finally for any nondecreasing φ, we have (6.1). Now under assumptions of Theorem 2 we may decompose φ as $\varphi = \psi + \gamma - \eta + \varphi_1 - \varphi_2$, where all components except ψ are nondecreasing, ψ has a bounded second derivative, γ and η are bounded and such that $\int_B(\mathrm{d}\gamma + \mathrm{d}\eta) = 0$, and φ_1 and φ_2 are arbitrarily small in mean square. However, that completes the proof of our assertion that the c_{Ni}s have to be bounded in the sense of (2.3), if the conditions of Theorem 2 are satisfied. □

7 PROOF OF THEOREMS 4 AND 5

Let us first suppose that the assumptions of Theorem 4 are satisfied, but let us use in the proof also the symbols L, M, λ, introduced in Theorem 5. Thus

$$\begin{aligned} L(t) &= F\,H^{-1}(t) = F\,H_0^{-1}(H_0\,H^{-1}(t)), \\ H_0 &= H + (\lambda_0 - \lambda)\,(F - G), \\ H_0\,H^{-1}(t) &= t + (\lambda_0 - \lambda)\,(L(t) - M(t)) = t' \text{ (say)}; \end{aligned}$$

i. e., for each $0 < t < 1$ there exists a $0 < t' < 1$ such that

$$L(t) = L_0(t') \quad \text{and} \quad |t - t'| \le K_{29}\,N^{-\frac{1}{2}}. \tag{7.1}$$

Further, from the identity $\lambda L(t) + (1 - \lambda)\,M(t) = t,\ 0 < t < 1$, and from the fact that the λs are bounded away from zero as well as from one, it follows that the functions L, M (and also L_0, M_0) satisfy the Lipschitz condition for some constant K_{30}, uniformly in $N \ge 1$. Hence also the validity of (2.12) follows for every $0 < v < 1$.

For each $v \notin B$ let us set

$$\begin{aligned} \ell_i(v) &= \lambda_0\,\lambda^{-1}\,L_0'(v), & 1 \le i \le m, \\ &= (1 - \lambda_0)\,(1 - \lambda)^{-1}\,M_0'(v), & M < i \le N. \end{aligned} \tag{7.2}$$

We shall show that (2.13) holds true, i. e. that for arbitrary $\epsilon > 0$ and $K > 0$

$$\left|(L(t) - L(v))/(t - v) - \lambda_0\,\lambda^{-1}\,L_0'(v)\right| < \epsilon\,N^{-\frac{1}{2}}|t - v|^{-1} \tag{7.3}$$

for all $N > N_0(\epsilon, K)$ and all $|t - v| \le K\,N^{-\frac{1}{2}}$ (and similarly for $M(t)$). For $|t - v| < \epsilon/2K_{30}\,N^{\frac{1}{2}}$, the inequality (7.3) is evident, because its right-hand side is larger than $2K_{30}$, whereas its left-hand side is less than $2K_{30}$ (provided that N is sufficiently large). So it suffices to prove (7.3) in the domain

$$\epsilon/2K_{30}\,N^{\frac{1}{2}} \le |t - v| \le K\,N^{-\frac{1}{2}}; \tag{7.4}$$

as the right-hand side of (7.3) is $\ge \epsilon/K$ in this domain, it actually suffices to prove

$$\left|(L(t) - L(v))/(t - v) - \lambda_0\,\lambda^{-1}\,L_0'(v)\right| < \epsilon/K \tag{7.5}$$

in the domain (7.4). We observe that

$$\begin{aligned}
& (L(t) - L(v))/(t - v) \qquad (7.6)\\
= \; & [(L_0(t') - L_0(v))/(t' - v)]\,(t' - v)/(t - v)\\
& \quad - [(L_0(v') - L_0(v))/(v' - v)]\,(v' - v)/(t - v),
\end{aligned}$$

where t' and v' are defined by (7.1); and note that

$$|(t' - v)/(t - v)| \leq 1 + 2K_{29}\,K_{30}/\epsilon$$

and

$$|(v' - v)/(t - v)| \leq 2K_{29}\,K_{30}/\epsilon$$

for all t from the domain (7.4). Furthermore,

$$|(t' - v')/(t - v) - 1| \leq K_{29}\,K_{30}\,N^{-\frac{1}{2}}$$

for all t. Finally, from (2.38) and from the existence of $L_0'(v)$ it follows that both the differences

$$\left|(L_0(t') - L_0(v))/(t' - v) - \lambda_0\,\lambda^{-1}\,L_0'(v)\right|$$

and

$$\left|(L_0(t') - L_0(v))/(v' - v) - \lambda_0\,\lambda^{-1}\,L_0'(v)\right|$$

are arbitrarily small (say $< \epsilon/3K(1 + 2K_{29}K_{30}\epsilon^{-1})$) for t from (7.4) and N large. Utilizing all these inequalities in (7.6), we obtain (7.5). Thus we have proved (2.13).

From (7.2) it immediately follows that (2.14) is fulfilled. (2.15) can also be easily proved. First, obviously, it cannot be that $L_0'(v) = 0$ and $M_0'(v) = 0$ simultaneously. So suppose $L_0'(v) > 0$ without loss of generality; hence $L_0(v)\,(1 - L_0(v))$ is positive, say equal to $\Delta > 0$. But $L(v) \to L_0(v)$ for $N \to \infty$ hence $L(v)\,(1 - L(v)) > \frac{1}{2}\Delta$ for large N; from the boundedness of λs away from 0 and 1, (2.15) then follows.

Now, evaluating (2.30) in our special case, we get

$$\begin{aligned}
\sigma^2 \;=\; & N\,\lambda\,\mathsf{var}\left(\int_{\frac{1}{2}}^{H(X_1)} (1 - \lambda_0)\,M_0'(v)\,\mathrm{d}\varphi(v)\right) \qquad (7.7)\\
& + N(1 - \lambda)\,\mathsf{var}\left(\int_{\frac{1}{2}}^{H(X_N)} \lambda_0\,L_0'(v)\,\mathrm{d}\varphi(v)\right).
\end{aligned}$$

It remains to pass to the limit (for $N \to \infty$). Let us denote

$$q(t) = \int_{\frac{1}{2}}^{t} M_0'(v)\,\mathrm{d}\varphi(v);$$

with this notation,

$$\text{var}\int_{\frac{1}{2}}^{H(X_1)} M_0'(v)\,\mathrm{d}\varphi(v) = \int_0^1 q^2(t)\,L'(t)\,\mathrm{d}t - \left(\int_0^1 q(t)\,L'(t)\,\mathrm{d}t\right)^2. \quad (7.8)$$

We may confine ourselves to a nondecreasing φ. The Lipschitz condition for M_0 entails the inequality

$$|q(t)| \le A|\varphi(t)| + B \quad \text{or} \quad q^2(t) \le C\,\varphi^2(t) + D$$

(with some constants A, B, C, D) and its uniform (in N) validity for the Ls entails further that the integrals

$$\left(\int_0^{\epsilon} + \int_{1-\epsilon}^{1}\right) q^2(t)\,L'(t)\,\mathrm{d}t$$

can be made arbitrarily small for all N and a suitable $\epsilon > 0$. Now, the integration by parts followed by the limit passage ($N \to \infty$) yields

$$\int_{\epsilon}^{1-\epsilon} q^2\,L'\,\mathrm{d}t \to \int_{\epsilon}^{1-\epsilon} q^2\,L_0'\,\mathrm{d}t,$$

and the limit is, in turn, arbitrarily close to $\int_0^1 q^2\,L_0'\,\mathrm{d}t$. As the same argument applies to the integral $\int_0^1 q(t)\,L'(t)\,\mathrm{d}t$, as well as to the second term on the right-hand side of (7.7), we get finally

$$\sigma^2 \approx N\,\lambda_0(1-\lambda_0)^2\,\text{var}\int_{\frac{1}{2}}^{H_0(X_1)} M_0'\,\mathrm{d}\varphi + N\,\lambda_0^2(1-\lambda_0)\,\text{var}\int_{\frac{1}{2}}^{H_0(X_N)} L_0'\,\mathrm{d}\varphi \quad (7.9)$$

(in the sense that the ratio of both sides tends to 1). But the right-hand side of (7.9) is exactly $N\,\tau_0^2$; and the postulate $\tau_0^2 > 0$ implies (2.23). An application of Theorem 2 then completes the proof of Theorem 4.

As to the proof of Theorem 5, it is easy to show that the assumptions (2.12) – (2.14) are satisfied with

$$\begin{aligned} \ell_i(v) &= L'(v) & & 1 \le i \le m, \\ &= M'(v) & & m < i \le N, \quad 0 < v < 1, \end{aligned}$$

and that (2.30) gives exactly $N\,\tau^2$. It remains to prove (2.15). Since f and g are continuous and $f(x) + g(x) > 0$ on (α, β) there is a point y such that $f(y)\,g(y) > 0$. Let

$$\epsilon = \min\left[v, \int_{\alpha}^{y} f(u)\,\mathrm{d}u, \int_{y}^{\beta} f(u)\,\mathrm{d}u, \int_{\alpha}^{y} g(u)\,\mathrm{d}u, \int_{y}^{\beta} g(u)\,\mathrm{d}u\right].$$

Now we shall show that

$$[G(H^{-1}(v)) < \epsilon \text{ or } G(H^{-1}(v)) > 1-\epsilon] \Rightarrow [\epsilon < F(H^{-1}(v)) < 1-\epsilon].$$

Assume that $G(H^{-1}(v)) < \epsilon$, for example. Then

$$m\,N^{-1}\,F(H^{-1}(v)) + n\,N^{-1}\,G(H^{-1}(v)) = v$$

entails $F(H^{-1}(v)) > \epsilon$. Further $H^{-1}(v) < y$, and

$$1 - F(H^{-1}(v)) = \int_{H^{-1}(v)}^{\beta} f(u)\,du > \int_{y}^{\beta} f(u)\,du \geq \epsilon.$$

Thus $F(H^{-1}(v)) < 1-\epsilon$. Similarly we may treat the case $G(H^{-1}(v)) > 1-\epsilon$.

Consequently,

$$N^{-1}\sum_{i=1}^{N} L_i(v)\,[1 - L_i(v)] \geq N^{-1}\min(m,n)\,\epsilon(1-\epsilon)$$

and (2.15) is entailed by (2.42).

Thus, again, an application of Theorem 2 completes the proof. □

ACKNOWLEDGEMENT

The authors wish to express their sincere thanks to the referee for his prompt and thorough reading of the manuscript and many valuable comments.

REFERENCES

Chernoff, H. and Savage, I. R. (1958). Asymptotic normality and efficiency of certain nonparametric test statistics. *Ann. Math. Statist. 29*, 972–994.

Cramér, H. (1946). *Mathematical Methods of Statistics.* Princeton Univ. Press.

Govingarajulu, Z., Le Cam, L. and Raghavachari, M. (1966). Generalizations of theorems of Chernoff and Savage on the asymptotic normality of test statistics. *Proc. Fifth Berkeley Symp. Math. Statist. Prob. 1*, 609–638.

Hájek, J. (1968). Asymptotic normality of simple linear rank statistics under alternatives. *Ann. Math. Statist. 39*, 325–346.

Loève, M. (1963). *Probability Theory* (3rd ed.). Van Nostrand, New York.

Petrov, V. V. (1957). A local theorem for lattice distribution. (In Russian) *Dokl. Akad. Nauk SSSR 115*, 49–52.

Pyke, R. and Shorack, G. R. (1968). Weak convergence of a two-sample empirical process and a new approach to Chernoff–Savage theorems. *Ann. Math. Statist.* *39*, 755–771.

CHAPTER 30

Miscellaneous Problems of Rank Test Theory

In: Nonparametric Techniques in Statistical Inference, 1970, (M. L. Puri, ed.), 3–19, Cambridge University Press.

1 INTRODUCTION AND SUMMARY

The paper deals with diverse problems, partly unpublished, partly published without proofs in Hájek (1969). Section 3 proposes three methods of density estimation which may be used in choosing proper scores for rank tests. The second of these methods is based on a theorem by Jurečková (1969) on asymptotic linearity of simple linear rank statistics, which is presented in a modified form in § 2. The same result by Jurečková is once again used in § 4 to prove that some statistics for testing scale (the symmetric case) may be adapted for the case when both medians are unknown without inflicting their asymptotic distribution. For a less general result of this kind, derived by a different method, see Raghavachari (1965). Section 5 provides a theorem to the effect that one partial ordering of permutations is finer than another. This theorem may be utilized in proving Jurečková's theorem, and also in proving unbiasedness of some nonparametric tests (see Lehmann (1966) and Hájek (1969)). The last section provides an explicit form for asymptotic normality of the conditional distribution of averaged rank statistics, when ties are present.

Collected Works of Jaroslav Hájek – With Commentary
Edited by M. Hušková, R. Beran and V. Dupač
Published in 1998 by John Wiley & Sons Ltd.

2 THE BEHAVIOUR OF LINEAR RANK STATISTICS UNDER A SHIFT

For each $N \geq 1$ let us consider a simple linear rank statistic

$$S_N = \sum_{i=1}^{N} c_{Ni}\, a_N(R_{Ni}), \tag{2.1}$$

where the ranks $R_{N1}, \ldots, R_{NN}$ are derived from a sequence of observations $Y_{N1}, \ldots, Y_{NN}$. Introducing

$$\begin{aligned} u(y) &= 1, \qquad x \geq 0, \\ &= 0, \qquad x < 0, \end{aligned} \tag{2.2}$$

and putting

$$s_N(y_1, \ldots, y_N) = \sum_{i=1}^{N} c_{Ni}\, a_N \left(\sum_{j=1}^{N} u(y_i - y_j) \right) \tag{2.3}$$

we may also write

$$S_N = s_N(Y_{N1}, \ldots, Y_{NN}). \tag{2.4}$$

We shall study the increment of S_N if the Y_{Ni}s are shifted by a random multiple of some numbers d_{Ni}. That is, considering simultaneously the statistic

$$S_N^* = s_N\left(Y_{N1} + \Delta_N\, d_{N1}, \ldots, Y_{NN} + \Delta_N\, d_{NN}\right) \tag{2.5}$$

we shall try to find an asymptotic expression for the difference $S_N^* - S_N$.

Assumptions A:

(A1) The regression constants satisfy the Noether condition:

$$\lim_{N\to\infty} \frac{\max_{1\leq i\leq N}(c_{Ni} - \bar{c}_N)^2}{\sum_{i=1}^{N}(c_{Ni} - \bar{c}_N)^2} = 0 \quad \left(\bar{c}_N = \frac{1}{N} \sum_{i=1}^{N} c_{Ni} \right). \tag{2.6}$$

(A2) The scores $a_N(i)$ are generated by a function $\phi(t)$, $0 < t < 1$, by either of the following two ways:

$$a_N(i) = \phi\left(\frac{i}{N+1}\right) \quad (1 \leq i \leq N), \tag{2.7}$$

$$a_N(i) = \mathsf{E}\,\phi\left(U_N^{(i)}\right) \quad (1 \leq i \leq N), \tag{2.8}$$

where $U_N^{(i)}$ denotes the ith order statistic in a sample of size N from the uniform distribution on $(0,1)$.

(A3) The score-generating function $\phi(t)$ is nonconstant and expressible as a difference of two *nondecreasing* and *square integrable* functions.

(A4) The constants d_{Ni} are such that

$$\sup_N \sum_{i=1}^{N} (d_{Ni} - \overline{d}_N)^2 < \infty \tag{2.9}$$

and

$$\lim_{N\to\infty} \max_{1\le i\le N} (d_{Ni} - \overline{d}_N)^2 = 0 \quad \left(\overline{d}_N = \frac{1}{N} \sum_{i=1}^{N} d_{Ni} \right). \tag{2.10}$$

(A5) The constants d_{Ni} and c_{Ni} are concordant in the following sense:

$$(c_{Ni} - c_{Nj})(d_{Ni} - d_{Nj}) \ge 0 \quad (1 \le i,\, j \le N). \tag{2.11}$$

(A6) The random variables $\Delta_N = \Delta_N(Y_{N1}, \ldots, Y_{NN})$ are *bounded in probability*.

(A7) $Y_{N1}, \ldots, Y_{NN}$ is a random sample from a distribution with a density f possessing finite Fisher information.

Put

$$\phi(t, f) = -\frac{f'(F^{-1}(t))}{f(F^{-1}(t))} \quad (0 < t < 1) \tag{2.12}$$

and

$$b_N = \sum_{i=1}^{N} d_{Ni}(c_{Ni} - \overline{c}_N) \int_0^1 \phi(t)\, \phi(t, f)\, \mathrm{d}t. \tag{2.13}$$

Theorem 2.1. *Under assumptions* (A1) *through* (A7),

$$\frac{S_N^* - S_N - \Delta_N\, b_N}{\sqrt{(\mathrm{var}\, S_N)}} \to 0, \tag{2.14}$$

holds in probability.

Proof. The theorem is an easy corollary of a result by J. Jurečková (1969). She showed that for nonrandom Δ_N (2.14) holds uniformly for $|\Delta_N| \le C$. Consequently (2.14) must also hold for Δ_N random and bounded in probability. □

Remark 2.1. Instead of (A4) and (A5) we could also assume that $d_{Ni} = d'_{Ni} - d''_{Ni}$, where both the constants d'_{Ni} and d''_{Ni} satisfy (A4) and (A5).

Remark 2.2. Assumption (A2) may be replaced by (A2*): the scores $a_N(i)$ satisfy

$$\frac{1}{N}\sum_{i=1}^{N}\left[a_N(i) - \mathsf{E}\,\phi(U_N^{(i)})\right]^2 \to 0. \tag{2.15}$$

3 THREE METHODS OF DENSITY TYPE ESTIMATION

Let $\mathcal{F}_1, \ldots, \mathcal{F}_k$ be k distinct density types generated by some one-dimensional densities $f_1, \ldots, f_k$:

$$\mathcal{F}_j = \{f : f(x) = \lambda\, f_j(\lambda x - u),\ -\infty < u < \infty,\ \lambda > 0\} \quad (1 \le j \le k). \tag{3.1}$$

Given a sample $(X_1, \ldots, X_N)$ governed by a density of the form

$$p(x_1, \ldots, x_N) = \prod_{i=1}^{N} f(x_i), \tag{3.2}$$

where f belongs to $\bigcup_{j=1}^{k} \mathcal{F}_j$ but otherwise is unknown, let us try to locate the type of f, i.e. find j such that $f \in \mathcal{F}_j$. Thus, a decision procedure for this problem will be a function $\delta(x_1, \ldots, x_N)$ taking its values in the set $\{1, 2, \ldots, k\}$. Assume that the loss incurred by $\delta(x) = d$ under $f \in \mathcal{F}_j$ is given by

$$\begin{aligned} L(j,d) &= 0 \quad \text{if} \quad j = d, \\ &= 1, \quad \text{if} \quad j \neq d,\ 1 \le j,\ d \le k. \end{aligned} \tag{3.3}$$

Restricting ourselves to procedures that are invariant relative to the group of positive linear transforms

$$((x_1, \ldots, x_N) \to (\lambda x_1 + u, \ldots, \lambda x_N + u),\ \lambda > 0)$$

the risk of a procedure given $f \in \mathcal{F}_j$ will equal

$$R(j, \delta) = 1 - \mathsf{P}\left(\delta(X_1, \ldots, X_N) = j \,|\, f \in \mathcal{F}_j\right). \tag{3.4}$$

Invariant Bayes solutions may be obtained from 'marginal' densities relative to the sub σ-field consisting of events that are invariant with respect to the group of positive linear transforms. These densities can be established for our types as follows:

$$\overline{p}_j(x_1, \ldots, x_N) = \int_0^\infty \int_{-\infty}^\infty \left[\prod_{i=1}^{N} f_j(\lambda x_i - u)\right] \lambda^{N-2}\, \mathrm{d}u\mathrm{d}\lambda \quad (1 \le j \le k) \tag{3.5}$$

(see Hájek and Šidák (1967), § II.2.2). If all types are considered *a priori* equiprobable, then the expected risk $(1/k)\sum_{j=1}^{k} R(j,\delta)$ is minimized by the procedure δ_0 such that

$$[\delta_0(x_1,\ldots,x_N)=j] \Rightarrow \left[\overline{p}_j(x_1,\ldots,x_N) = \max_{1\leq h\leq k} \overline{p}_h(x_1,\ldots,x_N)\right]. \tag{3.6}$$

Any such procedure satisfies

$$\sum_{j=1}^{k} R(j,\delta_0) \leq \sum_{j=1}^{k} R(j,\delta) \quad (\delta \text{ invariant}). \tag{3.7}$$

One should expect that

$$\sum_{j=1}^{k} R(j,\delta_0) \to 0 \quad \text{as } N\to\infty$$

under fairly general conditions concerning the types $\mathcal{F}_1,\ldots,\mathcal{F}_k$. If the ϕ-functions $\phi(t,f_j)$ given by (2.12) satisfy (A3), we shall be able to infer

$$\sum_{j=1}^{N} R(j,\delta_0) \to 0$$

from (3.7) and from the fact that

$$\sum_{j=1}^{N} R(j,\delta_1) \to 0$$

for some other invariant decision method δ_1 to be described next. We conclude the discussion of the procedure (3.6) by observing that the complexity of the formula for $\overline{p}_j$ makes its applicability doubtful.

Another selection procedure is yielded by Theorem 2.1. First introduce the functions

$$s_{Nj}(y_1,\ldots,y_N) = \sum_{i=1}^{[\frac{1}{2}N]} a_{Nj}\left(\sum_{h=1}^{N} u(y_i - y_h)\right) \quad (1\leq j\leq k), \tag{3.8}$$

where the scores $a_{Nj}(i)$ are generated either by (2.7) or by (2.8) with $\phi(t)$ replaced by $\phi(t,f_j) = -f_j'(F_j^{-1}(t))/f_j(F_j^{-1}(t))$. Further put

$$\begin{aligned} d_{Ni} &= 1/\sqrt{N} \qquad \left(1\leq i\leq [\tfrac{1}{2}N]\right), \\ &= 0 \qquad \left(\tfrac{1}{2}N < i\leq N\right), \end{aligned} \tag{3.9}$$

and denote by $\Delta_N = \Delta_N(y_1, \ldots, y_N)$ any statistic such that

$$\Delta_N(\lambda y_1 + u, \ldots, \lambda y_N + u) = \lambda \Delta_N(y_1, \ldots, y_N) \quad (-\infty < u < \infty,\ \lambda > 0) \tag{3.10}$$

and that

$$0 < p \lim_{f=f_j} \Delta_N(X_1, \ldots, X_N) = b_j < \infty \quad (1 \le j \le k). \tag{3.11}$$

For example, if $X_N^{(i)}$ denotes the ith order statistic we may put

$$\Delta_N = M\left[X_N^{(N+1-j_N)} - X_N^{(j_N)}\right] \quad \left(\frac{j_N}{N} \to \alpha,\ 0 < \alpha \le \frac{1}{2}\right). \tag{3.12}$$

Then, obviously,

$$b_j = M\left[F_j^{-1}(1-\alpha) - F_j^{-1}(\alpha)\right] \quad \left(F_j(\alpha) = \int_0^\alpha f_j(x)\,\mathrm{d}x\right). \tag{3.13}$$

Finally put

$$S_{Nj} = s_{Nj}(X_1, \ldots, X_N) \quad (1 \le j \le k) \tag{3.14}$$

and

$$S_{Nj}^* = s_{Nj}\left(X_1 + \Delta_N d_{N1}, \ldots, X_N + \Delta_N d_{NN}\right) \tag{3.15}$$

and compute the ratios

$$\ell_{Nj} = \frac{S_{Nj}^* - S_{Nj}}{\sqrt{(\operatorname{var} S_{Nj})}}. \tag{3.16}$$

The invariance of ℓ_{Nj} under positive linear transforms and a direct application of Theorem 2.1 provide

$$\begin{aligned} p \lim_{f \in \mathcal{F}_h} \ell_{Nj} &= p \lim_{f=f_h} \ell_{Nj} = p \lim_{f=f_h} \frac{\Delta_N \frac{1}{4} \sqrt{N} \int_0^1 \phi(t, f_h)\, \phi(t, f_j)\,\mathrm{d}t}{\left[\frac{1}{4} N \int_0^1 \phi^2(t, f_j)\,\mathrm{d}t\right]^{\frac{1}{2}}} \\ &= \frac{1}{2} b_h\, \rho_{jh} \sqrt{I_h}, \end{aligned} \tag{3.17}$$

where

$$I_h = \int_0^1 \phi^2(t, f_h)\,\mathrm{d}t$$

and

$$\rho_{jh} = [I_j I_h]^{-\frac{1}{2}} \int_0^1 \phi(t, f_h)\, \phi(t, f_j)\,\mathrm{d}t. \tag{3.18}$$

If the types $\mathcal{F}_1, \ldots, \mathcal{F}_k$ are distinct, we have

$$\rho_{jh} < 1 \quad (1 \le j \ne h \le k), \tag{3.19}$$

while $\rho_{jj} = 1,\ 1 \leq j \leq k$. Consequently the decision procedure δ_1 such that

$$[\delta_1(x_1, \ldots, x_N) = j] \Rightarrow \left[\ell_{Nj}(x_1, \ldots, x_N) = \max_{1 \leq h \leq k} \ell_{Nh}(x_1, \ldots, x_N)\right] \quad (3.20)$$

is consistent in the sense that $R(j, \delta_1) \to 0$ if $N \to \infty,\ 1 \leq j \leq k$. This entails

$$\sum_{j=1}^{k} R(j, \delta_1) \to 0,$$

and, in view of (3.7), also

$$\sum_{j=1}^{k} R(j, \delta_0) \to 0.$$

In order to apply the selection procedure δ_1 to the choice of proper scores in a rank statistic of form (2.1), we need δ_1 to be a function of the order statistic $X^{(\cdot)} = (X^{(1)}, \ldots, X^{(N)})$ only.

This goal may be easily achieved by introducing random variables

$$Y_i = X^{(Q_i)} \quad (1 \leq i \leq N), \quad (3.21)$$

where $Q = (Q_1, \ldots, Q_N)$ is independent of $(X_1, \ldots, X_N)$ and assumes each of $N!$ permutations of $(1, 2, \ldots, N)$ with probability $1/N!$. Then $(Y_1, \ldots, Y_N)$ has the same distribution as $(X_1, \ldots, X_N)$ under all densities of the form (3.2) and the selection procedure may equally well be applied to $Y_1, \ldots, Y_N$ instead of to $X_1, \ldots, X_N$. Thus we have the following theorem.

Theorem 3.1. *Under the above notations, if the functions $\phi(t, f_j)$ are expressible as differences of two nondecreasing and square integrable functions, $1 \leq j \leq k$, then*

$$\lim_{N \to \infty} P(\delta_1(Y_1, \ldots, Y_n) = j \mid q_{Nj}) = 1 \quad (3.22)$$

holds for any sequence of densities $q_{Nj}(x_1, \ldots, x_N)$ that are contiguous to densities of the form

$$p_{Nj}(x_1, \ldots, x_N) = \lambda_N^N \prod_{i=1}^{N} f_j(\lambda_N x_i + u_N) \quad (-\infty < u_N < \infty,\ \lambda_N > 0). \quad (3.23)$$

Proof. The above considerations yielded the proof for $q_{Nj} = p_{Nj}$, i. e. if f in (3.2) satisfied $f \in \mathcal{F}_j$ for every $N \geq 1$. However, if q_{Nj} are contiguous relative to p_{Nj} in the X-spaces, the induced distributions of $(Y_1, \ldots, Y_N)$ are also contiguous to p_{Nj} in the Y-spaces, as may be easily seen. Then it suffices to note that convergence to 1 under p_{Nj} entails the same under any contiguous alternative. □

Let us resume the discussion of how δ_1 may be applied in rank testing. Let us test the hypothesis of randomness against the alternative of two samples differing in location or against a general location shift alternative. Then the pertinent statistic is of form (2.1), where the scores $a_N(i)$ should correspond to the underlying density f in the well-known way. If the type of f is not known, we put forward a certain number of density types and compute for them the quantities ℓ_{Nj} applied to $Y_1, \ldots, Y_N$ generated as a random permutation of order statistics $X^{(1)}, \ldots, X^{(N)}$. The density type providing the largest ℓ_{Nj} is then chosen to generate the scores and the test is performed as if these scores were decided upon before knowing $X_1, \ldots, X_N$. Since under the hypothesis of randomness the vector of ranks $(R_1, \ldots, R_N)$ is independent of the vector of order statistics, the above selection of scores does not invalidate the significance level. For moderate N we can be deciding between three types corresponding to the following ϕ-functions; for example:

$$\begin{aligned} \phi_1(t) &= \Phi^{-1}(t) && \text{(normal type)}, \\ \phi_2(t) &= 2t-1 && \text{(logistic type)}, \\ \phi_3(t) &= -1 && \left(0 \le t \le \tfrac{1}{4}\right), \\ &= 4t-2 && \left(\tfrac{1}{4} \le t \le \tfrac{3}{4}\right), \\ &= 1 && \left(\tfrac{3}{4} \le t \le 1\right). \end{aligned} \tag{3.24}$$

(Φ^{-1} denotes the inverse normal distribution function.) It is advisable to use $\phi_3(t)$ instead of $\phi_4(t) = \operatorname{sign}(2t-1)$, which corresponds to the double-exponential type, since the discontinuity of ϕ_4 at $t = \frac{1}{2}$ increases unduly the variability of the corresponding ratio ℓ_{Nj}. Thus the method would consist in testing the sensitivity of the van der Waerden test, the Wilcoxon test, and a test lying 'between' the Wilcoxon test and the median test, to the shift of location of the sample $Y_1, \ldots, Y_{[\frac{1}{2}N]}$. The test which is most sensitive is then applied to the original problem, as far as the scores are concerned. If ϕ_3 is selected, we can apply the median test as well.

The third selection procedure is based on a partial ordering of ϕ-function and of corresponding types. Within the family of skew symmetric ϕ-functions, we shall say that ϕ increases more rapidly than ψ if

$$\phi(t) = b(t)\,\psi(t) \quad \left(\frac{1}{2} < t < 1\right), \tag{3.25}$$

where $b(t)$ is nondecreasing. Inspecting (3.24), it is easy to see that ϕ_1 increases more rapidly than ϕ_2, and ϕ_2 increases more rapidly than ϕ_3, and ϕ_3 increases more rapidly than $\phi_5(t) = \sqrt{2}\sin[\pi(2t-1)]$, which corresponds to the Cauchy distribution. Further, given two types $\mathcal{F}$ and $\mathcal{G}$ of symmetric densities, we shall say that the densities from $\mathcal{F}$ have shorter (lighter) tails than the densities from $\mathcal{G}$ if for any $f \in \mathcal{F}$ and $g \in \mathcal{G}$, such that their medians are zero, the corresponding quantile functions $F^{-1}(t)$ and $G^{-1}(t)$ satisfy

$$F^{-1}(t) = a(t)\, G^{-1}(t) \quad \left(\frac{1}{2} < t < 1\right), \tag{3.26}$$

with $a(t)$ nonincreasing.

In Hájek (1969), § 34, a theorem is given according to which (3.25) with $b(t)$ nondecreasing entails (3.26) with $a(t)$ nonincreasing, provided $\phi(t) = \phi(t, f)$ and $\psi(t) = \phi(t, g)$, using notation (2.12). Consequently, the normal type has shorter tails than the logistic type, the latter type has shorter tails than the type corresponding to ϕ_3, and the last type has shorter tails than the Cauchy type.

Since (3.26) with $a(t)$ nonincreasing entails that

$$F^{-1}(G(x)) \quad (x > 0), \quad \text{is concave,}$$

$$G^{-1}(F(x)) \quad (x > 0), \quad \text{is convex,}$$

we can test whether the true ϕ-function increases more rapidly than ϕ_0 as follows. We plot $\left[\frac{1}{2}N\right]$ points with co-ordinates

$$\left[x^{(N+1-i)} - x^{(i)},\ F_0^{-1}\left(-\frac{N+1-i}{N+1}\right)\right] \quad \left(1 \leq i \leq \left[\frac{1}{2}N\right]\right),$$

where F_0^{-1} is the quantile function corresponding to ϕ_0.

If the curve suggested by these $\left[\frac{1}{2}N\right]$ points is convex, than we may prefer a ϕ-function which is increasing more slowly than ϕ_0; if it is concave, we may prefer a ϕ which increases more rapidly than ϕ_0; if it is straight, we keep ϕ_0. The effectiveness of this method and the possibility of its formalization may be in doubt. On the other hand, it surely is the quickest of all three methods explained in this section.

4 TESTING SCALE WHEN BOTH MEDIANS ARE UNKNOWN

Jurečková's theorem may also be applied to an asymptotic treatment of nuissance medians in testing scale alternatives. First observe that the increment $(S_N^* - S_N)\big/\sqrt{(\operatorname{var} S_N)}$ will be asymptotically zero if

$$\int_0^1 \phi(t)\, \phi(t, f)\, \mathrm{d}t = 0. \tag{4.1}$$

Such a situation occurs if ϕ is symmetric,

$$\phi(t) = \phi(1-t) \quad (0 < t < 1), \tag{4.2}$$

and $\phi(t, f)$ is skew symmetric,

$$\phi(t, f) = -\phi(1-t, f) \quad (0 < t < 1). \tag{4.3}$$

Obviously, the skew symmetry of $\phi(t, f)$ is equivalent to the symmetry of f about its median.

Consider two samples $X_1, \ldots, X_m$ and $X_{m+1}, \ldots, X_{m+n}$ and assume that the respective densities are of the form $\sigma f[\sigma(x-\mu)]$ and $\lambda f[\lambda(x-\nu)]$. If $\mu = \nu$, proper rank tests for testing $\sigma = \lambda$ against $\sigma \neq \lambda$ (or $\sigma > \lambda$, or $\sigma < \lambda$) are based on statistics

$$S_N = s_N(X_1, \ldots, X_N), \tag{4.4}$$

where

$$s_N(x_1, \ldots, x_N) = \sum_{i=1}^{m_N} a_N\left(\sum_{j=1}^{N} u(x_i - x_j)\right) \tag{4.5}$$

and the scores $a_N(i)$ are generated by (2.7) or by (2.8). If $\mu \neq \nu$, but we have some estimates $\hat{\mu}_N$ and $\hat{\nu}_N$ for μ and ν, respectively, we may try to prove that the statistic

$$S_N^* = s_N\left(X_1 - \hat{\mu}_N, \ldots, X_m - \hat{\mu}_N, X_{m+1} - \hat{\nu}_N, \ldots, X_N - \hat{\nu}_N\right) \tag{4.6}$$

possesses the same limiting distribution as S_N.

Assumptions B:

(B1) The scores $a_N(i)$ are generated by (2.7) or (2.8), or satisfy (2.15), where $\phi(t)$ is square integrable, fulfils (4.2) and is nondecreasing for $t \in \left(\frac{1}{2}, 1\right)$.

(B2) Random vector $\boldsymbol{X}$ is governed by the density

$$p_N(x_1, \ldots, x_N) = \prod_{i=1}^{m_N} f(x_i - \mu) \prod_{j=1}^{n_N} f(x_{m+j} - \nu) \quad (m_N + n_N = N), \tag{4.7}$$

where μ and ν are arbitrary and f is a fixed density which is symmetric and possesses finite Fisher information.

(B3) The estimates $\hat{\mu}_m$ and $\hat{\nu}_n$ are square-root consistent, that is, random variables $\sqrt{m}\,(\hat{\mu}_m - \mu)$ and $\sqrt{n}\,(\hat{\nu}_n - \mu)$ are bounded in probability.

(B4) $$\min(m_N, n_N) \to \infty.$$

Theorem 4.1. *Let assumptions B be satisfied and denote by* $\mathsf{E}\, S_N$ *and* $\mathsf{var}\, S_N$ *the expectation and variance of* S_N *under the hypothesis of randomness.*

Then S_N^* *is asymptotically normal with parameters* $(\mathsf{E}\, S_N, \mathsf{var}\, S_N)$.

Moreover, if densities q_N *are contiguous to densities* p_N *of* (4.7), *and if* S_N *is asymptotically normal with parameters* $(a_N, \mathsf{var}\, S_N)$ *under* q_N, *then* S_N^* *is also asymptotically normal* $(a_N, \mathsf{var}\, S_N)$ *under* q_N.

Proof. Since S_N is a rank statistic, we may write equivalently

$$\begin{aligned} S_N &= s_N(X_1 - \mu + \Delta_N d_{N1}, \ldots, X_m - \mu + \Delta_N d_{Nm}, X_{m+1} - \nu \\ &\qquad + \Delta_N d_{N,m+1}, \ldots, X_N - \nu + \Delta_N d_{NN}), \end{aligned} \tag{4.8}$$

where

$$\Delta_N = (\mu - \hat{\mu}_n - \nu + \hat{\nu}_N)\left(\frac{1}{m_N} + \frac{1}{n_N}\right)^{-\frac{1}{2}} \tag{4.9}$$

$$(n_N = N - m_N,\ \hat{\mu}_n = \hat{\mu}_{mN},\ \hat{\nu}_N = \hat{\nu}_{nN})$$

and

$$\begin{aligned} d_{Ni} &= \left(\frac{1}{m_N} + \frac{1}{n_N}\right)^{\frac{1}{2}} \qquad (1 \le i \le m_N), \\ &= 0 \qquad\qquad\qquad\qquad (m_N < i \le N). \end{aligned} \tag{4.10}$$

Now we shall check if Assumptions A of Theorem 2.1 are satisfied. As for (A1), we have $c_{Ni} = 1,\ 1 \le i \le m_N$, and $c_{Ni} = 0,\ m_n < i \le N$, so that (B4) entails (2.6). Further, (B1) implies (A2), (A2*) and (A3). Next, definition (4.10) of the d_{Ni}'s entails (2.9), and (B4) entails (2.10); consequently (A4) holds. (A5) is obvious, and (A6) follows from (B3). Finally, (B2) entails the satisfaction of (A7) for

$$(Y_{N1}, \ldots, Y_{NN}) = (X_1 - \mu, \ldots, X_m - \mu,\ X_{m+1} - \nu, \ldots, X_N - \nu).$$

Now under (B1) and (B2) we have (4.2) and (4.3), and, consequently, also (4.1). Thus $(S_N^* - S_N)\big/\sqrt{(\mathsf{var}\, S_N)} \to 0$ under p_N. From this all conclusions of the theorem easily follow. □

Examples. The theorem is applicable among others to the following statistics. The Klotz statistic, the quartile statistic, the Mood statistic, the Capon statistics, the Ansari–Bradley statistic; these statistics are described in Hájek and Šidák (1967), § III.2.1. Further it also applies to the adapted Wilcoxon test (the Siegel–Tukey adaptation, for example) and the adapted van der Waerden test; see Hájek (1969), § 16.

5 PARTIAL ORDERINGS OF PERMUTATIONS

As is described in Lehmann (1966), § 7, one employs two distinct partial orderings of permutations. We shall call them Ordering 1 and Ordering 2, and instead of writing 'better ordered in the sense of Ordering 1' we shall write 'better ordered^1'; 'better ordered^2' will have a similar meaning.

Ordering 1. We say that a permutation $q = (q_1, \ldots, q_N)$ is better ordered^1 than $r = (r_1, \ldots, r_N)$ if q may be obtained from r by a number of steps, each of which consists in correcting an inversion. (To correct an inversion means to interchange two elements r_i and r_j such that $i < j$ and $r_i > r_j$.)

Ordering 2. We say that a permutation $q = (q_1, \ldots, q_N)$ is better ordered^2 than $r = (r_1, \ldots, r_N)$ if

$$[i < j,\ r_i < r_j] \Rightarrow [q_i < q_j]. \tag{5.1}$$

Ordering 1 does not entail Ordering 2, as is shown by example (a) in Lehmann (1966), p. 1149. On the other hand, we present below a theorem to the effect that Ordering 2 entails Ordering 1. This might seem to be in contradiction with example (b) in Lehmann (1966), p. 1149. However, it is not so, since 'correcting an inversion' in Lehmann's paper is understood as exchanging the positions of two *neighbouring* elements r_i and r_{i+1}, $r_i > r_{i+1}$.

Theorem 5.1. *If q is better ordered^2 than r in the sense of (5.1), then q may be obtained from r by a number of steps, each consisting in correcting an inversion such that the involved numbers differ by 1.*

Proof. If $r = q$ the proposition is trivial. Assume $r \neq q$ and denote by $d = (d_1, \ldots, d_N)$ the inverse permutation to r, that is, $r(d_i) = i$, $1 \leq i \leq N$. Certainly, we do not have $q(d_1) < q(d_2) < \cdots < q(d_N)$, since that would entail $q(d_i) = i$, and in turn, $q = r$. Consequently, there is a k, $1 \leq k \leq N$, such that

$$q(d_k) > q(d_{k+1}). \tag{5.2}$$

Since r is better ordered^2 than q,

$$[d_k < d_{k+1},\ r(d_k) = k < k+1 = r(d_{k+1})] \Rightarrow [q(d_k) < q(d_{k+1})]. \tag{5.3}$$

Since $q(d_k) < q(d_{k+1})$ is false, in view of (5.2), and $r(d_k) < r(d_{k+1})$ is true, we cannot have $d_k < d_{k+1}$. Consequently we have $d_{k+1} < d_k$, that is, $k+1$ precedes k in r. If we correct this inversion, the state of all other pairs will remain unchanged: the inversions will occur exactly where they were before. Thus the new permutation r' will be better ordered^2 than r, and q will be better ordered^2 than r', since for the pair where the inversion was corrected we have (5.2). Obviously, after a finite number of steps of this kind we must obtain a permutation which coincides with q. □

Example. Obviously, $q = (2, 1, 3, 4, 5)$ is better ordered^2 than

$$r = (3, 2, 5, 1, 4).$$

Correcting the inversion (2, 1) in r, we obtain $r^{(1)} = (3, 1, 5, 2, 4)$; correcting (3, 2) in $r^{(1)}$, we obtain $r^{(2)} = (2, 1, 5, 3, 4)$; correcting (5, 4) in $r^{(2)}$, we obtain $r^{(3)} = (2, 1, 4, 3, 5)$; finally, correcting (4, 3) in $r^{(3)}$, we obtain

$$r^{(4)} = (2, 1, 3, 4, 5) = q.$$

Theorem 5.1 may be used in the proof that the

$$s_N(y_1, \ldots, y_N, \Delta) = \sum_{i=1}^{N} c_i\, a\left(\sum_{j=1}^{N} u(y_i + \Delta d_i - y_j - \Delta d_j)\right)$$

is nondecreasing in Δ, if $a(1) \leq \cdots \leq a(N)$ and (2.11) hold (see Hájek (1969), § 7). A different proof is provided in Lehmann (1966), and still another in Jurečková (1969).

6 ASYMPTOTIC NORMALITY UNDER TIES

Let $X_1, \ldots, X_N$ be a random sample from an arbitrary distribution, possibly discrete. Then the set of order statistics $X^{(i)}$ decomposes into g groups of tied values:

$$X^{(1)} = \cdots = X^{(\tau_1)} < X^{(\tau_1+1)} = \cdots = X^{(\tau_1+\tau_2)} < \cdots$$
$$\cdots < X^{(\tau_1+\cdots+\tau_{g-1}+1)} = \cdots = X^{(N)}.$$

The occurrence of ties is characterized by the vector of tie-sizes $\tau = (\tau_1, \ldots, \tau_g)$, where g as well as the τ_is are random. Introducing averaged scores

$$a_N(i, \tau) = \frac{1}{\tau_k} \sum_{j=T_{k-1}+1}^{T_k} a(j) \tag{6.1}$$

if

$$T_{k-1} = \tau_1 + \cdots + \tau_{k-1} < i \leq \tau_1 + \cdots + \tau_k = T_k,$$

let us investigate the asymptotic distribution of the statistic

$$\overline{S}_N = \sum_{i=1}^{N} c_{Ni}\, a_N(R_{Ni}, \tau), \tag{6.2}$$

where $R_{Ni} = \sum_{j=1}^{N} u(X_i - X_j)$ as before. If the distribution of X_is is not continuous, then it is not true that $R = (R_{N1}, \dots, R_{NN})$ is a permutation of $(1, 2, \dots, N)$ with probability 1. However, the conditional distribution of $\overline{S}_N$ given τ is the same as if R were distributed uniformly over the space of permutations, in view of the special form of $a_N(i, \tau)$ (see Hájek (1969), § 29).

The statements concerning the asymptotic distribution may have three different forms:

Form 1. We assert that $\overline{S}_n$ is asymptotically normal with parameters $(\mathsf{E}\, S_N, \sigma_N^2)$, where σ_N^2 is independent of τ, but depends on the distribution function F of the X_is, if F is not continuous. (See Vorlíčková (1970), Theorem 3 of § 3).

Form 2. We assert that the *conditional distribution* of $\overline{S}_N$ given τ differs from $\mathcal{N}(\mathsf{E}\,\overline{S}_N, \mathsf{var}(\overline{S}_N \mid \tau))$ in the Kolmogorov distance by less than ϵ with probability greater than $1 - \epsilon$ if $N \geq N(\epsilon)$. (See Vorlíčková (1970), Theorem 4 of § 3).

Form 3. We explicitly establish regions W_N such that the Kolmogorov distance between $\mathcal{L}(\overline{S}_N \mid \tau)$ and $\mathcal{N}(\mathsf{E}\, S_N, \mathsf{var}(\overline{S}_N \mid, \tau))$ is smaller than ϵ if $\tau \in W_N$ and $N \geq N(\epsilon, \{W_N\})$.

Form 2 provides us with the knowledge that $\mathcal{L}(\overline{S}_N \mid \tau)$ should have been close to $\mathcal{N}(\mathsf{E}\,\overline{S}_N, \mathsf{var}(S_n \mid \tau))$ with great probability. Form 3 enables us to check whether this really happened, if τ is known.

Let us now prove a theorem in which the assertion takes on Form 3.

Theorem 6.1. *Let the scores $a_N(i)$ satisfy (2.15), where ϕ is nonconstant and square integrable. Let $X_1, \dots, X_N$ be independent and equidistributed.*

Then for every $\epsilon > 0$ and $\eta > 0$ there exists a $\delta = \delta(\epsilon, \eta)$ such that

$$\max_{1 \leq i \leq N} (c_{Ni} - \overline{c}_N)^2 < \delta \sum_{i=1}^{N} (c_{Ni} - \overline{c}_N)^2 \tag{6.3}$$

together with

$$\sum_{i=1}^{N} (a_N(i) - \overline{a}_N)^2 > \eta \sum_{i=1}^{N} (a_N(i, \tau) - \overline{a}_N)^2 \tag{6.4}$$

entails

$$\sup_{-\infty<x<\infty}\left|\mathsf{P}(\overline{S}_N \leq \mathsf{E}\,\overline{S}_N + x\sqrt{\mathsf{var}(\overline{S}_N\,|\,\tau)}\,|\,\tau) - \Phi(x)\right| < \epsilon. \tag{6.5}$$

Proof. Recall that we may assume that $\mathsf{P}(R = r) = 1/N!$ for all permutations r, without changing the conditional distribution of $\overline{S}_N$. Then keep η fixed and give the assertion the equivalent limiting form, which reads as follows. If (2.6) holds, then $\mathcal{L}\left[(\overline{S}_N - \mathsf{E}\,\overline{S}_N)\Big/\sqrt{\mathsf{var}(\overline{S}_N\,|\,\tau)}\Big|\,\tau\right] \to \mathcal{N}(0.1)$. According to Theorem 4.2 of Hájek (1962), it suffices to show that

$$\left[\frac{k_N}{N}\to 0\right] \Rightarrow \left[\frac{\max_{1\leq i_1<\cdots<i_{k_N}\leq N}\sum_{\alpha=1}^{k_N}[a_N(i_\alpha,\tau_N)-\overline{a}_N]^2}{\sum_{i=1}^{N}[a_N(i,\tau_N)-\overline{a}_N]^2}\to 0\right] \tag{6.6}$$

holds for any choice of $\tau = \tau_N$ such that (6.4) holds. Square integrability of ϕ and (2.15) entail that

$$\left[\frac{k_N}{N}\to 0\right] = \left[\frac{\max_{1\leq i_1<\cdots<i_{k_N}\leq N}\sum_{\alpha=1}^{k_N}[a_N(i_\alpha)-\overline{a}_N]^2}{\sum_{i=1}^{N}[a_N(i)-\overline{a}_N]^2}\to 0\right]. \tag{6.7}$$

On the other hand the averaging process (6.1) implies

$$\max_{1\leq i_1<\cdots<i_{k_N}\leq N}\sum_{\alpha=1}^{k_N}[a_N(i_\alpha,\tau_N)-\overline{a}_N]^2 \leq \max_{1\leq i_1<\cdots<i_{k_N}\leq N}\sum_{\alpha=1}^{k_N}[a_N(i_\alpha)-\overline{a}_N]^2, \tag{6.8}$$

which in conjunction with (6.4) yields (6.6). □

Form 3 may be given also to the limiting theorem concerning the chi-square-type statistics. Let $X_1,\ldots,X_N$ split into k samples and denote by $\overline{S}_j$ the sum of scores $a_N(i,\tau)$ over the jth sample.

Theorem 6.2. *Under conditions of Theorem 6.1 put*

$$\overline{Q}_N = (N-1)\left[\sum_{i=1}^{N}(a_N(i,\tau)-\overline{a}_N)^2\right]^{-1}\left[\sum_{j=1}^{k}\frac{\overline{S}_j^2}{n_j} - N\,\overline{a}_N^2\right], \tag{6.9}$$

where n_j denotes the size of the jth sample.

Then for every $\epsilon > 0$ and $\eta > 0$ there exist an $M = M(\epsilon,\eta)$ such that (6.4) and

$$\min(n_1,\ldots,n_k) > M \tag{6.10}$$

entail

$$\sup_{-\infty<x<\infty} \left|\mathsf{P}(\overline{Q}_N \le x \mid \tau) - \mathsf{P}(\chi^2_{k-1} \le x)\right| < \epsilon, \tag{6.11}$$

where χ^2_{k-1} possesses the chi-square distribution with $k-1$ degrees of freedom.

Proof. Theorem 6.2 is an easy consequence of Theorem 6.1. □

If testing the hypothesis of independence, we have a sequence of pairs $(X_1, Y_1), \ldots, (X_N, Y_N)$. Let τ_x and $(R_{N1}, \ldots, R_{NN})$ be the tie-sizes and ranks associated with $(X_1, \ldots, X_N)$; let τ_y and $(Q_{N1}, \ldots, Q_{NN})$ be the same quantities for $(Y_1, \ldots, Y_N)$. Consider the statistic

$$\overline{S}_N = \sum_{i=1}^{N} a_N(R_{Ni}, \tau_x)\, a_N(Q_{Ni}, \tau_y). \tag{6.12}$$

Theorem 6.3. *Let the scores $a_N(i)$ satisfy (2.15), where ϕ is nonconstant and square integrable. Let the samples $(X_1, \ldots, X_N)$ and $(Y_1, \ldots, Y_N)$ be independent and each equidistributed.*

Then for every $\epsilon > 0$ and $\eta > 0$ there exists an $N(\epsilon, \eta)$ such that $N > N(\epsilon, \eta)$ together with

$$\min\left\{\sum_{i=1}^{N}[a_N(i, \tau_x) - \overline{a}_N]^2, \sum_{i=1}^{N}[a_N(i, \tau_y) - \overline{a}_N]^2 > \eta \sum_{i=1}^{N}[a_N(i) - \overline{a}_N]^2\right\} \tag{6.13}$$

entails (6.5) with $\overline{S}_N$ given by (6.12) and $\tau = (\tau_x, \tau_y)$.

Proof. The proof may be based on Theorem 6.1 if we identify

$$c_{Ni} = a_N(i, \tau_y).$$

Since, under (6.13),

$$\frac{\max_{1\le i\le N}[a_N(i, \tau_y) - \overline{a}_N]^2}{\sum_{i=1}^{N}[a_N(i, \tau_y) - \overline{a}_N]^2} \le \frac{1}{\eta} \frac{\max_{1\le i\le N}[a_N(i) - \overline{a}_N]^2}{\sum_{i=1}^{N}[a_N(i) - \overline{a}_N]^2} \to 0, \tag{6.14}$$

we may apply Theorem 6.1. □

REFERENCES

Hájek, J. (1961). Some extensions of the Wald–Wolfowitz–Noether theorem. *Ann. Math. Statist. 32*, 506–523.

Hájek, J. (1969). *A Course in Nonparametric Statistics.* Holden–Day, San Francisco.

Hájek, J. and Šidák, Z. (1967). *Theory of Rank Test*. Academia, Prague and Academic Press, London.
Jurečková, J. (1969). Asymptotic linearity of a rank statistic. *Ann. Math. Statist. 40*, 1889–1900.
Lehmann, E. (1966). Some concepts of dependence. *Ann. Math. Statist. 37*, 1137–1153.
Raghavachari, M. (1965). The two-sample scale problem when locations are unknown. *Ann. Math. Statist. 36*, 1236–1242.
Vorlíčková, D. (1970). Asymptotic properties of rank tests under discrete distributions. *Z. Wahrscheinlichkeitstheorie verw. Geb. 14*, 275–289.

DISCUSSION ON HÁJEK'S PAPER W. HOEFFDING

Instead of elaborating on Professor Hájek's interesting and valuable paper, I will, in the spirit of the title of his paper, make a few methodological remarks on the problem of the asymptotic distribution of a linear rank statistic.

Let $X_1, \ldots, X_n$ be independent random variables with respective continuous distribution functions $F_1, \ldots, F_n$, and let $R_1, \ldots, R_n$ be the corresponding ranks. Consider the linear rank statistic

$$T_n = \sum_1^n c_i \, \phi\left(\frac{R_i}{n+1}\right),$$

where the c_i are constants and ϕ a function finite on $(0,1)$.

In his paper in *Ann. Math. Statist.* (1968), Professor Hájek proved the asymptotic normality of T_n under mild assumptions. He used his projection lemma to approximate T_n by a sum of independent random variables. With this approach the mean of the asymptotic distribution comes out as $\mathsf{E}\, T_n$. The following alternative approach is akin to von Mises' (1947) treatment of what he called differentiable statistical functions and also to the approach of Chernoff and Savage (1958).

Let $u(x) = 0$ or 1 according to whether $x < 0$ or $x \geq 0$;

$$S(x) = (n+1)^{-1} \sum_1^n u(x - X_j), \quad S^*(x) = \sum_1^n c_j \, u(x - X_j).$$

Then we can write

$$T_n = \int_{-\infty}^{\infty} \phi(S(x)) \, \mathrm{d}S^*(x).$$

Note that $\mathsf{E}\, S(x) = \overline{F}(x)$ and $\mathsf{E}\, S^*(x) = F^*(x)$, where

$$\overline{F}(x) = (n+1)^{-1} \sum_1^n F_j(x), \quad F^*(x) = \sum c_j \, F_j(x).$$

If ϕ is absolutely continuous inside $(0,1)$ and the integrals in what follows exist, we have

$$T_n - \int \phi(\overline{F})\,\mathrm{d}F^* = L_n + R_{1n} + R_{2n},$$

where

$$\begin{aligned} L_n &= \int \phi(\overline{F})\,\mathrm{d}(S^* - F^*) + \int \phi'(\overline{F})\,(S - \overline{F})\,\mathrm{d}F^*, \\ R_{1n} &= \int \{\phi(S) - \phi(\overline{F})\}\,\mathrm{d}(S^* - F^*), \\ R_{2n} &= \int \{\phi(S) - \phi(\overline{F}) - \phi'(\overline{F})\,(S - \overline{F})\}\,\mathrm{d}F^*. \end{aligned}$$

Here L_n is a sum of independent random variables with zero means and is almost identical with Hájek's projection approximation for T_n. Under the conditions of Theorem 2.1 of Hájek's (1968) paper, which include the assumption that ϕ has a bounded second derivative, it is rather straightforward to show that $\mathsf{E}\, R_{i,n}^2/\sigma_n^2 \to 0$ for $i = 1,\, 2$, where $\sigma_n^2 = \mathsf{var}\, L_n$. (Assume $\sum c_i = 0$; otherwise only a trivial modification is needed.) This provides an alternative proof of Hájek's Theorem 2.1. It is easy to show in this case that the centering constant $\mathsf{E}\, T_n$ of Hájek's theorem may be replaced by $\int \phi(\overline{F})\,\mathrm{d}F^*$.

However, under the conditions of Theorem 2.3 of Hájek's (1968) paper, which assumes much less about ϕ and somewhat more about the c_i, a direct proof of the asymptotic negligibility of $R_{1n} + R_{2n}$ seems to be much more difficult. Hájek's main tool is his powerful variance inequality.

REFERENCES

Chernoff, H. and Savage, I. R. (1958). Asymptotic normality and efficiency of certain nonparametric test statistics. *Ann. Math. Statist. 29*, 972–994.

Hájek, J. (1968). Asymptotic normality of simple linear rank statistics under alternatives. *Ann. Math. Statist. 39*, 325–346.

von Mises, R. (1947). On the asymptotic distribution of differentiable statistical functions. *Ann. Math. Statist. 18*, 309–348.

CHAPTER 31

A Characterization of Limiting Distributions of Regular Estimates

Z. Wahrscheinlichkeitstheorie und Verw. Gebiete 14 (1970), 323–330.

Reproduced by permission of Springer-Verlag GmbH & Co. KG.

Summary. We consider a sequence of estimates in a sequence of general estimation problems with a k-dimensional parameter. Under certain very general conditions we prove that the limiting distribution of the estimates, if properly normed, is a convolution of a certain normal distribution, which depends only on the underlying distributions, and of a further distribution, which depends on the choice of the estimate. As corollaries we obtain inequalities for asymptotic variance and for asymptotic probabilities of certain sets, thus generalizing some results of J. Wolfowitz (1965), S. Kaufman (1966), L. Schmetterer (1966) and G. G. Roussas (1968).

1 INTRODUCTION

In several papers there were established asymptotic lower bounds for variances and for probabilities of some sets provided the estimates are regular enough to avoid superefficiency. See Rao (1963), Bahadur (1964), Wolfowitz (1965), Kaufman (1966), Schmetterer (1966) and Roussas (1968), for example. In this chapter we obtain some of their results as corollaries of a representation theorem describing the class of all possible limiting laws. The present result refers to a general sequence of experiments with general norming matrices K_n. Our condition (2.2) below is implied by conditions imposed on the families of

Collected Works of Jaroslav Hájek – With Commentary
Edited by M. Hušková, R. Beran and V. Dupač
Published in 1998 by John Wiley & Sons Ltd.

distributions in the above mentioned papers. The same is true about regularity condition (2.3) concerning the limiting distribution of estimates. A comparison with Kaufman (1966) is given in Remark 1 below, and the same comparison could be made with Wolfowitz (1965). Schmetterer's (1966) condition, namely that the distribution law of normed estimates converges continuously in θ, also entails (2.3). See also Remark 2 for Bahadur's (1964) condition, which is weaker than (2.3).

Roughly speaking, condition (2.2) means that the likelihood functions may be locally asymptotically approximed by a family of normal densities differing in location only. Similarly, condition (2.3) expresses that the family of distributions of an estimate T_n may be approximated, in a local asymptotic sense, by a family of distributions differing again in location only.

The idea of the proof is based on considering the parameter as a random vector that has uniform distribution over certain cubes. In this respect the spirit of the proof is Bayesian, similarly to that in Weiss and Wolfowitz (1967). The mathematical technique of the present paper is borrowed from Le Cam (1960).

2 THE THEOREM

Consider a sequence of statistical problems $(\mathcal{X}_n, \mathcal{A}_n, \mathsf{P}_n(\cdot, \theta))$, $n \geq 1$, where θ runs through an open subset Θ of R^k. The nature of sets $\mathcal{X}_n$ may be completely arbitrary, and the only connection between individual statistical problems will be given by conditions (2.2) and (2.3). In particular, the nth problem may deal with n observations which may be independent, or may form a Markov chain, or whatever. A point from $\mathcal{X}_n$ will be denoted by x_n, and X_n will denote the abstract random variable defined by the identity function $X_n(x_n) = x_n$, $x_n \in \mathcal{X}_n$. The norm of a point $h = (h_1, \ldots, h_k)$ from R^k will be defined by $|h| = \max_{1 \leq i \leq k} |h_i|$; $h \leq v$ will mean that $h_i \leq v_i$, $1 \leq i \leq k$, and $h' v$ will denote the scalar product of two vectors h, $v \in R^k$.

Take a point $t \in \Theta$ and assume it is the true value of the parameter θ. We shall abbreviate $\mathsf{P}_n = \mathsf{P}_n(\cdot, t)$ and $\mathsf{P}_{nh} = \mathsf{P}_n(\cdot, t + K_n^{-1} h)$ where K_n^{-1} are inverses of some norming regular $(k \times k)$-matrices. Most usually the theorem below is applicable with K_n diagonal and having $n^{\frac{1}{2}}$ for its diagonal elements; then simply $K_n^{-1} h = n^{-\frac{1}{2}} h$. We shall need $\{K_n^{-1}\}$ to be a sequence of contracting transforms in the sense that for every $\varepsilon > 0$ and $a > 0$ there exists an n_0 such that $|h| < a$ entails $|K_n^{-1} h| < \varepsilon$ for all $n > n_0$. This property will be ensured by

$$\lim_{n \to \infty} (\text{the minimal eigenvalue of } K_n' K_n) = \infty, \tag{2.1}$$

where K_n' denotes the transpose of K_n. Since Θ is open, (2.1) ensures that for all $h \in R^k$ the points $t + K_n^{-1} h$ belong to Θ if n is sufficiently large. If

$\Theta = R^k$, the theorem below makes sense and is true without condition (2.1).

Given two probability measures P and Q, denote by $\mathrm{d}\mathsf{Q}/\mathrm{d}\mathsf{P}$ the Radon–Nikodym derivative of the *absolutely continuous part* of Q with respect to P. Thus, generally, we do not obtain $\int_A (\mathrm{d}\mathsf{Q}/\mathrm{d}\mathsf{P})\,\mathrm{d}\mathsf{P} = \mathsf{Q}(A)$, but only $\leq \mathsf{Q}(A)$. Introduce the family of likelihood ratios

$$r_n(h, x_n) = \frac{\mathrm{d}\mathsf{P}_{nh}}{\mathrm{d}\mathsf{P}_n}(x_n), \quad h \in R^k,\ n \geq n_h,\ x_n \in \mathcal{X}_n,$$

where n_h denotes the smallest n_0 such that $n \geq n_0$ entails $t + K_n^{-1}\,h \in \Theta$. In what follows the argument x_n will usually be omitted in order to stress the other arguments.

The distribution law of a random vector $Y_n = Y_n(x_n)$ under P_n will be denoted by $\mathcal{L}(Y_n \,|\, \mathsf{P}_n)$ and its weak convergence to L by $\mathcal{L}(Y_n \,|\, \mathsf{P}_n) \to L$. The k-dimensional normal distribution with zero expectation vector and covariance matrix Γ will be denoted by $\Phi(\cdot \,|\, \Gamma)$, and the corresponding law by $\mathcal{N}(0, \Gamma)$. The expectation with respect to P_n will be denoted by E_n, i. e. $\mathsf{E}_n(\cdot) = \int (\cdot)\,\mathrm{d}\mathsf{P}_n$.

Let $T_n = T_n(x_n)$ be a sequence of estimates of θ, i. e. for every n T_n is a measurable transform from $\mathcal{X}_n$ to R^k. (We shall say that the estimates T_n are regular if they satisfy condition (2.3) below.) Now we are prepared to formulate the following theorem.

Theorem. *In the above notation, let us assume that* (2.1) *holds and that*

$$r_n(h) = \exp\left\{h'\,\Delta_n - \frac{1}{2}\,h'\,\Gamma\,h + Z_n(h)\right\}, \quad h \in R^k,\ n \leq n_h, \tag{2.2}$$

where $\mathcal{N}(\Delta_n \,|\, \mathsf{P}_n) \to \mathcal{N}(0, \Gamma)$ *and* $Z_n(h) \to 0$ *in* P_n*-probability for every* $h \in R^k$. *Further, consider a sequence of estimates satisfying*

$$\mathsf{P}_{nh}(K_n(T_n - t) - h \leq v) \to L(v), \quad \text{for every } h \in R^k, \tag{2.3}$$

in continuity points of some distribution function $L(v)$, $v \in R^k$.

Then, if the matrix Γ *is regular, we have*

$$L(v) = \int \Phi\left(v - u \,|\, \Gamma^{-1}\right)\,\mathrm{d}G(u), \tag{2.4}$$

where $G(u)$ *is a certain distribution function in* R^k.

Remark. The dependence of Δ_n, Γ and $Z_n(h)$ on t, and of Δ_n and $Z_n(h)$ on x_n, is suppressed in our notation.

Proof. By Lemma 1 below we may also write

$$r_n(h) = \exp\left\{h'\,\Delta_n^* - \frac{1}{2}\,h'\,\Gamma\,h + Z_n^*(h)\right\}, \tag{2.5}$$

where the properties $\mathcal{L}(\Delta_n^* \mid \mathsf{P}_n) \to \mathcal{N}(0,\Gamma)$ and $Z_n^*(h) \to 0$ in P_n-probability are preserved, and, in addition

$$B_{nh} = \left[\mathsf{E}_n \exp\left(h'\,\Delta_n^* - \frac{1}{2}\,h'\,\Gamma\,h\right)\right]^{-1} \tag{2.6}$$

satisfies

$$\sup_{|h|<j} |B_{nh} - 1| \to 0 \quad \text{for every } j > 0, \text{ as } n \to \infty. \tag{2.7}$$

Put

$$r_n^*(h) = B_{nh} \exp\left(h'\,\Delta_n^* - \frac{1}{2}\,h'\,\Gamma\,h\right). \tag{2.8}$$

For our assumptions and from (2.7) it follows that $\mathcal{N}\left(-\frac{1}{2}\,h'\,\Gamma\,h,\, h'\,\Gamma\,h\right)$ is a limit for $\mathcal{L}(\log r_n(h) \mid \mathsf{P}_n)$ as well as for $\mathcal{L}(\log r_n^*(h) \mid \mathsf{P}_n)$. Let Y be a random variable such that $\mathcal{L}(Y) = \mathcal{N}\left(-\frac{1}{2}\,h'\,\Gamma\,h,\, h'\,\Gamma\,h\right)$. Then $\mathsf{E}\,e^Y = 1$. Further, by (2.8) and (2.6),

$$\mathsf{E}_n\, r_n^*(h) = 1.$$

On the other hand, since $r_n(h)$ is the Radon–Nikodym derivative of the absolutely continuous part of P_{nh} relative to P_n,

$$\limsup_n \mathsf{E}_n\, r_n(h) \le 1.$$

Since $r_n(h)$ is nonnegative and convergent in distribution to e^Y, we also have

$$\liminf_n \mathsf{E}\, r_n(h) \ge e^y = 1.$$

Consequently

$$\lim_{n\to\infty} \mathsf{E}_n\, r_n(h) = 1.$$

Since, furthermore, $[r_n(h) - r_n^*(h)] \to 0$ in P_n-probability, Lemma 2 below may be applied to the effect that

$$\int |r_n(h) - r_n^*(h)| \mathrm{d}\mathsf{P}_n \to 0, \quad h \in R^k. \tag{2.9}$$

Put

$$\mathsf{Q}_{nh}(A) = \int_A r_n^*(h)\,\mathrm{d}\mathsf{P}_n, \quad A \in \mathcal{A}_n. \tag{2.10}$$

Then (2.3) and (2.9) entail that

$$\mathsf{Q}_{nh}\left(K_n(T_n - t) - h \le v\right) \to L(v). \tag{2.11}$$

From (2.8) it follows that $\mathsf{Q}_{nh}(k_n(T_n - t) - h \le v)$ is a measurable function of h for every $v \in R^k$.

Let $\lambda_j(dh)$ be the uniform distribution over the cube $|h| \leq j$. It is easy to see that (2.11) entails for every natural j

$$\overline{L}_{nj}(v) \overset{\mathrm{Df}}{=} \int \mathsf{Q}_{nh}\left(K_n(T_n - t) - h \leq v\right) \mathrm{d}\lambda_j(h) \to L(v). \tag{2.12}$$

By Le Cam's lemma (Le Cam (1960) or Hájek–Šidák (1967), VI.1.4) we deduce

$$\mathcal{L}\left(\Gamma^{-1}\,\Delta_n^* \,|\, \mathsf{Q}_{nh}\right) \to \mathcal{N}(h,\,\Gamma^{-1}). \tag{2.13}$$

Put $a_j = (-j, \ldots, -j)$, $b_j = (j, \ldots, j)$ and $c_j = \left(\sqrt{j}, \ldots, \sqrt{j}\right)$. Let U be some random k-vector having a normal distribution with zero expectation and covariance matrix Γ^{-1}, i.e. $\mathcal{L}(U) = \mathcal{N}(0, \Gamma^{-1})$. Then, by (2.13),

$$\begin{aligned} &\int \mathsf{Q}_{nh}\left(a_j + c_j \leq \Gamma^{-1}\,\Delta_n^* \leq b_j - c_j\right) \mathrm{d}\lambda_j(h) \\ &\to \int \mathsf{P}\left(a_j + c_j \leq U + h \leq b_j - c_j\right) \mathrm{d}\lambda_j(h) \\ &\geq \int \mathsf{P}\left(a_j + 2c_j \leq h \leq b_j - 2c_j\right) \mathrm{d}\lambda_j(h) - \mathsf{P}\left(|U| > \sqrt{j}\right) \\ &= \left(1 - 2/\sqrt{j}\right)^k - \mathsf{P}\left(|U| > \sqrt{j}\right). \end{aligned} \tag{2.14}$$

For every natural j denote by m_j some integer such that

$$\sup_{|h|<j} |B_{nh} - 1| < \frac{1}{j}, \qquad n > m_j, \tag{2.15}$$

$$\rho(\overline{L}_{nj}, L) < \frac{1}{j}, \qquad n > m_j \tag{2.16}$$

(where $\overline{L}_{nj}$ is given by (2.12) and ρ denotes the Lèvy distance), and such that

$$\begin{aligned} &\int \mathsf{Q}_{nh}\left(a_j + c_j \leq \Gamma^{-1}\,\Delta_n^* \leq b_j - c_j\right) \mathrm{d}\lambda_j(h) \\ &> \left(1 - 2/\sqrt{j}\right)^k - \mathsf{P}\left(|U| > \sqrt{j}\right) - 1/j, \quad n > m_j. \end{aligned} \tag{2.17}$$

All this may be satisfied in view of (2.7), (2.12) and (2.14). We may assume that $m_1 < m_2 < \ldots$. Let $j(n)$ be defined by

$$m_{j(n)} \leq n < m_{j(n)+1},$$

and set

$$\overline{L}_n = \overline{L}_{nj(n)}, \quad \overline{\mathsf{Q}}_n = \int \mathsf{Q}_{nh}\,\mathrm{d}\lambda_{j(n)}(h). \tag{2.18}$$

From (2.16) it follows that

$$\overline{L}_n \to L. \tag{2.19}$$

On the other hand $\overline{L}_n(v)$ may also be interpreted as the probability of $\{K_n(T_n - t) - h \leq v\}$ if the joint distribution on (X_n, h) is given by the prior distribution $\lambda_{j(n)}$ and by the family Q_{nh} of conditional distributions of X_n given h. Denoting the posterior distribution function of h by $D_n(y \mid X_n)$, we may therefore write

$$\overline{L}_n(v) = \int [1 - D_n(K_n(T_n - t) - v - 0 \mid X_n)] \, \mathrm{d}\overline{\mathsf{Q}}_n. \tag{2.20}$$

Now, in view of (2.8), for $a_{j(n)} \leq n \leq b_{j(n)}$,

$$\begin{aligned} D_n(y \mid X_n) &= c(X_n) \int_{a_{j(n)}}^{y} B_{nh} \exp\left(h' \, \Delta_n^* - \frac{1}{2} h' \, \Gamma \, h\right) \mathrm{d}h \qquad (2.21) \\ &= c'(X_n) \int_{a_{j(n)}}^{y} B_{nh} \exp\left\{(h - \Gamma^{-1} \, \Delta_n^*)' \, \Gamma (h - \Gamma^{-1} \, \Delta_n^*)\right\} \mathrm{d}h, \end{aligned}$$

where $c(X_n)$ and $c'(X_n)$ are some constants depending on X_n only. Now, in view of (2.15),

$$\sup_{|h| < j(n)} |B_{nh} - 1| \to 1, \tag{2.22}$$

and in view of (2.17),

$$-\infty \leftarrow a_{j(n)} - \Gamma^{-1} \, \Delta_n^* \leq h - \Gamma^{-1} \, \Delta_n^* \leq b_{j(n)} - \Gamma^{-1} \, \Delta_n^* \to \infty, \tag{2.23}$$

where $-\infty = (-\infty, \ldots, -\infty)$, $\infty = (\infty, \ldots, \infty)$ and the convergence is in $\overline{\mathsf{Q}}_n$-probability. Consequently

$$D_n(y \mid X_n) \to \Phi\left(y - \Gamma^* \Delta_n^{-1} \mid \Gamma^{-1}\right) \tag{2.24}$$

in $\overline{\mathsf{Q}}_n$-probability, and, in turn

$$\left| \overline{L}_n(v) - \int \Phi(v - K_n(T_n - t) + \Gamma^{-1} \, \Delta_n^* \mid \Gamma^{-1}) \mathrm{d}\overline{\mathsf{Q}}_n \right| \to 0. \tag{2.25}$$

Denoting

$$G_n(u) = \overline{\mathsf{Q}}_n\left(K_n(T_n - t) - \Gamma^{-1} \, \Delta_n^* \leq u\right) \tag{2.26}$$

we may also write

$$\left| \overline{L}_n(v) - \int \Phi(v - u \mid \Gamma^{-1}) \, \mathrm{d}G_n(u) \right| \to 0. \tag{2.27}$$

Taking a subsequence $\{m\} \subset \{n\}$ such that $G_m \to G$, we obtain from (2.19) and (2.27)

$$L(v) \leftarrow \overline{L}_m(v) \to \int \Phi(v - u \mid \Gamma^{-1}) \, \mathrm{d}G(u).$$

Since $L(v)$ is a distribution function, $G(u)$ has to be a distribution function, too. This completes the proof. □

Lemma 1. *Let $\mathcal{L}(\Delta_n \,|\, \mathsf{P}_n) \to \mathcal{N}(0, \Gamma)$ hold. Then there exists a truncated version Δ_n^* of Δ_n such that $(\Delta_n - \Delta_n^*) \to 0$ in P_n-probability and*

$$\sup_{|h|<j} \left| \mathsf{E}_n \exp\left(h' \Delta_n^* - \frac{1}{2} h' \Gamma h \right) - 1 \right| \to 0, \quad j > 0. \tag{2.28}$$

Proof. For every natural i let us put

$$\begin{aligned} \Delta_{ni} &= \Delta_n, \quad \text{if } |\Delta_n| \le i, \\ &= 0, \quad \text{otherwise.} \end{aligned}$$

Let Y be some random vector such that $\mathcal{L}(Y) = \mathcal{N}(0, \Gamma)$ and let $Y_i = Y$, if $|Y| \le i$, and 0 otherwise. Let m_i be an integer such that

$$\sup_{|h|\le i} \left| \mathsf{E}_n \exp\left(h' \Delta_{ni} - \frac{1}{2} h' \Gamma h \right) - \mathsf{E} \exp\left(h' Y_i - \frac{1}{2} h' \Gamma h \right) \right| < \frac{1}{i}, \quad n > m_i.$$

Such an m_i exists, since $\mathcal{L}(\Delta_{ni} \,|\, \mathsf{P}_n) \to \mathcal{L}(Y_i)$ and the distributions are concentrated on a compact subset on which the system of functions $\exp(h' y - \frac{1}{2} h \Gamma h)$ is compact in the supremum metric. Note that further

$$\lim_{i\to\infty} \sup_{|h|<j} \left| \mathsf{E} \exp\left(h' Y_i - \frac{1}{2} h' \Gamma h \right) - 1 \right| = 0.$$

Assume that $m_1 < m_2 < \ldots$, define $i(n)$ by $m_{i(n)} \le n < m_{i(n)+1}$, and put $\Delta_n^* = \Delta_{ni(n)}$. Then the conclusion easily follows. □

Lemma 2. *Let $\{(U_n, V_n), n \ge 1\}$ be a sequence of pairs of nonnegative random variables such that*

(a) *U_n converges in distribution to U, $\mathsf{E}U < \infty$.*
(b) *$(U_n - V_n) \to 0$ in probability.*
(c) *$\mathsf{E}_n U_n \to \mathsf{E}U$, $\mathsf{E}_n V_n \to \mathsf{E}U$, where E_n and E denote expectations.*

Then

$$\mathsf{E}_n |U_n - V_n| \to 0 \quad \text{as } n \to \infty. \tag{2.29}$$

Proof. The proof follows from Loève (1963), Theorem 11.4.A. □

3 COROLLARIES

Corollary 1. *Let C be a convex symmetric set in R^k, and let Y be a random variable such that $\mathcal{L}(Y) = \mathcal{N}(0, \Gamma^{-1})$. Then, if the assumptions of the above theorem are satisfied for some $t \in \Theta$, we have for T_n from the same theorem*

$$\limsup_{n\to\infty} \mathsf{P}_n[K_n(T_n - t) \in C \,|\, t] \le \mathsf{P}(Y \in C). \tag{3.1}$$

Proof. Since the boundary of C has zero Lebesgue measure and since the limiting law of $K_n(T_n - t)$ is absolutely continuous, in view of (2.4), we have

$$\lim_{n\to\infty} \mathsf{P}_n[K_n(T_n - t) \in C \,|\, t] = \mathsf{P}(Z \in C),$$

where $\mathcal{L}(Z) = L$. Furthermore, (2.4) with the well-known lemma of Anderson (1955) entails

$$\mathsf{P}(Z \in C) \le \mathsf{P}(Y \in C).$$

Remark 1. Corollary 1 provides the essence of the main result by Kaufman (1966). His conditions, namely that the distribution of $K_n(T_n - \theta)$ converges uniformly in $R^k \times C$, where C is any compact subset of Θ, and his Lemma 5.1 entail our assumption (2.3). His regularity conditions for densities entail assumption (2.2). Contrary to Kaufman, we do not claim that the maximum likelihood estimate is after norming asymptotically normal $\mathcal{N}(0, \Gamma^{-1})$. For this conclusion we would need some global conditions concerning the distributions.

Under Le Cam's DN conditions in Le Cam (1960), there is an estimate, not necessarily the maximum likelihood one, which is after norming asymptotically normal $\mathcal{N}(0, \Gamma^{-1})$.

Remark 2. The theorem also entails

$$\liminf_{n\to\infty} \mathsf{E}_n \left(h' K_n(T_n - t)\right)^2 \ge h' \Gamma^{-1} h. \tag{3.2}$$

If we assume that $K_n(T_n - \theta)$ is asymptotically normal with zero variance, then by Bahadur (1964) we could derive (3.2) from weaker assumptions than (2.3). Actually, then (2.3) may be replaced by

$$\mathsf{P}_{nh} \left(h' K_n(T_n - t) \le h' h\right) \to \frac{1}{2}, \quad h \in R^k. \tag{3.3}$$

Remark 3. We have introduced K_n instead of $n^{\frac{1}{2}} I$ in order to cover such instances as

$$p_n(\theta) = \prod_{i=1}^{n} f\left(x_i - \sum_{j=1}^{k} \theta_j \, c_{nij}\right), \tag{3.4}$$

where we may put $K_n = \{\sum_{i=1}^n c_{nij}\, c_{nij'}\}_{j,j'=1}^k$. Alternatively, we may take for K_n a diagonal matrix coinciding on the diagonal with the previous one. Instances of this character occur in the asymptotic theory of linear regression.

Corollary 2. *Put $A_v = \{y : a'\, y \le v\}$, where $a \in R^k$ and $v \in R$. Assume that*

$$\int_{A_v} \mathrm{d}L(y) = \int_{A_{v^*}} \mathrm{d}\Phi(y \,|\, \Gamma^{-1}). \tag{3.5}$$

Then, under the assumptions of the above theorem,

$$\int_{A_{v+s} \div A_v} \mathrm{d}L(y) \le \int_{A_{v^*+s} \div A_{v^*}} \mathrm{d}\Phi(y \,|\, \Gamma^{-1}), \quad s \in R, \tag{3.6}$$

where $\div$ denotes the symmetric difference.

Proof. Consider two distributions of the pair $(Y, U) \in R^k \times R^k$, namely

$$\Phi(\cdot \,|\, \Gamma^{-1}) \times G \quad \text{and} \quad \Phi\left(\cdot + \lambda \Gamma^{-1} a \,|\, \Gamma^{-1}\right) \times G, \quad \lambda > 0.$$

Then (3.5) entails that the tests rejecting $\Phi(\cdot \,|\, \Gamma^{-1}) \times G$ if $a'Y \le v^*$ and $a'(Y + U) \le v$, respectively, have the same significance level, and (3.6) simply means that the former test has for $\lambda\, a'\, \Gamma^{-1}\, a = s > 0$ a larger power than the latter test, which easily follows from the Neyman–Pearson lemma. For $s < 0$ we proceed similarly.

Remark 4. Corollary 2 generalizes a result by Wolfowitz (1965) and Roussas (1968), if we note that

$$\mathsf{P}_n\left(a'\, K_n(T_n - t) \le v\right) \to \int_{A_v} \mathrm{d}L(v),$$

which follows again from the fact that the A_v has zero boundary.

Remark 5. If the loss incurred by T_n if θ is true equals $\ell(K_n(T_n - \theta))$, where $\ell(\cdot)$ may be represented as a mixture of indicators of complements of convex symmetric sets, then Corollary 1 entails that

$$\liminf_{n\to\infty} \mathsf{E}_n\, \ell_n[K_n(T_n - t)] \ge \mathsf{E}\, \ell(Y),$$

where $\mathcal{L}(Y) = N(0, \Gamma^{-1})$. In this sense $\mathcal{N}(0, \Gamma^{-1})$ may be regarded as a best possible limiting distribution. From (2.26) it is apparent that $\mathcal{L}(K_n(T_n - t) \,|\, \mathsf{P}_n) \to \mathcal{N}(0, \Gamma^{-1})$ if and only if $[K_n(T_n - t) - \Gamma^{-1}\, \Delta_n^*] \to 0$ in $\overline{\mathsf{Q}}_n$-probability.

REFERENCES

Anderson, T. W. (1955). The integral of a symmetric unimodal function. *Proc. Amer. Math. Soc. 6*, 170–176.

Bahadur, R. R. (1964). A note on Fisher's bound for asymptotic variance. *Ann. Math. Statist. 35*, 1545–1552.

Hájek, J. and Šidák, Z. (1967). *Theory of Rank Tests.* Academia, Prague and Academic Press, London.

Kaufman, S. (1966). Asymptotic efficiency of the maximum likelihood estimator. *Ann. Inst. Statist. Math. 18*, 155–178.

Le Cam, L. (1960). *Locally Asymptotically Normal Families of Distributions.* Univ. California Publ. Statist.

Loève, M. (1963). *Probability Theory.* 3rd ed. D. van Nostrand, Princeton – New Jersey – Toronto – New York – London.

Rao, C. R. (1963). Criteria of estimation in large samples. *Sankhyā 25a*, 189–206.

Roussas, G. G. (1968). Some applications of the asymptotic distribution of likelihood functions to the asymptotic efficiency of estimates. *Z. Wahrscheinlichkeitstheorie verw. Geb. 10*, 252–260.

Schmetterer, L. (1966). On asymptotic efficiency of estimates. *Research Papers in Statistics* (ed: F. N. David), Wiley, New York.

Weiss, L. and Wolfowitz, J. (1966). Generalized maximum likelihood estimators. *Teor. Veroyatn. Primen. 11*, 68–93.

Weiss, L. and Wolfowitz, J. (1967). Maximum probability estimators. *Ann. Inst. Statist. Math. 19*, 103–206.

Wolfowitz, J. (1965). Asymptotic efficiency of the maximum likelihood estimator. *Teor. Veroyatn. Primen. 10*, 267–281.

CHAPTER 32

Limiting Properties of Likelihoods and Inference

In: Foundations of Statistical Inference, 1971, (V. P. Godambe and D. A. Sprott, eds.), 142–162, Holt, Rinehart and Winston, Toronto.
Reproduced by permission of V.P. Godambe and D.A. Sprott

1 INTRODUCTION

This chapter supports to some extent the idea that the likelihood principle is applicable within a wide area. The sphere of applicability does not depend on the form of the sample space but on the form of likelihood functions (likelihoods, for short). If the likelihoods are exactly normal, then being given a likelihood function we may draw inference ignoring the statistical model in the background. This has been pointed out in Hájek (1968) and is resumed in Section 2. The main purpose of the present paper is to investigate how far the background model may be ignored if the likelihoods are only approximately normal. The investigation will be done in the framework of asymptotic theory, whose basic features the chapter discusses.

Although my initial aim was to explore only principles and to illustrate the points by examples, I could not resist answering some technical problems which occurred to me in the course of the writing. Thus the chapter contains some new propositions, which are fully stated, but given without proofs. The proofs will be published elsewhere.

My negative attitude towards the likelihood principle, if advocated as a *general* principle, will not be extended here, because the chapter is long enough already. The last section contains, however, a few remarks directed against a strict use of the likelihood principle even in cases where it is asymptotically justifiable.

Collected Works of Jaroslav Hájek – With Commentary
Edited by M. Hušková, R. Beran and V. Dupač
Published in 1998 by John Wiley & Sons Ltd.

1.1 Notation

The distribution law of random variable Y, if θ is the true parameter and value, will be denoted by $L(Y \mid \theta)$. The normal distribution with expectation vector a and covariance matrix Γ will be denoted by $N(a, \Gamma)$. The fact that a probability measure P has probability density p with respect to some dominating measure μ will be denoted by $\mathrm{d}\mathsf{P} = p\,\mathrm{d}\mu$. A randomization random variable, usually uniformly distributed over $(0,1)$, will be denoted by U. If $A - \Gamma$ is a positive semidefinite matrix, we shall write $A \geq \Gamma$. Given a one-dimensional density f, f' will denote its derivative, F the corresponding distribution function, and F^{-1} will be the inverse of F. For $x = (\nu_1, \ldots, \nu_k)$, $\|x\|$ will denote the usual norm $\left(\sum_{i=1}^k \nu_i^2\right)^{1/2}$. The k-dimensional Euclidean space will be denoted by R^k. $\mathsf{E}_{n\theta}$ will denote the expectation referring to the probability distribution $\mathsf{P}_{n\theta}$. For $x = (\nu_1, \ldots, \nu_k)$ and $y = (u_1, \ldots, u_k)$, $x'y$ will denote the scalar product $\sum_{i=1}^k \nu_i u_i$, and xy' will denote the matrix $\{\nu_i u_j\}_{i,j=1}^k$.

2 EXACTLY NORMAL LIKELIHOODS

Let $\{p_\theta,\, \theta \in R^k\}$ be a family of densities relative to a σ-finite measure on an arbitrary sample space (X, A). We shall say that the family has *exactly normal likelihood functions* if there exist a positive definite $(k \times k)$-matrix Γ, a function $c(x)$ and a vector statistic $T(x) = (T_1(x), \ldots, T_k(x))$ such that for all $x \in X$

$$p_\theta(x) = c(x) \exp\left[-\frac{1}{2}(T(x) - \theta)'\,\Gamma(T(x) - \theta)\right].$$

The statistic T will be called the *centering statistic.*

The following conclusions can be made for experiments with exactly normal likelihoods.

Proposition 1.

1. $T(x)$ is sufficient.
2. $L(T \mid \theta) = N(\theta, \Gamma^{-1})$.
3. In any decision problem to any randomized decision function $d(x, U)$ there exists another decision function $d^*(T(x), U)$ providing the same risks.
4. If for some estimate $\hat\theta$ the relation

$$\mathsf{E}[(\hat\theta - \theta)(\hat\theta - \theta)' \mid \theta] \leq \Gamma^{-1} \quad \text{for all } \theta \in R^k$$

holds, then $\hat\theta = T$.

5. The distribution of T given θ may be established from any single likelihood function.
6. For a uniform prior we have $L(\theta \mid x) = N(T(x), \Gamma^{-1})$.

Proof. Property 1 is obvious. Property 2 has been proved for $k = 1$ in Hájek (1968) on the basis of completeness of the normal family; the generalization to $k > 1$ presents no new problems. Property 3 follows from 1. Property 4 follows from the proof of admissibility of the usual estimator for $k = 1$. If 4 did not hold, then for some $(c_1, \ldots, c_k) \in R^k$ the estimate $\sum_{i=1}^{k} c_i T_i$ would not be admissible for $\sum_{i=1}^{k} c_i \theta_i$ under the usual mean square deviation risk. Property 5 simply means that any likelihood provides the information about Γ. Property 6 is obvious. □

The main topic of the present chapter is to discuss asymptotic versions of properties 1–6 if the likelihoods are normal only in an asymptotic sense.

3 ASYMPTOTICALLY NORMAL LIKELIHOODS

Consider a sequence of families of densities $\{p_n(x_n, \theta), \theta \in \Theta\}$ relative σ-finite measures μ_n on arbitrary sample spaces (X_n, A_n), $n = 1, 2, \ldots$. Assume that Θ is an open subset of R^k. Represent the densities in the form

$$p_n(x_n, \theta) = c_n(x_n) \exp\left\{-\frac{1}{2}\, n(T_n - \theta)'\, \Gamma_\theta (T_n - \theta) + Z_n(x_n, \theta)\right\}$$

where $T_n(x_n) = (T_{n1}(x_n), \ldots, T_{nk}(x_n))$ are some *centering* statistics, Γ_θ are $(k \times k)$-matrices, and $Z_n(x_n, \theta)$ are *remainders*.

By saying that the likelihoods are asymptotically normal, we understand that the following assumptions are satisfied.

Assumptions A. The following holds uniformly on every compact subset K of Θ:

A1. T_n is a square root consistent estimate of θ.

A2. The remainders $Z_n(x_n, \theta)$ are continuous in θ for every x_n, and, for every $a > 0$,

$$\max_t \left[|Z_n(x_n, t)| : \sqrt{n}|t - \theta| \leq a\right] \to 0 \text{ in } \mathsf{P}_{n\theta}\text{-probability}$$

where $\mathrm{d}\mathsf{P}_{n\theta} = p_n(x_n, \theta)\, \mathrm{d}\mu_n$.

A3. In $\Gamma_\theta = \{\gamma_{ij}(\theta)\}_{i,j=1}^{k}$ the functions $\gamma_{ij}(\theta)$ are continuous in θ and $\det \Gamma_\theta \geq \epsilon_k > 0$.

If the uniformity holds over all Θ, the Assumption A will be denoted by A*.

The spirit of the above assumptions is the same as that of Le Cam's (1960) definition of differentially asymptotically normal families: we do not restrict ourselves to n independent replications of a basic experiment and we also do not derive the existence of statistics T_n from some more elementary assumptions. It may be proved, in a non-trivial way, that Assumptions A are somewhat stronger than Le Cam's conditions DN1 – DN7 (see Le Cam, 1960, page 57) which are entailed if we put $\Delta_n(\theta) = \sqrt{n}(T_n - \theta)$ and $\mu(\theta) = 0$. Among other things it means that $\sqrt{n}(T_n - \theta)$ is asymptotically normal with parameters $(0, \Gamma_\theta^{-1})$, which will be formulated as a proposition in Section 5. On the other hand, Assumptions A are considerably simpler and, besides uniformities on compact subsets, the only part which is stronger than DN1 – DN7 is A2, where we assume the convergence of maxima instead of random variables $Z_n(x_n, t_n)$ for $\sqrt{n}(t_n - \theta) \to h$.

To summarize: Assumptions DN1 – DN7 are supremely economical, but drawing consequences from them is a very delicate matter; Assumptions A provide a simpler theory and have the advantage that no explicit assumption of normality of any random variables is made.

4 ASYMPTOTIC SUFFICIENCY

Following Le Cam, we say that the centering statistics T_n are asymptotically sufficient on compacts for the families $\{p_n(\cdot, \theta),\, \theta \in \Theta\}$, if they are sufficient for some families $\{q_n(\cdot, \theta),\, \theta \in \Theta\}$ such that for any compact $K \subset \Theta$

$$\sup_{\theta \in K} \int |p_n(x_n, \theta) - q_n(x_n, \theta)| \, \mathrm{d}\mu_n(x_n) \to 0 \quad \text{as } n \to \infty. \tag{4.1}$$

If (4.1) holds for K replaced by Θ, we say that T_n is asymptotically sufficient.

Asymptotic sufficiency of T_n under Assumptions A may be easily proved by the method of Le Cam (1956, p. 142). Choose a sequence of positive numbers $b_n \to \infty$ and $\epsilon_n \to 0$ and put

$$\chi_n(x_n) = \begin{cases} 1 & \text{if } \max_t \{|Z_n(x_n, t)| : \sqrt{n}\,|T_n(x_n) - t| < 2b_n\} < \epsilon_n \\ 0 & \text{otherwise, or if the above set of } t\text{-values is empty} \end{cases} \tag{4.2}$$

and

$$\nu_n(T_n - \theta) = \begin{cases} 1 & \text{if } \sqrt{n}\,|T_n - \theta| < b_n \\ 0 & \text{otherwise.} \end{cases} \tag{4.3}$$

It follows easily from Assumptions A that ϵ_n and b_n may be chosen so that we have

$$\mathsf{P}_{n\theta}\left(\chi_n(x_n)\, \nu_n(T_n(x_n) - \theta) = 1\right) \to 1 \tag{4.4}$$

uniformly on any compact $K \subset \Theta$. Then we put

$$q_n(x_n, \theta) = k_n(\theta)\, c_n(x_n)\, \chi_n(x_n)\, \nu_n(T_n - \theta) \exp\left\{-\frac{1}{2}\, n(T_n - \theta)'\, \Gamma_\theta(T_n - \theta)\right\} \tag{4.5}$$

where the functions $k_n(\theta)$ are introduced so as to obtain

$$\int q_n(x_n, \theta)\, \mathrm{d}\mu_n = 1, \quad \theta \in \Theta.$$

Proposition 2. Under Assumptions A the densities $q_n(x_n, \theta)$ of (4.5) satisfy (4.1) for a choice of numbers ϵ_n and b_n in (4.2) and (4.3). Furthermore, then

$$\sup_{\theta \in K} |k_n(\theta) - 1| \to 0 \quad \text{as } n \to \infty \tag{4.6}$$

for any compact $K \subset \Theta$.

Since T_n is evidently sufficient for $\{q_n(x_n, \theta),\ \theta \in \Theta\}$, it is asymptotically sufficient on compacts for $\{p_n(x_n, \theta),\ \theta \in \Theta\}$.

In view of the presence of the functions $k_n(\theta)$ and $\nu_n(T_n - \theta)$ the likelihoods of the densities $q_n(x_n, \theta)$ are only approximately normal. There are, however, a few cases when the likelihoods in $q_n(x_n, \theta)$ can be made exactly normal. To obtain such a result, we must obviously have $\Theta = R^k$.

Example 1. Put $x_n = (\nu_1, \ldots, \nu_n) \in R^n$, $\Theta = R$ and

$$p_n(x_n, \theta) = \prod_{i=1}^{n} f(\nu_i - \theta) \tag{4.7}$$

and assume that f has a finite Fisher's information I_f. Introducing $\phi(u) = -f'(F^{-1}(u))/f(F^{-1}(u))$, $0 < u < 1$, is means that

$$I_f = \int_0^1 \phi^2(u)\, \mathrm{d}u < \infty. \tag{4.8}$$

Then Assumptions A* are satisfied and there exist statistics $c_n^*(x_n)$ and $T_n(x_n)$ such that the densities

$$q_n(x_n, \theta) = c_n^*(x_n) \exp\left\{-\frac{1}{2}\, n(T_n - \theta)^2\, I_f\right\} \tag{4.9}$$

satisfy

$$\sup_{-\infty < \theta < \infty} \int \cdots \int |q_n(x_n, \theta) - p_n(x_n, \theta)|\, \mathrm{d}\nu_1 \cdots \mathrm{d}\nu_n \to 0. \tag{4.10}$$

In fact the integrals may be made constant in θ.

The centering statistic T_n may be obtained as *restricted* maximum likelihood estimates over the regions $\{\theta : |\theta - \hat{\theta}_n| \leq 1\}$, where $\hat{\theta}_n$ is some consistent location invariant estimate (see (Le Cam, 1969)). If

$$\int_{-\infty}^{\infty} f(x) \log f(x) \, dx > -\infty, \tag{4.11}$$

then T_n may be the maximum likelihood estimate over the whole space $-\infty < \theta < \infty$. If (4.11) does not hold, the behavior of the unrestricted maximum likelihood estimate is not known. Assuming that $\phi = \phi_1 - \phi_2$ where both ϕ_1 and ϕ_2 are nondecreasing and square integrable, which covers all reasonable cases, then we may take for T_n any estimate defined by

$$T_n = \tilde{\theta} - I_f^{-1} \, n^{-1} \sum_{i=1}^{n} \frac{f'(\nu_i - \tilde{\theta})}{f(\nu_i - \tilde{\theta})}, \tag{4.12}$$

where $\tilde{\theta}$ is an arbitrary location invariant and square root consistent estimate. For example, we may take for θ a properly shifted sample α-quantile for any α such that $f(F^{-1}(\alpha)) > 0$.

Since the likelihoods are *exactly* normal under q_n of (4.9), we have from Proposition 1, part 2,

$$L\left(\sqrt{n}(T_n - \theta) \,|\, q_{n\theta}\right) = N(0, \, I_f^{-1}). \tag{4.13}$$

In view of (4.10), it entails

$$L\left(\sqrt{n}(T_n - \theta) \,|\, p_{n\theta}\right) \to N(0, \, I_f^{-1}). \tag{4.14}$$

The cases when the approximating family $\{q_n\}$ can be made to have exactly normal likelihoods are relatively rare. In some instances we may get the result for a suitably transformed parameter.

Example 2. Let us consider the sequence of gamma densities:

$$p_n(x, \theta) = \theta^{-n}[\Gamma(n)]^{-1} x^{n-1} \exp(-x/\theta), \quad x > 0, \; \theta > 0. \tag{4.15}$$

Transforming the parameter by

$$\tau = \log \theta \tag{4.16}$$

we can approximate the family by

$$q_n(x, \theta) = c_n(x) \exp[-n(\log x - \tau)^2], \tag{4.17}$$

which has exactly normal likelihoods in terms of τ. We again have

$$\sup_{0<\theta<\infty} \int_0^{\infty} |p_n(x, \theta) - q_n(x, \theta)| \, dx \to 0 \quad \text{as } n \to \infty. \tag{4.18}$$

The reader recognizes in (4.16) the well-known variance-stabilizing transform. However, not every transform of this type provides an approximation family with exactly normal likelihoods. In the binomial and Poisson families, for example, the transformed parameter space does not coincide with R. In the normal correlation family again appropriate group structural properties are missing.

There are situations where the parameter space as well as its dimension may depend on n. This may be illustrated by the following.

Example 3. Fix some $b > 0$ and put

$$\Theta_n = \left\{ \theta_n = (d_1, \ldots, d_n) \in R^n : \sum_{i=1}^{n} d_i = 0, \sum_{i=1}^{n} d_i^2 \leq b \right\}. \tag{4.19}$$

Then take a density f with finite Fisher information and, denoting $x_n = (\nu_1, \ldots, \nu_n) \in R^n$, consider the families

$$p_n(x_n, \theta_n) = \prod_{i=1}^{n} f(\nu_i - d_i), \quad \theta_n \in \Theta_n. \tag{4.20}$$

Introducing the vector of ranks $R_n = (R_{n1}, \ldots, R_{nn})$, where $R_{ni}(x_n)$ denotes the number of coordinates of x_n that are smaller than or equal to ν_i, it may be derived (Hájek and Šidák 1967, VII 1.2) that the families may be approximated, uniformly on Θ_n, by families

$$q_n(x_n, \theta_n) = c_n(x_n) \, B_n(\theta_n) \exp\left\{ -\frac{1}{2} \sum_{i=1}^{n} [d_i - I_f^{-1} \, a_n(R_{ni})]^2 \right\},$$

where $a_n(i) = \mathsf{E}\,\phi(U_n^{(i)})$, with $\phi(u) = -f'(F^{-1}(u))/f(F^{-1}(u))$ and $U_n^{(i)}$ denoting the ith order statistic in a sample of size n from the uniform distribution over $(0, 1)$.

Of course, since the approximation of p_n by q_n holds uniformly on θ_n, it holds uniformly on any subsets of Θ_n that may be appropriate for applications. Since the vector of ranks R_n is evidently sufficient for the family $\{q_n(x_n, \theta_n), \theta_n \in \Theta_n\}$, it follows that it is asymptotically sufficient for $\{p_n(x_n, \theta_n), \theta_n \in \Theta_n\}$.

5 ASYMPTOTIC NORMALITY OF CENTERING STATISTICS

It is possible to prove the following.

Proposition 3. Under Assumptions A

$$L\left(\sqrt{n}(T_n - \theta) \,|\, \theta\right) \to N(0,\, \Gamma^{-1})$$

holds uniformly on compacts $K \subset \Theta$.

The proposition shows that asymptotic normality may be proved from square root consistency and certain properties of likelihood functions. The usefulness of this fact is exhibited by the following.

Example 4. Let $X_n = (Y_1, \ldots, Y_n)$ be a normal stationary autoregressive sequence of order one with zero expectations. Putting $\theta = (a, b)$ and $\Theta = \{(a, b) : a > 0,\ |b| < 1\}$, we obtain the following family of densities:

$$\begin{aligned} p_n(x_n, \theta) &= (2\pi)^{-n/2}\, a^n (1 - b^2)^{1/2} \\ &\quad \times \exp\left\{ -\frac{1}{2}\, a^2 (b - \hat{b}_n)^2 \sum_{i=2}^{n-1} y_i^2 - \frac{1}{2}\, a^2 \left(\sum_{i=1}^{n} y_i^2 - \hat{b}^2 \sum_{i=2}^{n-1} y_i^2 \right) \right\} \end{aligned} \tag{5.1}$$

where

$$\hat{b}_n = \frac{\sum_{i=1}^{n-1} y_i\, y_{i+1}}{\sum_{i=2}^{n-1} y_i^2} \quad \text{and} \quad x_n = (y_1, \ldots, y_n). \tag{5.2}$$

Now we may prove without difficulty that $\hat{b}$ is a square root consistent estimate of b uniformly on compacts $K \subset \Theta$, and $\frac{1}{n} \sum_{i=2}^{n-1} y_i^2$ is a square root consistent estimate of $a^{-2}(1 - b^2)^{-1}$. Consequently, we may write

$$\begin{aligned} p_n(x_n, \theta) &= c_n(x_n)\, a^n \\ &\quad \times \exp\left\{ -\frac{1}{2} n (b - \hat{b}_n)^2 / (1 - b^2) - \frac{1}{2}\, a^2 (1 - \hat{b}_n^2) \sum_{i=1}^{n} y_i^2 + Z_n(x_n, \theta) \right\}. \end{aligned} \tag{5.3}$$

Fixing a, we can see that Assumptions A are satisfied for $T_n = \hat{b}_n$ and $\Gamma_b = (1 - b^2)^{-1}$. Consequently whatever may be a, $\hat{b}_n$ is asymptotically normal $(b,\, n^{-1}(1 - b^2))$.

6 ASYMPTOTIC TESTING OF CONTIGUOUS ALTERNATIVES

Fix a simple hypothesis θ_0 and introduce local coordinates λ as follows:

$$\theta = \theta_0 + n^{-1/2} B\lambda, \tag{6.1}$$

where B is a matrix satisfying

$$B' B = \Gamma_{\theta_0}^{-1}. \tag{6.2}$$

Now choose a bounded subset L of R^k and consider alternatives K_n consisting of θ-points that correspond to $\lambda \in L$:

$$K_n = \theta_0 + n^{-1/2} BL, \quad n = 1, 2, \ldots. \tag{6.3}$$

We shall call a sequence of critical functions Φ_n asymptotically most powerful (in the maximin sense) on level α for testing θ_0 against K_n if

$$\limsup_{n\to\infty} \int \Phi_n(x_n)\, p_n(x_n, \theta_0)\, \mathrm{d}\mu_n \leq \alpha \tag{6.4}$$

and the minimum power

$$\beta_n = \inf_{\theta\in K_n} \int \Phi_n(x_n)\, p_n(x_n, \theta)\, \mathrm{d}\mu_n \tag{6.5}$$

has the largest possible limit.

For families satisfying Assumptions A the solution may be given in three steps:

1. We prove that Φ_n may depend on x_n through T_n.
2. We suggest a critical function which gives the best solution for testing $N(0, I)$ against the family $\{N(\lambda, I),\ \lambda \in L\}$, where I denotes the identity covariance matrix.
3. We prove that there is no better solution.

Step 1 depends on the sufficiency of T_n. Given any sequence of critical functions Φ_n let us put

$$\Phi_n^*(t) = \mathsf{E}\left[\Phi(x_n) \,|\, T_n = t,\ q_{n\theta} \text{ holds}\right]. \tag{6.6}$$

Since T_n is sufficient for $\{q_{n\theta}\}$, $\Phi_n^*(t)$ will be independent of θ. From (6.6) it follows that

$$\int \Phi_n^*(T_n(x_n))\, q_n(x_n, \theta)\, \mathrm{d}\mu_n = \int \Phi_n(x_n)\, q_n(x_n, \theta)\, \mathrm{d}\mu_n, \quad \theta \in \Theta.$$

This combined with (4.1) entails that

$$\int [\Phi_n(x_n) - \Phi^*(T_n(x_n))]\, p_n(x_n, \theta)\, \mathrm{d}\mu_n \to 0 \tag{6.7}$$

uniformly on compact subsets of Θ. Since there obviously is a compact set containing all the K_ns, the first step is proved by (6.7).

Step 2 is based on the fact that

$$L\left(n^{1/2} B^{-1}(T_n - \theta_0) \,|\, \theta\right) \to N(\lambda, I) \tag{6.8}$$

uniformly for $||\lambda|| < M < \infty$, if λ is related to θ, and B to Γ_{θ_0}, by (6.1) and (6.2). Statement (6.8) follows easily from Proposition 3 and continuity of Γ_θ. Now, if $\Phi^0(t)$ is the best critical function for testing $N(0, I)$ against $\{N(\lambda, I),\ \lambda \in L\}$, we obtain from (6.8)

$$\sup_{||\lambda||<M} \left| \int \phi^0 \left[n^{1/2} B^{-1}(T_n - \theta_0)\right] \mathrm{dP}_{n\theta} - \int \phi^0(t)\, (2\pi)^{-k/2} \right. \tag{6.9}$$
$$\left. \times \exp\left[-\frac{1}{2}\sum_{i=1}^{k}(t_i - \lambda_i)^2\right] \mathrm{d}t_1 \cdots \mathrm{d}t_k \right| \to 0$$

provided that $\Phi^0(t)$ is continuous almost everywhere (Lebesgue measure). This will be true for all reasonable alternatives L. However, even a completely general Φ^0 would be tractable, since it may be proved that there exist random variables Y_n of the form

$$Y_n = h_n(T_n, U), \tag{6.10}$$

where h_n are some functions, and U has uniform distribution over $(0,1)$ independently of T_n, such that $L(Y_n \,|\, \mathrm{P}_{n\theta})$ converges to $N(\lambda, I)$ in variation uniformly for $||\lambda|| < M$. Then (6.9) holds with $\Phi^0\left[n^{-1/2} B^{-1}(T_n - \theta_0)\right]$ replaced by $\int_0^1 \Phi^0[h_n(T_n, u)]\, \mathrm{d}u$. Now, if

$$\mathsf{E}[\Phi^0(T) \,|\, N(0,1)] \le \alpha, \quad \inf_{\lambda \in L} \mathsf{E}[\Phi^0(T) \,|\, N(\lambda, I)] = \beta, \tag{6.11}$$

then (6.5) entails

$$\limsup_{n\to\infty} \int \Phi^0 \left[n^{1/2} B^{-1}(T_n - \theta_0)\right] \mathrm{dP}_{n\theta_0} \le \alpha \tag{6.12}$$

$$\lim_{n\to\infty} \left[\inf_{\theta \in K_n} \int \Phi^0 \left[n^{1/2} B^{-1}(T_n - \theta_0)\right] \mathrm{dP}_{n\theta_0}\right] = \beta. \tag{6.13}$$

Example 5. Choose $c > 0$ and put $L = \{\lambda : ||\lambda||^2 = c\}$. The corresponding set in the Θ space will be, in view of (6.1) and (6.2),

$$K_n = \{\theta : n(\theta - \theta_0)' \Gamma_{\theta_0}(\theta - \theta_0) = c\}. \tag{6.14}$$

Now it is well known that the best critical function Φ^0 for testing $N(0, I)$ against $\{N(\lambda, I), \lambda \in L\}$ equals 1 if

$$||t||^2 = \sum_{i=1}^{n} t_i^2 \geq k_\alpha \tag{6.15}$$

and equals 0 otherwise. Since the Lebesgue measure of the set $||t||^2 = k_\alpha$, on which Φ^0 is discontinuous, equals 0, the above theory applies and the test which rejects θ_0 in favor of K_n if and only if

$$\left\| n^{-1/2} B^{-1}(T_n - \theta_0) \right\|^2 = n(T_n - \theta_0)' \Gamma_{\theta_0}(T_n - \theta_0) \geq k_\alpha \tag{6.16}$$

has the same asymptotic significance level and asymptotic minimal power as Φ^0 in testing $N(0, I)$ against $\{N(\lambda, I), ||\lambda||^2 = c\}$.

In order to complete step 3 it does not suffice to know that the distribution of T_n is asymptotically normal. We shall illustrate that by the following.

Example 6. Let ξ_n be a binomial random variable with parameters n and $p = 1/2$. Consider two distributions of T_n defined by means of ξ_n as follows:

$$\mathsf{P}_{n0}(T_n = k) = \mathsf{P}(\xi_n = k) \qquad k = 0, 1, \ldots, n, \tag{6.17}$$

$$\mathsf{P}_{n1}\left(T_n = k + \left[\sqrt{n}\right] + \frac{1}{2}\right) = \mathsf{P}(\xi_n = k) \qquad k = 0, 1, \ldots, n, \tag{6.18}$$

where $\lfloor\sqrt{n}\rfloor$ denotes the integer part of $\sqrt{n}$. Then, obviously, for every n the distributions P_{n0} and P_{n1} are disjoint and may be distinguished with no error (i. e. $\alpha = 0$, $\beta = 1$) by any critical function such that

$$\Phi(k) = 0, \quad \Phi\left(k + \frac{1}{2}\right) = 1, \quad \text{if } k \text{ is an integer.} \tag{6.19}$$

On the other hand the limiting distributions

$$L\left[n^{-1/2}\left(T_n - \frac{1}{2}n\right) \mid \mathsf{P}_{n0}\right] \to N\left(0, \frac{1}{4}\right) \tag{6.20}$$

$$L\left[n^{-1/2}\left(T_n - \frac{1}{2}n\right) \mid \mathsf{P}_{n1}\right] \to N\left(1, \frac{1}{4}\right) \tag{6.21}$$

lack this property and the sum of errors of both kinds cannot be smaller than a certain positive number.

Fortunately, in our case we know in addition that the likelihoods are approximately normal. From this we are able to derive, in a nontrivial way, that for every sequence of test functions $\Phi_n(x_n)$ there exist a subsequence Φ_{n_j} and a test function $\psi(t)$ in R^k such that

$$\int \Phi_{n_j}(x_n)\, p_{n_j}(x_n,\theta)\, \mathrm{d}\mu_n \to \int \psi(t)\,(2\pi)^{-k/2} \exp\left[-\frac{1}{2}\sum_{i=1}^{k}(t_i-\lambda_i)^2\right] \mathrm{d}t_1 \cdots \mathrm{d}t_k \tag{6.22}$$

uniformly for $||\lambda|| < M$.

All three steps yield the following.

Proposition 4. Under Assumptions A the best test for θ^0 against K_n of (6.3), where L is bounded, has in the limit the same power as the best test for testing $N(0,I)$ against $\{N(\lambda,I),\, \lambda \in L\}$.

Historical notes. Asymptotic hypothesis testing theory was founded by A. Wald (1943) in his very long and difficult paper. He considered n independent replications of a basic experiment and used assumptions that entail Assumptions A* with T_n being the maximum likelihood estimate. Using the *paving* method he proved that for any event $A_n \in \mathcal{A}_n$ there exists a Borel subset $B = B(A_n)$ of R^k such that

$$\left| \mathsf{P}_{n\theta}(A_n) - \int_{B(A_n)} (2\pi)^{-k/2} |\det \Gamma_\theta|^{1/2} \times \exp\left[-\frac{1}{2}\, n(t-\theta)'\, \Gamma_\theta (t-\theta)\right] \mathrm{d}t_1 \cdots \mathrm{d}t_k \right| < \epsilon$$

holds for $n > n_\epsilon$ uniformly in $A_n \in \mathcal{A}_n$ and $\theta \in \Theta$. From this Proposition 4 easily follows.

The most important subsequent development is due to L. Le Cam (1956, 1960). In his 1956 paper he relaxed Wald's conditions and replaced the maximum likelihood estimate by estimates obtainable from square-root consistent estimates by *improving*, as exemplified by (4.12) above. He proved the asymptotic sufficiency of centering statistics and then he established the existence of a function $\phi_n : R^k \to R^k$ such that $\sup_t |\phi_n(t) - t| \to 0$ as $n \to \infty$ and

$$\left| \mathsf{P}_{n\theta}(T_n \in B) - \int_{\phi_n^{-1}(B)} (2\pi)^{-k/2} |\det \Gamma_\theta|^{1/2} \times \exp\left[-\frac{1}{2}\, n(t-\theta)'\, \Gamma_\theta (t-\theta)\right] \mathrm{d}t_1 \cdots \mathrm{d}t_k \right| < \epsilon$$

for $n > n_\epsilon$ uniformly for Borel subsets B of R^k and θ belonging to compact subsets of K. This also yields Proposition 3 immediately. The point transforms ϕ_n established by Le Cam are more appropriate for estimating problems than the set transforms used by Wald. In his 1960 paper Le Cam elaborated consequences of the conditions of differential asymptotic normality DN1–DN7 mentioned in Section 3. Proposition 4 could also be derived under these extremely economical conditions.

In Hájek – Šidák (1967) asymptotic theory is developed for rank tests and an alternative approach to Proposition 4 is used.

In all four above sources not only are simple hypotheses considered, but also composite ones.

We shall conclude by explaining why we called the alternatives K_n contiguous in the title. Actually, Proposition 3 entails easily, by a Le Cam lemma (Hájek – Šidák 1967, VI. 1.2), that the sequences $\{\mathsf{P}_{n\theta_n}\}$, $\theta_n \in K_n$, are contiguous to the sequence $\{\mathsf{P}_{n\theta_0}\}$. Contiguity implies that convergence in $\mathsf{P}_{n\theta_0}$-probability entails the convergence in $\mathsf{P}_{n\theta_n}$-probability. Thus for $\alpha = 0$ the power of the best test converges to 0, for example.

7 ASYMPTOTIC ESTIMATION

Having obtained an asymptotic result we are not usually able to tell how far it applies to particular cases with finite n. In other words, given $\epsilon > 0$ we are not able to decide whether some n of interest is or is not larger than n_ϵ, whose existence the limit theorem ensures. Theorems providing a satisfactory answer to this question would be much more modest and much less elegant. Consequently, in applications we are guided by two epistemologically very different sources of knowledge: (i) we have limit theorems giving some hope, but no assurance, of practical sample sizes; (ii) we work with some numerical experience, which we extend to cases that seem to us to be *similar*.

Especially misinformative can be those limit results that are not uniform. Then the limit may exhibit some features that are not even approximately true for any finite n.

Example 7. In Hájek – Šidák (1967) a test is constructed that is asymptotically optimal for testing regression in location whatever may be the underlying density with finite Fisher information. The convergence of the power, however, is not uniform with respect to the choice of density. Thus no matter that the limiting power is the best possible one for all the densities, for any finite n; the actual power will be quite poor for most densities under consideration. A theorem, where convergence is replaced by uniform convergence, would pertain to much more modest classes of densities.

Example 8. Superefficient estimates produced by L. J. Hodges (see Le Cam (1953)) have their amazing properties only in the limit. For any finite n they behave quite poorly for some parameter values. These values, however, depend on n and disappear in the limit.

Fisher's program in the asymptotic theory of estimation was to prove that the best possible limiting distribution of an estimate is (under regularity conditions) $N(\theta,\, n^{-1}\,\Gamma_\theta)$ and that this optimum result is reached only by estimates that are asymptotically equivalent to the maximum likelihood estimate. If we interpret this conjecture in the sense of generally nonuniform convergence we come to paradoxes of Hodges' examples, and the conjecture is easily disproved. However, to come closer to what Fisher might have had in mind we must introduce some uniformity considerations. The requirements either impose some restrictions on the class of estimates that are allowed to compete (see Wolfowitz (1965) or Hájek (1970), for example), or they give a stricter meaning to the words *best possible*; see Huber (1966). Wolfowitz, while defending the first approach, argues that the limiting distribution of any *operational* estimate should be estimable. We may ask, however, why we should exclude from competition estimates whose limiting distribution is equal to or better than some estimable distribution? On the other hand, in Huber's approach we do not exclude any estimates, and only require that favourable behaviour of one parameter point should not be depreciated by poor behaviour at neighboring points.

The next proposition has been obtained by a different method than used by Huber (1966).

Proposition 5. Let Θ be an open subset of the line, i. e. $k = 1$. Consider a nondecreasing function $\ell(y),\ y \geq 0$; introduce

$$r(\sigma^2) = (2\pi)^{-1/2}\,\sigma^{-1} \int_{-\infty}^{\infty} \ell(|y|) \exp\left\{-\frac{1}{2}\,y^2/\sigma^2\right\} \mathrm{d}y \tag{7.1}$$

and assume that $r(\sigma^2) < \infty$ for any $\sigma^2 > 0$. Fix $\theta \in \Theta$.

Then, under Assumptions A, any sequence of estimates $\hat\theta_n$ either satisfies

$$\liminf_{n\to\infty} \mathsf{E}_{n\theta}\, \ell\left(\sqrt{n}|\hat\theta_n - \theta|\right) \geq r(\Gamma_\theta^{-1}) \tag{7.2}$$

or there exists a sequence of parameter points θ_n such that for some constant $M < \infty$

$$\sqrt{n}\ |\theta_n - \theta| < M \tag{7.3}$$

holds and

$$\limsup_{n\to\infty} \mathsf{E}_{n\theta_n} \ell\left(\sqrt{n}|\hat\theta_n - \theta_n|\right) > r(\Gamma_\theta^{-1}). \tag{7.4}$$

Example 9. For $\ell(y) = y^2$ we obtain that either the Cramér–Rao bound is not exceeded, i. e.

$$\liminf_{n\to\infty} \mathsf{E}_{n\theta}[n(\hat{\theta}_n - \theta)^2] \geq \Gamma_\theta^{-1}, \tag{7.5}$$

or it is exceeded by some positive number, independent of n, in the neighborhood of θ for some ns of arbitrary magnitude. Putting $\ell_b(y) = \min(y^2, b)$, we obtain either

$$\lim_{b\to\infty} \liminf_{n\to\infty} \mathsf{E}_{n\theta}\ell_b\left(\sqrt{n}\,|\hat{\theta}_n - \theta|\right) \geq \Gamma_\theta^{-1} \tag{7.6}$$

or uniformly sharp inequalities for a subsequence of a sequence θ_n satisfying (7.3).

Also, if $L[\sqrt{n}(\hat{\theta}_n - \theta_n) \mid \theta_n] \to G$ for any sequence $\theta_n \to \theta$, then

$$\int y^2 \,\mathrm{d}G(y) \geq \Gamma_\theta^{-1}$$

and, generally,

$$\int \ell(|y|) \,\mathrm{d}G(y) \geq r(\Gamma_\theta^{-1}).$$

Example 10. Another interesting loss function is

$$\lambda_a(y) = \begin{cases} 0 & \text{if} \quad 0 \leq y \leq a \\ 1 & \text{if} \quad a < y < \infty. \end{cases}$$

If we apply Proposition 5 to this function, we obtain that either

$$\liminf_{n\to\infty} \mathsf{P}_{n\theta}\left(\sqrt{n}|\hat{\theta}_n - \theta| \geq a\right) \geq 2\Phi\left(-a\,\Gamma_\theta^{1/2}\right),$$

where $\Phi(x) = (2\pi)^{-1/2} \int_{-\infty}^{x} \exp\left(\ \frac{1}{2}\,y^2\right)\,\mathrm{d}y$, or that there is a uniformly sharp inequality for a subsequence of a sequence satisfying (7.3).

In order to obtain analogues of Proposition 5 for a multidimensional parameter space, we must consider the loss function also multidimensional (i. e. vector-valued or matrix-valued).

Example 11. If $\theta \in \Theta \subset R^k$, k arbitrary, then either

$$\liminf_{n\to\infty} \mathsf{E}_{n\theta}(\hat{\theta}_n - \theta)\,(\hat{\theta}_n - \theta)' \geq \Gamma_\theta^{-1}$$

or there exists a vector $c \in R^k$, a constant $M < \infty$, and a sequence θ_n such that $\sqrt{n}||\theta_n - \theta|| < M$, for which

$$\limsup_{n\to\infty} \mathsf{E}_{n\theta_n}\left[c'(\hat{\theta}_n - \theta_n)'\,(\hat{\theta}_n - \theta_n)\,c\right] > c'\,\Gamma_\theta^{-1}\,c.$$

As is known from nonasymptotic consideration (Stein (1956)), Proposition 5 (with $|\hat{\theta}_n - \theta|$ replaced by $||\theta_n - \theta||$) is not correct for $k > 2$. It may be correct for $k = 2$, but this has not yet been proved. The explanation of failure for $k > 2$ may be found in the fact that estimating a multidimensional parameter with a one-dimensional loss function includes as a special case repeated estimation of a one-dimensional parameter that varies from one experiment to another. If k is sufficiently large we may successfully estimate and utilize prior distribution of this parameter if we are interested in minimizing the average risk. Proposition 5 allows us to say that Fisher's program has been essentially sound.

8 ESTIMATION OF THE INFORMATION MATRIX

If we know the family $\{p_n(x_n, \theta)\}$ and a centering statistic we can estimate Γ_θ by simply substituting T_n for θ. This gives a consistent estimate, uniformly on compacts. If we are, however, adherents of the likelihood principle, we must seek for estimation methods that are independent of our knowledge of $\{p_n(x_n, \theta)\}$. One possibility is to estimate Γ_θ by

$$\left\{ -\frac{1}{n} \frac{\partial^2}{\partial\theta_i \, \partial\theta_j} \log p_n(x_n, \theta) \bigg|_{\theta = T_n \, \hat{\theta}_n} \right\}_{i,j=1}^{k}$$

where $\hat{\theta}_n$ is the maximum likelihood estimate.

To prove consistency of this estimate would require conditions much stronger than our Assumptions A (namely in the point A2), but these alternative conditions are often satisfied in practical models. A more general method is given in the following.

Proposition 6. Under Assumptions A there exists a sequence $b_n \to \infty$ such that the matrix

$$c_n(x_n) \int \cdots \int_{\sqrt{n}||t - T_n|| < b_n} (t - T_n)(t - T_n)' \, p_n(x_n, t) \, \mathrm{d}t_1 \cdots \mathrm{d}t_k,$$

where $t = (t_1, \ldots, t_k)$ and $c_n(x_n)$, defined by

$$c_n(x_n) \int \cdots \int_{\sqrt{n}||t - T_n|| < b_n} p_n(x_n, t) \, \mathrm{d}t_1 \cdots \mathrm{d}t_k = 1,$$

is a consistent estimate of Γ_θ^{-1}, uniformly on compact subsets of Θ.

The above proposition is not exactly in the likelihood principle spirit, since the speed of convergence of $b_n \to \infty$ and the choice of T_n depend on the sequence of families $\{p_n(x_n, \theta)\}$. However, it shows that methods based on the knowledge of likelihood functions only, with ignorance of the distribution family in the background, may often be successful.

9 BAYES ESTIMATES

Consistently using the likelihood principle, we also have to define the centering statistics as functionals on the likelihoods, without reference to the family of probability densities in the background. One way is to take for the centering statistic the maximum likelihood estimate. Of course, satisfaction of Assumptions A is not sufficient for proving even consistency of the maximum likelihood estimate. However, if we know that the ML estimate is square-root consistent, uniformly on compacts, then it is easy to show that it has all the properties of a centering statistic under A.

Another possibility is to take for the centering statistic the conditional expectation of θ given X_n with respect to some prior distribution, possibly depending on X_n.

Proposition 7. Under Assumptions A, there exists a sequence $b_n \to \infty$ such that

$$T_n^* = c_n(x_n) \int_{\sqrt{n}\|t-T_n\| \le b_n} \cdots \int t\, p_n(x_n, t)\, \mathrm{d}t_1 \cdots \mathrm{d}t_k \tag{9.1}$$

is also a centering statistic. Consequently, $L\,[\sqrt{n}(T_n^* - \theta) \,|\, \theta] \to N(0, \Gamma_\theta^{-1})$ as $n \to \infty$, uniformly on compacts.

Of course, T_n^* will be a proper likelihood principle estimate if T_n has the same property and b_n is defined independently of the underlying family of distributions. Both T_n and b_n may be removed from consideration if the integration in (9.1) could be extended over all Θ. A partial result of this kind is given in the following.

Example 12. Let $f(x)$ have a finite Fisher's information I_f and let there exist $\alpha > 0$ such that

$$\sup_{-\infty < x < \infty} f(x)\,|x|^\alpha < \infty. \tag{9.2}$$

Then, for the family considered in Example 1, the statistics

$$T_n^* = \left[\int_{-\infty}^{\infty}\left(\prod_{i=1}^{n} f(\nu_i - t)\right)\mathrm{d}t\right]^{-1}\int_{-\infty}^{\infty} t\prod_{i=1}^{n} f(\nu_i - t)\,\mathrm{d}t \tag{9.3}$$

are well defined for $n > 1+1/\alpha$ and have the properties of centering statistics. Consequently, $L\,(\sqrt{n}(T_n^* - \theta)\,|\,\theta) \to N(0,\, I_f^{-1})$ uniformly for $\theta \in R$.

For the Cauchy density, for example, (9.2) is satisfied for any $0 < \alpha < 1$.

For more general results see Bickel (1969).

10 LOCAL THEORY AND LARGE DEVIATIONS THEORY

The theory exposed in preceding sections has local character. Its results do not depend on the behavior of likelihoods at joint points (θ, x_n) such that $|T_n(x_n) - \theta| > \epsilon$, however positive ϵ may be. For this reason we were able to truncate the likelihoods as severely as was done in (4.5).

The local type of asymptotic theory is supported by the following fact: any two families of densities satisfying Assumptions A such that their likelihoods coincide over regions $|T_n(x_n) - \theta| < b_n/\sqrt{n}$, $b_n \to \infty$, are asymptotically undistinguishable in the sense that any test designed to separate them has the sum of errors of both kinds tending to 1. Therefore, it is desirable not to let the procedures depend on features that cannot be tested empirically.

However reasonable this consideration may sound, it has been recently shown that *tails* of the likelihoods are very well worth studying. This is being done under the name of large deviation theory. The parts of the structure of a sequence of density families $\{p_n(x_n, \theta)\}$ studied by each of these two theories are essentially independent: without changing the local results we can vary the tails of the likelihoods so as to obtain any large deviation results we wish, and the opposite is also true. Practically, however, for well-behaved families there is a smooth transition from large deviation results to local results. Combining both theories we may hope to get a better fit of the asymptotic results to finite sample sizes.

It is very urgent for the applications of large deviation results to know the model precisely. The same is true to a lesser extent with respect to local theory. If we are vague about the model we can be quite confused by the mess of mutually inconsistent recommendations. Theoretical answers to this kind of difficulty are robustness and uniform optimality.

11 CONCLUDING REMARKS

The above discussion shows that not much harm may be caused, at least for sufficiently large n, if we base our inferences on certain functionals of likelihood that are independent of the family of densities (models, for short) in the background. On the other hand, it is also true that in order to prove it we would need much stronger assumptions than our Assumptions A. Furthermore, it is notoriously known that maximum likelihood estimates and Bayes estimates are often very difficult to obtain explicitly. This is well illustrated by Example 4, where the maximum likelihood estimate for $\theta = (a, b)$ is a difficult function of observations. On the other hand, utilizing the knowledge of the model, we may modify the likelihoods [by neglecting the term $Z_n(x_n, \theta)$ in (5.3), e. g.] so as to obtain very simple asymptotically sufficient statistics as $\hat{b}_n$ of (5.2). In Example 3 again the analysis of the model provides asymptotic sufficiency of the vector of ranks – a kind of information, which is not apparent from the likelihood function. Also, if the likelihood functions are very wild, we can replace the observations X by a statistic $T(X)$ for which the likelihoods are smoothened enough to be a basis for inference. To do that, however, we again need to know the model.

For these and similar reasons, I believe that strict adherence to the likelihood principle would complicate and impoverish the statistical theory.

REFERENCES

Bickel, P. J. and Yahav, J. A. (1969). Some contributions to the asymptotic theory of Bayes solutions. *Z. Wahrscheinlichkeitstheorie Verw. Geb. 11*, 257–276.

Hájek, J. (1968). On basic concepts of statistics. *Proceedings of the 5th Berkeley Symposium on Math. Statist. and Probability, Vol. 1*, University of California Press, 139–162.

Hájek, J. (1970). A characterization of limiting distributions of regular estimates. *Z. Wahrscheinlichkeitstheorie Verw. Gebiete 14*, 323–330.

Hájek, J. and Šidák, Z. (1967). *Theory of Rank Tests.* Academic Press, London.

Huber, P. (1966). Strict efficiency excludes super efficiency. Unpublished. Abstract in *Annals of Mathematical Statistics 37*, 1425.

Le Cam, L. (1956). On the asymptotic theory of estimation and testing hypotheses. *Proc. 3rd Berkeley Symp. on Math. Statist. and Probability, Vol. 1*, University of California Press, 129–156.

Le Cam, L. (1960). Locally asymptotically normal families of distributions. *University of California Publications in Statistics, 3,(2)*, 37–98.

Le Cam, L. (1969). On the assumptions used to prove asymptotic normality of maximum likelihood estimates. Preprint.

Stein, C. (1956). Inadmissibility of the usual estimator of the mean of a multivariate normal distribution. *Proc. 3rd Berkeley Symposium on Math. Statist. and Probability*. University of California Press, 197–206.

Wald, A. (1943). Tests of statistical hypotheses concerning several parameters when the number of observations is large. *Trans. Amer. Math. Soc. 54*, 426–482.

Wolfowitz, J. (1965). Asymptotic efficiency of the maximum likelihood estimator. *Teorie Veroyatn. Primen. 10*, 267–281.

COMMENTS

V. P. Godambe:

Though I do not share the view often expressed that, asymptotically, answers to the questions are independent of the questions asked, I believe that a considerable portion of the asymptotics in statistics has no stronger relevance to the central problem of statistical inference. This is particularly borne out by the recent literature on nonparametric statistics which many times consists of meaningless investigations of the asymptotic properties of some arbitrarily chosen estimators or test criteria. (Refer to papers in *Ann. Math. Statist.* during the last few years.) I am glad to say that a happy exception to my foregoing remarks is Professor Hájek's work in this field.

To me it looks as though asymptotic investigation is necessitated occasionally by too narrow a formulation of statistical concepts. For instance, an optimum property of the maximum likelihood estimator obtained by carrying the criterion of unbiasedness and minimum variance to its logical ends (mostly ignored by statisticians) is as follows. Let X be any abstract sample space, Ω the parameter space (real interval) and for every $x \in X$ and $\theta \in \Omega$, $f(x \mid \theta)$ the frequency function. Then it can be very easily shown that under mild regularity conditions, of all the real functions $g(x, \theta)$ on $X \times \Omega$ such that $\mathsf{E}(g \mid \theta) = 0$ the variance $g\left(\mathsf{E}\left(\frac{\partial g}{\partial \theta}\middle|\theta\right)\right)^{-1}$ is minimized for every $\theta \in \Omega$, for $g = \partial \log f(x \mid \theta)/\partial\theta$ (Godambe, 1960, *Ann. Math. Statist.*). Surely this property of the m. l. estimator which has its appeal for physicists (see, for instance, Janossy's book, 1963, on statistics) and for anyone who has a feeling for statistical concepts has no appeal whatsoever to many mathematical statisticians who take on mathematical formalism of statistics without its conceptual statistical content. [Professor Barnard here added that the property of m. l. estimation referred to by Godambe was an important one which deserved more publicity than it had received.]

C. R. Rao:

Le Cam–Hájek formulation of the problem providing an asymptotically sufficient statistic seems to involve undue assumptions, and the method of finding such a statistic also seems to be somewhat involved. An alternative approach which I have advocated (see, for instance, the book, *Linear Statistical Inference and its Applications*) is as follows. If $L_n(x_n, \theta)$ denotes the log likelihood based on sample size n, then it is well known that

$$\frac{\mathrm{d}}{\mathrm{d}\theta} L_n(x_n, \theta) = S_n$$

has very important properties, especially when n is large, in drawing inference on θ. It is shown that under mild regularity conditions s_n is asymptotically equivalent to $T_n - \theta$ uniformly in compact intervals of θ, where T_n is the maximum likelihood estimator. This shows that for purposes of inference T_n is as good as the sample in large samples. The argument can be extended to many parameters and also to the case of dependent observations.

H. Rubin:

Professor Hájek's approach is inadequate for the decision-theoretic view-point. For example

$$\text{let } T_n(x) = \begin{cases} x_1 \text{ if } x_1 \text{ is in the largest } n^{1/2} \text{ observations,} \\ \overline{x} \text{ otherwise.} \end{cases}$$

If the loss ρ is approximately the squared error for small deviations and monotone, the risk of $\overline{x}$ is σ^2/n but that of T_n is greater than $n^{-1/2}\,\mathsf{E}(\rho(x_1 - \theta) \mid x_1$ is in the largest $n^{1/2}$ observations) which is asymptotically infinitely greater. However T_n satisfies Hájek's conditions.

Professor Hájek's discussion of large deviations, brief though it is, conveys the impression that all deviations of the type he has not considered can be called large. *Large* deviations is a technical term in the literature, as are *excessive*, and *moderate*. The appropriate term for this whole class is *extraordinary*.

REPLY

To Professor Rubin:

Proposition 5 gives a lower bound and generally does not claim that T_n satisfies it. If $\ell(y) = C|y|^\alpha$ then $\ell(\sqrt{n}|T_n - \theta|) = n^{\alpha/2\,\alpha}\ell(|T_n - \theta|)$, otherwise there may, of course, be a difference. The former choice of passing to the limit makes the theory more feasible.

To Professor Godambe:

As is well known, many asymptotic results have been tested numerically with great success, namely in the nonparametric field. As I stressed in my paper, in general, we do not have a more satisfactory substitute for them. Professor Godambe's suggestion about how to prove the 'optimality' of the maximum likelihood estimate for any finite n is, for me, not convincing enough. This estimate fails even to be admissible in many instances, for example, in estimating the location when the density is not normal.

To Professor Rao:

I welcome Professor Rao's comment. The following is an attempt to clarify the relation between two approaches.

On page 285 of his book, *Linear Statistical Inference*, Professor Rao introduces the concept of asymptotic efficiency of a sequence of estimates. Such estimates T_n must be square-root consistent and satisfy

$$\sqrt{n}\left|T_n - \theta - \beta(\theta)\,\frac{1}{n}\,\frac{\partial}{\partial\theta}\log p_n(x_n,\theta)\right| \to 0 \qquad (*)$$

in probability. This condition for T_n is very close to what in the language of my paper would be described as 'the densities have asymptotically normal likelihoods with T_n serving for the centering statistic'. Actually, if $\log p_n(x_n\theta)$ is differentiable, we have

$$p_n(x_n,\theta) = p_n(x_n,T_n)\exp\left\{-\frac{1}{2}\,n(T_n-\theta)^2\,\Gamma_n + \int_{T_n}^{\theta}\left[\frac{\partial}{\partial t}\log p_n(x_n,t) - n(T_n-t)\,\Gamma_t\right]dt\right\}$$

so that conditions A of my paper correspond roughly to $(*)$ with

$$c_n(x_n) = p_n(x_n,T_n), \quad \Gamma_\theta = [\beta(\theta)]^{-1}$$

and

$$Z_n(x_n,\theta) = \int_{T_n}^{\theta}\left[\frac{\partial}{\partial t}\log p_n(x_n,t) - n(T_n-t)\,\Gamma_t\right]dt.$$

Now since $|T_n-\theta|$ is of order $1/\sqrt{n}$, the smallness of $(*)$ and of $Z_n(x_n,\theta)$ (see A2) are related conditions, even if neither implies the other. In $(*)$ one stresses the closeness of $T_n-\theta$ to $\frac{1}{n}[\partial/\partial\theta]\log p_n(x_n,\theta)$;in Assumptions A I prefer to stress the approximately normal form of likelihoods, because just this fact is crucial for deriving the propositions of my paper. I do not see that Assumptions A would be much more 'undue' than the assumptions that there exists an estimate satisfying $(*)$. I also do not see how $(*)$ could be used to prove more refined theorems corresponding to Propositions 4 and 5 of my paper. Let me also reiterate that under $(*)$ we derive the asymptotic normality of T_n via the asymptotic normality of $[\partial/\partial\theta] \log p_n(x_n, \theta)$, whereas in Proposition 2 of my paper a different way not using the central limit theorem is used.

One source of misunderstanding may be the fact that the theory of my paper starts at the point where the theory in Professor Rao's book ends. Professor Rao, introducing traditional differentiability and uniformity conditions (page 299 of his book) proves that the maximum likelihood estimator, if consistent, satisfies $(*)$. In my paper, assuming that there exists an estimate T_n satisfying Assumptions A (which are a variant of $(*)$), we prove certain properties that are not even considered in Rao's book. True enough, the proofs of these properties are too difficult to be embodied in a textbook, but the results themselves could and should be included easily, and only these properties justify introduction of such terms as asymptotically efficient or sufficient.

Not being concerned with the construction of T_n from more basic assumptions, I do not see why Professor Rao mentions that methods of finding T_n are 'involved'. If he means that the maximum likelihood estimates are always easiest to find, there are plenty of examples showing the opposite. Let me refer to Example 4 of my paper in that connection.

CHAPTER 33

Local Asymptotic Minimax and Admissibility in Estimation

In: Proc. Sixth Berkeley Symp. Math. Statist. Prob. 1 (1972), 175–194, Univ. of California Press.

Reproduced by permission of Lucien Le Cam, editor.

1 INTRODUCTION

In their vigorous search for an adequate asymptotic theory of estimation, statisticians have tried almost all their methodological tools: prior densities, minimax, admissibility, large deviations, restricted classes of estimates (invariant, unbiased), contiguity, and so forth. The resulting body of knowledge is somewhat atomized and a certain synthetic work seems to be needed. In this section, let us try to single out several pieces of the jigsaw puzzle and combine them into a logically connected theory. Along with this we shall criticize some other approaches and make a few historical remarks.

Consider a fixed parametric space θ and a sequence of experiments described by families of densities $p_n(x_n, \theta)$, say $p_n(x_n, \theta) = \prod_{i=1}^{n} f(y_i, \theta)$, where $x_n = (y_1, \ldots, y_n)$. First of all, it is necessary to single out 'regular cases'. This should not be done only formally, for example, only in terms of θ derivatives of $f(y, \theta)$. The statistical essence of regularity consists in the possibility of replacing the family of distributions by a normal family in a local asymptotic sense. Loosely speaking, given a point $t \in \theta$ and a small vicinity V_t of t, the quantity

$$\Delta_{n,t} = n^{-1/2} \frac{\partial}{\partial \theta} \log p_n(x_n, \theta) | \theta = t \tag{1.1}$$

should be approximately sufficient and normal (with constant covariance and expectation linear in θ) for $\theta \in V_t$ and n large. See Section 3 for a precise definition. The idea of approximating a general family by a normal family

Collected Works of Jaroslav Hájek – With Commentary
Edited by M. Hušková, R. Beran and V. Dupač
Published in 1998 by John Wiley & Sons Ltd.

was first formulated by A. Wald (1943), and then sophisticatedly developed by L. Le Cam (1953, 1956, 1960). In spite of its importance, the idea has not yet found its way into current textbooks.

The next step is to get rid of ill behaved estimates and to characterize optimum ones. This may be achieved by scrutinizing an arbitrary sequence of estimates T_n from the point of view of minimax and admissibility, again in a *local* asymptotic sense. Theorem 4.1 below entails that there is a lower bound for asymptotic local maximum risk and that this bound may be achieved only if

$$\left[\sqrt{n}(T_n - t) - \frac{\Delta_{n,t}}{\Gamma_t}\right] \to 0, \quad [p_{n,t}], \tag{1.2}$$

where Γ_t does not depend on the observation nor on the choice of $\{T_n\}$.

Once the condition (1.2) has been established, we can develop particular methods providing estimates satisfying (1.2) for every $t \in \theta$. These methods would include Bayes estimates with respect to diffuse priors, maximum likelihood estimates, maximum probability estimates, RBAN estimates, 'nonparametric' estimates, and so forth. The justification for a diffuse prior in Bayes estimation – and probably the only sound one – is that the resulting estimates satisfy (1.2), and, under additional conditions, have satisfactory local asymptotic minimax properties. Minimax scrutiny of Bayes estimates seems to be even more important in cases when we estimate a scalar parametric function $\tau = \tau(\theta)$ of a parameter whose dimensionality increases with n. Then it may be far from obvious what a 'diffuse' prior means. Unfortunately local asymptotic minimax results are not systematically available for this case.

In maximum likelihood and Bayes estimation, we are troubled by the behavior of likelihood tails. This may be avoided by considering only some vicinity of a consistent estimate. This leads to consistent estimates considered as a starting point for arriving at better estimates.

If the families $p_n(\cdot, \theta)$ are well behaved, then $\mathcal{L}_\theta(\Delta_{n,\theta})$ will be smooth in θ. In turn, the distribution of any estimate T_n for which (1.2) is satisfied will be smooth in θ. Assuming that the laws $\mathcal{L}_\theta\,[\sqrt{n}(T_n - \theta)]$ converge continuously in θ to some laws L_θ, it was proved in Hájek (1970) that L_t may be decomposed as a convolution

$$L_t = \phi_t * G_t, \tag{1.3}$$

where ϕ_t is a normal distribution and G_t is a distribution depending on the choice of T_n. It also may be shown that the best possible G_t is degenerate at the point zero. If G_t is normal, the L_t is also normal with larger covariance matrix than ϕ_t. In such a case, the ratio of the generalized variances of ϕ_t and L_t may serve as a measure of the efficiency of T_n. A simple short proof of (1.3) has been recently suggested by P. Bickel (personal communication). Le Cam (1971) provided still another proof and extended the result to families which may not be locally asymptotically normal. Then, of course, ϕ_t will not be a normal distribution, but, for example, the one sided exponential distribution.

Some points in the above exposition need comment. First, let us explain what we understand by local asymptotic minimax and admissibility. Roughly speaking, by saying that $\{T_n\}$ is locally asymptotically minimax (admissible) we shall mean that for any open interval $V \subset \theta$ the estimate is approximately minimax (admissible) on V if n is large. Actually, for n large there is little justification for relating the minimax criterion and admissibility to the whole parametric space, since after having obtained our observations, we are able to locate the unknown parameter with considerable precision.

It is important to note that an estimate may be exactly minimax for every n, but may fail to be *locally* asymptotically minimax. For example, this is true about the minimax estimator of the binomial p from X successes in n trials:

$$t_n(X) = \frac{X + \sqrt{n}/2}{n + \sqrt{n}}. \tag{1.4}$$

This estimate is obviously not approximately minimax relative to any open interval V not containing $p = 1/2$, and consequently it is not locally asymptotically minimax. Actually, the maximum mean square error of t_n on V is $\left[4n(1 + n^{-1/2})^2\right]^{-1}$, whereas the minimax on V is smaller than $n^{-1} \max_{p \in V}\{p(1-p)\}$.

Also admissibility for all n does not entail local asymptotic admissibility. Estimate (1.4) may illustrate this point as well. Another example may be given from the field of sample surveys, where Joshi and Godambe (see Joshi (1965)) proved that the sample average is an admissible estimate for the population average no matter what the sample design is. Again this estimate fails to be locally asymptotically admissible if the concept is properly defined for this situation.

According to Chernoff (1956) the idea of local asymptotic minimax is due to C. Stein and H. Rubin, who showed that Fisher's programs could be rescued by it. The proof that local asymptotic minimax implies local asymptotic admissibility was first given by Le Cam (1953, Theorem 14). There, he proved that superefficiency at one point entails bad risk values in the vicinity of this point, that is, that superefficiency excludes the local asymptotic minimax property. Apparently not many people have studied Le Cam's paper so far as to read this very last theorem, and the present author is indebted to Professor Le Cam for giving him the reference. Theorems 4.1 and 4.2 below may be regarded as extensions of the above mentioned result by Le Cam and a related theorem by Huber. Huber (1966), in addition to Le Cam's statement, proves that a locally asymptotically minimax estimate must be asymptotically normal with a given variance. This is made more precise by (1.2), since (1.2) entails the asymptotic normality of $\sqrt{n}(T_n - t)$ on the basis of the assumed asymptotic normality of $\Delta_{n,t}$. From the methodological point of view, Le Cam's and the present paper both use the Blyth (1951) approach to admissibility based on a normal prior, whereas Huber used the Hodges–

Lehmann (1950) approach based on the Cramér–Rao inequality.

In his subsequent papers (1956, 1960) Le Cam perfected the idea of a locally asymptotically normal family. We shall base our proof on these papers rather than on his 1953 paper, which contains some omissions. When approximating a general family by a normal one, the basic point is that the distance should be expressed in terms of the L_1 norm (variation) given by (3.5) below. The Lèvy or Prohorov distance would not do in proving (4.8) below. Of course, we may not expect that, for example, the distribution of $\Delta_{n,t}$ approaches its normal limit in the L_1 norm. However, as was shown by Le Cam (1956) and as is proved here under different assumptions in Lemmas 3.2 and 3.3, it approaches in the L_1 norm a slightly deformed normal law with slightly deformed likelihood function, which is sufficient for our purposes. Another important point provided in Le Cam (1960) was a general definition of locally asymptotically normal families, which is not restricted to partial sequences in an infinite sequence of independent replications of some basic experiment. We take this attitude also, and our local asymptotic normality (LAN) conditions of Section 3 represent a selection from Le Cam's (1960) conditions DN1–7.

Another comment is invited by (1.2) in connection with the approach suggested by C. R. Rao (1965, p. 285) in his admirable book. He uses property (1.2) directly as a definition of asymptotic efficiency, which seems to be justified by the fact that the ratio of Fisher's information for T_n and for the whole observation, respectively, approaches unity under (1.2) and some additional assumptions. However, since Fisher's information is invariant under one-to-one transformations, I doubt whether giving it a central position in estimation problems can be rationally explained, even if used jointly with consistency. For example, in estimating the location parameter from n independent observations if the density is otherwise known, the maximum likelihood estimator contains all of the Fisher information (the apparently lost part of it may be 'recovered'), but it is not the best estimate, not even the best location invariant estimate. The extent to which the maximum likelihood estimate lags behind the best one depends on how much the likelihoods are irregularly shaped, asymmetric, for example. Theorem 4.1 below provides a minimax justification for (1.2), which is somewhat less mystical.

To be more precise, Rao calls an estimate efficient, if (1.2) holds with Γ_t^{-1} replaced by any function β_t not involving the observations, but possibly depending on the estimate T_n. However, Theorem 4.1 implies that T_n cannot be locally asymptotically minimax if $\beta_t \neq \Gamma_t^{-1}$. Rao is aware of bad consequences of $\beta_t \neq \Gamma_t^{-1}$ and proves in Rao (1963) that then the law of $\sqrt{n}(T_n - \theta)$ does not converge uniformly in θ, so that the limiting laws cannot be used to provide approximate confidence intervals.

Continuing our comments on approaches not embodied in the above exposition, let us mention another paper by Le Cam (1958). There he proved that under certain conditions Bayes estimates provide asymptotic risks that can be improved on a set of measure zero only. Thus, if the prior is diffuse and the asymptotic risk of a Bayes estimate is continuous in θ, as it usually is, then it dominates the asymptotic risk of any other estimate having continuous asymptotic risk. A disadvantage of this approach is that it defines lower bounds in terms of certain estimates – Bayes estimates – and not in terms of the intrinsic properties of the families of distributions involved. Also, a verification of assumptions entails difficult consistency investigations. On the other hand, it is the first paper of a very general scope; for example, the estimates are not assumed to be necessarily asymptotically normal. In regular cases, on which the present paper is focused, Bayes estimates for diffuse priors will satisfy (1.2), so that we get a good agreement with our previous considerations.

Investigations of estimates which are not necessarily asymptotically normal were started from a different angle by J. Wolfowitz (1965). He again defined the best possible asymptotic behavior in terms of certain special estimates – generalized maximum likelihood estimates. In order to be comparable in his sense, the estimates must be 'regular', so that his approach cannot be applied to arbitrary estimates. However, for regular estimates we can under LAN conditions establish the decomposition (1.3), which provides the conclusions obtained by J. Wolfowitz more directly. As was shown by Le Cam (1971) similar decomposition as (1.3) may be obtained also for the 'nonregular case' investigated by Weiss and Wolfowitz (1966).

The property of being a generalized maximum likelihood estimate is implied by (1.2) and in the usual situation also the converse is true. The same holds about maximum probability estimators.

All of the above approaches are based on 'ordinary' deviations of the estimate from the parameter, which are typically of order $n^{-1/2}$. A concept of asymptotic efficiency based on large deviations has been suggested by D. Basu (1956) and R. R. Bahadur (1960, 1967). Fixing θ and some $\varepsilon > 0$, we define $\tau_n(\varepsilon, \theta)$ as the standard deviation of a normal distribution under which $|T_n - \theta| \geq \varepsilon$ would have the same probability as under $\mathsf{P}_{n,\theta}$. Formally

$$\mathsf{P}_\theta(|T_n - \theta| \geq \varepsilon) = 2\Phi\left[-\frac{\varepsilon}{\tau_n(\varepsilon, \theta)}\right]. \tag{1.5}$$

Now, under regularity conditions somewhat stronger than LAN below it may be proved that for any consistent estimate

$$\lim_{\varepsilon \to 0} \lim_{n \to \infty} \{n\, \tau_n^2(\varepsilon, \theta)\} \geq \Gamma_\theta^{-1} \tag{1.6}$$

and that the equality holds for the maximum likelihood estimator. It will generally hold also for estimates satisfying (1.2). The only trouble with this approach is that (1.2) is no longer necessary and that too many estimates satisfy equality in (1.6), for example, the 'superefficient' Hodges estimator, which must clearly be rejected by the minimax and ordinary deviations point of view. However, equality in (1.6) certainly may be used as an additional requirement for good estimates. Our main proposition is partially proved by a different method in Le Cam (1971).

2 NORMAL FAMILIES OF DISTRIBUTIONS

Consider the estimation of $\theta \in R$ by a random variable Z which has the normal distribution $N(\theta, 1)$ given θ. We shall assume that the loss function is $\ell(\hat{\theta} - \theta)$, where ℓ satisfies the following conditions:

$$\ell(y) = \ell(|y|), \tag{2.1}$$

$$\ell(y) \leq \ell(z) \quad |y| \leq |z|, \tag{2.2}$$

$$\int_{-\infty}^{\infty} \ell(y) \exp\left\{-\frac{1}{2}\lambda y^2\right\} dy < \infty \quad \lambda > 0, \tag{2.3}$$

$$\ell(0) = 0. \tag{2.4}$$

Condition (2.3) entails

$$\int_{-\infty}^{\infty} \ell(y)\, y^2 \exp\left\{-\frac{1}{2}\lambda y^2\right\} dy < \infty, \quad \lambda > 0. \tag{2.5}$$

We shall also introduce a truncated version of ℓ:

$$\ell_a(y) = \min(\ell(y),\, a), \quad 0 < a \leq +\infty, \tag{2.6}$$

$$r = (2\pi)^{-1/2} \int_{-\infty}^{\infty} \ell(y) \exp\left\{-\frac{1}{2} y^2\right\} dy, \tag{2.7}$$

$$r_a = (2\pi)^{-1/2} \int_{-\infty}^{\infty} \ell_a(y) \exp\left\{-\frac{1}{2} y^2\right\} dy, \tag{2.8}$$

and

$$r_a(b) = (2\pi)^{-1/2} \int_{-\sqrt{b}}^{\sqrt{b}} \ell_a(y) \exp\left\{-\frac{1}{2} y^2\right\} dy. \tag{2.9}$$

The estimator will be considered randomized, that is,

$$\hat{\theta} = \xi(Z, U) \tag{2.10}$$

where U is a randomized variable. For convenience, we shall assume that U is uniformly distributed on $(0,1)$. As always, U is independent of Z and θ. The introduction of randomized estimates is justified since our loss function $\ell(y)$ may not be convex.

The risk function corresponding to ℓ_a and ξ will be denoted as follows:

$$R_a(\theta; \xi) = (2\pi)^{-1/2} \int_{-\infty}^{\infty} \int_0^1 \ell_a[\xi(z,u) - \theta] \exp\left\{-\frac{1}{2}(z-\theta)^2\right\} du dz. \tag{2.11}$$

The following is an extension of a result of Blyth (1951) to situations involving truncation and randomization. The important feature in the lemma is the independence of the numbers a, α, b on ξ if (2.12) holds. Since the lemma follows the pattern of Blyth (1951), the proof will be condensed.

Lemma 2.1. *Under the above notations and assumptions, for any $\varepsilon > 0$ there exist positive numbers, a, b, α and a prior density $\pi(\theta)$, all depending on ε only, with the following property.*

For any randomized estimator $\xi(Z, U)$ such that

$$\mathsf{P}\left(|\xi(Z,U) - Z| > \varepsilon \,|\, \theta = 0\right) > \varepsilon \tag{2.12}$$

then

$$\int_{-b}^{b} \pi(\theta)\, R_a(\theta;\xi)\, \mathrm{d}\theta > r + \alpha. \tag{2.13}$$

Proof. It suffices to show that

$$\int_{-b}^{b} \pi(\theta)\, R_a(\theta;\xi)\, \mathrm{d}\theta > r_a(b) + 2\alpha \tag{2.14}$$

for a, b sufficiently large and for an α which is independent of a, b. Note that

$$\begin{aligned}
&(2\pi)^{-1/2}\int_{-\sqrt{b}}^{\sqrt{b}} \ell_a(y-\beta)\exp\left\{-\frac{1}{2}\,y^2\right\}\mathrm{d}y && (2.15)\\
\geq\;& r_a(b) + (2\pi)^{-1/2}\int_0^{\sqrt{b}}\left[\int_{-x}^{x} - \int_{-x+\beta}^{x+\beta}\right]\exp\left\{-\frac{1}{2}\,y^2\right\}\mathrm{d}y\mathrm{d}\ell_a(x)\\
\geq\;& r_a(b) + \delta, \quad \text{if } |\beta| > \frac{1}{2}\,\varepsilon,
\end{aligned}$$

where $\delta > 0$ depends only on ε, but not on a, b when they are sufficiently large.

Next, following the idea of Blyth (1951), we shall assume θ to be distributed with density

$$\pi(\theta) = \frac{1}{\sigma(2\pi)^{1/2}}\exp\left\{-\frac{\theta^2}{2\sigma^2}\right\}, \tag{2.16}$$

where σ will be appropriately chosen to depend on ε later. Then the conditional density of θ given $Z = z$ is

$$\psi(\theta \,|\, z) = \frac{(1+\sigma^2)^{1/2}}{\sigma(2\pi)^{1/2}}\exp\left\{-\frac{1+\sigma^2}{2\sigma^2}\left(\theta - \frac{z\sigma^2}{1+\sigma^2}\right)^2\right\} \tag{2.17}$$

and the overall density of Z will be

$$f(z) = \frac{1}{[(1+\sigma^2)(2\pi)]^{1/2}} \exp\left\{-\frac{z^2}{2(1+\sigma^2)}\right\}. \tag{2.18}$$

In what follows we shall assume that

$$|z| \leq b - \sqrt{b}. \tag{2.19}$$

Then

$$\begin{aligned} &\int_{-b}^{b} \ell_a[\xi(z,u)-\theta]\,\psi(\theta\,|\,z)\mathrm{d}\theta \qquad (2.20)\\ \geq\ &(2\pi)^{-1/2}\int_{-\sqrt{b}}^{\sqrt{b}} \ell_a(y)\exp\left\{-\frac{y^2(1+\sigma^2)}{2\sigma^2}\right\}\mathrm{d}y \\ \geq\ &r_a(b) - \frac{K}{\sigma^2}, \end{aligned}$$

where K does not depend on a, b, σ^2. We have used the inequality

$$\exp\left\{-\frac{y^2(1+\sigma^2)}{2\sigma^2}\right\} > \left[1-\frac{y^2}{\sigma^2}\right]\exp\left\{-\frac{1}{2}y^2\right\}. \tag{2.21}$$

On the other hand, if

$$|\xi(z,u)-z| > \varepsilon, \quad |z| < M, \tag{2.22}$$

we shall have

$$\left|\xi(z,u) - \frac{z\sigma^2}{1+\sigma^2}\right| > \frac{1}{2}\varepsilon \quad \text{for } (1+\sigma^2) > \frac{2M}{\varepsilon}. \tag{2.23}$$

Consequently, in view of (2.15), we have

$$\begin{aligned} &(2\pi)^{-1/2}\int_{-b}^{b} \ell_a[\xi(z,u)-\theta]\,\psi(\theta\,|\,z)\,\mathrm{d}\theta \qquad (2.24)\\ \geq\ &(2\pi)^{-1/2}\int_{-\sqrt{b}}^{\sqrt{b}} \ell_a\left(y+\frac{1}{2}\varepsilon\right)\exp\left\{-\frac{y^2(1+\sigma^2)}{2\sigma^2}\right\}\mathrm{d}y \\ \geq\ &(2\pi)^{-1/2}\int_{-\sqrt{b}}^{\sqrt{b}} \ell_a\left(y+\frac{1}{2}\varepsilon\right)\exp\left\{-\frac{1}{2}y^2\right\}\mathrm{d}y - \frac{K}{\sigma^2} \\ \geq\ &r_a(b) + \delta - \frac{K}{\sigma^2}, \end{aligned}$$

if (2.19) and (2.23) hold.

Altogether, we have

$$\int_{-b}^{b} \pi(\theta)\, R_a(\theta;\xi)\,\mathrm{d}\theta \tag{2.25}$$
$$= \int_0^1 \int_{-\infty}^{\infty} \int_{-b}^{b} \ell_a[\xi(z,u)-\theta]\psi(\theta\,|\,z)\, f(z)\,\mathrm{d}\theta\mathrm{d}z\mathrm{d}u$$
$$\geq r_a(b)\,\mathsf{P}\left(|Z|<b-\sqrt{b}\right)-\frac{K}{\sigma^2}+\delta\,\mathsf{P}\left(|\xi(Z,U)-Z|>\varepsilon,\ |Z|<M\right).$$

From (2.12) we see that

$$\mathsf{P}\left(|\xi(Z,U)-Z|>\varepsilon,\ |Z|<M|\theta=0\right)>\frac{1}{2}\,\varepsilon \tag{2.26}$$

for M sufficiently large. Now under $\theta = 0$ the density of Z is $(2\pi)^{-1/2} \times \exp\left\{-\frac{1}{2}\,z^2\right\}$, whereas the overall density is given by $f(z)$ of (2.18). The likelihood ratio for the two densities is for $|z| < M$ greater than

$$\frac{1}{(1+\sigma^2)^{1/2}}\exp\left\{-\frac{1}{2}\,M^2\right\}. \tag{2.27}$$

Thus

$$\mathsf{P}\left(|\xi(Z,U)-Z|>\varepsilon,\ |Z|<M\right)>\frac{\varepsilon}{2(1+\sigma^2)^{1/2}}\exp\left\{-\frac{1}{2}\,M^2\right\}. \tag{2.28}$$

Now it suffices to put

$$3\alpha=\frac{\delta\varepsilon}{2(1+\sigma^2)^{1/2}}\exp\left\{-\frac{1}{2}\,M^2\right\}-\frac{K}{\sigma^2} \tag{2.29}$$

which is positive for σ sufficiently large. In the last step we choose a and b in such a way that

$$r_a(b)>r-\frac{1}{2}\,\alpha \tag{2.30}$$

and given the above chosen σ^2,

$$r\,\mathsf{P}\left(|Z|>b-\sqrt{b}\right)<\frac{1}{2}\,\alpha. \tag{2.31}$$

Now (2.25) to (2.31) yield (2.14). □

Of course, usually there will be nuisance parameters, and then the following k-dimensional version of the preceding lemma is useful.

Lemma 2.2. *Let $Z = (Z_1, \dots, Z_k)$ be a normal random vector with expectation $\theta = (\theta_1, \dots, \theta_k) \in R^k$ and fixed positive definite covariance matrix $\Sigma = \{\sigma_{i,j}\}_{i,j=1}^k$. Consider a function ℓ satisfying (2.1) to (2.4). Then for every $\varepsilon > 0$ there exist positive numbers a, b, α and a one-dimensional density $\pi(\cdot)$ with the following property.*

For any randomized estimator $\xi(Z_1, \dots, Z_k, U)$ of θ_1 such that

$$\mathsf{P}\left(|\xi(Z_1, \dots, Z_k, U) - Z_1| > \varepsilon | \theta_1 = \dots = \theta_k = 0\right) > \varepsilon \tag{2.32}$$

then

$$\int_{-b}^{b} \pi(\theta_1) \mathsf{E}\left\{\ell_a[\xi(Z_1, \dots, Z_k, U) - \theta_1] | \theta_i = \theta_1 \sigma_{1,i} \sigma_{1,1}^{-1}, 1 \le i \le k\right\} \mathrm{d}\theta_1 \tag{2.33}$$
$$\ge (2\pi)^{-1/2} \int_{-\infty}^{\infty} \ell[y\sigma_1] \exp\left\{-\frac{1}{2} y^2\right\} \mathrm{d}y + \alpha,$$

where $\sigma_1 = (\sigma_{1,1})^{1/2}$.

Proof. For the submodel $\theta_i = \theta_1 \sigma_{1,i} \sigma_{1,1}^{-1}$, $1 \le i \le k$, $-\infty < \theta_1 < \infty$, Z_1 is a sufficient statistic, and the result follows from the preceeding lemma. □

Lemma 2.3. *Let Z be the same vector as in Lemma 2.2. Then for every $\delta > 0$ there exist positive numbers a, b and a density $\pi(\cdot)$ such that for any randomized estimator $\xi(Z_1, \dots, Z_k, U)$:*

$$\int_{-b}^{b} \pi(\theta_1) \mathsf{E}\left\{\ell_a[\xi(Z_1, \dots, Z_k, U) - \theta_1] | \theta_i = \theta_1 \sigma_{1,i} \sigma_{1,1}^{-1}, 1 \le i \le k\right\} \mathrm{d}\theta_1 \tag{2.34}$$
$$\ge (2\pi)^{-1/2} \int_{-\infty}^{\infty} \ell[y\sigma_1] \exp\left\{-\frac{1}{2} y^2\right\} \mathrm{d}y - \delta,$$

where σ_1 has the same meaning as in Lemma 2.2.

Proof. The proof follows the same lines as the proofs of the previous lemmas. □

3 LOCALLY ASYMPTOTICALLY NORMAL FAMILIES OF DISTRIBUTIONS

The essence of the definition of a locally asymptotically normal family is that the log likelihood ratio is asymptotically normally distributed with a covariance matrix, which is locally constant, and with an expectation which is locally a linear function of θ. For our purpose – to obtain lower bounds for risks and necessary conditions for allowing equality in corresponding inequalities – we need the following version, employed already in Hájek (1970): consider a sequence of statistical experiments $(\mathcal{X}_n, \mathcal{A}_n, \mathsf{P}_n(\cdot, \theta))$, $n \geq 1$, where θ runs through an open subset Θ of R^k. Take a point $t \in \Theta$ and assume it to be the true value of the parameter θ. We shall abbreviate $\mathsf{P}_n = \mathsf{P}_n(\cdot, t)$ and $\mathsf{P}_{n,h} = \mathsf{P}_n\left(\cdot, t + n^{-1/2} h\right)$. If appropriate, we could use in place of $\sqrt{n}$ some more general norming numbers $k(n) \to \infty$, or even matrices as in Hájek (1970).

The norm of a point $h = (h_1, \ldots, h_k)$ from R^k will be denoted by $|h| = \max_{1 \leq i \leq k} |h_i|$, and $h' v$ will denote the scalar product of two vectors.

Given two probability measures P and Q, denote by $\mathrm{d}\mathsf{Q}/\mathrm{d}\mathsf{P}$ the Radon–Nikodym derivative of the *absolutely continuous* part of Q with respect to P. Introduce the family of likelihood ratios

$$r_n(h, x_n) = \frac{\mathrm{d}\mathsf{P}_{n,h}}{\mathrm{d}\mathsf{P}_n}(x_n), \quad h \in R^k, \ n \geq n_h, \ x_n \in \mathcal{X}_n, \tag{3.1}$$

where n_h denotes the smallest integer such that $n \geq n_h$ entails $t + n^{-1/2} h \in \Theta$. In what follows the argument x_n will usually be omitted.

Assumption 3.1. LAN *(local asymptotic normality) at $\theta = t$. Assume that*

$$r_n(h) = \exp\left\{h' \Delta_{n,t} - \frac{1}{2} h' \Gamma_t h + Z_n(h, t)\right\}, \quad h \in R^k, \ n \geq n_h, \tag{3.2}$$

where the random vector $\Delta_{n,t}$ satisfies $\mathcal{L}(\Delta_{n,t} \,|\, \mathsf{P}_n) \to N(0, \Gamma_t)$, and $Z_n(h,t) \to 0$ in P_n probability for every $h \in R^k$. Further assume that $\det \Gamma_t > 0$.

Example. Consider the case $k = 1$, $x_n = (y_1, \ldots, y_n) \in R^n$ and $p_n(x_n, \theta) = \prod_{i=1}^n f(y_i, \theta)$.

Then the existence of Fisher's information, its positivity and continuity at $\theta = t$, entails LAN. Of course, in order for Fisher's information to be well defined, $f(y, \theta)$ must be absolutely continuous in θ in a vicinity of t, and the derivative $\dot{f}(y, t) = (\partial/\partial\theta) f(y, \theta)|_{\theta = t}$ must exist for almost all y. Then the information equals

$$I_\theta = \int_{-\infty}^{\infty} \left\{\frac{[\dot{f}(y, \theta)]^2}{f(y, \theta)}\right\} \mathrm{d}y. \tag{3.3}$$

Satisfaction of LAN for this case may be proved by methods developed in Hájek and Šidák (1967) as is shown in the Appendix below. Specifically (3.2) will be satisfied for

$$\Delta_{n,t} = n^{-1/2} \sum_{i=1}^n \frac{\dot{f}(y_i, t)}{f(y_i, t)}, \quad \Gamma_t = I_t. \tag{3.4}$$

Let $||\mathsf{P}-\mathsf{Q}||$ be the L_1 norm of two probability measures. If p and q are densities of P and Q with respect to μ, then

$$||\mathsf{P}-\mathsf{Q}|| = \int |p-q|\,\mathrm{d}\mu. \tag{3.5}$$

The following lemmas are essentially contained in Le Cam (1956), and form a bridge between exactly normal models and locally asymptotically normal models.

Lemma 3.1. *For any sequence of statistics $\{S_n\}$ put*

$$s_n(x,u) = \inf\{y : \mathsf{P}_n[S_n \le y \,|\, \Delta_{n,t} = x] \ge u\}, \quad x \in R,\ 0 < u < 1, \tag{3.6}$$

and denote by $F_{n,h}$ the distribution of S_n under $\mathsf{P}_{n,h}$ and by $F^*_{n,h}$ the distribution of $s_n(\Delta_{n,t}, U)$ also under $\mathsf{P}_{n,h}$, if U is uniformly distributed on $(0,1)$ and independent of $\Delta_{n,t}$.

Then, under Assumption LAN,

$$\lim_{n\to\infty} ||F_{n,h} - F^*_{n,h}|| = 0, \quad h \in R. \tag{3.7}$$

Proof. We see from (3.6) that $F_{n,h} = F^*_{n,h}$ if $h=0$. Now, as shown in the proof of the theorem in Hájek (1970), $\mathsf{P}_{n,h}$ may be approximated by $\mathsf{Q}_{n,h}$ such that $\Delta_{n,t}$ is a sufficient statistic for the pair $(\mathsf{P}_n,\ \mathsf{Q}_{n,h})$ and $||\mathsf{Q}_{n,h} - \mathsf{P}_{n,h}|| \to 0$. That yields (3.7). □

In the next two lemmas we shall treat the cases $k=1$ and $k>1$ separately, because for $k=1$ we are able to reach a greater degree of explicitness.

Lemma 3.2. *Assume $k=1$. Denote $G_{n,h}(x) = \mathsf{P}_{n,h}(\Delta_{n,t} \le x)$ and $\Phi(x) = (2\pi)\int_{-\infty}^{x} \exp\{-\frac{1}{2}y^2\}\,\mathrm{d}y$. Let $G^*_{n,h}$ be the distribution of $G^{-1}_{n,0}\,\Phi(Z\,\Gamma_t^{-1/2})$ if Z is normal $(h\Gamma_t,\ \Gamma_t)$.*

Then, under Assumption LAN,

$$\lim_{n\to\infty} ||G_{n,h} - G^*_{n,h}|| = 0, \quad h \in R. \tag{3.8}$$

Proof. We again have $G_{n,h} = G^*_{n,h}$ for $h=0$. Since $G_{n,0}(x) \to \Phi(x\Gamma_t^{-1/2})$ uniformly under Assumptions LAN we have

$$G^{-1}_{n,0}\,\Phi\left(x\Gamma_t^{-1/2}\right) \to x \quad \text{uniformly on compacts.} \tag{3.9}$$

Further, referring again to the proof of the theorem of Hájek (1970), we may approximate $G_{n,h}$ by $\overline{G}_{n,h}(x) = \mathsf{Q}_{n,h}(\Delta_{n,t} \le x)$ satisfying $||G_{n,h} - \overline{G}_{n,h}|| \to 0$. The rest follows by showing that

$$\frac{\mathrm{d}\overline{G}_{n,h}}{\mathrm{d}G_{n,0}}(x) \quad \to \quad \exp\left\{-hx - \frac{1}{2}h^2\Gamma_t\right\}, \tag{3.10}$$

$$\frac{\mathrm{d}G^*_{n,h}}{\mathrm{d}G_{n,0}}(x) \quad \to \quad \exp\left\{-hx - \frac{1}{2}h^2\Gamma_t\right\}. \tag{3.11}$$

□

Lemma 3.3. *Assume $k \ge 1$. Denote $G_{n,h}(x) = \mathsf{P}_{n,h}(\Delta_{n,t} \le x)$, $x \in R^k$ and denote by $Z = (Z_1, \ldots, Z_k)$ a random vector such that $\mathcal{L}(Z) = N(\Gamma_t h, \Gamma_t)$ if $\theta = t + n^{-1/2}h$. Then there exists a sequence of functions $\phi_n(x)$ such that*

$$\lim_{n\to\infty} \sup_{x \in R^k} |\phi_n(x) - x| = 0 \tag{3.12}$$

and the distribution of $\phi_n(Z)$ under $\theta = t + n^{-1/2}h$, say $G^*_{n,h}$, satisfies

$$\lim_{n\to\infty} ||G^*_{n,h} - G_{n,h}|| = 0. \tag{3.13}$$

Proof. Consider a sequence of cubes

$$C_j = \{(x_1, \ldots, x_k) : |x_i| \le j,\ 1 \le i \le k\}, \quad j = 1, 2, \ldots. \tag{3.14}$$

Partition each cube C_j into j^{2k} subcubes $C_{j,i}$, $1 \le i \le j^{2k}$, of all equal volume $(2/j)^k$. Since $G_{n,0} \to N(0, \Gamma_t)$, we have

$$\int_{C_{j,i}} \mathrm{d}G_{n,0} \to \int_{C_{j,i}} \mathrm{d}N(0, \Gamma_t). \tag{3.15}$$

Let $\phi_{n,j}$ be a function mapping $C_{j,i}$ into $C_{j,i}$,

$$\phi_{n,j} : C_{j,i} \to C_{j,i}, \quad 1 \le j \le \infty,\ 1 \le i \le j^{2k} \tag{3.16}$$

and such that

$$\phi_{n,j}(x) = x \quad \text{for } x \notin C_j. \tag{3.17}$$

Furthermore, we choose $\phi_{n,j}$ so that the difference between the $G^*_{n,j,0} = \mathcal{L}(\phi_{n,j}(Z) \mid N(0, \Gamma_t))$ and $G_{n,0}$ tends to zero in the L_1 norm, that is,

$$||G_{n,j,0} - G_{n,0}|| < \frac{1}{j} \quad \text{as } n \ge n_j. \tag{3.18}$$

This is possible in view of (3.15). Now define $j(n)$ by $n_{j(n)} \le n < n_{j(n)+1}$ and put

$$\phi_n(x) = \phi_{n,j(n)}(x). \tag{3.19}$$

Then $\phi_n(x)$ satisfies (3.12) because of (3.17) and (3.16) and the fact that diameters of the cubes $C_{j,i}$ converge to 0 as $j \to \infty$. Furthermore, (3.18) entails $\|G^*_{n,0} - G_{n,0}\| \to 0$ since $G^*_{n,0} = G^*_{n,j(n),0}$.

The proof may be concluded by showing that (3.10) holds, referring again to the theorem of Hájek (1970), and that (3.11) is true. The last statement follows from (3.12). □

4 THE MAIN PROPOSITION

We shall generalize and modify theorems by Le Cam (1953) and P. Huber (1966), treating the cases $k = 1$ and $k \ge 1$ separately, again.

Theorem 4.1. *Assume $k = 1$. Under Assumptions* LAN *of Section 3, any sequence of estimates $\{T_n\}$ for θ satisfies for ℓ of* (2.1) *to* (2.4)

$$\begin{aligned} &\lim_{\delta\to 0}\liminf_{n\to\infty} \sup_{|\theta-t|<\delta} \mathsf{E}_\theta \left\{\ell\left[\sqrt{n}(T_n-\theta)\right]\right\} \\ \ge\ & (2\pi)^{-1/2}\int_{-\infty}^{\infty} \ell(y\,\Gamma_t^{-1/2})\exp\left\{-\frac{1}{2}y^2\right\}\mathrm{d}y. \end{aligned} \tag{4.1}$$

Furthermore, we can have for a nonconstant ℓ

$$\begin{aligned} &\lim_{\delta\to 0}\liminf_{n\to\infty} \sup_{|\theta-t|<\delta} \mathsf{E}_\theta \left\{\ell\left[\sqrt{n}(T_n-\theta)\right]\right\} \\ =\ & (2\pi)^{-1/2}\int_{-\infty}^{\infty} \ell(y\,\Gamma_t^{-1/2})\exp\left\{-\frac{1}{2}y^2\right\}\mathrm{d}y \end{aligned} \tag{4.2}$$

only if

$$\sqrt{n}(T_n - t) - \Gamma_t^{-1}\,\Delta_{n,t} \to 0 \tag{4.3}$$

in P_n probability.

Proof. We shall first prove (4.1). Introducing local coordinates by $\theta = t + h\,n^{-1/2}$ and using $\mathsf{E}_\theta(\cdot)$ and $\mathsf{E}(\cdot\,|\,\theta)$ interchangeably, we may write for n sufficiently large

$$\begin{aligned} &\sup_{|\theta-t|<\delta} \mathsf{E}_\theta \left\{\ell\left[\sqrt{n}(T_n-\theta)\right]\right\} \\ \ge\ & \int_{-b}^{b} \pi(h)\,\mathsf{E}\left\{\ell_a\left[\sqrt{n}(T_n-t)-h\right]\,|\,t+n^{-1/2}h\right\}\mathrm{d}h, \end{aligned} \tag{4.4}$$

whatever the constants a, b and density $\pi(\cdot)$ may be. We fix some $\overline{\delta} > 0$ and choose a, b and π in such a way that

$$\int_{-b}^{b} \pi(h)\, \mathsf{E}\left\{\ell_a[\xi(Z,U) - h] \,|\, t + n^{-1/2}\, h\right\} \mathrm{d}h \tag{4.5}$$
$$\geq \; (2\pi)^{-1/2} \int_{-\infty}^{\infty} \ell\left(y\, \Gamma_t^{-1/2}\right) \exp\left\{-\frac{1}{2}\, y^2\right\} \mathrm{d}y - \overline{\delta}$$

for any estimator $\xi(Z,U)$, provided that $\mathcal{L}\left(Z\,|\, t + n^{-1/2}\, h\right) = N(h\, \Gamma_t,\, h\, \Gamma_t)$. This is possible according to Lemma 2.3, if applied to $\Gamma_t^{-1}\, Z$ which is normal $(h,\, \Gamma_t^{-1})$.

Next we identify $S_n = \lambda_n(T_n - t)$ in Lemma 3.1 and conclude (ℓ_a is bounded!) that for every $h \in R$

$$\left|\mathsf{E}\left\{\ell_a\left[\sqrt{n}(T_n - t) - h\right] \,|\, t + n^{-1/2}\, h\right\}\right. \tag{4.6}$$
$$\left. - \mathsf{E}\left\{\ell_a[s_n(\Delta_{n,t},\, U) - h] \,|\, t + n^{-1/2}\, h\right\}\right| \to 0.$$

Furthermore, by Lemma 3.2, putting

$$\xi_n(Z,U) = s_n\left(G_{n,0}^{-1}\, \phi(Z\Gamma_t^{-1/2}),\, U\right) \tag{4.7}$$

we obtain for every $h \in R$

$$\left|\mathsf{E}\left\{\ell_a\left[s_n(\Delta_{n,t}, U) - h\right] \,|\, t + n^{-1/2}\, h\right\}\right. \tag{4.8}$$
$$\left. - \mathsf{E}\left\{\ell_a[\xi_n(Z,U) - h] \,|\, t + n^{-1/2}\, h\right\}\right| \to 0.$$

Consequently

$$\int_{-b}^{b} \pi(h)\, \mathsf{E}\left\{\ell_a\left[\sqrt{n}(T_n - t) - h\right] \,|\, t + n^{-1/2}\, h\right\} \mathrm{d}h \tag{4.9}$$
$$\geq \; \int_{-b}^{b} \pi(h)\, \mathsf{E}\left\{\ell_a\left[\xi_n(Z,U) - h\right] \,|\, t + n^{-1/2}\, h\right\} \mathrm{d}h - \overline{\delta}, \quad n > n(a, b, \pi, \overline{\delta}).$$

Combining (4.4), (4.9), and (4.5), we obtain (4.1).

Now we shall prove the necessity of (4.3) for (4.2). Assuming that (4.3) does not hold we shall contradict (4.2). Again putting $S_n = \lambda_n(T_n - t)$, and recalling Lemma 3.1, we can see that $\lambda_n(T_n - t) - \Gamma_t^{-1}\Delta_{n,t}$ has the same distribution under P_n as

$$s_n(\Delta_{n,t}, U) - \Gamma_t^{-1}\, \Delta_{n,t}. \tag{4.10}$$

Furthermore, in view of (3.9), if expression (4.10) fails to converge to zero in probability, then also

$$\xi_n(Z,U) - \Gamma_t^{-1}\, Z \tag{4.11}$$

fails to do so. But then, there is an $\varepsilon > 0$ such that for every n there exists an $m > n$ such that

$$\mathsf{P}_m\left(|\xi_m(Z,U) - \Gamma_t^{-1}U| > \varepsilon\right) > \varepsilon. \tag{4.12}$$

Therefore, according to Lemma 2.2, applied to $\Gamma_t^{-1}Z$ and $k = 1$, we choose a, b, α, and $\pi(\cdot)$ such that

$$\int_{-b}^{b} \pi(h)\,\mathsf{E}\left\{\ell_a[\xi_m(Z,U) - h]\,|\,t + n^{-1/2}h\right\}\mathrm{d}h \tag{4.13}$$
$$> (2\pi)^{-1/2}\int_{-\infty}^{\infty} \ell\left(y\,\Gamma_t^{-1/2}\right)\exp\left\{-\frac{1}{2}y^2\right\}\mathrm{d}y + \alpha.$$

This in connection with (4.4) and (4.9) contradicts (4.2), since $\overline{\delta}$ can be made smaller than α. □

Without proof let us also formulate a k-dimensional version of the previous theorem. The parametric function of interest will be the first coordinate, θ_1, while the other coordinates will be regarded as nuisance parameters. The proof could be based on Lemma 2.2.

Theorem 4.2. *Under Assumption* LAN *above, any sequence of estimates T_n for the first coordinate θ_1 of θ satisfies*

$$\lim_{\delta\to 0}\liminf_{n\to\infty}\sup_{|\theta-t|<\delta}\mathsf{E}_\theta\left\{\ell[\sqrt{n}(T_n - \theta_1)]\right\} \tag{4.14}$$
$$\geq (2\pi)^{-1/2}\int_{-\infty}^{\infty}\ell(\sigma_1^t y)\exp\left\{-\frac{1}{2}y^2\right\}\mathrm{d}y,$$

where $\sigma_1^t = (\sigma_{t,1,1})^{1/2}$ and $\{\sigma_{t,i,j}\}_{i,j=1}^k = \Gamma_t^{-1}$.

We can have

$$\lim_{\delta\to 0}\lim_{n\to\infty}\sup_{|\theta-t|<\delta}\mathsf{E}_\theta\left\{\ell\left[\sqrt{n}(T_n - \theta_1)\right]\right\} \tag{4.15}$$
$$= (2\pi)^{-1/2}\int_{-\infty}^{\infty}\ell(\sigma_1^t y)\exp\left\{-\frac{1}{2}y^2\right\}\mathrm{d}y$$

only if

$$\left\{\sqrt{n}(T_n - t_1) - (\Gamma_t^{-1}\Delta_{n,t})_1\right\} \to 0 \tag{4.16}$$

in P_n probability, where $(\cdot)_1$ denotes the first coordinate in the corresponding vector.

Remark 1. In (4.1) we could replace the left side by

$$\lim_{a\to\infty}\liminf_{n\to\infty}\sup_{\sqrt{n}|\theta-t|<a}\mathsf{E}_\theta\left\{\ell\left[\sqrt{n}(T_n - \theta)\right]\right\}. \tag{4.17}$$

Remark 2. (4.16) in connection with asymptotic normality entails that $\sqrt{n}(T_n - t_1)$ is asymptotically normal $N(0, \sigma_{t,1,1})$, as is proved in Huber (1966).

Remark 3. Condition (4.16) is only necessary. In order that there exists an estimate for which (4.15) holds for all $t \in \theta$, additional global as well as local conditions on the underlying family of distributions are necessary. If ℓ is not bounded, we also need to know something about how fast probabilities of moderate deviations of T_n from t approach zero.

Remark 4. If we are interested in a general scalar function $\sigma = \tau(\theta)$, we may reduce the problem to one considered in Theorem 4.2 by introducing new local coordinates $(\tau_1, \ldots, \tau_k)$ such that $\tau_1 \equiv \tau$.

APPENDIX

We shall here prove that the LAN conditions are satisfied under the assumptions of the Example in Section 3. To this end it will be sufficient to adapt Theorem VI.2.1 of Hájek and Šidák (1967) for the present situation (see also Problem 7 of Chapter VI 1.c).

Let us restate our conditions carefully:

1. In some vicinity of $\theta = t$ the functions $f(y, \theta)$ are absolutely continuous in θ for all $y \in R$.
2. For every θ in some vicinity of t the θ derivative $\dot{f}(y, \theta) = (\partial / \partial \theta) f(y, \theta)$ exists for almost all (Lebesgue measure) $y \in R$.
3. The Fisher information

$$I_\theta = \int_{-\infty}^{\infty} \left\{ \frac{[\dot{f}(y, \theta)]^2}{f(y, \theta)} \right\} dy \tag{A.1}$$

exists, is continuous at $\theta = t$ and $I_t > 0$. (The integrand in (A.1) is to be interpreted as zero if $f(y, \theta) = 0$.)

Our preliminary goal is to show that

$$s(y, \theta) = [f(y, \theta)]^{1/2} \tag{A.2}$$

is also absolutely continuous and has a mean square derivative at $\theta = t$.

Lemma A.1. *If $g(\theta) \geq 0$ is absolutely continuous on (a, b) and its derivative $\dot{g}(\theta)$ satisfies*

$$\int_a^b \frac{|\dot{g}(\theta)|}{[g(\theta)]^{1/2}} \, d\theta < \infty, \tag{A.3}$$

then $[g(\theta)]^{1/2}$ is also absolutely continuous on (a, b).

Proof. If $g(\theta) > 0$ and $\dot{g}(\theta)$ exists for some point θ, then it is well known from calculus that

$$\frac{\partial}{\partial\theta}[g(\theta)]^{1/2} = \frac{\dot{g}(\theta)}{2[g(\theta)]^{1/2}}. \tag{A.4}$$

Furthermore, if $a \leq \alpha < \beta \leq b$ and g is positive on $[\alpha, \beta]$, it is easy to see that $[g(\alpha)]^{1/2}$ is absolutely continuous on $[\alpha, \beta]$ and

$$[g(\beta)]^{1/2} - [g(\alpha)]^{1/2} = \frac{1}{2}\int_{\alpha}^{\beta} \frac{\dot{g}(u)}{[g(u)]^{1/2}}\,\mathrm{d}u. \tag{A.5}$$

In view of the continuity of $g(\theta)$ and in view of (A.3), (A.5) extends to $\alpha < \beta$ such that g is positive on (α, β), that is, even if possibly $g(\alpha) = 0$ or $g(\beta) = 0$. Now for any c, $a < c \leq b$ the interval (a, c) may be decomposed as follows:

$$(a, c) = \left[\bigcup_{i=1}^{\infty}(\alpha_i, \beta_i)\right] \cup A, \tag{A.6}$$

where (α_i, β_i) are disjoint intervals such that $g(\theta)$ is positive on them, $g(\alpha_i) = 0$ if $\alpha_i \neq a$ and $g(\beta_i) = 0$ if $\beta_i \neq c$, and $g(\theta) = 0$ for $\theta \in A$. Then, interpreting $\dot{g}(\theta)\,[g(\theta)]^{-1/2}$ as zero when $g(\theta) = 0$, we may write, in view of (A.3),

$$\int_a^c \frac{\dot{g}(\theta)}{[g(\theta)]^{1/2}}\,\mathrm{d}\theta = \sum_{i=1}^{\infty}\int_{\alpha_i}^{\beta} \frac{\dot{g}(\theta)}{[g(\theta)]^{1/2}}\,\mathrm{d}\theta = [g(c)]^{1/2} - [g(a)]^{1/2}, \tag{A.7}$$

because the summands with endpoints satisfying $0 = g(\alpha_i) = g(\beta_i)$ vanish according to (A.5). Relation (A.7) holding for all $c \in (a, b)$ proves absolute continuity. □

Lemma A.2. *Under conditions 1–3 the functions $s(y, \theta)$ are absolutely continuous in some vicinity of $\theta = t$ for almost all y.*

Proof. Continuity of I_θ and the Fubini theorem imply for some $\varepsilon > 0$

$$\infty > \int_{t-\varepsilon}^{t+\varepsilon} I_\theta\,\mathrm{d}\theta = \int_{-\infty}^{\infty}\int_{t-\varepsilon}^{t+\varepsilon}\left(\frac{[\dot{f}(y,\theta)]^2}{f(y,\theta)}\right)\mathrm{d}\theta\mathrm{d}y. \tag{A.8}$$

Consequently, for almost all y,

$$\int_{t-\varepsilon}^{t+\varepsilon}\left(\frac{[\dot{f}(y,\theta)]^2}{f(y,\theta)}\right)\mathrm{d}\theta < \infty \tag{A.9}$$

and, in turn, for almost all y,

$$\int_{t-\varepsilon}^{t+\varepsilon}\left(\frac{|\dot{f}(y,\theta)|}{[f(y,\theta)]^{1/2}}\right)\mathrm{d}\theta<\infty. \tag{A.10}$$

Thus it suffices to apply Lemma A.1 for $g(\theta)=f(y,\theta)$, for those y that satisfy (A.10). □

Lemma A.3 *Under conditions 1–3 the function $\dot{s}(y,t)$ defined by*

$$\dot{s}(y,t)=\begin{cases}\dfrac{\dot{f}(y,t)}{2[f(y,t)]^{1/2}} & \text{if } f(y,t)>0 \text{ and } \dot{f}(y,t) \text{ exists}\\ 0 & \text{otherwise}\end{cases} \tag{A.11}$$

is the mean square derivative of $s(y,\theta)$ at $\theta=t$, that is

$$\lim_{\Delta\to0}\int_{-\infty}^{\infty}\left\{\frac{1}{\Delta}[s(y,t+\Delta)-s(y,t)]-\dot{s}(y,t)\right\}^2\mathrm{d}y=0. \tag{A.12}$$

Proof. Lemma A.2 entails

$$\begin{aligned}&\left\{\frac{1}{\Delta}[s(y,t+\Delta)-s(y,t)]\right\}^2 \qquad (A.13)\\ =\;&\left(\frac{1}{\Delta}\right)^2\left(\int_0^{\Delta}\dot{s}(y,t+\lambda)\,\mathrm{d}\lambda\right)^2\\ \le\;&\frac{1}{\Delta}\int_0^{\Delta}[\dot{s}(y,t+\lambda)]^2\,\mathrm{d}\lambda.\end{aligned}$$

Consequently, in view of the continuity of I_θ,

$$\begin{aligned}&\int_{-\infty}^{\infty}\left\{\frac{1}{\Delta}[s(y,t+\Delta)-s(y,t)]\right\}^2\mathrm{d}y \qquad (A.14)\\ \le\;&\frac{1}{\Delta}\int_{-\infty}^{\infty}\int_0^{\Delta}[\dot{s}(y,t+\lambda)^2]\,\mathrm{d}\lambda\mathrm{d}y=\frac{1}{4}\frac{1}{\Delta}\int_0^{\Delta}I_{t+\lambda}\,\mathrm{d}\lambda\\ \to\;&\frac{1}{4}I_t=\int_{-\infty}^{\infty}[\dot{s}(y,t)]^2\,\mathrm{d}y,\quad\text{as }\Delta\to0.\end{aligned}$$

Put $M=\{y:s(y,t)>0\}$. Then $(1/\Delta)[s(y,t+\Delta)-s(y,t)]$ converges to $\dot{s}(y,t)$ almost everywhere (Lebesgue measure) on M, and (A.14) entails

$$\begin{aligned}&\limsup_{\Delta\to0}\int_M\left\{\frac{1}{\Delta}[s(y,t+\Delta)-s(y,t)]\right\}^2\mathrm{d}y \qquad (A.15)\\ &\le\int_{-\infty}^{\infty}[\dot{s}(y,t)]^2\,\mathrm{d}y=\int_M[\dot{s}(y,t)]^2\,\mathrm{d}y.\end{aligned}$$

Utilizing Theorem V.I.3 of Hájek and Šidák (1967), we conclude that

$$\lim_{\Delta\to 0}\int_M \left\{\frac{1}{\Delta}[s(y,t+\Delta)-s(y,t)]-\dot{s}(y,t)\right\}^2 \mathrm{d}y = 0 \tag{A.16}$$

and

$$\lim_{\Delta\to 0}\int_M \left\{\frac{1}{\Delta}[s(y,t+\Delta)-s(y,t)]\right\}^2 \mathrm{d}y = \int_{-\infty}^{\infty}[\dot{s}(y,t)]^2\,\mathrm{d}y. \tag{A.17}$$

However (A.17) and (A.14) are compatible only if

$$\lim_{\Delta\to 0}\int_{M^c} \left\{\frac{1}{\Delta}[s(y,t+\Delta)-s(y,t)]\right\}^2 \mathrm{d}y = 0, \tag{A.18}$$

where $M^c = R - M$. Now (A.17) and (A.18) together are equivalent to (A.12) if $\dot{s}(y,t)$ is defined by (A.11). □

Remark A.1. Since $s(y,t) = 0$ on M^c, (A.18) is equivalent to

$$\int_{\{y:f(y,t)=0\}} f(y,\theta)\,\mathrm{d}y = o[(\theta-t)^2] \tag{A.19}$$

which describes how large the singular part of $f(y,\theta)$ relative to $f(y,t)$ may be. Satisfaction of (A.18) is necessary for $q_n(x) = \prod_{i=1}^n f\left(y_i, t+n^{-1/2}h\right)$ to be contiguous with respect to $p_n(x) = \prod_{i=1}^n f(y_i,t)$.

Lemma A.4. *Under conditions* 1–3 *the* LAN *conditions defined in Section 3 are satisfied at* $\theta = t$ *with* $\Gamma_t = I_t$ *and* $\Delta_{n,t}$ *of* (3.4).

Proof. Introduce

$$\begin{aligned} L_{n,h} &= \sum_{i=1}^n \log\frac{f(Y_i, t+n^{-1/2}h)}{f(Y_i,t)} \\ W_{n,h} &= 2\sum_{i=1}^n\left\{\frac{s(Y_i, t+n^{-1/2}h)}{s(Y_i,t)} - 1\right\}. \end{aligned} \tag{A.20}$$

We need to prove that for every $h \in R$

$$\left(L_{n,h} - h\,\Delta_{n,t} + \frac{1}{2}h^2 I_t\right) \xrightarrow{\mathsf{P}_n} 0 \tag{A.21}$$

in P_n probability, P_n referring to $\theta = t$. According to Le Cam's second lemma in Hájek and Šidák (1967) (A.21) is equivalent to

$$\left(W_{n,h} - h\,\Delta_{n,t} + \frac{1}{4}h^2 I_t\right) \xrightarrow{\mathsf{P}_n} 0 \tag{A.22}$$

if we show that $\mathcal{L}(W_{n,h} \mid \mathsf{P}_n) \rightarrow N\left(-\frac{1}{4}h^2 I_t, h^2 I_t\right)$. All this will be accomplished if we prove the following relations:

$$\mathsf{E}(\Delta_{n,t} \mid \mathsf{P}_n) = 0, \tag{A.23}$$

$$\mathsf{E}(W_{n,h} \mid \mathsf{P}_n) \rightarrow -\frac{1}{4} h^2 I_t, \tag{A.24}$$

$$\mathsf{var}(W_{n,h} - h\,\Delta_{n,t} \mid \mathsf{P}_n) \rightarrow 0, \tag{A.25}$$

$$\mathcal{L}(\Delta_{n,t} \mid \mathsf{P}_n) \rightarrow N(0, I_t). \tag{A.26}$$

We shall start with (A.23). We have

$$\begin{aligned} 0 &= \int_{-\infty}^{\infty} \frac{1}{\Delta}[f(y, t+\Delta) - f(y,t)]\,dy \qquad\qquad (A.27) \\ &= \int_{-\infty}^{\infty} \frac{1}{\Delta}[s^2(y, t+\Delta) - s^2(y,t)]\,dy \\ &= \int_{-\infty}^{\infty} \frac{1}{\Delta}[s(y, t+\Delta) - s(y,t)]^2\,dy \\ &\quad +2\int_{-\infty}^{\infty} \frac{1}{\Delta}[s(y, t+\Delta) - s(y,t)]\,s(y,t)\,dy \\ &\rightarrow 0 + 2\int_{-\infty}^{\infty} s(y,t)\,\dot{s}(y,t)\,dy = \int_{-\infty}^{\infty} \dot{f}(y,t)\,dy \quad \text{as } \Delta \rightarrow 0. \end{aligned}$$

The last statement follows from (A.11) and (A.12). Consequently

$$\begin{aligned} \mathsf{E}(\Delta_{n,t} \mid \mathsf{P}_n) &= n^{-1/2} \prod_{i=1}^{n} \mathsf{E}\left[\left.\frac{\dot{f}(Y_i,t)}{f(Y_i,t)}\right| \mathsf{P}_n\right] \qquad (A.28) \\ &= n^{1/2} \int_{-\infty}^{\infty} \dot{f}(y,t)\,dy = 0. \end{aligned}$$

In order to prove (A.24) and (A.25) we employ the same idea as in Lemmas VI.2.1a and VI.2.1b in Hájek and Šidák (1967):

$$\begin{aligned} \mathsf{E}(W_{n,h} \mid \mathsf{P}_n) &= 2\sum_{i=1}^{n} \mathsf{E}\left[\frac{s(Y_i, t+n^{-1/2}h)}{s(y_i,t)} - 1\right] \qquad (A.29) \\ &= -h^2 \int_{-\infty}^{\infty} \left[\frac{s(y,t+n^{-1/2}h) - s(y,t)}{n^{-1/2}\,h}\right]^2 dy \\ &\rightarrow -h^2 \int_{-\infty}^{\infty} [\dot{s}(y,t)]^2\,dy = -\frac{1}{4} h^2 I_t. \end{aligned}$$

Furthermore,

$$\mathsf{var}(W_{n,h} - h\,\Delta_{n,t}) \tag{A.30}$$

$$
\begin{aligned}
&= 4\sum_{i=1}^{n} \mathsf{var}\left[\frac{s(Y_i,\, t+n^{-1/2}\,h)}{s(Y_i,t)} - 1 - \frac{1}{2}\,n^{-1/2}\,h\,\frac{\dot f(Y_i,t)}{f(Y_i,t)}\right] \\
&= 4\sum_{i=1}^{n} \mathsf{E}\left[\frac{s(Y_i,\, t+n^{-1/2}\,h)}{s(Y_i,t)} - 1 - \frac{1}{2}\,n^{-1/2}\,h\,\frac{\dot f(Y_i,t)}{f(Y_i,t)}\right]^2 \\
&\le 4\,h^2 \int_{-\infty}^{\infty} \left[\frac{s(y,\, t+n^{-1/2}\,h) - s(y,t)}{n^{-1/2}\,h} - \dot s(y,t)\right]^2 dy \to 0.
\end{aligned}
$$

Finally, (A.26) follows easily from (3.4) by the central limit theorem. □

5 ACKNOWLEDGEMENT

The research was sponsored in part by National Science Foundation Grant # Gu 2612 to the Florida State University and in part by the Army, Navy and Air Force under Office of Naval Research Contract Number NONR–988 (08), Task Order NR 042–004.

REFERENCES

Bahadur, R. R. (1960). On the asymptotic efficiency of tests and estimates. *Sankhyā 22*, 229–252.

Bahadur, R. R. (1967). Rates of convergence of estimates and test statistics. Ann. Math. Statist. 38, 303–324.

Basu, D. (1956). The concept of asymptotic efficiency. *Sankhyā 17*, 193–196.

Blyth, C. R. (1951). On minimax statistical decision procedures and their admissibility. *Ann. Math. Statist. 22*, 22–42.

Chernoff, H. (1956). Large sample theory: Parametric case. *Ann. Math. Statist. 27*, 1–22.

Hájek, J. (1970). A characterization of limiting distributions of regular estimates. *Z. Wahrscheinlichkeitsheorie und Verw. Gebiete 14*, 323–330.

Hájek, J. (1971). Limiting properties of likelihoods and inference. *Foundations of Statistical Inference* (eds: V. P. Godambe and D. A. Sprott). Holt, Rinehart and Winston, Toronto.

Hájek, and Šidák, Z. (1967). *Theory of Rank Tests.* Academic Press, New York.

Hodges, J. L. and Lehmann, E. L. (1950). Some applications of the Cramér–Rao inequality. *Proceedings of the Second Berkeley Symposium on Mathematical Statistics and Probability.* University of California Press, Berkeley and Los Angeles, 13–22.

Huber, P. (1966). Strict efficiency excludes superefficiency. (Abstract.) *Ann. Math. Statist. 37*, 1425. Proof unpublished but available.

Joshi, V. M. (1965). Admissibility and Bayes estimation in sampling finite finite population II. *Ann. Math. Statist. 36* (1965), 1723–1729.

Le Cam, L. (1953). On some asymptotic properties of maximum likelihood estimates and related Bayes' estimates. *Univ. California Publ. Statist. 1*, 277–330.

Le Cam, L. (1956). On the asymptotic theory of estimation and testing hypotheses. *Proceedings of the Third Berkeley Symposium on Mathematical Statistics and Probability.* University of California Press, Berkeley and Los Angeles, Vol. 1, 129–156.

Le Cam, L. (1958). Les propriétés asymptotiques des solutions de Bayes. *Publ. Inst. Statist. Univ. Paris 7, fasc. 3–4*, 17–35.

Le Cam, L. (1960). Locally asymptotically normal families of distributions. *Univ. California Publ. Statist. 3*, 27–98.

Le Cam, L. (1971). Limits of experiments. *Proceedings of the Sixth Berkeley Symposium of Mathematical Statistics and Probability*, University of California Press, Berkeley and Los Angeles, 245–261.

Rao, C. R. (1963). Criteria of estimation in large samples. *Sankhyā Ser. A 25*, 189–206.

Rao, C. R. (1965). *Linear Statistical Inference and Its Applications.* Wiley, New York.

Wald, A. (1943). Tests of statistical hypothesis concerning several parameters when the number of observations is large. *Trans. Amer. Math. Soc. 54*, 426–482.

Weiss, L. and Wolfowitz, J. (1966). Generalized maximum likelihood estimators. *Teor. Veroyatnost. i Primenen. 11*, 68–93.

Weiss, L. and Wolfowitz, J. (1967). Maximum probability estimators. *Ann. Inst. Statist. Math. 19*, 193–206.

Wolfowitz, J. (1965). Asymptotic efficiency of the maximum likelihood estimator. *Teor. Veroyatnost. i Primenen. 10*, 267–281.

CHAPTER 34

Asymptotic Sufficiency of the Vector of Ranks in the Bahadur Sense

Ann. Statist. 2 (1974), 75–83.
Reproduced by permission of The Institute of Mathematical Statistics.

We shall consider the hypothesis of randomness under which two samples $X_1, \ldots, X_n$ and $Y_1, \ldots, Y_m$ have an identical but arbitrary continuous distribution. The vector of ranks $(R_1, \ldots, R_{n+m})$ will be shown to be asymptotically sufficient in the Bahadur sense for testing randomness against a general class of two-sample alternatives, simple ones as well as composite ones. In other words, the best exact slope will be attainable by rank statistics, uniformly throughout the alternative.

1 INTRODUCTION

We shall consider the hypothesis of randomness under which two samples $X_1, \ldots, X_n$ and $Y_1, \ldots, Y_m$ have an identical but arbitrary continuous distribution. The vector of ranks $(R_1, \ldots, R_{n+m})$ will be shown to be asymptotically sufficient in the Bahadur sense for testing randomness against a general class of two-sample alternatives, simple ones as well as composite ones. In other words, the best exact slope will be attainable by rank statistics, uniformly throughout the alternative.

Our basic tool will be a law of large numbers for simple linear statistics and a large deviation theorem for those statistics due to G. Woodworth (1970). Established properties of simple linear statistics will have important consequences for the rank likelihood ratio statistic, without which the composite alternative could not be treated.

We shall also provide a specialized version of the Berk–Savage theorem from

Collected Works of Jaroslav Hájek – With Commentary
Edited by M. Hušková, R. Beran and V. Dupač
Published in 1998 by John Wiley & Sons Ltd.

Berk and Savage (1968).

The reader not acquainted with Bahadur's theory may consult papers by Bahadur (1967) and Bahadur and Raghavachari (1970).

2 THE TWO-SAMPLE CASE

Fix two densities f and g defined on R, and a number λ, $0 < \lambda < 1$. Put $F(x) = \int_{-\infty}^{x} f(y)\,\mathrm{d}y$, $G(x) = \int_{-\infty}^{x} g(y)\,\mathrm{d}y$ and

$$H(x) = \lambda F(x) + (1-\lambda)\,G(x) \quad -\infty < x < \infty.$$

Introduce new densities

$$\overline{f}(u) = \frac{\mathrm{d}}{\mathrm{d}u} F(H^{-1}(u)), \quad \overline{g}(u) = \frac{\mathrm{d}}{\mathrm{d}u} G(H^{-1}(u)), \quad 0 < u < 1.$$

Obviously, if f and g correspond to X and Y, then $\overline{f}$ and $\overline{g}$ correspond to $H(X)$ and $H(Y)$, respectively. Since $\lambda F(H^{-1}(u)) + (1-\lambda)\,G(H^{-1}(u)) = u$, we have

$$\lambda\overline{f}(u) + (1-\lambda)\,\overline{g}(u) = 1, \quad 0 < u < 1. \tag{2.1}$$

Now given a vector of ranks $(R_1, \ldots, R_N)$ and a double sequence of scores $a_N(i)$, $1 \leq i \leq N < \infty$, let us consider simple linear rank statistic

$$S_N = \sum_{i=1}^{N} a_N(R_i). \tag{2.2}$$

We shall assume that for some integrable function $\phi(u)$, $0 < u < 1$, the following holds:

$$\lim_{N\to\infty} \int_0^1 |a_N(1 + [uN]) - \phi(u)|\,\mathrm{d}u = 0, \tag{2.3}$$

where $[uN]$ denotes the integral part of uN.

Statistics S_N satisfy the following law of large numbers.

Theorem 1. *Assume that the functions $a_N(1 + [uN])$ have uniformly bounded variation on closed subintervals of $(0,1)$. Let $X_1, X_2, \ldots, Y_1, Y_2, \ldots$ be independent random variables, the first sequence having density f and the other density g. Let $R_1, \ldots, R_N$ be the ranks corresponding to $(X_1, \ldots, X_n, Y_1, \ldots,$
$Y_m)$, where $n + m = N$ and*

$$\frac{n}{N} \to \lambda, \quad 0 < \lambda < 1. \tag{2.4}$$

Then, under (2.3),

$$\frac{1}{N} S_N \to \lambda \int_0^1 \phi(u)\,\overline{f}(u)\,\mathrm{d}u$$

holds with probability 1, *with* $\overline{f}(u)$ *defined above.*

Proof. For every $\delta > 0$ we can find $K > 0$ such that truncated scores

$$\begin{aligned} \overline{a}_N(i) &= a_N(i), \qquad \text{if } |a_N(i)| \le K, \\ &= 0, \qquad \text{otherwise} \end{aligned}$$

satisfy

$$\frac{1}{N} \sum_{i=1}^{N} |a_N(R_i) - \overline{a}_N(R_i)| < \delta \tag{2.5}$$

for all $N > N_0$ and $(R_1, \ldots, R_N)$. Similarly

$$\begin{aligned} \overline{\phi}(u) &= \phi(u), \qquad \text{if } |\phi(u)| \le K, \\ &= 0, \qquad \text{otherwise} \end{aligned}$$

will satisfy

$$\int_0^1 |\phi(u) - \overline{\phi}(u)|\,\mathrm{d}u < \delta.$$

Relation (2.1) entails $\lambda \overline{f}(u) \le 1$, so that we also have

$$\left| \lambda \int_0^1 \phi(u)\,\overline{f}(u)\,\mathrm{d}u - \lambda \int_0^1 \overline{\phi}(u)\,\overline{f}(u)\,\mathrm{d}u \right| < \delta. \tag{2.6}$$

From (2.5) and (2.6) it follows that it is sufficient to prove the theorem for the case when ϕ and $a_N(\cdot)$ have bounded variations over all $(0, 1)$.

Denoting by F_n the empirical distribution function corresponding to $(X_1, \ldots, X_n)$ and by H_N the empirical distribution corresponding to $(X_1, \ldots, X_n, Y_1, \ldots, Y_m)$, we can write

$$\begin{aligned} \frac{1}{N} S_N &= \frac{n}{N} \int_0^1 a_N(1 + [uN])\,\mathrm{d}F_n(H_N^{-1}(u)) \\ &= \frac{n}{N} \int_0^1 a_N(1 + [uN])\,\mathrm{d}\overline{F}_n(u) \\ &= \frac{n}{N} \int_0^1 \phi\,\overline{f}\,\mathrm{d}u + \frac{n}{N} \int_0^1 (a_N - \phi)\,\overline{f}\,\mathrm{d}u + \frac{n}{N} \int_0^1 a_N\,\mathrm{d}(\overline{F}_n - \overline{F}), \end{aligned} \tag{2.7}$$

where

$$\overline{F}(u) = F(H^{-1}(u)), \quad \overline{F}_n(u) = F_n(H_N^{-1}(u)).$$

Now (2.3) and (2.4) entail

$$\left|\frac{n}{N}\int_0^1 (a_N - \phi)\,\overline{f}\,\mathrm{d}u\right| \to 0 \quad \text{as } N \to \infty. \tag{2.8}$$

Moreover, by the Glivenko–Cantelli theorem

$$H_N(x) \to H(x), \quad F_n(y) \to F(y), \quad \overline{F}_n(u) \to \overline{F}(u), \quad \text{as } N \to \infty,$$

the convergence being uniform. Thus, if the variations of $a_N(\cdot)$ are bounded by V, we have

$$\begin{aligned} \left|\frac{n}{N}\int_0^1 a_N\,\mathrm{d}(\overline{F}_n - \overline{F})\right| &= \frac{n}{N}\left|\int (\overline{F}_n - \overline{F})\,\mathrm{d}a_N(\cdot)\right| \\ &= \max_{0<u<1} |\overline{F}_n(u) - \overline{F}(u)|\, V \to 0. \end{aligned} \tag{2.9}$$

Substituting relations (2.8) and (2.9) into (2.7) we can see that the theorem is proved. □

Our next aim is to establish the Bahadur exact slope for S_N-tests used for discriminating the hypothesis of randomness from the alternative considered in Theorem 1. The necessary (extra) large deviation result is contained in the paper by G. Woodworth (1970). For any ρ satisfying

$$\lambda \int_0^1 \phi\,\mathrm{d}u < \rho < \sup_A \left\{ \int_A \phi\,\mathrm{d}u : \int_A \mathrm{d}u = \lambda \right\}$$

we have, under the hypothesis of randomness (the distribution of $(R_1, \ldots, R_N)$ is uniform over the space of all permutations), that

$$\lim_{N\to\infty} \frac{1}{N} \log \mathsf{P}\left(\frac{1}{N} S_N > \rho\right) = -b(\lambda, \rho),$$

where

$$\begin{aligned} b(\lambda,\rho) &= \rho H + (1-\lambda)\log R - \int_0^1 \log\left(e^{H\phi(u)} + R\right)\mathrm{d}u \\ &\quad -\lambda\log\lambda - (1-\lambda)\log(1-\lambda) \end{aligned} \tag{2.10}$$

with (H, R) being the unique solution of the following two equations:

$$\int_0^1 [1 + R\exp(-H\,\phi(u))]^{-1}\,\mathrm{d}u = \lambda, \tag{2.11}$$

$$\int_0^1 \phi(u)\,[1 + R\exp(-H\,\phi(u))]^{-1}\,\mathrm{d}u = \rho. \tag{2.12}$$

It is obvious that $-b(\lambda, \rho)$ is continuous in ρ.

Remark 1. G. Woodworth, in applying his Theorem 1 to the two-sample case, assumes that $a_N(\cdot) \to \phi$ in L_2. But it is possible to show that convergence in L_1 is also sufficient for establishing Property A of his article.

The above rather complicated result simplifies tremendously if applied to the ϕ and ρ used here. We shall put

$$\phi(u) = \log \frac{\overline{f}(u)}{\overline{g}(u)}, \tag{2.13}$$

and

$$\rho = \lambda \int_0^1 \overline{f}(u) \log \frac{\overline{f}(u)}{\overline{g}(u)} \, \mathrm{d}u = \lambda \, K(\overline{f}, \overline{g}), \tag{2.14}$$

$K(\cdot,\cdot)$ denoting the Kullback–Leibler information number.
Note that $(1/N)\, S_N \to \rho$ under the alternative (f, g), and also $(\overline{f}, \overline{g})$, according to Theorem 1, if ϕ is given by (2.13). Introduce

$$\overline{J}(f, g, \lambda) = \lambda \int_0^1 \overline{f} \log \overline{f} \, \mathrm{d}u + (1-\lambda) \int_0^1 \overline{g} \log \overline{g} \, \mathrm{d}u. \tag{2.15}$$

Theorem 2. *Let* (2.3) *be satisfied for ϕ given by* (2.13). *Then, under the hypothesis of randomness,*

$$\lim_{N\to\infty} \frac{1}{N} \log \mathsf{P}\left(\frac{1}{N} S_N > \lambda \, K(\overline{f}, \overline{g}) \right) = -\overline{J}(f, g, \lambda), \tag{2.16}$$

where $\overline{J}(f, g, \lambda)$ is given by (2.15), *and $K(\overline{f}, \overline{g})$ by* (2.14).

Proof. For ϕ and ρ given by (2.13) and (2.14), the equations (2.11) and (2.12) become

$$\int_0^1 \left[1 + R(\overline{g}/\overline{f})^H\right]^{-1} \mathrm{d}u \;=\; \lambda, \tag{2.17}$$

$$\int_0^1 \log \frac{\overline{f}}{\overline{g}} \left[1 + R(\overline{g}/\overline{f})^H\right]^{-1} \mathrm{d}u \;=\; \lambda(K(\overline{f}, \overline{g})), \tag{2.18}$$

which is solved by $H = 1$ and $R = (1-\lambda)/\lambda$, as may be easily seen. Substituting for ρ, ϕ, H and R into (2.1), we obtain

$$\begin{aligned} b(\lambda, \rho) &= \lambda K(f, g) + (1-\lambda) \log[(1-\lambda)/\lambda] \\ &\quad - \int_0^1 \log[\overline{f}/\overline{g} + (1-\lambda)/\lambda] \, \mathrm{d}u - \lambda \log \lambda - (1-\lambda) \log(1-\lambda) \\ &= \lambda K(\overline{f}, \overline{g}) + \int_0^1 \log \overline{g} \, \mathrm{d}u = \overline{J}(f, g, \lambda). \end{aligned}$$

This completes the proof of Theorem 2. □

Putting

$$L_N(t) = \mathsf{P}\left(\frac{1}{N} S_N \geq t\right)$$

we are now able to compute the exact slope of the S_N-test for the (f, g, λ) alternative. See Bahadur (1967) for details.

Corollary 1. *Under conditions of Theorem 2,*

$$\lim_{N\to\infty} \frac{1}{N} \log L_N\left(\frac{1}{N} S_N\right) = -J(f, g, \lambda) \tag{2.19}$$

with probability 1 *under the* (f, g, λ)*-alternative described in Theorem 1.*

Proof. We know that $(1/N) \log L_N(t) \to -b(\lambda, t)$ which is continuous in t. Further, $(1/N) S_N \to K(\overline{f}, \overline{g})$ with probability 1 under the alternative and $b(\lambda, \lambda K(\overline{f}, \overline{g})) = \overline{J}(f, g, \lambda)$. This completes the proof of Corollary 1. □

The most striking feature of Corollary 1 is that the exact slope for S_N is the best possible slope of all.

Corollary 2. *Under conditions of Theorem 2,* $c = 2J(f, g, \lambda)$ *is the best exact slope for testing the hypothesis of randomness against the alternative*

$$q(x_1, \ldots, x_n, y_1, \ldots, y_m) = \prod_{i=1}^{n} f(x_i) \prod_{j=1}^{m} g(y_j).$$

Proof. The hypothesis of randomness H_0 is composite and is satisfied whenever $X_1, X_2, \ldots, X_n, Y_1, \ldots, Y_m$ are independent and have a common distribution, which may be arbitrary but continuous. The best exact slope cannot be larger for H_0 against q than for p against q, where p is a particular member of H_0. Since the best exact slope for testing

$$p(x_1, \ldots, x_n, y_1, \ldots, y_m) = \prod_{i=1}^{n} f_0(x_i) \prod_{j=1}^{m} f_0(y_j)$$

against q is known to satisfy

$$c \leq 2\left[\lambda \int_{-\infty}^{\infty} f \log \frac{f}{g_0}\, dx + (1-\lambda) \int_{-\infty}^{\infty} g \log \frac{g}{f_0}\, dx\right] \tag{2.20}$$

(Raghavachari (1970), Theorem 1), the least favorable p will correspond to f_0 that minimizes the right side of (2.20). It is easy to see that the minimum occurs for

$$f_0 = \lambda f + (1-\lambda) g$$

and that for this choice of f_0 the right side of (2.20) equals $2\overline{J}(f, g, \lambda)$. Thus $2\overline{J}(f, g, \lambda)$ actually is the best possible exact slope. □

Remark 2. The result in Raghavachari (1970) corresponds to a 'one-sample' situation. However, it extends easily to a two-sample situation as well.

3 THE RANK-LIKELIHOOD RATIO STATISTIC

We intend to prove that the rank-likelihood statistic also provides the best possible exact slope. If $(R_1, \ldots, R_N)$ is the vector of ranks and $r = (r_1, \ldots, r_N)$ is a particular permutation, we have under H_0

$$\mathsf{P}(R = r) = 1/N!.$$

If the density q is true, let us denote the probability of the same event as $Q(R = r)$ and put

$$T_N(r) = \log[N!\, Q(R = r)]. \tag{3.1}$$

Net $U_N^{(i)}$ be the ith order statistic from a uniform sample over $(0, 1)$ of size N. Consider again densities f, g and their 'normed' counterparts $\overline{f}$ and $\overline{g}$. Put

$$a_N(i) = E\left\{\log\left[\overline{f}(U_N^{(i)})/\overline{g}(U_N^{(i)})\right]\right\}. \tag{3.2}$$

Theorem 3. *If the scores are given by* (3.2) *and* T_N *is defined by* (3.1), *then for every* $R = (R_1, \ldots, R_N)$

$$\sum_{i=1}^{n} a_N(R_i) + N\int_0^1 \log\overline{g}(u)\,du \le T_N(R). \tag{3.3}$$

Proof. Introducing

$$\Lambda_N = \sum_{j=1}^{n} \log\overline{f}(X_i) + \sum_{j=1}^{m} \log\overline{g}(Y_j)$$

we have

$$T_N = \log \mathsf{E}_1\left(e^{\Lambda_N} \mid R\right),$$

where $\mathsf{E}_1(\cdot \mid R)$ refers to the conditioning given R via the uniform density $p(x_1, \ldots, x_n, y_1, \ldots, y_m) = 1,\ 0 \le x_1, \ldots, y_m \le 1$. Now

$$\begin{aligned}
\log \mathsf{E}_1\left(e^{\Lambda_N} R\right) &\ge \mathsf{E}_1(\Lambda_N \mid R)\\
&= \sum_{i=1}^{n} \mathsf{E}_1\left\{\log\overline{f}(X_i) \mid R\right\} + \sum_{j=1}^{m} \mathsf{E}_1\left\{\log\overline{g}(Y_j) \mid R\right\}\\
&= \sum_{i=1}^{n} a_N(R_i) + N\int_0^1 \log\overline{g}(u)\,du.
\end{aligned}$$

This completes the proof of Theorem 3. □

Theorem 3 entails Theorem 4.

Theorem 4. *The exact slope corresponding to the rank-likelihood ratio statistic* T_N *of* (3.1) *equals* $2\overline{J}(f, g, \lambda)$, *if testing the hypothesis of randomness against the alternative described in Theorem 1, and if* $\log(\overline{f}(u)/\overline{g}(u))$ *is integrable and of bounded variation on every closed subinterval of* $(0, 1)$.

Proof. Let $\widehat{L}_N$ be the level attained by T_N. Since T_N is a likelihood ratio statistic, we have

$$-\frac{1}{N}\log \widehat{L}_N \geq \frac{1}{N} T_N. \tag{3.4}$$

Now, by Theorem 3,

$$\frac{1}{N} T_N \geq \frac{1}{N} S_N + \int_0^1 \log \overline{g}\, du, \tag{3.5}$$

where S_N is given by (2.2) with scores of (3.2). From Theorem 1 we conclude that

$$\frac{1}{N} S_N \to \lambda \int_0^1 \phi(u)\, \overline{f}(u)\, du = \lambda \int_0^1 \overline{f}(u) \log \frac{\overline{f}(u)}{\overline{g}(u)}\, du \tag{3.6}$$

since (2.3) is satisfied for $\phi(u) = \log[\overline{f}(u)/\overline{g}(u)]$ in view of (3.2). (In order to prove it we may use the martingale argument of Lemma 6.1 in Hájek (1961).) Putting (3.4) through (3.6) together, we obtain

$$\liminf_{N\to\infty} \left[-\frac{1}{N}\log \widehat{L}_N\right] \geq \lambda \int_0^1 \overline{f}(u) \log \frac{\overline{f}(u)}{\overline{g}(u)}\, du + \int_0^1 \log \overline{g}(u)\, du = \overline{J}(f, g, \lambda).$$

Again, Theorem 1 in Raghavachari (1970) implies that

$$\limsup_{N\to\infty} \left[-\frac{1}{N}\log \widehat{L}_N\right] \leq \overline{J}(f, g, \lambda).$$

This completes the proof of Theorem 4. □

Thus it was easier to establish the exact slope for the rank-likelihood statistic than for the linear rank statistic S_N; the only tools used were the strong law of large numbers for S_N, the Raghavachari theorem and inequality (3.3).

4 COMPOSITE ALTERNATIVES

We shall now show that a rank statistic can serve efficiently any finite number of alternatives as tested against the hypothesis of randomness.

Theorem 5. *Let $\log \overline{f}(u)/\overline{g}(u)$ be integrable and have bounded variation on every closed subinterval of $(0,1)$. Let $X_1, X_2, \ldots$ have density f_1 and $Y_1, Y_2, \ldots$ density g_1, and $n/N \to \lambda \in (0,1)$.*

Then

$$\liminf_{N\to\infty} \frac{1}{N} T_N \geq \lambda \int_0^1 \overline{f}_1 \log \overline{f}\, du + (1-\lambda) \int_0^1 \overline{g}_1 \log \overline{g}\, du \tag{4.1}$$

with probability 1, where again $\overline{f}_1 = [F_1(H_1^{-1})]'$, $\overline{g}_1 = [G_1(H_1^{-1})]'$.

Proof. Theorem 1 entails

$$\frac{1}{N}\sum_{i=1}^{n} a_N(R_i) + \int_0^1 \overline{f}_1 \log \overline{g}\, du \to \lambda \int_0^1 \overline{f}_1 \log[\overline{f}/\overline{g}]\, du + \int_0^1 \log \overline{g}\, du$$

$$= \lambda \int_0^1 \overline{f}_1 \log \overline{f}\, du + \int_0^1 [1-\lambda \overline{f}_1] \log \overline{g}\, du.$$

Theorem 5 now follows by noting that $1 - \lambda \overline{f}_1 = (1-\lambda)\overline{g}_1$ and by using inequality (3.3).

Consider a composite alternative given by k pairs

$$A = \{(f_j, g_j),\ 1 \leq j \leq k\}. \tag{4.2}$$

For each member of the alternative, let us establish the rank likelihood ratio statistic

$$T_{Nj}(r) = \log\left[N!\, Q_j(R=r)\right], \quad 1 \leq j \leq k,$$

where Q_j corresponds to $X_1, \ldots, X_n$ having density f_j, and $Y_1, \ldots, Y_m$ density g_j. Put

$$U_N(r) = \max_{1\leq j\leq k} T_{Nj}(r).$$

□

Theorem 6. *Let $\log \overline{f}_j(u)/\overline{g}_j(u)$, $1 \leq j \leq k$, be integrable and have bounded variation on every closed subinterval of $(0,1)$, and $n/N \to \lambda \in (0,1)$. The exact slope of U_N equals*

$$\begin{aligned} c &= 2\lambda \int_0^1 \overline{f}_j \log \overline{f}_j\, du + 2(1-\lambda) \int_0^1 \overline{g}_j \log \overline{g}_j\, du \\ &= 2\overline{J}(f_j, g_j, \lambda) \end{aligned}$$

for the jth member of the alternative A of (4.2), and it is the best attainable exact slope.

Proof. We have, under the hypothesis of randomness,

$$\mathsf{P}\left(\frac{1}{N} U_N \geq t\right) \leq k \max_{1\leq j\leq k} \mathsf{P}\left(\frac{1}{N} T_{Nj} \geq t\right) \leq k\, e^{-Nt},$$

since T_{Nj} are log-rank likelihood ratios. Thus the level obtained by U_N, say $\widehat{M}_N$, satisfies

$$\frac{1}{N}\log \widehat{M}_N \leq \frac{1}{N}\log k - \frac{1}{N}U_N.$$

On the other hand, if (f_j, g_j) obtains, then

$$\limsup_{N\to\infty}\left(-\frac{1}{N}U_N\right) \leq \limsup_{N\to\infty}\left(-\frac{1}{N}T_{Nj}\right) \leq -\overline{J}(\overline{f}_j, \overline{g}_j, \lambda)$$

according to (3.3) and Theorem 1. Consequently,

$$\limsup_{N\to\infty}\left\{\frac{1}{N}\log \widehat{M}_N\right\} \leq -\overline{J}(f_j, g_j, \lambda).$$

However, the sharp inequality can hold only with probability 0, since $-\overline{J}(f_j, g_j, \lambda)$ is a lower bound according to the above mentioned result by Raghavachari (1970). This completes the proof of Theorem 6. □

As a side product we obtain the following inequalities for the log-rank-likelihood ratio statistic.

Theorem 7. *If under conditions of Theorem 6, T_{Nj} refers to (f_j, g_j) and (f_i, g_i) obtains, then with probability* 1

$$\begin{aligned} &\lambda\int_0^1 \overline{f}_i \log \overline{f}_j\,du + (1-\lambda)\int_0^1 \overline{g}_i \log \overline{g}_j\,du \\ \leq\ &\liminf_{N\to\infty}\left[\frac{1}{N}T_{Nj}\right] \leq \limsup_{N\to\infty}\left[\frac{1}{N}T_{Nj}\right] \\ \leq\ &\lambda\int_0^1 \overline{f}_i \log \overline{f}_i\,du + (1-\lambda)\int_0^1 \overline{g}_i \log \overline{g}_i\,du. \end{aligned} \tag{4.3}$$

Proof. The first inequality is obtained from Theorem 5, and the violation of the last inequality would conflict with the Raghavachari result, since for (f_i, g_i) the exact slope of U_N would exceed $2\overline{J}(f_i, g_i, \lambda)$. □

Remark 3. If $f_i = f_j$, $g_i = g_j$, the inequalities become equalities and we obtain a special case of the Berk–Savage theorem in Berk and Savage (1968) under somewhat relaxed conditions.

Remark 4. The above result for the finite alternative A could be extended to infinite but compactable alternatives in a standard way. This is not done here.

ACKNOWLEDGEMENTS

My interest in the topic was initiated and catalysed by inspiring discussion and letter exchanges I had with Professor R. R. Bahadur, I. R. Savage and J. Sethuraman during my stay at the Florida State University. I also appreciate Professor Raghavachari's sending me a preprint of his paper (1970).

The research was supported by the Army, Navy and Air Force under Office of Naval Research Contract No. NONR 988(08), Task Order NR 042-004. Reproduction in whole or in part is permitted for any purpose by the United States Government.

REFERENCES

Bahadur, R. R. (1967). Rates of convergence of estimates and test statistics. *Ann. Math. Statist. 38*, 303–324.

Bahadur, R. R. and Raghavachari, M. (1970). Some asymptotic properties of likelihood ratios on general sample spaces. *Proc. Sixth Berkeley Symp. Math. Statist. Prob. I*, 129–152.

Berk, R. and Savage, I. R. (1968). The information in a rank order and the stopping time of some associated SPRT's. *Ann. Math. Statist. 39*, 1661–1674.

Hájek, J. (1961). Some extensions of the Wald–Wolfowitz–Noether theorem. *Ann. Math. Statist. 35*, 506–523.

Raghavachari, M. (1970). On a theorem of Bahadur on the rate of convergence of test statistics. *Ann. Math. Statist. 41*, 1695–1699.

Woodworth, G. (1970). Large deviations and Bahadur efficiency in linear rank statistics. *Ann. Math. Statist. 41*, 251–283.

CHAPTER 35

Regression Designs in Autoregressive Stochastic Processes

Co-author: George Kimeldorf, Department of Statistics,
Florida State University, Tallahassee, Florida 32 306.
Ann. Statist. 2 (1974), 520–527.
Reproduced by permission of The Institute of Mathematical Statistics.

This chapter extends some recent results on asymptotically optimal sequences of experimental designs for regression problems in stochastic processes. In the regression model $Y(t) = \beta f(t) + X(t)$, $0 \le t \le 1$, the constant β is to be estimated based on observations of $Y(t)$ and its first $m-1$ derivatives at each of a set T_n of n distinct points. The function f is assumed known, as is the covariance kernel of $X(t)$, a zero-mean mth order autoregressive process. Under certain conditions, we derive a sequence $\{T_n\}$ of experimental designs which are asymptotically optimal for estimating β.

1 INTRODUCTION AND SUMMARY

Let $\{Y(t) : 0 \le t \le 1\}$ be a stochastic process of the form

$$Y(t) = \beta f(t) + X(t), \tag{1.1}$$

where β is an unknown constant, $\{X(t)\}$ is a zero-mean, real stochastic process with known covariance kernel $k(s,t) = \mathsf{E}\, X(s)\, X(t)$, and f is a known function of the form

$$f(t) = \int_0^1 k(s,t)\,\phi(s)\,\mathrm{d}s. \tag{1.2}$$

Given a subset T of $[0,1]$, we observe $Y(t)$ and its first $m-1$ derivatives at each point $t \in T$. The parameter β is to be estimated by an unbiased estimator

Collected Works of Jaroslav Hájek – With Commentary
Edited by M. Hušková, R. Beran and V. Dupač
Published in 1998 by John Wiley & Sons Ltd.

$\hat{\beta}_T$ which is linear in the observation set $\{Y^{(j)}(t) : j = 0, 1, \ldots, m-1; t \in T\}$ and which has minimum variance among all such estimators. Given any n, let $\mathcal{D}_n$ be the class of all subsets T_n of $[0, 1]$ containing exactly n points. Then the problem of optimal design is to find a set $\hat{T}_n \in \mathcal{D}_n$ for which $\mathsf{var}\, \hat{\beta}_{\hat{T}_n} \leq \mathsf{var}\, \hat{\beta}_{T_n}$ for all $T_n \in \mathcal{D}_n$. Because optimal designs when they exist are generally very difficult to compute, Sacks and Ylvisaker (1966) introduced the concept of an asymptotically optimal sequence of designs.

Definition. A sequence $\{\hat{T}_n\}$ of designs is said to be *asymptotically optimal* (in the sense of Sacks and Ylvisaker) if

$$\lim_{n\to\infty} \frac{\mathsf{var}\, \hat{\beta}_{\hat{T}_n} - \mathsf{var}\, \hat{\beta}_T}{\inf \mathsf{var}\, \hat{\beta}_{T_n} - \mathsf{var}\, \hat{\beta}_T} = 1, \tag{1.3}$$

where $T = [0, 1]$ and the infimum is taken over all $T_n \in \mathcal{D}_n$.

Sacks and Ylvisaker (1969) considered processes $\{X(t)\}$ of the form

$$X(t) = \int_0^t \int_0^{t_{m-1}} \cdots \int_0^{t_2} V(t_1)\, \mathrm{d}t_1 \cdots \mathrm{d}t_{m-1},$$

where

$$\lim_{s\to t-} \frac{\partial}{\partial s} \mathsf{E}[V(s)\, V(t)] - \lim_{s\to t+} \frac{\partial}{\partial s} \mathsf{E}[V(s)\, V(t)] = c,$$

where c is a positive constant. Under certain assumptions, they showed that the deigns $T_n = \{t_{in} : i = 1, 2, \ldots, n\}$ defined by

$$\int_0^{t_{in}} [\phi(t)]^{2/(2m+1)} \mathrm{d}t = \frac{i}{n} \int_0^1 [\phi(t)]^{2/(2m+1)} \mathrm{d}t$$

form an asymptotically optimal sequence. Wahba (1971) recently considered processes $\{X(t)\}$ satisfying the stochastic differential equation

$$L\, X(t) = \mathrm{d}W(t)/\mathrm{d}t, \tag{1.4}$$

where $W(t)$ is Brownian motion and L is an mth order linear differential operator of the special form

$$L = D \frac{1}{\omega_m(t)}\, D \frac{1}{\omega_{m-1}(t)} \cdots D \frac{1}{\omega_1(t)}, \tag{1.5}$$

where $\omega_j(t) > 0$ and $\omega_j(t) \in C^{m-j}$. (The symbol D denotes differentiation.) A differential operator L of the form (1.5) has the property that its null space is spanned by an extended complete Tchebyscheff system.

This chapter considers a much broader class of processes $\{X(t)\}$, namely those processes which satisfy a linear stochastic differential equation (1.4), where L is any mth order linear differential operator of the form

$$L = \sum_{j=0}^{m} a_j(t)\, D^j, \quad a_j(t) \in C^j, \ a_m(t) \neq 0; \tag{1.6}$$

thus, $\{X(t)\}$ is an mth order autoregressive process. The goal of this chapter is to prove, under certain mild conditions, that an asymptotically optimal sequence of designs for estimating β in (1.1) is given by

$$\int_0^{t_{im}} [\phi(t)/a_m(t)]^{2/(2m+1)} \mathrm{d}t = \frac{i}{n} \int_0^1 [\phi(t)/a_m(t)]^{2/(2m+1)} \mathrm{d}t.$$

This result, with a form of error estimate, is stated precisely at the beginning of Section 3.

There are many examples of processes for which the results of Sacks and Ylvisaker (1966, 1968, 1969) and of Wahba (1971) are not applicable, but for which the present results are. One such process $\{X(t)\}$ is given by

$$X(t) = \frac{1}{2\pi} \int_0^t \frac{\sin 2\pi(t-s)}{1+s}\, \mathrm{d}W(s),$$

which satisfies (1.4) when $L = (1+t)\, D^2 + 4\pi^2(1+t)$.

2 SOME LEMMAS

We restrict our attention to an interval $[0, u]$, $0 < u \leq 1$, where u is fixed. For any t, $0 \leq t < u$, let $\psi(\cdot, t)$ be the unique solution on the interval $[t, u]$ of the differential equation $L\psi(\cdot, t) = 0$ subject to the initial conditions

$$\psi_{j0}(t^+, t) = \delta_{j,m-1}/a_m(t), \quad j = 0, 1, \ldots, m-1,$$

where δ is the Kronecker delta and the notation $\psi_{pq}(a, b)$ means

$$\left.\frac{\partial^{p+q}}{\partial s^p\, \partial t^q}\, \psi(s, t)\right|_{s=a,\, t=b}$$

Let $g(s, t) = \psi(s, t)$ if $s \geq t$, and $g(s, t) = 0$ otherwise. Then g is the Green function for the differential equation $Lf = h$ subject to the initial conditions $f^{(j)}(0) = 0$, $j = 0, 1, \ldots, m-1$. Thus, for fixed t, $g(\cdot, t) \in C^m$ on $(0, t)$ and on (t, u) while $g_{m-1,0}(t^+, t) - g_{m-1,0}(t^-, t) = 1/a_m(t)$. Similarly, if L^* is the operator $L^* f = \sum_{j=0}^{m} (-1)^j\, D^j\, [a_j f]$, which is adjoint to L, then the Green function for the differential equation $L^* f = h$ subject to the initial conditions $f^{(j)}(u) = 0$, $j = 0, 1, \ldots, m-1$, is $g^*(s, t) = g(t, s)$.

Let $\{X(t)\}$ be the process defined by

$$X(t) = \int_0^u g(t,s)\,\mathrm{d}W(s). \tag{2.1}$$

We state some well-known results about the process $\{X(t) : 0 \le t \le u\}$. The process $\{X(t)\}$ satisfies (1.4) and the initial conditions

$$X^{(j)}(0) = 0, \quad j = 0, 1, \ldots, m-1. \tag{2.2}$$

Moreover, $\{X(t)\}$ is an mth order autoregressive process so that for $0 < s_1 < t < s_2 < u$, $X(s_1)$ and $X(s_2)$ are conditionally uncorrelated given the set $M = \{X^{(j)}(t) : j = 0, 1, \ldots, m-1\}$, and hence

$$\mathsf{E}\,\{X(s_1) - \mathsf{E}[X(s_1)\,|\,M]\}\,\{X(s_2) - \mathsf{E}[X(s_2)\,|\,M]\} = 0. \tag{2.3}$$

We associate with the process $\{X(t)\}$ two Hilbert spaces. The first is the space $\mathcal{X}$ which is the closure of the vector space spanned by $\{X(t) : 0 \le t \le u\}$ with an inner product determined by $\langle X(s),\, X(t)\rangle = \mathsf{E}\,X(s)\,X(t) = k(s,t)$. The second is the space $\mathcal{H}$ of all real functions h on $[0,u]$ such that (i): For $j = 0, 1, \ldots, m-1$, $h^{(j)}$ is absolutely continuous and $h^{(j)}(0) = 0$, and (ii): Lh is square integrable. The inner product in $\mathcal{H}$ is given by $\langle h_1, h_2\rangle = \int (Lh_1)\,(Lh_2)$ and $\mathcal{H}$ is a reproducing kernel Hilbert space (RKHS) with kernel k. There is an isomorphism between $\mathcal{X}$ and $\mathcal{H}$ which preserves inner products. The image under this isomorphism of any random variable $Z \in \mathcal{X}$ is the function $h(t) = \mathsf{E}[Z\,X(t)]$. For details on the above facts, the reader is referred to Hájek (1962), Parzen (1961) and Dolph and Woodbury (1952).

Using the foregoing theory with $u = 1$ it can be shown that for any design T, $\mathsf{var}\,\hat{\beta}_T = ||P_T\,f||^{-2}$, where the norm is taken in $\mathcal{H}$ and P_T is the projection onto the subspace spanned by $\{k(\cdot,t) : t \in T\}$. Hence the definition (1.3) of asymptotic optimality is equivalent to

$$\lim_{n\to\infty} \frac{||f - P_{\hat{T}_n} f||^2}{\inf ||f - P_{T_n} f||^2} = 1, \tag{2.4}$$

where the infimum is taken over all $T_n \in \mathcal{D}_n$.

For fixed u, $0 \le u \le 1$, let $\mathcal{U} = \{X^{(j)}(u) : j = 0, 1, \ldots, m-1\}$ and P_u be the projection operator in $\mathcal{X}$ onto the subspace spanned by $\mathcal{U}$. Consider the stochastic process $\{\overline{X}(t) : 0 \le t \le u\}$, where $\overline{X}(t) = X(t) - P_u\,X(t)$. The RKHS $\overline{\mathcal{H}}$ associated with the process $\{\overline{X}(t)\}$ can easily be shown to consist of the subspace of all functions $h \in \mathcal{H}$ for which $h^{(j)}(u) = 0$ for all $j = 0, 1, \ldots, m-1$.

Lemma 1. *Let ϕ be a continuous function and let*

$$\begin{aligned} w(t) &= \mathsf{E}\left[\int_0^u \phi(s)\,\overline{X}(s)\,\mathrm{d}s\right]\overline{X}t \\ &= \mathsf{E}\left[\int_0^u \phi(s)\,X(s)\,\mathrm{d}s - P_u \int_0^u \phi(s)\,X(s)\,\mathrm{d}s\right]X(t). \end{aligned} \tag{2.5}$$

Then $w(t)$ is the unique solution to the 2mth order linear differential equation:

$$L^* \, Lw(t) = \phi(t) \tag{2.6}$$

subject to the boundary conditions

$$w^{(j)}(0) = w^{(j)}(u) = 0, \quad j = 0, 1, \ldots, m-1. \tag{2.7}$$

Proof. To show that the solution to (2.6) and (2.7) is unique, we note that if $L^* \, Lv(t) = 0$, then $0 = \int (L^* \, Lv) \, v = \int (Lv)^2$ and hence $Lv = 0$ so that by (2.7) we have $v = 0$. Now $w(t)$ defined by (2.5) clearly satisfies (2.7). To show it satisfies (2.6) let $\overline{h}_t$ be the function in $\overline{\mathcal{H}}$ corresponding to $\overline{X}(t)$. From m suitable functions $\alpha_j(t)$,

$$\begin{aligned} \overline{X}(t) &= X(t) - \sum_{j=0}^{m-1} \alpha_j(t) \, X^{(j)}(u) \\ &= \int_0^u \left[g(t,s) - \sum \alpha_j(t) \, g_{j0}(u,s) \right] \mathrm{d}W(s). \end{aligned}$$

Hence $L\overline{h}_t(s) = g(t,s) - \sum \alpha_j(t) \, g_{j0}(u,s)$ and

$$Lw(s) = \int_0^u \phi(t) \left[g(t,s) - \sum \alpha_j(t) \, g_{j0}(u,s) \right] \mathrm{d}t.$$

Now $L^* g_{j0}(u,s) = 0$ and $L^* \int_0^u \phi(t) \, g(t,s) \, \mathrm{d}t = \phi(s)$; hence $L^* \, Lw(s) = \phi(s)$ and the lemma is proved. □

Lemma 2. $\operatorname{var} \int_0^u \phi(t) \, \overline{X}(t) \mathrm{d}t = \int_0^u \phi(t) \, w(t) \, \mathrm{d}t$, *where w is defined by (2.5).*

Proof.

$$\begin{aligned} \operatorname{var} \int_0^u \phi(t) \, \overline{X}(t) \, \mathrm{d}t &= \mathsf{E}\left[\int_0^u \phi(t) \int_0^u \phi(s) \, \overline{X}(s) \, \overline{X}(t) \, \mathrm{d}s\mathrm{d}t \right] \\ &= \int_0^u \phi(t) \, \mathsf{E}\left[\int_0^u \phi(s) \, \overline{X}(s) \, \overline{X}(t) \, \mathrm{d}s \right] \mathrm{d}t \\ &= \int_0^u \phi(t) \, w(t) \, \mathrm{d}t. \end{aligned}$$

□

In the preceding discussion u has been fixed. We now allow u to vary and indicate the dependence on u by a subscript.

Lemma 3. *For* $j = 0, 1, \ldots, m$,

$$\lim_{u \to 0+} u^{j-m} w_u^{(m+j)}(0) = \frac{(-1)^j \binom{m}{j} (m+j)!}{(2m)!\, a_m^2(0)} \phi(0). \tag{2.8}$$

Proof. By (2.6) we have

$$w_u^{(2m)}(t) = (-1)^m a_m^{-2}(t) \left[\phi(t) - \sum_{j=0}^{2m-1} b_j(t)\, w_u^{(j)}(t) \right], \tag{2.9}$$

where the functions $b_j(t)$, which are expressible in terms of the coefficients $a_j(t)$, are of no present interest. Because of the boundary conditions at 0, the Taylor expansions for $w_u(t)$ and its first $m-1$ derivatives take the form

$$w_u^{(i)}(t) = \sum_{j=0}^{m} \frac{t^{m+j-i}\, w_u^{(m+j)}(0)}{(m+j-i)!} + o(t^{2m-i}), \tag{2.10}$$

which, by (2.9), equals

$$\sum_{j=0}^{m-1} \frac{t^{m+j-i}\, w_u^{(m+j)}(0)}{(m+j-i)!} + (-1)^m t^{2m-i} \frac{\phi(0) - \sum_{j=0}^{2m-1} b_j(0)\, w_u^{(j)}(0)}{a_m^2(0)\, (2m-i)!} + o(t^{2m-i}).$$

Hence the boundary conditions imply

$$0 = \sum_{j=0}^{m-1} \frac{u^{m+j-i}\, w_u^{(m+j)}(0)}{(m+j-i)!} + (-1)^m u^{2m-i} \frac{\phi(0) - \sum_{j=m}^{2m-1} b_j(0)\, w_u^{(j)}(0)}{a_m^2(0)\, (2m-i)!} + o(u^{2m-i}).$$

Multiplying by u^{i-2m} and denoting $u^{j-m} w_u^{(m+j)}(0)$ by $x_j(u)$, we derive

$$\frac{(-1)^{(m+1)} \left[\phi(0) - \sum_{j=0}^{m-1} u^{m-j}\, b_{m+j}(u)\, x_j(u) \right]}{(2m-i)!\, a_m^2(0)} = \sum_{j=0}^{m-1} \frac{x_j(u)}{(m+j-i)!} + o(1), \tag{2.11}$$

a system of m identities in u. The system (2.11) implies that for $j = 0, 1, \ldots, m-1$, $x_j = \lim_{u \to 0+} x_j(u)$ exist and satisfy

$$\frac{(-1)^{m+1} \phi(0)}{(2m-i)!\, a_m^2(0)} = \sum_{j=0}^{m-1} \frac{x_j}{(m+j-i)!}, \quad i = 0, 1, \ldots, m-1. \tag{2.12}$$

But equation (8) of Riordan (1968, p. 10) shows that the solution to the system (2.12) is

$$x_j = \frac{(-1)^j \binom{m}{j} (m+j)!\, \phi(0)}{(2m)!\, a_m^2(0)}. \tag{2.13}$$

The proof of the lemma is complete because (2.9) shows that (2.8) also holds for $j = m$. □

Lemma 4.

$$\lim_{u \to 0+} u^{-2m-1} \operatorname{var} \int_0^u \phi(t)\, \overline{X}(t)\, dt = \frac{\phi^2(0)\,(m!)^2}{a_m^2(0)\,(2m)!\,(2m+1)!}.$$

Proof. From (2.10) with $i = 0$, we get

$$\begin{aligned}
&\lim_{u \to 0+} u^{-2m-1} \int_0^u \phi(t)\, w_u(t)\, dt \\
&= \lim_{u \to 0+} \sum_{j=0}^{m} \frac{u^{j-m}\, w_u^{(m+j)}(0)\, \phi(u_j)}{(m+j+1)!}, \quad \text{where } 0 \le u_j \le u \\
&= \frac{\phi^2(0)}{(2m)!\, a_m^2(0)} \sum_{j=0}^{m} \frac{(-1)^j \binom{m}{j}}{m+j+1} \\
&= \frac{\phi^2(0)\, B(m+1,\, m+1)}{(2m)!\, a_m^2(0)},
\end{aligned}$$

where B is the (complete) Beta function. □

In the next section we shall need the following lemma, whose proof follows from Lemma 4 by letting $Z(t) = X(t - t_1)$.

Lemma 5. *Suppose $0 \le t_1 < t_2 \le 1$. Let $\Delta = t_2 - t_1$ and P be the projection operator in $\mathcal{X}$ onto the subspace spanned by $\{X^{(j)}(t_1),\, X^{(j)}(t_2) : j = 0, 1, \ldots, m-1\}$. Then*

$$\operatorname{var} \int_{t_1}^{t_2} \phi(t)\, [X(t) - P\, X(t)]\, dt = \frac{(m!)^2\, \Delta^{2m+1}\, \phi^2(t_1)}{a_m^2(t_1)\,(2m)!\,(2m+1)!} + o(\Delta^{2m+1}).$$

3 THE PRINCIPAL RESULT

We shall prove the following theorem.

Theorem. *Let $\{X(t) : 0 \le t \le 1\}$ be a real stochastic process satisfying the stochastic differential equation $L\,X(t) = \mathrm{d}W(t)/\mathrm{d}t$, where W is the Wiener process and L is any mth order linear differential operator,*

$$L = \sum_{j=0}^{m} a_j(t)\, D^j, \quad a_j \in C^j, \tag{3.1}$$

whose leading coefficient $a_m(t)$ is never zero. Let $f(t) = \int k(s,t)\,\phi(s)\,\mathrm{d}s$, where ϕ is non-vanishing and continuous on $[0,1]$ and where $k(s,t) = \mathsf{E}[X(s)\,X(t)]$. If, in the regression model $Y(t) = \beta\, f(t) + X(t)$, β is to be estimated based on observations $\{Y^{(j)}(t_{in}) : j = 0, 1, \ldots, m-1; t_{in} \in T_n\}$, then an asymptotically optimal sequence of designs is $\{\hat{T}_n\} = \{\{\hat{t}_{in} : i = 1, 2, \ldots, n\}\}$, where

$$\int_0^{\hat{t}_{in}} [\phi(t)/a_m(t)]^{2/(2m+1)}\mathrm{d}t = \frac{i}{n}\int_0^1 [\phi(t)/a_m(t)]^{2/(2m+1)}\mathrm{d}t. \tag{3.2}$$

Moreover,

$$\lim_{n\to\infty} n^{2m}\|f - \hat{P}f\|^2 = \frac{(m!)^2}{(2m)!\,(2m+1)!}\left[\int_0^1 [\phi(t)/a_m(t)]^{2/(2m+1)}\mathrm{d}t\right]^{2m+1}, \tag{3.3}$$

where $\hat{P}$ is the projection in $\mathcal{H}$ onto the space spanned by $\{k^{(j)}(\hat{t}_{in}, \cdot)\}$.

Proof. Without loss of generality we can assume $\{X(t)\}$ satisfies (2.2). (Otherwise we append the point $t_{0n} = 0$ to the design without affecting its asymptotic optimality.) Similarly, we can assume $t_{nn} = 1$ for every design. Also, we note that any candidate $\{T_n\}$ for an asymptotically optimal sequence must have

$$\lim_n \max_i\,(t_{i+1,n} - t_{in}) = 0. \tag{3.4}$$

Now, let $\{T_n\} = \{\{t_{in} : i = 1, 2, \ldots, n\}\}$ be any sequence of designs satisfying (3.4) and $t_{nn} = 1$. For notational convenience we suppress the second subscript of t_{in}, we denote the function $[\phi(t)/a_m(t)]^2$ by $h(t)$, and we denote the constants $2m+1$, $(2m+1)/(2m)$ and $(m!)^2\,[(2m)!\,(2m+1)!]^{-1}$ by p, q, and c, respectively. By (2.3)

$$\begin{aligned}\|f - Pf\|^2 &= \mathsf{var}\int_0^1 \phi(t)\,[X(t) - P\,X(t)]\,\mathrm{d}t\\ &= \mathsf{var}\sum_{i=0}^{n-1}\int_{t_i}^{t_{i+1}} \phi(t)\,[X(t) - P\,X(t)]\,\mathrm{d}t\\ &= \sum_{i=0}^{n-1}\mathsf{var}\int_{t_i}^{t_{i+1}} \phi(t)\,[X(t) - P\,X(t)]\,\mathrm{d}t,\end{aligned}$$

where $t_0 = 0$ and P is the projection in $\mathcal{H}$ onto the subspace spanned by $\{k^{(j)}(t_i, \cdot)\}$. Since (3.4) holds, Lemma 5 implies

$$\|f - Pf\|^2 = c\sum_{i=0}^{n-1} h(t_i)\,\left[(t_{i+1} - t_i)^p + o(t_{i+1} - t_i)^p\right]. \tag{3.5}$$

Having established (3.5) we complete the proof in a manner similar to that of Sacks and Ylvisaker (1966). Hölder's inequality implies

$$\begin{aligned} \|f - Pf\|^2 &\geq c\,n^{-p/q}\left[\sum_{i=0}^{n-1} h^{1/p}(t_i)\,(t_{i+1} - t_i)\right]^p + o(n^{-2m}) \qquad (3.6) \\ &= c\,n^{-2m}\left[\int_0^1 h^{1/p}(t)\,\mathrm{d}t\right]^p + o(n^{-2m}). \end{aligned}$$

To evaluate $\|f - \hat{P}f\|^2$ we apply a mean value theorem to (3.2), which yields

$$(\hat{t}_{i+1} - \hat{t}_i)\; h^{1/p}(t_i) = n^{-1}\int_0^1 h^{1/p}(t)\,\mathrm{d}t + o(n^{-1}),$$

and apply (3.5) which yields

$$\begin{aligned} \|f - \hat{P}f\|^2 &= c\sum_{i=0}^{n-1} n^{-p}\left[\int_0^1 h^{1/p}(t)\,\mathrm{d}t\right]^p + o(n^{-2m}) \qquad (3.7) \\ &= c\,n^{-2m}\left[\int_0^1 h^{1/p}(t)\,\mathrm{d}t\right]^p + o(n^{-2m}). \end{aligned}$$

This proves (3.3), and the asymptotic optimality of $\{\hat{T}_n\}$ follows from (3.6) and (3.7). □

4 REMARKS

Remark 1. The conditions of the theorem can be weakened somewhat. First, the condition that ϕ never vanish can be replaced by a condition which restricts the behavior of ϕ in the neighborhood of any zero. (For example, see Sacks and Ylvisaker (1969).) Second, although we assume that $\{X(t)\}$ satisfies the stochastic differential equation (1.4), we use only properties of the first two moments of $\{X(t)\}$, namely $\mathsf{E}X(t) = 0$ and $\mathsf{E}[X(s)\,X(t)] = k(s,t) = \int_0^1 g(t,\cdot)\,g(s,\cdot)$, where g is the Green function corresponding to L. Hence the results of this chapter are applicable to all processes of the form $X(t) = \int_0^1 g(t,s)\,\mathrm{d}U(s)$, where $\{U(t)\}$ is a random orthogonal process with structural measure equal to the Lebesgue measure.

Remark 2. There is a close relation between the statistical design problem discussed here and certain integral approximation problems. The interested reader is referred to Sacks and Ylvisaker (1970). (Also, see Karlin (1972).)

Remark 3. Many of the results of this paper can be extended to the more general regression model

$$Y(t) = \sum \beta_j\, f_j(t) + X(t),$$

where $\mathsf{var}\,\hat{\beta}$ is replaced by a suitable norm on the matrix $\mathsf{var}\,\hat{\beta}$. (See Sacks and Ylvisaker (1968).)

ACKNOWLEDGEMENTS

We thank Professor T. Hallam and J. Sethuraman for helpful conversation with one of the authors, and Professor J. Sacks for correcting an error in an earlier version.

This research was sponsored in part by Grant No. Gu 2612 from the National Science Foundation to the Florida State University.

REFERENCES

Dolph, C. L. and Woodbury, M. A. (1952). On the relation between Green's functions and covariances of certain stochastic processes and its application to unbiased linear prediction. *Trans. Amer. Math. Soc.* *72*, 519–550.

Hájek, J. (1962). On linear statistical problems in stochastic processes. *Czechoslovak Math. J.* *12*, 404–444.

Karlin, S. (1972). On a class of best nonlinear approximation problems. *Bull. Amer. Math. Soc.* *78*, 43–49.

Parzen, E. (1961). An approach to time series analysis. *Ann. Math. Statist.* *32*, 951–989.

Riordan, J. (1968). *Combinatorial Identities.* Wiley, New York.

Sacks, J. and Ylvisaker, D. (1966). Designs for regression with correlated errors. *Ann. Math. Statist.* *37*, 66–89.

Sacks, J. and Ylvisaker, D. (1968). Design for regression with correlated errors; many parameters. *Ann. Math. Statist.* *39*, 49–69.

Sacks, J. and Ylvisaker, D. (1969). Design for regression problems with correlated errors, III. *Ann. Math. Statist.* *41*, 2057–2074.

Sacks, J. and Ylvisaker, D. (1970). Statistical designs and integral approximation. In: R. Pyke, ed. *Seminar of the Canadian Mathematical Congress on Time Series and Stochastic Processes.*

Wahba, G. (1971). On the regression design problem of Sacks and Ylvisaker. *Ann. Math. Statist.* *42*, 1035–1053.

CHAPTER 36

Asymptotic Theories of Sampling with Varying Probabilities without Replacement

In: Proc. Prague Symp. on Asymptotic Statistics 1973 (J. Hájek, ed.), 1, 127–138.
Reproduced by permission of Hájek family: Al. Hájková, H. Kazárová and Al. Slámová.

A finite population will be identified as the set of natural numbers $\{1, 2, \ldots, N\}$ and a sample s will be an arbitrary subset of it. A sampling design (plan) will be defined by probabilities $p(s)$ associated with all possible samples. The probability of inclusion for the unit i will be the probability of obtaining a sample which includes the unit

$$\pi_i = \mathsf{P}(s \ni i) = \sum_{s \ni i} p(s), \quad 1 \le i \le N.$$

Similarly, for a pair of units (i, j) we may define probabilities of inclusion of second order:

$$\pi_{ij} = \mathsf{P}(s \ni i,\, j), \quad 1 \le i \ne j \le N.$$

The purpose of sampling is to estimate a total

$$Y = \sum_{i=1}^{N} y_i$$

of a variable 'y' under issue.

The simple estimate (Horwitz–Thompson est.) is defined as follows:

$$\hat{Y} = \sum_{i \in s} \frac{y_i}{\pi_i}.$$

Collected Works of Jaroslav Hájek – With Commentary
Edited by M. Hušková, R. Beran and V. Dupač
Published in 1998 by John Wiley & Sons Ltd.

Given some explanatory variables 'x_j' whose totals X_j are a priori known, we may estimate Y by a regression estimate

$$\tilde{Y} = \hat{Y} - \sum_{j=1}^{m} \hat{\beta}_j (\hat{X}_j - X_j),$$

where $\hat{\beta}_j$ is the solution of

$$\sum_{i \in s} \left(y_i - \sum_{j=1}^{m} \beta_j \, x_{ji} \right)^2 \frac{t_i}{\pi_i} = \text{minimum}$$

for some weights t_i. If $m = 1$ and $t_i = 1/x_i$, the regression estimate turns into the ratio estimate:

$$\breve{Y} = X \, \frac{\hat{Y}}{\hat{X}}.$$

Now we shall ask which sampling with varying probabilities without replacement is a most natural extension of simple random sampling. We shall consider the following four designs:

(a) rejective sampling;
(b) Sampford–Durbin sampling;
(c) successive sampling;
(d) randomized systematic sampling.

Rejective sampling may be defined as sampling with replacement with drawing probabilities α_i at each draw, conditioned by the requirement that all drawn units are distinct. As soon as we obtain a replication, we reject the whole partially built up sample and start completely anew. Required inclusion probabilities π_i can be obtained by a proper choice of drawing probabilities α_i. (See Hájek (1964).) Sampford–Durbin sampling is defined similarly, but the first unit is drawn with drawing probabilities

$$\alpha_i^{\mathrm{I}} = \frac{1}{n} \, \pi_i, \quad 1 \leq i \leq N,$$

and in the next $(n-1)$ draws we utilize drawing probabilities

$$\alpha_i^{\mathrm{II}} = \lambda \frac{\pi_i}{1 - \pi_i}, \quad 1 \leq i \leq N,$$

where λ is a proper constant such that $\sum_{i=1}^{N} \alpha_i^{\mathrm{II}} = 1$. Again in Sampford–Durbin sampling a sample is accepted only if all selected units are distinct. (See Sampford (1967).)

In successive sampling we draw units one by one with drawing probabilities α_i and, if a replication occurs, we reject only this replication and continue until we have n distinct units in the sample. (See Rosén (1972).)

Randomized systematic sampling is systematic sampling applied to a randomly permuted population. We shall not be able to say much about this sampling, since the only known results obtained by Hartley–Rao (1962) seem to be incorrect under their conditions.

Let $I_i = 1$ if s includes i, and $I_i = 0$ in the opposite case. Thus I_i is the indicator of inclusion for the unit i. We have

$$-\mathsf{cov}(I_i, I_j) = \pi_i \pi_j - \pi_{ij}, \quad 1 \le i \ne j \le N.$$

Under simple random sampling

$$-\mathsf{cov}(I_i, I_j) = \left(\frac{n}{N}\right)^2 \frac{1}{n} \frac{N-n}{N-1} > 0,$$

i. e. the inclusions are negatively correlated for all pairs (i, j). This property is enjoyed also for designs (a) and (b). For (a) it has been shown by the present author and also by J. Lanke (not published) and for (b) by Sampford (1967). For (c) it holds for $n = 2$ but generally not for $n > 3$ as was shown by D. Singh (1954). The status for (d) is probably not known.

Another property of simple random sampling is that it maximizes the entropy $-\sum p(\Delta) \log p(\Delta)$ in the family of all designs of size n and with given π_is, $1 \le i \le N$. It has been shown (Hájek (1959)) that this property is enjoyed by (a).

A further property of simple random sampling is the simple form of the negative covariances $\pi_i \pi_j - \pi_{ij}$: they are constant. Let us ask which of the above designs shares this property. First of all we must define what we mean under simplicity. It may be easily shown that for varying π_is there is no design of fixed size such that

$$\pi_i \pi_j - \pi_{ij} = \begin{array}{ll} c, & \text{or} \\ c\,\pi_i \pi_j, & \text{or} \\ c\left[\pi_i(1-\pi_i)\,\pi_j(1-\pi_j)\right]^{1/2}. & \end{array}$$

However, in nonpathological situations there will be a design such that for some constants $c_1, \ldots, c_N$ the relation

$$\pi_i \pi_j - \pi_{ij} = c_i c_j, \quad 1 \le i \ne j \le N. \tag{1.1}$$

Then, however, the constants c_i will be rather complicated functions of the π_is; the present author is able to construct such a design only for $n = 2$, provided that $\sum_{i=1}^{N} \pi_i(1-\pi_i) > 0.75$.

From among the designs (a), (b), (c) the Sampford–Durbin sampling seems to approximate (1.1) most closely.

The last point of view we shall consider is the simplicity of the formula for the variance of the simple estimate $\hat{Y}$. Generally, we have

$$\operatorname{var}\hat{Y} = \sum_{i=1}^{n} y_i^2 \left(\frac{1}{\pi_i} - 1\right) + \sum_{i \neq j}^{N}\sum^{N} y_i\, y_j \left(\frac{\pi_{ij}}{\pi_i\, \pi_j} - 1\right).$$

What we would like to obtain is

$$\operatorname{var}\hat{Y} = \sum_{i=1}^{N} \left(y_i - \frac{Y}{n}\pi_i\right)^2 \frac{1-\lambda_i}{\pi_i},$$

where the λ_is are some constants equalling approximately π_i. We cannot have $\lambda_i = \pi_i$ exactly, as can be shown. For the π_is proportionate to x_is we should like to have

$$\operatorname{var}\hat{Y} = \sum_{i=1}^{N} \left(y_i - \frac{Y}{X}x_i\right)^2 \frac{X}{n\,x_i}\,\frac{X - n\,x_i}{X - x_i} \tag{1.2}$$

at least with a high degree of approximation. This ideal is most closely satisfied by successive sampling. Since the last criterion seems to be most important, we may consider successive sampling as a most suitable generalization of simple random sampling to varying probabilities in the family of fixed-size designs. For $x_i \equiv 1$, (1.2) becomes the well-known formula for simple random sampling.

Let us now support the previous statements by some asymptotic results. There exist two asymptotic theories. In the first theory n is fixed, $N \to \infty$ and $N \max \pi_i$ is bounded. Then $\pi_i\,\pi_j - \pi_{ij} = O(N^{-3})$ and up to this order we have

$$\pi_i\,\pi_j - \pi_{ij} = \frac{\pi_i\left(1 - \frac{n-1}{n}\pi_i\right)\pi_j\left(1 - \frac{n-1}{n}\pi_j\right)}{\sum_{k=1}^{N}\pi_k\left(1 - \frac{n-1}{n}\pi_k\right)} + O(N^{-4}), \quad 1 \leq i \neq j \leq N$$

for all designs (a), (b), (c). If we go one step further, it may be shown that

$$\pi_i\,\pi_j - \pi_{ij} = \frac{\pi_i(1-\lambda_i)\,\pi_j(1-\lambda_j)}{\sum_{k=1}^{N}\pi_k(1-\lambda_k)}\,[1 - a_n(\pi_i - \bar{\pi})\,(\pi_j - \bar{\pi})] + O(N^{-5}), \quad 1 \leq i \neq j \leq N, \tag{1.3}$$

where the λ_is are solutions of the system of equations

$$(1-\lambda_i)\left[1 - \frac{\pi_i(1-\lambda_i)}{\sum_{k=1}^{N}\pi_k(1-\lambda_k)}\right] = 1 - \pi_i, \quad 1 \leq i \leq N, \tag{1.4}$$

$$\bar{\pi} = \sum_{i=1}^{N}\pi_i^2(1-\lambda_i) \Big/ \sum_{i=1}^{N}\pi_i(1-\lambda_i) \quad \text{and}$$

$$a_n = \begin{cases} \dfrac{n-1}{n^2} & \text{for Sampford–Durbin sampling} \\ 2\,\dfrac{n-1}{n^2} & \text{for rejective sampling} \\ \dfrac{1}{4}\,\dfrac{(n+1)\,(n+5)}{n^2} & \text{for successive sampling.} \end{cases}$$

The approximate formula for $\mathsf{var}\,\hat{Y} = \mathsf{E}(\hat{Y} - Y)^2$ based on (1.3) gives for $\pi_i = n\,x_i/X,\ 1 \le i \le N$,

$$\overline{\mathsf{E}}(\hat{Y} - Y)^2 = \sum_{i=1}^{N}(y_i - \tilde{\beta}\,x_i)^2\,\frac{1-\lambda_i}{\pi_i}\left[1 + a_n\left(\rho^2 - \frac{1}{N}\right)\delta^2\right], \tag{1.5}$$

where

$$\tilde{\beta} = \frac{\sum_{i=1}^{N} y_i(1-\lambda_i)}{\sum_{i=1}^{N} x_i(1-\lambda_i)},$$

ρ is the correlation coefficient of y_i/x_i and x_i with weights $\pi_i(1-\lambda_i)$, and

$$\delta^2 = \sum_{i=1}^{N}(\pi_i - \overline{\pi})^2\,\pi_i(1-\lambda_i)\,\Big/\sum_{i=1}^{N}\pi_i(1-\lambda_i). \tag{1.6}$$

We can see that the smaller a_n is, the less the variance suffers from large values of ρ^2, which occur if the regression of y_i on x_i is not linear but rather quadratic. The smallest a_n is obtained for Sampford–Durbin sampling; for rejective sampling a_n is twice as large, but for both designs $a_n \to 0$ as $n \to \infty$. For successive sampling a_n is smaller than for rejective sampling for $n = 2$, but it is larger for $n > 3$, and converges to $1/4$ as $n \to \infty$. Formula (1.3) for designs (a) and (c) and $n = 2$ has been obtained by J. N. K. Rao (1963), and the rest by the present author and J. Á. Víšek (unpublished).

Asymptotic theory with n fixed does not seem very promising at first sight, since the indicators of inclusion become independent asymptotically as $N \to \infty$. However, numerical investigations show that results obtained in this way provide very good approximations even for realistic situations.

Another asymptotic theory assumes that $N \to \infty$ as well as $n \to \infty$. For rejective and Sampford–Durbin sampling results have been obtained under the sole condition that

$$\sum_{i=1}^{N}\pi_i(1-\pi_i) \to \infty, \tag{1.7}$$

and for successive sampling under a more restrictive condition that

$$N \to \infty, \quad \frac{n}{N} \to \lambda, \quad 0 < \lambda < 1, \quad 0 < \varepsilon \le \frac{\min \alpha_i}{\max \alpha_i}. \tag{1.8}$$

For rejective and Sampford–Durbin sampling it has been shown under (1.7) that

$$\pi_i \pi_j - \pi_{ij} \sim \frac{\pi_i(1-\pi_i)\,\pi_j(1-\pi_j)}{\sum_{k=1}^N \pi_k(1-\pi_k)}, \quad 1 \le i \ne j \le N, \tag{1.9}$$

where $\sim$ means that the ratio of both sides converges to 1.

For successive sampling we have under (1.8)

$$\pi_i \pi_j - \pi_{ij} \sim \frac{\pi_i(1-\pi_i)\,\pi_j(1-\pi_j)}{\sum_{k=1}^N \pi_k(1-\pi_k)} \left[1 - \left(1 - \frac{\overline{\pi}\alpha_i}{\overline{\alpha}\pi_i}\right)\left(1 - \frac{\overline{\pi}\alpha_j}{\overline{\alpha}\pi_j}\right)\right], \tag{1.10}$$

where $\overline{\pi}/\overline{\alpha} = \sum_{i=1}^N \pi_i(1-\pi_i) \Big/ \sum_{i=1}^N \alpha_i(1-\pi_i)$. Under normal conditions the difference between (1.9) and (1.10) will be very small. The variance formula for $\hat{Y}$ yielded by (1.9) and (1.10) is

$$\overline{\mathsf{E}}(\hat{Y} - Y)^2 = \sum_{i=1}^N (y_i - \tilde{\beta}\,\pi_i)^2 \left(\frac{1}{\pi_i} - 1\right), \tag{1.11}$$

where

$$\tilde{\beta} = \frac{\sum_{i=1}^N y_i\, t_i}{\sum_{i=1}^N x_i\, t_i}$$

and

$$t_i = \begin{cases} 1 - \pi_i & \text{for rejective and Sampford–Durbin sampling} \\ \dfrac{\alpha_i}{\pi_i}(1-\pi_i) \doteq \lambda\left(1 - \dfrac{1}{2}\pi_i\right) & \text{for successive sampling.} \end{cases}$$

Thus successive sampling is closest to the ideal $t_i = 1$. Under $n \to \infty$ more precise results for $\pi_i \pi_j - \pi_{ij}$ are not available.

The author hopes to present the unpublished results in a forthcoming monograph.

Example. Consider a sample for $n = 2$ from a population of $N = 21$, with probabilities π_i proportional to x_i. The values x_i arranged in ascending order were as follows: 1, 2, 2, 2, 3, 3, 3, 3, 3, 4, 4, 4, 4, 5, 5, 5, 6, 6, 6, 7, 9. The quantity (1.6) equals $\delta^2 = 0.0020$ for this population. Now we considered populations of the values y_i with different squared correlation coefficient ρ^2 between y_i/x_i and x_i weighted by $\pi_i(1-\lambda_i)$. Table 1 gives exact variance values and their approximations (1.5), provided that the y_is had been chosen so that formula (1.11) gave $\overline{\mathsf{E}}(\hat{Y} - Y)^2 = 1$.

Table 1

Sampling	Formula	ρ^2		
		0.005	0.184	1.000
Rejective	exact	0.99994	1.00014	1.00126
	(1.5)	0.99996	1.00014	1.00095
Sampford	exact	0.99997	1.00007	1.00065
-Durbin	(1.5)	0.99998	1.00007	1.00048
Successive	exact	0.99995	1.00012	1.00109
	(1.5)	0.99996	1.00012	1.00083

We can see that the departures of the exact values from the value 1 obtained by (1.11) depend on ρ^2 and the kind of sampling in accordance with our asymptotic theory. Moreover, a large portion of the deviations between the exact value and (1.11) is explained by (1.5).

REFERENCES

Hájek, J. (1959). Optimum strategy and other problems in probability sampling. *Časopis Pěst. Mat. 84*, 387–423.

Hájek, J. (1964). Asymptotic theory of rejective sampling. *Ann. Math. Statist. 35*, 1491–1523.

Hartley, H.O. and Rao, J.N.K. (1962). Sampling with unequal probabilities and without replacement. *Ann. Math. Statist. 33*, 350–374.

Lanke, J. On non-rejective variance estimators in survey sampling. (Unpublished.)

Rao, J. N. K. (1963). On three procedures of unequal probability sampling without replacement. *J. Amer. Statist. Assoc. 58*, 202–215.

Rosén, B. (1972). Asymptotic theory for succesive sampling with varying probabilities without replacement I, II. *Ann. Math. Statist. 43*, 373–397, 748–776.

Sampford, M. R. (1967). On sampling without replacement with unequal probabilities of selection. *Biometrika 54*, 499–513.

Singh, Daroga (1954). On efficiency of the sampling with varying probabilities without replacement. *J. Ind. Stat. Agr. Soc. 6*, 48–57.

WILEY SERIES IN PROBABILITY
AND STATISTICS

Probability and Statistics

ANDERSON • An Introduction to Multivariate Statistical Analysis, *Second Edition*
*ANDERSON • The Statistical Analysis of Time Series
ARNOLD, BALAKRISHNAN and NAGARAJA • A First Course in Order Statistics
BACCELLI, COHEN, OLSDER and QUADRAT • Synchronization and Linearity: An Algebra for Discrete Event Systems
BARTOSZYNSKI and NIEWIADOMSKA-BUGAJ • Probability and Statistical Inference
BASILEVSKY • Statistical Factor Analysis and Related Methods
BERNARDO and SMITH • Bayesian Statistical Concepts and Theory
BHATTACHARYYA and JOHNSON • Statistical Concepts and Methods
BILLINGSLEY • Convergence of Probability Measures
BILLINGSLEY • Probability and Measure, *Third Edition*
BOROVKOV • Asymptotic Methods in Queueing Theory
BRANDT, FRANKEN and LISEK • Stationary Stochastic Models
CAINES • Linear Stochastic Systems
CAIROLI and DALANG • Sequential Stochastic Optimization
CHEN • Recursive Estimation and Control for Stochastic Systems
CONSTANTINE • Combinatorial Theory and Statistical Design
COOK and WEISBERG • An Introduction to Regression Graphics
COVER and THOMAS • Elements of Information Theory
CSÖRGŐ and HORVÁTH • Weighted Approximations in Probability and Statistics
CSÖRGŐ and HORVÁTH • Limit Theorems in Change-Point Analysis
*DOOB • Stochastic Processes
DRYDEN and MARDIA • Statistical Analysis of Shape
DUDEWICZ and MISHRA • Modern Mathematical Statistics
DUPUIS and ELLIS • A Weak Convergence Approach to the Theory of Large Deviations
ENDERS • Applied Econometric Time Series
ETHIER and KURTZ • Markov Processes: Characterization and Convergence
FELLER • An Introduction to Probability Theory and Its Applications, Volume 1, *Third Edition*, Revised; Volume II, *Second Edition*
FREEMAN and SMITH • Aspects of Uncertainty: A Tribute to D.V. Lindley
FULLER • Introduction to Statistical Time Series, *Second Edition*
FULLER • Measurement Error Models
GELFAND and SMITH • Bayesian Computation
GHOSH, MUKHOPADHYAY and SEN • Sequential Estimation
GIFI • Nonlinear Multivariate Analysis
GUTTORP • Statistical Inference for Branching Processes
HALD • A History of Probability and Statistics and Their Applications before 1750
HALL • Introduction to the Theory of Coverage Processes

*Now available in a lower priced paperback edition in the Wiley Classics Library

HAND • Construction and Assessment of Classification Rules
HANNAN and DEISTLER • The Statistical Theory of Linear Systems
*HANSEN, HURWITZ and MADOW • Sample Survey Methods and Theory, 2 Vol Set
HEDAYAT and SINHA • Design and Inference in Finite Population Sampling
HOEL • Introduction to Mathematical Statistics, *Fifth Edition*
HUBER • Robust Statistics
IMAN and CONOVER • A Modern Approach to Statistics
JOHNSON, KOTZ and KEMP • Univariate Discrete Distribution
JUPP and MARDIA • Statistics of Directional Data
JUREK and MASON • Operator-Limit Distributions in Probability Theory
KASS • The Geometrical Foundations of Asymptotic Inference
KAUFMAN and ROUSSEEUW • Finding Groups in Data: An Introduction to Cluster Analysis
KELLY • Probability, Statistics and Optimization
LAMPERTI • Probability: A Survey of the Mathematical Theory, *Second Edition*
LARSON • Introduction to Probability Theory and Statistical Inference, *Third Edition*
LESSLER and KALSBEEK • Nonsampling Error in Surveys
LINDVALL • Lectures on the Coupling Method
McLACHLAN • Discriminant Analysis and Statistical Pattern Recognition
McLACHLAN and KRISHNAN • The EM Algorithm
McNEIL • Epidemiological Research Methods
MANTON, WOODBURY and TOLLEY • Statistical Applications Using Fuzzy Sets
MARDIA • The Art of Statistical Science: A Tribute to G.S. Watson
MOLCHANOV • Statistics of the Boolean Model for Practitioners and Mathematicians
MORGENTHALER and TUKEY • Configural Polysampling: A Route to Practical Robustness
MUIRHEAD • Aspects of Multivariate Statistical Theory
OLIVER and SMITH • Inference Diagrams, Belief Nets and Decision Analysis
*PARZEN • Modern Probability Theory and Its Applications
PRESS • Bayesian Statistics: Principles, Models, and Applications
PUKELSHEIM • Optimal Experimental Design
PURI and SEN • Nonparametric Methods in General Linear Models
PURI, VILAPLANA and WERTZ • New Perspectives in Theoretical and Applied Statistics
RAO • Asymptotic Theory of Statistical Inference
RAO • Linear Statistical Inference and Its Applications, *Second Edition*
RAO and SHANBHAG • Choquet-Deny Type Functional Equations and Applications to Stochastic Models
RENCHER • Methods of Multivariate Analysis
ROBERTSON, WRIGHT and DYKSTRA • Order Restricted Statistical Inference
ROGERS and WILLIAMS • Diffusions, Markov Processes, and Martingales, Volume I: Foundations, *Second Edition*, Volume II: Itô Calculus
ROHATGI • An Introduction to Probability Theory and Mathematical Statistics
ROSS • Stochastic Processes
RUBINSTEIN • Simulation and the Monte Carlo Method
RUBINSTEIN and SHAPIRO • Discrete Event Systems: Sensitivity Analysis and Stochastic Optimization by the Score Function Method
RUZSA and SZEKELY • Algebraic Probability Theory
SCHEFFE • The Analysis of Variance

*Now available in a lower priced paperback edition in the Wiley Classics Library

SEBER • Linear Regression Analysis
SEBER • Multivariate Observations
SEBER and WILD • Nonlinear Regression
SERFLING • Approximation Theorems of Mathematical Statistics
SHORACK and WELLNER • Empirical Processes with Applications to Statistics
SMALL and McLEISH • Hilbert Space Methods in Probability and Statistical Inference
STAPLETON • Linear Statistical Models
STAUDTE and SHEATHER • Robust Estimation and Testing
STOYANOV • Counterexamples in Probability, *Second Edition*
STYAN • The Collected Papers of T.W. Anderson 1943–1985
TANAKA • Time Series Analysis. Nonstationary and Noninvertible Distribution Theory
THOMPSON anD SEBER • Adaptive Sampling
WELSH • Aspects of Statistical Inference
WHITTAKER • Graphic Models in Applied Multivariate Statistics
YANG • The Construction Theory of Denumerable Markov Processes

Applied Probability and Statistics

ABRAHAM and LEDOLTER • Statistical Methods for Forecasting
AGRESTI • Analysis of Ordinal Categorical Data
AGRESTI • Categorical Data Analysis
AGRESTI • An Introduction to Categorical Data Analysis
ANDERSON and LOYNES • The Teaching of Practical Statistics
ANDERSON, AUQUIER, HAUCH, OAKES, VANDAELE and WEISBERG • Statistical Methods for Comparative Studies
ARMITAGE and DAVID (editors) • Advances in Biometry
*ARTHANARI and DODGE • Mathematical Programming in Statistics
ASMUSSEN • Applied Probability and Queues
*BAILEY • The Elements of Stochastic Processes with Applications to the Natural Sciences
BARNETT and LEWIS • Outliers in Statistical Data, *Third Edition*
BARTHOLOMEW, FORBES, and McLEAN • Statistical Techniques for Manpower Planning, *Second Edition*
BATES and WATTS • Nonlinear Regression Analysis and Its Applications
BECHOFER, SANTNER and GOLDSMAN • Design and Analysis of Experiments for Statistical Selection, Screening and Multiple Comparisons
BELSLEY • Conditioning Diagnostics: Collinearity and Weak Data in Regression
BELSLEY, KUH and WELSCH • Regression Diagnostics: Identifying Influential Data and Sources of Collinearity•
BERRY • Bayesian Analysis in Statistics and Econometrics: Essays in Honor of Arnold Zellner
BERRY, CHALONER and GEWEKE • Bayesian Analysis in Statistics and Econometrics Essays in Honor of Arnold Zellner
BHAT • Elements of Applied Stochastic Processes, *Second Edition*
BHATTACHARYA and WAYMIRE • Stochastic Processes with Applications
BIEMER, GROVES, LYBERG, MATHIOWETZ and SUDMAN • Measurement Errors in Surveys
BIRKES and DODGE • Alternative Methods of Regression
BLOOMFIELD • Fourier Analysis of Time Series: An Introduction
BOLLEN • Structural Equations with Latent Variables

*Now available in a lower priced paperback edition in the Wiley Classics Library

BOULEAU • Numerical Methods for Stochastic Processes
BOX • R.A. Fisher, the Life of a Scientist
BOX and DRAPER • Empirical Model-Building and Response Surfaces
BOX and DRAPER • Evolutionary Operation: A Statistical Method for Process Improvement
BOX, HUNTER and HUNTER • Statistics for Experimenters: An Introduction to Design, Data Analysis, and Model Building
BOX and LUCENO • Statistical Control by Monitoring and Feedback Adjustment
BROWN and HOLLANDER • Statistics: A Biomedical Introduction
BUCKLEW • Large Deviation Techniques in Decision, Simulation, and Estimation
BUNKE and BUNKE • Non-linear Regression, Functional Relations and Robust Methods: Statistical Methods of Model Building
CHATTERJEE and HADI • Sensitivity Analysis in Linear Regression
CHATTERJEE and PRICE • Regression Analysis by Example, *Second Edition*
CLARKE and DISNEY • Probability and Random Processes: A First Course with Applications, *Second Edition*
COCHRAN • Sampling Techniques, *Third Edition*
*COCHRAN and COX • Experimental Designs, *Second Edition*
CONOVER • Practical Nonparametric Statistics, *Second Edition*
CONOVER and IMAN • Introduction to Modern Business Statistics
CORNELL • Experiments with Mixtures, Designs, Models, and the Analysis of Mixture Data, *Second Edition*
COX • A Handbook of Introductory Statistical Methods
*COX • Planning of Experiments
COX, BINDER, CHINNAPPA, CHRISTIANSON, COLLEDGE, and KOTT • Business Survey Methods
CRESSIE • Statistics for Spatial Data, *Revised Edition*
DANIEL • Applications of Statistics to Industrial Experimentation
DANIEL • Biostatistics: A Foundation for Analysis in the Health Sciences, *Sixth Edition*
DANIEL Fitting Equations into Data: Computer Analysis of Multifactor Data, *Second Edition*
DAVID • Order Statistics, *Second Edition*
*DEGROOT, FIENBERG and KADANE • Statistics and the Law
*DEMING • Sample Design in Business Research
DETTE and STUDDEN • The Theory of Canonical Moments with Applications in Statistics, Probability and Analysis
DILLON and GOLDSTEIN • Multivariate Analysis: Methods and Applications
DODGE and ROMIG • Sampling Inspection Tables, *Second Edition*
DOWDY and WEARDEN • Statistics for Research, *Second Edition*
DRAPER and SMITH • Applied Regression Analysis, *Second Edition*
DUNN • Basic Statistics: A Primer for the Biomedical Sciences, *Second Edition*
DUNN and CLARK • Applied Statistics: Analysis of Variance and Regression, *Second Edition*
DUPUIS • Large Deviations
ELANDT-JOHNSON and JOHNSON • Survival Models and Data Analysis
EVANS, PEACOCK and HASTINGS • Statistical Distributions, *Second Edition*
FISHER and VAN BELLE • Biostatistics: A Methodology for the Health Sciences
FLEISS • The Design and Analysis of Clinical Experiments
FLEISS • Statistical Methods for Rates and Proportion, *Second Edition*
FLEMING and HARRINGTON • Counting Processes and Survival Analysis

*Now available in a lower priced paperback edition in the Wiley Classics Library

FLURY • Common Principal Components and Related Multivariate Models
GALLANT • Nonlinear Statistical Models
GHOSH • Estimation
GLASSERMAN and YAO • Monotone Structure in Discrete-Event Systems
GNANADESIKAN • Analysis, *Second Edition*
GOLDSTEIN and LEWIS • Assessment: Problems, Developments and Statistical Issues
GOLDSTEIN and WOOFF • Bayes Linear Statistics
GREENWOOD and NIKULIN • A Guide to Chi-squared Testing
GROSS and HARRIS • Fundamentals of Queuing Theory, *Second Edition*
GROVES • Survey Errors and Survey Costs
GROVES, BIEMER, LYBERG, MASSEY, NICHOLLS and WAKSBERG • Telephone Survey Methodology
HAHN • and MEEKER • Statistical Intervals: A Guide for Practitioners
HAND • Construction and Assessment of Classification Rules
HAND • Discrimination and Classification
*HANSEN, HURWITZ and MADOW • Sample Survey Methods and Theory, Volume 1: Methods and Applications
*HANSEN, HURWITZ and MADOW • Sample Survey Methods and Theory, Volume II: Theory
HEIBERGER • Computation for the Analysis of Designed Experiments
HELLER • MACSYMA for Statisticians
HINKELMAN and KEMPTHORNE • Design and Analysis of Experiments, Volume 1: Introduction to Experimental Design
HOAGLIN, MOSTELLER and TUKEY • Exploratory Approach to Analysis of Variance
HOAGLIN, MOSTELLER and TUKEY • Exploring Data Tables, Trends and Shapes
HOAGLIN, MOSTELLER and TUKEY • Understanding Robust and Exploratory Data Analysis
HOCHBERG and TAMHANE • Multiple Comparison Procedures
HOCKING • Methods and Applications of Linear Models: Regression and the Analysis of Variance
HOEL • Elementary Statistics, *Fifth Edition*
HOGG and KLUGMAN • Loss Distributions
HOLLANDER and WOLFE • Nonparametric Statistical Methods
HOSMER and LEMESHOW • Applied Logistic Regression
HØYLAND and RAUSAND • System Reliability Theory: Models and Statistical Methods
HUBERTY • Applied Discriminant Analysis
IMAN and CONOVER • Modern Business Statistics
JACKSON • A User's Guide to Principle Components'
JOHN • Statistical Methods in Engineering and Quality Assurance
JOHNSON • Multivariate Statistical Simulation
JOHNSON & KOTZ • Distributions in Statistics
• Continuous Multivariate Distributions
JOHNSON, KOTZ and BALAKRISHNAN • Continuous Univariate Distributions, Volume 1, *Second Edition*; Volume 2, *Second Edition*
JOHNSON, KOTZ, and BALAKRISHNAN • Discrete Multivariate Distributions
JOHNSON, KOTZ and KEMP • Univariate Discrete Distribution, *Second Edition*
JUDGE, GRIFFITHS, HILL, LÜTKEPOHL, and LEE • The Theory and Practice of Econometrics, *Second Edition*

*Now available in a lower priced paperback edition in the Wiley Classics Library

JUDGE, HILL, GRIFFITHS, LÜTKEPOHL, and LEE • Introduction to the Theory and Practice of Econometrics, *Second Edition*
JURECKOVÁ and SEN • Robust Statistical Procedures: Asymptotics and Interrelations
KADANE • Bayesian Methods and Ethics in a Clinical Trial Design
KADANE and SCHUM • A Probabilistic Analysis of the Sacco and Vanzetti Evidence
KALBFLEISCH and PRENTICE • The Statistical Analysis of Failure Time Data
KASPRZYK, DUNCAN, KALTON and SINGH • Panel Surveys
KAURI • Advanced Calculus with Applications in Statistics
KISH • Statistical Design for Research
*KISH • Survey Sampling
KOTZ • Personalities
KOVALENKO, KUZNETZOV and PEGG • Mathematical Theory of Reliability of Time-dependent Systems with Practical Applications
LAD • Operational Subjective Statistical Methods: A Mathematical, Philosophical and Historical Introduction
LANGE, RYAN, BILLARD, BRILLINGER, CONQUEST, and GREENHOUSE • Case Studies in Biometry
LAWLESS • Statistical Models and Methods for Lifetime Data
LEBART, MORINEAU and WARWICK • Multivariate Descriptive Statistical Analysis: Correspondence Analysis and Related Techniques for Large Matrices
LEE • Statistical Methods for Survival Data Analysis, *Second Edition*
LePAGE and BILLARD • Exploring the Limits of Bootstrap
LESSLER and KALSBEEK • Nonsampling Error in Surveys
LEVY and LEMESHOW • Sampling of Populations: Methods and Applications
LINHART and ZUCCHINI • Model Selection
LITTLE and RUBIN Statistical Analysis with Missing Data
LYBERG • Survey Measurement
McLACHLAN • Discriminant Analysis and Statistical Pattern Recognition
McLACHLAN and KRISHNAN • The EM Algorithm and Extensions
McNEIL • Epidemiological Research Methods
MAGNUS and NEUDECKER • Matrix Differential Calculus with Applications in Statistics and Econometrics
MALLER and ZHOU • Survival Analysis with Long Term Survivors
MALLOWS • Design, Data, and Analysis by Some Friends of Cuthbert Daniel
MANN, SCHAFER, and SINPURWALLA • Methods for Statistical Analysis of Reliability and Life Data
MASON, GUNST, and HESS • Statistical Design and Analysis of Experiments with Applications to Engineering and Science
MILLER • Survival Analysis
MONTGOMERY and MYERS • Response Surface Methodology: Process and Product in Optimization Using Designed Experiments
MONTGOMERY and PECK • Introduction to Linear Regression Analysis, *Second Edition*
MORGENTHALER and TUKEY • Configural Polysampling
MYERS and MONTGOMERY • Response Surface Methodology
NELSON • Accelerated Testing, Statistical Models, Test Plans, and Data Analyses
NELSON • Applied Life Data Analysis
OCHI • Applied Probability and Stochastic Processed in Engineering and Physical Sciences

*Now available in a lower priced paperback edition in the Wiley Classics Library

OKABE, BOOTS, and SUGIHARA • Spatial Tesselations: Concepts and Applications of Voronoi Diagrams
PANKRATZ • Forecasting with Dynamic Regression Models
PANKRATZ • Forecasting with Univariate Box-Jenkins Models: Concepts and Cases
PIANTADOSI • Clinical Trials: A Methodological Perspective
PORT • Theoretical Probability for Applications
PUKELSHEIM • Optimal Design of Experiments
PUTERMAN • Markov Decision Processes: Discrete Stochastic Dynamic Programming
RACHEV • Probability Metrics and the Stability of Stochastic Models
RADHAKRISHNA RAO and SHANBHAG • Choquet-Deny Type Functional Equations with Applications to Stochastic Models
RÉNYI • A Diary on Information Theory
RIPLEY • Spatial Statistics
RIPLEY • Stochastic Simulation
ROSS • Introduction to Probability and Statistics for Engineers and Scientists
ROUSSEEUW and LEROY • Robust Regression and Outlier Detection
RUBIN • Multiple Imputation for Nonresponse in Surveys
RUBINSTEIN and SHAPIRO Discrete Event Systems: Sensitivity Analysis and Stochastic Optimization by the Score
RYAN • Modern Regression Methods
RYAN • Statistical Methods for Quality Improvement
SCHOTT • Matrix
SCOTT • Multivariate Density Estimation: Theory, Practice, and Visualization
SEARLE • Linear Models
SEARLE • Linear Models for Unbalanced Data
SEARLE • Matrix Algebra Useful for Statistics
SEARLE, CASELLA and McCULLOCH • Variance Components
SKINNER, HOLT, and SMITH • Analysis of Complex Surveys
STOYAN, KENDALL, and MECKE • Stochastic Geometry and Its Applications, *Second Edition*
STOYAN and STOYAN • Fractals, Random Shapes and Point Fields: Methods of Geometrical Statistics
THOMPSON • Empirical Model Building
THOMPSON • Sampling
TIERNEY • LISP-STAT: An Object-Oriented Environment for Statistical Computing and Dynamic Graphics
TIJMS • Stochastic Models: An Algorithmic Approach
TITTERINGTON, SMITH and MARKOV • Statistical Analysis of Finite Mixture Distributions
UPTON and FINGLETON • Spatial Data Analysis by Example, Volume 1: Point Pattern and Quantitative Data
UPTON and FINGLETON • Spatial Data Analysis by Example, Volume II: Categorical and Directional Data
VAN RIJKEVORSEL and DE LEEUW • Component and Correspondence Analysis
WEISBERG • Applied Linear Regression, *Second Edition*
WESTFALL and YOUNG • Resampling-Based Multiple Testing: Examples and Methods for *p*-Value Adjustment
WHITTLE • Optimization Over Time: Dynamic Programming and Stochastic Control, Volume 1 and Volume II
WHITTLE • Systems in Stochastic Equilibrium
WONNACOTT and WONNACOTT • Econometrics, *Second Edition*
WONNACOTT and WONNACOTT • Introductory Statistics, *Fifth Edition*

WONNACOTT and WONNACOTT • Introductory Statistics for Business and Economics, *Fourth Edition*
WOODING • Planning Pharmaceutical Clinical Trials: Basic Statistical Principles
WOOLSON • Statistical Methods for the Analysis of Biomedical Data
*ZELLNER • An Introduction to Bayesian Inference in Econometrics

Tracts on Probability and Statistics
BILLINGSLEY • Convergence of Probability Measures
KELLY • Reversibility and Stochastic Networks

*Now available in a lower priced paperback edition in the Wiley Classics Library